Anatomy and Human Movement
Structure and Function

SIXTH EDITION

Nigel Palastanga MA BA FCSP DMS DipTP

Pro-Vice-Chancellor, University of Wales; formerly Director of Physiotherapy Education,
Cardiff University, Cardiff, UK

Roger Soames BSc PhD

Centre for Anatomy and Human Identification, College of Life Sciences,
University of Dundee, Dundee, UK

CHURCHILL
LIVINGSTONE

ELSEVIER

Edinburgh London New York Oxford Philadelphia St Louis Sydney Toronto 2012

ELSEVIER
CHURCHILL
LIVINGSTONE

First edition 1989
First published as a paperback edition 1990
Second edition 1994
Third edition 1998
Fourth edition 2002
Fifth edition 2006
Sixth edition 2012

Formerly 978 0 7020 3553 1
ISBN 978 0 7020 5308 5

International ISBN 978 0 7020 4053 5

British Library Cataloguing in Publication Data
A catalogue record for this book is available from the British Library

Library of Congress Cataloging in Publication Data
A catalog record for this book is available from the Library of Congress

ELSEVIER your source for books, journals and multimedia in the health sciences
www.elsevierhealth.com

Working together to grow libraries in developing countries
www.elsevier.com | www.bookaid.org | www.sabre.org

ELSEVIER   BOOK AID International   Sabre Foundation

The Publisher's policy is to use **paper manufactured from sustainable forests**

Printed in China

Contents

Preface
to the first edition

We have designed and written this book for the student of anatomy who is concerned with the study of the living body and who wishes to use this knowledge functionally for a greater understanding of the mechanisms which allow movement to take place. Traditional anatomy texts are written as an adjunct to the study of the human body in the dissecting room, but only the surgeon has the advantage of directly viewing living musculoskeletal structures. The vast majority of students interested in musculoskeletal anatomy as well as those involved with human movement and its disorders are confronted by an intact skin and therefore must visualize the structures involved by palpation and analysis of movement. *Anatomy and Human Movement* presents the musculoskeletal structures as a living dynamic system – an approach lacking in many existing textbooks. The applied anatomy of the musculoskeletal system occupies the greater part of the book and is built up from a study of the bones and muscles (which are grouped according to their major functions, rather than as seen in the dissecting room) to a consideration of joints and their biomechanics. Anatomical descriptions of each joint are given with a detailed explanation of how it functions, the forces generated across it and how it might fail. We have placed great emphasis on the joints, as these are of major concern to those interested in active movement and passive manipulation, and we give examples of common traumatic or pathological problems affecting the structures described. Where possible, we describe palpation and analyse movement with respect to the joints and muscles involved, as well as any accessory movements. The course and distribution of the major peripheral nerves and blood vessels, together with the lymphatic drainage of the region, are given at the end of each relevant section. There are separate chapters on embryology and the skin and its appendages, and we have included, in the introduction, a section on the terminology used in the book. There is also an account of the structure and function of the nervous system written by Nikolai Bogduk whose contribution has been extremely valuable.

The format of the book matches a page of text to a page of illustrations, whenever possible, and we hope that this will allow the reader to confirm his or her understanding of the text with the visual information provided. The book is extensively illustrated with large, clear, fully labelled diagrams, all of which have been specially prepared. In the sections covering the joints and biomechanics, the illustrations have been drawn by Roger Soames, and these are particularly detailed as they pull together the anatomy from the previous parts of that chapter.

We hope that this new approach to the teaching of anatomy will serve to fill the gap which has always existed for those who have to learn their anatomy on a living subject and eventually have to determine their diagnoses and apply their treatments through an intact skin.

Nigel Palastanga
Derek Field
Roger Soames
1989

Preface
to the sixth edition

In this sixth edition of *Anatomy and Human Movement*, we have made a number of small but significant changes, mainly to the text, which are designed to further improve the usefulness of this already successful book. In this edition, there have been further improvements to the illustrations and we have added some new illustrations. Where possible, structures such as ligaments or cartilage have been coloured the same throughout the book, in an attempt to further enhance the clarity of illustrations. The book has progressed a long way since the first edition in 1989 when only black and white was used throughout the text.

As regards the content, we have continued to engage with the community of users of the book, namely, students, teachers, practitioners, and have responded to yet more suggestions. As a result of this very positive feedback, there has been some minor reordering within the text and expansion of some sections, namely those on joint replacement.

The use of Summary Boxes at the end of various sections has been retained: they are intended to be used as a quick aid to revision when the majority of the subject area has been learned, but feedback suggests that some will use them to get a general overview before starting to tackle the detailed text. Helping individuals to understand, remember and apply anatomy has always been the prime purpose of this book.

In response to the advances in the range of electronic aids to learning, the sixth edition also gives automatic access (unlocked by the unique PIN found on the inside front cover) to the e-book via Pageburst, allowing users to quickly search the entire book, make notes, add highlights and study more efficiently. In addition to this, the authors have helped create a separate five-hour modular e-learning course in functional human anatomy. At present, this is only available for purchase by universities and other institutions. This is presented in short lesson chunks following the body's regional structure – ideal for the student to study at their own pace and time and the lecturer to assign accordingly. Learning is delivered via outcome measures, animation, video, quizzes, activity analyses and exams.

Whether the book is being used by undergraduates, postgraduates, lecturers or practitioners, the changes and additions made to this edition are designed to make it even more useful to your learning, teaching or practice. Changes to the way education is managed and delivered require students and educators to be flexible in their learning. It is recognized that most learning takes place outside the classroom, and this requires the provision of high-quality material, both electronic and hard copy, to support learning wherever it is taking place. This edition of *Anatomy and Human Movement* and its accompanying electronic resources are designed to facilitate study in this new environment of varied learning spaces. The range of health professional groups using *Anatomy and Human Movement* has grown extensively over the past 22 years, with its use spreading to other groups interested in human movement, such as sports scientists.

Anatomy and Human Movement has been written by authors experienced in teaching living anatomy and human movement. This remains the prime reason for producing the book, and whilst the authors visit dissecting rooms regularly, the appreciation of anatomy through the intact skin of a living person is fundamental to the practice of a number of professions. We hope that this sixth edition will provide the reader with stimulation that will aid their understanding and learning of anatomy and their ability to apply this knowledge to human movement.

Nigel Palastanga
Roger Soames
2012

Acknowledgements

Nigel Palastanga and Roger Soames would like to acknowledge the tremendous contribution to the first four editions made by Derek Field.

In addition, we are very grateful to Dot Palastanga who is a lecturer teaching Occupational Therapy students at Cardiff University. She has been very much involved in the proofreading of every edition, including this the sixth: her attention to detail, as well as gentle reminders, has kept things on schedule and allowed us to meet deadlines. This is a very belated acknowledgement for her contribution over the past 20 years.

About the authors

Nigel Palastanga has been involved in higher education for nearly 40 years and for most of those has taught undergraduate and postgraduate physiotherapists. He has been the head of physiotherapy education at Cardiff University and was a pro-vice chancellor both there and in the University of Wales College of Medicine. He was awarded a Fellowship of the Chartered Society of Physiotherapy in 2001 and is still an external examiner for a number of physiotherapy programmes. He is currently pro-vice chancellor of the University of Wales where his role is concerned with the quality and standards of their awards.

Roger Soames moved from James Cook University in Townsville, Queensland, Australia in 2007 to become Principal Anatomist at the University of Dundee. In 2009 he was awarded a personal chair in Functional and Applied Anatomy and later the same year was appointed to the Cox Chair of Anatomy at the University of Dundee. He has spent the whole of his career within the discipline of anatomy, teaching and examining on a wide range of undergraduate and postgraduate degree programmes and courses. He has research interests in the musculoskeletal system and is also involved in developing surgical and clinical training opportunities utilizing Thiel preserved cadaveric material.

Part | 1 |

Introduction

CONTENTS

TERMINOLOGY

It is essential for students beginning their study of anatomy to become familiar with an internationally accepted vocabulary, allowing communication and understanding between all members of the medical and paramedical professions throughout the world. Perhaps the single, most important descriptive feature of this vocabulary is the adoption of an unequivocal position of the human body. This is known as *the anatomical position*. It is described as follows: the body is standing erect and facing forwards; the legs are together with the feet parallel so that the toes point forwards; the arms hang loosely by the sides with the palm of the hand facing forwards so that the thumb is lateral (Fig. 1.1). All positional terminology is used with reference to this position, irrespective of the actual position of the body when performing an activity.

The following is a list of more commonly used terms which describe the position of anatomical structures:

Anterior (ventral) To the front or in front, e.g. the patella lies anterior to the knee joint.

Posterior (dorsal) To the rear or behind, e.g. gluteus maximus lies posterior to the hip joint. (Ventral and dorsal are used more commonly in quadrupeds.)

Superior (cephalic) Above, e.g. the head is superior to the trunk.

Inferior (caudal) Below, e.g. the knee is inferior to the hip.

Cephalic (the head) **Caudal** (the tail) May be used in relation to the trunk.

Lateral Away from the median plane or midline, e.g. the thumb lies lateral to the index finger.

Medial Towards the median plane or midline, e.g. the index finger lies medial to the thumb.

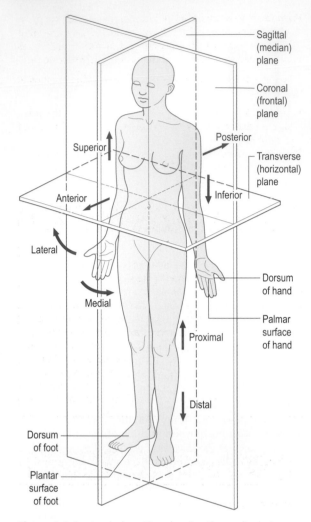

Figure 1.1 labels:
Sagittal (median) plane
Coronal (frontal) plane
Posterior
Transverse (horizontal) plane
Inferior
Superior
Anterior
Lateral
Medial
Dorsum of hand
Palmar surface of hand
Proximal
Distal
Dorsum of foot
Plantar surface of foot

Figure 1.1 Anatomical position showing the cardinal planes and directional terminology.

Distal Further away from the trunk or root of the limb, e.g. the foot is distal to the knee.

Proximal Closer to the trunk or root of the limb, e.g. the elbow is proximal to the hand.

Superficial Closer to the surface of the body or skin, e.g. the ulnar nerve passes superficial to the flexor retinaculum of the wrist.

Deep Further away from the body surface or skin, e.g. the tendon of tibialis posterior passes deep to the flexor retinaculum at the ankle.

To facilitate the understanding of the relation of structures one to another and the movement of one segment with respect to another, imaginary reference planes pass through the body in such a way that they are mutually perpendicular to each other (Fig. 1.1). Passing through the body from front to back and dividing it into two symmetrical right and left halves is the *sagittal* (*median*) *plane*. Any plane parallel to this is also known as a *parasagittal* (*paramedian*) *plane*.

A plane passing through the body from top to bottom and lying at right angles to the sagittal plane is the *coronal* (*frontal*) *plane*. This divides the body into anterior and posterior parts. All planes that divide the body in this way are known as coronal planes. Finally, a plane passing through the body at right angles to both the sagittal and coronal planes dividing it into upper and lower parts is known as a *transverse* (*horizontal*) *plane*. A whole family of parallel transverse planes exist; it is therefore usual when presenting a particular transverse section to specify the level at which it is taken. This may be done by specifying the vertebral level or the position within the limb, e.g. C6 or mid-shaft of humerus, respectively.

Within each plane a single axis can be identified, usually in association with a particular joint, about which movement takes place. An anteroposteriorly directed axis in the sagittal (or a paramedian) plane allows movement in a coronal plane. Similarly, a vertical axis in a coronal plane allows movement in a transverse plane. Lastly a transverse (right to left) axis in a coronal plane provides movement in a paramedian plane.

By arranging that these various axes intersect at the centre of joints, the movements possible at the joint can be broken down into simple components. It also becomes easier to understand how specific muscle groups produce particular movements, as well as to determine the resultant movement of combined muscle actions.

> **Section summary**
>
> **Terminology**
> - Specific terms are used to describe the relationship of one body part/segment/region to another and are considered in relation to the anatomical position of the body.
> - The anatomical position is: standing erect facing forwards, legs together toes pointing forwards, arms at the side and palms facing forwards.

TERMS USED IN DESCRIBING MOVEMENT

Rarely do movements of one body segment with respect to another take place in a single plane. They almost invariably occur in two or three planes simultaneously, producing a complex pattern of movement. However, it is convenient to consider movements about each of the three defined axes separately. Movement about a transverse axis occurring in the paramedian plane is referred to as *flexion* and *extension;* that about an anteroposterior axis in a coronal plane is termed *abduction* and *adduction;* and finally, that about a vertical axis in a transverse plane is termed *medial* and *lateral rotation.*

All movements are described, unless otherwise stated, with respect to the anatomical position, this being the position of reference. In this position joints are often referred to as being in a 'neutral position'.

Flexion The bending of adjacent body segments in a paramedian plane so that their two anterior/posterior surfaces are brought together, e.g. bending the elbow so that the anterior surfaces of the forearm and arm move towards each other. (For flexion of the knee joint the posterior surfaces of the leg and thigh move towards each other.)

Extension The moving apart of two opposing surfaces in a paramedian plane, e.g. the straightening of the flexed knee or elbow. Extension also refers to movement beyond the neutral position in a direction opposite to flexion, e.g. extension at the wrist occurs when the posterior surfaces of the hand and forearm move towards each other.

Flexion and extension of the foot at the ankle joint may be referred to as plantarflexion and dorsiflexion respectively.

Plantarflexion Moving the top (dorsum) of the foot away from the anterior surface of the leg.

Dorsiflexion Bringing the dorsum of the foot towards the anterior surface of the leg.

Abduction The movement of a body segment in a coronal plane such that it moves away from the midline of the body, e.g. movement of the upper limb away from the side of the trunk.

Adduction The movement of a body segment in a coronal plane such that it moves towards the midline of the body, e.g. movement of the upper limb back towards the side of the trunk.

Lateral flexion (bending) A term used to denote bending of the trunk (vertebral column) to one side, e.g. lateral bending of the trunk to the left. The movement occurs in the coronal plane.

Medial rotation Rotation of a limb segment about its longitudinal axis such that the anterior surface comes to face towards the midline of the body, e.g. turning the lower limb inwards so that the toes point towards the midline.

Lateral rotation Rotation of a limb segment about its longitudinal axis so that its anterior surface faces away from the midline plane, e.g. turning the lower limb so that the toes point away from the midline.

Supination and *pronation* are terms used in conjunction with the movements of the forearm and foot.

Supination Movement of the forearm so that the palm of the hand faces forwards. In the foot it is the movement whereby the forefoot is turned so that the sole faces medially; it is always accompanied by adduction of the forefoot.

Pronation Movement of the forearm that makes the palm of the hand face backwards. In the foot it is a movement of the forefoot which causes the sole to face laterally; it is always accompanied by abduction of the forefoot.

Inversion and *eversion* are terms used to describe composite movements of the foot.

Inversion Movement of the whole foot to make the sole face medially. It consists of supination and adduction of the forefoot.

Eversion Movement of the whole foot so that the sole comes to face laterally. It consists of pronation and abduction of the forefoot.

Section summary

Movements

- Specific terms refer to different types of movement between body parts/segments and/or regions.
- Flexion/extension occur about a transverse axis in a paramedian plane; abduction/adduction occur about an anteroposterior axis in a coronal plane; medial/lateral rotation occur about a vertical axis in a transverse plane.
- More specific terms are used for movements associated with some segments/regions: plantarflexion/dorsiflexion of the foot at the ankle joint; supination/pronation of the forearm; supination/pronation/inversion/eversion within the foot; lateral flexion of the vertebral column.

NERVOUS SYSTEM

Introduction

The nervous system consists of highly specialized cells designed to transmit information rapidly between various parts of the body. Topographically it can be divided into two major parts: the central nervous system (CNS) and the peripheral nervous system (PNS). The brain and spinal cord constitute the CNS, which lies within the skull and vertebral canal (see p. 572 and p. 489), while nerves in the PNS connect the CNS with all other parts of the body (see p. 7).

The CNS is a massive collection of nerve cells connected in an intricate and complex fashion to subserve the higher order functions of the nervous system, such as: thought, language, emotion, control of movement and the analysis of sensation. It is isolated from the rest of the body; its cardinal characteristic is that it is located wholly within the skull and vertebral canal.

The PNS consists of cells that connect the CNS with the other tissues of the body. These cells are aggregated into a large number of cable-like structures called nerves, which are threaded like wires throughout the tissues of the body.

Cellular structure

The basic cellular unit of the nervous system is the nerve cell or *neuron*, which differ in size and shape according to their function and location within the nervous system. All neurons have three characteristic components: a *cell body*, an *axon*, and *dendrites* (Fig. 1.2A).

The *cell body* is an expanded part of the cell containing the nucleus and apparatus necessary to sustain the metabolic activities of the cell. The *axon* is a longitudinal, tubular extension (process) of the cell membrane and cytoplasm which transmits information away from the cell body. The cell membrane surrounding the axon is referred to as the *axolemma*. Dendrites are extensions of the cell membrane that radiate from the cell body in various directions, and are responsible for receiving information and transmitting it to the cell body.

Structurally, dendrites differ from axons as they typically undergo extensive branching close to the cell body, whereas axons remain singular for most of their course only branching at their terminal ends. A neuron has only one axon, but may have several dendrites; the lengths and calibre of axons and dendrites vary, depending on the particular function of the neuron.

Interneural connections

Individual neurons convey information by conducting electrical action potentials along their cell membrane, with communication between separate neurons occurring chemically at a specialized structure called a *synapse*.

Synapses are formed by the close approximation of a small discrete area of the cell membrane of one neuron to a reciprocal area of the membrane of a second neuron (Fig. 1.2B). The apposed membranes are specially modified and are separated by a narrow gap about 0.02 μm in width, called the *synaptic cleft*. Across the synapse one cell communicates with the next, with communication being unidirectional. The membrane of the cell transmitting the information is the *presynaptic membrane*, and that of the cell receiving the information the *postsynaptic membrane*.

Near the presynaptic membrane, the cytoplasm of the transmitting neuron contains numerous small vesicles filled with chemicals known as neurotransmitters. The nature of the neurotransmitters varies according to the function of the neuron, but in all cases, when an electrical signal arrives at the terminal end of an axon, the neurotransmitter is released from the vesicles through the presynaptic membrane into the synaptic cleft. There it flows across to the postsynaptic membrane where it exerts its effect. This effect can be excitatory or inhibitory depending on the nature of the neurotransmitter and the nature of the receptors on the postsynaptic membrane with which it interacts. Excitatory substances generate an action potential in the postsynaptic neuron which then propagates this potential to its other end. Inhibitory substances alter the electrical potential of the postsynaptic neuron temporarily reducing its capacity to be stimulated by other neurons.

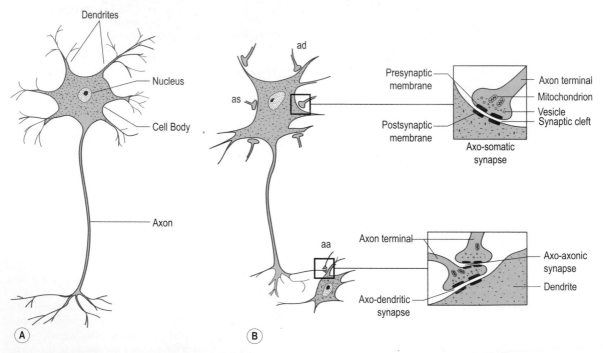

Figure 1.2 (A) Microscopic appearance of a neuron and its component parts, (B) neuron with multiple axo-somatic (as), axo-dendritic (ad) and axo-axonic (aa) synapses. The insets show the detailed structure of the synapse.

Synapses typically occur between the axon of one neuron and the dendrite of another. However, they may also occur between axons and cell bodies, axons and axons, and even between dendrites and dendrites. Some neurons receive only a few synapses while others may receive thousands.

The purpose of synapses is not to relieve one neuron of information and simply pass it on to the next, but to allow the interaction of information from perhaps several diverse sources on a single neuron. The activity of a given neuron may be influenced by many others and, conversely, by having several terminal branches to its axon, a single neuron may influence many other neurons.

By being connected to one another in diverse ways, groups of neurons are organized to subserve different functions of the nervous system. The patterns of connections, or circuits, vary in complexity and, in general, the more sophisticated the function being subserved, the more complex the circuitry.

Myelination

Myelination is a process whereby individual axons are wrapped in a sheath of lipid called *myelin*. The myelin sheath serves as a protective and insulating coating for the axon, and enhances the speed of conduction of electrical impulses along it.

In the PNS, myelination is by cells called *Schwann cells*: it is achieved by a Schwann cell curling an extension of its cell membrane around the shaft of the axon in a spiral manner (Fig. 1.3A). As the Schwann cell extension wraps around

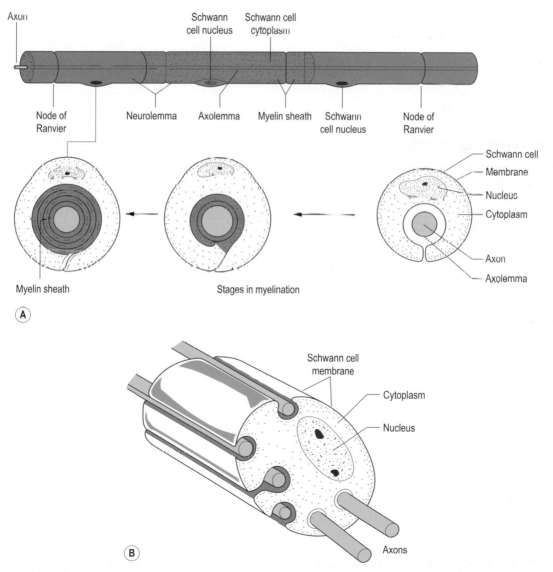

Figure 1.3 (A) Myelinated axon in the peripheral nervous system, (B) relationship between unmyelinated axons and Schwann cells.

the axon, the cytoplasm in it is squeezed out until only a double layer of the cell membrane of the Schwann cell remains spiralling around the axon (Fig. 1.3A). In this way the myelin sheath is formed by the lipid and protein of the cell membrane of the Schwann cell that surrounds the axon. Peripheral to the myelin sheath, the axon is surrounded by the cytoplasm of the Schwann cell, with the outermost cell membrane of the Schwann cell acting as a second membrane to the axon; this forms the *neurolemma* (Fig. 1.3A).

Along its length, a given axon is ensheathed in series by a large number of Schwann cells, but each Schwann cell is related only to a single axon (Fig. 1.3A). Junction points where the myelin sheath of one Schwann cell ends and the sheath of the next begins are known as *Nodes of Ranvier*. An axon together with the myelin sheath and the Schwann cells that surround it is referred to as a *nerve fibre*.

While many axons in the PNS are myelinated a large number remain unmyelinated. Instead of being enveloped by a tightly spiralling sheath of myelin, they run embedded in invaginations of Schwann cell membranes, so that a single Schwann cell may envelop several axons (Fig. 1.3B). Unmyelinated axons are afforded less physical protection than myelinated ones, but the foremost difference between them is that unmyelinated axons conduct impulses at much slower velocities.

In the CNS, myelination is performed by specialized cells called *oligodendrocytes*. The process is basically similar to that in the PNS, except that a given oligodendrocyte is usually involved in the simultaneous myelination of several separate axons (Fig. 1.4). Unmyelinated axons in the CNS run embedded in cytoplasmic extensions, or processes, of oligodendrocytes.

Several diseases may affect myelinating cells, including toxic and metabolic diseases, and most notably multiple sclerosis. In these diseases, neurons are not necessarily directly affected, but the loss of their myelin covering affects the way they conduct action potentials resulting in disordered neural function.

Structure of a peripheral nerve

A peripheral nerve is formed by the parallel aggregation of a variable number of myelinated and unmyelinated axons: the greater the number of axons, the larger the size of the nerve. Microscopically, within a peripheral nerve, myelinated axons are surrounded by their individual Schwann cell sheaths and unmyelinated axons run embedded in invaginations of the Schwann cell membrane. The axons are held together by sheaths of fibrous tissue which constitute additional coatings that protect them from external mechanical and chemical insults.

Individual myelinated axons are surrounded by a tubular sheath of fibrous tissue, the *endoneurium*, whereas clusters of axons are held together by a larger fibrous sheath, the *perineurium* (Fig. 1.5). Unmyelinated axons are not enclosed by endoneurium but run in isolated bundles

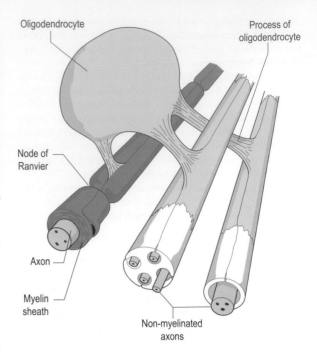

Figure 1.4 Relationship between an oligodendrocyte and the several myelinated and unmyelinated axons it ensheathes.

parallel to myelinated axons and are enclosed with them in the perineurial sheath.

A bundle of axons enclosed within a single perineurial sheath is referred to as a nerve *fascicle* (Fig. 1.5). Axons within a fascicle largely remain within that same fascicle

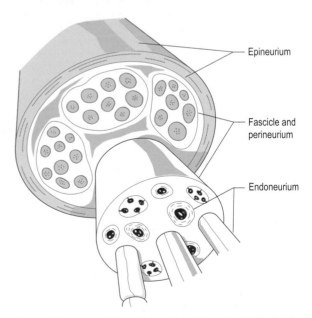

Figure 1.5 Composition of a peripheral nerve.

throughout the course of the nerve until their peripheral distribution. However, along the course of a peripheral nerve axons may at times leave one fascicle to enter and continue within an adjacent fascicle.

The fascicles within a peripheral nerve are bound together by an external sheath of fibrous tissue, the *epineurium* (Fig. 1.5), which forms the external surface of the macroscopic nerve. As a peripheral nerve passes through the tissues of the body, it gives off branches composed of one or more fascicles of the parent nerve, which leave it to reach their particular destination. Along the course of a nerve, this process is repeated until all the fascicles and axons in the nerve have been distributed to their target tissues.

The peripheral nervous system

Nerves that supply the structural tissues, such as bone, muscle and skin, are referred to as *somatic nerves*. Groups of somatic nerves innervate specific areas or regions and are covered in detail in the respective region.

The nerves that supply the viscera, such as the heart, lungs and digestive tract, are referred to as *visceral nerves*. However, because the nervous functions concerning viscera are largely automatic and subconscious, that part of the nervous system that innervates viscera is referred to as the *autonomic nervous system* (ANS). It has components in both the central and PNSs: it is described in detail on page 500.

Constituents of peripheral nerves

Peripheral nerves are composed of different types of axons that are classified according to their size, function or physiological characteristics. The broadest classification of axons recognizes *afferent* (or *sensory*) *fibres* and *efferent* (or *motor*) *fibres*. The terms 'afferent' and 'efferent' refer to the direction in which axons conduct information: afferent fibres conduct towards, while efferent fibres conduct away from the CNS. The term 'sensory' refers to axons that convey information to the CNS about events that occur in the periphery; 'motor' fibres cause events in the periphery, usually in the form of contraction of voluntary or smooth muscle.

Axons may also be classified according to their conduction velocities, which in turn are proportional to their sizes (Fig. 1.6). If a peripheral nerve is stimulated electrically and recordings of the evoked activity are taken some distance along the nerve, a wave of electrical activity can be recorded. The wave is generated by the summation of the electrical activity in each of the axons in the nerve, its size being proportional to the number of axons in the nerve, while its shape reflects the type of axons present.

The waveform obtained from a nerve containing every known type of axon is represented in Fig. 1.6. It has three principal peaks, named the A, B and C waves. The A wave

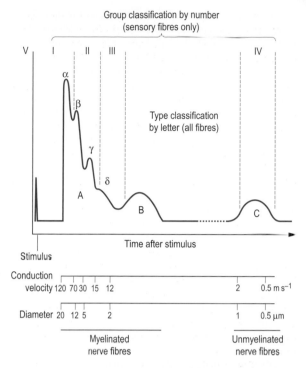

Figure 1.6 Neurogram of an idealized peripheral nerve. Activity in different types of nerve fibres is reflected in depolarizations that occur at different times after the delivery of an electrical stimulus to the nerve.

occurs very soon after the triggering stimulus, as it is produced by rapidly conducting axons with velocities in the range 12–120 ms^{-1}; these are classified as *A fibres*. The A wave is broken down into several secondary peaks called Aα, Aβ, Aγ and Aδ waves, each being produced by a particular subgroup of rapidly conducting axons.

The B wave is produced by more slowly conducting, but nevertheless fast, fibres classified as *B fibres*. The C wave arrives at the recording electrode much later as it is produced by slowly conducting axons classified as *C fibres* with conduction velocities in the range 0.5–2 ms^{-1}.

Not all peripheral nerves contain every type of fibre, therefore not all nerves will exhibit A, B and C waves. For example, B fibres are not regularly present in all nerves; B waves are therefore not always recordable.

Myelinated axons conduct impulses faster than unmyelinated ones, and among myelinated axons those with larger diameters conduct impulses faster than those with smaller diameters. The conduction velocity of a myelinated axon is directly proportional to the total diameter of the axon and its myelin sheath. A and B fibres are myelinated axons of various diameters and fast conduction velocities, while C fibres are unmyelinated axons with small diameters and slow conduction velocities.

Functionally, Aα and Aγ fibres represent motor fibres connected to voluntary muscles and also include certain sensory fibres that transmit position sensation from skeletal muscles. Aβ fibres mediate the sensations of touch, vibration and pressure from skin; Aδ fibres are sensory fibres that mediate pressure, pain and temperature sensations from skin, pain and pressure from muscle, and pain, pressure and position sensation from ligaments and joints. B fibres are preganglionic sympathetic efferent fibres. C fibres are largely unmyelinated sensory fibres that arise in virtually all tissues of the body and transmit pain, temperature and pressure sensations, however, some are postganglionic sympathetic neurons. Because Aβ fibres have large diameters they are sometimes referred to as large diameter afferent fibres, while Aδ and C fibres are referred to collectively as small diameter afferent fibres.

Another system of axon classification relates specifically to sensory fibres. In this system, sensory fibres are classified according to their conduction velocities into groups I, II, III and IV. These groups have the same conduction velocities as type Aα, Aβ, Aδ and C fibres respectively, as shown in Fig. 1.6, but do not include the Aα and Aβ motor fibres. The fibres of groups III and IV largely mediate pain and temperature sensations. Group II constitutes fibres that mediate pressure and touch, and fibres that form spray endings in muscle spindles (p. 15). Group I is divided into groups Ia and Ib. Group Ia fibres are slightly larger and innervate muscle spindles, while group Ib fibres innervate Golgi tendon organs (p. 15).

Nerve endings

The terminals of axons in peripheral nerves have unique structures depending on their function. Motor axons have terminals designed to deliver a stimulus to muscle cells (see p. 16), while sensory axons have terminals designed to detect particular types of stimuli. These terminals are called receptors, and a diversity of morphological types can be found (see p. 26).

Section summary

Nervous system

- Consists of specialized cells (neurons) concerned with the transmission of information throughout the body.
- Neurons have a cell body (expanded part of the cell), an axon (single long process arising from the cell body) and dendrites (multi-branching radiating processes arising from the cell body).
- Two parts: CNS comprises brain, spinal cord and spinal nerves; PNS comprises nerves (peripheral and cranial) and sensory receptors.

- Connections (synapses) between neurons allow integration of information from several sources. Resultant effect may be excitatory or inhibitory depending on the type of neurotransmitter and type of receptors on the postsynaptic membrane.
- Axons may be myelinated or unmyelinated. In CNS myelination is by oligodendrocytes, in PNS by Schwann cells. Myelinated axons have faster conduction velocities.

Peripheral nerves

- Parallel aggregations of individual axons (fibres).
- Afferent (sensory) fibres transmit information towards and efferent (motor) away from the CNS.
- The larger the fibre diameter the faster the nerve conduction velocity.

COMPONENTS OF THE MUSCULOSKELETAL SYSTEM

As this book is concerned essentially with the musculoskeletal system a brief account of the major tissues of the system, i.e. connective, skeletal and muscular tissue, and of the type of joints which enable varying degrees of movement to occur, is given as it will aid in understanding the mobility and inherent stability of various segments. The initiation and coordination of movement is the responsibility of the nervous system.

Connective tissue

Connective tissue is of mesodermal origin and in the adult has many forms; the character of the tissue depends on the organization of its constituent cells and fibres.

Fat

Fat is a packing and insulating material; however, in some circumstances it can act as a shock absorber, an important function as far as the musculoskeletal system is concerned. Under the heel, in the buttock and palm of the hand, the fat is divided into lobules by fibrous tissue septa, thereby stiffening it for the demands placed upon it.

Fibrous tissue

Fibrous tissue is of two types. In *white fibrous tissue* there is an abundance of collagen bundles, whereas in *yellow fibrous tissue* there is a preponderance of elastic fibres.

White fibrous tissue is dense, providing considerable strength without being rigid or elastic. It forms: (1) ligaments, which pass from one bone to another in the region of joints, uniting the bones and limiting joint movement; (2) tendons for attaching muscles to bones; and (3) protective membranes around muscle (*perimysium*), bone (*periosteum*) and many other structures.

Yellow fibrous tissue, on the other hand, is highly specialized, being capable of considerable deformation yet returning to its original shape. It is found in the ligamenta flava associated with the vertebral column as well as in the walls of arteries.

Skeletal tissue

Skeletal tissues are modified connective tissues, in which the cells and fibres have a particular organization and become condensed so that the tissue is rigid.

Cartilage

Cartilage is supplementary to bone, forming wherever strength, rigidity and some elasticity are required. In fetal development, cartilage is often a temporary tissue, being later replaced by bone. However, in many places cartilage persists throughout life. Although a rigid tissue, it is not as hard or strong as bone. It is also relatively non-vascular, being nourished by tissue fluids. A vascular invasion of cartilage often results in the death of the cells during the process of ossification of the cartilage and its eventual replacement by bone. Except for the articular cartilage of synovial joints, cartilage possesses a fibrous covering layer, the *perichondrium*.

There are three main types of cartilage: *hyaline cartilage*, *white fibrocartilage* and *yellow fibrocartilage*.

Hyaline cartilage

Hyaline cartilage forms the temporary skeleton of the fetus from which many bones develop. Its remnants can be seen as the articular cartilages of synovial joints, the epiphyseal growth plates between parts of an ossifying bone during growth, and the costal cartilages of the ribs. At joint surfaces it provides a limited degree of elasticity to offset and absorb shocks, as well as providing a relatively smooth surface permitting free movement to occur. With increasing age, hyaline cartilage tends to become calcified and sometimes ossified.

White fibrocartilage

White fibrocartilage contains bundles of white fibrous tissue which give it great tensile strength combined with some elasticity so that it is able to resist considerable pressure. It is found at many sites within the musculoskeletal system: (1) within the intervertebral discs between adjacent vertebrae; (2) in the menisci of the knee joint; (3) in the labrum surrounding and deepening the glenoid fossa of the shoulder joint and the acetabulum of the hip joint; (4) in the articular discs of the radiocarpal (wrist), sternoclavicular, acromioclavicular and temporomandibular joints; and (5) as the articular covering of bones which ossify in membrane, e.g. the clavicle and mandible.

White fibrocartilage may calcify and ossify.

Yellow fibrocartilage

Yellow fibrocartilage contains bundles of elastic fibres with little or no white fibrous tissue. It does not calcify or ossify and is not found within the musculoskeletal system.

Bone

Bone is extremely hard with a certain amount of resilience. It is essentially an organic matrix of fibrous connective tissue impregnated with mineral salts. The connective tissue gives the bone its toughness and elasticity, while the mineral salts provide hardness and rigidity, the two being skilfully blended together. The mineral component provides a ready store of calcium, which is continuously exchanged with that in body fluids, with the rate of exchange and overall balance of these mineral ions being influenced by several factors including hormones.

Each bone is enclosed in a dense layer of fibrous tissue, the *periosteum*, with its form and structure adapted to the functions of support and resistance of mechanical stresses. Being a living tissue, bone is continually remodelled to meet these demands; this is particularly so during growth. The structure of any bone cannot be satisfactorily considered in isolation; it is dependent upon its relationship to adjacent bones and the type of articulation between them, as well as the attachment of muscles, tendons and ligaments to it.

The internal architecture of bone reveals a system of struts and plates (*trabeculae*) running in many directions (Fig. 1.7), which are organized to resist compressive, tensile and shearing stresses. Surrounding these trabecular systems, which tend to be found at the ends of long bones, is a thin layer of condensed or compact bone (Fig. 1.7). The network of the trabeculae, because of its appearance, is known as *cancellous* or *spongy bone*. In the shaft of a long bone there is an outer, relatively thick ring of *compact bone* surrounding a cavity, which in life contains bone marrow.

Red and white blood cells are formed in red bone marrow, which after birth is the only source of red blood cells and the main source of white blood cells. In infants, the cavities of all bones contain red marrow. However, this gradually becomes replaced by yellow fat marrow, so that at puberty red marrow is only found in cavities associated with cancellous bone. With increasing age many of these red marrow containing regions are replaced by yellow marrow. Nevertheless, red marrow tends to persist throughout life in the vertebrae, ribs and sternum, and the proximal ends of the femur and humerus.

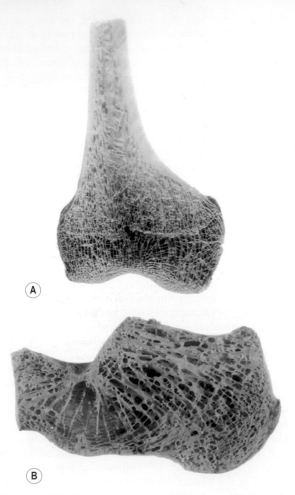

Figure 1.7 Trabecular arrangement within bone: (A) coronal section through the lower end of the femur, (B) sagittal section through the calcaneus.

For descriptive purposes bones can be classified according to their shape:

1. *Long bones* are found within the limbs: each consists of a shaft (*diaphysis*) and two expanded ends (*epiphyses*).
2. *Short bones* are the bones of the wrist and part of the foot, the carpal and tarsal bones respectively.
3. *Flat bones* are thin and tend to be curved in spite of their classification; they include the bones of the skull vault and ribs. Structurally, they consist of two layers of compact bone enclosing cancellous bone.
4. *Irregular bones* are those which fit none of the above categories, and include the vertebrae and many of the bones of the skull and face.

Both irregular and short bones consist of a thin layer of compact bone surrounding cancellous bone.

Bone development

Bone develops either directly in the mesoderm by the deposition of mineral salts, or in a previously formed cartilage model. When the process of calcification and then ossification takes place without an intervening cartilage model, it is known as *intramembranous ossification*, with the bone being referred to as *membrane bone*. However, if there is an intervening cartilage model, the process is known as *endochondral ossification*, with the bone being referred to as *cartilage bone*. This latter process is by far the most common.

Intramembranous ossification

The site of bone formation is initially indicated by a condensation of cells and collagen fibres accompanied by the laying down of organic bone matrix, which becomes impregnated with mineral salts. The formation of new bone continues in a manner similar to bone developed in cartilage (p. 11). Intramembranous ossification occurs in certain bones of the skull, the mandible and clavicle.

Endochondral ossification

Again the first step in the process is the accumulation of mesodermal cells in the region where the bone is to develop. A cartilage model of the future bone develops from these mesodermal cells (Fig. 1.8A). In long bones the cartilage model grows principally at its ends, so that the oldest part of the model is near the middle. As time progresses, the cartilage matrix in this older region is impregnated with mineral salts so that it becomes calcified. Consequently the cartilage cells, being cut off from their nutrient supply, die. The greater part of the calcified cartilage is subsequently removed and bone is formed around its few remaining spicules (Fig. 1.8A). Ultimately, the continual process of excavation of calcified cartilage and the deposition of bone leads to the complete removal of the calcified cartilage (Fig. 1.8A).

The cartilage at the ends of the bone continues to grow due to multiplication of its cells. However, the deeper layers gradually become calcified and replaced by bone. Therefore, the increase in length of a long bone is due to active cartilage at its ends, while an increase in width is by deposition of new bone on that already existing.

When first laid down, bone is cancellous in appearance, having no particular organization: it is referred to as woven bone. In the repair of fractures, the newly formed bone also has this woven appearance. However, in response to stresses applied to the bone by muscles, tendons, ligaments as well as the forces transmitted across joints, the woven bone gradually assumes a specific pattern in response to these stresses.

Growth and remodelling of bone

During growth there is an obvious change in the shape of a bone. However, it should be remembered that even in the adult, bone is being continuously remodelled, principally under the direct control of hormones to stabilize blood

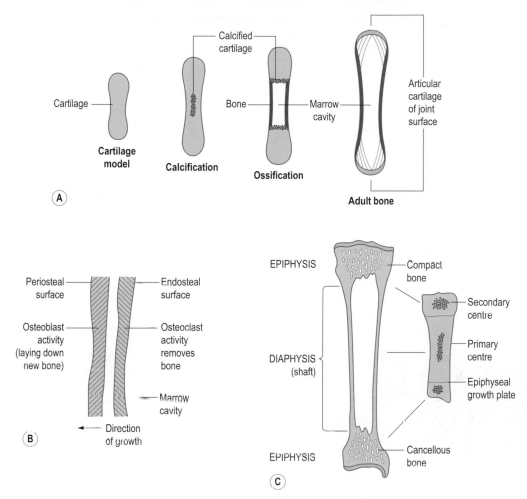

Figure 1.8 (A) Stages in calcification and ossification of bone from a cartilage model, (B) schematic representation of osteoblast and osteoclast activity, (C) sites of ossification centres in long bones, and the parts of the bone in each form.

calcium levels, but also in response to long-term changes in the pattern of forces applied to the bone.

Both growth and remodelling depend on the balanced activity of two cell types, one of which removes bone tissue (*osteoclasts*) and one which lays down new bone (*osteoblasts*). In a growing bone, for example, new bone is laid down around the circumference of the shaft in order to increase its diameter. At the same time the deepest layers of bone are being removed, thereby maintaining a reasonable thickness of cortical bone and enlarging the marrow cavity (Fig. 1.8B). Should the combined process of deposition and removal fail to match, then either a very thick or a very thin shaft results.

Ossification centres

The regions where bone begins to be laid down are known as ossification centres. It is from these centres that the

process of ossification spreads. The earliest, and usually the principal, centre of ossification in a bone is referred to as the *primary ossification centre*. Primary centres of ossification appear at different times in different bones, but are relatively constant between individuals: they also appear in an orderly sequence. The majority of such centres appear between the seventh and twelfth week of intrauterine life: virtually all are present before birth. In long bones, the primary ossification centre appears in the shaft of the bone (Fig. 1.8C).

Secondary ossification centres appear much later than primary centres, usually after birth, being formed in parts of the cartilage model into which ossification from the primary centre has not spread (Fig. 1.8C). All of the long bones in the body, and many others, have secondary centres of ossification: the bone formed is almost entirely cancellous.

The part of a long bone which ossifies from the primary centre is the *diaphysis*, while that from the secondary centres is an *epiphysis*. The plate of cartilage between these two regions is where the diaphysis continues to grow in length. Consequently, it is referred to as the *epiphyseal growth plate* (Fig. 1.8C). When this growth plate disappears the diaphysis and epiphysis become fused and growth of the bone ceases.

Section summary

Skeletal tissues

Cartilage

- Supplementary to bone providing strength, rigidity and some elasticity.
- Hyaline cartilage forms the temporary skeleton of bone and the articular surfaces of synovial joints.
- White fibrocartilage has great tensile strength and is able to resist compressive stresses.
- Yellow fibrocartilage contains elastic fibres.

Bone

- An organic matrix of cells and fibres impregnated with mineral salts surrounded by periosteum.
- Compact bone is found in the shaft (diaphysis) of long bones and as a thin shell covering cancellous bone.
- Cancellous bone is found in the ends of long bones (epiphyses), short and flat bones.
- Bone develops by either intramembranous or endochondral ossification initiated from a series of primary and secondary ossification centres.

Muscular tissue

Within the body there are three varieties of muscle: (1) *smooth muscle*, also referred to as involuntary or non-striated muscle; (2) *cardiac muscle*, and (3) *skeletal muscle*, also known as *voluntary* or *striated* muscle. Smooth muscle forms the muscular layer in the walls of blood vessels and of hollow organs such as the stomach. It is not under voluntary control, contracting more slowly and less powerfully than skeletal muscle. However, it is able to maintain its contraction longer. Cardiac muscle is also not under voluntary control; although it exhibits striations, it is considered to be different from skeletal muscle.

Skeletal muscle

Skeletal muscle constitutes over one-third of the total human body mass. It consists of non-branching striated muscle fibres, bound together by loose areolar tissue. Muscles have various forms; some are flat and sheet-like, some short and thick, while others are long and slender. The length of a muscle, exclusive of tendons, is closely related to the distance through which it is required to contract. Experiments have shown that muscle fibres

have the ability to shorten to almost half their resting length. Consequently, the arrangement of fibres within a muscle determines how much it can shorten when it contracts. However, irrespective of muscle fibre arrangement, all movement is brought about by muscle contraction (shortening), with the consequent action of pulling across joints changing the relative positions of the bones involved.

Muscle forms

The arrangement of the individual fibres within a muscle can be in one of two ways only; either parallel or oblique to the line of pull of the whole muscle. When the fibres are parallel to the line of pull they are arranged as a discrete bundle giving a *fusiform muscle* (Fig. 1.9A) (e.g. biceps brachii) or spread out as a broad, thin sheet (Fig. 1.9B) (e.g. external oblique of the abdomen). When contraction occurs it does so through the maximal distance allowed by the length of the muscle fibres. However, the muscle has limited power.

Muscles whose fibres are oblique to the line of pull cannot shorten to the same extent, but because of the increased number of fibres packed into the same unit area they are much more powerful. Such arrangements of fibres are known as *pennate*, of which there are three main patterns (Fig. 1.9C). In *unipennate muscles* the fibres attach to one side of the tendon only (e.g. flexor pollicis longus). *Bipennate muscles* have a central septum with the muscle fibres attaching to both sides and to its continuous central tendon (e.g. rectus femoris). Finally, some muscles possess several intermediate septa, each of which has associated with it a bipennate arrangement of fibres: the whole is known as a *multipennate muscle* (e.g. deltoid (Fig. 1.9C)).

Muscle structure

Muscle consists of many individual fibres, each being a long, cylindrical, multinucleated cell of varying length and width. Each fibre has a delicate connective tissue covering, the *endomysium*, separating it from its neighbours, yet connecting them together. Bundles of parallel fibres (*fasciculi*) are bound together by a more dense connective tissue covering, the *perimysium*. Groups of fasciculi are bound together to form whole muscles (Fig. 1.10) and are enclosed in a fibrous covering, the *epimysium*, which may be thick and strong or thin and relatively weak.

Muscle attachments

The attachment of muscle to bone or some other tissue is always via its connective tissue elements. Sometimes the perimysium and epimysium unite directly with the periosteum of bone or with the joint capsule. Where this connective tissue element cannot readily be seen, the muscle has a fleshy attachment and leaves no mark on the bone, although the area is often flattened or depressed. In many instances the connective tissue elements of the muscle fuse

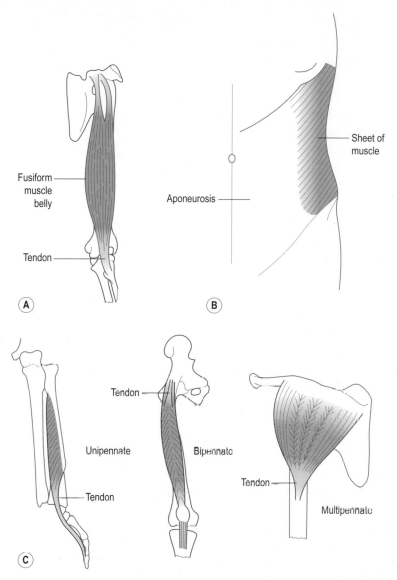

Figure 1.9 Arrangement of fibres within muscles: (A) fusiform, (B) sheet, (C) pennate.

together to form a tendon, consisting of bundles of collagen fibres. There is, however, no direct continuity between the fibres of the muscle and those of the tendon. Tendons can take various forms, all of which are generally strong. They can be round cords, flattened bands or thin sheets, the latter being known as an *aponeurosis*. Attachments of tendon to bone nearly always leave a smooth mark; it is only when the attachment is by a mixture of fleshy and tendinous fibres, or when the attachment is via a long aponeurosis, that the bone surface is roughened.

Where a muscle or tendon passes over or around the edge of a bone it is usually separated from it by a *bursa*, which serves to reduce friction during movement. Bursae are sac-like dilatations which may communicate directly with an adjacent joint cavity or exist independently: they contain a fluid similar to synovial fluid.

When a tendon is subjected to friction it may develop a *sesamoid bone*. Once formed these have the general effect of increasing the lever arm of the muscle, and act as a pulley enabling a slight change in the direction of pull of the muscle, e.g. the patella and the quadriceps tendon (Fig. 1.11A).

Because each end of a muscle attaches to different bones, observation of its principal action led to the designation of one end being the origin and the other the insertion: the insertion being to the bone that showed the freest movement. Such a designation is, however, misleading, since

13

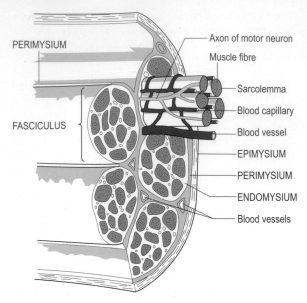

PERIMYSIUM

Axon of motor neuron

Muscle fibre

Sarcolemma

Blood capillary

FASCICULUS

Blood vessel

EPIMYSIUM

PERIMYSIUM

ENDOMYSIUM

Blood vessels

Figure 1.10 Organization of individual muscle fibres into whole muscles, together with their investing connective layers.

the muscle can cause either of the two attachments to move relatively freely. The term *attachment* is therefore preferred.

Muscle action

When stimulated, a muscle contracts bringing its two ends closer together. If this is allowed to happen the length of the muscle obviously changes, although the tension generated remains more or less constant; such contractions are termed *isotonic*. If, however, the length of the muscle remains unaltered (due, for example, to an externally applied force) then the tension it develops usually increases in an attempt to overcome the resistance; such contractions are termed *isometric*.

Isotonic contractions can be of two types: *concentric*, in which the muscle shortens, or *eccentric*, in which the muscle lengthens. Eccentric contraction occurs when the muscle is being used to control the movement of a body segment against an applied force.

When a muscle, or group of muscles, contracts to produce a specific movement, it is termed a *prime mover*. Muscles which directly oppose this action are called *antagonists*. Muscles which prevent unwanted movements associated with the action of the prime movers are known as *synergists*.

In all actions, part (often the larger part) of the muscle activity is directed across the joint, thereby stabilizing it by pulling the two articular surfaces together (Fig. 1.11B).

When testing the action of a muscle to determine whether it is weakened or paralysed, the subject is usually asked to perform the principal action of the muscle against resistance. This may be insufficient to confirm the integrity of the muscle. The only infallible guide is to palpate the muscle belly or its tendons to determine whether it is contracting during the manoeuvre.

Sensory receptors in muscle

Muscles contain free nerve endings and two types of specialized receptors: Golgi tendon organs and muscle spindles. The free nerve endings are responsible for mediating pain; they are sparsely scattered throughout the muscle belly occurring more densely at the myotendinous junction.

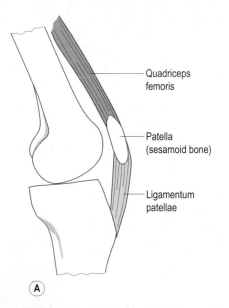

Quadriceps femoris

Patella (sesamoid bone)

Ligamentum patellae

(A)

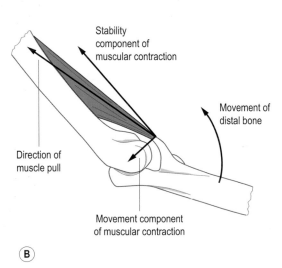

Stability component of muscular contraction

Movement of distal bone

Direction of muscle pull

Movement component of muscular contraction

(B)

Figure 1.11 (A) Arrangement of sesamoid bone within a tendon, (B) the components of muscle action across a joint.

Golgi tendon organs are formed by multiple, terminal branches of an axon (nerve fibre) weaving between the collagen fibres of the tendon, with the region encompassed by these branches surrounded by a fibrous capsule (Fig. 1.12A). When a muscle contracts, the collagen fibres in its tendon are stretched and are brought closer together compressing the nerve terminals and triggering the stimulus. In this way, the Golgi tendon organ monitors the extent of muscle contraction and the force exerted by the muscle.

Muscle spindles are highly elaborate structures essentially consisting of two types of modified muscle fibres (Fig. 1.12B). Muscle spindles occur throughout the muscle belly and are surrounded by a fibrous capsule.

The muscle fibres within a spindle monitor changes in muscle length. Because of this difference in function, and because they are located within the fusiform capsule of the spindle, these muscle fibres are referred to as *intrafusal* muscle fibres. The muscle fibres that produce movement upon muscle contraction are known as the *extrafusal* fibres.

There are two types of intrafusal muscle fibres: one type (nuclear bag fibres) has its nuclei grouped into an expanded region in the middle of the fibre, while the other type (nuclear chain fibres) has its nuclei spread along its length. At their middle, both nuclear bag and nuclear chain fibres are surrounded, in a spiral fashion, by branches of a group Ia sensory neuron. Group II neurons form similar spiral endings around nuclear chain fibres, but spray-like endings on nuclear bag fibres (Fig. 1.12B).

As a muscle lengthens or shortens, the degree of stretching or relaxation of the intrafusal muscle fibres alters the activity in the Ia and II fibres that innervate them. This activity is relayed to the CNS where it is used to determine the length of the muscle and its rate of lengthening or shortening. Indirectly, and in combination with other receptors, joint position is thereby determined. Various subgroups of bag and chain fibres have been identified, and each is responsible for detecting a different component of the change in muscle length, or the rate at which it occurs.

Muscle spindles also receive a motor innervation; this is from γ-motor neurons. Terminals of γ-motor neurons end on intrafusal fibres either side of the central regions surrounded by the sensory neurons. Activity in the γ-motor neurons causes the peripheral ends of the intrafusal fibres to contract, thereby stretching the central sensory region. This stretch alters the sensitivity or setting of the central

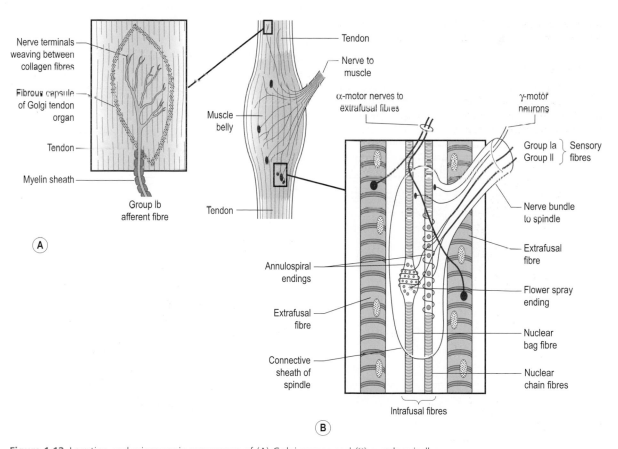

Figure 1.12 Location and microscopic appearance of (A) Golgi organs and (B) muscle spindles.

region and allows the sensory function of the spindle to keep in phase with the overall lengthening or shortening of the muscle as a whole as it relaxes or contracts.

Without γ activity, muscle spindles would respond to stretch of the muscle only when it was fully extended. Shortening the muscle would relieve the stretch, with the muscle spindle ceasing to signal muscle length. By contracting, under the control of γ-motor neurons, intrafusal fibres keep the central sensory region of the spindle taut at all times, allowing it to keep monitoring muscle length throughout the total range of movement.

Motor nerve endings

α-motor neurons form endings designed to deliver stimuli to extrafusal muscle fibres; similar but smaller endings are formed by γ-motor neurons on intrafusal fibres. The junction between a motor neuron and a muscle cell involves elaboration of both the neuron and muscle cell and of certain surrounding tissues, the entire complex being known as a *motor end plate* (Fig. 1.13).

Structurally and functionally, neuromuscular junctions resemble synapses (Fig. 1.2B). As a motor axon approaches its target muscle cell it loses its myelin sheath and forms a

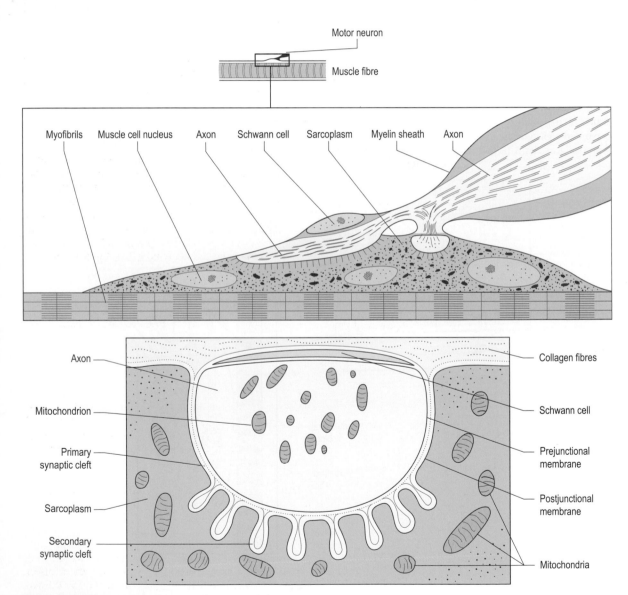

Figure 1.13 Microscopic appearance of motor end plates.

flattened expansion applied to the surface of the muscle membrane: the expansion is covered by a Schwann cell sheath insulating the neuromuscular junction from the external environment. That portion of the muscle cell to which the nerve is apposed is modified by forming a flattened bump on its surface, formed by the focal accumulation of cytoplasm (sarcoplasm) and organelles within the muscle cell that raises its membrane underneath the nerve terminal. That part of the muscle cell membrane applied to the nerve cell is known as the *postjunctional membrane* (Fig. 1.13). Reciprocally, that part of the nerve cell membrane applied to the muscle cell is known as the *prejunctional membrane.*

The postjunctional membrane forms troughs which lodge reciprocal folds of the prejunctional membrane. Within the troughs, the membranes of the nerve cell and muscle cell are separated by a synaptic cleft measuring 20–50 nm which is filled with an amorphous ground substance. In the floor of the trough, the postjunctional membrane is thrown into further folds, known as *secondary synaptic clefts*, which increases the surface area of the receptive muscle membrane (Fig. 1.13).

The terminal expansion of the motor neuron contains vesicles filled with the transmitter substance acetylcholine. When an action potential arrives at the nerve terminal, the vesicles release the acetylcholine through the presynaptic membrane into the synaptic cleft. The transmitter substance then flows across the cleft to react with molecular receptors on the postjunctional membrane, which generates an action potential in the muscle membrane. This potential is then propagated into the muscle cell causing contraction of the muscle fibres.

Section summary

Muscular tissue

- There are three varieties of muscle: smooth, associated with viscera and blood vessels; cardiac, associated with the heart; and skeletal.
- Muscle fibres are either aligned with (fusiform) or oblique to (pennate) the direction of pull.
- Fusiform muscles can shorten quickly over a large distance, but have limited power; pennate muscles are more powerful, but cannot shorten to the same extent.
- Muscles attach either directly to bones or via a tendon.
- Isotonic muscle contraction involves a change in muscle length often under constant force generation: in concentric contraction the muscle shortens, while in eccentric contraction it lengthens.
- Isometric contraction involves no change in muscle length but is usually associated with a change in the tension generated.

- Prime movers are responsible for producing a specific movement, antagonists oppose a movement and synergists help prevent unwanted movements.
- Muscles contain free nerve endings (which mediate pain), Golgi tendon organs (which monitor the extent and force of contraction) and muscle spindles (which monitor changes in length).
- Muscle spindles (intrafusal fibres) are of two types (nuclear bag and nuclear chain fibres).
- Innervation of muscle spindles by γ-motor neurons maintains their length in relation to the overall muscle length, keeping them taut throughout the full range of movement.
- Motor end plates are specialized endings between the nerve and muscle.
- Transmission of the neural impulse to the muscle occurs at the motor end plate.

Joints

The bones of the body come together to form joints: it is through these articulations that movement occurs. However, the type and extent of the movement possible depend on the structure and function of the joint; these latter can, and do, vary considerably. Nevertheless, the variation that exists in the form and function of the many joints of the body allows them to be grouped into well-defined classes: *fibrous, cartilaginous* and *synovial*, with the degree of mobility gradually increasing from fibrous to synovial.

Fibrous joints

Fibrous joints are of three types: *suture, gomphosis* and *syndesmosis.*

Suture

A form of fibrous joint that exists between the bones of the skull. It permits no movement as the edges of the articulating bones are often highly serrated, as well as being united by an intermediate layer of fibrous tissue (Fig. 1.14A). Either side of this fibrous tissue the inner and outer periosteal layers of the bones are continuous, and in fact constitute the main bond between them.

Sutures are not permanent joints, as they usually become partially obliterated after 30 years of age.

Gomphosis

In this type of fibrous joint a peg fits into a socket, being held in place by a fibrous ligament or band; the roots of the teeth held within their sockets in the maxilla and mandible are such examples (Fig. 1.14B): the fibrous band

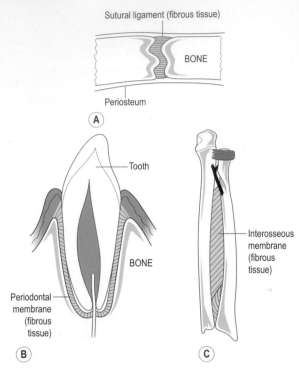

Figure 1.14 Fibrous joints: (A) suture, (B) gomphosis, (C) syndesmosis.

connecting tooth and bone being the periodontal ligament (membrane).

Syndesmosis

In a syndesmosis the uniting fibrous tissue is greater in amount than in a suture, constituting a ligament or an interosseous membrane (Fig. 1.14C). In adults, examples are the inferior tibiofibular joint where the two bones are held together by an interosseous ligament, and the interosseous membrane between the radius and ulna. Flexibility of the membrane or twisting and stretching of the ligament permit movement at the joint; however, the movement allowed is restricted and controlled.

Cartilaginous joints

In cartilaginous joints the two bones are united by a continuous pad of cartilage. There are two types of cartilaginous joint, *primary* and *secondary* (*synchondrosis* and *symphysis* respectively).

Primary cartilaginous

Between the ends of the bones involved is a continuous layer of hyaline cartilage (Fig. 1.15A). These joints occur at the epiphyseal growth plates of growing and developing

bone; these obviously become obliterated with fusion of the two parts (diaphysis and epiphysis). Because the plate of hyaline cartilage is relatively rigid, such joints exhibit no movement. However, there is one such joint in the adult, which is slightly modified because, by virtue of its structure, it enables slight movement to occur. This is the first sterno-costal joint.

Secondary cartilaginous

These occur in the midline of the body and are slightly more specialized. Furthermore, their structure enables a small amount of controlled movement to take place. Hyaline cartilage covers the articular surfaces of the bones involved, but interposed between the hyaline cartilage coverings is a pad of fibrocartilage. Examples are the joints between the bodies of adjacent vertebrae (Fig. 1.15B), where the fibrocartilaginous pad is in fact the intervertebral disc, and the joint between the bodies of the pubic bones (pubic symphysis).

Synovial joints

Synovial joints are a class of freely mobile joints, with movement being limited by associated joint capsules, ligaments and muscles crossing the joint. The majority of the joints of the limbs are synovial. In synovial joints the articular surfaces of the bones involved are covered

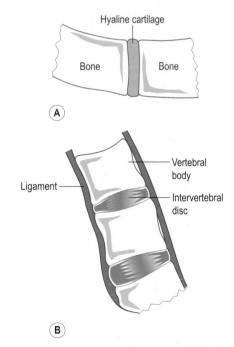

Figure 1.15 Cartilaginous joints: (A) primary cartilaginous joint, (B) secondary cartilaginous joint.

with *articular (hyaline) cartilage*, which because of their hardness and smoothness enable the bones to move against each other with minimum friction. Surrounding the joint, attaching either at or away from the articular margins, is a *fibrous articular capsule*, which may be strengthened by ligaments or the deeper parts of muscles crossing the joint. Lining the deep surface of the capsule is the *synovial membrane*, which covers all non-articular surfaces within the capsule (Fig. 1.16A). The synovial membrane secretes synovial fluid into the *joint space (cavity)* enclosed by the capsule, and serves to lubricate and nourish the articular cartilage as well as the opposing joint surfaces. During movement the joint surfaces either glide or roll past each other.

If the bones involved in the articulation originally ossified in membrane (p. 10), then the articular cartilage has a large fibrous element. In addition, an intra-articular disc,

which may not be complete separates the two articular surfaces enclosed by the capsule (Fig. 1.16B).

Bursae are often associated with synovial joints, sometimes communicating directly with the joint space.

Because of the large number of synovial joints within the body and their differing forms they can be subdivided according to the shape of their articular surfaces and the movements possible at the joint.

Plane joint

The joint surfaces are flat, or at least relatively flat, and of approximately equal extent. The movement possible is either a single gliding or twisting of one bone against the other, usually within narrow limits. An example is the acromioclavicular joint.

Saddle joint

The two surfaces are reciprocally concavoconvex, as a rider sitting on a saddle. The principal movements possible at the joint occur about two mutually perpendicular axes. However, because of the nature of the joint surfaces there is usually a small amount of movement about a third axis. The best example in the body is the carpometacarpal joint of the thumb.

Hinge joint

The surfaces are so arranged as to allow movement about one axis only. Consequently, the 'fit' of the two articular surfaces is usually good; in addition the joint is supported by strong collateral ligaments. The elbow is a typical hinge joint. The knee joint is considered to be a modified hinge joint, as it permits some movement about a second axis. In this case the movement is possible because of the poor fit of the articular surfaces.

Pivot joint

Again movement occurs about a single axis, with the articular surfaces arranged so that one bone rotates within a fibro-osseous ring. The atlantoaxial joint is a good example of a pivot joint.

Ball and socket joint

As the name suggests the 'ball' of one bone fits into the 'socket' of the other. This type of joint allows movement about three principal mutually perpendicular axes. The hip joint is a ball and socket joint.

Condyloid joint

This is a modified form of a ball and socket joint, which only allows active movement to occur about two perpendicular axes. However, passive movement may occur about the third axis. The metacarpophalangeal joints are examples of such joints.

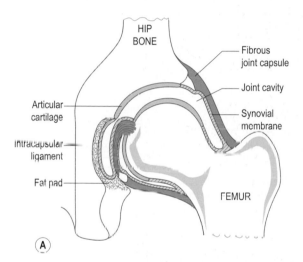

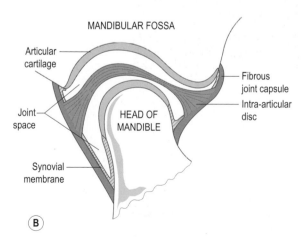

Figure 1.16 Structure of the synovial joints: (A) without the presence of an intra-articular disc, (B) with the presence of an intra-articular disc.

Ellipsoid joint

This is another form of a ball and socket joint, although in this case the surfaces are ellipsoid in nature. Consequently, movement is only possible about two perpendicular axes. The radiocarpal joint is an ellipsoid joint.

Receptors in joints and ligaments

Joints and ligaments typically have three types of receptors: free nerve endings believed to be responsible for mediating pain, nerve endings that resemble the Ruffini corpuscles and Pacinian corpuscles found in skin. The latter are not as well developed as the Pacinian corpuscles in the skin, and are referred to as paciniform endings. Ruffini-type and paciniform endings are responsible for detecting stretch and pressure applied to joint capsules and/or ligaments and so are involved in position sense.

Cartilaginous joints

- Permit some limited and controlled movement: there are two types – primary and secondary cartilaginous.
- In primary cartilaginous joints a layer of hyaline cartilage separates the bones.
- In secondary cartilaginous joints a fibrocartilage pad separates the hyaline cartilage layers.

Synovial joints

- No direct connection between the articular surfaces, therefore freely mobile; movement is limited by the joint capsule, ligaments and muscles crossing the joint.
- Surrounded by fibrous capsule lined with synovial membrane, which secretes synovial fluid to lubricate and nourish the articulating surfaces.
- Different surface shapes permit different types of movement.
- Subdivisions are: plane, saddle, hinge, pivot, ball and socket, condyloid and ellipsoid.

Section summary

Joints

The articulation between two or more bones: there are three classes – fibrous, cartilaginous and synovial.

Fibrous joints

- Generally permit little or no movement: there are three types – sutures, gomphoses and syndesmoses.

Spin, roll and slide

The movements actually occurring between articular surfaces can be complex; the terms spin, roll and slide are used to help explain them. Spin, in which one surface spins relative to the other, occurs about a fixed central axis (Fig. 1.17A). Roll is where one surface rolls across the other so that new

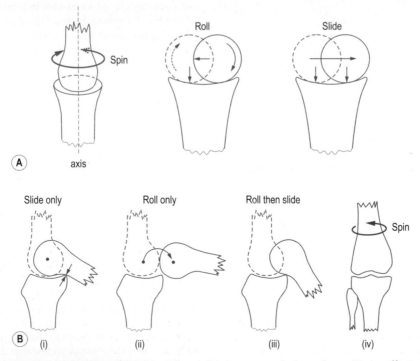

Figure 1.17 Diagrammatic representation of (A) spin, roll and slide between articular surfaces, (B) the effect of pure sliding (i) or rolling (ii) and a combination of rolling then sliding at the knee (iii), together with the spin that accompanies full extension (iv).

parts of both surfaces are continually coming into contact with each other, as in a wheel rolling along the ground (Fig. 1.17A). Slide occurs when one surface slides over the other so that new points on one surface continually make contact with the same point on the other surface, as in a wheel sliding across an icy surface (Fig. 1.17A).

Normally, spin, roll and slide do not occur separately as they complement one another to facilitate the complex movements available at joints. Combinations of spin, roll and slide are, therefore the basic components underlying movement at all joints. This concept can best be illustrated at the knee joint, a modified hinge joint, because of the type of movement available: a true hinge joint would permit slide only as one surface moves past the other about a fixed axis. For example, if only sliding movements were possible at the knee, movement would soon be restricted because of contact of the popliteal surface of the femur with the posterior part of the tibial condyle (Fig. 1.17B(i)). Similarly, if the femoral condyles only rolled over the tibial plateaux a situation would soon be reached where the femur would hypothetically roll off the tibia, because the profile of the femoral articular condyle is much longer than that of the tibia (Fig. 1.17B(ii)). The actual movement at the knee joint is a combination of both rolling and sliding between the two articular surfaces, under the control of ligaments associated with the joint, allowing a greater range of movement to be achieved (Fig. 1.17B(iii)). Spin also occurs at the knee joint as full extension is approached, as the femur spins on the tibia about its longitudinal axis so that the medial femoral condyle moves backwards (Fig 1.17B (iv)). The resultant effect is to place the knee into its close-packed position of maximum congruity between the joint surfaces. As with the combination of rolling and sliding, the ligaments of the joint are primarily responsible for bringing about spin at the knee. See also page 323, where movements of the knee are described in more detail.

Levers

An understanding of the action and principle of levers is important when considering the application of the forces applied to bones. The following is a simplified description of the mechanics of levers and how they are applied in the human body.

A lever may be considered to be a simple rigid bar, with no account taken of its shape or structure. Most long bones appear as rigid bars but although many bones, such as those of the skull, are far from the usual concept of a lever they can, nevertheless, still act in this way.

The fulcrum is the point around which the lever rotates. That part of the lever between the fulcrum and point of application of the force is known as the *force arm*, and that between the fulcrum and the point of application of the load is known as the *load arm*. This concept is easy to understand when applied to a child's see-saw (Fig. 1.18A). Different arrangements of the fulcrum, load and force arms produce different *classes of lever*. There are three possible

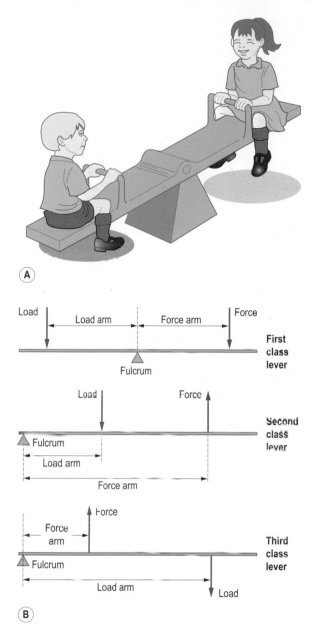

Figure 1.18 Levers: (A) a child's see-saw, (B) classes of lever.

arrangements: a *first class lever* has the fulcrum between the load and force arms; a *second class lever* has the fulcrum at one end and the applied force at the other, with the load situated between them; a *third class lever* again has the fulcrum at one end but the load at the other with the applied force between (Fig. 1.18B).

All three classes of lever are found within the human body: the fulcra are usually situated at the joints; the load may be body weight or some external resistance; the force is usually produced by muscular effort. It is the complex arrangement of all three classes of lever within the human body that produces movement.

A *first class lever* is used in balancing weight and/or changing the direction of pull. There is usually no gain in mechanical advantage, e.g. when standing on the right lower limb the fulcrum is the right hip joint, the load being body weight to the left of the hip, while the force is provided by the contraction of the right gluteus medius and minimus muscles.

A *second class lever* (the principle on which weight is lifted in a wheelbarrow) gains mechanical advantage, thereby allowing large loads to be moved, but with a loss of speed. Rising up onto the toes is a good example of such a system; the metatarsal heads act as the fulcrum, the weight of the body acting down through the tibia is the load while the calf muscles contract to produce the required force. The load arm is thus the distance from the tibia to the metatarsal heads, while the force arm is the distance between the attachment of the calf muscles to the calcaneus and the metatarsal heads.

A *third class lever* is the most commonly found within the body. It works at a mechanical disadvantage moving less weight but often at great speed. Biceps brachii acting across the elbow joint is a good example of this class of lever. The elbow is the fulcrum, the weight is the forearm and hand being supported, with the force being provided by the contraction of biceps. In this example the load arm is the distance between the elbow and centre of mass of the forearm and hand, whereas the force arm is the distance between the elbow joint and the attachment of biceps.

All movements of the human body are dependent on the interaction of these three classes of lever. It is as well to remember when studying the structure of the human body the relationship between the joints, the attachment of relevant muscles and the load to be moved. This will lead to an understanding of functional anatomy and with it, human movement.

Section summary

Levers

- Three classes of lever are found within the body, with the third class being the most common.
- First class levers have the load and muscle action on opposite sides of the joint (fulcrum); second class levers have the load between the joint and the muscle attachment; third class levers have the muscle attachment between the load and the joint.

SKIN AND ITS APPENDAGES

Introduction

The skin is a tough, pliable waterproof covering of the body, blending with the more delicate lining membranes of the body at the mouth, nose, eyelids, urogenital and anal openings. It is the largest single organ in the body. Not only does it provide a surface covering, it is also a sensory organ endowed with a host of nerve endings which provide sensitivity to touch and pressure, changes in temperature and painful stimuli. As far as general sensations are concerned, the skin is their principal source. The waterproofing function of the skin is essentially concerned with the prevention of fluid loss from the body. To this end, fatty secretions from sebaceous glands help to maintain this waterproofing, as well as being acted upon to produce vitamin D. However, the efficient waterproofing mechanism does not prevent the skin having an absorptive function when certain drugs, vitamins and hormones are applied to it in a suitable form. Nor does it prevent the excretion of certain crystalloids through sweating; if sweating is copious as much as 1 g of non-protein nitrogen may be eliminated in an hour. Because human beings are warm-blooded, body temperature must be kept within relatively narrow limits despite enormous variations in environmental temperature. Reduction of body temperature is a special function of the skin; because of the variability in its blood supply and the presence of sweat glands heat is lost through radiation, convection and evaporation. Together with the lungs, the skin accounts for over 90% of total heat loss from the body. As well as the ability of the blood vessels to 'open up' to promote heat loss, they can also be 'closed down' in an attempt to conserve body heat in cool environments.

The metabolic functions of the skin require a large surface area for effective functioning. In adults, the area is approximately 1.8 m^2, being seven times greater than at birth. Skin thickness also varies, not only with age but also from region to region. It is thinnest over the eyelids (0.5 mm) and thickest over the back of the neck and upper trunk, the palm of the hand and the sole of the foot. It generally tends to be thicker over posterior and extensor surfaces than over anterior and flexor surfaces, usually being between 1 and 2 mm thick.

The total thickness of the skin depends on the thickness of both the epidermis and the dermis. On the palms of the hands and soles of the feet the epidermis is responsible for the thickness of the skin, the dermis being relatively thin. This arrangement provides protection for the underlying dermis, as the palms and soles are regions of great wear and tear. The character of flexor and extensor skin differs in more respects than just thickness. The extensor skin of the limbs tends to be more hairy than the flexor skin, while the flexor skin is usually far more sensitive as it has a rich nerve supply.

The skin is loosely applied to underlying tissues so that it is easily displaced. However, in some regions it may be firmly attached to the underlying structures, e.g. the cartilage of the ear and nose, the subcutaneous periosteal surface of the tibia, and the deep fascia surrounding joints. In response to continued friction, skin reacts by increasing the thickness of its superficial layers. When wounded it responds by increased growth and repair.

The skin of young individuals is extremely elastic, rapidly returning to its original shape and position. However, this elasticity is increasingly lost with increasing age so that unless it is firmly attached to the underlying tissues it stretches. The stretching tends to occur in one direction because of the orientation of the collagen fibres in the deeper layers, which runs predominantly at right angles to the direction of stretch, therefore being parallel to the communicating grooves present on the skin surface. In some places the skin is bound down to the underlying deep fascia to allow freedom of movement without interference from subcutaneous fat and otherwise highly mobile skin. For example, at flexion creases of joints the skin is bound down to the underlying fibrous tissue. Where the skin has to be pulled around a joint when it is flexed, it is bound down in loose folds which are taken up in flexion: the joints of the fingers clearly show this arrangement.

In adjusting to allow movement, the skin follows the contours of the body. Although this is enabled by its intrinsic elasticity the skin is nevertheless subjected to internal stresses, which vary from region to region. These stress lines are often referred to as Langer's lines or cleavage lines (Fig. 1.19). They are important because incisions along these lines heal with a minimum of scarring, whereas in wounds across them the scar may become thicker, with the possible risk of scar contraction, because the wound edges are being pulled apart by the internal stresses of the skin. Langer's lines do not always correspond with the stress lines of life; they merely reflect the stresses within the skin at rest.

Skin colour depends on the presence of pigment (melanin) and the vascularity of the dermis. When hot the skin appears reddened due to the reflection of large quantities of blood through the epidermis. Similarly, when cold the skin appears paler due to the reduction of blood flow to it. The degree of oxygenation of the blood also influences skin colour; anaemic individuals generally appear pale. Individual and racial variations in skin colour are dependent upon the presence of melanin in the deepest layers of the epidermis. In darker skinned races, the melanin is distributed throughout the layers of the epidermis. In response to sunlight and heat the skin increases its degree of pigmentation, thereby making it appear darker. This physiological increase in pigment formation, i.e. tanning from exposure to sunlight, is widely sought after in white populations. Some areas of the body show a constant deeper pigmentation, these being the external genital and perianal regions, the axilla and the areola of the breast.

On the pads of the fingers and toes, and extending over the palm of the hand and sole of the foot, are a series of alternating ridges and depressions. The arrangement of these ridges and depressions is highly individual, so much so that even identical twins have different patterns. It is this patterning which forms the basis for identification through finger prints. They are due to the specific arrangement of the large dermal papillae under the epidermis and act to improve the grip and prevent slippage. Along the summits of these ridges sweat glands open; sebaceous glands and hair are absent on these surfaces.

The skin and subcutaneous tissues camouflage the deeper structures of the body. Nevertheless, it is often necessary to identify and manipulate these deeper structures through the skin, and also to test their function, effectiveness and efficiency. To do this the examiner relies heavily on sensory information provided by their own skin, particularly that of the digits and hands. Indeed, the skin of this region is so richly endowed with sensory nerve endings that it allows objects to be identified by touch alone, culminating in the ability of the blind to read with their fingers.

As well as the functions of the skin referred to above, nails, hair, sebaceous and sweat glands are all derived from the epidermis. The delicate creases extending in all directions across the skin form irregular diamond-shaped regions. It is at the intersections of these creases that hairs typically emerge.

Structure

The skin consists of a superficial layer of ectodermal origin known as the *epidermis*, and a deeper mesodermal-derived layer known as the *dermis* (Fig. 1.20A).

Epidermis

The epidermis is a layer of stratified squamous epithelium of varying thickness (0.3–1.0 mm), being composed of many layers of cells. The deeper cells are living and actively

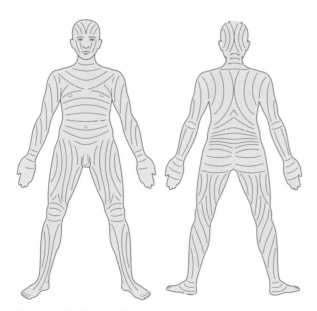

Figure 1.19 Cleavage line orientation.

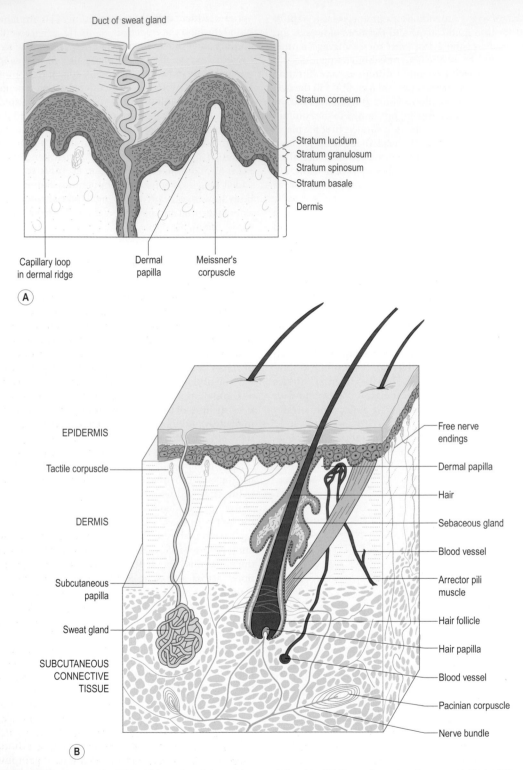

Duct of sweat gland

Stratum corneum

Stratum lucidum
Stratum granulosum
Stratum spinosum
Stratum basale

Dermis

Capillary loop
in dermal ridge

Dermal
papilla

Meissner's
corpuscle

(A)

EPIDERMIS

DERMIS

SUBCUTANEOUS
CONNECTIVE
TISSUE

Tactile corpuscle

Subcutaneous
papilla

Sweat gland

Free nerve
endings

Dermal papilla

Hair

Sebaceous gland

Blood vessel

Arrector pili
muscle

Hair follicle

Hair papilla

Blood vessel

Pacinian corpuscle

Nerve bundle

(B)

Figure 1.20 (A) Arrangement of the epidermal and dermal layers of the skin, (B) three-dimensional representation of the skin and subcutaneous connective tissue layer showing the arrangement of hair, glands and blood vessels.

proliferating, with the cells produced gradually passing toward the surface; as they do so they become cornified (*keratinized*). They are ultimately shed as the skin rubs against the clothing and other surfaces. The epidermis is avascular but is penetrated by sensory nerve endings. Its deep surface is firmly locked to the underlying dermis by projections into it known as *epidermal pegs*, with the reciprocal projections from the dermis being known as *dermal papillae* (Fig. 1.20A).

It is convenient to consider the epidermis as being divided into a number of layers, particularly in the so-called thick skin of the palm or sole of the foot. These layers are from within outwards the *stratum basale, stratum spinosum, stratum granulosum, stratum lucidum* and finally the *stratum corneum* (Fig. 1.20A).

The stratum basale consists of a single layer of cells adjacent to the dermis. It is in this layer, as well as in the stratum spinosum that new cells are produced to replace those lost from the surface. The stratum spinosum itself consists of several layers of irregularly shaped cells, which become flattened as they approach the stratum granulosum. The stratum basale and stratum spinosum together are often referred to as the germinal zone, because of their role in new cell production.

Collectively, the remaining layers of the epidermis (granulosum, lucidum and corneum) are often referred to as the horny layer. In the stratum granulosum the cells become increasingly flattened and the process of keratinization begins: the cells in this layer are in the process of dying. A relatively thin transparent layer (the stratum lucidum) lies between the granulosum and the superficial stratum corneum. It is from the stratum corneum that the cells are shed: it is also mainly responsible for the thickness of the skin.

The *epidermal melanocytes*, which are responsible for the pigmentation of the skin, lie within the deepest layers of the epidermis.

Dermis

The dermis is the deeper interlacing feltwork of collagen and elastic fibres, which generally comprises the greater part of total skin thickness. It can be divided into a superficial finely-textured *papillary layer*, which, although clearly separated from it, interdigitates with the epidermis, and a deeper coarser *reticular layer*, which gradually blends into the underlying subcutaneous connective tissue.

The projecting dermal papillae usually contain capillary networks which bring the blood into close association with the epidermis (Fig. 1.20A). The ability to open up or close down these networks is responsible for the regulation of heat loss through the skin, as well as causing individuals to blush in moments of embarrassment. Some papillae contain tactile receptors, which are obviously more numerous in regions of high tactile sensitivity (e.g. fingers, lips) and less so in other regions (e.g. back).

The reticular layer of dermis consists of a dense mass of interweaving collagen and elastic connective tissue fibres. It is this layer which gives the skin its toughness and strength. The fibres run in all directions, but are generally tangential to the surface. However, there is a predominant orientation of fibre bundles, with respect to the skin surface, which varies in different regions of the body. It is this orientation which gives rise to the cleavage lines of the skin (Fig. 1.19).

The dermis contains numerous blood vessels and lymphatic channels, nerves and sensory nerve endings as well as a small amount of fat. In addition, it also contains hair follicles, sweat and sebaceous glands, and smooth muscle (*arrector pili*). The deep surface of the dermis is invaginated by projections of subcutaneous connective tissue, which serve partly for the entrance of the nerves and blood vessels into the skin (Fig. 1.20B).

Subcutaneous connective tissue

This is a layer of loosely arranged connective tissue containing fat and some elastic fibres. The amount of subcutaneous fat varies in different parts of the body, being completely absent in only a few regions (eyelid, scrotum, penis, nipple and areola). The distribution of subcutaneous fat also differs between men and women, constituting a secondary sexual characteristic in women, e.g. the breast as well as the rounded contour of the hips. The subcutaneous connective tissue contains blood and lymph vessels, the roots of hair follicles, the secretory parts of sweat glands, cutaneous nerves, and sensory endings (particularly Pacinian (*pressure*) corpuscles) (Figs. 1.20B and 1.21).

In the subcutaneous tissue overlying joints, subcutaneous bursae exist, which contain a small amount of fluid thereby facilitating movement of the skin in these regions.

Cutaneous sensory receptors

The nerve endings in skin vary in structure from simple to complex formations. Simple nerve endings are those in which the axon terminates without undergoing any branching or other elaboration. Complex endings are those in which the axons undergo any of a variety of changes, such as forming expansions, assuming a tangled appearance, or becoming surrounded by additional specialized tissue. When an axon terminates by losing its myelin sheath ending as a naked axon, it is known as a *free nerve ending*.

The types of nerve endings in skin are shown in Fig. 1.21. Simple, free nerve endings occur within the epidermis and the dermis. In the epidermis they run as naked axons between the epidermal cells and may undergo branching; they are generally orientated perpendicular to the skin surface. In contrast, in the dermis free nerve endings run parallel to the skin surface. Free nerve endings are thought to detect stimuli that generate the sensations of pain and temperature, but many are also capable of responding to mechanical stimuli like touch and pressure which deform

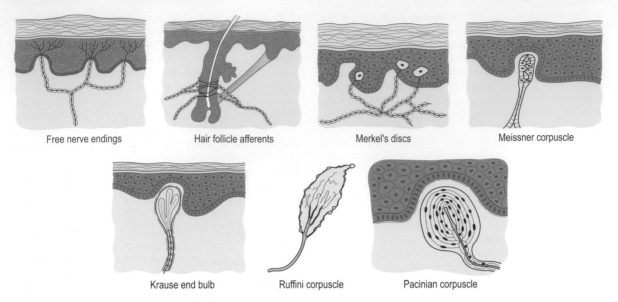

| Free nerve endings | Hair follicle afferents | Merkel's discs | Meissner corpuscle |

| Krause end bulb | Ruffini corpuscle | Pacinian corpuscle |

Figure 1.21 Types of receptors found in skin.

the skin. Free nerve endings are also found surrounding the roots of hairs. These endings are stimulated by deformation of the hair, and are involved in detecting coarse stimuli causing the hairs on the skin to bend.

Merkel's discs are complex endings formed when a free nerve ending terminates as a disc-like expansion under a specialized cell in the epidermis called a Merkel cell or a tactile cell. They are found in hairless skin, particularly in the fingertips, and are thought to mediate the sensation of touch.

Meissner's corpuscles are complex encapsulated receptors formed by a spiralling axon surrounded by flattened Schwann cells in turn surrounded by fibrous tissue continuous with the endoneurium of the axon. Meissner's corpuscles are believed to be involved in the sensation of touch. *Krause's end bulbs* consist of an axon that forms a cluster of multiple short branches surrounded by a fibrous capsule. Simpler varieties occur in which the axon does not undergo branching, forming only a bulbous ending surrounded by a poorly developed capsule. The function of these nerve endings is unclear, but they occur in the dermis and are believed to respond to mechanical stimulation of the skin.

Ruffini corpuscles consist of an axon that forms a flattened tangle of branches embedded in a bundle of collagen fibres. These nerve endings occur in the dermis and respond to stretching of the collagen fibres when the skin is deformed by pressure. *Pacinian corpuscles* are the most complex of sensory nerve endings and consist of a single axon surrounded by several concentric laminae of modified Schwann cells all enclosed in a fibrous capsule. These structures are located in the dermis and are designed to respond to pressure stimuli.

Appendages of the skin

These are nails, hairs, sebaceous, sweat and mammary glands, and are all derived from the epidermis.

Nails

A nail consists of an approximately rectangular plate of horny tissue found on the dorsum of the terminal phalanx of the fingers, thumb and the toes (Fig. 1.22). They are a special modification of the two most superficial layers of the epidermis, particularly the stratum lucidum. Its transparency allows the pinkness of the underlying highly vascular nail bed to show through. The nail is partly surrounded by a fold of skin, the nail wall, which is firmly adherent to the underlying *nail bed* with some fibres ending in the periosteum of the distal phalanx. It is this firm attachment

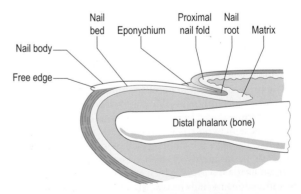

Figure 1.22 Relationship of the nail to the skin.

which enables the nails to be used for scratching and as instruments for prizing open various objects.

The distal end of the nail is free, while the proximal covered part constitutes the *nail root*. There is an abundant supply of sensory nerve endings and blood vessels to the nail bed. The nails grow at approximately 1 mm per week, being faster in summer than in winter.

Hairs

Hairs are widely distributed over the body surface, notable exceptions being the palm of the hand and the sole of the foot. Hairs vary in thickness and length; most are extremely fine so that the skin may appear hairless. There is a marked sexual difference in the distribution of coarse hair, particularly on the face and trunk, and in its loss from the scalp. This coarse hair tends to become more prominent after puberty, particularly in the axilla, over the pubes, and on the face in males.

Except for the eyelashes, all hairs emerge obliquely from the skin surface, with the hairs in any one region doing so in the same direction. The part of the hair which projects from the skin surface is the shaft and that under the skin is the root, which is ensheathed in a sleeve of epidermis known as the *follicle* extending into the subcutaneous tissue (Fig. 1.20B). The shaft appears circular in cross-section. Throughout most of its length the hair consists of the keratinized remains of cells. Hair colour is due to pigment in the hair cells (melanin and a subtle red pigment), and to air within the shaft of the hair. The hair of the head has a life span of between 2 and 4 years, while that of the eyelashes is only 3–5 months. All hairs are intermittently shed and replaced.

In the growing hair, the deepest part of the hair follicle expands to form a cap, known as the bulb of the hair, which almost completely surrounds some loose, vascular connective tissue, known as the papilla. The cells of the follicle around the papilla proliferate to form the various layers of the hair. In the resting hair follicle the bulb and papilla shrink, with the deepest part of the follicle being irregular in shape.

Associated with each hair are one or more sebaceous glands, which lie in the angle between the slanting hair follicle and the skin surface with their ducts opening into the neck of the follicle. Bundles of smooth muscle fibres (the arrector pili) attach to the sheath of the hair follicle, deep to the sebaceous gland, and pass to the papillary layers of the dermis on the side towards which the hair slopes (Fig. 1.20B). Contraction of the muscle fibres causes the hair to stand away from the skin, elevating the skin around the opening of the hair follicles, producing 'goose flesh'. This action also compresses the sebaceous glands causing them to empty their secretions onto the skin surface. Elevation of the hairs traps a layer of air against the skin surface in an attempt to produce an insulating layer to reduce heat loss, while the sebaceous secretions are important in 'waterproofing' the skin surface and in aiding the absorption of fat-soluble substances through the skin.

Glands

Sebaceous glands

These are associated with all hairs and hair follicles, there being between one and four associated with each hair. They may also exist where there is no hair, such as the corner of the mouth and adjacent mucosa, the lips, the areola and the nipple, opening directly onto the skin surface. However, they are absent from the skin of the palm and sole and the dorsum of the distal segments of the digits. The glands vary in size between 0.2 and 2.0 mm in diameter. The cells of the glands are continuously destroyed in the production of the oily secretions, known as sebum. This mode of secretion production is known as *holocrine secretion*.

Inflammation and accumulation of secretion within the sebaceous glands give rise to acne. If plugging of the outlet is permanent, a sebaceous cyst may be formed in the duct and follicles. These may become so enlarged that they require surgical removal. Sebaceous glands do not appear to be under nervous control.

Sweat glands

These have a wide distribution throughout the body (Fig. 1.23), being more numerous on its exposed parts, especially on the palms, soles and flexor surfaces of the

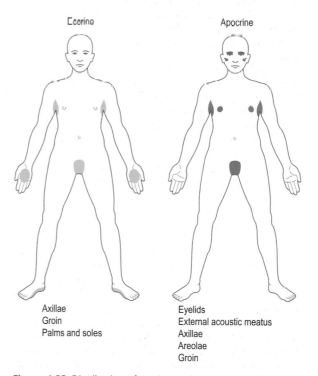

Eccrine	Apocrine
Axillae	Eyelids
Groin	External acoustic meatus
Palms and soles	Axillae
	Areolae
	Groin

Figure 1.23 Distribution of eccrine and apocrine sweat glands.

digits. Here the ducts open onto the summits of the epidermal ridges. Each gland has a long tube extending into the subcutaneous tissue, where it becomes coiled, forming the secretory body of the gland (Fig. 1.20B). The glands produce sweat, which is a clear fluid without any cellular elements, for secretion (*eccrine secretion*). The production of sweat is important in temperature regulation, as its evaporation from the skin surface promotes heat loss. These eccrine sweat glands are innervated by sympathetic nerves. Consequently, any disturbance in the sympathetic nervous system will result in a dry warm skin (anhydrosis) either locally or extensively.

In the axilla, groin and around the anus there are large modified sweat glands between 3 and 5 mm in diameter lying deeply in the subcutaneous layer. Their ducts may be associated with a hair follicle or they may open directly onto the skin surface. The secretions of the glands include some disintegration products of the gland cells (*apocrine secretions*). The odour associated with these glands is not from the secretion itself, but is due to bacterial invasion and contamination from the skin. Pigment granules associated with axillary glands produce a slight coloration of the secretion. The apocrine glands vary with sexual development, enlarging at puberty. In females they show cyclical changes associated with the menstrual cycle.

The glands which open at the margins of the eyelid (*ciliary glands*) are modified, uncoiled sweat glands, as are the glands of the external auditory meatus (*ceruminous glands*). The cells of the ceruminous glands contain a yellowish pigment which colours the wax secretion (*cerumen*).

Mammary gland (breast)

The mammary glands are modified sweat glands, being accessory to the reproductive function in females, secreting milk (lactation) for the nourishment of the infant. In children prior to puberty and in adult males, the glands are rudimentary and functionless.

Blood supply and lymphatic drainage of the skin

The arterial supply of the skin is derived from vessels in the subcutaneous connective tissue layer, which form a network at the boundary between the dermis and subcutaneous tissue (Fig. 1.20B). Branches from the network supply the fat, sweat glands and deep parts of the hair follicles. Branches within the dermis form a subpapillary plexus. The epidermis is avascular. Abundant arteriovenous anastomoses occur within the skin. Lymphatics of the skin begin in the dermal papillae as networks or blind outgrowths which form a dense mesh of lymphatic capillaries in the papillary layer. Larger lymphatic vessels pass deeply to the boundary between the dermis and subcutaneous tissue to accompany the arteries as they pass centrally.

Nerves of the skin

The nerves of the skin are of two types, afferent somatic fibres mediating pain, touch, pressure, heat and cold (general sensations), and efferent autonomic (sympathetic) fibres supplying blood vessels, arrector pili and sweat glands. The sensory (afferent) endings have several forms. Free nerve endings extend between cells of the basal layer of the epidermis, terminating around and adjacent to hair follicles. They are receptive to general tactile sensation as well as painful stimuli. Enclosed tactile corpuscles lie in the dermal papillae, being sensitive to touch. Pacinian corpuscles (Figs. 1.20B and 1.21) exist in the subcutaneous tissue, being particularly plentiful along the sides of the digits, and act as pressure receptors. Specific endings for heat and cold have been described, although general agreement as to their identity has not been reached. Details of all of the above receptors are given on pages 25–26.

Application

The majority of physiotherapy techniques are applied either directly or indirectly via the patient's skin. Manual manipulations, such as massage, and thermal treatments both have an effect on the skin. The skin provides an extremely important barrier as it restricts the penetration of damaging electromagnetic radiations in the ultraviolet spectrum. All but the very longest ultraviolet wavelengths are absorbed by the skin, and if sufficiently high levels of ultraviolet have been absorbed then the characteristic effects of erythema, thickening of the epidermis, increased pigmentation and finally peeling will all be seen.

The general dryness and natural greasiness of the skin surface mean that it has a high electrical resistance. Consequently, if electrical currents are to be applied directly to body tissues this resistance must be reduced. This is usually successfully achieved by the application of moist pads or conducting gels to the skin below the site of electrode attachment.

Section summary

Skin

- A tough pliable waterproof covering of the body endowed with sensory receptors for pain, pressure, touch and temperature.
- On average it is between 1 and 2 mm thick, being thicker over posterior and extensor surfaces.
- Consists of a superficial epidermis (0.3–1.0 mm thick) and an intimately related deeper dermis overlying a variable amount of subcutaneous connective tissue.
- Nails are modifications of the two superficial layers of the epidermis.

- Hairs are widely distributed over the skin, emerging at an angle from the surface.
- Sebaceous glands release oily secretions onto the skin surface; sweat glands are important in temperature regulation; the mammary gland (breast) is a modified sweat gland.

EARLY EMBRYOLOGY

Stages in development

The process of development begins with the penetration of the zona pellucida of the ovum (egg) by the head of the sperm – this is fertilization. It is this event which activates the ovum biochemically and leads, some 40 weeks later, to the birth of an infant. Between fertilization and birth, however, a series of complicated changes, involving both differentiation and reorganization, occur which gradually build up different tissues, organs and organ systems to create a viable individual. This account is intended as an introduction designed to help in the understanding of early human development. Unfortunately many new, and sometimes confusing, terms are used particularly when describing early development; these new terms, wherever possible, are kept to a minimum

Fertilization and cleavage

Fertilization usually occurs within the Fallopian (uterine) tubes, and it is not until some 6 days later that the resulting cell mass implants itself in the uterine wall. However, during this time considerable differentiation and organization has already taken place. Within 12 h of fertilization the male and female pronuclei have met and fused near the centre of the ovum, forming a *zygote*. Within another 12 h cleavage occurs; this consists of repeated mitotic divisions of the zygote (still within the zona pellucida) with the consequent formation of an increasing number of cells (now called *blastomeres*) without an increase in total cytoplasmic mass, which obviously becomes partitioned among the blastomeres. By the time there are 16 cells, i.e. after four divisions, the mass of cells is known as a *morula* (Fig. 1.24A).

The morula

Evidence suggests that the nuclei of individual cells come to lie in quantitatively and qualitatively different cytoplasmic environments; the initial circumstances of cellular differentiation have therefore been created. Furthermore, the blastomeres of the future inner cell mass are able to move with respect to one another. Consequently, the basis for cell movement during gastrulation and the possibility of inductive cellular interaction, resulting from the acquisition of a new microenvironment by an individual cell, are established.

At the 16-cell stage, the morula enters the uterus and the process of compaction occurs, as a result of which individual blastomeres become less distinct. Cells on the outside of the morula adhere to each other and a topographical difference is established between these surface cells and those inside. The outer cells, under the influence of the *inner cell mass*, form the *trophoectoderm*, which eventually forms the *fetal membranes*. The inner cell mass forms the *embryonic cells*, and may also contribute to the *extra-embryonic membranes*.

The blastocyst

Some 4–6 days after fertilization the morula takes in uterine fluid, which passes through the zona pellucida, eventually forming a *blastocyst cavity* (Fig. 1.24B). This separates the inner cell mass from the trophoblast except in the region of the *polar trophoblast*, which overlies the inner cell mass (Fig. 1.24B). During blastocyst formation the zona pellucida thins and is eventually shed, exposing the cells. The polar trophoblast adheres to the uterine wall and the process of implantation begins.

Implantation

During the early stages of implantation (6–8 days) the inner cell mass begins to differentiate and will eventually form the complete embryo. First, those cells facing the blastocyst cavity form a single layer of *primary embryonic endoderm*. The remaining cells form another layer which constitutes the precursor of the *embryonic ectoderm*, also giving rise later to the *embryonic mesoderm* (Fig. 1.24C). The *amniotic cavity* appears between these cells and an overlying layer of cells which form the *amniotic ectoderm*, derived from the deep aspect of the polar trophoblast (Fig. 1.24C). The trophoblast has formed the *cytotrophoblast* surrounding the *blastocyst cavity*, and the *syncytiotrophoblast* which penetrates the endometrial lining of the uterus (Fig. 1.24C). Thus the inner cell mass, from which the cells of the future embryo arise, forms no more than a *bilaminar disc* while the remainder of the blastocyst forms the fetal membranes.

Development of the yolk sac

While the amniotic cavity is being formed, the blastocyst cavity becomes lined by cells of *extra-embryonic endoderm*. The blastocyst wall now consists of three layers: an outer trophoblast (the extra-embryonic ectoderm) consisting of cytotrophoblast and syncytiotrophoblast, a loose reticular layer of *extra-embryonic mesoderm* and an inner cell layer of *extra-embryonic endoderm*. The space remaining is the *yolk sac* (Fig. 1.24D).

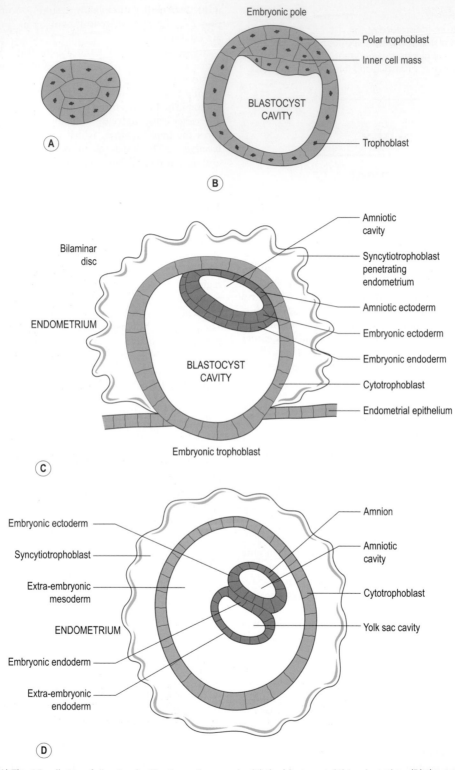

Figure 1.24 (A) The 16-cell stage following fertilization – the morula, (B) the blastocyst, (C) implantation, (D) the appearance of the yolk sac.

Prochordal plate

In a localized area in the roof of the yolk sac the *prochordal plate* is formed. It occurs at the cranial end of the future embryo and gives the bilaminar disc bilateral symmetry. At the same time the extra-embryonic mesoderm begins to develop fluid-filled spaces, which join together forming a large cavity surrounding the whole of the yolk sac and the amnion, except for the mesodermal *connecting stalk* (Fig. 1.25A). This space is the *extra-embryonic coelom*, which splits the extra-embryonic mesoderm into visceral and parietal layers and separates the amniotic and yolk cavities from the outer wall of the conceptus (Fig. 1.25A).

The primary germ layers

The primary germ layers as well as the supporting membranes have now been established. From the embryonic ectoderm the outer covering of the embryo is formed. This includes the outer layers of the skin and its derivatives (hair, nails); the mucous membrane of the cranial and caudal ends of the alimentary canal; the central and peripheral nervous systems, including the retina; and part of the iris of the eye. In general terms, the embryonic endoderm forms epithelial tissues in the adult, these being the epithelial lining of the alimentary canal, the parenchyma of its associated glands (liver, pancreas), the lining of the respiratory

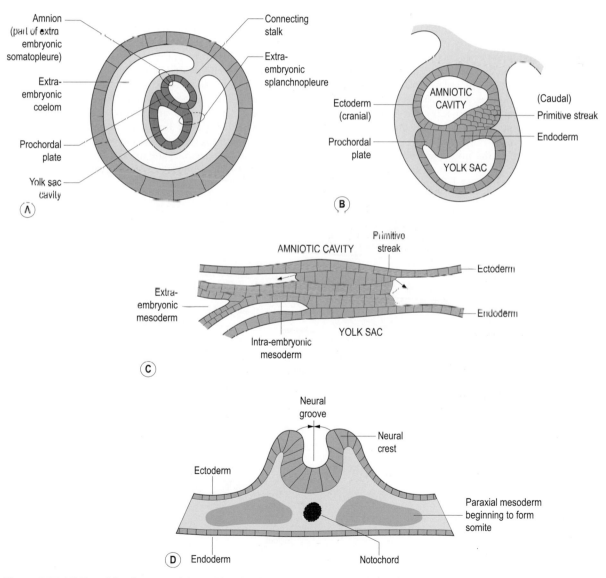

Figure 1.25 (A) Site of development of the prochordal plate, (B) further development of the prochordal plate and appearance of the primitive streak, (C) the trilaminar disc, (D) formation of the neural groove.

system, and most of the epithelium of the bladder and urethra.

About 15 days after fertilization, the trilaminar disc begins to form as a heaping up of cells in the upper layer of the bilaminar disc towards the posterior part of the midline, forming the *primitive streak* (Fig. 1.25B). The heaping up is mainly due to a medial and backward migration of actively proliferating ectodermal cells which spread laterally and forward between the ectoderm and endoderm layers as *intra-embryonic mesoderm* (Fig. 1.25C). At the lateral extremes of this migration, the mesodermal cells become continuous with the *extra-embryonic mesoderm* (splanchnopleuric) covering the yolk sac and that covering the amnion (somatopleuric). Anteriorly, the embryonic mesodermal cells become continuous across the midline in front of the prochordal plate.

Soon after the appearance of the primitive streak, which forms as a line on the surface, a further heaping of cells occurs at the anterior end. This is the *primitive knot* from which the notochordal process extends forwards to the posterior edge of the prochordal plate. The ectoderm overlying the notochordal process, as well as that immediately anterior, becomes thickened to form the *neural plate*, from which the neural tube, and eventually the brain and spinal cord, develop.

On either side of the notochord the mesoderm forms two longitudinal strips known as the *paraxial mesoderm* (Fig. 1.25D). Each of these becomes segmented, forming approximately 44 blocks of mesoderm – *somites*, none of which are formed anterior to the notochord. Lateral to the paraxial mesoderm is a thinner layer, the *lateral plate mesoderm*, which is continuous at its edges with the extra-embryonic mesoderm. Connecting the edge of the paraxial mesoderm to the lateral plate mesoderm is a longitudinal tract, the *intermediate mesoderm*, from which arises the *nephrogenic cord*.

Within the lateral plate mesoderm small, fluid-filled spaces appear which join together to form the intra-embryonic mesoderm, continuous across the midline (Fig. 1.26A). The intra-embryonic mesoderm forms the pericardial (heart), pleural (lung) and peritoneal (abdominal) cavities.

During the fourth week following fertilization the trilaminar disc bulges further into the amniotic cavity. Under the cranial and caudal parts of the disc a head and tail fold appear; at the same time marked folding occurs along the lateral margins of the embryo (Fig. 1.26B).

Further development is considered in the appropriate section as necessary. Development of the ear and eye are outlined on pages 565 and 568 respectively, the musculoskeletal system and limbs on pages 36 (upper limb) and 203 (lower limb), the cardiovascular system on page 507, the respiratory system on page 515, the digestive system on page 522, the urinary system on page 532, the genital system on page 535, and the nervous system on page 490.

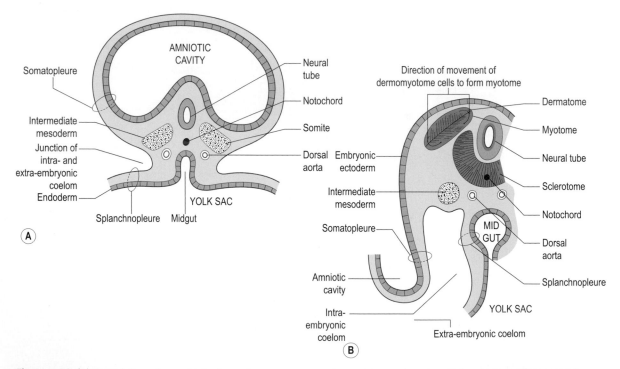

Figure 1.26 (A) Completion of neural tube formation, appearance of paraxial mesoderm (somite) and intermediate mesoderm, (B) differentiation and migration of the somite mesoderm.

Section summary

Stages in development

Time from ovulation		Event
Within	24 h	Fertilization
	72 h	Passage of conceptus through isthmus of Fallopian tubes
	80 h	Entry of conceptus into uterus
	4–6 days	Blastocyst formation
	7 days	Implantation
	9–13 days	Bilaminar embryonic disc
	14–15 days	Beginning of primitive streak; appearance of extra-embryonic coelom
	16 days	Beginning of notochord process
	17–18 days	Neural plate appears
	19–20 days	Intra-embryonic coelom begins to appear
	20–30 days	Formation of somites
	24 days	Head, tail and lateral body folds have established basic embryonic shape
	24–26 days	Limb buds appear
	5 weeks	Hands and feet begin to develop
	8 weeks	Primary ossification centres appear in long bones
	12 weeks	Formation of definitive body wall complete
	3 months	Nails appear
	4 months	Hair (lanugo) appears
	5 months	Whitish slippery coating (vernix caseosa) begins to develop – probably makes skin more waterproof
	7 months	Scalp and eyebrow hair develops
	Birth	Vertebrae in three parts (centrum and two neural arches); nails have grown to end of digits; shafts of long bones completely ossified; secondary ossification beginning to appear (distal end femur, proximal end tibia)

Part | 2 |

The upper limb

CONTENTS

INTRODUCTION

The human upper limb has almost no locomotor function, instead it is an organ for grasping and manipulating. With the evolutionary adaptation of bipedalism the upper limb acquired a great degree of freedom of movement. However, the upper limb has still retained its ability to act as a locomotor prop, as when grasping an immobile object and pulling the body towards the hand. Alternatively, it may be used in conjunction with a walking aid to support the body during gait. Nevertheless, the bones of the upper limb are not as robust as their counterparts in the lower limb.

The upper limb is attached to the trunk by the pectoral girdle, which consists of the scapula and the clavicle, the only point of articulation with the axial skeleton is at the sternoclavicular joint. The scapula rides in a sea of muscles attaching it to the head, neck and thorax, while the clavicle acts as a strut holding the upper limb away from the trunk. Between the trunk and the hand are a series of highly mobile joints and a system of levers, which enable the hand to be brought to any point in space and to hold it there steadily and securely while it performs its task. However, it is the development of the hand as a sensitive instrument of precision, power and delicacy which is the acme of human evolution. The importance of the opposability of the thumb in providing effective grasping and manipulating skills makes the hand the most efficient tool in the animal kingdom. In grasping, the thumb is equal in value to the other four digits; loss of the thumb is as disabling as the loss of all four fingers. For these skills the hand has a rich motor and sensory nerve supply. It is no coincidence that the hand has large representations in both the motor and sensory regions of the cerebral cortex. The adoption of a bipedal gait during human evolution freed the upper limbs for functions other than locomotion; this is one reason why the brain developed and enlarged to its present form. Not to be overlooked in the functional effectiveness of the hand is the important contribution made by the extensive vascular network in supporting its metabolic requirements.

As the upper limb is also used for carrying loads and supporting the body, the question arises as to how these forces are transmitted to the axial skeleton. The usual means is by tension developed in the muscles and ligaments crossing the various joints. In addition, because the upper limb itself is heavy, every movement that it makes has to be accompanied by postural contractions of the muscles of the trunk and lower limb to compensate for shifts in the body's centre of gravity.

DEVELOPMENT OF THE MUSCULOSKELETAL SYSTEM

Mesodermal somites

By the end of the third week after fertilization the paraxial mesoderm begins to divide into *mesodermal somites* which are easily recognizable during the fourth and fifth weeks (Fig. 2.1A). Eventually some 44 pairs of somites develop, although not all are present at the same time; however, the paraxial mesoderm at the cranial end of the embryo remains unsegmented. There are 4 occipital somites, followed by 8 cervical, 12 thoracic, 5 lumbar, 5 sacral and 8–10 coccygeal somites. The growth and migration of these somitic cells are responsible for the thickening of the body wall, as well as the development of bone and muscle. The deeper layers of the skin are also of somitic origin. Somite-derived tissue spreads medially to form the vertebrae, dorsally to form the musculature of the back, and ventrally into the body wall to form the ribs, and the intercostal and abdominal muscles.

Soon after its formation each somite becomes differentiated into three parts. The ventromedial part forms the *sclerotome*, which migrates medially towards the notochord and neural tube to take part in the formation of the vertebrae and ribs (Fig. 2.1A). The remainder of the somite is known as the *dermomyotome*. The cells of the dorsal and ventral edges proliferate and move medially to form the *myotome*, whose cells migrate widely and become differentiated into myoblasts (primitive muscle cells). The remaining thin layer of cells forms the *dermatome*, which spreads out to form the dermis of the skin.

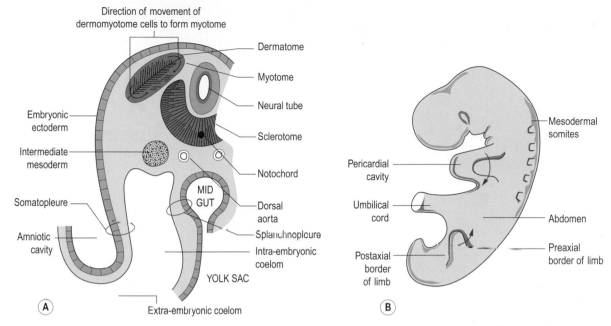

Figure 2.1 (A) Somites in the developing embryo, (B) the development of the upper limb showing limb buds and the direction they will rotate.

The myotome of each somite receives a single spinal nerve which innervates all the muscle derived from that myotome, no matter how far it eventually migrates. The dorsal aortae lie adjacent to the somites and give off a series of intersegmental arteries which lie between them.

Development of the limbs

The limbs initially appear as flipper-like projections (the limb buds), with the forelimbs appearing first, between 24 and 26 days, each bud consisting of a mass of mesenchyme covered by ectoderm with a thickened ectodermal ridge at the tip; the ectodermal ridge controls normal development of the limb. Consequently, damage to it will result in some trauma to the limb. At the beginning of the second month the elbow and knee prominences project laterally and backwards. At about the same time, the hand and foot plates appear as flattened expansions at the end of the limb bud. Between 36 and 38 days, five radiating thickenings forming the fingers and toes can be distinguished, the webs between the thickenings disappear freeing the digits. Appropriate spinal nerves grow into the limbs in association with migration of the myotomes: C5, 6, 7, 8 and T1 for the upper limb, and L4, 5, S1, 2 and 3 for the lower limb. The limb bones differentiate from the mesenchyme of the bud. The limbs grow in such a way that they rotate in opposite directions, the upper limb laterally and lower limb medially (Fig. 2.1B). Consequently, the thumb becomes the lateral digit of the hand, while the great toe is the medial digit of the foot.

During development the upper limb bud appears as a swelling from the body wall (Fig. 2.2(i)) at the level of the lower cervical and first thoracic segments. At first the limb buds project at right angles to the surface of the body, having ventral and dorsal surfaces and cephalic (preaxial) and caudal (postaxial) borders (Fig. 2.2(ii)). As the limb increases in length it becomes differentiated, during which time it is folded ventrally so that the ventral surface becomes medial (Fig. 2.2(iii)), with the convexity of the elbow directed laterally (Fig. 2.2(iv)). At a later stage the upper and lower limbs rotate in opposite directions so that the convexity of the elbow is directed towards the caudal end of the body (Fig. 2.2(v)). As the limb bud develops, the primitive muscle mass becomes compartmentalized, foreshadowing the adult pattern. Intermuscular septa, extending outwards from the periosteum of the humerus, divide the arm into anterior and posterior compartments. As in the lower limb, some of the anterior compartment musculature has become separated into an adductor group. However, in the upper limb this muscle mass has degenerated phylogenetically so that all that remains is coracobrachialis. Adduction of the upper limb is a powerful action in humans, being served by great sheets of muscle that have migrated into it; these are latissimus dorsi posteriorly and pectoralis major anteriorly. In the forearm, the radius (preaxial bone) and the ulna (postaxial bone) are connected by an interosseous membrane, and to the investing fascia by intermuscular septa. The compartments so formed enclose muscles of similar or related functions.

With the upper limb in the anatomical position the anterior preaxial compartments are in a continuous plane with the muscles being supplied by branches from the

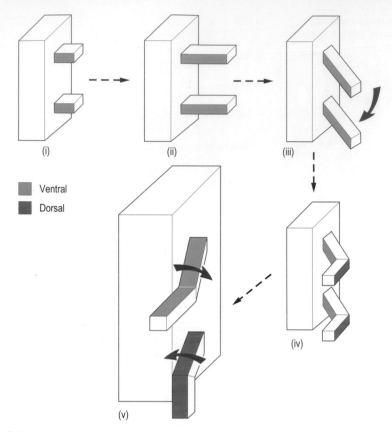

(i)　(ii)　(iii)

Ventral

Dorsal

(iv)

(v)

Figure 2.2 Rotation of the limbs during development.

lateral and medial cords of the brachial plexus, which are all derived from the anterior divisions of the nerve trunks. Similarly, the posterior postaxial compartment muscles are all supplied by branches of the posterior cord, derived from the posterior divisions of the nerve trunks.

The median, musculocutaneous and ulnar nerves are responsible for preaxial innervation, whereas the radial nerve supplies all of the postaxial musculature of the upper limb below the shoulder. Within the pectoral girdle the clavicle is the anterior preaxial bone and the scapula, with the exception of the coracoid process (which is also an anterior bone), is the posterior postaxial bone. The distinction with respect to the coracoid process is that phylogenetically it is a separate bone; its fusion with the scapula is secondary. Consequently, muscles arising from the clavicle or coracoid process belong to the preaxial group, and are therefore supplied by preaxial branches of the brachial plexus. Similarly, muscles arising from the remainder of the scapula are part of the postaxial group and are innervated by postaxial branches of the plexus.

Furthermore, there is a serial arrangement of the nerves in the brachial plexus with respect to both their motor innervation and their sensory supply. The order is retained from the primitive serial morphology of the embryo.

Remembering that the skin has essentially been stretched over the developing limb, the fifth cervical nerve in the adult is sensory to the cranial part of the limb, and the first thoracic to its caudal part, with the seventh cervical nerve lying in the middle of the limb. The pattern of motor innervation, in simple terms, progresses from C5 for shoulder movements to T1 for intrinsic hand movements, with the elbow being served by C5 and 6, the forearm by C6, the wrist by C6 and 7, and the fingers and thumb by C7 and 8.

As in the lower limb, many muscles cross two or more joints, and can therefore act on all of them. Consequently, a complex system of synergists and fixators is required in order to prevent or restrict unwanted movements. Procedures for testing for loss of muscle action, as in paralysis, can thus be quite complicated.

The upper limit of the upper limb is not so easily defined as in the lower limb. In spite of muscular attachments to the head, neck and thorax, the upper limits can be conveniently considered as the superior surface of the clavicle anteriorly and the superior border of the scapula posteriorly. The free upper limb is divided into the arm between the shoulder and elbow, the forearm between the elbow and the wrist, and the hand beyond the wrist. The hand has an anterior or palmar surface, and a posterior or dorsal surface (Fig. 2.3).

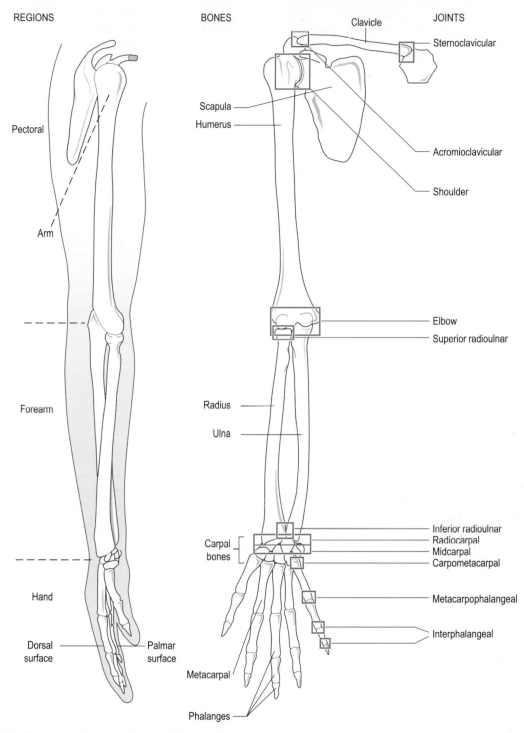

REGIONS

Pectoral

Arm

Forearm

Hand

Dorsal
surface

Palmar
surface

BONES

Scapula
Humerus

Radius

Ulna

Carpal
bones

Metacarpal

Phalanges

JOINTS

Clavicle

Sternoclavicular

Acromioclavicular

Shoulder

Elbow
Superior radioulnar

Inferior radioulnar
Radiocarpal
Midcarpal
Carpometacarpal

Metacarpophalangeal

Interphalangeal

Figure 2.3 The regions, bones and joints of the upper limb.

The bones of the upper limb are the *clavicle* and *scapula* of the pectoral girdle, the *humerus* in the arm, the lateral *radius* and medial *ulna* in the forearm, the eight *carpal* bones of the wrist, the five *metacarpals* of the hand and the *phalanges* of the digits – two in the thumb and three in each finger (Fig. 2.3).

BONES

The scapula

This is a large, flat, triangular plate of bone on the posterolateral aspect of the thorax, overlying the second to the seventh ribs. Suspended in muscles, the scapula is held in position by the strut-like clavicle, but retains great mobility relative to the thorax. Being a triangular bone it has three angles, three borders and two surfaces which support three bony processes (Figs. 2.4 and 2.5A).

The costal surface which faces the ribs is slightly hollowed and ridged with a smooth, narrow strip along its entire medial border. It is also known as the *subscapular fossa*.

The dorsal surface faces posterolaterally and is divided by the *spine* of the scapula into a smaller *supraspinous fossa* above and a larger *infraspinous fossa* below. The supraspinous and infraspinous fossae communicate via the *spinoglenoid notch* between the lateral end of the spine and the neck of the scapula. The spine of the scapula has upper and lower free borders which diverge laterally enclosing the *acromion*.

The thin *medial border* lies between the inferior and superior angles, being slightly angled at the medial end of the spine. The *lateral border* is thicker, being deeply invested in muscles, and runs down from the *infraglenoid tubercle* below the *glenoid fossa* to meet the medial border at the *inferior angle*. The *superior border*, which is thin and sharp, is the shortest, and has the *suprascapular notch* at the junction with the root of the *coracoid process*.

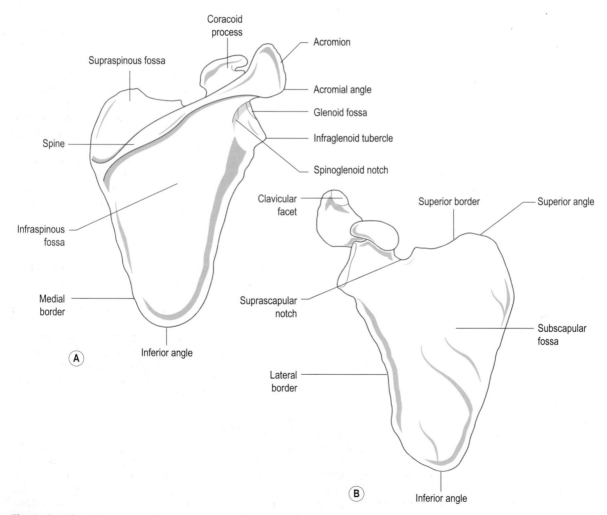

Figure 2.4 The right scapula: (A) posterior view, (B) anterior view.

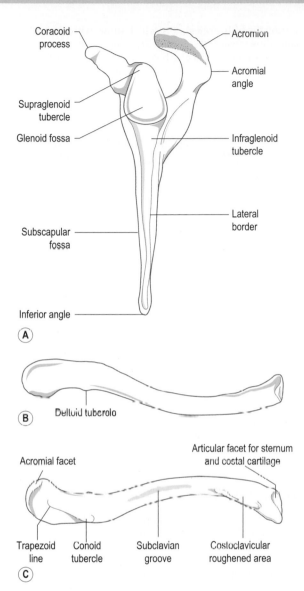

Figure 2.5 (A) Left scapula, lateral view, (B) right clavicle, superior view, (C) right clavicle, inferior view.

Inferiorly the thick inferior angle lies over the seventh rib and is easily palpated. The *superior angle* lies at the junction of the medial and superior borders, whilst the *lateral angle* is truncated and broadened to support the head and *glenoid fossa* of the scapula.

The head of the scapula is an expanded part of the bone joined to a flat blade by a short inconspicuous neck. The glenoid fossa (or cavity) is found on the head as a shallow, pear-shaped concavity, facing anterolaterally. The glenoid fossa is broader below and articulates with the head of the humerus forming the shoulder (glenohumeral) joint. Immediately above the glenoid fossa is the *supraglenoid tubercle*.

The *acromion*, which is the expanded lateral end of the spine, is large and quadrilateral, projecting forwards at right angles to the spine. The lower border of the crest of the spine continues as the lateral border of the acromion; the junction of these two borders forms the palpable *acromial angle*. The upper border of the crest becomes continuous with the medial border of the acromion and presents an oval facet for articulation with the clavicle at the acromioclavicular joint. The superior surface of the acromion is flattened and subcutaneous.

The *coracoid process* is a hook-like projection with a broad base directed upwards and forwards from the upper part of the head, and a narrow more horizontal part which passes anterolaterally from the upper edge of the base. The tip lies below the junction of the middle and lateral thirds of the clavicle.

Palpation

Starting at the lowest point, the inferior angle can readily be gripped between thumb and index finger and, if the subject relaxes sufficiently, can be lifted away from the thorax. The medial border can be followed along its whole length from inferior to superior angles. The spine of the scapula can be palpated as a small triangular area medially increasing in size as the fingers move laterally along it. The flat crest with its upper and lower borders can be identified. Continuing along the lower border of the crest to its most lateral point, the sharp 90° acromial angle can be felt; this continues as the palpable lateral border of the acromion. Running onto this lateral border, the flat upper surface of the acromion can be felt above the shoulder joint. The coracoid process can be palpated as an anterior projection below the junction of the middle and lateral thirds of the clavicle, and therefore is a useful reference point for surface marking the shoulder joint as it lies just medial to the joint line.

Ossification

The scapula ossifies from a number of centres. The primary ossification centre appears in the region of the neck by the eighth week *in utero*, so that at birth the coracoid process, acromion, glenoid cavity, medial border and inferior angle are still cartilaginous. Secondary centres appear in each of these regions, except the coracoid, between the ages of 12 and 14 years, fusing with the body between 20 and 25 years. The secondary centre for the coracoid process, however, appears during the first year and fuses with the body between 12 and 14 years.

The clavicle

This is a subcutaneous bone running horizontally from the sternum to the acromion (Fig. 2.5B,C). It acts as a strut holding the scapula laterally, thus enabling the arm to be clear of the trunk – an essential feature in primates. The scapula and clavicle together form the pectoral (shoulder) girdle, transmitting the weight of the upper limb to the axial skeleton and facilitating a wide range of movement of the upper limb.

The medial two-thirds of the clavicle is convex forwards and is roughly triangular in cross-section. The lateral third is concave forwards and flattened from above downwards. The medial convexity conforms to the curvature of the superior thoracic aperture, the lateral concavity to the shape of the shoulder.

The lateral (acromial) end of the clavicle is the most flattened part and has a small *deltoid tubercle* on its anterior border. Inferiorly the rounded *conoid tubercle* is present at the posterior edge, with the rough *trapezoid line* (Fig. 2.5C) running forwards and laterally away from it. The conoid tubercle and trapezoid line give attachment to the conoid and trapezoid parts of the coracoclavicular ligament binding the clavicle and scapula together. Laterally is a small oval facet for the acromion: it faces obliquely downwards and laterally.

The medial (sternal) end of the clavicle is enlarged and faces downwards and medially. The lower three-quarters is bevelled and articulates with the clavicular notch of the manubrium and the costal cartilage of the first rib, forming the sternoclavicular joint. The cylindrical clavicle projects above the shallow notch on the sternum; this can be confirmed by palpation. The superior quarter of the sternal end is roughened for the attachment of the intra-articular disc and ligaments. Between the lateral and medial ends, the superior surface is smooth, while the inferior surface is marked by a rough *subclavian groove* centrally and a large oval roughened area for the *costoclavicular* ligament medially. The anterior and posterior borders are roughened by muscle attachments.

The clavicle is often fractured by the direct violence of a blow, or by indirect forces transmitted up the limb following a fall on the outstretched arm. The fracture usually occurs at the junction of the two curvatures, and the resultant fracture appearance is caused by the weight of the arm pulling the shoulder downwards and medially so that the medial fragment of the clavicle overrides the lateral at the fracture site.

Palpation

In a slender subject the whole length of the clavicle can often be seen directly beneath the skin. Initially, the enlarged medial end of the clavicle can be palpated with the fingers, with the line of the sternoclavicular joint also being identified. Moving laterally, almost the whole length of the shaft of the clavicle can be gripped between finger and thumb. At the lateral end, the bulk of deltoid may require deeper pressure; nevertheless the line of the acromioclavicular joint should be palpable, particularly from above.

Ossification

The clavicle ossifies in membrane, being the first bone in the body to begin ossification. Two primary centres appear during the fifth week *in utero* which unite, with ossification spreading towards the ends of the bone. A secondary centre appears in the medial end between 14 and 18 years, fusing with the main part of the bone as early as 18–20 years in females and 23–25 years in males. An additional centre may appear in the lateral end at puberty; however, it soon fuses with the main bone.

Section summary

Bones of the pectoral girdle

Scapula

- Triangular bone located on the posterolateral aspect of the thorax.
- Has superior, lateral and inferior angles; medial, lateral and superior borders; posterior spine; laterally projecting acromion; anteriorly projecting coracoid process.
- Articulates with head of humerus at glenoid fossa forming the shoulder (glenohumeral) joint; clavicle at acromion forming acromioclavicular joint.

Clavicle

- Curved bone – medially convex anteriorly, laterally concave anteriorly.
- Has smooth superior surface; roughened inferior surface; expanded medial and flattened lateral ends.
- Articulates with acromion of scapula laterally forming acromioclavicular joint; sternum medially forming sternoclavicular joint.

The humerus

Largest bone in the upper limb (Fig. 2.6), being a typical long bone with a shaft (body) and two extremities (ends). Proximally, it articulates with the glenoid fossa of the scapula forming the shoulder joint, and distally with the radius and ulna forming the elbow joint.

Proximally, the major feature is the almost hemispherical *head* of the humerus with its smooth, rounded articular surface facing upwards, medially and backwards; it is considerably larger than the socket formed by the glenoid fossa. The head is joined to the upper end of the shaft by the *anatomical neck*, a slightly constricted region encircling the bone at the articular margin, separating it from the tubercles.

The *greater tubercle* is a prominence on the upper lateral part of the bone, next to the head. It merges with the shaft below and is marked by three distinct impressions for muscular attachment. The greater tubercle projects laterally past the margin of the acromion and is the most lateral bony point at the shoulder.

The smaller *lesser tubercle* is a distinct prominence on the anterior aspect below the anatomical neck. It has a

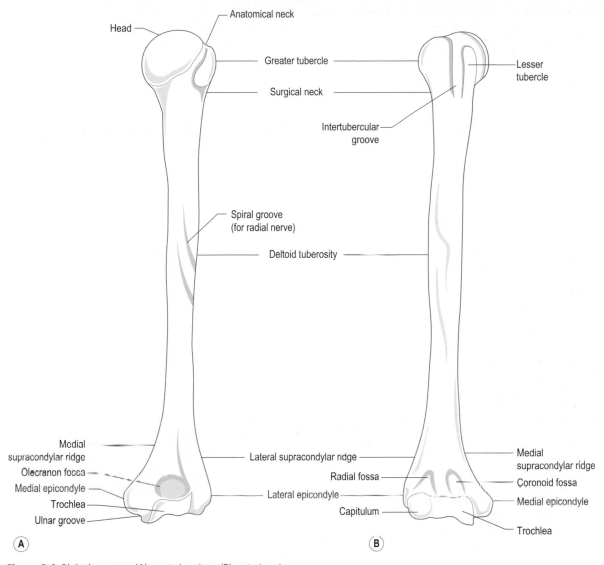

Figure 2.6 Right humerus: (A) posterior view, (B) anterior view.

well-marked impression on its medial side for muscular attachment. Between these two tubercles, and passing onto the shaft of the humerus, is the deep *intertubercular groove (sulcus)*. The crests of the greater and lesser tubercles continue down from the anterior borders of the tubercles to form the lateral and medial lips of the groove. Between the two lips is the floor of the groove.

Below where the head and tubercles join the shaft there is a definite constriction. This region is termed the *surgical neck* because fractures often occur here, particularly in the elderly.

The shaft of the humerus is almost cylindrical above, becoming triangular in its lower part with distinct medial and lateral borders. It presents three borders (anterior, medial,

lateral), although they are frequently rounded and indistinct, between which are the three surfaces (anteromedial, anterolateral, posterior) of the shaft. The intertubercular groove is continuous with the anteromedial surface, the medial border beginning as the crest of the lesser tubercle and ending by curving towards the medial epicondyle. The smooth anterolateral surface is marked about its middle by the *deltoid tuberosity*. The posterior surface is crossed obliquely from superomedial to inferolateral by the *spiral (radial) groove*, which reaches the lateral border below the deltoid tuberosity, but is often poorly marked.

The lower end of the humerus is expanded laterally, flattened anteroposteriorly, and curves slightly forwards. It presents two articular surfaces separated by a ridge. The lateral articular surface, the *capitulum*, is situated anteroinferiorly and is a rounded, convex surface, being less than a hemisphere in size. The capitulum articulates with the radius, making its greatest contact with the radius when the elbow is fully flexed.

Medial to the capitulum is the *trochlea*, the articular surface for the ulna. The trochlea is a grooved surface rather like a pulley, the medial edge projecting further distally and anteriorly than the lateral. This causes the ulna also to project laterally and results in a carrying angle between the humerus and ulna (see Fig. 2.79B).

On the medial side of the trochlea is the large *medial epicondyle*; its posterior surface is smooth with a shallow groove for the ulnar nerve. The sharp *medial supracondylar ridge*, comprising the lower third of the medial border, runs upwards onto the shaft. On the lateral side of the capitulum is the *lateral epicondyle* with the *lateral supracondylar ridge*, comprising the lower third of the lateral border, also running upwards onto the shaft.

Just above the articular surfaces, the lower end of the humerus presents three fossae for the bony processes of the radius and ulna. Posteriorly is the deep *olecranon fossa*, which on full extension of the elbow receives the olecranon process of the ulna. Anteriorly, there are two fossae, the lateral *radial* and medial *coronoid fossae*, which on full flexion of the elbow receive the head of the radius and coronoid process of the ulna respectively. Many of the bony features previously described can be seen in Figs. 2.7 and 2.9.

Palpation

At the upper end of the humerus the most lateral bony point at the shoulder is the greater tubercle, whose quadrilateral superior, anterior and posterior surfaces can be felt. Further differentiation can be made by palpating the lateral margin of the acromion (p. 41) and then running the fingers off its edge onto the greater tubercle. The rounded lesser tubercle can be felt through the deltoid, and is just lateral to the tip of the coracoid process. To the lateral side of the lesser tubercle the impression of the intertubercular sulcus can usually be felt. The shaft of the humerus is covered with thick muscle, but can be palpated on its medial and lateral sides. At the lower end, the prominent medial epicondyle is the most obvious bony landmark. The ulnar nerve can be rolled in the groove behind it (the 'funny bone'). Running upwards from the medial epicondyle the sharp medial supracondylar ridge can be palpated. The lateral epicondyle can be palpated at the base of a dimple on the lateral aspect of the elbow, as can the lateral supracondylar ridge running upwards from it. Posteriorly, the olecranon fossa can be felt through the triceps tendon, if the relaxed elbow is flexed.

Ossification

A primary ossification centre appears in the shaft in the eighth week *in utero* and spreads until, at birth, only the ends are cartilaginous. Secondary centres appear in

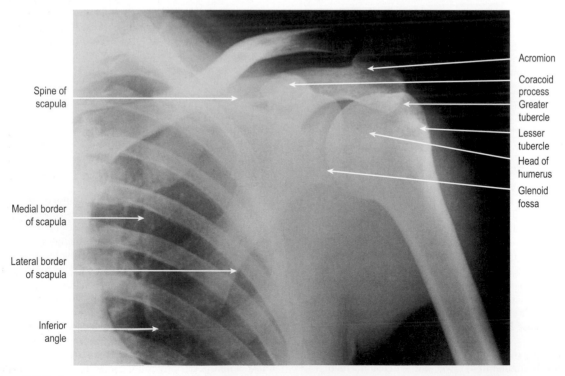

Spine of scapula

Medial border of scapula

Lateral border of scapula

Inferior angle

Acromion

Coracoid process

Greater tubercle

Lesser tubercle

Head of humerus

Glenoid fossa

Figure 2.7 Radiograph of left shoulder, anteroposterior view with the arm laterally rotated and slightly abducted.

the head early in the first year, in the greater tubercle at about 3 years and in the lesser tubercle at about 5 years. These fuse to form a single cap of bone between the ages of 6 and 8 years, finally fusing with the shaft between 18 and 20 years in females and 20 and 22 years in males.

At the lower end of the humerus, secondary centres appear for the capitulum during the second year, for the trochlea between 9 and 10 years, and for the lateral epicondyle between 12 and 14 years. These join together at about 14 years, fusing with the shaft at 15 years in females and 18 years in males. A separate centre for the medial epicondyle appears between 6 and 8 years and fuses between 15 and 18 years with a spicule of bone projecting down from the shaft medial to the trochlea. This latter ossification centre lies entirely outside the joint capsule. Most of the growth in length of the humerus occurs at its upper end.

The bones of the forearm are the *radius* laterally and the *ulna* medially, which articulate proximally with the humerus at the elbow joint and contribute to the wrist joint distally. They are connected by a strong interosseous membrane between their shafts, and synovial pivot joints at each end. Both are long bones, the ulna being expanded proximally and the radius distally (Fig. 2.8A,B). The shaft of the radius is convex laterally, allowing it to move around the ulna carrying the hand with it in pronation of the forearm.

The radius

The radius lies lateral to the ulna and is the shorter of the two bones. It articulates proximally with the capitulum of the humerus, distally with the scaphoid and lunate bones of the proximal row of the carpus, and at each end with the ulna. It has a shaft and two ends, the inferior being the larger.

The *head* is a thick disc with a concave superior surface for articulation with the capitulum. The outer, articular surface of the head is flattened, articulating with a fibro-osseous ring formed by the radial notch of the ulna and the annular ligament. Below the head is the constricted *neck*, which slopes medially as it approaches the shaft. Where the shaft joins the neck it is round, but it becomes triangular lower down. Together with the neck, the shaft has a slight medial convexity in its upper quarter, with a lateral convexity in its remaining lower part. The *radial tuberosity* lies anteromedially on the upper part of the shaft at the maximum convexity of the medial curve. The majority of the shaft presents three borders (anterior, posterior, interosseous) and three surfaces (lateral, anterior, posterior).

The sharp interosseous border, to which the interosseous membrane attaches, faces medially. It extends from just below the radial tuberosity to the medial side of the lower end of the radius, splitting into two ridges which become continuous with the anterior and posterior margins of the ulnar notch. The anterior and posterior borders pass obliquely downwards and laterally from either side of the radial

tuberosity to the roughened area for pronator teres lower down. The anterior border becomes distinct lower down, while the posterior border becomes more rounded. These borders enclose the lateral, anterior and flatter posterior surfaces.

The inferior end of the radius is expanded having five distinct surfaces. The lateral surface, which extends down to the *styloid process*, has a shallow groove anteriorly for the tendons of the abductor pollicis longus and extensor pollicis brevis. The medial surface forms the concave ulnar notch for articulation with the head of the ulna: it has a roughened triangular area superiorly. The posterior surface is convex and grooved by tendons. The prominent ridge in the middle of this surface is the *dorsal (Lister's) tubercle*. The lateral half of this surface continues down onto the styloid process. The anterior surface is smooth and curves forward to a distinct anterior margin. The distal articular surface is concave and extends onto the styloid process; it is divided by a ridge into two areas, a lateral triangular area for articulation with the scaphoid and a medial quadrilateral area for the lunate (Fig. 2.8C).

Palpation

The head of the radius can be palpated in a 'dimple' on the posterolateral aspect of the elbow, particularly when the elbow joint is extended as it overhangs the capitulum; it can be felt rotating during pronation and supination. The shaft of the radius can be palpated on the lateral side in the lower half of the forearm. Distally, on the posterior aspect, the dorsal tubercle can be identified above the wrist, as can the styloid process laterally between the extensor tendons of the thumb within the 'anatomical snuffbox'.

Ossification

A primary ossification centre appears in the shaft during the eighth week *in utero*, so that at birth only the head, inferior end and radial tuberosity are cartilaginous. The first secondary centre appears in the inferior end during the first year of life, fusing with the shaft between the ages of 20 and 22 years. The secondary centre for the head appears at about 6 years and fuses with the shaft between 15 and 17 years. A secondary centre usually appears in the radial tuberosity between 14 and 15 years, but soon fuses with the shaft.

The ulna

The ulna lies medial to the radius and is the longer of the two bones. It has a shaft and two ends, of which the superior is the larger presenting as a hook-like projection for articulation with the trochlea of the humerus. The smaller rounded distal end is the *head* of the ulna (Fig. 2.8B): it does not articulate directly with the carpus. The ulna articulates laterally at each end with the radius.

The large upper end of the ulna has two projecting processes, enclosing a concavity. The *olecranon process* is

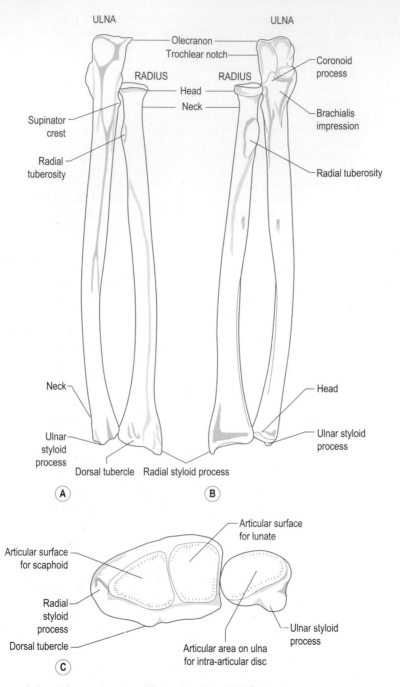

Figure 2.8 Right radius and ulna: (A) posterior view, (B) anterior view, (C) inferior view.

the larger of the two processes and forms the proximal part of the bone. It is beak-shaped and is directed forwards, being continuous inferiorly with the shaft. Posteriorly, it is smooth and subcutaneous while anteriorly it is concave, forming the upper part of the articular surface of the *trochlear notch*. The borders of the olecranon are thickened and rough.

The *coronoid process* projects from the front of the shaft and has an upper articular surface which completes the trochlear notch (Figs. 2.8B and 2.9). These two surfaces

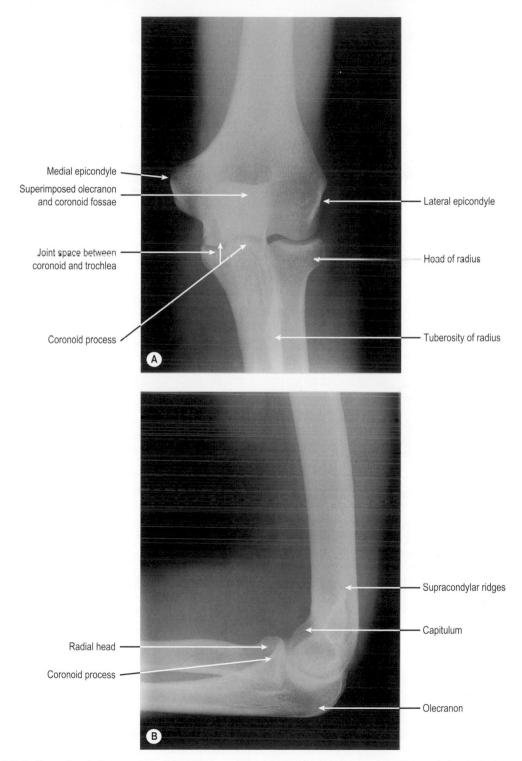

Medial epicondyle

Superimposed olecranon
and coronoid fossae

Lateral epicondyle

Joint space between
coronoid and trochlea

Head of radius

Coronoid process

Tuberosity of radius

A

Supracondylar ridges

Capitulum

Radial head

Coronoid process

Olecranon

B

Figure 2.9 Radiographs of elbow region of the right arm: (A) posterior view in full extension, (B) lateral view in flexion.

are often separated by a roughened non-articular area running horizontally across the notch. The trochlear notch is divided by a vertical ridge into a larger medial part and a smaller lateral part, the latter being continuous over its outer edge with the articular surface of the radial notch on the lateral side of the coronoid process. There is a small tubercle where the medial and anterior edges of the articular surface of the coronoid process meet; this gives attachment to the anterior part of the ulnar collateral ligament. The irregular, anterior surface of the coronoid ends inferiorly as the rough tuberosity of the ulna. Both this surface and the tuberosity give attachment to brachialis. At the upper medial part of the coronoid is the small sublime tubercle from which the pronator ridge runs downwards and laterally.

On the lateral side of the coronoid process, the concave radial notch receives the head of the radius. Below this and extending onto the shaft is the triangular supinator fossa bound posteriorly by the distinct *supinator crest*. The medial border of this area forms a prominent ridge which has a small tubercle at its upper end.

The prominent interosseous border, to which the interosseous membrane attaches, runs down from the apex of the supinator fossa. The anterior border runs down from the medial margin of the coronoid process but is indistinct. The sinuous, subcutaneous posterior border, prominent in its upper part, is continuous with the subcutaneous region of the olecranon and upper part of the shaft. Between these borders are three surfaces, the anterior and medial being continuous at the rounded anterior border. The lower quarter of the anterior surface is marked by an oblique ridge running downwards and medially. On the posterior surface, an oblique ridge runs downwards and backwards from the radial notch to the posterior border. The remaining posterior surface has faint ridges laterally and is smooth medially.

The lower end of the ulna has a narrowed neck which expands into a small, rounded head. From the posteromedial part of the head the conical *styloid process* projects downwards. The head has a smooth articular surface for the radius on its anterior and lateral aspects. The distal surface of the head is smooth and almost flat, and articulates with an articular disc which intervenes between it and the triquetral (one of the carpal bones).

Palpation

At the upper and posterior aspect of the elbow the outline of the olecranon can be identified; it forms the 'point' of the elbow seen in flexion. Running downwards from this point the posterior border can be palpated throughout its length. At the lower end, the neck, head and styloid process can all be palpated, with the styloid process being the most posterior. When the forearm is fully pronated the rounded head of the ulna is prominent on the back of the wrist.

Ossification

A primary ossification centre appears in the shaft during the eighth week *in utero*. The body, coronoid process and major part of the olecranon ossify from this primary centre. A secondary centre appears in the head during the fifth year and fuses with the shaft between 20 and 22 years. The secondary centre for the remainder of the olecranon appears at about 11 years, with fusion occurring between 16 and 19 years. There may be several secondary centres for the olecranon.

Section summary

Bones of the arm and forearm

Humerus

- Long bone of arm having a proximal head and greater and lesser tubercles; rounded shaft; flattened distal end with medial and lateral epicondyles, capitulum and trochlea.
- Head articulates with glenoid fossa of scapula forming shoulder (glenohumeral) joint.
- Capitulum articulates with head of radius; trochlea with trochlear notch of ulna forming elbow joint.

Radius

- Lateral bone of forearm having proximal disc-like head; shaft with radial tuberosity; expanded distal end with radial styloid process laterally and ulnar notch medially.
- Head articulates with capitulum forming lateral part of elbow joint and with radial notch of ulna forming superior radioulnar joint.
- Distally articulates with scaphoid and lunate as part of radiocarpal joint of wrist and with head of ulna with ulnar notch of radius forming inferior radioulnar joint.

Ulna

- Medial bone of forearm having olecranon (expanded hook-like projection) posterior and coronoid process anterior; shaft with sharp lateral interosseous border; distal head with ulnar styloid process medially.
- Trochlear notch articulates with trochlea of humerus forming medial part of elbow joint.
- Radial notch articulates with head of radius forming superior radioulnar joint.
- Head articulates with ulnar notch of radius forming inferior radioulnar joint and with intra-articular disc proximal to radiocarpal joint.

The carpus

The carpus consists of eight separate bones arranged around the capitate, but commonly described as being in two rows each of four bones. Three of the bones in the proximal row articulate proximally with the radius or articular disc at the radiocarpal joint, whilst distally they articulate with the distal row of bones forming the midcarpal joint. The four carpal bones of the distal row articulate with the bases of the five metacarpal bones via the carpometacarpal joints. There are intercarpal joints between the adjacent carpal bones in each of the rows.

The bones are bound together by ligaments forming a compact mass, which is curved giving a posterior convexity and a pronounced anterior concavity (the carpal sulcus). The sulcus is converted into a canal (carpal tunnel) by the flexor retinaculum.

The individual carpal bones are clinically important because they are often injured, especially the scaphoid and lunate, and because they provide recognizable bony landmarks in the wrist region.

From lateral to medial the proximal and distal rows are arranged as follows (Figs. 2.10 and 2.11):

Proximal: scaphoid, lunate, triquetral, pisiform
Distal: trapezium, trapezoid, capitate, hamate.

The three lateral bones of the proximal row form a convex articular surface facing proximally to fit into the concavity formed by the radius and the articular disc. Individually, each bone has a characteristic shape and its own set of articular surfaces.

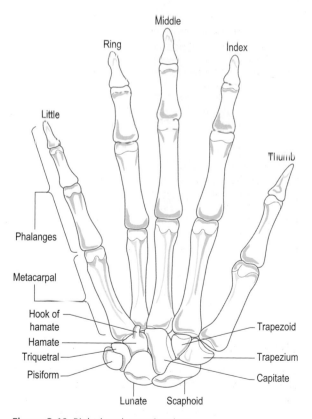

Figure 2.10 Right hand, anterior view.

Scaphoid

Marked anteriorly by a prominent palpable tubercle and a narrowed waist around its centre. Articular surfaces are present proximally for the radius, medially for the lunate and more distally for the head of the capitate, and lateral to the tubercle for the trapezium and trapezoid. The small, non-articular surface of the tubercle is the only region available for the entry of blood vessels. It is a common site of fracture.

Lunate

Has a smooth convex palmar surface which is larger than its dorsal surface. On its medial side is a square articular surface for the triquetral, and laterally a crescent-shaped area for the scaphoid. Distally, there is a deep concavity for the head of the capitate, while proximally the bone is convex where it articulates with the radius and articular disc.

Triquetral

Lies in the angle between the lunate and hamate, with which it articulates via a sinuous surface. The square lateral articular surface is for the lunate. The triquetral is distinguished by a circular articular surface for the pisiform. The proximal part enters the radiocarpal joint during adduction of the hand.

Pisiform

A small round sesamoid bone found in the tendon of flexor carpi ulnaris. It articulates with the palmar surface of the triquetral. The anterior surface projects distally and laterally forming the medial part of the carpal tunnel.

The distal row of carpal bones presents a more complex proximal articular surface, being flat laterally and convex medially. Individually, the bones all have a characteristic shape.

Trapezium

The most irregular of the carpal bones, with a palpable tubercle and groove medially on its anterior surface. It has articular surfaces proximally for the scaphoid and trapezoid, which are set at an angle to each other. Its main feature is the articular surface for the base of the first metacarpal, which is saddle-shaped and faces distally, laterally and slightly forwards; it contributes greatly to the mobility of the carpometacarpal joint of the thumb.

Trapezoid

A small irregular bone, which articulates with the second metacarpal. It lies in the space bounded by the metacarpal, scaphoid, capitate and trapezium, articulating with each.

Capitate

The largest of the carpal bones being centrally placed with a rounded head articulating with the concavities of the lunate and scaphoid. Medially and laterally are flatter articular surfaces for the hamate and trapezoid respectively. The dorsal surface is flat, but the palmar aspect is roughened by ligamentous attachments. The distal surface articulates mainly

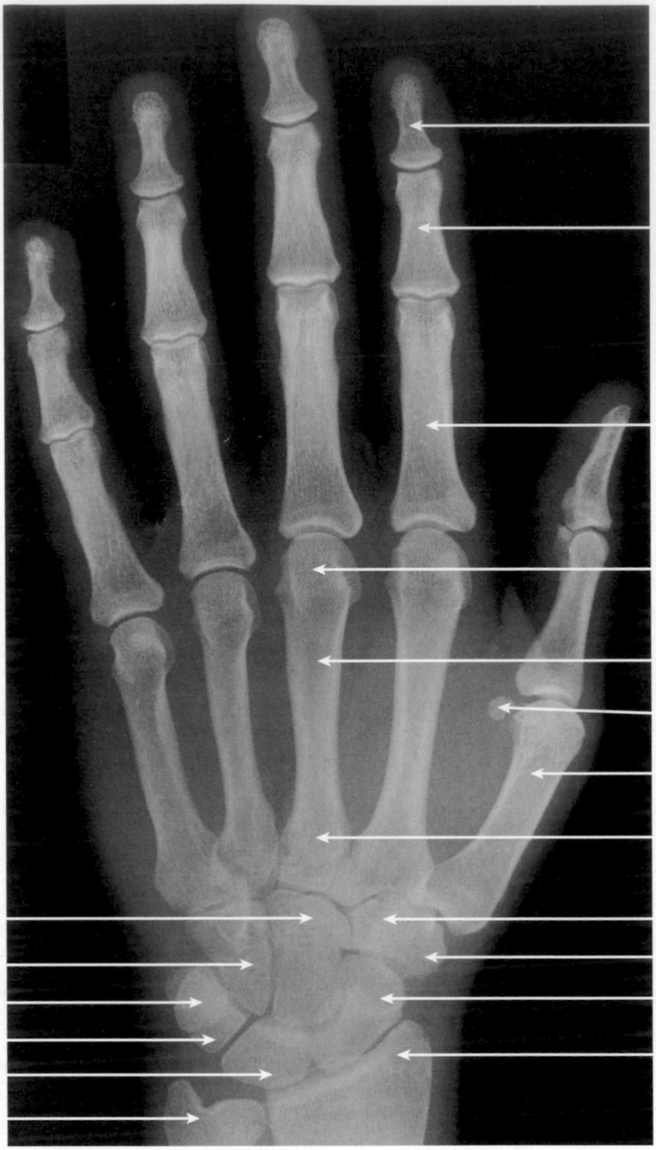

Figure 2.11 Radiograph of the right hand and wrist, anterior view.

with the base of the third metacarpal, but also by narrow surfaces with the bases of the second and fourth metacarpals.

Hamate

Wedge-shaped with a large curved palpable hook projecting from its palmar surface near the base of the fifth metacarpal. The hook is concave medially forming part of the carpal tunnel. The distal base of the wedge articulates with the bases of the fourth and fifth metacarpals. The wedge passes up between the capitate and triquetral to reach the lunate. The articular surface for the capitate is flat while that for the triquetral is sinuous.

Overall the carpus presents a deep transverse concavity on the palmar surface of the wrist. The flexor retinaculum bridges the concavity, attaching to the tubercles of the scaphoid and trapezium laterally, and the pisiform and hook of hamate medially, forming the roof of the carpal tunnel.

Palpation

Starting on the medial side of the palmar aspect of the wrist at the proximal part of the hypothenar eminence, the pisiform can be easily distinguished with the tendon of flexor carpi ulnaris running proximally from it. Immediately distal and slightly lateral to the pisiform, the hook of the hamate can be palpated if sufficient pressure is applied through the hypothenar muscles.

On the lateral side of the carpus just proximal to the distal wrist crease, the prominent tubercle of the scaphoid can be palpated, immediately beyond which is the tubercle of the trapezium. The scaphoid can be 'pinched' between the palpating thumb and index finger if these are placed on the tubercle and in the 'anatomical snuffbox' at the base of the thumb on its dorsal surface.

Ossification

Each carpal bone ossifies from a single centre, all of which appear after birth. During the first year of life the centres for the capitate and hamate appear. These are followed by centres for the triquetral between 2 and 4 years, the lunate between 3 and 5 years, the scaphoid, trapezium and trapezoid all between 4 and 6 years, and finally the pisiform between 9 and 14 years. Ossification is not complete until between 20 and 25 years. The hook of hamate may remain separate. Small additional nodules may also be present. The shape of the individual carpal bones, and not their size, can be used to determine the age of an individual.

The metacarpus

The metacarpus consists of five bones, the metacarpals, one corresponding to each digit and numbered in sequence from the lateral side. Each is a long bone with a proximal quadrilateral *base*, a shaft (body) and a distal rounded *head* (Figs. 2.10 and 2.11). Variations in the shape of the bases provide a means by which each can be distinguished. The base of the first metacarpal has a saddle-shaped articular surface which fits a corresponding surface on the trapezium. The base of the second metacarpal articulates with the trapezium, trapezoid and capitate. The base of the third has a single articulation with the capitate. The bases of the fourth and fifth metacarpals articulate with the hamate. The bases of the second to fifth also articulate with the adjacent metacarpals, having articular facets in appropriate positions.

The heads of the metacarpals are smooth and rounded, extending further onto the palmar surface. The palmar articular margin is notched in the midline. The head of the first metacarpal is wider than the others having two *sesamoid bones*, usually found in the short tendons crossing the joint, which articulate with the palmar part of the joint surface, occasionally grooving it. The heads fit into a concavity on the base of the proximal phalanx at the metacarpophalangeal joints. The shaft of the metacarpals is slightly curved with a longitudinal palmar concavity. The shaft of the first metacarpal is nearly as wide as the base and has a rounded dorsal surface. The palmar surface is divided by a blunt ridge into a larger lateral part and a smaller medial part.

Palpation

If the fingers are flexed to form a fist, the heads of the metacarpals can easily be palpated as the knuckles. Running proximally on the dorsal surface of the hand the shafts can also be distinguished. At the proximal end of the shaft the gap between the base of the metacarpal and the carpus can be palpated as the line of the carpometacarpal joint.

Ossification

Primary ossification centres appear in the shaft in the ninth week *in utero*, so that the bones are well ossified at birth. Secondary centres appear in the heads of the second to fifth metacarpals between 2 and 3 years. The secondary centre for the base of the first metacarpal appears slightly later. Fusion of the epiphysis with the shaft occurs between 17 and 19 years for all metacarpals. Occasionally, a secondary centre may appear in the head of the first metacarpal.

The phalanges

There are 14 phalanges in each hand, three for each finger (2nd to 5th digits) and two for the thumb (1st digit). As they are long bones, each phalanx has a shaft, a large proximal end and a smaller distal end, the head (Figs. 2.10 and 2.11). The phalanges of the thumb are shorter and broader than those of the fingers.

The proximal phalanx has a concave oval facet on its base for articulation with the head of the metacarpal. The rounded head, which extends further onto the palmar surface, has a wide, pulley-shaped articular surface for the base of the next phalanx. The shaft is curved along its length, being convex dorsally. It is convex from side to side on its dorsal surface and flat on the palmar surface. The middle and distal phalanges are similar to the proximal phalanx. However, the base of the distal phalanx is large, and the head is expanded to support the pulp pad of the digits.

By convention, the digits are described by name rather than by number, and are from lateral to medial, the thumb, index, middle, ring and little fingers.

Palpation

By flexing the fingers into a fist, the heads of the proximal and middle phalanges can be palpated. The shafts of the phalanges are also easily followed throughout their length, especially on their dorsal surface.

Ossification

Primary ossification centres appear in the shafts of the phalanges between the eighth and twelfth week *in utero*, with the distal phalanges ossifying first. Secondary centres appear in the bases of the phalanges during the second and

51

third year, fusing with the shaft between 17 and 19 years. Occasionally, a secondary centre may appear in the head as well as in the base.

Section summary

Bones of the wrist and hand

Carpus

- Eight small bones arranged as two rows between the forearm and hand. Proximal row (lateral to medial): scaphoid, lunate, triquetral, pisiform (sesamoid bone). Distal row (lateral to medial) trapezium, trapezoid, capitate, hamate.
- Proximal row articulates with distal end of the radius and intra-articular disc forming the radiocarpal joint.
- Distal row articulates with proximal row (except pisiform) forming the midcarpal joint and with bases of metacarpals forming the carpometacarpal joints.

Metacarpus

- Five long bones (metacarpals) in hand each with quadrilateral base proximally; shaft, rounded head distally.
- Bases articulate with distal row of carpal bones forming the carpometacarpal joints and with adjacent metacarpals forming intercarpal joints.
- Heads articulate with base of corresponding proximal phalanx forming the metacarpophalangeal joints.

Phalanges

- Fourteen individual bones with two in thumb (proximal, distal) and three in each finger (proximal, middle, distal) each having base; shaft; head (flattened pad in distal phalanges).
- Bases of proximal phalanges articulate with heads of metacarpals forming the metacarpophalangeal joints.
- Interphalangeal joints formed between heads and bases of adjacent phalanges.

MUSCLES

Movements of the pectoral (shoulder) girdle

The pectoral girdle consists of the scapula and clavicle, articulating with each other at the acromioclavicular joint. It provides the link between the upper limb and the axial skeleton via the shoulder (glenohumeral) and sternoclavicular joints respectively. The flattened triangular scapula provides attachment for many muscles, some of which anchor the pectoral girdle to the thorax while others control the position of the upper limb. The clavicle, however, acts primarily as a strut holding the upper limb away from the trunk. Because of the connections (both muscular and ligamentous) between the scapula and clavicle, movements of the pectoral girdle, either independently or in association with the upper limb, mean that both bones are always involved. The position and movement of the scapula are determined by the activity of the muscles attached to it. An individual muscle when acting in concert with various combinations of other muscles may be involved in producing several different movements of the pectoral girdle. Consequently, in the following account individual muscles will be described in detail according to their major action. Movements between the scapula and thorax are allowed because the fascia covering adjacent layers of muscles facilitates gliding and sliding movements.

The movements of the shoulder girdle are described as taking place from the anatomical position where the scapula lies obliquely over the second to seventh ribs on the posterior wall of the thorax with the coracoid process pointing anteriorly. The movements described are:

Retraction – movement of the scapula, whilst maintaining its vertical position, such that its medial border approaches the vertebral column, as in bracing the shoulder. The glenoid fossa thus comes to face more directly lateral.

Protraction – movement of the scapula forwards around the chest wall, as in rounding the shoulders. There may be some associated lateral rotation in this movement. The glenoid fossa comes to face more directly forwards.

Elevation – where the pectoral girdle is lifted upwards as in shrugging the shoulders.

Depression – where the pectoral girdle is pulled downwards.

Lateral (forward) rotation of the pectoral girdle is a complex movement whereby the inferior angle of the scapula moves laterally around the chest wall, while the strut-like action of the clavicle results in a concomitant upward movement of the scapula, thus causing the glenoid fossa to be turned increasingly upwards.

Medial (backward) rotation of the pectoral girdle returns the scapula to its resting position from lateral rotation.

Muscles retracting the pectoral (shoulder) girdle

Rhomboid minor
Rhomboid major
Trapezius

Rhomboid minor

A small quadrilateral muscle (Fig. 2.12) whose fibres run obliquely downwards and laterally from the *spinous processes* of C7 and T1 and the *supraspinous ligament* between them and the lower part of the *ligamentum nuchae*, to attach to the *medial border* of the smooth *triangular area* at the base of the *spine of the scapula*.

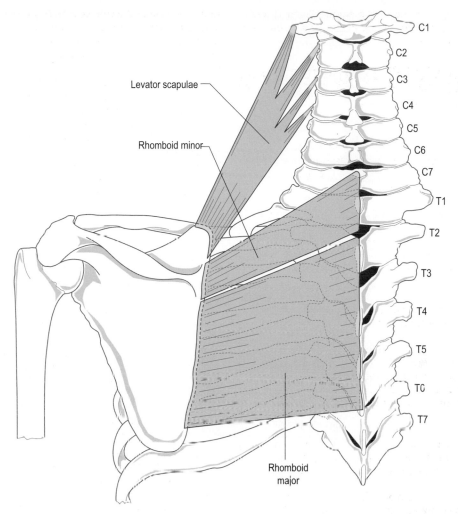

Levator scapulae

Rhomboid minor

C1
C2
C3
C4
C5
C6
C7
T1
T2
T3
T4
T5
T6
T7

Rhomboid major

Figure 2.12 The attachments of levator scapulae, rhomboid minor and rhomboid major shown on the posterior aspect of the thorax.

Rhomboid major

Although larger than rhomboid minor, rhomboid major may be continuous with it (Fig. 2.12). It arises by tendinous slips from the *spinous processes* of *T2* to *T5* inclusive and the intervening *supraspinous ligament*. The muscle fibres run obliquely downwards and laterally to attach to the *medial border* of the *scapula* between the base of the spine and the inferior angle.

Both rhomboids lie superficial to the long back muscles, being themselves covered by trapezius, except for the lower border of rhomboid major which forms the floor of the 'triangle of auscultation'.

Nerve supply

Both rhomboids are supplied by the *dorsal scapular nerve* (root value C5).

Action

They act principally to retract the scapula but are also active in medial rotation of the pectoral girdle. In addition, they also act as important stabilizers of the scapula when other muscle groups are active.

Palpation

With the subject's hand placed in the small of the back (to relax trapezius), the rhomboids can be palpated through trapezius when the hand is moved backwards. Contraction of the rhomboids can be felt (and occasionally seen) between the medial border of the scapula and the vertebral column.

Trapezius

A large, flat triangular sheet of muscle extending from the skull and spine medially to the pectoral girdle laterally

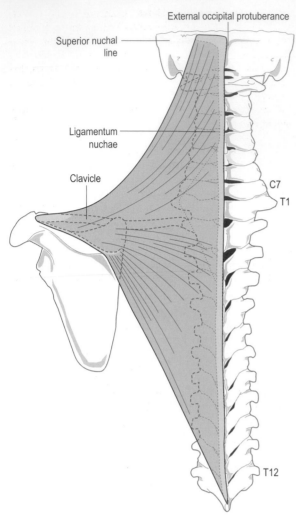

Figure 2.13 The attachments of trapezius shown on the posterior aspect of the thorax.

(Fig. 2.13). It is the most superficial muscle in the upper back and with its fellow of the opposite side it forms a trapezium, hence its name.

The medial attachment runs from the medial third of the *superior nuchal line* and *external occipital protuberance* of the *occipital bone*, the *ligamentum nuchae*, the *spinous processes* of *C7* to *T12* inclusive and the intervening *supraspinous ligament*. The majority of this attachment is by direct muscular slips; however, a triangular aponeurosis exists in the trapezius between C6 and T3 which corresponds to a hollow seen in the living subject.

From this extensive medial attachment there is a continuous line of attachment to the clavicle and scapula; the upper fibres run downwards and laterally, the middle fibres almost horizontally, and the lower fibres upwards and laterally. The upper fibres descend to the *posterior border* of the *lateral third*

of the *clavicle*, while the middle fibres pass to the *medial border* of the *acromion* and *upper border* of the *crest* of the *spine of the scapula*, being separated from the smooth area on the medial part of the spine by a small bursa. The lowermost fibres converge to a tendon which attaches to the *tubercle* on the inferior edge at the medial end of the *spine of the scapula*.

The upper free edge of trapezius forms the posterior border of the posterior triangle of the neck, while the lower free border forms the medial boundary of the triangle of auscultation. This latter triangle is an area of the chest wall free of bony obstruction by the scapula and thinly covered by muscle. Its other boundaries are inferiorly the upper border of latissimus dorsi and laterally the medial border of the scapula.

Nerve supply

The motor supply is via the spinal part of the *accessory nerve (XI)* which enters it from the posterior triangle. It also receives sensory fibres from the *ventral rami* of C3 and 4 via the cervical plexus. The skin over trapezius is supplied by the *dorsal rami* of C3 to T12.

Action

Trapezius has an important function in stabilizing the scapula during movements of the upper limb. The middle horizontal fibres pull the scapula towards the midline, that is retraction, and may be aided by the upper and lower fibres contracting together to produce a 'resolved' force towards the midline. The upper fibres of trapezius elevate the pectoral girdle and maintain the level of the shoulders against the effect of gravity, or when a weight is being carried in the hand. When both left and right muscles contract they can extend the neck, but when acting singularly the upper fibres produce lateral (side) flexion of the neck. The lower fibres pull down the medial part of the scapula and thus lower the shoulder, especially against resistance, for example when using the arms to get out of a chair. The upper and lower fibres working together produce lateral rotation of the scapula about a point towards the base of the spine. Trapezius is thus important in the overall function of the upper limb as its action increases the possible range of movement.

Paralysis of trapezius, particularly its upper part, results in the scapula moving forwards around the chest wall with the inferior angle moving medially. The usually smooth curve of its upper border between the occiput and the acromion may become markedly angulated.

Palpation

To demonstrate and palpate all three parts of trapezius, the subject should abduct both arms to 90°, flex the elbows to 90° and then rotate them laterally so that the fingers are pointing upwards.

In this position the three sets of fibres can be readily palpated; in a lean subject contraction of the various parts of the muscle can be seen. For the lower fibres the contraction can be further enhanced by asking the subject to clasp his or her hands together above the head and pull hard.

Soft tissue techniques are often applied to the upper muscular fibres of trapezius in the presence of muscle spasm secondary to neck pain, with the aim of inducing relaxation. Deep transverse frictions can also be applied to the tendinous attachment of trapezius on the superior nuchal line when this is the site of a lesion causing pain in the neck or occipital region.

Muscles protracting the pectoral (shoulder) girdle

Serratus anterior
Pectoralis minor

Serratus anterior

A large flat muscular sheet covering the side of the thorax sandwiched between the ribs and the scapula (Fig. 2.14). Loose fascia exists between the deep surface of the muscle and the ribs or intercostal fascia, and also between its superficial surface and subscapularis in order to facilitate free movement of the scapula. Serratus anterior forms the medial wall of the axilla, and is partly covered by the breast inferolaterally. The upper digitations are behind the clavicle while latissimus dorsi crosses its lower border.

It attaches by fleshy digitations just anterior to the mid-axillary line to the *outer surfaces* of the *upper eight* or *nine ribs* and the intervening *intercostal fascia*. The upper digitation arises from ribs one and two, whereas each of the remaining digitations arises from a single rib. The lower four digitations interdigitate with the costal attachment of external oblique of the abdomen.

From this extensive attachment the muscle fibres run backwards to insert into the *costal surface* of the *medial border* of the *scapula* between the superior and inferior angles. The digitations are not, however, evenly distributed in their attachment to the scapula. The first passes almost horizontally to the superior angle, while the lower four condense to

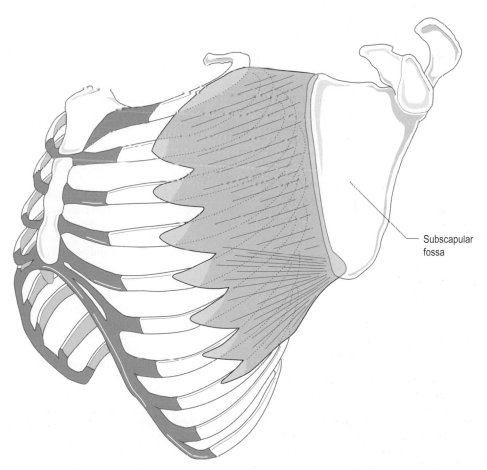

Subscapular fossa

Figure 2.14 The attachments of serratus anterior shown on the lateral thoracic wall with the scapula lifted posteriorly.

attach to the inferior angle, with the intervening ones spread along the medial border.

Nerve supply

By the *long thoracic nerve* (root value C5, 6 and 7), with the first two digitations supplied by C5, the next two by C6 and the remaining four by C7. The nerve enters the muscle on its superficial aspect. The skin over the accessible parts of the muscle is supplied by nerves with root values T3 to T7.

Action

Serratus anterior is a major protractor of the pectoral girdle and as such is involved in all thrusting, pushing and punching movements where the scapula is driven forwards carrying the upper limb with it. Note the massive development of this muscle in boxers.

It plays a vital role in stabilizing the scapula during movements of the upper limb and contracts strongly to hold the medial border of the scapula against the chest wall when the arm is flexed or when a weight is carried in front of the body. Failure to perform this action, as when paralysed, results in 'winging' of the scapula in which the medial border stands away from the chest wall, thus severely affecting the function and mobility of the upper limb.

The lower digitations of the muscle work together with trapezius to rotate the scapula laterally, so that the glenoid fossa looks upwards and forwards. When paralysed, loss of the rotating action of serratus anterior means that the upper limb cannot be abducted by more than approximately 90°, thereby seriously limiting the functional capacity of the upper limb. There is some controversy as to whether serratus anterior acts as an accessory muscle of inspiration during respiratory distress. The line of action of the muscle fibres, except perhaps for the first two digitations and maybe the last, are not directed to cause elevation of the ribs. Indeed, they are more likely to cause depression of the ribs.

Palpation

In a muscular subject the digitations of serratus anterior can be felt and often seen running forwards in the region of the midaxillary line, especially when performing 'press-ups'.

Pectoralis minor

A thin, flat, triangular muscle situated on the anterior chest wall deep to pectoralis major (Fig. 2.15). Inferiorly it attaches to the *outer surfaces* of the *third*, *fourth* and *fifth ribs* close to their costal cartilages and the intervening *intercostal fascia*. There may be additional attachments to the second or sixth rib, or more rarely, to both. The fibres converge to a short, flat tendon as they pass superolaterally to attach to the *upper surface* and *medial border* of the *coracoid process* of the *scapula*.

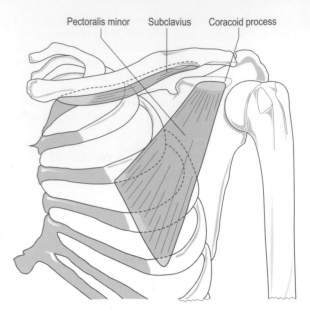

Figure 2.15 The attachments of subclavius and pectoralis minor shown on the anterior aspect of the thorax.

Nerve supply

By the *medial pectoral nerve* which pierces it. However, within the axilla the medial and lateral pectoral nerves communicate, thereby ensuring that pectoralis minor is supplied by both nerves. The segmental supply is by C6, 7 and 8.

Action

By exerting a strong pull on the coracoid process, the scapula can be pulled forwards and downwards during pushing and punching movements. When leaning on the hands it helps to transfer the weight of the trunk to the upper limb. Its attachment to the coracoid process allows pectoralis minor to help produce medial rotation of the scapula against resistance (Fig. 2.16B). With the scapula and upper limb fixed, pectoralis minor may be used as an accessory muscle of inspiration during respiratory distress.

Palpation

Contraction of the pectoralis minor, lying deep to the great bulk of the pectoralis major, is difficult to palpate.

Muscles elevating the pectoral (shoulder) girdle

Trapezius (upper fibres) (p. 53)
Levator scapulae

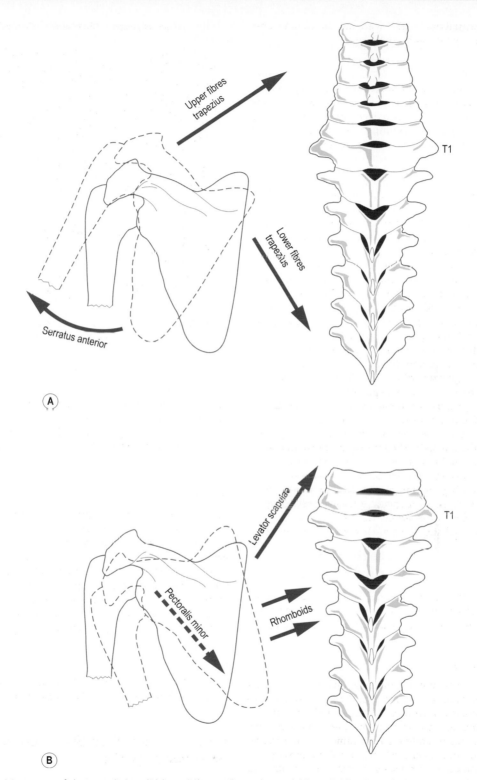

Figure 2.16 Movement of the scapula into (A) lateral (forward) rotation and (B) medial (backward) rotation, represented in diagrammatic form indicating the direction of pull of the principal muscles involved.

Levator scapulae

Situated in the posterior part of the neck, the upper part of levator scapulae is under cover of sternomastoid and the lower part deep to trapezius (Fig. 2.12). Its middle portion forms part of the floor of the posterior triangle. It lies superficial to the extensor muscles of the neck and attaches by tendinous slips to the *transverse processes* of the *upper three or four cervical vertebrae* (attaching to the posterior tubercles of the lower two) behind the attachment of scalenus medius. From here the fibres run downwards and laterally to attach to the *medial margin* of the *scapula* between the superior angle and base of the spine.

Nerve supply

Partly from the *dorsal scapular nerve* (C5) and directly from the *ventral rami* of C3 and 4.

Action

Working with trapezius, levator scapulae can produce elevation and retraction of the pectoral girdle or resist its downward movement, as when carrying a load in the hand. Again when working with trapezius, contraction of both sides produces extension of the neck, while one side produces lateral flexion of the neck. Levator scapulae also helps to stabilize the scapula and is active in producing medial rotation of the scapula.

Palpation

Levator scapulae can be palpated when trapezius is not contracting (as for the rhomboids), with the subject upright and the hands in the small of the back. Levator scapulae can be felt anterior to trapezius in the posterolateral part of the neck when the subject's hand is moved backwards with the elbow flexed.

Muscles laterally rotating the pectoral (shoulder) girdle

Trapezius
Serratus anterior

The detailed attachments of both muscles have already been described (pp. 53–56).

As can be seen in Figs. 2.13 and 2.14, both muscles are well positioned to cause the inferior angle of the scapula to move laterally around the thoracic wall. The clavicle acting as a strut restricts movement at the acromion, so that the overall effect of their action is to elevate the acromion and move the inferior angle laterally, thereby enabling the glenoid fossa to face more directly upwards. This movement of the pectoral girdle is extremely important for increasing the range of movement possible, particularly in terms of abduction and flexion of the upper limb at the shoulder (p. 124).

Trapezius contributes to the rotation by contraction of its upper fibres which lift the lateral end of the clavicle and acromion upwards while at the same time its lower fibres pull downwards on the medial end of the spine of the scapula.

Serratus anterior, the more important of the two muscles in this movement, pulls strongly on the inferior angle of the scapula, where the majority of its muscle fibres insert, to pull it laterally around the chest wall. The notional axis about which this rotation takes place is just below the spine of the scapula towards the base. The resultant movements are shown in Fig. 2.16.

Muscles medially rotating the pectoral (shoulder) girdle

Rhomboid major
Rhomboid minor
Pectoralis minor
Levator scapulae

The detailed attachments of the above muscles have already been described on pages 52–58, as all are active in other movements of the pectoral girdle.

Movement of the inferior angle of the scapula towards the vertebral column is frequently produced by the action of gravity, being controlled by the eccentric activity of the lateral rotators trapezius and serratus anterior. However, the above muscles contract strongly if the pectoral girdle is to be medially rotated against resistance, as when moving the weight of the body from a position of hanging from a beam to full chin-up. The notional axis of rotation is just below the spine of the scapula, towards the base. Pectoralis minor exerts a downward pull on the lateral side of this axis via its attachment to the coracoid process, while the rhomboids and levator scapulae pull upwards on the medial side. The resultant movements are shown in Fig. 2.16. Details of the movements of the joints of the pectoral girdle can be found in the section on joints (pp. 101–112).

Muscles stabilizing the clavicle

Subclavius

Subclavius

The subclavius (Fig. 2.15) lies entirely beneath the clavicle under cover of pectoralis major. The small, fleshy belly attaches to the *floor* of the *subclavian groove* on the undersurface of the *clavicle*. The fibres converge and pass medially, becoming tendinous, to attach to the *first rib* near its junction with the costal cartilage.

Nerve supply

By the *nerve to the subclavius* (root value C5 and 6), which arises from the upper trunk of the brachial plexus.

Action

The principal action of subclavius is to steady the clavicle, by pulling it towards the disc of the sternoclavicular joint, and the sternum during movements of the pectoral girdle. This action tends to depress the lateral end of the clavicle. Paralysis of subclavius has no demonstrable effect.

Section summary

Movements at pectoral (shoulder) girdle

The pectoral (shoulder) girdle consists of the clavicle and scapula. It is capable of a range of independent movements produced by the following muscles:

Movement	Muscles
Retraction	Rhomboid major and minor
	Trapezius
Protraction	Serratus anterior
	Pectoralis minor
Elevation	Trapezius (upper fibres)
	Levator scapulae
Depression	Pectoralis minor
	Trapezius (lower fibres)
Lateral rotation	Trapezius
	Serratus anterior
Medial rotation	Rhomboid major and minor
	Levator scapulae
	Pectoralis minor

- It is important to consider the effect of gravity or resistance on these movements and muscle work – for example, the elevators may work concentrically against gravity or resistance to raise the pectoral girdle but eccentrically to lower it back down.
- The muscles, especially serratus anterior, act to hold the scapula against the thorax.
- Movements of the pectoral girdle increase the range of movement of the upper limb by combining with that available at the shoulder joint during functional activities.

Movements of the shoulder joint

The movements of the shoulder joint will be considered as follows: abduction and adduction in the coronal plane, flexion and extension in the sagittal plane, and medial and lateral rotation about the long axis of the arm.

Reference will be made to functional movements which combine some of the above.

Muscles abducting the arm at the shoulder joint

Supraspinatus
Deltoid

Supraspinatus

Supraspinatus (Fig. 2.17) arises from the *medial two-thirds* of the *supraspinous fossa* and the deep surface of the dense fascia which covers the muscle. The muscle and the tendon which form within it pass laterally below trapezius, the acromion and the coracoacromial ligament to cross over the top of the shoulder joint. The tendon of supraspinatus blends on its deep surface with the capsule of the shoulder joint prior to inserting into the *upper facet* on the *greater tubercle* of the *humerus*.

Nerve supply

By the *suprascapular nerve* (root value C5 and 6), a branch from the upper trunk of the brachial plexus. The skin over the muscle is supplied from roots C4 and T2.

Action

The supraspinatus initiates abduction at the shoulder joint, being more important during the early part of the movement than later when deltoid takes over. Its role is probably twofold during this movement; it braces the head of the humerus firmly against the glenoid fossa to prevent an upward shearing of the humeral head (this has been likened to a 'foot on the ladder' where a small force applied at one end will produce a rotatory rather than a shearing movement) while at the same time producing abduction. After the initial 20° of abduction, when the stronger deltoid takes over, supraspinatus acts to hold the humeral head against the glenoid fossa.

Functional activity

Supraspinatus is one of the four muscles which form a musculotendinous cuff (or rotator cuff) around the head of the humerus. They function to keep the head of the humerus in the glenoid fossa during movements of the shoulder joint.

Palpation

Contraction of supraspinatus can be felt through the trapezius if the examiner's fingers are pressed into the medial part of the supraspinous fossa when the subject initiates abduction at the shoulder joint. In the anatomical position, the tendon of supraspinatus is covered by the acromion but it can be palpated if the subject medially rotates the shoulder with the hand resting passively in the small of the back. During this manoeuvre, the greater tubercle moves anteriorly so that the tendon can now be rolled against the bone by a medial to lateral pressure of the examiner's finger against the tubercle. The tendon of supraspinatus is the most frequently damaged soft tissue in the shoulder region and techniques such as transverse frictions, injection and ultrasound are often applied to this exact location. In severe cases the tendon may be sufficiently eroded to cause its rupture, which then affects the

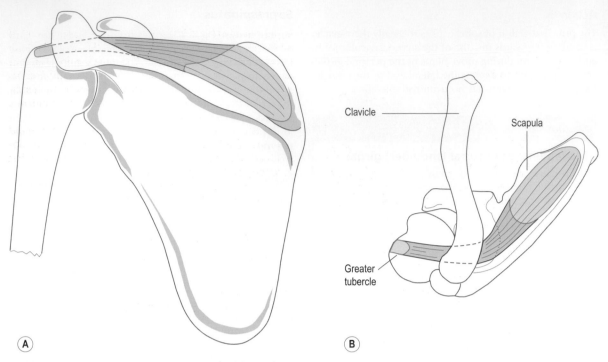

Figure 2.17 The attachments of supraspinatus: (A) posterior view, the broken lines of the muscle indicating its passage below the coracoacromial arch, (B) superior view.

ease with which abduction can occur. In such cases, or when supraspinatus is paralysed, the patient can still initiate abduction by leaning slightly to the side, so using gravity. Alternatively, the patient may use the opposite arm to push the affected limb away from the side, or jerk the hips to 'kick' the elbow out. By each of these actions a small yet sufficient degree of abduction occurs to enable the powerful deltoid to take over.

Deltoid

A coarse, thick, triangular muscle (Fig. 2.18) giving the shoulder its rounded contour. Functionally it can be divided into three parts, anterior, posterior and middle, of which only the middle portion is multipennate. It has an extensive attachment to the pectoral girdle. Anteriorly, the fibres attach to the *anterior border* of the *lateral third* of the *clavicle*, whilst posteriorly they attach to the *lower lip* of the *crest* of the *spine* of the *scapula*. The most anterior and posterior fibres both run obliquely, in an uninterrupted manner, to the *deltoid tuberosity* on the lateral surface of the shaft of the *humerus*.

The middle muscle fibres are more complex because of their multipennate arrangement (Fig. 2.18). These shorter oblique fibres run from four tendinous slips attached to the *lateral margin* of the *acromion* to join three intersecting tendinous slips which ultimately run to the *deltoid tuberosity*

of the *humerus.* Consequently, these shorter, more numerous middle fibres, working under considerable mechanical disadvantage when active, give this part of the muscle great strength.

Deltoid is separated from the coracoacromial arch and the upper and lateral aspects of the shoulder joint (and the tendons lying on it) by the subacromial bursa.

Nerve supply

By the *axillary nerve* (root value C5 and 6) from the posterior cord of the brachial plexus. The skin covering deltoid is supplied by roots C4 and 5.

Action

Deltoid is the principal abductor of the arm at the shoulder joint, the movement being produced by its middle, multipennate fibres. However, deltoid can only produce this movement efficiently after it has been initiated by supraspinatus.

The true plane of abduction is in line with the blade of the scapula and for this the anterior and posterior fibres are active in order to maintain the plane of abduction by acting as 'guy ropes'. The tendency for deltoid to produce an upward shearing of the head of the humerus is resisted by the rotator cuff muscles, that is by

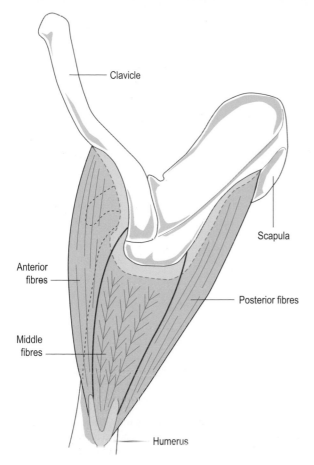

Clavicle

Anterior
fibres

Middle
fibres

Scapula

Posterior fibres

Humerus

Figure 2.18 Deltoid, viewed from above, with its attachments.

subscapularis anteriorly, teres minor and infraspinatus posteriorly, and supraspinatus superiorly.

The anterior part of deltoid is a strong flexor and medial rotator of the humerus, while the posterior part is a strong extensor and lateral rotator, and can help in the transfer of the strain of heavy weights carried in the hand to the pectoral girdle. The posterior part of deltoid is also active during adduction of the arm, to counteract the medial rotation produced by pectoralis major and latissimus dorsi.

Functional activity

Deltoid is active in abduction when the middle fibres contract concentrically, but the massive development and multipennate nature of the muscle are probably due to the fact that many activities of the upper limb require that it is maintained or 'held' in this position for long periods of time. Consequently, the middle fibres contract statically when performing activities with the arms in front of the trunk; they then lower the arm back to the side by working eccentrically.

Palpation

If the seated subject is asked to raise the arm to 60° of abduction in the plane of the scapula, the triangular bulk of deltoid can be felt and seen. Palpating the upper surface of the acromion and moving the fingers laterally from its edge, the depressions in the muscle caused by the tendinous intersections can be felt if anteroposterior pressure of the fingers is applied.

The anterior and posterior fibres can be made to stand out more clearly if, in the same position as above, the subject is asked to maintain the position against resistance first anteriorly and then posteriorly.

Paralysis of deltoid severely affects the functioning of the shoulder joint and therefore of the upper limb.

Muscles flexing the arm at the shoulder joint

Pectoralis major
Deltoid (anterior fibres) (p. 60)
Biceps brachii – long head (p. 68)
Coracobrachialis (p. 64)

Pectoralis major

A thick trianglar muscle located on the upper half of the anterior surface of the thoracic wall (Fig. 2.19). It has clavicular and sternocostal parts, which may be separated by a groove, although they are usually continuous with each other. As the fibres pass towards the humerus they twist forming the rounded anterior fold of the axilla.

The smaller, clavicular attachment is from the *medial half of the anterior surface of the clavicle*; the larger, sternocostal attachment comes from the *anterior surface* of the *manubrium* and *body* of the *sternum*, the anterior aspects of the *upper six costal cartilages*, the anterior part of the *sixth rib* as well as the *aponeurosis* of the *external oblique muscle* of the abdomen.

From this large medial attachment, the muscle narrows and inserts via a laminated tendon into the *lateral lip* of the *intertubercular groove* of the *humerus*. The anterior lamina, which is the clavicular part of the muscle, runs to the lower part of the humeral attachment. The sternocostal part forms the posterior lamina, which passes upwards behind the anterior lamina to the upper part of the attachment to the humerus. In this way the tendon comes to resemble a U in cross-section. The posterior part blends with the shoulder joint capsule, while the anterior, clavicular fibres blend with the attachment of deltoid.

As the most superficial muscle of the anterior thoracic wall, pectoralis major lies on top of pectoralis minor, the ribs and serratus anterior. In females, the muscle is covered by the breast; indeed the fibrous septa of the breast are attached to the deep fascia overlying pectoralis major.

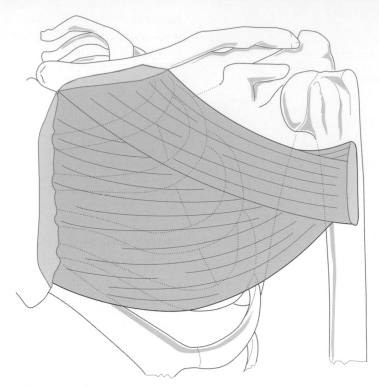

Figure 2.19 The attachments of pectoralis major, anterior view.

Pectoralis major is separated from the deltoid by the delto-pectoral groove (the infraclavicular fossa) in which lie the cephalic vein and branches from the thoracoacromial artery.

Nerve supply

By the *medial* (C8, T1) and the *lateral* (C5, 6, 7) *pectoral nerves;* the clavicular part by roots C5 and 6, and the sternocostal part by C7, 8 and T1. The skin over pectoralis major is supplied by roots T2 to T6.

Action

Pectoralis major as a whole is a powerful adductor and medial rotator of the humerus at the shoulder joint. In addition, the clavicular part can flex the humerus to the horizontal, while the sternocostal fibres, because of their direction, can extend the flexed humerus, particularly against resistance in the anatomical position. With the humerus fixed, as in gripping a bed, table or chair back, pectoralis major pulls on the upper ribs to assist inspiration during respiratory distress.

Functional activity

Pectoralis major is one of the major climbing muscles, so that if the arms are fixed above the head, the power of the muscle can be used to pull the trunk upwards. It is assisted in this activity by latissimus dorsi. In pushing, punching and throwing movements, pectoralis major acts to move the humerus forcefully forwards, while serratus anterior and pectoralis minor simultaneously protract the pectoral girdle.

In exercises, such as the 'press-up', pectoralis major contracts concentrically on the upward movement raising the body and eccentrically on the downward movement when lowering the body.

Palpation

The clavicular part of pectoralis major can be readily palpated if the arm is flexed to 60° and held against downward pressure. The sternocostal part is best palpated if this same position is maintained against an upward pressure. The integrity of the muscle can be tested by adduction of the arm against resistance.

Muscles extending the arm at the shoulder joint

Latissimus dorsi
Teres major
Pectoralis major (p. 61)
Deltoid (posterior fibres) (p. 60)
Triceps (long head) (p. 71)

Latissimus dorsi

A large flat triangular sheet of muscle running between the trunk and the humerus (Fig. 2.20); consequently, it acts on the shoulder joint. The superior border forms the lower border of the triangle of auscultation, while its lateral border forms the medial border of the lumbar triangle.

Latissimus dorsi arises from the posterior layer of the *thoracolumbar fascia*, which attaches to the *spinous processes* of the *lower six thoracic* and *all* of the *lumbar* and *sacral vertebrae*, as well as to the intervening *supraspinous* and *interspinous ligaments*.

That part arising from the lower six thoracic vertebrae is covered by trapezius. In addition to this vertebral attachment, latissimus dorsi arises from the posterior part of the *outer lip* of the *iliac crest*, most laterally by direct muscular slips. As the muscle fibres sweep upwards and laterally across the lower part of the thorax, they attach to the *outer surfaces* of the *lower three* or *four ribs* and via fascia to the *inferior angle* of the *scapula*. From this widespread origin the fibres converge as they pass to the humerus forming a thin flattened tendon, which winds around the lower border of teres major, and inserts into the *floor* of the *inter tubercular groove* of the *humerus* anterior to the tendon of teres major, being separated from it by a bursa. The effect of twisting of the muscle through 180° means that the anterior surface of the tendon is continuous with the posterior surface of the rest of the muscle. Consequently, the fibres with the lowest origin on the trunk gain the highest attachment on the humerus.

Nerve supply

By the *thoracodorsal nerve* (root value C6, 7 and 8) from the posterior cord of the brachial plexus, which enters the muscle on its deep surface. The skin covering the muscle is supplied by roots T4 to T12 inclusive, by both ventral and dorsal rami, and by the dorsal rami of L1 to L3.

Action

The latissimus dorsi is a strong extensor of the flexed arm; however, if the humerus is fixed relative to the scapula it retracts the pectoral girdle. It is also a strong adductor and medial rotator of the humerus at the shoulder joint.

Functional activity

Functionally, latissimus dorsi is a climbing muscle, and with the arms fixed above the head it can raise the trunk upwards, in conjunction with pectoralis major. Latissimus dorsi has an important function in rowing and during the downstroke in swimming. Attachment of the muscle to the ribs means that it is active in violent expiration, and can be felt pressing forcibly inwards during a cough or sneeze, as it acts to compress the thorax and abdomen.

The attachment to the inferior angle of the scapula allows latissimus dorsi to assist in holding it against the thorax during movements of the upper limb.

If the humerus becomes the fixed point when standing, as for example when using crutches, latissimus dorsi is able to pull the trunk forwards relative to the arms; associated with this is a lifting of the pelvis. In patients with paralysis of the lower half of the body, the fact that latissimus dorsi attaches to the pelvis and is still innervated allows it to be used to produce movement of the pelvis and trunk. Consequently, patients wearing calipers and using crutches can produce a modified gait by fixing the arms and hitching the hips by the alternate contraction of each latissimus dorsi.

Palpation

In a lean subject, latissimus dorsi can be made to stand out relative to the thorax by asking the subject to raise the arm to 90° flexion and to hold it steady against an upwardly directed pressure. The muscle can be felt contracting if the posterior axillary fold is held between the finger and thumb

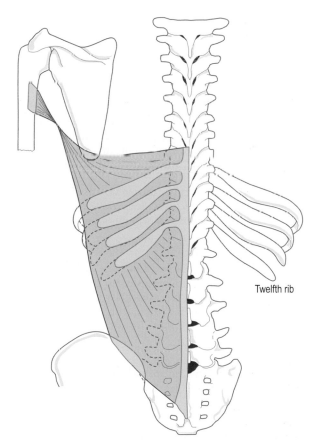

Twelfth rib

Figure 2.20 The attachments of latissimus dorsi, posterior view.

while the subject coughs. Adduction of the abducted arm against resistance also enables latissimus dorsi to be seen and felt.

Teres major

In the posterior part of the axilla, teres major (Fig. 2.21) forms the lower boundary of both the upper triangular and quadrangular spaces. A thick, chunky muscle, with latissimus dorsi, it forms the posterior fold of the axilla. It arises from an oval area on the *dorsal surface* of the *scapula* near the *inferior angle*, and from the fascia between it and adjacent muscles. The muscle fibres adjacent to latissimus dorsi run upwards and laterally forming a broad, flat tendon which attaches along the *medial lip* of the *intertubercular groove* of the *humerus*. The tendon is separated from that of latissimus dorsi by a bursa, with the latter muscle virtually covering the whole of teres major.

Nerve supply

By the *lower subscapular nerve* (root value C6 and 7) from the posterior cord of the brachial plexus.

Action

Teres major adducts and medially rotates the humerus at the shoulder joint. In addition it can help to extend the flexed arm.

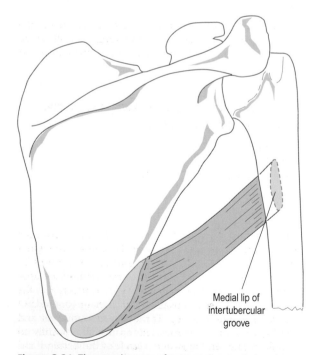

Medial lip of intertubercular groove

Figure 2.21 The attachments of teres major, posterior view.

Functional activity

Teres major, like latissimus dorsi, is a climbing muscle and works with the latter and pectoralis major to pull the trunk upwards when the arms are fixed. In conjunction with the latissimus dorsi and pectoralis major, teres major is important in stabilizing the shoulder joint.

Palpation

Teres major is covered by latissimus dorsi, and as these two muscles have similar actions, considerable care must be exercised when it is tested. The inferior angle of the scapula must first be found, the fingers are then moved upwards and laterally into the posterior wall of the axilla. The subject should abduct the arm to 90° and then adduct against an upwardly directed resistance. The rounded contour of teres major should now be palpable. During this same manoeuvre the flattened tendon of latissimus dorsi, as it twists around the teres major, may also be felt.

Muscles adducting the arm at the shoulder joint

Coracobrachialis
Pectoralis major (p. 61)
Latissimus dorsi (p. 63)
Teres major

Coracobrachialis

The only true representative in the arm of the adductor group of muscles found on the leg, coracobrachialis (Fig. 2.22) arises via a rounded tendon, in conjunction with the short head of biceps brachii, from the *apex* of the *coracoid process* of the *scapula* and attaches by a flat tendon to the *medial side* of the *shaft* of the *humerus* at its midpoint, between triceps and brachialis. Some fibres may continue into the medial intermuscular septum of the arm.

Nerve supply

By the *musculocutaneous nerve* (root value C6 and 7) from the lateral cord of the brachial plexus. However, as it pierces the muscle, the nerve to coracobrachialis may arise directly from the lateral cord of the brachial plexus. The skin over the muscle is supplied by roots T1 and T2.

Action

Coracobrachialis is an adductor and weak flexor of the arm at the shoulder joint.

Palpation

Coracobrachialis can be seen and felt as a rounded muscular ridge on the medial side of the arm when it is fully abducted and then adducted against resistance.

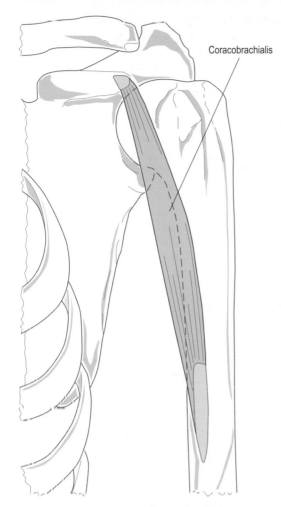

Coracobrachialis

Figure 2.22 The attachments of coracobrachialis, anterior view.

Muscles medially rotating the arm at the shoulder joint

Subscapularis
Teres major (p. 64)
Latissimus dorsi (p. 63)
Pectoralis major (p. 61)
Deltoid (anterior fibres) (p. 60)

Subscapularis

Forming the greater part of the posterior wall of the axilla, subscapularis (Fig. 2.23) lies close to teres major and latissimus dorsi. The anterior surface of the muscle lies on serratus anterior. When seen from the front it forms the upper boundary of the triangular and quadrangular spaces (see Fig. 2.71).

Subscapularis is a multipennate muscle which arises from the *medial two-thirds* of the *subscapular fossa* and from the tendinous *septa*, which reinforce the muscle, attached to bony ridges in the fossa. There is also an attachment to the fascia covering the muscle. The muscle fibres narrow and form a broad, thick tendon which attaches to the *lesser tubercle* of the *humerus*, the capsule of the shoulder joint and the front of the humerus below the tuberosity. A bursa, which communicates directly with the shoulder joint, separates the tendon from the neck of the scapula.

Nerve supply

By the *upper* and *lower subscapular nerves* (root value C5, 6 and 7) from the posterior cord of the brachial plexus.

Action

Subscapularis is a strong medial rotator of the arm at the shoulder joint; it may also assist in adduction of the arm.

Functional activity

As part of the 'rotator cuff', subscapularis plays an important role in maintaining the integrity of the shoulder joint during movement, by keeping the head of the humerus within the glenoid fossa. It also resists upward displacement of the humeral head when deltoid, biceps brachii and the long head of triceps are active.

Palpation

The muscle belly cannot be palpated as it lies deep to the scapula. However, careful deep palpation may allow the tendon to be felt just before its insertion onto the lesser tuberosity.

Muscles laterally rotating the arm at the shoulder joint

Teres minor
Infraspinatus
Deltoid (posterior fibres) (p. 60)

Teres minor

When seen from the back, teres minor (Fig. 2.24) forms the upper boundary of both the triangular and quadrangular spaces (see Fig. 2.71). It is a thin muscle which arises by two heads, separated by a groove for the circumflex scapular artery, from the *upper two-thirds* of the *lateral border* of the *scapula*, and the fascia between it and teres major (below) and infraspinatus (above). The fibres run upwards and laterally forming a narrow tendon which attaches to the *inferior facet* on the *greater tubercle* of the *humerus* and to the bone immediately below. The tendon reinforces

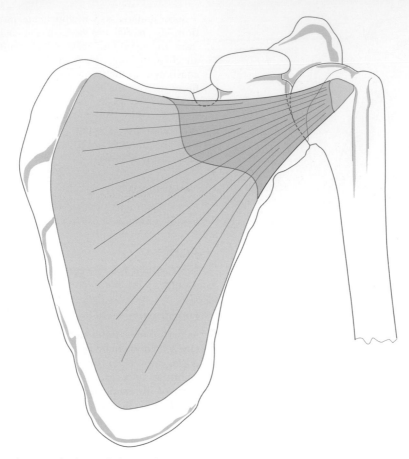

Figure 2.23 The attachments of subscapularis, anterior view.

and blends with the lower posterior part of the shoulder joint capsule.

Nerve supply

By the *axillary nerve* (root value C5 and 6) from the posterior cord of the brachial plexus. The skin over the muscle is supplied by roots T1, 2 and 3.

Action

In the anatomical position teres minor is a lateral rotator, but when the arm is abducted it laterally rotates and adducts.

Palpation

Teres minor can be felt contracting if the examiner's fingers are placed halfway up the lateral border of the scapula and the arm is then actively laterally rotated. The tendon is found just below that of infraspinatus as determined above.

Infraspinatus

A thick, triangular muscle, infraspinatus (Fig. 2.24) arises from the *medial two-thirds* of the *infraspinous fossa* of the *scapula*, tendinous intersections attached to ridges in this fossa, and the thick fascia covering the muscle. The fibres converge to a narrow tendon which inserts onto the *middle facet* on the *greater tubercle* of the *humerus*, and into the posterior part of the shoulder joint capsule. A bursa, which occasionally communicates with the shoulder joint, separates the muscle from the neck of the scapula. The upper part of the muscle lies deep to trapezius, deltoid and the acromion process; however, the lower part is superficial.

Nerve supply

By the *suprascapular nerve* (root value C5 and 6) from the upper trunk of the brachial plexus. The skin over the muscle is supplied by the dorsal rami of T1 to T6.

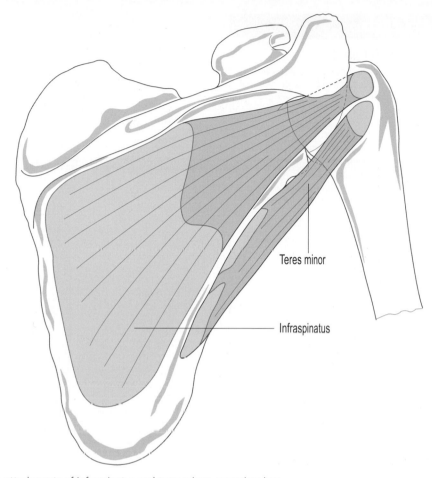

Teres minor

Infraspinatus

Figure 2.24 The attachments of infraspinatus and teres minor, posterior view.

Action

Infraspinatus is a lateral rotator of the arm at the shoulder joint.

Palpation

When the arm is laterally rotated, contraction of infraspinatus can be felt in the medial part of the infraspinous fossa. Its tendon can be palpated if the greater tubercle is moved from below the acromion. To accomplish this, the subject lies prone supporting themselves on the elbows and forearms. The arm is then laterally rotated some 25° and slightly adducted. The tendon can now be palpated just below the acromial angle. It is at this point that soft tissue techniques, such as transverse frictions or electrical treatments, are applied if the tendon becomes inflamed.

Functional activity

Infraspinatus and teres minor are of importance during the sequence of movements which occur when the arm is fully abducted. During the latter part of this movement the humerus is laterally rotated so that the greater tubercle moves clear of the coracoacromial arch, thereby enabling the remaining part of the humeral head to come into contact with the glenoid fossa, and full abduction to occur (p. 124).

Teres minor, infraspinatus, supraspinatus and subscapularis, the musculotendinous 'rotator cuff' of extensible ligaments around the shoulder joint, are all concerned with its stability; the proximity of their tendons to the joint enhances their effect. During movements of the head of the humerus on the glenoid fossa, interplay between these muscles reduces the sliding and shearing movements which tend to occur. When carrying a weight in the hand these same four muscles brace the head of the humerus against the glenoid fossa.

Movements at shoulder joint

The shoulder joint consists of the head of humerus and glenoid fossa of the scapula. It is capable of a wide range of independent movements produced by the following muscles:

Movement	Muscles
Abduction	Supraspinatus
	Deltoid
Adduction	Coracobrachialis
	Pectoralis major
	Latissimus dorsi
	Teres major
Flexion	Pectoralis major
	Deltoid (anterior fibres)
	Coracobrachialis
	Biceps (long head)
Extension	Latissimus dorsi
	Teres major
	Pectoralis major (to midline)
	Deltoid (posterior fibres)
	Triceps (long head)
Medial rotation	Subscapularis
	Teres major
	Latissimus dorsi
	Pectoralis major
	Deltoid (anterior fibres)
Lateral rotation	Teres minor
	Infraspinatus
	Deltoid (posterior fibres)

- All the muscles listed contribute to the stability of the shoulder joint.
- It is important to consider the effects of gravity or resistance on the above movements and muscle work – for example, adduction against resistance is produced by the muscles listed, whereas when lowering the arm (adduction) with gravity producing the movement, the abductors will work eccentrically to control the rate of descent.

Muscles flexing the elbow joint

Biceps brachii
Brachialis
Brachioradialis
Pronator teres (p. 75)

Biceps brachii

A prominent fusiform muscle on the anterior aspect of the arm (Fig. 2.25). It arises by two tendinous heads (short and long) at its upper end, and attaches by one tendinous insertion and one aponeurotic insertion at its lower end.

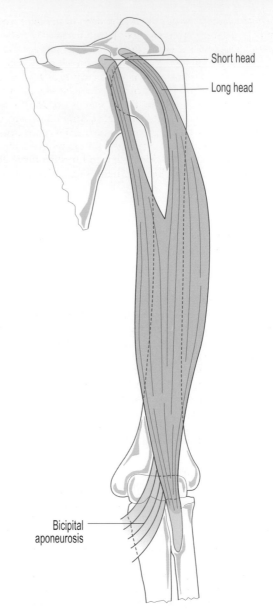

Figure 2.25 The attachments of biceps brachii, anterior view.

The upper end is covered by deltoid and pectoralis major, but the main part of the muscle is only covered by skin and subcutaneous fat.

The short head arises by a flat tendon, shared with cora-cobrachialis, from the *apex* of the *coracoid process* of the *scapula*. The long head arises from the *supraglenoid tubercle* of the *scapula* and adjacent *glenoid labrum* of the shoulder joint. The tendon of the long head runs within the shoulder joint enclosed within a synovial sleeve and leaves to enter the *intertubercular groove* by passing deep to the transverse humeral ligament (see Fig. 2.66). The two fleshy bellies continue towards the elbow fusing to form a single muscle just below the middle of the arm. At the elbow, a single flattened tendon is formed which twists through 90° before attaching to the *posterior part* of the *radial tuberosity*. A bursa separates the tendon from the remainder of the radial tuberosity. The prominent *bicipital aponeurosis*, a strong membranous band arising from the lateral side of the main tendon, runs downwards and medially across the cubital fossa, in front of the brachial artery and median nerve, to attach to the deep fascia on the ulnar side of the forearm (Fig. 2.25).

Nerve supply

By the *musculocutaneous nerve* (root value C5 and 6) from the lateral cord of the brachial plexus. The skin over biceps is supplied by the roots C5, 6, T2 and 3.

Action

Biceps brachii is not only an important flexor of the elbow joint, but also a powerful supinator of the forearm. Often these two actions are performed together with any unwanted actions being cancelled by antagonists. Maximum power is achieved for both flexion and supination with the elbow at 90°. When the elbow is fully extended the supinating action of biceps is lost. Biceps is also a flexor of the shoulder joint, and the fact that the long head crosses the superior part of the joint means that it has an important stabilizing role.

Functional activity

Biceps may use its supinatory and flexing actions sequentially in an activity, as, for example, in inserting a corkscrew and pulling out the cork. During this activity the head of the ulna may move medially due to the force of the biceps contraction transmitted to its posterior border via the bicipital aponeurosis.

When deltoid is paralysed, the long head of biceps can be re-educated to abduct the shoulder. This is achieved by laterally rotating the humerus at the shoulder joint in order to put the long head into a more appropriate position.

Palpation

With the elbow flexed to 90° and the forearm pronated, the muscle can be felt contracting in the middle of the arm when supination against resistance is attempted.

The lower part of the muscle is easily palpated through the skin. Proximally, each tendon may be palpated but with some degree of difficulty. The tendon of the long head lies between the greater and lesser tubercles of the humerus. Having determined these, firm deep pressure between them is needed to locate the tendon. This is the point at which

deep transverse frictions or electrical treatments are applied when the tendon becomes inflamed.

The short head can be found by first palpating the apex of the coracoid process, and then placing the fingers just below it. As the elbow is flexed, the tendon can be felt to stand out.

At the elbow, the tendon of insertion is best palpated with the elbow flexed to 20°. In this position it can be easily gripped between the index finger and thumb. If, in this same position, the subject is asked to resist a strong downward pressure on the forearm, the upper border of the bicipital aponeurosis can be seen and felt as a crescentic border running downwards and medially from the main tendon.

The tendon of biceps is the point at which the biceps reflex is tested, often by the examiner placing their thumb over the tendon and then tapping it with a patella hammer. The resultant reflex contraction can be felt below the thumb and biceps may be seen contracting if the reflex is brisk enough.

Brachialis

Brachialis (Fig. 2.26) lies under cover of biceps brachii in the lower half of the anterior aspect of the arm. It arises from the *distal two-thirds* of the *anterior surface* of the *shaft* of the *humerus* extending outwards onto the *medial and lateral intermuscular septa*. The muscle fibres are separated from the lower part of the lateral intermuscular septum by brachioradialis, with which it may be partly fused, and extensor carpi radialis longus. The fibres converge to a thick tendon which forms the floor of the cubital fossa and attaches to the rough *triangular brachialis impression* on the *inferior* part of the *coronoid process* and *tuberosity of the ulna*. Some deeper fibres of brachialis insert into the capsule of the elbow joint serving to pull it away from the moving bones during flexion preventing it from becoming trapped.

Nerve supply

Mainly mainly by the *musculocutaneous nerve* (root value C5 and 6) from the lateral cord of the brachial plexus, but also by a branch from the *radial nerve* from the posterior cord of the brachial plexus (root value C5 and 6), as it runs along the lateral border of the muscle. However, this latter branch is thought to be almost entirely sensory.

Action

Brachialis is the main flexor of the elbow joint.

Functional activity

Although brachialis flexes the elbow, it is important in controlling the extension produced by gravity. In this situation, the flexors of the elbow control the movement by an eccentric contraction.

Brachioradialis

Brachioradialis is a superficial muscle on the lateral side of the forearm extending almost as far as the wrist (Fig. 2.27). Brachioradialis forms the lateral border of the cubital fossa, and is covered in its upper part by brachialis, with which it may be partly fused. Its proximal attachment is to the *upper two-thirds* of the *front* of the *lateral supracondylar ridge* of the *humerus* and the adjacent part of the *lateral intermuscular septum*. From here, the fibres run downwards forming a long, narrow, flat tendon in the middle of the lateral side of the forearm. The tendon is crossed by those of abductor pollicis longus and extensor pollicis brevis before it attaches to the *lateral surface* of the *radius* just above the styloid process.

Nerve supply

By a branch from the *radial nerve* (root value C5 and 6) from the posterior cord of the brachial plexus, which enters its medial side above the elbow. The skin over the muscle is also supplied by roots C5 and 6.

Action

Brachioradialis flexes the elbow joint, particularly when the forearm is midway between pronation and supination. It also helps to return the forearm to this mid position from the extremes of either pronation or supination; this can be confirmed by palpation.

Functional activity

The brachioradialis acts primarily to maintain the integrity of the elbow joint since its fibres run more or less parallel to the radius. It also works eccentrically as an extensor of the elbow joint in activities such as hammering.

Palpation

With the elbow flexed to 90° and the forearm in a midpronated position, brachioradialis can be felt along the top of the forearm when the position is maintained against resistance. Using firm pressure, the tendon can be palpated proximal to the radial styloid process. The brachioradialis reflex can be elicited by firmly tapping its tendon just above the wrist.

Muscles extending the elbow joint

Triceps brachii
Anconeus

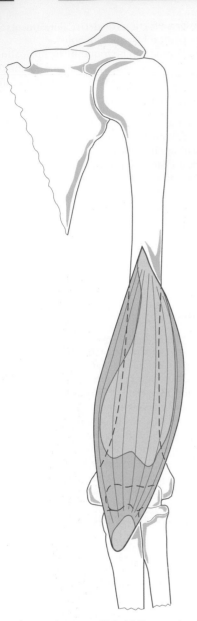

Figure 2.26 The attachments of brachialis, anterior view.

Palpation

When biceps brachii has been identified (p. 69), brachialis can be felt extending either side of its belly with the elbow flexed. The tendon can be palpated by applying deep pressure just above the coronoid process of the ulna.

Triceps brachii

Situated on the posterior aspect of the arm, triceps (Fig. 2.28), as its name suggests, arises by three heads. Two of the heads arise from the humerus, being separated by the spiral groove, and the third from the scapula; the three heads are referred to as the long head, lateral head and medial head. The muscle attaches via a tendon to the olecranon of the ulna.

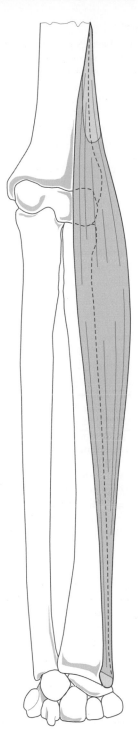

Figure 2.27 The attachments of brachioradialis, anterior view.

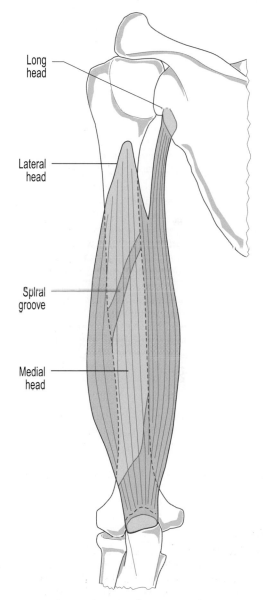

Figure 2.28 The attachments of triceps, posterior view.

Long head The tendinous long head comes from the *infraglenoid tubercle* of the *scapula* and the adjacent *glenoid labrum*, where it blends with the lower part of the shoulder joint capsule. It is the most medial of the three heads, the fibres running downwards superficial to the medial head before joining the tendon of insertion. As the long head descends from the infraglenoid tubercle, it passes anterior to teres minor and posterior to teres major. In its course it forms the medial border of the quadrangular space and triangular interval, and the lateral border of the triangular space (see Fig. 2.71).

Lateral head The fleshy lateral head arises *above* and *lateral* to the *spiral groove* on the *posterior surface* of the *humerus* between the attachments of the teres minor and deltoid. As the fibres pass to join with those of the medial head, they cover the spiral groove.

Medial head The large, fleshy medial head lies deep to the other two, and arises from the *posterior surface* of the *humerus*, *below* and *medial* to the *spiral groove* as far distally as the olecranon fossa. It has an additional attachment to the posterior aspect of the *medial* and *lateral intermuscular septa*.

The three heads of triceps come together to form a broad, laminated tendon, the superficial part of which covers the posterior aspect of the lower third of the muscle, while the deeper part arises from within the substance of the muscle. This arrangement provides a larger surface area for the attachment of the muscle fibres. Both laminae blend to form a single tendon which attaches to the *posterior part* of the *proximal surface* of the *olecranon* of the *ulna*, and to the *deep fascia* of the *forearm* on either side. Some muscle fibres from the medial head attach to the posterior part of the capsule of the elbow joint serving to pull it clear of the moving bones preventing it becoming trapped during extension of the joint.

Nerve supply

All three parts of the muscle are supplied separately by branches from the *radial nerve* from the posterior cord of the brachial plexus. The branch to the lateral head is derived from C6, 7 and 8, while those to the long and medial heads come from C7 and 8. Of these, the medial head receives two branches, one of which accompanies the ulnar nerve for some distance before entering the distal part of the muscle. The other branch enters more proximally, continuing through the muscle to end in, and supply, anconeus. The skin over the muscle is supplied by roots C5, 7, T1 and 2.

Action

Triceps brachii is the extensor of the elbow joint. The long head can also adduct the arm and extend it from a flexed position.

Functional activity

Once the elbow has been flexed, gravity often provides the necessary force for extension, with the elbow flexors working eccentrically to control the movement. Triceps only becomes active in this form of extension when the speed of the movement becomes important as in executing a karate chop. Triceps works strongly in pushing and punching activities, and when performing 'press-ups'. In the latter it is working concentrically in the upward movement and eccentrically in the downward movement. It works in a similar manner when using the arms to get out of or to lower oneself into a chair with arms, or when using crutches or parallel bars to relieve body-weight from the legs during walking. When using a wheelchair, triceps brachii works strongly to push the wheel around and so propel the chair forwards.

Triceps brachii is also an important extensile 'ligament' on the undersurface of the shoulder joint capsule during abduction of the arm.

Palpation

The bulk of the triceps is easy to see and feel on the posterior aspect of the arm (Fig. 2.29). All three heads can be felt contracting if the subject flexes the elbow to 90° with the hand resting on a table, and then alternately presses downwards and relaxes. The long head can be felt high up on the back of the arm almost at the axilla; careful palpation enables it to be traced almost to its insertion on the scapula. The lateral head can be felt on the upper lateral part of the arm, extending as far round as the biceps brachii, while the medial head, covered by the other two heads, can be felt contracting just above the olecranon.

The thick tendon of the triceps can be easily gripped between the thumb and index finger of the examiner's hand, just above the olecranon of the ulna. The triceps reflex is elicited by tapping the tendon just above its insertion, with the elbow slightly flexed.

Anconeus

A small triangular muscle situated immediately behind the elbow joint, anconeus (Fig. 2.30) appears almost to be part of triceps brachii. It arises from the *posterior surface* of the *lateral epicondyle* of the *humerus* and the adjacent part of the elbow joint capsule. The fibres pass medially and distally to attach to the *lateral surface* of the *olecranon* and *upper quarter* of the *posterior surface* of the *ulna* and to the fascia which covers it.

Nerve supply

By a branch of the *radial nerve* (root value C7 and 8) to the medial head of triceps. The skin over anconeus is supplied by root T1.

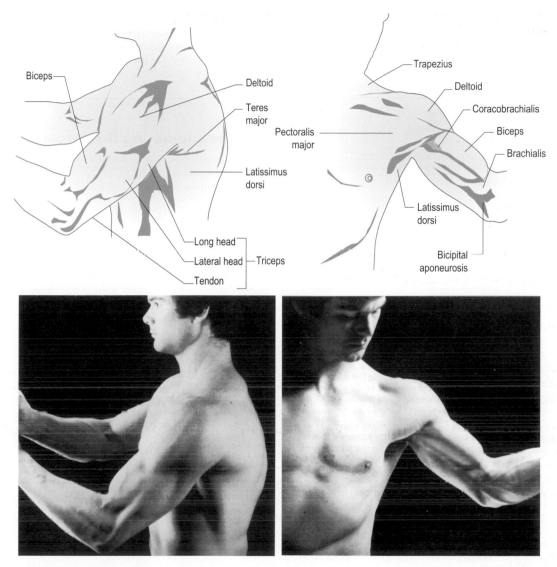

Figure 2.29 Muscles of the shoulder region and arm. *(Lower figures, reproduced with permission from Keogh B, Ebbs S (1984) Normal Surface Anatomy, William Heinemann Medical Books Ltd.)*

Action

Anconeus assists in extension of the elbow joint.

Functional activity

By virtue of its long attachment on the ulna, it is thought that anconeus produces lateral movement (abduction of the ulna) and extension of the bone at its distal end. These movements occur during pronation and are essential if a tool, such as a screwdriver, is being used. The movement of the ulna with respect to the radius allows the axis of pronation and supination to be altered so that the forearm rotates about a single axis and so does not describe an arc. This can be seen on an articulated skeleton. This action allows the rotatory movement of the forearm to be transmitted along the screwdriver into the head of the screw.

Palpation

Anconeus can be palpated between the lateral epicondyle of the humerus and the upper part of the ulna during pronation and supination, particularly if the axis of rotation is maintained through the extended index finger. As a practical exercise, it can be demonstrated that anconeus alters the axis of pronation and supination by making each of the fingers, in turn, the axis of rotation.

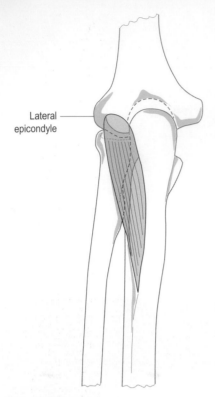

Lateral epicondyle

Figure 2.30 The attachments of anconeus, posterior view.

Section summary

Movements at elbow joint

The elbow joint consists of the distal end of the humerus and upper ends of the radius and ulna. It is capable of only two major movements produced by the following muscles:

Movement	Muscles
Flexion	Biceps brachii
	Brachialis
	Brachioradialis
	Pronator teres
	Superficial forearm flexors:
	Flexor digitorum superficialis
	Flexors carpi radialis and ulnaris
	Palmaris longus
Extension	Triceps
	Anconeus
	Superficial forearm extensors:
	Extensors carpi radialis longus and brevis
	Extensor carpi ulnaris
	Extensors digitorum and digiti minimi

- The muscles shown in italics have their primary action at the wrist or in the digits. Once this has been achieved they can also act to aid movement at the elbow joint.

- During many functional activities extension of the elbow is produced by gravity and is controlled by eccentric work of the flexors.
- Anconeus can move the ulna into slight abduction to alter the axis for pronation/supination during some fine movements.

Muscles supinating the forearm

Supinator
Biceps brachii (p. 68)
Brachioradialis (p. 70)

Supinator

Deep in the upper part of the forearm, supinator (Fig. 2.31) lies concealed by the superficial muscles as it surrounds the upper end of the radius. Its two heads arise in a continuous manner from the *inferior aspect* of the *lateral epicondyle* of the *humerus*, the *radial collateral ligament*, the *annular ligament*, the *supinator crest* and *fossa* of the *ulna*. It is often convenient, however, to think of supinator as arising by two heads, humeral and ulnar, between which passes the posterior interosseous nerve to gain access to the extensor compartment of the forearm. From this extensive origin, the muscle fibres pass downwards and laterally to wrap around the upper third of the radius. They insert into the *posterior*, *lateral* and *anterior aspects* of the *radius*, as far forwards as the anterior margin between the neck and the attachment of pronator teres.

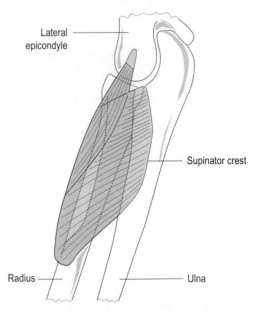

Lateral epicondyle

Supinator crest

Radius

Ulna

Figure 2.31 The attachments of supinator of the left arm, lateral view.

Nerve supply

By the *posterior interosseous branch* of the *radial nerve* (root value C5 and 6). However, the skin overlying supinator is supplied by roots C5, 6 and T1.

Action

As its name suggests, supinator supinates the forearm, in which there is an anterolateral movement of the distal end of the radius around the ulna causing the two bones to lie parallel to each other. Unless a particularly powerful supinatory action is required, supinator is probably the prime mover. However, if a powerful movement is required, biceps brachii is also recruited. It must be remembered, however, that biceps brachii cannot function as a supinator with the elbow fully extended, and consequently powerful supinatory movements are performed with the elbow flexed to about 120°.

Palpation

With the arm fully extended at the elbow, and the forearm in a midpronated position, supinator can be felt contracting over the posterior part of the upper third of the radius when the arm is supinated against resistance.

Muscles pronating the forearm

Pronator teres
Pronator quadratus
Brachioradialis (p. 70)

Pronator teres

Forming the medial border of the cubital fossa at the elbow, pronator teres (Fig. 2.32) is the most lateral of the superficial muscles in the flexor compartment of the forearm. It arises by two heads – the humeral head and the ulnar head. The humeral head arises from the lower part of the *medial supracondylar ridge* and adjacent *intermuscular septum*, as well as from the common flexor origin on the *medial epicondyle* of the *humerus* and the covering fascia. The ulnar head arises from the *pronator ridge* on the *ulna*, which runs downwards from the medial part of the *coronoid process*, joining the humeral head on its deep surface. Between these two heads passes the median nerve. The muscle fibres pass downwards and laterally to attach via a flattened tendon into a *roughened oval area* on the middle of the *lateral surface* of the *radius*.

Nerve supply

By the *median nerve* (root value C6 and 7) from the medial and lateral cords of the brachial plexus. The overlying skin is supplied by roots C6 and T1.

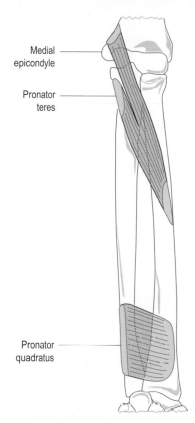

Figure 2.32 The attachments of pronator teres and pronator quadratus of the left arm, anterior view.

Action

Pronator teres pronates the forearm by producing an anteromedial movement of the lower end of the radius across the ulna, carrying the hand with it. Pronator teres is also a weak flexor of the elbow.

Palpation

The muscle can be palpated running along the medial border of the cubital fossa between the medial epicondyle of the humerus and the middle of the radius. Pronator teres can be most easily felt, and occasionally seen, when resisting pronation of the forearm.

Pronator quadratus

Pronator quadratus is a fleshy, quadrangular muscle lying within the flexor compartment of the forearm (Fig. 2.32), passing transversely from the *lower quarter* of the *anterior surface* of the *ulna* to the *lower quarter* of the *anterior surface* of the *radius*. Some of its deeper fibres attach to the triangular area above the ulnar notch of the radius.

Nerve supply

By the *anterior interosseous branch* of the *median nerve* (root value C8 and T1). The skin overlying the muscle is supplied by roots C6, 7 and 8.

Action

Pronator quadratus initiates pronation of the forearm. The transverse nature of its fibres allows the lower ends of the radius and ulna to be held together when upward pressure is applied, for example when the hand is weight-bearing. It therefore protects the inferior radioulnar joint.

Palpation

Pronator quadratus is difficult to palpate because of its deep position, but if firm pressure is applied between the long flexor tendons in the lower part of the forearm, contraction of the muscle may be felt when it acts against resistance.

Section summary

Movements at radioulnar joints

The radius and ulna articulate at their proximal and distal ends at the superior and inferior radioulnar joints. The lower end of the radius is moved around the ulna by the following muscles:

Movement	Muscles
Supination	Supinator
	Biceps brachii
	Brachioradialis
Pronation	Pronators teres and quadratus
	Brachioradialis

- Movement of the radius around the ulna carries the hand with it and is therefore an important component of hand function.
- The proximal end of the radius remains lateral to the ulna whilst the lower end crosses to the medial side during pronation and returns during supination.

Muscles flexing the wrist

Flexor carpi ulnaris
Flexor carpi radialis
Palmaris longus
Flexor digitorum superficialis (p. 81)
Flexor digitorum profundus (p. 82)
Flexor pollicis longus (p. 84)

Flexor carpi ulnaris

Lying along the medial border of the forearm flexor carpi ulnaris (Fig. 2.33A) is the most medial of the superficial flexor group of muscles. It arises from the humerus and

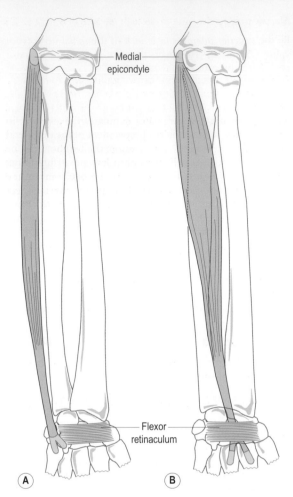

Figure 2.33 The attachments of (A) flexor carpi ulnaris and (B) flexor carpi radialis, anterior view.

ulna. The humeral head arises from the *common flexor origin* on the *medial epicondyle* of the *humerus* and the adjacent fascia. The ulnar head arises from the *medial border* of the *olecranon* and, by an aponeurotic attachment, from the *upper two-thirds* of the *posterior border* of the *ulna*. Between these two heads passes the ulnar nerve to gain the medial side of the flexor compartment of the forearm.

The muscle forms a long tendon about halfway down the forearm, which attaches to and invests the *pisiform*. The tendon is prolonged to reach the *hook of the hamate* and *base* of the *fifth metacarpal* by the pisohamate and pisometacarpal ligaments respectively. Occasionally, some fibres may be prolonged into abductor digiti minimi. Lateral to the tendon are the ulnar nerve and blood vessels.

Nerve supply

By several branches from the *ulnar nerve* (root value C7 and 8) from the medial cord of the brachial plexus.

The skin overlying the muscle is supplied by roots C8 and T1.

Action

In conjunction with flexor carpi radialis, and to some extent palmaris longus, flexor carpi ulnaris flexes the hand at the wrist. However, when working with extensor carpi ulnaris, it produces adduction or ulnar deviation of the hand at the wrist. It also plays an important role in stabilizing the pisiform during abduction of the little finger, so that abductor digiti minimi has a firm base from which to work. As with flexor carpi radialis, flexor carpi ulnaris is an important synergist in extension of the fingers, preventing unwanted extension of the wrist.

Palpation

The tendon of flexor carpi ulnaris can easily be identified running proximally from the pisiform where it can be pinched between the thumb and index finger. In the upper medial part of the forearm, the muscle belly can be palpated when flexion of the wrist is performed against resistance.

Flexor carpi radialis

A fusiform muscle, flexor carpi radialis (Fig. 2.33B) is the most lateral of the superficial flexor muscles in the lower half of the forearm. It arises from the *medial epicondyle* of the *humerus* via the *common flexor tendon* and the adjacent fascia. Halfway down the forearm, the muscle fibres condense to form a long tendon which passes beneath the flexor retinaculum, where it lies in its own lateral compartment within the carpal tunnel. Here the tendon is surrounded by its own synovial sheath as it grooves the trapezium. Distally the tendon inserts into the *palmar surface* of the *bases* of the *second* and *third metacarpals*. Its course in the forearm is oblique, running from medial to lateral and from above downwards. At the wrist, the tendon lies between the radial vessels laterally and median nerve medially.

Nerve supply

By the *median nerve* (root value C6 and 7) from the medial and lateral cords of the brachial plexus. The skin over the muscle is supplied by roots C6 and T1.

Action

Working with palmaris longus and flexor carpi ulnaris, flexor carpi radialis acts to flex the wrist. Abduction or radial deviation of the wrist is produced by the combined action of flexor carpi radialis and extensors carpi radialis longus and brevis. Because of its oblique course in the forearm, flexor carpi radialis may aid pronation. It can also help to flex the elbow. It works in a similar way to flexor carpi ulnaris in preventing unwanted extension of the wrist when extending the fingers.

Palpation

When the wrist joint is flexed and abducted, the tendon of flexor carpi radialis can be palpated as the most lateral of the tendons on the anterior aspect of the wrist, at the level of the radial styloid process.

Palmaris longus

A small vestigial muscle palmaris longus (Figs. 2.34 and 2.35) is absent in about 10% of the population. Lying centrally among the superficial flexor muscles of the forearm, palmaris longus arises from the front of the *medial epicondyle* of the *humerus* via the *common flexor origin*. The short muscle fibres soon form a long and slender tendon which passes distally to attach to the superficial surface of the *flexor retinaculum* and inserts into the *apex* of the *palmar aponeurosis*. At the wrist, the tendon lies on top of the median nerve.

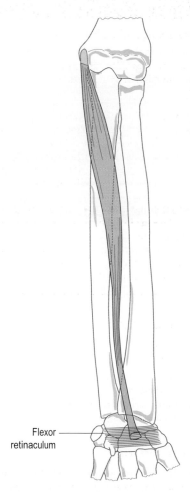

Flexor retinaculum

Figure 2.34 The attachments of palmaris longus, anterior view.

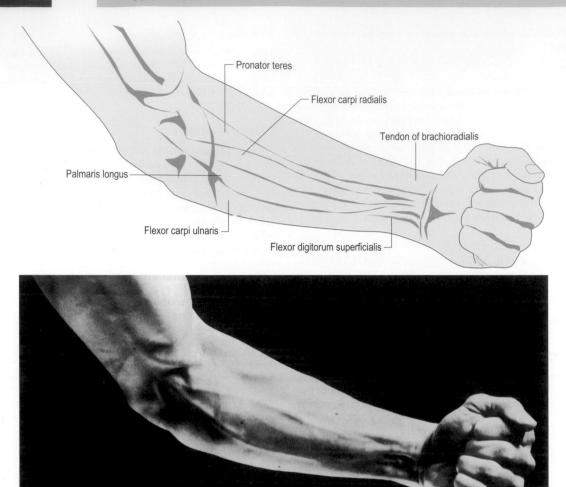

Figure 2.35 Flexor surface of the left forearm. *(Lower part, reproduced with permission from Keogh B, Ebbs S (1984) Normal Surface Anatomy, William Heinemann Medical Books Ltd.)*

Nerve supply

By the *median nerve* (root value C8) from the medial and lateral cords of the brachial plexus. The skin over the muscle is supplied by roots C7 and T1.

Action

Palmaris longus is a weak flexor of the wrist. However, because of its attachment to the palmar aponeurosis, it may have some slight action in flexing the metacarpophalangeal joints as it tightens the palmar fascia.

Palpation

The tendon of palmaris longus can be identified just proximal to the wrist, where it is the most central structure when flexion of the wrist is resisted. The tendon lies on the medial side of that of flexor carpi radialis.

Muscles extending the wrist

Extensor carpi radialis longus
Extensor carpi radialis brevis
Extensor carpi ulnaris
Extensor digitorum (p. 85)
Extensor indicis (p. 87)
Extensor digiti minimi (p. 87)
Extensor pollicis longus (p. 88)
Extensor pollicis brevis (p. 89)

Extensor carpi radialis longus

Lying on the lateral side of the posterior compartment of the forearm, extensor carpi radialis longus (Fig. 2.36A) is partly covered by brachioradialis. It arises from the *anterior part* of the *lower third* of the *lateral supracondylar ridge* of the *humerus* and adjacent *intermuscular septum*. Occasionally,

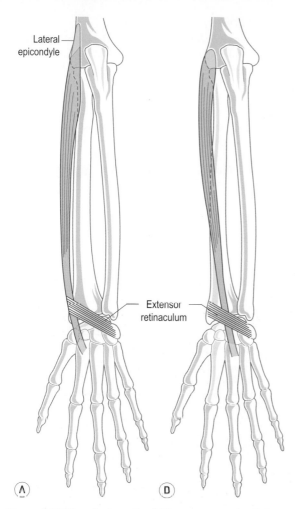

Figure 2.36 The attachments of (A) extensor carpi radialis longus and (B) extensor carpi radialis brevis, posterior view.

there may be an attachment to the lateral epicondyle by the common extensor tendon. Approximately in the middle of the forearm, the muscle forms a flattened tendon which runs distally over the lateral surface of the radius. In the lower third of the forearm, the tendon, together with that of extensor carpi radialis brevis, is crossed by the tendons of abductor pollicis longus and extensor pollicis brevis. The tendons of extensor carpi radialis longus and brevis pass deep to the extensor retinaculum in a common synovial sheath (see Fig. 2.109B). Together they groove the posterior surface of the styloid process of the radius. The tendon of extensor carpi radialis longus attaches to the *posterior surface* of the *base* of the *second metacarpal*.

Nerve supply

By the *radial nerve* (root value C6 and 7) from the posterior cord of the brachial plexus, which enters the muscle above the elbow. The skin over the muscle is supplied by roots C5 and 6.

Action and palpation

These are considered with extensor carpi radialis brevis.

Extensor carpi radialis brevis

Lying adjacent to and partly covered by extensor carpi radialis longus, to which it may be partly fused, is extensor carpi radialis brevis (Fig. 2.36B). It arises from the *lateral epicondyle* of the *humerus* via the *common extensor tendon*, the *lateral ligament* of the *elbow* and the adjacent fascia. The tendon of extensor carpi radialis brevis forms halfway down the forearm and runs with that of extensor carpi radialis longus deep to abductor pollicis longus and extensor pollicis brevis. It passes below the extensor retinaculum, in a common synovial sheath, to attach to the *posterior surface* of the *base* of the *third metacarpal*.

Nerve supply

By the *posterior interosseous* branch of the *radial nerve* (root value C6 and 7). The skin over the muscle is supplied by roots C5, 6 and 7.

Action

Working with extensor carpi ulnaris, extensors carpi radialis longus and brevis produce extension of the wrist. Working with flexor carpi radialis, however, they will produce abduction (radial deviation) of the wrist. In addition, extensor carpi radialis longus may help to flex the elbow joint.

Functional activity

Functionally, the wrist extensors work strongly in the action of gripping, where they have a synergistic role. The synergy of extensors carpi radialis longus and brevis and carpi ulnaris is a vital factor in the gripping action. By maintaining the wrist in an extended position, flexion of the wrist under the action of flexors digitorum superficialis and profundus is prevented, with the result that these muscles act on the fingers. If the wrist is then allowed to flex the flexor tendons cannot shorten sufficiently to produce effective movement at the interphalangeal joints. This therefore becomes a state of active insufficiency.

If the radial nerve is damaged the patient is unable to produce an effective grip because of paralysis of the wrist extensors. However, with the wrist splinted in extension, the tendons of flexors digitorum superficialis and profundus act on the fingers and a functional grip can be obtained. (see also p. 191)

Palpation

When the wrist is extended and abducted against resistance, both extensors carpi radialis longus and brevis can be palpated in the upper lateral aspect of the posterior part of the forearm. The tendons, particularly longus, can be palpated in the floor of the 'anatomical snuffbox' if the same movement of extension and abduction is carried out.

Extensor carpi ulnaris

Extensor carpi ulnaris (Figs. 2.37 and 2.38) arises from the *lateral epicondyle* of the *humerus* via the *common extensor tendon* and the adjacent fascia. There is also a strong attachment, via a common *aponeurosis* shared with flexor digitorum profundus and flexor carpi ulnaris, from the *posterior border* of the *ulna*. The muscle forms a tendon near the wrist which passes below the extensor retinaculum in its own synovial sheath and compartment (see Fig. 2.109B) in a groove next to the ulnar styloid process. The tendon attaches to a *tubercle* on the *medial side* of the *base* of the *fifth metacarpal*.

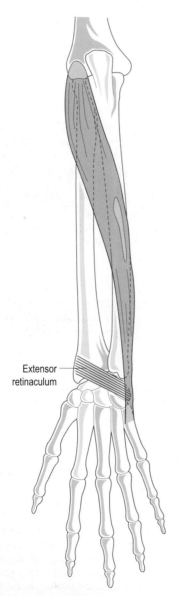

Extensor retinaculum

Figure 2.37 The attachments of the extensor carpi ulnaris, posterior view.

Nerve supply

By the *posterior interosseous* branch of the *radial nerve* (root value C7 and 8). The skin over the muscles is supplied by roots C6, 7 and 8.

Action

Working with extensors carpi radialis longus and brevis extensor carpi ulnaris produces extension of the wrist. The functional significance of this has been described in the actions of extensor carpi radialis muscles. Working with flexor carpi ulnaris, extensor carpi ulnaris produces adduction (ulnar deviation) at the wrist.

Palpation

The tendon of extensor carpi ulnaris can be identified on the dorsum of the wrist, when the wrist is extended and adducted against resistance. It lies on the lateral side of the ulnar styloid process.

Section summary

Movements at wrist joint

The wrist joint consists of the distal ends of the radius and ulna (via an intra-articular disc) and the proximal row of carpal bones. In addition to the prime movers working on the joint, other muscles crossing the wrist (to the fingers and thumb) also contribute to the movements.

Movement	Muscles
Flexion	Flexors carpi ulnaris and radialis
	Palmaris longus
	Flexors digitorum superficialis and profundus
	Flexor pollicis longus
Extension	Extensors carpi ulnaris and radialis longus and brevis
	Extensors digitorum, indicis and digiti minimi
	Extensors pollicis longus and brevis
Abduction (radial deviation)	Flexor carpi radialis
	Extensors carpi radialis longus and brevis
Adduction (ulnar deviation)	Flexor carpi ulnaris
	Extensor carpi ulnaris

- The muscles shown in italics have a primary function of flexing or extending the digits. Once this action has been achieved they act to move the wrist in continued action.
- The extensors of the wrist have an important functional action during gripping. Acting as synergists, they serve to prevent unwanted continued action of the finger flexors at the wrist. Holding the wrist in extension during gripping also prevents active insufficiency of the finger flexors.

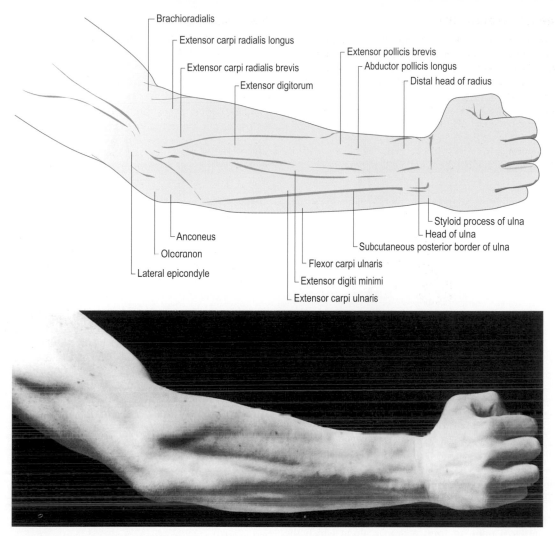

Figure 2.38 Extensor surface of the right forearm and hand. *(Lower part, reproduced with permission from Keogh B, Ebbs S (1984) Normal Surface Anatomy, William Heinemann Medical Books Ltd.)*

Muscles flexing the fingers

Flexor digitorum superficialis
Flexor digitorum profundus
Lumbricals
Flexor digiti minimi brevis

Flexor digitorum superficialis

A large muscle (Fig. 2.39A) lying in the anterior compartment of the forearm deep to pronator teres, palmaris longus and flexors carpi radialis and ulnaris and superficial to flexor digitorum profundus and flexor pollicis longus. It has a long, linear origin but may be considered to arise by two heads. The medial or humeroulnar head arises from the *medial epicondyle* of the *humerus* via the *common flexor tendon*, the anterior part of the *ulnar collateral ligament* and the *sublime tubercle* at the upper medial part of the

coronoid process of the *ulna*. The lateral (radial) head arises from the *upper two-thirds* of the *anterior border* of the *radius*, which runs downwards and laterally from the radial tuberosity.

About halfway down the forearm, the muscle narrows to form four separate tendons which pass deep to the flexor retinaculum where they are arranged in two pairs to enter the hand. The superficial pair pass to the middle and ring fingers, while the deep pair pass to the index and little fingers. Within the carpal tunnel the tendons of flexor digitorum superficialis are superficial to those of flexor digitorum profundus, with which they share a common synovial sheath (see Figs. 2.51A and 2.109B). In the palm, the tendons separate and pass towards their respective fingers, still lying superficial to the profundus tendons. At the level of the metacarpophalangeal joint, each superficialis tendon splits longitudinally into two parts. The two parts

81

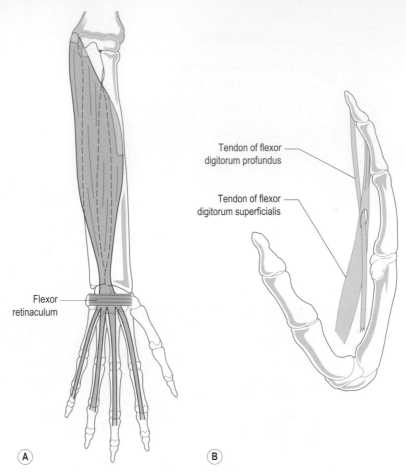

Figure 2.39 (A) The position and attachments of flexor digitorum superficialis, (B) the relationships and insertions of the long flexor tendons.

pass around the profundus tendon, twisting so that their outer surfaces unite to form a groove along which the tendon of flexor digitorum profundus passes. Prior to attaching to either side of the *palmar surface* of the *base* of the *middle phalanx*, the tendon splits again.

This peculiar arrangement of the superficialis tendons provides a tunnel which allows the profundus tendon to become superficial (Fig. 2.39B). The effect, as far as the tendon of flexor digitorum superficialis is concerned, is to increase its lever arm at the proximal interphalangeal joint, thereby enabling a powerful grip of the fingers to be exerted. As well as their main attachment to the middle phalanx, tendons of flexor digitorum superficialis also provide attachments for the vincula tendinum which convey blood vessels to the tendon.

Nerve supply

By the *median nerve* (root value C7, 8 and T1) from the medial and lateral cords of the brachial plexus. The skin overlying the muscle and its tendons are supplied by roots C6, 7, 8 and T1.

Action

Flexor digitorum superficialis is primarily a flexor of the metacarpophalangeal and proximal interphalangeal joints. Because it crosses the wrist joint, it also aids flexion of the wrist if its action is continued.

Palpation

Contraction of flexor digitorum superficialis can be felt by applying deep pressure through the superficial flexor muscles in the upper part of the forearm whilst the fingers are flexed. The tendons can be palpated in a similar manner proximal to the flexor retinaculum. The muscle can be tested specifically by asking the subject to flex the proximal interphalangeal joint without flexing the distal interphalangeal joint.

Flexor digitorum profundus

The flexor digitorum profundus (Fig. 2.40A) lies deep to flexor digitorum superficialis on the medial side of the forearm. It arises from the *medial side* of the *coronoid process* of

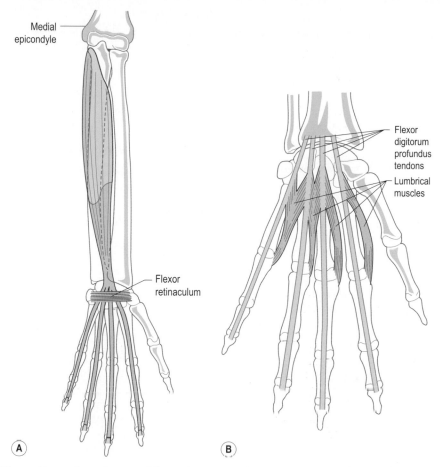

Figure 2.40 (A) The position and attachments of flexor digitorum profundus, anterior view, (B) the four lumbrical muscles, anterior view.

the *ulna*, the *upper three-quarters* of the *anterior* and *medial surfaces* of the *ulna*, and from the *medial, middle-third* of the *anterior surface* of the adjacent *interosseous membrane*. It also arises from the *aponeurosis* attaching flexor carpi ulnaris to the *posterior border* of the *ulna*.

The part of the muscle arising from the interosseous membrane forms a separate tendon about halfway down the forearm which passes to the index finger. The remaining tendons are not usually formed until just proximal to the flexor retinaculum. The separate tendons, however, pass below the flexor retinaculum where they lie side by side, deep to those of flexor digitorum superficialis, but within the same synovial sheath (see Figs. 2.51A and 2.109B). In the palm, the four tendons pass to their respective fingers. At first, they travel deep to superficialis, but then pass superficially as they go through the tunnel formed by the superficialis tendon at the level of the meta-carpophalangeal joint (Fig. 2.39B).

The tendon of flexor digitorum profundus eventually inserts into the *base* of the *palmar surface* of the *distal phalanx* having passed through a fibro-osseous tunnel

(see Fig. 2.126). Like the tendons of flexor digitorum superficialis, the profundus tendons are provided with vincula tendinum.

Nerve supply

Flexor digitorum profundus has a dual nerve supply. The lateral part of the muscle, which gives tendons to the index and middle fingers, is supplied by the *anterior interosseous* branch of the *median nerve* (root value C7, 8 and T1). The medial part of the muscle, which gives tendons to the ring and little fingers, is supplied by the *ulnar nerve* (root value C8 and T1) from the medial cord of the brachial plexus. The skin over the muscle is supplied by roots C7, 8 and T1.

Action

The primary action of flexor digitorum profundus is flexion of the distal interphalangeal joint. However, because it crosses several other joints during its course, it also aids in flexion of the proximal interphalangeal, metacarpopha-langeal and wrist joints.

Palpation

The muscular part of flexor digitorum profundus can be palpated immediately medial to the posterior border of the ulna, where its contraction can be felt as the fingers are fully flexed from a position of extension.

Lumbricals

Four small, round muscles found in the palm in association with the tendons of flexor digitorum profundus. The lateral two lumbricals are frequently unipennate arising from the *lateral side* of the *flexor digitorum profundus tendons*, while the medial two muscles are bipennate arising from the adjacent sides of the *tendons* of the middle and ring fingers (Fig. 2.40B).

From this proximal attachment the muscles pass distally to their respective fingers, anterior to the deep transverse metacarpal ligament and then obliquely on the lateral side of the metacarpophalangeal joint. They then attach to the *lateral edge* of the *dorsal digital expansion* at the side of the proximal phalanx (see Fig. 2.123C). A few fibres make their way to the bone of the middle phalanx, but the majority can be traced to the base of the distal phalanx via the extensor expansion.

Nerve supply

The nerve supply varies, but the most common arrangement is for the lateral two lumbricals to be supplied by the *median nerve* (root value T1) from the medial and lateral cords of the brachial plexus, and the medial two by the *ulnar nerve* (root value T1) from the medial cord of the brachial plexus.

Action

The lumbricals in both hands and feet are unique muscles as they pass between the flexor and extensor tendons. Their anatomical position means that in the hand they flex the metacarpophalangeal joint, and extend both interphalangeal joints of the corresponding finger, the 'lumbrical action'. In theory the lumbricals should be able to rotate their respective fingers at the metacarpophalangeal joint, however this appears to be well marked only in the index finger so that the pad faces medially. The first dorsal interosseous muscle is also involved in producing this movement. Functionally, the lumbricals are involved in the coordination of complex activities of the fingers which involve movements of both flexion and extension, for example as in writing. As a group they have major functional significance in the dexterity of the hand, which is further enhanced by their rich sensory innervation.

Flexor digiti minimi brevis

Flexor digiti minimi brevis is not always present, but when it is it lies on the lateral side of abductor digiti minimi, arising from the *hook of the hamate* and the adjacent *flexor retinaculum*. It inserts, together with abductor digiti minimi, into the *base* of the *proximal phalanx* of the *little finger* on its ulnar side (see Fig. 2.42).

Nerve supply

By the deep branch of the *ulnar nerve* (root value T1) from the medial cord of the brachial plexus.

Action

Flexor digiti minimi brevis flexes the metacarpophalangeal joint of the little finger.

Muscles flexing the thumb

Flexor pollicis longus
Flexor pollicis brevis

Flexor pollicis longus

Flexor pollicis longus (Fig. 2.41) lies on the lateral side of flexor digitorum profundus. It arises from the *anterior surface* of the *radius* between the radial tuberosity above and

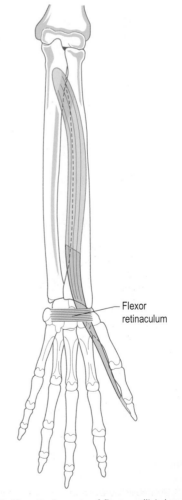

Flexor retinaculum

Figure 2.41 The attachments of flexor pollicis longus.

pronator quadratus below, and the adjacent *anterior surface* of the *interosseous membrane*. Occasionally, it also arises by a small slip from the medial border of the coronoid process of the ulna. The fibres pass almost to the wrist before a single tendon is formed.

The tendon of flexor pollicis longus passes below the flexor retinaculum, in its own synovial sheath (see Fig. 2.109B), to insert into the *palmar surface* of the *base* of the *distal phalanx* of the *thumb*.

Nerve supply

By the *anterior interosseous* branch of the *median nerve* (root value C8 and T1).

Action

The flexor pollicis longus is the only flexor of the interphalangeal joint of the thumb, and is thus vital for all gripping activities of the hand. It also flexes the metacarpophalangeal joint of the thumb and the wrist joint.

Palpation

When only the thumb is flexed contraction of flexor pollicis longus can be felt in the lower third of the forearm, immediately lateral to the superficial flexor tendons.

Flexor pollicis brevis

The most medial of the three thenar muscles, flexor pollicis brevis (Fig. 2.42) is usually partly fused with opponens pollicis. It arises from the distal border of the *flexor retinaculum* and *tubercle* of the *trapezium* (the superficial part), with a deeper attachment to the *capitate* and *trapezoid*. Both parts form a single tendon containing a sesamoid bone which inserts into the *radial side* of the *base* of the *proximal phalanx* of the *thumb*.

Nerve supply

By the *median nerve* (root value T1) from the medial and lateral cords of the brachial plexus, but frequently the deep part is supplied by the *ulnar nerve* (root value T1) from the medial cord of the brachial plexus. The skin over the muscle is supplied by root C6.

Action

Flexor pollicis brevis produces flexion of the thumb at the metacarpophalangeal and carpometacarpal joints. Its continued action tends to produce medial rotation of the thumb, since flexion of the carpometacarpal joint automatically involves this movement.

Palpation

Flexor pollicis brevis can be palpated in the medial part of the thenar eminence if flexion of the thumb is resisted.

Muscles extending the fingers

Extensor digitorum
Extensor digiti minimi
Extensor indicis
Interossei (p. 92)
Lumbricals (p. 84)

Extensor digitorum

Extensor digitorum (Fig. 2.43A) is centrally placed within the posterior compartment of the forearm. It arises from the *lateral epicondyle* of the *humerus* via the *common extensor tendon*, the covering fascia and the *intermuscular septa* at its sides. In the lower part of the forearm it forms four tendons which pass deep to the extensor retinaculum in a synovial sheath shared with the tendon of extensor indicis (see Fig. 2.109B). On the dorsum of the hand the tendons diverge towards the medial four digits, being interconnected by obliquely placed fibrous bands. The arrangement of these bands is variable; however, the tendons for the ring and little fingers are attached usually until just proximal to the metacarpophalangeal joints. The tendon to the index finger may not be attached to that of the middle finger.

The distal attachment of extensor digitorum is complex in that each tendon helps to form an aponeurosis over the dorsum of the fingers known as the *dorsal digital expansion* or *extensor hood*. In its simplest form it is best thought of as a movable triangular hood, the base of which lies proximally over the metacarpophalangeal joint. From here the sides of the hood wrap around the phalanx with the apex pointing distally. The extensor hood forms the dorsal part

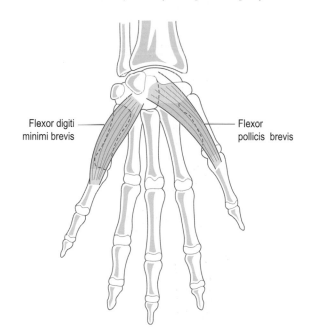

Flexor digiti minimi brevis

Flexor pollicis brevis

Figure 2.42 Flexor pollicis brevis and flexor digiti minimi brevis, anterior view.

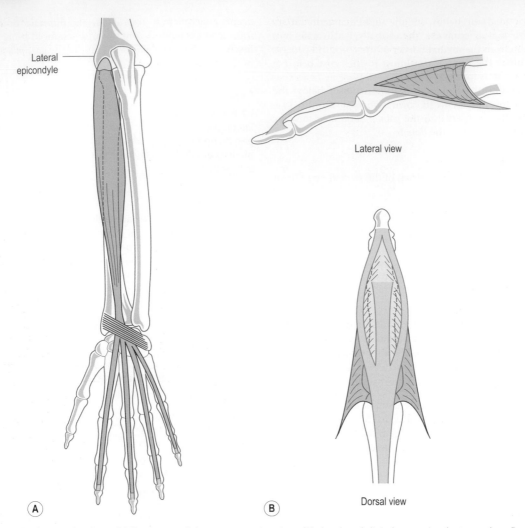

Lateral epicondyle

Lateral view

Dorsal view

(A)　　　(B)

Figure 2.43 The attachments of (A) extensor digitorum, posterior view, (B) the dorsal digital expansion (extensor hood).

of the capsule of the metacarpophalangeal joint and extends forwards either side of the metacarpal head to fuse with the deep transverse metacarpal ligament. As the extensor hood approaches the proximal interphalangeal joint, it narrows and is reinforced on either side by the interosseous and lumbrical muscles associated with that finger. At the distal end of the proximal phalanx, the extensor hood divides into three parts (Fig. 2.43B). The central part, which is directly continuous with the extensor tendon, inserts into the *base* of the *middle phalanx* on its *dorsal aspect*. The two collateral parts, which are continuous with the tendons of the interossei and lumbricals, reunite to insert into the *base* of the *distal phalanx* on its *dorsal aspect*. There may be an attachment to the base of the proximal phalanx but this is unusual.

The way in which the extensor hood wraps around the phalanges towards the palm facilitates the attachment of the lumbricals and interossei. These enable the complex co-ordinated movements of the fingers to take place, involving flexion of some joints and extension of others, as for example in writing. The overall arrangement of this complex structure is shown in Fig. 2.43B. Detachment or rupture of the dorsal digital expansion from the distal phalanx gives rise to 'mallet finger'.

Nerve supply

By the *posterior interosseous* branch of the *radial nerve* (root value C7 and 8). The skin over the muscle is supplied by roots C6, 7 and 8.

Action

The primary action of extensor digitorum is to produce extension of the metacarpophalangeal joint. It also helps to extend both interphalangeal joints: however, the main extensors of these joints are the interossei and lumbricals, which also help to prevent hyperextension of the metacarpophalangeal joint. When these small muscles are paralysed, extensor digitorum hyperextends the metacarpophalangeal joint and is then unable to extend the interphalangeal joints. Extensor digitorum also extends the wrist.

Palpation

The tendons of extensor digitorum can easily be palpated on the dorsum of the hand when the fingers are extended. It may also be possible to identify the fibrous interconnections between some of these tendons. Contraction of the muscle belly can be palpated in the posterior central part of the upper forearm during the same movement.

Extensor digiti minimi

A small muscle lying on the medial side of the extensor digitorum (Fig. 2.44A). It arises from the *lateral epicondyle* of the *humerus* via the *common extensor tendon* and the surrounding fascia. In the lower part of the forearm, extensor digiti minimi forms a single tendon which passes deep to the extensor retinaculum in a separate compartment in its own synovial sheath (see Fig. 2.109B). Below the extensor retinaculum it lies immediately behind the inferior radioulnar joint. On the dorsum of the hand, the tendon splits into two with both parts inserting into the *dorsal digital expansion* of the *little finger*. The double tendon of extensor digiti minimi lies medial to that of the extensor digitorum.

Nerve supply

By the *posterior interosseous* branch of the *radial nerve* (root value C7 and 8). The skin over the muscle is supplied by roots C8 and T1.

Action

Extensor digiti minimi assists extensor digitorum in extending the metacarpophalangeal joint of the little finger, and via the extensor hood, the interphalangeal joints. It can also cause ulnar deviation of the little finger. The muscle also aids extension of the wrist.

Palpation

The tendon of extensor digiti minimi can be palpated just distal to the inferior radioulnar joint when the little finger is extended. The muscle belly can be palpated medial to that of extensor digitorum when the same movement is performed.

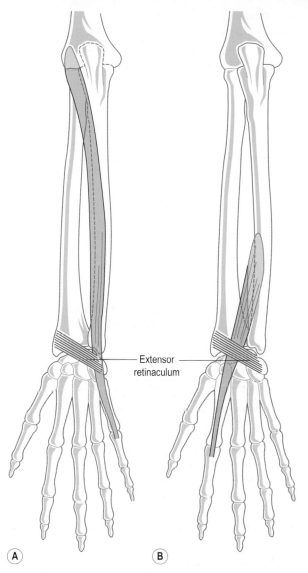

Extensor retinaculum

(A)   (B)

Figure 2.44 The attachments of (A) extensor digiti minimi and (B) extensor indicis, posterior view.

Extensor indicis

Lying deep to extensor digitorum, extensor indicis (Fig. 2.44B) arises from the lower part of the *posterior surface* of the *ulna* and the adjacent *interosseous membrane*. It forms a single tendon which passes deep to the extensor retinaculum in the same synovial sheath as extensor digitorum. On the dorsum of the hand the tendon lies on the medial side of the extensor digitorum tendon. It eventually inserts into the *dorsal digital expansion* on the back of the proximal phalanx of the *index finger*.

Nerve supply

By the *posterior interosseous* branch of the *radial nerve* (root value C7 and 8). The skin over the muscle is supplied by roots C6 and 7.

Action

Extensor indicis assists extensor digitorum in its actions with respect to the joints of the index finger. It does, however, enable the index finger to be used independently, and also aids extension of the wrist.

Palpation

The tendon of extensor indicis, lying medial to that of extensor digitorum, can be palpated on the dorsum of the hand when the index finger is extended. The muscle belly can be palpated by deep pressure over the lower part of the ulna during the same movement.

Muscles extending the thumb

Extensor pollicis longus
Extensor pollicis brevis

Extensor pollicis longus

The extensor pollicis longus (Fig. 2.45A) lies deep to extensor digitorum in the posterior compartment of the forearm. It arises from the *lateral part* of the *middle third* of the *posterior surface* of the *ulna* and the adjacent *interosseous membrane* above extensor indicis. Above the wrist it forms a single tendon which passes in its own synovial sheath deep to the extensor retinaculum (see Fig. 2.109B), lying in a groove on the back of the radius medial to the dorsal tubercle. Winding around the dorsal tubercle, the tendon changes direction and crosses the tendons of extensor carpi radialis longus and brevis (see Fig. 2.110A) and the radial artery. As the tendon crosses the metacarpophalangeal joint it forms the dorsal part of the joint capsule, where it is joined by slips from abductor pollicis brevis laterally and adductor pollicis medially. The whole arrangement forms a triangular expansion, not unlike that of extensor digitorum. The expansion may also be joined by extensor pollicis brevis. The tendon of extensor pollicis longus attaches to the *dorsal surface* of the *base* of the *distal phalanx* of the *thumb*. As it passes across the back of the hand, the tendon forms the medial boundary of a region known as the 'anatomical snuffbox'.

Nerve supply

By the *posterior interosseous* branch of the *radial nerve* (root value C7 and 8). The skin over the muscle is supplied by roots C6 and 7.

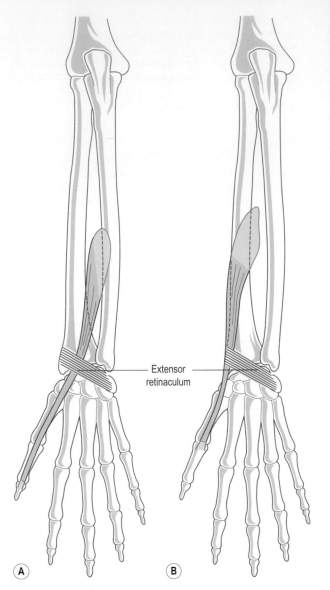

Extensor retinaculum

(A)   (B)

Figure 2.45 The attachments of (A) extensor pollicis longus and (B) extensor pollicis brevis, posterior view.

Action

Extensor pollicis longus extends all joints of the thumb, and assists in extension and abduction of the wrist. Once this has been achieved, the obliquity of the tendon pulls the first metacarpal into a laterally rotated and abducted position. In full abduction or extension of the thumb, extensor pollicis longus can act as an adductor. In addition, it may contribute slightly to supination, due to its oblique course across the lower part of the forearm.

Palpation

The tendon of extensor pollicis longus is easily identified with the thumb extended as the medial border of the 'anatomical snuffbox'. The tendon frequently ruptures in conditions such as rheumatoid arthritis or post Colles' fracture. As a result, the patient is incapable of extending the interphalangeal joint of the thumb.

Extensor pollicis brevis

Extensor pollicis brevis (Fig. 2.45B) lies on the lateral side of and is adjacent to extensor pollicis longus, and distal to abductor pollicis longus to which it closely adheres. It arises from the *middle part* of the *posterior surface* of the *radius* and adjacent *interosseous membrane*. The tendon is formed above the wrist, and runs with that of abductor pollicis longus deep to the extensor retinaculum in a common synovial sheath (see Fig. 2.109B), in a groove on the lateral surface of the radial styloid process. Below this point, the two tendons form the lateral boundary of the 'anatomical snuffbox'. The tendon of extensor pollicis brevis partially replaces the dorsal part of the capsule of the metacarpophalangeal joint of the thumb and inserts into the *dorsal surface* of the *base* of the *proximal phalanx*. Occasionally the tendon may be prolonged, so that it runs with and attaches to the tendon of extensor pollicis longus.

Nerve supply

By the *posterior interosseous* branch of the *radial nerve* (root value C7 and 8). The skin over the muscle is supplied by roots C6 and 7.

Action

Extensor pollicis brevis extends both the carpometacarpal and metacarpophalangeal joints of the thumb. It may also help in extending and abducting the wrist, particularly against resistance.

Palpation

The tendons of extensor pollicis brevis and abductor pollicis longus run together from the lower lateral aspect of the radius where they form the lateral boundary of the 'anatomical snuffbox'. Both tendons can be palpated when the thumb is extended, that of abductor pollicis longus being the most anteriorly placed.

Application

The synovial sheath of extensor pollicis brevis and abductor pollicis longus frequently becomes inflamed in the region of the radial styloid process (de Quervain's syndrome). Techniques such as transverse frictions, ultrasound or injection can be usefully applied to this point.

Muscles abducting/adducting/ opposing the thumb

Abductor pollicis longus
Abductor pollicis brevis
Opponens pollicis
Adductor pollicis
Palmaris brevis

Abductor pollicis longus

Abductor pollicis longus (Fig. 2.46) lies deep to extensor digitorum in the posterior compartment of the forearm. It arises from the *upper part of the posterior surface* of the *ulna* below anconeus, the *middle third* of the *posterior surface* of the *radius* below supinator and intervening *interosseous membrane*.

As it passes distally, abductor pollicis longus emerges from its deep position to lie superficially in the lower part of the forearm. The tendon forms above the wrist and passes, with that of extensor pollicis brevis, in the same synovial sheath below the extensor retinaculum (see Fig. 2.109B), where it lies in a groove on the lateral surface of the radial styloid process. The tendon inserts primarily into the *radial side* of the *base* of the *first metacarpal*, with a slip which passes to the trapezium, and another which passes to the abductor pollicis brevis and the fascia over the thenar eminence.

Nerve supply

By the *posterior interosseous* branch of the *radial nerve* (root value C7 and 8). The skin over the muscle is supplied by roots C6 and 7.

Action

By itself, the muscle puts the thumb into a mid-extended and abducted position. Working with the extensors, abductor pollicis longus helps to extend the thumb at the carpometacarpal joint, while with abductor pollicis brevis it abducts the thumb.

Palpation

Palpation of this muscle is described with that of extensor pollicis brevis on page 89.

Abductor pollicis brevis

Abductor pollicis brevis (Fig. 2.47) is the most lateral and superficial of the three muscles forming the thenar eminence. It takes its origin mainly from the front of the *flexor retinaculum*, extending onto the *tubercles* of the *scaphoid* and *trapezium* with an occasional contribution from the tendon of abductor pollicis longus. The muscle forms a short tendon which attaches to the *radial side* of the *base* of the

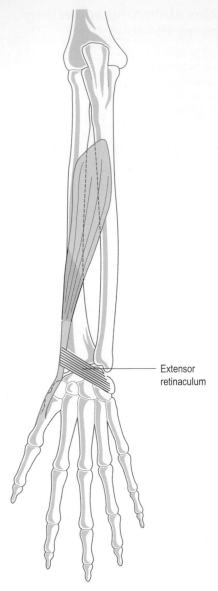

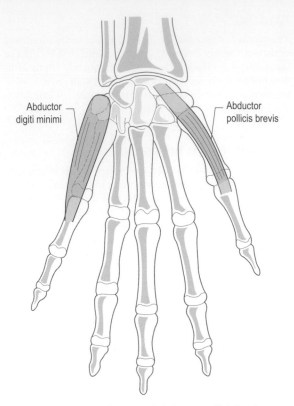

Figure 2.47 The attachments of abductor pollicis brevis and abductor digiti minimi, anterior view.

it to move anteriorly at right angles to the palm. In order to achieve this position there must be some medial rotation at the carpometacarpal joint, the remainder occurring at the metacarpophalangeal joint. This movement is of great significance in terms of the function of the hand as the thumb can be moved towards the fingertips (opposition) where it can carry out precision tasks that require a pincer grip. Because of its partial insertion into the long extensor tendon of the thumb, abductor pollicis brevis can aid in flexion of the metacarpophalangeal joint and extension of the interphalangeal joint.

Opponens pollicis

Opponens pollicis (Fig. 2.48A) is covered by abductor pollicis brevis. Arising from the *flexor retinaculum* and *tubercle* of the *trapezium*, it inserts into the whole length of the *lateral half of* the *anterior surface* of the *first metacarpal*.

Nerve supply

By the *median nerve* (root value T1) from the medial and lateral cords of the brachial plexus. Occasionally, it may be supplied by the ulnar nerve. The skin over the muscle is supplied by root C6.

Figure 2.46 The attachments of abductor pollicis longus, posterior view.

proximal phalanx of the *thumb*, with some fibres reaching the expansion of the extensor pollicis longus tendon.

Nerve supply

By the *median nerve* (root value T1) from the medial and lateral cords of the brachial plexus. The skin over the muscle is supplied by root C6.

Action

Abductor pollicis brevis abducts the thumb at both the carpometacarpal and metacarpophalangeal joints, causing

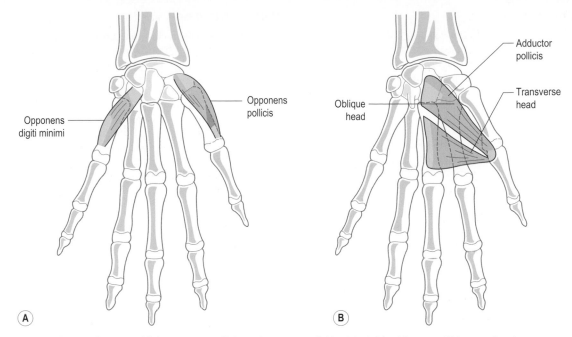

Figure 2.48 The attachments of (A) opponens pollicis and opponens digiti minimi, (B) adductor pollicis, anterior view.

Action

Opponens pollicis produces the complex movements of the thumb called opposition where the first metacarpal is drawn forwards and medially in an arc towards the fingers. The movement involves, in order of action, abduction, medial rotation, and finally flexion and adduction at the carpometacarpal joint of the thumb. The importance of this action is that the tip of the thumb can be brought into contact with the tip of any finger, thus allowing for very precise action of the hand.

Palpation

The three muscles forming the thenar eminence of the thumb lie closely together and are covered with tough fascia which makes identification of individual muscles difficult. However, if the thenar eminence is carefully palpated during resisted movements, all but opponens pollicis can be identified. If abduction of the thumb is resisted (that is movement away from the palm in a plane at 90° to the palm) abductor pollicis brevis can be identified in the lateral part of the thenar eminence.

Adductor pollicis

Adductor pollicis (Fig. 2.48B) is found in the web space of the thumb on its palmar aspect, and has oblique and transverse heads. The oblique head arises from the *sheath* of the *tendon of the flexor carpi radialis*, the *bases* of the *second, third* and *fourth metacarpals*, and the *trapezoid* and *capitate*. The transverse head arises from the *longitudinal ridge* on the *anterior surface* of the *shaft* of the *third metacarpal*. The radial artery passes between the two heads to gain access to the palmar aspect of the hand. Both heads are inserted into the *medial side* of the *base* of the *proximal phalanx* of the *thumb* by a tendon containing a sesamoid bone.

Nerve supply

By the *deep* branch of the *ulnar nerve* (root value C8 and T1). The skin over the muscle is supplied by roots C6 and 7.

Action

Adductor pollicis is a strong muscle that brings the thumb back to the palm from a position of abduction. It is also active in the later stages of opposition. Functionally, the strength of this muscle can be demonstrated when the tip of the index finger and thumb are held together in a pincer grip and an attempt is made to pull them apart. Consequently, adductor pollicis is an important muscle in maintaining the precision grip of the hand.

Palpation

Adductor pollicis is found in the web between the thumb and index finger, and can be felt on the palmar aspect when the movement of adduction is resisted.

91

Palmaris brevis

A superficial muscle found covering the hypothenar eminence of the little finger (see Fig. 2.49). It arises from the *medial border* of the *palmar aponeurosis* and the front of the *flexor retinaculum* and inserts into the *skin* of the *medial border* of the *hand*.

Nerve supply

By the *superficial* branch of the *ulnar nerve* (root value T1). The skin over the muscle is supplied by root C8.

Action

Contraction of palmaris brevis wrinkles the skin on the ulnar side of the hand. It is included in this section as its main function is to assist the thumb in producing a good grip.

Muscles abducting/adducting/opposing the fingers

Palmar interossei
Dorsal interossei
Abductor digiti minimi
Opponens digiti minimi

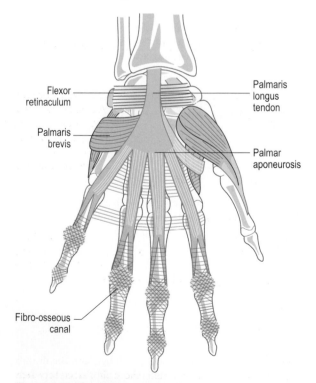

Figure 2.49 The palmar aponeurosis.

Palmar interossei

Situated between the metacarpals are the four palmar interossei (Fig. 2.50A), with one running to the thumb, index, ring and little fingers. Each muscle arises from the *shaft* of the *metacarpal* of the digit on which it acts, and inserts into the *dorsal digital expansion* and the *base* of the *proximal phalanx* of the same digit.

The first palmar interosseous lies on the medial side of the thumb and passes between the first dorsal interosseous and the oblique head of adductor pollicis to insert into the medial side of the base of its proximal phalanx with adductor pollicis. Of the remaining palmar interossei, the second lies on the medial side of the index finger, while the third and fourth lie on the lateral side of the ring finger and little finger respectively. The tendons of all but the first palmar interosseous pass posterior to the deep transverse metacarpal ligament on their way to the extensor expansion.

Nerve supply

All are supplied by the *deep* branch of the *ulnar nerve* (root value T1).

Action

The palmar interossei adduct the thumb, index, ring and little finger towards the middle finger (the midline of the hand). The first palmar interosseous also assists flexor pollicis brevis in flexing the thumb at the metacarpophalangeal joint. The three remaining interossei, by their attachment to the dorsal digital expansion, assist the lumbricals in flexing the metacarpophalangeal joint and extending both interphalangeal joints.

Dorsal interossei

Lying superficially in the spaces between the metacarpals on the dorsum of the hand are the four dorsal interossei (Fig. 2.50B). Two attach to the middle finger, and one each to the index and ring fingers. Each is a bipennate muscle arising from the *sides* of *adjacent metacarpals*. The first and second dorsal interossei lie on the lateral side of the index and middle finger respectively, while the third and fourth lie on the medial side of the middle and ring fingers respectively. All of the dorsal interossei insert into the *proximal phalanx* and *dorsal digital expansion* of the appropriate finger, having passed posterior to the deep transverse metacarpal ligament.

Nerve supply

All are supplied by the *deep* branch of the *ulnar nerve* (root value T1). The skin overlying the muscles is supplied by roots C6, 7 and 8.

Action

The dorsal interossei are abductors of the index, middle and ring fingers. In addition, the first dorsal interosseous

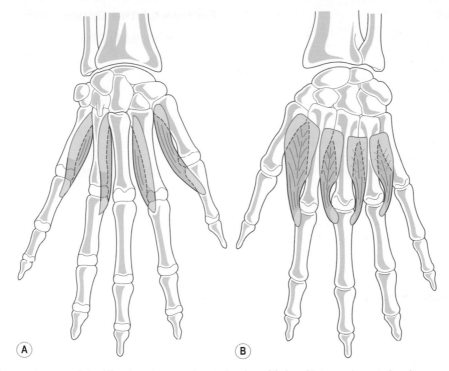

Figure 2.50 The attachments of the (A) palmar interossei, anterior view, (B) dorsal interossei, posterior view.

can rotate the index finger at the metacarpophalangeal joint, and may assist adductor pollicis in adduction of the thumb. Their attachment to the dorsal digital expansion means that, like the palmar interossei, they assist the lumbricals in producing flexion and extension of the metacarpophalangeal and interphalangeal joints respectively.

Palpation

During resisted abduction of the index, middle or ring fingers the dorsal interossei can be palpated on the dorsum of the hand between the metacarpals. The first dorsal interosseous may be seen contracting against resistance in the thumb web. The palmar interossei and lumbricals are too deep to palpate but their actions can be demonstrated by accurate electrical stimulation.

Abductor digiti minimi

Abductor digiti minimi (Fig. 2.47) is the most superficial of the hypothenar muscles, lying in series with the dorsal interossei. It arises from the *pisiform*, the *pisohamate* and *pisometacarpal ligaments*, and the tendon of flexor carpi ulnaris, to insert by a tendon into the *ulnar side* of the *proximal phalanx* of the *little finger* and its *dorsal digital expansion*.

Nerve supply

By the *deep* branch of the *ulnar nerve* (root value T1).

Action

Abductor digiti minimi pulls the little finger away from the ring finger into a position of abduction. It also helps to flex the metacarpophalangeal joint. By its attachment to the dorsal digital expansion, the muscle may help in extending the interphalangeal joints. Abductor digiti minimi is a powerful muscle playing an important role in grasping a large object with outspread fingers.

Opponens digiti minimi

Lying deep to abductor digiti minimi, opponens digiti minimi (Fig. 2.48A) arises from the *hook of the hamate* and adjacent *flexor retinaculum*, to insert into the *medial half of* the *palmar surface* of the *fifth metacarpal*.

Nerve supply

By the *deep* branch of the *ulnar nerve* (root value T1).

Action

Opponens digiti minimi pulls the little finger forwards towards the palm and rotates it laterally at the carpometacarpal joint. This movement deepens the hollow of the hand, and is a necessary part of opposition of the little finger.

Palpation

The hypothenar muscles are closely related to one another making individual palpation difficult. Resistance to abduction of the little finger enables abductor digiti minimi to be palpated on the ulnar border of the hand. For the movements of flexion and opposition of the little finger it is difficult to localize the action of the remaining hypothenar muscles, so accurate palpation is hard to achieve.

Section summary

Movements at joints of fingers and thumb

Each finger comprises three joints, the distal (DIP) and proximal (PIP) interphalangeal joints and the metacarpophalangeal (MCP) joint. The thumb has one interphalangeal (IP) joint, a metacarpophalangeal (MCP) joint and a carpometacarpal (CMJ) joint. The joints that the muscles work on are given in brackets.

Movement (fingers)	Muscles
Flexion	Flexor digitorum profundus (DIP, PIP, MCP)
	Flexor digitorum superficialis (PIP, MCP)
	Lumbricals and interossei (MCP)
Extension	Extensor digitorum (DIP, PIP, MCP)
	Extensors indicis and digiti minimi (DIP, PIP, MCP for named digit)

Movement (fingers)	Muscles
Abduction	Dorsal interossei (MCP) Abductor digiti minimi (MCP)
Adduction	Palmar interossei (MCP)
Opposition	Opponens digiti minimi (CMJ) (little finger only)

Movement (thumb)	Muscles
Flexion	Flexor pollicis longus (IP, MCP) Flexor pollicis brevis (MCP)
Extension	Extensor pollicis longus (IP, MCP)
	Extensor pollicis brevis (MCP)
Abduction	Abductors pollicis longus and brevis (CMJ)
Adduction	Adductor pollicis (CMJ)
Opposition	Opponens pollicis (CMJ)

Many of the above muscles cross several joints and are involved in complex combined movements; therefore considering individual muscles is somewhat artificial.

Interaction of the extrinsic and intrinsic muscles of the fingers and thumb allows the wide range of complex grips and functions possible in the human hand.

Fasciae of the upper limb

The superficial fascia

The superficial fascia of the upper limb shows regional differences between, for example, the shoulder and the hand. In the shoulder region and arm, it contains a variable amount of fat. In females there is deposition of fat in this region (a secondary sexual characteristic), the amount of which tends to increase after middle age. At the elbow, a subcutaneous bursa is present between the skin and the olecranon process. This may become enlarged in individuals who often tend to lean on their elbows, giving rise to a condition known as 'student's elbow'.

There is nothing particularly noteworthy about the superficial fascia in the forearm. However, in the hand there are several specializations, most of which enhance the hand's tactile or prehensile capabilities. On the dorsum of the hand the fascia is loose and thin, and can be readily lifted away from the underlying tissue. It is in the palm of the hand, as well as the palmar surface of the digits, where specializations of the fascia can be seen. In the centre of the palm, strong bands of connective tissue connect the skin to the palmar aponeurosis, a thickening of the deep fascia. Overlying the thenar and hypothenar regions fixation of the skin to the deep fascia is less marked, but here the superficial fascia is thicker and less fibrous to facilitate the gripping action of the hand. This is because it can adapt to the contours of the object being held. Palmaris brevis lies in the superficial fascia over the hypothenar eminence; by wrinkling the skin, it improves the grip. Similar less fibrous pads of tissue are also found opposite the metacarpophalangeal joints, where the superficial transverse metacarpal ligament (a band of transverse fibres) connects to the palmar surfaces of the fibrous flexor sheaths of the fingers.

The pads on the palmar surfaces of the distal phalanges are highly specialized regions which house numerous tactile nerve endings. Here the skin is firmly attached to the distal two-thirds of the distal phalanx. The blood supply to the distal phalanx passes through this highly specialized pad; if the pad becomes infected there may be compression of the artery with death of this part of the bone. On the dorsum of the distal phalanx is the nail; there is no superficial fascia deep to it.

The deep fascia

The deep fascia of the upper limb is continuous with that of the upper back, and consequently can be traced superiorly to the superior nuchal line on the occipital bone, to the ligamentum nuchae in the midline in the cervical region, and to the supraspinous and interspinous ligaments in the thoracic region. In the shoulder region, the deep fascia is extremely strong over infraspinatus and teres minor, being firmly attached to the medial and lateral borders of the scapula. Superiorly, a sheath is formed for deltoid, which attaches to the clavicle, acromion and spine of the scapula.

The deep fascia covering pectoralis major attaches above to the clavicle, and may be traced, via the clavicle, to the neck. Inferiorly it is continuous with the fascia of the anterior abdominal wall. Medially, the fascia is firmly attached to the sternum, whereas laterally it becomes thickened as the axillary fascia, which forms the floor of the axilla. Further laterally it becomes continuous with the deep fascia of the arm.

Deep to pectoralis major is the clavipectoral fascia. Medially, it attaches to the first costal cartilage and passes to the coracoid process and coracoclavicular ligament laterally. The clavipectoral fascia splits to surround subclavius superiorly to attach to the undersurface of the clavicle. It also splits to enclose pectoralis minor inferiorly. An extension of the fascia from the lateral border of pectoralis minor passes into the axilla and attaches to the axillary floor. This is often referred to as the suspensory ligament of the axilla. The deep surface of the clavipectoral fascia is connected to the axillary sheath surrounding the axillary vessels and the brachial plexus.

In the arm, the deep fascia forms an investing layer around the muscles. At the elbow it attaches to the medial and lateral epicondyles of the humerus and the olecranon process, becoming continuous with the deep fascia of the forearm. Two intermuscular septa arise from the deep surface of this investing layer and pass to attach to the supracondylar ridges of the humerus. Both the medial and lateral intermuscular septa are found only in the lower half of the arm. Besides separating the arm into flexor and extensor compartments the septa also give attachment to muscles in each compartment. Of the two, the medial intermuscular septum is said to be the stronger.

In the forearm, the deep fascia of the elbow is very strong because many of the muscles arising from either the common flexor or extensor origins also arise from the overlying fascia. The bicipital aponeurosis helps to strengthen the fascia anteriorly, while the triceps insertion does so posteriorly. The deep fascia is also strong and thick where it attaches to the posterior border of the ulna, because it gives attachment to flexor digitorum profundus and flexor and extensor carpi ulnaris. However, the fascia becomes thinner as it approaches the wrist, although at the wrist there are two thickenings of the transverse fibres forming the flexor and extensor retinacula, which serve to hold the tendons entering the hand in place preventing 'bowstringing'.

The flexor retinaculum

Found anterior to the carpus the flexor retinaculum (Fig. 2.51A) acts as a strong band for retention of the long

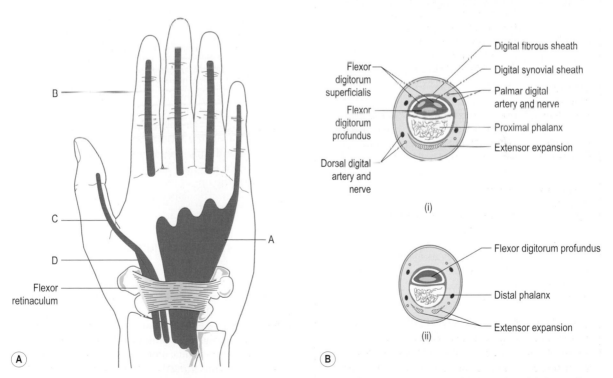

Figure 2.51 (A) Diagrammatic representation of the synovial sheaths of: A, flexors digitorum superficialis and profundus; B, flexor tendons in the fingers; C, flexor pollicis longus; D, flexor carpi radialis. (B) Cross-section at the level of the proximal (i) and distal (ii) phalanx to show the arrangement of the tendons and synovial sheath.

flexor tendons, converting the carpal sulcus into a tunnel. It attaches to the tubercle of the scaphoid and to both lips of the groove on the trapezium laterally, and to the pisiform and the hook of the hamate medially. Flexor carpi radialis passes below the flexor retinaculum in its own lateral compartment surrounded by its synovial sheath. Medially the single tendon of flexor pollicis longus lies within its synovial sheath. In a common synovial sheath all eight tendons of flexors digitorum superficialis and profundus pass deep to the retinaculum. The median nerve also enters the hand by passing below the flexor retinaculum, where it lies in front of the superficialis tendons. It is here that it may become compressed if the synovial sheaths become inflamed, giving rise to the 'carpal tunnel syndrome'.

There are two layers of fascia in the palm of the hand. The deeper layer covers the interossei and encloses adductor pollicis, while the more superficial layer is strong in its central part, forming the palmar aponeurosis. On each side of the palmar aponeurosis the fascia thins out to cover the thenar and hypothenar muscles. The palmar aponeurosis strengthens the hand for gripping, yet also protects the underlying vessels and nerves. It is a dense, thick triangular structure bound to the overlying superficial fascia. The apex is at the wrist and its base at the webs of the fingers. From the base four slips pass into the fingers to become continuous with the digital sheaths of the flexor tendons. Each slip further divides and has attachments to the deep transverse metacarpal ligament, the capsule of the metacarpophalangeal joint and the sides of the proximal phalanx. The slips cross in front of the lumbricals, and the digital vessels and nerves.

In some individuals, often the elderly, the medial part of the palmar aponeurosis and the fibrous flexor sheaths of the little and ring fingers may become shortened. There is a progressive flexion of the fingers at the metacarpophalangeal and proximal interphalangeal joints pulling the fingers towards the palm. The distal phalanx is not flexed because the tip of the finger is pressed against the palm. Indeed, in severe cases, the distal interphalangeal joint may become hyperextended. There is an inability to extend the fingers even passively. Treatment usually involves the removal of the offending fibrous tissue; however, there is a tendency for it to reform and contract again. The condition is known as *Dupuytren's contracture* (Fig. 2.52).

Fibro-osseous canals, retinacula and synovial sheaths of the flexors of the wrist and fingers

The tendons of the digital flexors are held close to the phalanges by fibrous sheaths (Fig. 2.51B), which act to prevent 'bowstringing' of the tendons ensuring that their pull produces immediate movement at the interphalangeal joints. The canals are formed by a shallow groove on the anterior surface of the phalanges and by a fibrous sheath attaching

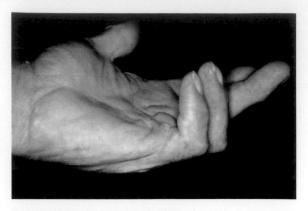

Figure 2.52 Dupuytren's contracture.

to the raised lateral and medial edges of the palmar surfaces of the proximal and middle phalanges and the palmar surface of the distal phalanx. The canal is closed distally by attaching to the distal phalanx, but is open proximally deep to the palmar aponeurosis. Most of the fibres of this sheath are arranged transversely, but at the interphalangeal joints they have a criss-cross arrangement to allow flexion to occur (see Fig. 2.126). All five fibro-osseous canals are lined with a synovial sheath which surrounds the enclosed tendons. In the fingers, the synovial sheath surrounds the tendons of flexors digitorum superficialis and profundus and is connected to them by the vincula. The sheath of the thumb contains only the tendon of flexor pollicis longus within a synovial covering.

In the condition of 'trigger finger', a swelling of the flexor tendons proximal to the fibro-osseous canals means that when the finger (or thumb) is flexed, the swelling moves into the canal and difficulty is thus experienced on attempting extension. In Dupuytren's contracture, the medial portion of the palmar aponeurosis and the fibrous sheaths of the ring and little fingers may become shortened, in which case these fingers are pulled towards the palm. In severe cases, the tips of these fingers may become pressed into the palm. The lateral two fibrous sheaths are less commonly affected.

The extensor retinaculum and synovial sheaths

On the back of the forearm and wrist is a thickening of the deep fascia, the extensor retinaculum. It attaches laterally to the distal part of the anterior surface of the radius, and medially to the distal end of the ulna, pisiform, triquetral and ulnar collateral ligament of the wrist. Septa run from the deep surface of the retinaculum to the distal ends of the radius and ulna, converting the grooves on the dorsum of the wrist into six separate tunnels, each of which is lined by a synovial sheath. From medial to lateral, the compartments transmit extensor carpi ulnaris; extensor digiti minimi;

extensor digitorum and extensor indicis; extensor pollicis longus; extensor carpi radialis longus and brevis, and finally abductor pollicis longus and extensor pollicis brevis. The synovial sheaths extend just proximal to the extensor retinaculum. All except that surrounding extensor pollicis longus, which extends almost as far as the metacarpophalangeal joint of the thumb, end at the middle of the dorsum of the hand (see Fig. 2.110B).

The retinaculum functions to retain the extensor tendons in their position preventing 'bowstringing'.

Palmar regions, compartments and spaces

From the medial and lateral borders of the palmar aponeurosis, septa pass down to fuse with the fascia covering the interossei, dividing the palm into three compartments. Of these, the medial and lateral compartments contain the hypothenar and thenar muscles respectively. The intermediate compartment contains the long flexor tendons and the lumbricals, surrounded by loose connective tissue. A further septum, although often incomplete, runs from the deep aspect of this connective tissue to the shaft of the third metacarpal, further dividing this compartment into two potential spaces – the lateral and medial midpalmar spaces. Proximally they communicate, behind the flexor tendons, with a similar potential space in front of pronator quadratus. Distally, they are prolonged around the lumbrical muscles as the lumbrical canals which pass into the web of the fingers. The web space in turn is in free communication with the sub-aponeurotic space on the back of the hand deep to the extensor tendons, and with the dorsal subcutaneous space. Accumulations of fluid in the midpalmar spaces, following injury or infection of the hand, can therefore track back in the deep aspect of the flexor compartment, or via the webs of the fingers up the dorsum of the forearm in the loose superficial fascia.

The common synovial sheath for the flexor digitorum superficialis and profundus tendons extends from 2 cm above the flexor retinaculum and downwards through the carpal tunnel to end in the midpalmar region. Only the synovial sheath of the little finger is continuous with the common synovial sheath in the palm; the synovial sheaths lining the fibro-osseous canals of the other three fingers are separated from it (Fig. 2.51A). The single synovial sheath of flexor pollicis longus has the same upward limit but continues downwards through the carpal tunnel and along the fibro-osseous canal of the thumb (Fig. 2.51A).

Simple activities of the upper limb

Introduction

The purpose of this section is to provide a simple analysis of some common activities to show how different muscle groups cooperate in producing a desired movement. The analysis describes the starting position, omitting muscle work, and then goes on to describe the sequence of movements involved, giving the joints and the muscles responsible, and ends with a description of how the muscles work with respect to joint movement. These general principles can then be applied to any movement.

Abduction of the arm through to full elevation

Starting position

This movement begins with the individual standing in the anatomical position.

Sequence of movements

Abduction of the humerus   Abduction is initiated by *supraspinatus*, which raises the upper limb by approximately 10°, thereby putting it into a position that allows *deltoid* to produce a rotatory force (torque) to continue the movement. Upward shearing of the head of the humerus against the glenoid fossa is prevented by the *rotator cuff muscles*, which all act isometrically as stabilizers with the exception of supraspinatus, which shortens concentrically. Deltoid also works concentrically, the principal force being exerted by its middle multipennate fibres, with the anterior and posterior fibres acting as guides to control the plane of abduction.

Lateral rotation of the scapula   On reaching 30° of abduction lateral rotation of the scapula begins, turning the glenoid fossa to face upwards increasing the total range of movement. For every 15° of abduction from this point onwards, approximately 10° occurs at the shoulder joint and 5° by lateral rotation of the scapula. Lateral rotation of the scapula is produced by concentric contraction of the upper and lower fibres of *trapezius*, and the lower half of *serratus anterior*. Other muscles assist by holding the scapula against the chest wall.

Lateral rotation of the humerus   As shoulder joint movement approaches 90° (i.e. 120° abduction), there is lateral rotation of the humerus to prevent it: (1) coming into contact with the coracoacromial arch and (2) stopping any further movement – a locked position. This lateral rotation brings the lower part of the articular surface of the humeral head into contact with the glenoid fossa and is brought about by concentric contraction of *teres minor* and *infraspinatus*.

As the clavicle is firmly anchored to the acromion, lateral scapular rotation elevates the lateral end of the clavicle, producing a downward movement of the clavicle at the sternoclavicular joint. There is also some rotation of the clavicle about its long axis, which takes place at both the acromioclavicular and sternoclavicular joints.

Throwing a ball overarm

Techniques for throwing a ball overarm vary enormously depending on the distance, height and direction of the throw. The type, weight and size of the ball will influence the grip used, while the surface of the ball may affect its trajectory and resultant movement as it hits the ground.

For this particular analysis a cricket ball is being thrown with some force from the boundary of a cricket field towards the wicket keeper.

Starting position

Drawing the arm backwards prior to the throw is an important factor in determining the velocity of the ball. The starting position of the activity is with the ball in the right hand and the arm at the side of the body. As this position has little to do with the actual throw, the muscle work and joint position are not described. However, it is from this position that the activity begins.

Sequence of movements

Stretch phase: To achieve maximum power in the throw the muscles involved in the activity must be stretched as much as possible. The pectoral girdle is retracted and depressed by the *middle* and *lower fibres* of *trapezius*, thus stretching the protractors and elevators. The shoulder joint is extended by the *posterior fibres* of *deltoid*, *latissimus dorsi*, *teres major* and possibly also the *long head of triceps* stretching the flexors. The shoulder joint is also abducted, by the *middle fibres* of *deltoid*, and laterally rotated by the *posterior fibres* of *deltoid* and *infraspinatus*. The elbow is extended by *triceps* and *anconeus*. The wrist is extended by *extensors carpi radialis longus* and *brevis*, *extensor carpi ulnaris*, and the *long digital extensors*, and adducted by *extensor* and *flexor carpi ulnaris*. All the muscles producing these movements are working concentrically.

The ball is usually held in the ulnar side of the hand in a type of power grip, with the fingers flexed around it by the *long digital flexors*. The distal and proximal interphalangeal and metacarpophalangeal joints are flexed by *flexor digitorum profundus*, *flexor digitorum superficialis* and the *lumbricals* respectively. To obtain a better grip the little finger is also often opposed to the ball by *opponens digiti minimi*. The thumb is flexed at the interphalangeal and at the metacarpophalangeal and carpometacapral joints by *flexor pollicis longus* and *flexor pollicis brevis*. The thumb is also abducted and opposed; abduction is due to the shape of the ball with the adductors working to grip the ball, while opposition is brought about by *opponens pollicis*. These muscles of the hand are all working isometrically.

These movements of the upper right limb are accompanied by rotation of the trunk to the right with right lateral flexion and extension, brought about by the concentric contraction of the *right internal oblique* and *left external oblique* and *right quadratus lumborum*, and the *postvertebral muscles* respectively. Movements of the lower limbs will not be covered in this analysis but are considerable.

The trajectory phase: All muscles producing movement in this phase of the activity are working concentrically unless otherwise stated. The pectoral girdle is rapidly protracted by *serratus anterior* and *pectoralis minor* and elevated by the *upper fibres* of *trapezius* and *levator scapulae*. The shoulder joint is flexed and elevated by the *anterior fibres of deltoid* and *pectoralis major* (clavicular head), and medially rotated by the *anterior fibres* of *deltoid* and *pectoralis major*, aided perhaps, by *latissimus dorsi* and *teres major* acting as a 'drag' on the medial side of the arm. During this movement the shoulder passes through abduction, produced by *deltoid*, initially the *posterior* and *middle fibres* then the *middle fibres*, and finally by the *middle* and *anterior fibres*.

The elbow moves from full extension into a variable degree of flexion according to technique, brought about by *biceps brachii* and *brachialis*. At the point when the upper limb moves in front of the trunk the elbow is fully extended by *triceps* and *anconeus*.

The precise trajectory of the movement of the upper limb is determined by the required elevation the ball. At the top of the trajectory the ball is released by extension of the fingers and extension by the action of *extensor digitorum* aided by the *lumbricals* and *interossei*, and by extension and abduction of the thumb by the action of *extensors pollicis longus* and *brevis* and *abductors longus* and *brevis*.

At the same time as the trajectory phase of the upper limb, the trunk powerfully flexes with lateral flexion and rotation to the left. Flexion is brought about by the *oblique muscles* and *rectus abdominis* of the left side. Left rotation is brought about by the *left internal oblique* and *right external oblique*, lateral flexion by the *left quadratus lumborum*. All of these movements are accompanied by powerful actions of the lower limbs, with body weight being transferred from the back of the right foot to the front of the left foot. There is also considerable movement of the neck to maintain the eyes in the direction in which the ball is being thrown.

Pushing or punching

Starting position.

The shoulder joint is abducted 90° and fully extended and the elbow joint fully flexed. The hand is either open, for pushing, or closed, for punching. For this activity the muscle work and joint positions of the hand will not be considered.

Sequence of movements

Abduction and flexion of the shoulder   The shoulder is maintained in abduction by isometric contraction of the *middle fibres* of *deltoid*, and is then rapidly flexed by the *anterior fibres* of *deltoid* and *pectoralis major*, assisted by *coracobrachialis* and the *long head* of *biceps*, all of which work concentrically.

Protraction of the scapula and extension of the forearm   To increase the force and range of the movement, the scapula is protracted by a strong concentric contraction of *serratus anterior* and *pectoralis minor*. At the same time there is a forceful contraction of *triceps* to produce rapid extension of the forearm at the elbow joint.

The major difference between pushing and punching is the velocity of the movement, with the latter being a more rapid and explosive movement.

Press-ups

Starting position

The subject lies prone with the palmar surfaces of the hands on the ground below the shoulders, which are abducted 90° and in slight extension; the elbows are fully flexed. It will be assumed that static muscle work in the legs and trunk will maintain their neutral position throughout

Sequence of movements

Upward movement: There is simultaneous elbow extension by *triceps* and *anconeus*, and shoulder flexion by *pectoralis major* and the *anterior fibres* of *deltoid*, assisted by *coracobrachialis* and the *long head* of *biceps*. The pectoral girdle is strongly protracted by *serratus anterior* and *pectoralis minor*, the arms are straight and the shoulders are protracted. All the muscles work concentrically.

From this position, the downward movement can now be considered.

Downward movement: In the downward phase there is simultaneous movement at both the elbow and shoulder joints under the influence of gravity. The elbows are flexed under the eccentric contraction of *triceps* and *anconeus* working to control the rate of descent. At the shoulder joint, extension is controlled by eccentric contraction of *pectoralis major* and the *anterior fibres* of *deltoid*, with the *long head* of *biceps* and *coracobrachialis* assisting these muscles. The pectoral girdle retracts under the eccentric control of *serratus anterior* and *pectoralis minor*.

It is worth remembering that the same muscles are involved in producing both movements, upwards by shortening and downwards by lengthening.

Pulling

Starting position

The subject stands with the shoulder flexed to 90°, the elbow in full extension and the hand firmly gripping the object to be pulled.

Sequence of movements

The elbow is flexed by concentric contraction of *biceps* and *brachialis*, while the shoulder is extended by the strong concentric action of *teres major*, *latissimus dorsi* and the *posterior fibres* of *deltoid*. The pectoral girdle is retracted by concentric contraction of *trapezius* and *rhomboids major* and *minor*.

Rowing

Rowing a boat is one of life's most pleasurable activities whether it takes place on a lake, river or the sea. The type of rowing, however, varies considerably according to its purpose. Rowing machines are also frequently used in homes and sports centres as a means of exercising.

This brief outline of the joint movement and muscle work involved in rowing is based on an amateur oarsman in an ordinary rowing boat which has a sliding seat and rowlocks to extend the fulcrum of the oars. As styles vary considerably no attempt is made to recommend any particular technique.

The description begins in the fully forward position with the arms straight, the grip tightening on the oar, the back fully flexed, the neck slightly flexed and the head slightly extended on the neck. The moveable seat is fully forward with the ankles dorsiflexed and the knees and hips fully flexed. Description of activities in the lower limb are considered in the section on the lower limb (p. 275).

Starting position

The blades of the oars are in the water at right angles to the surface. The hands are placed on the ends of the handles of the oars with the fingers and palms over the top and the thumbs below the oars. The interphalangeal and metacarpophalangeal joints of all fingers are in a semiflexed position around the oars, being held by the static muscle work of *flexor digitorum profundus*, and all other joints by *flexors digitorum superficialis* and *profundus*.

The thumb grips the under side of the oars due to the static action of *adductors pollicis longus* and *brevis*. Semiflexion of the interphalangeal joints is due to the static muscle work of *flexor pollicis longus*, and semiflexion of the metacarpophalangeal joint by the action of *flexors pollicis longus* and *brevis*. There is also abduction of the carpometacarpal joint due to the thickness of the oar.

The wrists are in a neutral position with both the flexors and extensors working statically. The elbows are in full extension with *triceps* working statically; the shoulders are flexed and in almost full elevation due to the static muscle work of the *anterior* and *middle fibres* of *deltoid* and the *clavicular head* of *pectoralis major*. The trunk is held in full flexion by the abdominal muscles particularly *rectus abdominis*, aided by the *abdominal obliques*.

Sequence of movements

Stroke phase: At the beginning of this phase the grip is tightened on the oars as they are pulled backwards by flexion of the elbows, brought about by powerful concentric action of *biceps brachii*, *brachialis* and *brachioradialis*. The shoulders are extended through abduction by the concentric action of *latissimus dorsi* and *teres major* aided by the *middle* and *posterior fibres* of the *deltoid*. The pectoral girdle is retracted by the concentric activity of the *rhomboid major* and *minor* and the *middle fibres* of the *trapezius* and depressed by the *lower fibres* of the *trapezius*.

The trunk is almost fully extended by the *erectores spinae* (*iliocostalis*, *longissimus* and *spinalis*) aided by *quadratus lumborum* and *latissimus dorsi* all working concentrically. The neck is extended to the neutral position by concentric contraction of the *upper fibres of trapezius*, *splenius cervicis* and *semispinalis*, while the head is stabilized in a neutral position.

This powerful stroke is arrested by the oar coming into contact with the upper abdomen. The oars are then released from the water by a strong extension of the wrists brought about by the concentric action of *extensors carpi radialis longus* and *brevis* and *extensor carpi ulnaris*. During the stroke phase the lower limbs are powerfully extended (p. 275).

Recovery phase: The grip on the oars is maintained, but relaxed a little. Extension of the wrists is maintained until the end of the recovery phase and is brought about by static muscle work of the wrist extensors.

The elbows are extended fully by the concentric contraction of *triceps*. The shoulders are flexed into almost full elevation by the *anterior fibres* of *deltoid*, the *clavicular head* of *pectoralis major*, *coracobrachialis* and the *long head* of *biceps brachii*. Both pectoral girdles are protracted by the concentric action of *pectoralis minor* and *serratus anterior* and partially elevated by the *upper fibres* of *trapezius* and *levator scapulae*.

The trunk is flexed by the powerful concentric action of the abdominal muscles, particularly *rectus abdominis*. The neck is partially flexed by concentric contraction of *sternomastoid* and *prevertebral muscles*. The head is slightly extended by the posterior suboccipital muscles (*rectus capitis posterior major* and *minor*) and the *upper fibres* of *trapezius*.

During this recovery phase the lower limbs are flexing and drawing the seat forward on the slide ready for the next stroke (p. 275).

Finally, when the individual has come as far forward as possible without leaving the seat, the wrists are flexed to the neutral position and the blade of the oar put cleanly into the water ready for the next stroke phase.

Gripping

A number of different types of grip can be described depending upon the activity being undertaken.

Power grip

A power grip (see p. 180) is the position adopted by the fingers, thumb and wrist when a strong grip on an object, such as a hammer, is required. There is little movement involved and so the joint position and muscle work only will be described.

The distal and proximal interphalangeal joints of the fingers are flexed by *flexor digitorum profundus* and *flexor digitorum superficialis* respectively. The metacarpophalangeal joints are flexed by the *lumbricals* as well as by the long flexors. The thumb is pressed strongly against the flexed fingers by *flexor pollicis longus* and the *intrinsic* thumb muscles. The wrist is held in extension by the synergistic action of *extensors carpi radialis longus* and *brevis* and *extensor carpi ulnaris*. This wrist position prevents the continued contraction of the long flexors pulling the wrist into flexion, and thus prevents a loss of power in the gripping fingers by 'active insufficiency' of the flexor muscles.

Once in the gripping position, all of the muscles named are working isometrically, with the joint positions determined by the size of the object being held.

Hook grip

In this grip, as in that described earlier, the wrist is maintained in a similar position of extension. The fingers are hooked through some sort of handle to allow carrying, as when holding a suitcase (see p. 180). The distal and proximal interphalangeal joints are flexed by *flexor digitorum profundus* and *flexor digitorum superficialis* respectively, which are working isometrically to maintain the grip. The metacarpophalangeal joints are held in a neutral position.

Precision grip

The position of the wrist and the muscle work involved are the same as for the power grip. Depending upon the activity being undertaken there will be varying degrees of flexion and extension in the joints of the fingers and thumb. In writing, for example, the pen is held between the thumb, index and middle fingers. The muscle work involved is similar to that for the power grip, but the strength of contraction is much less. In addition to the long flexors of the fingers and thumb holding the pen, the *lumbricals* and *interossei* produce and control the very fine movements of the fingers, which, together with those produced in the thumb, allow the pen to be moved in order to write. See also page 180.

Hand movements are extremely precise and the ability to manipulate small objects and carry out very intricate tasks is one of the characteristics of human development.

JOINTS

The pectoral girdle

Introduction

The upper limb has become highly specialized in its functions of prehension and manipulation. Evolution has produced a limb which is extremely mobile without losing the stability required to give this acquired mobility, force and precision. The result of this evolutionary development is an upper limb which has no locomotor function, except in infants and in those who, of necessity, use walking aids.

All vertebrates possess four limbs of one form or another, these being connected to the axial skeleton by the pectoral and pelvic girdles. In humans the pelvic girdle is firmly anchored to the vertebral column, thereby providing the stability necessary for bipedal locomotion. The pectoral girdle, however, does not articulate with the vertebral column; it articulates only with the thoracic cage (Fig. 2.53), which while providing a mechanism whereby forces generated in the upper limb can be partially transferred to the axial skeleton, does not unduly restrict movement of the pectoral girdle as a whole.

The shoulder blade (scapula) and clavicle are the bones of the pectoral girdle. The shoulder blade is usually referred to as the scapula in descriptive anatomy; morphologically the term has a slightly more restricted meaning (see later). Although movements of the shoulder joint accompany nearly all movements of the joints of the pectoral girdle, it is not part of the pectoral girdle. The shoulder blade is slung in muscle from the thoracic cage, while the clavicle interposes itself between shoulder blade and thorax providing a strut which steadies and braces the pectoral girdle during movements, particularly adduction. The clavicle articulates with the thorax by the sternoclavicular joint (Fig. 2.53), and with the shoulder blade by the acromioclavicular joint (Fig. 2.53). The humerus of the upper limb joins the shoulder blade at the glenohumeral (shoulder) joint. It is the summation of the mobility of these three individual, yet mutually interdependent, joints which gives the upper limb its freedom of movement; consequently, the ultimate range of movement of the shoulder complex is much greater than that of the hip. The clavicle moves with respect to the sternum, the shoulder blade with respect to the clavicle and the humerus with respect to the shoulder blade. In addition, the shoulder blade moves with respect to the thoracic wall. This arrangement favours mobility of the shoulder–arm complex; however, it makes stabilization of the upper limb against the axial skeleton much more difficult. Stability is achieved by the powerful musculature which attaches the pectoral girdle to the thorax, vertebral column, head and neck. Furthermore, this musculature acts as a shock absorber when body-weight is received by the upper limbs.

Just as the innominate bone consists of three parts which meet and fuse at the acetabulum, so there are three parts of the shoulder blade (Fig. 2.54). The broad, flat, dorsal part of the shoulder blade is the scapula; the ventral part is the coracoid process, the two joining together in the upper part of the glenoid fossa. (These are the counterparts of the ilium and ischium respectively.) A small piece of bone, the precoracoid, ossifying separately at the tip of the coracoid process, is the counterpart of the pubis.

The clavicle has no counterpart in the pelvic girdle. Forces from the upper limb are transmitted by trapezius to the cervical spine and by the clavicle to the axial skeleton by the coracoclavicular and costoclavicular ligaments, so that normally neither end of the clavicle transmits much force.

The sternoclavicular joint

The synovial sternoclavicular joint provides the only point of bony connection between the pectoral girdle and upper limb, and the trunk. Although the joint is functionally a ball and socket joint, it does not have the form of such a joint.

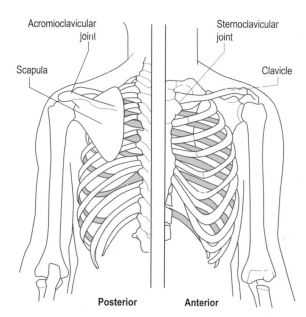

Acromioclavicular joint

Sternoclavicular joint

Scapula

Clavicle

Posterior   Anterior

Figure 2.53 Relation of the pectoral girdle to the thorax.

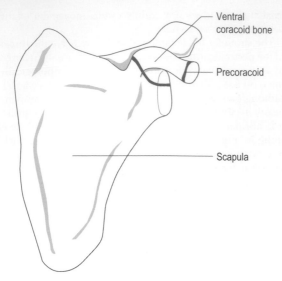

Figure 2.54 The parts of the scapula blade equivalent to the innominate bone.

Articular surfaces

The medial end of the clavicle articulates with the clavicular notch at the superolateral angle of the sternum and the adjacent upper medial surface of the first costal cartilage. The clavicular articular surface tends to be larger than that on the sternum; consequently the medial end of the clavicle projects above the upper margin of the manubrium sterni.

The articular surfaces are reciprocally concavoconvex (Fig. 2.55A), although they do not usually have similar radii of curvature. The joint, therefore, is not particularly congruent; congruence is partly provided by an intra-articular fibrocartilaginous disc. The articular surface on the manubrium sterni is set at approximately 45° to the vertical. It is markedly concave from above downwards, and convex from behind forwards, being covered with hyaline cartilage. The clavicular articular surface is convex vertically and flattened or slightly concave horizontally (Fig. 2.55A), with the concavity being continued over the inferior surface of the shaft for articulation with the costal cartilage of the first rib. The greater horizontal articular surface of the clavicle overlaps the sternocostal surface anteriorly and particularly posteriorly, the whole being covered with fibrocartilage rather than hyaline cartilage.

Joint capsule and synovial membrane

A fibrous capsule surrounds the whole joint like a sleeve attaching to the articular margins of both the clavicle and the sternum, with its inferior part passing between the clavicle and the upper surface of the first costal cartilage (Fig. 2.55B). Except for this inferior part, which is weak, the joint capsule is relatively strong, being strengthened anteriorly, posteriorly and superiorly by capsular thickenings known as the anterior and posterior sternoclavicular and interclavicular ligaments respectively.

Because there are two separate cavities associated with the joint (see below), there are two synovial membranes. A relatively loose lateral membrane lines the capsule, being reflected from the articular margin of the medial end of the clavicle to the margins of the articular disc. Similarly, the medial membrane attaches to the articular margins on the sternum and to the margins of the disc.

Intra-articular structures

A complete, intra-articular, fibrocartilaginous disc divides the joint into two separate synovial cavities (Fig. 2.55B). The disc is flat and round, being thinner centrally, where it may occasionally be perforated and permit communication between the two cavities, than around the periphery. It is attached at its circumference to the joint capsule, particularly in front and behind. More importantly, however, the disc is firmly attached superiorly and posteriorly to the upper border of the medial end of the clavicle, and inferiorly to the first costal cartilage near its sternal end (Fig. 2.55B). Consequently, as well as providing some cushioning between the articular surfaces from forces transmitted from the upper limb and compensating for incongruity of the joint surfaces, the disc also has an important ligamentous action. Although mainly fibrocartilaginous, it is fibrous or ligamentous at its circumference, and holds the medial end of the clavicle against the sternum. It prevents the clavicle moving upwards and medially along the sloping sternochondral surface under the influence of strong, thrusting forces transmitted from the limb, or when the lateral clavicle is depressed as by a heavy weight carried in the hand.

Ligaments

The joint capsule is strengthened anteriorly, posteriorly and superiorly by the anterior and posterior sternoclavicular and interclavicular ligaments respectively. In addition, an accessory ligament, the costoclavicular ligament, binds the clavicle to the first costal cartilage just lateral to the joint.

Anterior sternoclavicular ligament

This is a strong, broad band of fibres attaching above to the superior and anterior parts of the medial end of the clavicle, passing obliquely downwards and medially to the front of the upper part of the manubrium sterni (Fig. 2.56). It is reinforced by the tendinous origin of sternomastoid.

Posterior sternoclavicular ligament

Although not as strong as the anterior ligament the posterior sternoclavicular ligament (Fig. 2.56B) is also a broad

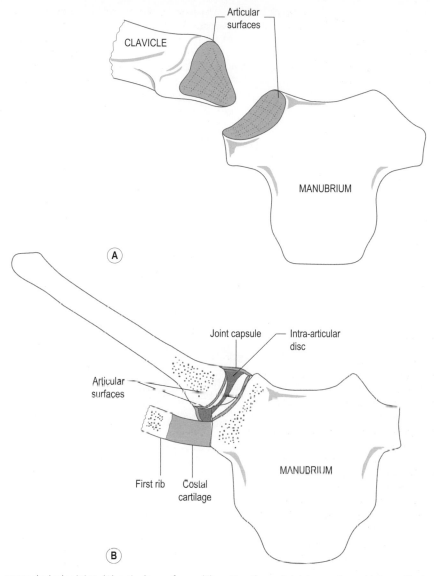

Figure 2.55 The sternoclavicular joint: (A) articular surfaces, (B) section through joint capsule and intra-articular disc.

band running obliquely downwards and medially. Laterally it attaches to the superior and posterior parts of the medial end of the clavicle, while medially it is attached to the back of the upper part of the manubrium sterni. The sternal attachment of sternohyoid extends across, and reinforces part of, the posterior ligament.

Interclavicular ligament

This strengthens the capsule superiorly (Fig. 2.56): it is formed by fibres attaching to the upper aspect of the sternal end of one clavicle passing across the jugular notch to join similar fibres from the opposite side. Some of these fibres attach to the floor of the jugular notch.

Costoclavicular ligament

This is an extremely strong, extracapsular, short, dense band of fibres (Fig. 2.56). It is attached to the upper surface of the first costal cartilage near its lateral end, and to a roughened area on the posterior aspect of the inferior surface of the medial end of the clavicle. The ligament has two laminae, usually separated by a bursa, which are attached to the anterior and posterior lips of the clavicular rhomboid impression. The anterior fibres run upwards and laterally, while those of the posterior lamina run upwards and medially; thus the fibres have a cruciate arrangement. The direction of fibres in the two laminae is the same as those in the external and internal intercostal muscles respectively.

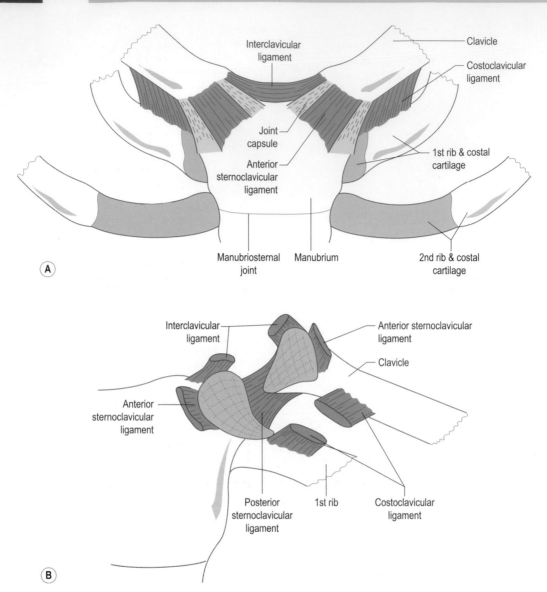

Figure 2.56 Ligaments associated with the sternoclavicular joint (A) seen from the front and (B) with the joint opened out and capsule removed.

The costoclavicular ligament essentially limits elevation of the clavicle; however, it is also active in preventing excessive anterior or posterior movements of the medial end of the clavicle. Its position and strength compensate for the weakness of the adjacent inferior part of the joint capsule.

Blood and nerve supply

The arterial supply of the sternoclavicular joint is from branches of the internal thoracic artery, the superior thoracic branch of the axillary artery, the clavicular branch of the thoracoacromial trunk, and the suprascapular artery.

Venous drainage is to the axillary and external jugular veins. Lymphatics from the joint pass to the lower deep cervical group of nodes, sometimes called the supraclavicular nodes, and thence to the jugular trunk. A few lymphatics may pass to the apical group of axillary nodes.

The nerve supply of the joint is by twigs from the medial supraclavicular nerve (C3 and 4) and the nerve to subclavius (C5 and 6).

Relations

Overlying the joint anteriorly is the tendinous attachment of the sternal head of sternomastoid (Fig. 2.57A).

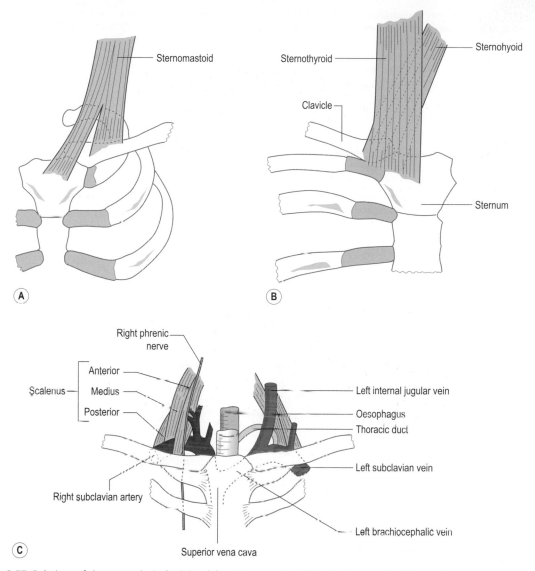

Figure 2.57 Relations of the sternoclavicular joint: (A) anterior muscles, (B) posterior muscles, (C) vessels.

Posteriorly, the sternoclavicular joint is separated from the brachiocephalic vein and common carotid artery on the left and the brachiocephalic trunk on the right, by sternohyoid and sternothyroid (Fig. 2.57B). The superior vena cava is formed on the right hand side, by the union of the two brachiocephalic veins, just below the joint at the lower border of the first costal cartilage (Fig. 2.57C).

On the right hand side, both the phrenic and vagus nerves lie lateral to the sternoclavicular joint as they enter the thorax from the neck. However, on the left hand side the vagus may pass behind the joint as it descends between the common carotid and subclavian arteries.

Stability

The shape of the articular surfaces and the surrounding musculature provide only a limited amount of security for the joint. Joint stability is primarily dependent on the strength and integrity of its ligaments, particularly the costoclavicular ligament. Unfortunately, when dislocation of the joint takes place it is liable to recur.

Movements

Although the articular surfaces do not conform to those of a ball and socket joint, the sternoclavicular joint nevertheless has three degrees of freedom of movement; that is elevation and depression, protraction and retraction, and axial rotation. The fulcrum of these movements, except axial rotation, is not at the joint centre but through the costoclavicular ligament. Consequently, the movements of elevation and depression, and protraction and retraction involve a gliding (i) between the clavicle and the intra-articular disc, and (ii) between the disc and the sternum.

Elevation and depression

The axis of rotation for elevation and depression runs horizontally and slightly obliquely anterolaterally through the costoclavicular ligament (Fig. 2.58A). Some authorities have suggested that two axes of rotation can be identified

for elevation and depression, one for gliding of the clavicle with respect to the disc, and the other for gliding of the disc against the sternum. Nevertheless, functionally the combined axis of movement runs through the costoclavicular ligament.

Because the axis of movement is somewhat removed from the joint centre, as the lateral end of the clavicle moves in one direction its medial end moves in the opposite direction (Fig. 2.58A). Consequently, elevation of the lateral end of the clavicle causes the medial end to move downwards and laterally. The range of movement of the lateral end of the clavicle is approximately 10 cm of elevation and 3 cm of depression, giving a total angular range of movement of some 60°. Elevation is limited by tension in the

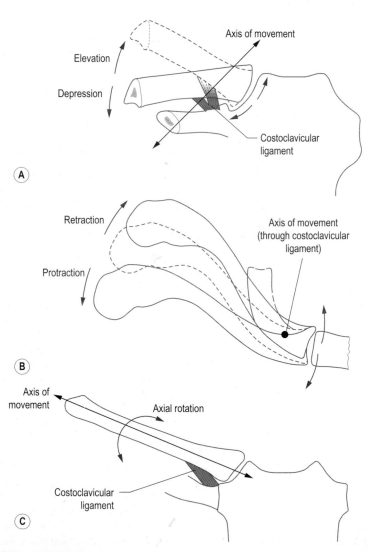

Figure 2.58 Movements of the clavicle at the sternoclavicular joint: (A) elevation and depression, (B) protraction and retraction (viewed from above), (C) axial rotation.

costoclavicular ligament and by tone in subclavius. Depression of the clavicle, in which the medial end moves upwards and medially, is limited by tension in the interclavicular ligament as well as by the intra-articular disc. If these two mechanisms fail, then movement is eventually limited by contact between the clavicle and upper surface of the first rib.

Protraction and retraction

The axis of movement for protraction and retraction lies in a vertical plane running obliquely inferolaterally through the middle part of the costoclavicular ligament (Fig. 2.58B). Again the two ends of the clavicle move in opposite directions because of the position of the fulcrum about which movement takes place (Fig. 2.58B), so that in protraction of the lateral end, the medial end moves posteriorly and vice versa. In these movements, the medial end of the clavicle and the intra-articular disc tend to move as one unit against the sternum. The range of movement of the lateral end of the clavicle is approximately 5 cm of protraction (anterior movement) and 2 cm of retraction (posterior movement), giving a total angular range of movement of about 35°. Movement anteriorly is limited by tension in the anterior sternoclavicular and costoclavicular ligaments, while posterior movement is limited by the posterior sternoclavicular and costoclavicular ligaments.

Axial rotation

Whereas elevation, depression, protraction and retraction of the clavicle are active movements brought about by direct muscle action, axial rotation (Fig. 2.58C) is entirely passive, being produced by rotation of the scapula transmitted to the clavicle by the coracoclavicular ligament. Pure axial rotation of the clavicle is not possible in the living subject; it always accompanies movements in other planes. The axis about which rotation occurs passes through the centre of the articular surfaces of the sternoclavicular and acromioclavicular joints.

The range of movement is small when the clavicle is in the frontal plane, but increases considerably when the lateral end of the clavicle is carried backwards. The degree of axial rotation possible is between 20° and 40° depending on the position of the clavicle.

That there should be any axial rotation possible at the sternoclavicular joint is due to (i) the relative incongruity of the articular surfaces, (ii) the presence of an intra-articular disc and (iii) the relative laxness of the capsular thickenings.

Accessory movements

With the subject lying supine, a downward pressure by the thumb on the medial end of the clavicle produces a posterior gliding of the clavicle against the sternum.

Palpation

The line of the sternoclavicular joint can be easily identified through the skin and subcutaneous tissues at the medial end of the clavicle. The projection of the medial end of the clavicle above the sternum can also be palpated.

Section summary

Sternoclavicular joint

Type	Saddle-shaped synovial, but functionally a ball and socket joint
Articular surfaces	Medial end of clavicle, clavicular notch of sternum and first costal cartilage. A complete intra-articular disc divides the joint into two separate compartments
Capsule	Complete fibrous capsule
Ligaments	Anterior and posterior sternoclavicular: interclavicular: costoclavicular
Stability	Mainly provided by costoclavicular ligament
Movements	Active elevation and depression, and protraction and retraction; passive axial rotation

---

The acromioclavicular joint

The plane synovial acromioclavicular joint connects the clavicle with the shoulder blade. The role it plays in movements of the pectoral girdle is considered by some to be greater than that of the sternoclavicular joint, particularly for movements in or close to the sagittal plane.

Articular surfaces

The articulation is between an oval flat or slightly convex facet on the lateral end of the clavicle, and a similarly shaped flat or slightly concave facet on the anteromedial border of the acromion (Fig. 2.59A). Both joint surfaces are covered with fibrocartilage. The major axis of both facets runs from anterolateral to posteromedial, so that the clavicular facet faces laterally and posteriorly, with that on the acromion facing medially and anteriorly. Consequently, the lateral end of the clavicle tends to over-ride the acromion, which together with the slope of their articulating surfaces favours displacement of the acromion downwards and under the clavicle in dislocations.

Joint capsule and synovial membrane

A relatively loose fibrous capsule surrounds the joint attaching to the articular margins. Its strong coarse fibres run in parallel fasciculi from one bone to the other. The capsule is thickest and strongest above where it is reinforced by the fibres of trapezius. Some authorities contend that the joint capsule is reinforced by two strong ligaments, the superior and inferior acromioclavicular ligaments,

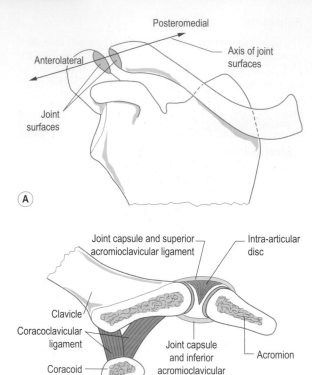

Figure 2.59 (A) Acromioclavicular joint open showing surfaces and (B) acromioclavicular joint frontal section, showing intra-articular disc.

passing between the adjacent surfaces of the two bones (Fig. 2.59B). In reality these are no more than capsular thickenings, which will show varying degrees of thickening in different individuals.

The synovial membrane lines the inner surface of the capsule attaching to the margins of the articular surfaces.

Intra-articular structures

A wedge-shaped, fibrocartilaginous articular disc partially divides the cavity in most joints (Fig. 2.59B). When present, the disc is attached to the upper inner part of the capsule extending down between the two articulating surfaces. Only rarely does the disc form a complete partition within the joint. The presence of the articular disc partially compensates for the small degree of incongruity between the two joint surfaces.

Ligaments

Apart from the capsular thickening referred to above, the strength of the acromioclavicular joint is provided by an extracapsular accessory ligament, the coracoclavicular ligament.

Coracoclavicular ligament

An extremely powerful ligament situated medial to the acromioclavicular joint anchoring the lateral end of the clavicle to the coracoid process (Fig. 2.60A) and thus stabilizing the clavicle with respect to the acromion. It is in two parts named according to their shapes, these being the posteromedial *conoid* and anterolateral *trapezoid* ligaments (Fig. 2.60B). The two parts tend to be continuous with each other posteriorly but are separated anteriorly by a small gap in which is found a synovial bursa.

The apex of the fan-shaped conoid ligament is attached posteromedially to the 'elbow' of the coracoid process (Fig. 2.60B). From here the ligament widens as it passes upwards, more or less in the frontal plane, to attach to the conoid tubercle on the undersurface of the clavicle.

The stronger and more powerful trapezoid ligament is a flat quadrilateral band attached inferiorly to a roughened ridge on the upper surface of the coracoid process (Fig. 2.60B). The wider superior surface of the ligament is attached to the trapezoid line on the undersurface of the clavicle, which runs anterolaterally from the conoid tubercle. Although the two surfaces of the trapezoid ligament are set obliquely, the ligament lies more or less in the sagittal plane, being more nearly horizontal than vertical.

Because the conoid and trapezoid ligaments lie in different planes, which are more or less at right angles to each other, and because the posterior edge of the trapezoid ligament is usually in contact with the lateral edge of the conoid ligament, a solid angle facing anteromedially is formed between them.

The two parts of the coracoclavicular ligament are set so as to restrain opposite movements of the scapula with respect to the clavicle. The conoid ligament limits forward movement of the scapula, while the trapezoid limits backward movement. The importance of these limiting movements is discussed more fully in the section on movements of the pectoral girdle (p. 110). Both ligaments, but especially the trapezoid ligament, prevent the acromion being carried medially under the lateral end of the clavicle when laterally directed forces are applied to the shoulder.

Blood and nerve supply

The arterial supply to the joint is by branches from the suprascapular branch of the subclavian artery and the acromial branch of the thoracoacromial trunk. Venous drainage is to the external jugular and axillary veins. Lymphatic drainage is to the apical group of axillary nodes.

The nerve supply to the joint is by twigs from the lateral supraclavicular, lateral pectoral, suprascapular and axillary nerves, from roots C4, 5 and 6.

Relations

The attachments of trapezius and deltoid cover the posterosuperior and anterosuperior aspects of the joint

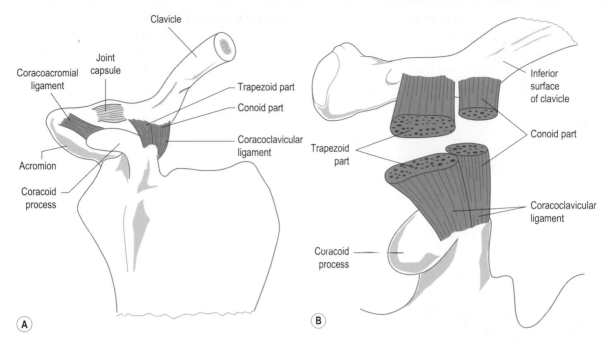

Figure 2.60 The coracoclavicular ligament (A) from the front and (B) sectioned to show its attachments to the coracoid process and clavicle.

respectively. Medial to the coracoclavicular ligament the transverse superior scapular ligament converts the scapula notch into a foramen, through which passes the suprascapular nerve; the suprascapular vessels pass above the transverse ligament. The lateral supraclavicular nerve crosses the clavicle medial to the acromioclavicular joint.

Although not directly associated with the joint, the coracoacromial ligament, as its name suggests, is attached to both the coracoid and acromion processes. Full details are given on page 117.

Stability

The stability of the joint is essentially provided by the coracoclavicular ligament. Trapezius and deltoid because they cross the joint will also provide some stability during movements of the joint.

Movement

The movements of the joint are entirely passive as there are no muscles connecting the bones which could cause one to move with respect to the other. Muscles which move the shoulder blade cause it to move on the clavicle. Indeed, all movements of the shoulder blade involve movement at both the acromioclavicular and sternoclavicular joints. All movements at the acromioclavicular joint, except axial rotation, are gliding movements with the coracoclavicular ligament acting to limit the movements.

The acromioclavicular joint has three degrees of freedom of motion about three axes. These movements are probably best described in terms of their relation to the shoulder blade rather than with respect to the cardinal axes of the body, since the joint constantly changes its relation to the trunk. The most important function of the joint is that it provides an additional range of movement for the pectoral girdle after the range of movement at the sternoclavicular joint has been exhausted.

Movement about a vertical axis

This is associated with protraction and retraction of the shoulder blade. The axis of movement passes vertically through the lateral end of the clavicle midway between the joint and the coracoclavicular ligament (Fig. 2.61A). As the acromion glides backwards with respect to the clavicle, the angle between the clavicle and shoulder blade increases; similarly as the acromion glides forwards this angle decreases (Fig. 2.61A). Backward movement of the acromion is checked by the anterior joint capsule and is actively limited by the trapezoid ligament as it becomes stretched. Forward movement is checked by the posterior joint capsule and limited by the conoid ligament. Towards the end of forward movement of the acromion, the trapezoid ligament may also be put under tension and therefore help to limit the movement. Compensatory movements of the clavicle at the sternoclavicular joint accompany those at the acromioclavicular joint.

109

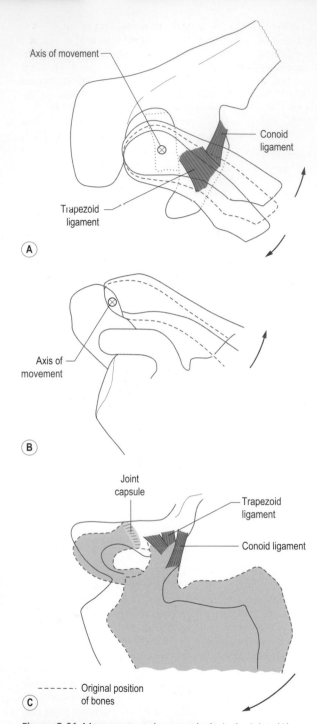

Figure 2.61 Movements at the acromioclavicular joint: (A) about a vertical axis associated with protraction/retraction of scapula, (B) about a sagittal axis associated with elevation/depression of scapula, (C) axial rotation of scapula.

Movement about a sagittal axis

Movement about a sagittal axis (Fig. 2.59B) occurs when the shoulder blade is elevated or depressed, with the total range of movement being no more than 15°. Elevation is limited by tension developed in both parts of the coracoclavicular ligament, with the conoid ligament coming into play first; depression is checked by the coracoid process coming into contact with the undersurface of the clavicle.

Axial rotation

This is associated with medial and lateral rotation of the shoulder blade; that is when the glenoid fossa faces inferiorly or superiorly respectively. The range of rotation of the shoulder blade with respect to the clavicle is of the order of 30°, and occurs about an axis that passes through the conoid ligament and the acromioclavicular joint (Fig. 2.61C). It allows the flexed arm to be fully elevated. Restraints to rotation are provided by both parts of the coracoclavicular ligament.

Accessory movements

With the subject lying supine, downward pressure applied with the thumb on the lateral end of the clavicle causes it to glide backward against the acromion.

Palpation

The line of the acromioclavicular joint can be palpated from above by applying a downward pressure to the lateral end of the clavicle.

Section summary	
Acromioclavicular joint	
Type	Synovial plane joint
Articular surfaces	Lateral end of clavicle and acromion of scapula. An incomplete intra-articular disc is placed between the two bones
Capsule	Fibrous capsule surrounds articular margin
Ligaments	Coracoclavicular (conoid and trapezoid parts)
Stability	Mainly from coracoclavicular ligament
Movements	Passive gliding during movements of pectoral girdle

Movements of the pectoral girdle as a whole

Movements of the pectoral girdle serve to increase the range of movement of the shoulder joint, principally by changing the relative position of the glenoid fossa with respect to the chest wall. In all of these movements the clavicle, acting as a strut, holds the shoulder away from the trunk thereby securing greater freedom of movement of the upper limb. It should be remembered that movements of the pectoral girdle accompany virtually all movements of the shoulder joint (pp. 123–125).

In movements of the pectoral girdle, the glenoid fossa travels in an arc of a circle whose radius is the clavicle, while the medial border of the scapula, held against the chest wall, travels in a curve of shorter radius. Consequently, the relative positions of the clavicle and shoulder blade must be capable of changing. This occurs at the acromioclavicular joint. It now becomes obvious that a rigid union between the clavicle and shoulder blade would severely limit the mobility of the upper limb.

Lateral movement of the shoulder blade around the chest wall brings it to lie more in a sagittal plane so that the glenoid fossa faces more directly forwards (Fig. 2.62A,B). Medial movement towards the vertebral column brings the shoulder blade to lie more in the frontal plane, with the glenoid fossa facing more directly laterally. These two extreme

positions of the scapula form a solid angle of 40°–45° (Fig. 2.62A). Furthermore, the angle between the clavicle and shoulder blade decreases to approximately 60° on full lateral movement of the shoulder blade, and increases to approximately 70° on full medial movement (Fig. 2.62A). The total range of linear translation of the shoulder blade around the chest wall is about 15 cm.

Elevation and depression of the shoulder blade has a linear range of some 10–12 cm (Fig. 2.62C). However, it is usually accompanied by some degree of rotation, so that in elevation the glenoid fossa comes to face increasingly upwards. In depression, the fossa points increasingly downwards. The rotation of the shoulder blade that occurs with respect to the chest wall does so about an axis perpendicular to the plane of the scapula, and is situated a little below the spine close to the superomedial angle. The total range of angular rotation of the scapula amounts to 60° (Fig. 2.62D). This involves a displacement of the inferior angle of the scapula of 10–12 cm, and that of the superolateral angle of 5–6 cm.

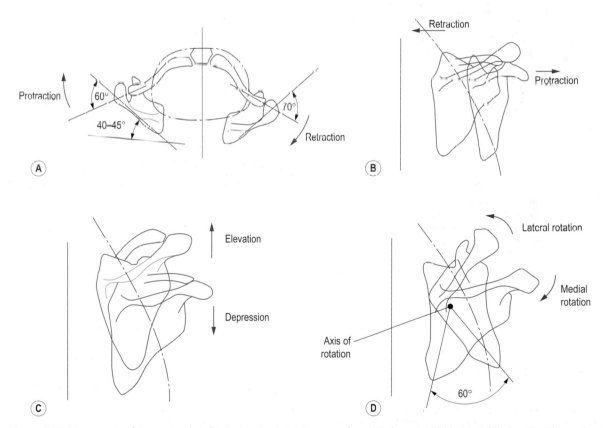

Figure 2.62 Movements of the pectoral girdle. Protraction/retraction seen from (A) above and (B) behind, (C) elevation/depression, (D) axial rotation.

During abduction or flexion of the arm, the clavicle rotates about its long axis so that its superior surface is increasingly directed posteriorly. Towards the end of the range of rotation of the scapula against the clavicle, the coracoclavicular ligament becomes taut and transmits the rotating force to the clavicle, whose rotation then accounts for the scapular rotation on the chest wall. Any impairment to clavicular rotation at either the sternoclavicular or acromioclavicular joints will interfere with the free movement of the shoulder blade and upper limb as a whole.

Although movements of medial and lateral translation (retraction and protraction) of the scapula, elevation and depression, and its rotation with respect to the chest wall have been discussed, it is important to remember that they do not occur as pure movements. All movements of the pectoral girdle will involve some degree of each of the above pure movements.

Biomechanics

Stresses on the clavicle

The clavicle is subjected to both compression and tension stresses, which under normal conditions are absorbed within it. They only become apparent when the integrity of the pectoral girdle is compromised, as by fracture, dislocation or muscular imbalance. Unlike the pelvic girdle, because the pectoral girdle is not a complete bony ring, the intrinsic stresses require the cooperation of muscles attached to it for there to be equilibrium.

A compressive stress along the length of the clavicle directed towards the sternoclavicular joint is produced by the action of trapezius and pectoralis minor as they pull the clavicle towards the sternum. Forces transmitted medially from the upper limb to the glenoid fossa are transmitted from the scapula to the clavicle by the trapezoid ligament, and from the clavicle to the first rib by the costoclavicular ligament. Consequently, falling on an outstretched hand or elbow puts no strain on the joints at either end of the clavicle. If the clavicle fractures as a result, it does so between these two ligaments. In such fractures, the two fragments tend to over-ride one another. These compressive stresses are increased by lying on one shoulder.

Tension stresses within the clavicle are produced, under the action of the deltoid, when the upper limb is abducted. Hanging and swinging forwards by the arms increases these tension stresses. If sufficiently large they may lead to some discontinuity of the pectoral girdle; however, a joint dislocation is much more likely to occur than a fracture.

A rotational force transmitted through the scapula to the clavicle tends to damage the acromioclavicular and/or sternoclavicular joints and their associated ligaments rather than causing a fracture of the bone. Downward forces applied to the lateral end of the clavicle create bending stresses within it, which, if sufficiently large or if the clavicle comes into contact with the coracoid process, may lead to fracture of the bone.

The shoulder joint

Introduction

The shoulder (glenohumeral) joint is the articulation between the head of the humerus and the glenoid fossa of the scapula (Fig. 2.63). The intracapsular presence of the epiphyseal line between the ventral coracoid and dorsal scapula in the upper part of the glenoid fossa facilitates adjustments of the joint surfaces during the growth of the bone. Similar arrangements are also to be found in the hip and elbow joints.

It is a synovial joint of the ball and socket variety, with the head of the humerus forming the ball and the glenoid fossa the socket, in which freedom of movement has been developed at the expense of stability. The mobility of the upper limb is partly due to changes in the shoulder joint which have occurred with the freeing of the upper limbs from locomotor activity, and partly due to mobility of the pectoral girdle linking the upper limb to the trunk. Comparison of the shoulder and hip joints, equivalent joints in the upper and lower limbs, together with their mode of attachment to the axial skeleton, reveals significant and important differences between them even though their basic features are similar.

In the frontal plane, the axis of the head and neck of the humerus forms an angle of 135° to 140° with the long axis of the shaft; this is the angle of inclination (Fig. 2.64A). Because of this angulation, the centre of the humeral head lies about 1 cm medial to the long axis of the humerus. Although the anatomical and mechanical axes of the humerus do not exactly coincide, unlike the femur they both lie inside the bone (Fig. 2.64A). Consequently, the action of muscle groups producing movement at the shoulder joint, especially medial and lateral rotation, is more easily understood.

As well as being set at an angle to the shaft of the humerus, the axis of the head and neck is rotated backwards against the shaft 30° to 40°; this is the angle of retroversion (Fig. 2.64B). The magnitude of the angle of retroversion is said to vary both with age and race. It is thought that this angle has increased with the attainment of bipedalism, in which there has also been flattening of the thoracic cage anteroposteriorly, and a backward displacement of the scapula, the result being that as the glenoid fossa came to be directed more laterally, the head and neck of the humerus became more twisted in an attempt to maintain maximum joint contact between the two articulating surfaces. Structurally this adaptation was not entirely successful, which to some extent was fortuitous for in humans the use of the upper limb gradually changed from one of support to one of manipulation.

Articular surfaces

The articular surfaces of the joint are the rounded head of the humerus and the rather shallow glenoid fossa. The adaptation of these two surfaces contributes very little,

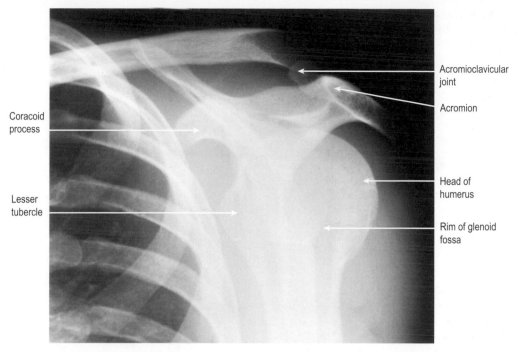

Acromioclavicular joint

Acromion

Coracoid process

Head of humerus

Lesser tubercle

Rim of glenoid fossa

Figure 2.63 Radiograph of the shoulder in the anatomical position, oblique view.

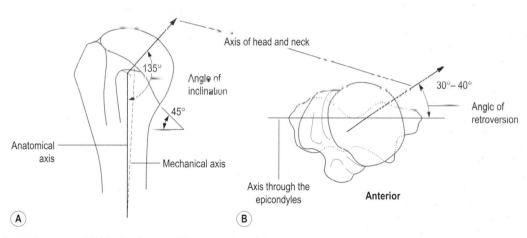

Axis of head and neck

135°

Angle of inclination

45°

30°–40°

Angle of retroversion

Anatomical axis

Mechanical axis

Axis through the epicondyles

Anterior

(A)

(B)

Figure 2.64 The angle of (A) inclination and (B) retroversion of the humerus.

if at all, to the stability and security of the joint. As in the majority of synovial joints, the two articular surfaces are covered by hyaline cartilage.

Glenoid fossa

Situated at the superolateral angle of the scapula it faces laterally, anteriorly and slightly superiorly. It is pear-shaped in outline, with the narrower region superiorly, and concave both vertically and transversely (Fig. 2.65A). However, the concavity of the joint is irregular and less deep than the convexity of the head of the humerus. In the plane of the axis of the head and neck of the humerus it has been estimated that the curvature of the glenoid fossa, with its larger radius, subtends an angle of some 75°. The articular surface of the fossa is little more than one-third of that of the humeral head. It is deepened to some extent by the presence of the glenoid labrum (p. 117).

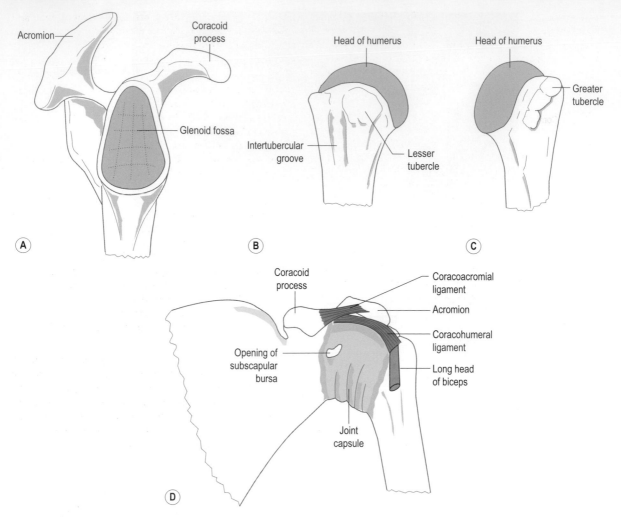

Figure 2.65 Articular surfaces of the shoulder joint: (A) glenoid fossa, (B) head of the humerus viewed anteriorly, (C) head of the humerus viewed posteriorly, (D) anterior aspect of the lax joint capsule.

Head of the humerus

This represents two-fifths of a sphere and faces superiorly, medially and posteriorly (Fig. 2.65B,C). With its smaller radius of curvature, the articular surface subtends an angle of approximately 150° in the plane of the axis of the head and neck. Regardless of the position of the joint, only one-third of the humeral head is in contact with the glenoid fossa at any time. It is essentially the mismatch in coaptation of the articular surfaces which gives the joint its mobility.

Joint capsule and synovial membrane

The fibrous joint capsule forms a loose cylindrical sleeve between the two bones (Fig. 2.65D). The majority of the capsular fibres pass horizontally between scapula and humerus, but some oblique and transverse fibres are also present. Although it is thick and strong in parts, particularly anteriorly, because of its laxness the capsule conveys little stability to the joint. On the scapula, the capsule attaches just outside the glenoid labrum anteriorly and inferiorly, and to the labrum superiorly and posteriorly. Recesses formed between the anterior capsular attachment and the glenoid labrum may have pathological significance in shoulder joint trauma.

On the humerus, the capsule attaches to the anatomical neck, around the articular margins of the head, medial to the greater and lesser tubercles, except inferiorly where it joins the medial surface of the shaft about 1 cm below the articular margin (Fig. 2.65D). The downward extension of the capsular attachment medially causes the medial end of the upper epiphyseal line of the humerus to become intracapsular.

The anterior part of the capsule is thickened and strengthened by the presence of the three glenohumeral ligaments, which can only be seen on its inner aspect. The superoposterior part is strengthened near its humeral attachment by the coracohumeral ligament. The tendons of the 'rotator cuff' (p. 67) muscles spread out over the capsule, blending with it near their humeral attachments. These short scapular muscles act as extensible ligaments, and are extremely important in maintaining the integrity of the joint.

With the arm hanging in the anatomical position, the lower part of the joint capsule is lax and forms a redundant fold. When the arm is abducted this part of the capsule becomes increasingly taut.

There are two openings in the fibrous capsule; occasionally a third may be present. One is at the upper end of the intertubercular groove to allow the long head of biceps to pass into the arm (Fig. 2.65D). This part of the capsule is thickened, forming the transverse humeral ligament, arching over the tendon as it emerges from the capsule. The second opening is in front of the capsule, between the superior and middle glenohumeral ligaments, and communicates with the subscapular bursa deep to the tendon of subscapularis (Fig. 2.65D). A third opening is often present posteriorly, allowing a communication of the joint cavity with the infraspinatus bursa

Capsular ligaments

The anterior part of the joint capsule is reinforced by three longitudinal bands of fibres called the glenohumeral ligaments (Fig. 2.66). They are seldom prominent and when present radiate from the anterior glenoid margin, extending downwards from the supraglenoid tubercle.

Superior glenohumeral ligament: A slender ligament arising from the upper part of the glenoid margin and adjacent labrum immediately anterior to the attachment of the tendon of the long head of biceps. It runs laterally parallel to the biceps tendon to the upper surface of the lesser tubercle.

Middle glenohumeral ligament: This arises below the superior ligament and attaches to the humerus on the front of the lesser tubercle below the insertion of subscapularis.

Inferior glenohumeral ligament: This is usually the most well developed of the three ligaments, although it is occasionally absent. It arises from the glenoid margin below the notch in its anterior border and the adjacent anterior border of the glenoid labrum. It descends slightly obliquely to the humerus to attach to the anteroinferior part of the anatomical neck. As it passes from scapula to humerus, the upper part of the inferior ligament may merge with the lower part of the middle glenohumeral ligament.

Although they have no real stabilizing function, certain movements of the shoulder joint will tend to increase the tension in some or all of the glenohumeral ligaments. Lateral rotation of the humerus will put all three ligaments under tension, whereas medial rotation relaxes them. In abduction, only the middle and inferior ligaments become taut, while the superior becomes relaxed.

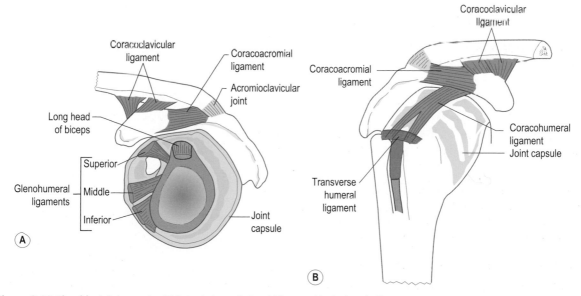

Figure 2.66 Shoulder joint capsule: (A) lateral view of glenoid fossa with the head of humerus removed to show the glenohumeral ligaments, (B) transverse humeral ligament.

Transverse humeral ligament: At the upper end of the intertubercular groove, the transverse humeral ligament bridges the gap between the greater and lesser tubercles (Fig. 2.66B). It is formed by some of the transverse fibres of the capsule and serves to hold the biceps tendon in the intertubercular groove as it leaves the joint.

Synovial membrane

Lines the capsule and therefore also extends downwards as a pouch when the arm is hanging by the side (Fig. 2.67A). It attaches to the articular margins of both bones, and is therefore reflected upwards on the medial side of the humeral shaft to its attachment. Consequently, although the medial part of the epiphyseal line is intracapsular, it is extrasynovial (Fig. 2.67A). The membrane extends through the anterior opening of the capsule forming the subscapularis bursa, which may be limited in its extent to the posterior surface of the tendon of subscapularis. However, it may be sufficiently large to extend above the upper border of the tendon and so come to lie below the coracoid process. This upward extension may be replaced by a

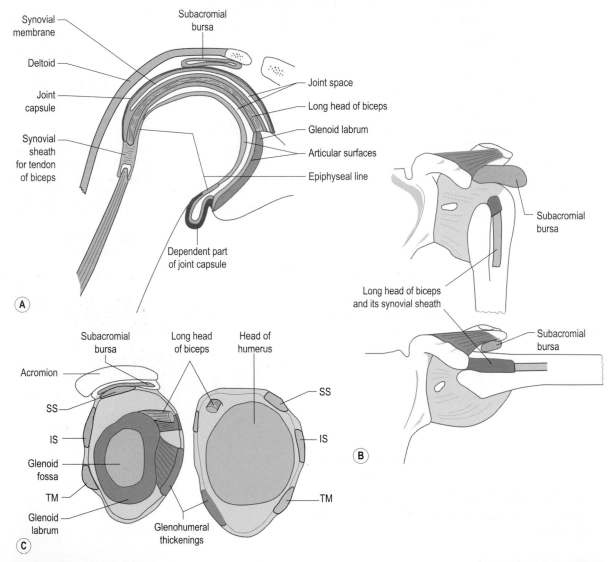

Figure 2.67 (A) Coronal section of the shoulder joint showing the reflection of the synovial membrane around the long head of biceps, (B) withdrawal of the subacromial bursa, and protrusion of the biceps synovial sheath from the joint capsule when the arm is abducted, (C) the joint opened out, also showing the blending of some of the rotator cuff muscles to the capsule. IS, infraspinatus; SS, supraspinatus; TM, teres minor.

separate subcoracoid bursa. The posterior extension of the membrane through the joint capsule forms the infraspinatus bursa.

The intracapsular part of the long head of biceps is enclosed within a double-layered tubular sheath of synovial membrane continuous with that of the joint at its glenoid attachment (Fig. 2.65A). This sheath surrounds the biceps tendon as it passes beneath the transverse humeral ligament into the intertubercular groove, and extends some 2 cm into the arm (Fig. 2.67B).

An important, but non-communicating, bursa associated with the shoulder joint is the subacromial bursa (Fig. 2.67). It lies between, and separates, the coracoacromial arch and deltoid from the superolateral part of the shoulder joint. That part of the bursa which extends laterally under the deltoid is usually referred to as the subdeltoid bursa.

Some of these bursae are of clinical significance as adhesions may form, preventing free gliding movements. This is particularly true for the bicipital sheath and the subdeltoid bursa. Indeed, the subdeltoid bursa may become inflamed (bursitis) and affect the underlying tendon of supraspinatus, leading to its rupture in a small number of cases.

Intra-articular structures

Glenoid labrum

The glenoid fossa is deepened by the presence of a fibrocartilaginous rim, the glenoid labrum (Fig. 2.67C), which is triangular in cross-section; it has a thin free edge and is about 4 mm deep. The base of the labrum attaches to the margin of the glenoid fossa; the outer surface gives attachment to the joint capsule posteriorly and superiorly, while the inner (joint) surface is in contact with the head of the humerus and is lined by cartilage continuous with that of the glenoid fossa. The upper part of the labrum may not be completely fixed to the bone so that its inner edge may project into the joint like a meniscus.

The outer margin of the labrum gives attachment to the tendon of the long head of biceps superiorly, while inferiorly the tendon of the long head of triceps partly arises from it.

Long head of the biceps

The tendon of the long head of biceps runs intracapsularly from its attachment to the supraglenoid tubercle and adjacent superior margin of the glenoid labrum until it emerges from the joint deep to the transverse humeral ligament (Fig. 2.66). During its intracapsular course, and for some 2 cm beyond, the tendon is ensheathed in a synovial sleeve.

Accessory ligaments

Apart from the capsular ligaments, there are two further ligaments associated with the shoulder joint. Both are considered to be accessory ligaments, although one, the coracohumeral ligament, blends with the joint capsule. The other, the coracoacromial ligament, completes a fibro-osseous arch above the joint.

Coracohumeral ligament

A fairly strong, broad band arising from the lateral border of the coracoid process near its root. It becomes flattened as it passes laterally, with its two margins diverging above the intertubercular groove to attach to the upper part of the anatomical neck in the region of the greater and lesser tubercles and to the intervening transverse humeral ligament (Fig. 2.68).

The anterior border of the medial part of the ligament is free, but as it passes laterally it fuses with the tendon of subscapularis as it blends with the capsule prior to its insertion on the lesser tubercle. The posterior part of the ligament blends with the tendon of supraspinatus as it attaches to the superior facet on the greater tubercle of the humerus.

Coracoacromial ligament

This is not directly associated with the joint but forms, together with the coracoid and acromion processes, a fibro-osseous arch above the head of the humerus. It is a strong, triangular ligament, with the anterior and posterior borders tending to be thicker than the intermediate part.

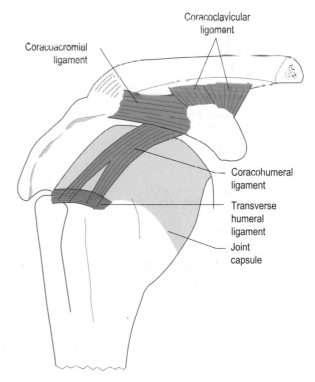

Figure 2.68 The coracohumeral and coracoacromial ligaments.

Occasionally, the tendon of pectoralis minor is prolonged and pierces the base of the ligament to become continuous with the coracohumeral ligament. The coracoacromial ligament is attached by a broad base to the lateral border of the horizontal part of the coracoid process, with the blunt apex attached to the apex of the acromion in front of the acromioclavicular joint (Fig. 2.68). Superiorly are the clavicle and deltoid, while inferiorly it is separated from the tendon of supraspinatus and the shoulder joint by the subacromial bursa.

The arch formed by the ligament and bony processes serves to increase the surface upon which the head of the humerus is supported when force is transmitted upwards along the humerus.

Blood and nerve supply

The arterial supply is from numerous sources as there is an important anastomosis around the scapula involving vessels from the subclavian and axillary arteries and the descending aorta (Fig. 2.69). The supply to the shoulder joint is by branches from the suprascapular branch of the subclavian artery, the acromial branch of the thoracoacromial artery, and branches from the anterior and posterior circumflex humeral arteries. The latter three are all branches of the axillary artery. The venous drainage is by similarly named veins which drain into the external jugular and axillary veins.

Lymphatic drainage of the joint is to lymph nodes within the axilla, eventually passing by the apical group of nodes into the subclavian lymph trunk.

The nerve supply to the shoulder joint is by twigs from the suprascapular, axillary, subscapular, lateral pectoral and musculocutaneous nerves, and have a root value of C5, 6 and 7.

Relations

The shoulder joint is almost completely surrounded by muscles passing between the pectoral girdle and the humerus (Fig. 2.70). These serve to protect the joint by helping to suspend the upper limb from the pectoral girdle and so convey some degree of stability to the joint. Some muscles are more important in this respect than others. The anterior, superior and posterior parts of the joint are directly related to the tendons of subscapularis, supraspinatus, and infraspinatus and teres minor, respectively. These tendons all blend with the humeral part of the joint capsule, and because of their action on the shoulder joint (see pp. 59, 65–68), have become known as the *rotator cuff*. Covering the superolateral part of the joint is deltoid, giving the shoulder its rounded appearance (Fig. 2.70). The greater and lesser tubercles of the humerus can be palpated through deltoid, as can the long head of biceps as it emerges from within the capsule. The tendon of the long head of biceps passes directly above the joint within the capsule. Superiorly, and separated from the joint by the tendon of supraspinatus, is the coracoacromial arch. Inferiorly, arising from the infraglenoid tubercle, is the long head of triceps as it passes into the arm almost parallel to the humerus.

Immediately below the shoulder joint is the quadrangular space, bounded superiorly by teres minor, inferiorly by teres major, medially by the long head of triceps and laterally by the shaft of the humerus (Fig. 2.71). Passing

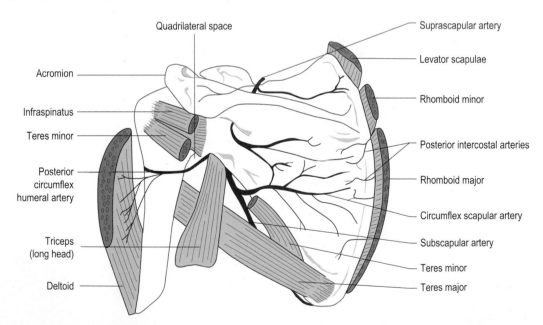

Figure 2.69 Vessels involved in supplying the shoulder region, together with those involved in the scapular collateral anastomosis.

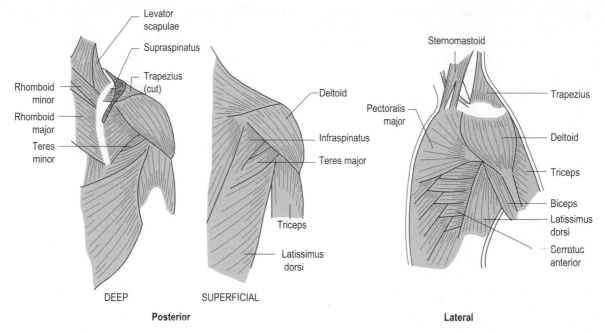

Figure 2.70 Muscles of the shoulder region.

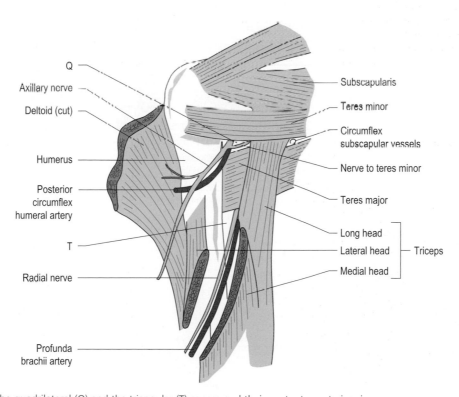

Figure 2.71 The quadrilateral (Q) and the triangular (T) spaces and their contents, posterior view.

through this space from anterior to posterior are the axillary nerve and posterior circumflex humeral artery. Downward dislocation of the head of the humerus or prolonged upwardly applied pressure, for example as when falling asleep with the arm hanging over the back of a chair, may cause temporary or permanent damage to the nerve and consequent loss of function of deltoid and teres minor. Immediately medial to the quadrangular space is the triangular space, bounded by teres minor and major and the long head of triceps, through which pass the circumflex scapular vessels (Fig. 2.71). The triangular interval lies below the quadrilateral space and has the long head of triceps as its medial border, the shaft of the humerus laterally and the lower border of teres major as its base (Fig. 2.71). It is an important region as the radial nerve and the profunda brachii vessels pass through it to the posterior compartment of the arm. Fractures of the shaft of the humerus, or pressure from the axillary pad of an incorrectly used crutch, may involve the radial nerve, resulting in radial nerve palsy, i.e. wrist drop, which will affect the functional use of the hand.

The axilla

Inferomedial to the shoulder joint is the pyramid-shaped axilla – the space between the arm and the thorax which enables vessels and nerves to pass between the neck and the upper limb (Fig. 2.72A). The apex of the axilla is formed by the clavicle anteriorly, scapula posteriorly and outer border of the first rib medially. Its concave base (floor) is formed by deep fascia extending from that over serratus anterior to the deep fascia of the arm, attached to the margins of the axillary folds anteriorly and posteriorly and supported by the suspensory ligament of the axilla, itself a downward extension of the clavipectoral fascia below pectoralis minor (Fig. 2.72B). The anterior axillary wall is formed by pectoralis major, the lower rounded fold formed by twisting of the muscle fibres as they pass from the chest wall to the humerus (p. 61). The posterior wall, which extends lower than the anterior, is formed by subscapularis and teres major, with the tendon of latissimus dorsi twisting around teres major (Fig. 2.72B). The medial wall is formed by serratus anterior, and the

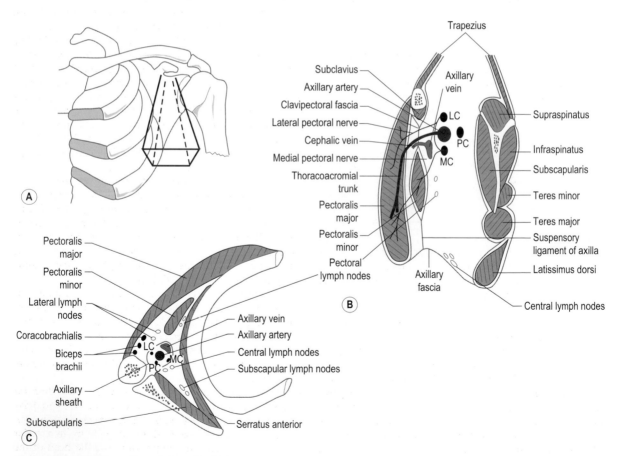

Figure 2.72 (A) The position of the axilla, (B) longitudinal section through the axilla showing its boundaries and contents, (C) transverse section. LC, lateral cord; MC, medial cord; PC, posterior cord of the brachial plexus.

lateral wall by the floor of the intertubercular groove (Fig. 2.72C). Both the anterior and posterior axillary folds can be easily palpated. A vertical line midway between them, running down the thoracic wall, is the *midaxillary line.*

When the arm is fully abducted the axillary folds virtually disappear as the muscles forming them run almost parallel to the humerus. The axillary hollow may be replaced by a bulge.

The principal contents of the axilla are the blood vessels and nerves passing between the neck and upper limb (Fig. 2.72B,C). These are the axillary artery and its branches, the axillary vein and its tributaries, and the brachial plexus and its terminal branches. Together, with the various groups of lymph nodes, these structures are surrounded by fat and loose areolar tissue. The tendon of the long head of biceps runs in the intertubercular groove, and so lies just within the axilla. Also within the axilla are the short head of biceps and coracobrachialis. The major vessels and nerve trunks are enclosed within the axillary sheath, a fascial extension of the prevertebral layer of cervical fascia. The axillary sheath is adherent to the clavipectoral fascia behind pectoralis minor, while just beyond the second part of the axillary artery it blends with the tunica adventitia of the vessels.

The axillary artery runs through the axilla, being posterior and superior to the vein; it can be indicated on the surface of the abducted arm by a straight line running from the middle of the clavicle to the medial prominence of coracobrachialis. With the arm hanging by the side, the artery describes a gentle curve with the concavity facing inferomedially.

The axillary artery is crossed anteriorly by the tendon of pectoralis minor, dividing it into three parts. Above the first part of the artery are the lateral and posterior cords of the brachial plexus and behind is the medial cord – it is crossed by a communicating loop between the medial and lateral pectoral nerves. The second part of the axillary artery, behind pectoralis minor, has the various cords of the brachial plexus in their named positions. The third part of the artery, lying laterally against coracobrachialis, has the musculocutaneous nerve laterally, the median nerve anteriorly, the ulnar and medial cutaneous nerves of the arm and forearm medially, and the axillary and radial nerves posteriorly.

The axillary lymph nodes are widely distributed within the axillary fat, but may be conveniently divided into five groups (Fig. 2.72B,C). The lateral nodes lie along and above the axillary vein and receive the majority of the lymphatic drainage of the upper limb. The subscapular (posterior) nodes lie along the subscapular artery and receive lymph from the scapular region and back above the level of the umbilicus. The pectoral (anterior) nodes, alongside the lateral thoracic artery, receive lymph from the anterior chest wall including the breast. These groups of nodes drain to a central group, which lie above the axillary floor. From here efferents pass to the apical group (the only group lying above the tendon of the pectoralis minor) and then to the subclavian lymph trunk.

Because of the involvement of the axillary nodes in the lymphatic drainage of the breast, they may be subjected to radiotherapy treatment in an attempt to limit the secondary spread of cancer from the breast. It is important to remember that the lateral group of nodes lie above the axillary vein so that they can be excluded from treatment programmes; otherwise severe problems with the lymphatic drainage of the upper limb may result.

Stability

The incongruity of the articular surfaces, together with the laxness of the joint capsule, suggests that the shoulder joint is not very stable. Although dislocation of the shoulder is common it is by no means an everyday occurrence. What factors are responsible for conferring stability on the joint? The glenoid labrum, as well as deepening the fossa, also improves joint congruency, and thus becomes a significant stabilizing factor. Fracture of the glenoid or tearing of the labrum often results in dislocation.

The most important factor, however, is the tone in the short scapular (rotator cuff) muscles (supraspinatus, infraspinatus, teres minor and subscapularis). Not only do they attach close to the joint, but they fuse with the lateral part of the capsule (Fig. 2.73). In this way they act as ligaments of variable length and tension, and also prevent the lax capsule and its synovial lining from becoming trapped between the articulating surfaces. The inferior part of the capsule is the weakest, being relatively unsupported by muscles. However, as the arm is gradually abducted, the long head of triceps and teres major become increasingly applied to this aspect of the joint.

In addition to the rotator cuff muscles, all the muscles passing between the pectoral girdle and the humerus assist in maintaining the stability of the joint (Fig. 2.74). Particularly important are the long heads of biceps and triceps. The tendon of the long head of biceps, being partly intracapsular, acts as a strong support over the superior part of the joint. The long head of triceps gives support below the joint when the arm is abducted.

An upward displacement of the head of the humerus is resisted by the coracoacromial arch. Although not part of the joint, this arch, separated from the joint by the subacromial bursa, functions mechanically as an articular surface. The arch is so strong that an upward thrust on the humerus will fracture either the clavicle or the humerus first before compromising the arch.

Dislocation of the shoulder is more common than for many joints, being a consequence of the need to have the joint as mobile as possible. In addition, the long humerus has great leverage in dislocating forces. In anterior dislocation, which is the more common, the head of the humerus comes to lie under the coracoid process, producing a bulge in the region of the clavipectoral groove. At the

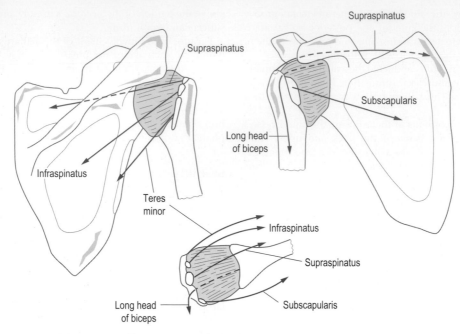

Figure 2.73 The action of the rotator cuff muscles in stabilizing the shoulder joint.

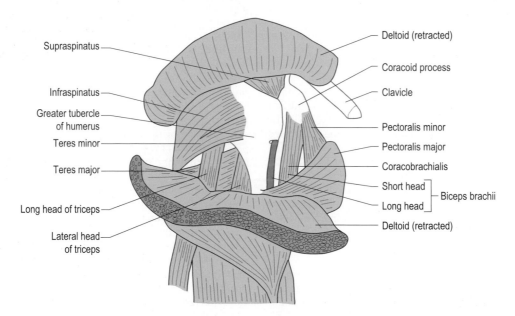

Figure 2.74 Lateral view of muscles involved in stabilizing the shoulder joint.

same time the roundness of the shoulder is lost. In such dislocations, the humeral head usually comes through the joint capsule between the long head of triceps and the inferior glenohumeral ligament.

Because the glenoid fossa faces anterolaterally, it is better situated to resist posteriorly directed forces. The presence of infraspinatus and teres minor reinforces the capsule posteriorly. Posterior dislocation may result when a large force is applied to the long axis of the humerus when the arm is medially rotated and abducted. The joint capsule tears in the region of teres minor with the head of the humerus coming to lie below the spinous process of the scapula.

Movements

The architecture of the shoulder joint gives it a greater range of movement than at any other joint within the body. Its ball and socket shape means that movement can take place around an infinite number of axes intersecting at the centre of the head of the humerus. For descriptive purposes, the movements of which the shoulder joint is capable are flexion and extension, abduction and adduction, and medial and lateral rotation. However, the axes about which they occur have to be carefully defined as the plane of the glenoid fossa does not coincide with any of the cardinal planes of the body, being inclined approximately 45° to both the frontal and sagittal planes. It is therefore possible to define two sets of axes about which movements occur, one with respect to the cardinal planes of the body (Fig. 2.75A) and another with respect to the plane of the glenoid fossa (Fig. 2.75B). If the cardinal planes are used, flexion and extension occur about a transverse axis, abduction and adduction occur about an anteroposterior axis, and medial and lateral rotation occur about the longitudinal axis of the humerus, passing between the centre of the head and the centre of the capitulum (Fig. 2.75A). If, however, movements are considered with respect to the plane of the scapula, then flexion and extension take place about an axis perpendicular to the plane of the glenoid fossa, abduction and adduction about an axis parallel to the plane of the glenoid fossa, with medial and lateral rotation occurring about the longitudinal axis of the humerus (Fig. 2.75B). While the presentation of these two sets of axes may appear

to be initially confusing, the importance of those with respect to the plane of the glenoid fossa is that in the treatment of some injuries of the shoulder, the position of the joint which will not create asymmetric tension on the joint capsule is when the joint is abducted to 90° in the plane of the glenoid fossa.

Irrespective of the orientation of these axes, the incongruity of the joint surfaces means that all movements, except axial rotation, are a combination of gliding and rolling of the articular surfaces against each other. However, unlike the knee, it is not possible to define the extent of each type of motion in each of the various movements. Although the range of movement at the shoulder joint is relatively large, the mobility of the upper limb against the trunk is increased by movements of the pectoral girdle. Indeed, flexion and extension, and abduction and adduction, may be considered to be always accompanied by scapular and clavicular movements, except perhaps for the initial stages. Shoulder joint movement is more concerned with bringing the arm to the horizontal position, while pectoral girdle movements, principally of the scapula, are more concerned in bringing the arm into a vertical position.

The association of shoulder and pectoral girdle movement also increases the power of the movement. The rotator cuff muscles, being attached close to the axes of movement, have a poor mechanical advantage compared with muscles acting on the scapula, which are generally more powerful as well as having considerable leverage,

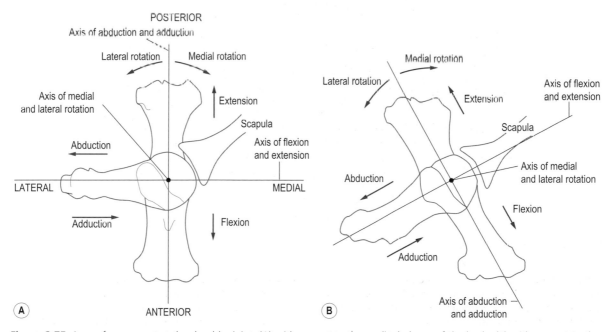

Figure 2.75 Axes of movement at the shoulder joint: (A) with respect to the cardinal planes of the body, (B) with respect to the plane of the glenoid fossa.

In patients with fused or fixed shoulder joints a large degree of upper limb mobility with respect to the trunk is still possible because of pectoral girdle movements.

In the following account shoulder joint movements are considered initially with respect to the plane of the glenoid fossa. How these are related to movements of the upper limb in the cardinal planes is discussed at the end of this section.

Flexion and extension

These occur about an axis perpendicular to the plane of the glenoid fossa, so that in flexion the arm moves forwards and medially at an angle of approximately 45° to the sagittal plane (Fig. 2.76A). In extension it is carried backwards and laterally (Fig. 2.74A). The range of flexion is approximately 110° and that of extension 70°. Both of these ranges may be extended by movements of the pectoral girdle so that flexion of the upper limb with respect to the trunk reaches 180° and extension just exceeds 90°. Extension is limited by the greater tubercle of the humerus coming into contact with the coracoacromial arch.

Flexion is produced by the anterior fibres of deltoid, clavicular head of pectoralis major, coracobrachialis and biceps. Passive extension from the flexed position is essentially due to the eccentric contraction of these same muscles. Beyond the neutral position, however, extension is produced by the posterior fibres of deltoid, teres major and latissimus dorsi; to these may be added the long head of triceps and the sternal fibres of pectoralis major when active extension is performed from a flexed position.

Abduction and adduction

These occur about an oblique horizontal axis in the same plane as the glenoid fossa. In abduction the arm moves anterolaterally away from the trunk (Fig. 2.76B). The total range of movement at the shoulder joint is 120°; however, only the first 25° occurs without concomitant rotation of the scapula, so that between 30° and 180° scapula rotation augments shoulder abduction in the ratio of 1:2. The terminal part of shoulder joint abduction is accompanied by lateral rotation of the humerus. This occurs not to prevent bony contact between the greater tubercle and the acromion, but to provide further articular surface on the head of the humerus for the glenoid fossa. Abduction of the medially rotated humerus is limited by tension in the posterior capsule and the lateral rotators. Adduction beyond the neutral position of the joint is not possible because of the presence of the trunk.

Abduction is initiated by supraspinatus which, although nowhere near as strong as deltoid, is better placed to act on the humerus. With the arm hanging at the side, the fibres of deltoid, especially the middle fibres, run almost parallel to the humerus, so that on contraction they pull the humerus upwards. Once the arm has been pulled away from the side then deltoid takes over and

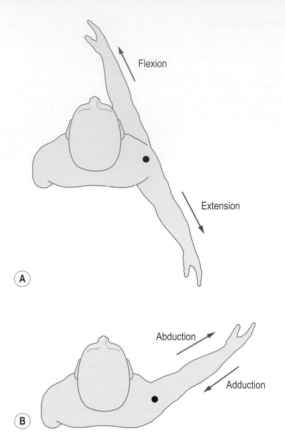

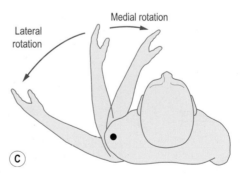

Figure 2.76 Movements at the shoulder joint with respect to the plane of the glenoid fossa, (A) flexion and extension, (B) adduction and abduction, (C) medial and lateral rotation.

continues the movement. Supraspinatus is required for the first 20° of abduction. If deltoid is paralysed, supraspinatus is not strong enough to abduct the upper limb fully. If, however, supraspinatus is paralysed, deltoid cannot initiate abduction. A passive abduction of some 20°, or leaning to the affected side so that the limb hangs away from the body, will enable deltoid to continue the movement. In some circumstances biceps may be re-educated to take over the initiating role of a paralysed supraspinatus. As abduction proceeds, teres minor and major hold the head of the humerus down against the pull of deltoid, while together with subscapularis and infraspinatus, they stabilize the humeral head against the glenoid fossa.

Lateral rotation of the scapula accompanying abduction of the humerus is produced by the force-couple of the lower part of serratus anterior, acting on the inferior angle of the scapula, and the upper fibres of trapezius pulling on acromion.

Adduction from an abducted position of the upper limb is produced by the eccentric contraction of serratus anterior, trapezius, deltoid and supraspinatus, under the action of gravity. If adduction is resisted then a forceful movement is produced by pectoralis major, teres major, latissimus dorsi and coracobrachialis.

Medial and lateral rotation

Rotation takes place about the longitudinal axis of the humerus described earlier. In lateral rotation the anterior surface of the humerus is turned laterally (Fig. 2.76C). It is produced by infraspinatus, teres minor and the posterior fibres of deltoid, and has a maximum range of 80°. In medial rotation the anterior surface of the humerus is turned medially (Fig. 2.76C). The maximum range of medial rotation is in excess of 90°; however, to reach this value with the elbow flexed the forearm has to be pulled behind the trunk, otherwise contact between the trunk and forearm limits the movements. Medial rotation is produced by subscapularis, pectoralis major, latissimus dorsi, teres major and the anterior fibres of deltoid. The combined range of rotation varies with the position of the arm, being greatest when the arm is by the side, decreasing to 90° with the arm horizontal, and being negligible as the arm approaches the vertical.

Rotation is limited by the extent of the articular surfaces, and tension in the appropriate part of the joint capsule and opposing musculature. Furthermore, it is the movement most commonly affected by pathology or injury to the shoulder joint. When assessing the range of rotation possible at the joint, the elbow must be flexed so as to exclude the possibility of any pronatory or supinatory action of the forearm.

Movements of the shoulder joint with respect to the cardinal planes of the body

Although movements of the shoulder joint have been considered with respect to the plane of the glenoid fossa, it is often more convenient to test the range of movement possible with respect to the cardinal planes of the body.

Movements of the arm about a transverse axis through the humeral head produce what are termed 'flexion' and 'extension'. Strictly speaking these movements are a combination of flexion and abduction, and extension and adduction – the degree of each component depending on the position of the scapula on the chest wall. The forward 'flexion' movement has a range of 180°, while the backward 'extension' movement is limited to 50°.

In relation to an anteroposterior axis, the movements produced are called 'abduction' and 'adduction'. 'Abduction' is a combination of abduction and extension, and has a range of 180° with scapula rotation. 'Adduction' is the combined movement of adduction and flexion. Again the movement is limited by the trunk so that adduction beyond the neutral position of the joint is not possible. However, with protraction of the pectoral girdle some 30° of 'adduction' is possible as the arm is brought across the front of the chest. Similarly, retraction of the pectoral girdle allows a minimal amount of 'adduction' to occur behind the back.

Although the terminology used to describe these various movements of the arm at the shoulder joint is of little practical significance, it is important to understand the context in which it is being used. It is also important to be fully aware of which movements are being tested when asking individuals to perform certain actions. Two simple activities that demonstrate the mobility of the shoulder joint and pectoral girdle are (i) combing the hair and (ii) putting on a coat or jacket.

With respect to the cardinal planes when the arm is flexed at 45° and abducted 60° and neither medially nor laterally rotated, it is said to be in the position of function of the shoulder. This corresponds to the position of equilibrium of the short scapular muscles; hence its use when immobilizing fractures of the humeral shaft.

Accessory movements

When the subject is lying supine the muscles around the shoulder are relatively relaxed. In this position the relative laxity of the ligaments and joint capsule allows an appreciable range of accessory movements. By placing the hand high up in the axilla and applying a lateral force to the upper medial aspect of the arm, the head of the humerus can be lifted away from the glenoid fossa by as much as 1 cm.

Proximal and distal gliding movements of the head of the humerus against the glenoid fossa can be produced by forces applied along the shaft of the humerus. Similarly, anterior and posterior gliding movements can be produced by applying pressure in an appropriate direction, to the region of the surgical neck.

Palpation

The line of the shoulder joint cannot be directly palpated due to the mass of muscles surrounding it. However, the surface projection of the joint line can be estimated first by identifying the surface projection of the midpoint of the joint, which is approximately 1 cm lateral to the apex of the coracoid process. A vertical line, slightly concave laterally, through this point gives an indication of the joint line.

Biomechanics

The rotator cuff muscles are active during abduction and lateral rotation, providing stability at the shoulder joint. However, they are probably also involved in the pathogenesis of dislocation of the shoulder. In any equilibrium analysis of the joint certain assumptions have to be made. The following is based on an account given by Morrey and Chao (1981).[1] The assumptions made were that:

1. each muscle contributing to the equilibrium acts with a force proportional to its cross-sectional area, this being 6.2 kg/cm^2
2. each muscle is equally active
3. the active muscle contracts along a straight line connecting the centres of its two areas of attachment.

While none of these assumptions is necessarily true, they do provide a framework within which to work. When the unloaded arm is laterally rotated and abducted to 90° there is a compressive force of approximately 70 kg between the articular surfaces, and anterior and inferior shear forces of 12 and 14 kg respectively. These forces are produced by muscles actively resisting the weight of the arm; the resultant force is directed 12° anteriorly with a magnitude of 72 kg.

If, as well as being abducted and laterally rotated, the arm is also extended by 30° and loaded so that the muscles are contracting maximally, then the magnitude of the various forces across the joint increases dramatically. The compressive force across the joint is now of the order of 210 kg, while the anterior and inferior shear forces have increased to 42 and 58 kg respectively. The resultant force is now directed 36° anteriorly and has a magnitude of 222 kg. To prevent anterior dislocation the shearing forces must be balanced by the joint capsule and its associated ligaments, because the glenoid fossa is too shallow to provide much constraint. As the tensile strength of the capsule and ligaments is of the order of 50 kg, an imbalance of forces may occur leading to dislocation at the joint. Once the anterior part of the capsule has been torn then less force is required for subsequent dislocations to occur.

The above force analysis is comparable to the situation when an individual slips when walking on ice and puts out their hand and arm to break a backward fall.

Velocity of movement

With the shoulder joint being extremely mobile, some of its movements are performed at fairly high velocities. In many instances, for example when studying natural or artificial joints and their lubrication, knowledge of the sliding velocities at the articulating surfaces is of importance.

Using cine film techniques and by suitable trigonometric relationships the maximum sliding velocities at the shoulder joint in various common activities have been determined. These are for hanging clothes 100 mm/s, sweeping 34 mm/s, bedmaking 40 mm/s, arm swing during walking 30 mm/s, eating 13 mm/s and dressing 25 mm/s. Obviously in activities requiring fast and forceful movement at the shoulder, such as the tennis serve, then the sliding velocities will be much greater. The demands made upon the lubricating fluid and articular surfaces in such situations are high. It is not surprising therefore that sometimes the system breaks down and some form of joint trauma results.

Shoulder joint replacement

Pathologies of the shoulder joint do not all require surgical intervention; arthroplasty is only considered as a last resort for those who have exhausted conservative treatments and have demonstrated unsatisfactory improvement in pain relief. Replacement of a damaged or arthritic humeral head may offer the immediate relief of pain, however, an intact rotator cuff and a normal glenoid fossa are often prerequisites for hemi-arthroplasty. The replacement, when used following a severe fracture of the humeral head, should be done as soon as possible, and certainly not later than 4 weeks following the injury, because of the extensive development of scar tissue and the subsequent limitation of motion.

Total shoulder arthroplasty replaces both articular surfaces with prosthetic components which mimic normal shoulder anatomy, except for reverse shoulder arthroplasty in which the glenoid fossa is replaced with a glenosphere and the head of the humerus is changed into a concave humerosocket (Fig. 2.77). Constrained prostheses do not rely on the surrounding tissues to provide support as the two components are physically linked, however there is limited joint movement. They are rarely used except as a last resort. Semi-constrained prostheses, which have a hooded glenoid component to restrict upward migration of the humeral head, are also rarely used. Unconstrained total arthroplasty is used extensively in the treatment of primary avascular necrosis, fractures of the humeral head and especially glenohumeral osteoarthritis; however, the primary cause of failure in such prostheses is glenoid component loosening. Interpositional arthroplasty, although not a new idea, has been developed during the last few years. It involves resurfacing the glenoid fossa with meniscus, tendon or fascia lata allograft, and can be peformed with hemi-arthroplasty of the humeral head or independently via arthroscopy to improve joint congruence.

All forms of total shoulder arthroplasty have proved to be inefficient at restoring function in patients with rotator

---

[1]Morrey BF, Chao EYS (1981) Recurrent anterior dislocation of the shoulder. In: Black J, Dumbleton JH (eds) Clinical Biomechanics: a case history approach, pp. 24–46. Edinburgh: Churchill Livingstone.

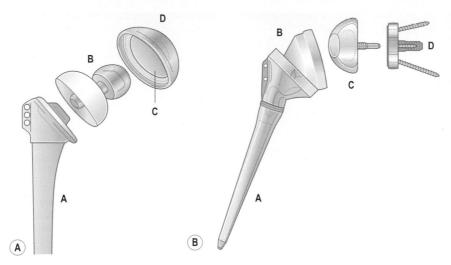

Figure 2.77 (A) Bipolar hemi-arthroplasty shoulder prosthesis: A, humeral stem; B, humeral head; C, polyethylene cup liner; D, humeral head cup. (B) Reverse shoulder prosthesis: A, humeral stem; B, the humeral socket; C, the glenoid sphere; D, the glenoid base plate.

cuff tears or arthropathy. However, interpositional arthoplasty appears to be more promising, with some reports stating that it exceeds the functional results of conventional hemi-arthroplasty. Nevertheless, superior migration and subluxation of the humeral head remain a problem.

Section summary

Shoulder joint

Type	Synovial ball and socket joint
Articular surfaces	Head of humerus, glenoid fossa of scapula. Glenoid fossa is deepened by the glenoid labrum
Capsule	Loose and attaches to articular margins of both bones but on humerus attaches a short way down onto shaft
Ligaments	Superior, middle and inferior glenohumeral (reinforce capsule)
	Transverse humeral (across intertubercular groove)
	Coracohumeral; coracoacromial (forms an arch above the joint)
Stability	Effectively provided by rotator cuff muscles (supraspinatus, infraspinatus, subscapularis and teres minor)
Movements	Can be described with respect to the plane of the scapula or cardinal planes of the body: flexion and extension, abduction and adduction, medial and lateral rotation

• These movements are usually accompanied by movements of the pectoral girdle.

The elbow joint

Introduction

The elbow joint is the intermediate joint of the upper limb, being between the arm and the forearm. It can be considered to be subservient to the hand in the sense that it enables the hand and fingers to be properly placed in space. The elbow joint is responsible for shortening and lengthening the upper limb; the ability to carry food to the mouth is due to flexion at the elbow. If situations arise in which the hand and forearm are not able to move, then the arm and trunk can move towards the hand.

A synovial joint of the hinge variety, the elbow joint shares its joint capsule with the superior radioulnar joint. The superior radioulnar joint is considered in detail on page 141, but it is important to remember that it plays no part and has no function at the elbow. The elbow joint shows the fundamental characteristics of all hinge joints. The articular surfaces are reciprocally shaped (Fig. 2.78); it has strong collateral ligaments with the forearm muscles grouped at the sides of the joint where they do not interfere with movement.

When viewed laterally, the distal end of the humerus projects anteriorly and inferiorly at an angle of 45° so that the trochlea lies anterior to the axis of the shaft (Fig. 2.79B). In a similar way the trochlear notch of the ulna projects anteriorly and superiorly at an angle of 45°, and so lies anterior to the axis of the shaft of the ulna (Fig. 2.79A). It is these projections of articular surfaces that facilitate and promote a large range of flexion at the elbow. It delays contact between the humerus and ulna, in addition to which it provides a space between them to accommodate the musculature until the bones are almost parallel. Without

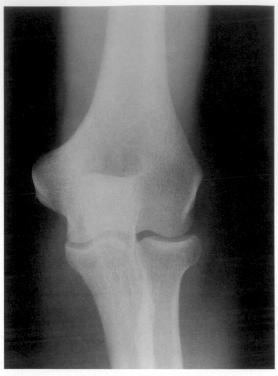

Posterior view

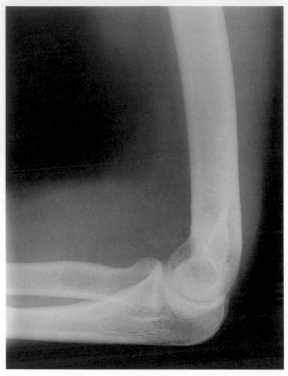
Lateral view

Figure 2.78 Radiographs of the elbow joint.

these two features, particularly the first, flexion beyond 90° would be severely limited.

In spite of the anterior projections of the humerus and ulna, the long axes of the two bones coincide when viewed laterally. However, when seen from the front, the ulnar axis deviates laterally from that of the humerus (Fig. 2.79B). This deviation is referred to as the carrying angle, and is said to be approximately 10°–15° in men and 20°–25° in women. Normally, the transverse axis of the elbow joint bisects this angle so that when the elbow is fully flexed the forearm overlies the arm and the hand covers the shoulder joint. If, however, the bisected parts of the carrying angle (A and B, Fig. 2.79B) are not equal then the hand will be lateral (A < B) or medial (A > B) to the shoulder on full flexion of the joint.

The transverse axis of the elbow joint runs from inferior posteromedially to superior anterolaterally, passing approximately through the middle of the trochlea. Because of this slight obliquity there has been some debate as to whether the joint exhibits a pure hinge movement, especially as this axis also oscillates slightly during flexion and extension. However, for practical purposes it can be considered as a pure hinge joint.

Articular surfaces

Three bones are involved in the articulation at the elbow joint; the distal end of the humerus and the proximal ends of the radius and ulna. The distal end of the humerus shows two joined articular regions, the grooved trochlea medially and rounded capitulum laterally, separated by a groove of variable depth (Fig. 2.80A). The whole of this composite surface is covered by a continuous layer of hyaline cartilage. The trochlea articulates with the trochlear notch of the ulna and the capitulum with the cupped head of the radius. Both of the latter surfaces are also covered with hyaline cartilage.

Trochlea of the humerus

The pulley-shaped trochlea with its groove presents a concave surface in the frontal plane and is convex sagittally. It almost forms a complete circle, being separated by a thin wall of bone, which itself may be perforated, so that 320°–330° of the surface is cartilage covered (Fig. 2.80A,B). The medial free border of the trochlea is not circular but describes part of a helix with a slant directed radially. The groove of the trochlea is limited by a sharp and prominent ridge medially and a lower and blunter ridge laterally

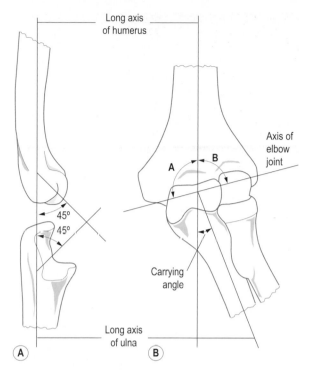

Figure 2.79 (A) Relation of the axes of the elbow joint, (B) the carrying angle.

which blends with the articular surface of the capitulum (Fig. 2.80A). The tilt on the trochlea partly accounts for the carrying angle of the elbow.

Although the groove of the trochlea appears to lie in the sagittal plane it does in fact run obliquely. This obliquity shows individual variation; however, the most common form is with the anterior part of the groove being vertical and the posterior part running obliquely distally and laterally. As a whole the groove runs in a spiral around the axis of the bone (Fig. 2.80C(i)). Occasionally, the groove runs obliquely proximally and laterally at the front and distally and laterally at the back, and so as a whole forms a true spiral around the axis of the bone (Fig. 2.80C(ii)). Rarely, the groove may run obliquely proximally and medially anteriorly, and distally and laterally posteriorly, so that as a whole it forms a circle (Fig. 2.80C(iii)). The functional significance of these variations in the angulation of the trochlea is minimal. The only observable differences are in the degree of the carrying angle and the relative positions of the arm and forearm in full flexion at the elbow.

Immediately above the trochlea anteriorly is the concave coronoid fossa (Fig. 2.80A), which receives the coronoid process of the ulna during flexion. Posteriorly, in a similar position, is the olecranon fossa which receives the olecranon process during extension (Fig. 2.80B). If these two fossae are particularly deep the intervening thin plate

of bone may be perforated, allowing them to communicate with each other.

Capitulum

This is not a complete sphere but a hemisphere on the anterior and inferior surface of the humerus (Fig. 2.80A): it does not extend posteriorly like the trochlea. Although described as hemispherical, its radius of curvature is not constant, increasing slightly from proximally to distally. The cartilage covering the capitulum is thickest in its central region, where it may be as much as 5 mm thick. The medial border of the capitulum is truncated, forming the capitulotrochlear groove. Above the capitulum anteriorly is the radial fossa, which receives the rim of the head of the radius during flexion (Fig. 2.80A).

Trochlear notch of the ulna

The proximal end of the ulna has the deep trochlear notch (Fig. 2.81A,B) which articulates with the trochlea of the humerus. It has a rounded, curved longitudinal ridge extending from the tip of the olecranon process superiorly to the tip of the coronoid process inferiorly. The ridge snugly fits the groove of the trochlea, on either side of which is a concave surface for the lips of the trochlea. The cartilage of the trochlear notch is interrupted by a transverse line across its deepest part, providing two separate surfaces, one on the olecranon and the other on the coronoid process.

The obliquity of the shaft of the ulna to the ridge accounts for the majority of the carrying angle.

Head of the radius

The superior surface of the head of the radius is concave for articulation with the capitulum, with the raised margin articulating with the capitulotrochlear groove (Fig. 2.81C,D). The cartilage of this surface is continuous with that around the sides of the head; it is thickest in the middle of the concavity.

Because of the articulations between the radius and ulna, their proximal surfaces may be considered as constituting a single articular surface. However, because of the movements between these two bones, they do not maintain the same relative positions with respect to the humerus, nor does the radius always maintain contact with the humerus (p. 137).

As stated earlier, the articular surface of the trochlea has an angular value of 330°, while that of the capitulum is 180° (Fig. 2.80A,B). The angular values of the articular surfaces of the ulna and radius are much smaller, leaving a large part of the humeral surfaces exposed at all positions of the joint. The angular value of the trochlear notch is 190°, while that of the head of the radius is only 40°. The difference in angular values between corresponding parts of the elbow is therefore 140°, a value very close to the range of flexion–extension possible at the joint.

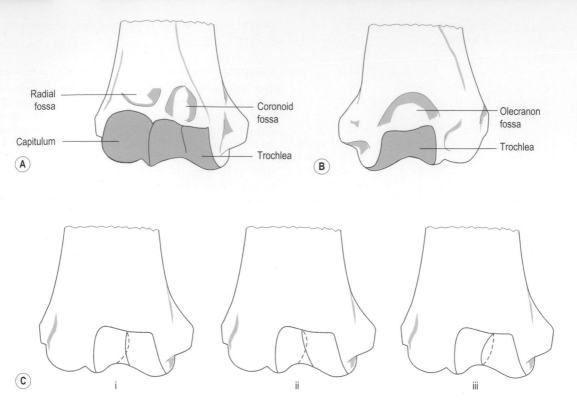

Figure 2.80 Articular surfaces of the humerus participating in the elbow joint: (A) anterior view, (B) posterior view, (C) the nature of the trochlear groove, anterior view.

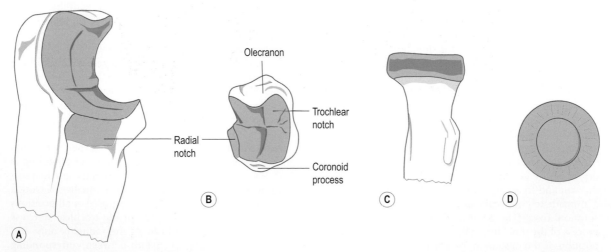

Figure 2.81 Articular surfaces of the ulna: (A) lateral view, (B) from above. Articular surfaces of the radius: (C) anterior view, (D) from above.

Joint capsule and synovial membrane

A single fibrous capsule completely encloses the elbow and superior radioulnar joints. It has no openings in it, but slight pouching of the synovial membrane may occur beneath the edge of the capsule in one or two areas.

Anteriorly the capsule arises from the medial epicondyle away from the articular surface of the trochlea. It arches upwards and laterally, attaching to the margins of the coronoid and radial fossae, and to the articular margin of the capitulum as it reaches the lateral epicondyle. Posteriorly the capsule follows the lateral margins of the capitulum and arches upwards around the olecranon fossa, returning to the medial epicondyle some distance from the edge of the trochlear surface (Fig. 2.82).

Distally the capsule attaches to the margins of the trochlear notch around the olecranon and coronoid processes. As it reaches the region of the radial notch the capsule passes onto and attaches to the annular ligament of the radius. Medially and laterally it blends with the collateral ligaments of the joint (Fig. 2.82). The joint capsule has no direct attachment to the radius. If this were the case, then the movements possible between the radius and ulna would be severely limited.

Because it blends with the collateral ligaments at the sides, the capsule is strengthened in these regions; however, it is relatively weak in front and behind. Anteriorly the capsule consists mainly of longitudinal fibres running from above the coronoid and radial fossae on the humerus to the anterior border of the coronoid process and in front of the annular ligament (Fig. 2.82A). Some bundles among these longitudinal fibres run obliquely and transversely (Fig. 2.82A). Consequently, this part of the capsule is thicker in its middle region than at the sides, a feature which has led to it being referred to as the capsular ligament. Some of the deep fibres of brachialis insert into the front of the capsule as the muscle passes anteriorly across the joint. This serves to pull the capsule and underlying synovial membrane upwards when the joint is flexed, preventing them becoming trapped between the two moving bones.

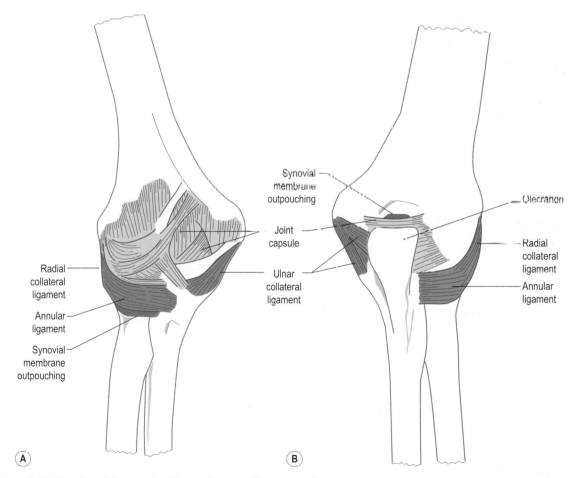

Figure 2.82 The elbow joint capsule: (A) anterior view, (B) posterior view.

The posterior part of the capsule is thin and membranous, being composed mainly of transverse fibres extending loosely between the margins of the olecranon and the edges of the olecranon fossa. A few fibres stretch across the fossa as a transverse band with a free upper border, which does not reach as high as the upper margin of the fossa, without attaching to the olecranon (Fig. 2.82B). Posteriorly the capsule also passes laterally from the lateral epicondyle to the posterior border of the radial notch and posterior part of the annular ligament. The weakest part of the capsule posteriorly is in the midline of the joint. However, here it is attached to the tendon of triceps which supports it, and performs a similar function to the deep part of the brachialis during extension of the joint.

Synovial membrane

The synovial membrane of the joint is extensive, attaching to the articular margins of the humerus and ulna. It lines the joint capsule and is reflected onto the humerus to cover the coronoid and radial fossae anteriorly and the olecranon fossa posteriorly. Distally it is prolonged onto the upper part of the deep surface of the annular ligament. The membrane continues into the superior radioulnar articulation covering the lower part of the annular ligament, and is then reflected onto the neck of the radius. Below the lower border of the annular ligament, the membrane emerges as a redundant fold to give freedom of movement to the head of the radius (Fig. 2.82A). This downward reflection is supported by a few loose fibres which pass from the lower border of the annular ligament to the neck of the radius. The quadrate ligament (p. 142) supports the synovial membrane as it passes from the medial side of the neck of the radius to the lower border of the radial notch, so preventing its herniation between the anterior and posterior free edges of the annular ligament.

Various synovial folds project into the recesses of the joint between the edges of the articular surfaces. An especially constant fold is one which forms an almost complete ring overlying the periphery of the head of the radius, projecting into the crevice between it and the capitulum. Slight pouching of the synovial membrane may occur below the lower borders of the annular ligament and the transverse band of the ulnar collateral ligament, as well as above the transverse capsular fibres across the upper part of the olecranon fossa (Fig. 2.82B).

Well-marked extrasynovial fat pads lie adjacent to the articular fossae. In extension they fill the radial and coronoid fossae and in flexion the olecranon fossa. They are displaced when the appropriate parts of the ulna or radius occupy the fossae.

Ligaments

The collateral ligaments associated with the elbow joint are strong triangular bands which blend with the sides of the joint capsule. Their location is such that they lie across the axis of movement in all positions of the joint. Consequently, they are relatively tense in all positions of flexion and extension, and impose strict limitations on abduction and adduction and axial rotation.

Ulnar collateral ligament

This fans out from the medial epicondyle and has thick anterior and posterior bands united by a thinner intermediate portion (Fig. 2.83A). The anterior band passes from the front of the medial epicondyle to the medial edge of the coronoid process. It is intimately associated with the common tendon of the superficial forearm flexors, giving rise to some of the fibres of flexor digitorum superficialis. The posterior band runs from the back of the medial epicondyle to the medial edge of the olecranon. The apex of the thinner intermediate part of the ligament is attached to the undersurface of the medial epicondyle, while its base is attached to the transverse band passing between the attachments of

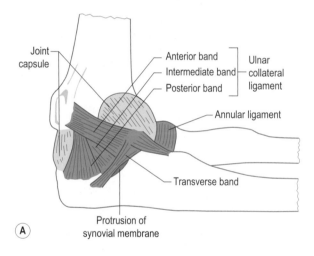

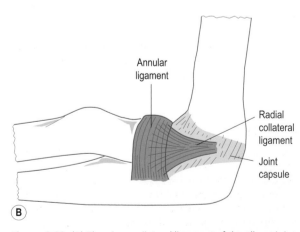

Figure 2.83 (A) The ulnar collateral ligament of the elbow joint and (B) the radial collateral ligament of the elbow joint.

the anterior and posterior bands to the coronoid process and olecranon (Fig. 2.83A). The synovial membrane protrudes below the free edge of the transverse ligament during movement at the joint. The intermediate grooved part of the ligament is crossed by the ulnar nerve as it passes behind the medial epicondyle to enter the forearm.

Radial collateral ligament

This is a strong, triangular band attaching above to a depression on the anteroinferior aspect of the lateral epicondyle deep to the overlying common extensor tendon (Fig. 2.83B). Below, it blends with the annular ligament of the radius, the slightly thicker anterior and posterior margins passing forwards and backwards to attach to the margins of the radial notch on the ulna (Fig. 2.83B). The ligament is less distinct than the ulnar collateral ligament.

Blood and nerve supply

The arterial supply to the joint is from an extensive anastomosis around the elbow involving the brachial artery and its terminal branches. Descending from above are the superior and inferior ulnar collateral branches of the brachial artery, and the radial and middle collateral branches of the profunda brachii artery. These anastomose on the surface of the joint capsule with one another, and with the anterior and posterior recurrent branches of the ulnar artery, the radial recurrent branch of the radial artery and the interosseous recurrent branch of the common interosseous artery (Fig. 2.84).

Venous drainage, by vessels accompanying the above arteries, is into the radial, ulnar and brachial veins. Lymphatic drainage of the elbow joint is predominantly to the deep cubital nodes at the bifurcation of the brachial artery, the efferents of which pass to the lateral group of nodes in the axilla. Some of the lymphatics from the joint may pass to small nodes situated along the interosseous, ulnar, radial or brachial arteries and then to the lateral axillary group.

The joint is innervated by twigs derived anteriorly from the musculocutaneous, median and radial nerves, and posteriorly from the ulnar nerve and radial nerve by its branch to anconeus. The root value of these nerves is C5, 6, 7 and 8.

Palpation

Palpation of the joint line anteriorly is not possible because of the muscles crossing the joint and its deepness within the cubital fossa. Nevertheless, it can be approximated by drawing a line joining the points 1 cm below the lateral epicondyle and 2 cm below the medial epicondyle. Posteriorly the gap between the head of the radius and the capitulum can be palpated in the large dimple present at the back of the extended elbow.

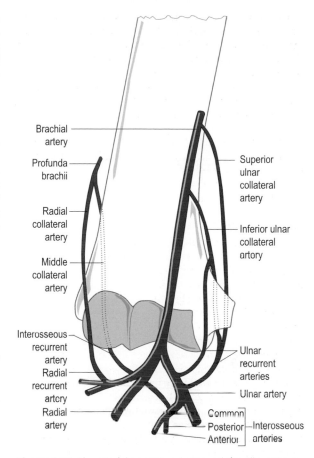

Figure 2.84 The arterial anastomosis around the elbow joint, anterior view

Posteriorly the olecranon is subcutaneous and can be readily palpated, either side of which the medial and lateral epicondyles form easily recognizable bony landmarks.

Relations

Brachialis lies anteriorly forming the majority of the floor of the cubital fossa, a hollow in front of the elbow through which pass many of the vessels and nerves entering or leaving the forearm.

The cubital fossa is a triangular space bounded above by an imaginary line between the medial and lateral epicondyles, and at the sides by the converging medial borders of brachioradialis laterally and pronator teres medially (Fig. 2.85A). The floor of the fossa is formed mainly by brachialis with supinator inferolaterally (Fig. 2.85A). It is roofed over by the deep fascia of the forearm, reinforced medially by the bicipital aponeurosis passing from the tendon of biceps downwards and medially to the deep fascia of the forearm (Fig. 2.85B). The deep fascia separates the

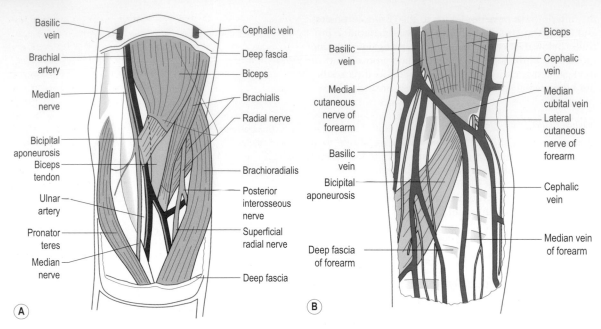

Figure 2.85 The cubital fossa: (A) deep structures, (B) superficial structures.

superficial veins and nerves from the deeper more important structures (Fig. 2.85B).

This region is of considerable importance because the large superficial veins are frequently used for venepuncture, while the deeper brachial artery is used for determining blood pressure. The main superficial veins are the cephalic laterally, the basilic medially and the median cubital passing obliquely upwards and medially between them (Fig. 2.85B). It is not unusual for the median cubital vein to lie towards the lateral side of the fossa, or to be joined by the median vein of the forearm. Occasionally, the median cubital vein is absent and the median vein of the forearm divides into lateral and medial branches to join with the cephalic and basilic veins respectively. Lateral to the cephalic vein runs the lateral cutaneous nerve of the forearm, the terminal branch of the musculocutaneous nerve. Crossing the median cubital vein and running with the basilic vein and its tributaries are branches of the medial cutaneous nerve of the forearm (Fig. 2.85B).

Lying within the cubital fossa deep to the deep fascia are several structures passing into the forearm. The most medial is the median nerve as it passes downwards through the fossa to emerge between the two heads of pronator teres to enter the forearm (Figs. 2.85A and 2.138). While in the fossa it gives off a branch to pronator teres, and the anterior interosseous branch as it passes through pronator teres. Lateral to the median nerve is the brachial artery, which bifurcates at the neck of the radius in the lower part of the fossa into the ulnar and radial arteries (Fig. 2.85A; see also Fig. 2.84). The ulnar artery passes inferomedially

deep to pronator teres, giving off recurrent branches to the elbow joint and the common interosseous artery. The radial artery passes inferolaterally over the tendon of biceps brachii deep to brachioradialis, giving off its recurrent branch to the elbow joint. Running through the central region of the fossa, lateral to both the brachial artery and the median nerve, is the tendon of biceps brachii towards its insertion on the radial tuberosity (Fig. 2.85). As it passes through the fossa the tendon twists on itself so that its anterior surface faces laterally. The most lateral structure passing through the fossa is the radial nerve. In the upper part of the fossa it lies between brachialis and brachioradialis, supplying both muscles, and then divides into its terminal branches, the superficial radial and posterior interosseous (deep radial) nerves. The superficial branch continues downwards into the forearm under cover of brachioradialis. The posterior interosseous nerve passes posteriorly around the lateral side of the radius to enter the forearm between the two heads of supinator.

The ulnar nerve passes behind the medial epicondyle of the humerus on the intermediate part of the ulnar collateral ligament posteromedial to the elbow joint. It does not pass through the cubital fossa.

Stability

The shape of the articular surfaces of the trochlea and capitulum of the humerus, and of the trochlear notch of the ulna and head of the radius confer some stability on the elbow joint. Nevertheless, without strong collateral ligaments and the muscular cuff of triceps, biceps, brachialis, brachioradialis, and the common tendons of the superficial flexors and extensors arising from the medial and lateral epicondyles of the humerus, the elbow joint cannot be considered as an inherently stable joint. The bony surfaces are in closest contact with the forearm flexed to 90° in a position of mid pronation–supination. This, therefore, is the position of greatest stability of the joint, and is the position naturally assumed when fine manipulation of the hand and fingers is required, as for example when writing.

In spite of ligaments and muscles crossing the joint, dislocations of the elbow can and do occur. In young children, because the head of the radius is small relative to the annular ligament, it is commonly dislocated by traction forces applied to the forearm and hand. In older individuals, a fall on the hand with the forearm extended may tear the annular ligament with a consequent anterior displacement of the head of the radius (Fig. 2.86). The head of the radius may also be dislocated by tearing the annular ligament in extreme pronation. In either case it can be palpated in the cubital fossa.

The majority of elbow dislocations involve backward movement of the ulna through the relatively weak posterior capsule, often associated with a fracture of the coronoid process (Fig. 2.86B). Both the radius and ulna may be displaced together due to their connections at the superior radioulnar joint. This backward displacement can lead to pressure on the brachial artery, which may go into spasm and reduce the blood supply to the forearm and hand. Pressure on the brachial artery can also arise in supracondylar fractures as the lower fragment moves forwards. Either of these events can also lead to injury to the median nerve

with a consequent loss of pronation and reduced function of the hand. Both dislocations and supracondylar fractures result in considerable swelling in the region of the elbow. The alignment of the humeral epicondyles and the olecranon can be used to determine the nature of the trauma in an individual with a swollen elbow. The alignment remains unchanged in supracondylar fractures (Fig. 2.86C), while in dislocations it is changed (Fig. 2.86D). When an apparently dislocated joint cannot be reduced, fracture of the olecranon must be considered, particularly if the joint is extremely unstable.

A forceful abduction applied to the forearm may be sufficient to rupture the ulnar collateral ligament, or more commonly result in avulsion of the medial epicondyle. The ulnar nerve is especially liable to damage at the time of the injury. If the fracture does not unite or the ligament heal, the forearm tends to become more and more abducted with a consequent stretching of the ulnar nerve, leading to sensory disturbances and muscle weakness or paralysis.

Movements

Flexion and extension are possible at the elbow joint, which take place about a transverse axis through the humeral epicondyles (Fig. 2.87). The axis is not at right angles to the long axis of the humerus or of the forearm, as it bisects the carrying angle at the elbow (Fig. 2.79B). Consequently, its medial end is slightly lower than the lateral. Except at the extremes of flexion and extension, movement between the humerus and the radius and ulna is one of sliding. It is only at the extremes of movement, when the axis

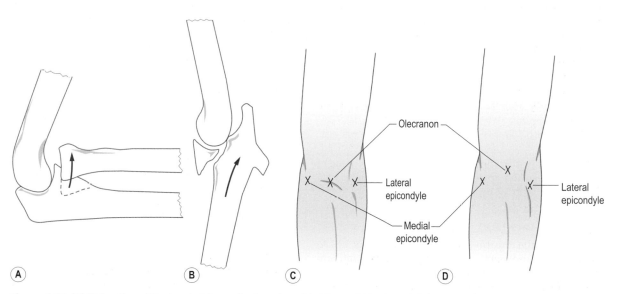

Figure 2.86 (A) Dislocation of the head of the radius into the cubital fossa, (B) backward dislocation of the ulna with fracture of coronoid process, (C) normal alignment of the olecranon and epicondyles of the humerus, (D) alignment when ulna is dislocated.

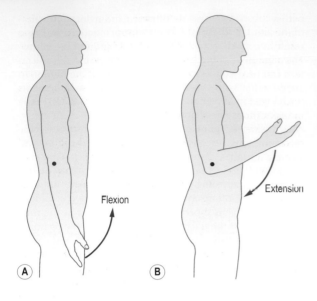

Flexion

Extension

(A)

(B)

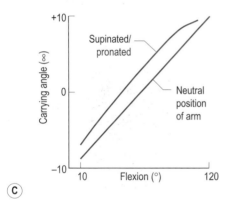

(C)

Figure 2.87 (A, B) Movements at the elbow joint, (C) changes in the carrying angle with flexion at the joint. *(Adapted from Chao EY, Morrey BF (1978) Three-dimensional rotation of the elbow. Journal of Biomechanics, 11, 57–74).*

Flexion is brought about by brachialis, biceps brachii and brachioradialis. In slow gentle movements, brachialis is the main muscle used, whereas rapid and forceful movements will use all three muscles as well as the superficial forearm flexors arising from the medial epicondyle. Flexion is primarily limited by apposition of the anterior muscles of the arm and forearm, with tension in the posterior part of the capsule and triceps; the impact of bony surfaces is insignificant.

Extension is movement of the forearm posteriorly, and is best defined as returning the forearm to the anatomical position (Fig. 2.87B). Strictly speaking the range of extension possible at the joint is zero since full extension corresponds to the anatomical position. However, relative extension is always possible from any position of the joint. Active extension is brought about by triceps and anconeus, while passive extension is due to gravity controlled by the eccentric contraction of the elbow flexors, particularly when a weight is being lowered. Extension is usually limited by tension in the anterior joint capsule and flexor muscles, and to some extent in the anterior parts of the collateral ligaments.

Limitation of movement, either flexion or extension, is rarely due to bony contact, although the presence of small cartilage-covered facets at the bottom of the coronoid fossa and at the sides of the olecranon fossa in some individuals suggests that bony contact occurs during life.

Because the axis of flexion–extension bisects the carrying angle it is only to be expected that there is a linear change in the carrying angle during flexion of the elbow. This occurs irrespective of whether the forearm is fully pronated or supinated or is in some position between (Fig. 2.87C). The small changes in the position of the axis of movement at the extremes of the range result in a small degree of axial rotation, which is due to the configuration of the humeroulnar articulation and ligamentous constraints. During initial flexion, the forearm may medially rotate up to 5°, and may laterally rotate up to 5° during terminal flexion. Although not much, the movement nevertheless occurs.

Abduction and adduction

Being a hinge joint, the only movements expected at the elbow have already been described. However, during pronation and supination of the forearm there is a small degree of abduction and adduction respectively, between the trochlear notch of the ulna and the trochlea of the humerus. Full details of these movements are given on page 146.

Accessory movements

With the elbow almost fully extended a small degree of abduction and adduction at the joint is possible. This can best be demonstrated with the subject lying supine. Holding the lower part of the arm steady and applying alternate medial and lateral pressure to the distal end of the forearm will produce these accessory movements.

changes slightly, that the sliding motion changes to one of rolling between the articular surfaces. The collateral ligaments are tense in all positions of the joint.

Flexion and extension

Movement of the forearm anteriorly is flexion (Fig. 2.87A) and continues until contact between the forearm and arm prevents further movement. The active range of flexion is 145°; passively, 160° of flexion can be attained. Because of the obliquity of the axis about which flexion occurs, the hand moves medially to come to lie over the shoulder.

Biomechanics

Contact areas

Because of the rounded ridge extending from the tip of the olecranon to the tip of the coronoid process, and the transverse ridge observable on the articular cartilage, the trochlear notch can be conveniently divided into four quadrants (Fig. 2.88A). Direct observation reveals that the contact area on the humerus changes during flexion and extension. In general, the humeroulnar contact area increases from full extension to full flexion, and the head of the radius establishes more and more contact with the capitulum. The increasing area of contact supports the view that stability at the joint increases with flexion. As far as the radius is concerned, it means that it must move proximally during flexion.

In full extension the contact areas of the elbow are in the lower part of the trochlear notch, with concentrations on the medial aspects (Fig. 2.88A); there is no contact between the radius and capitulum. At 90° flexion, the contact area is a diagonal strip running from the lower medial to the upper lateral compartment. Small parts of the superior surface of the coronoid process and the inferior surface of the olecranon also show contact. There is some contact between the capitulum and the head of the radius (Fig. 2.88B). It is only when the elbow is fully flexed that a definite area of contact between the radius and capitulum can be identified (Fig. 2.88B). Trochlear notch contact areas, although of a similar diagonal orientation, are larger and extend into the upper medial quadrant (Fig. 2.88A).

This general increase in contact areas across the joint serves to reduce the pressure applied to the cartilage, and so help protect it, particularly when loading is being supported with the elbow flexed, or when fine manipulative movements are being performed in which many muscles crossing the joint may be active.

Muscular action and its efficiency

As a group, the flexor muscles are more powerful than the extensors. Consequently, with the arm hanging loosely by the side, the elbow tends to be slightly flexed. Not only does the power of the muscle groups vary with the position of the shoulder, because of the attachment of biceps

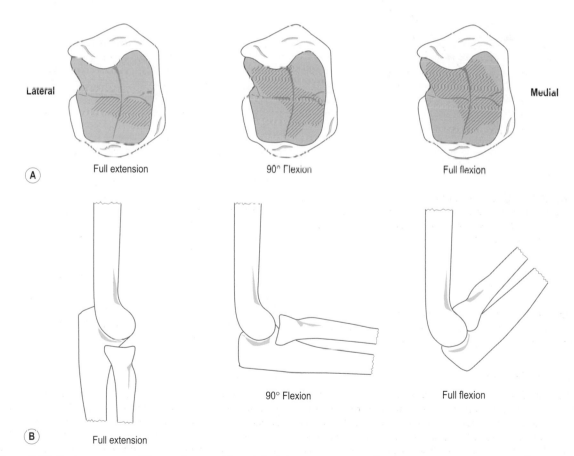

Lateral   Medial

(A)   Full extension   90° Flexion   Full flexion

(B)   Full extension   90° Flexion   Full flexion

Figure 2.88 Contact areas at different elbow positions: (A) of the trochlear notch, (B) of head of the radius.

and triceps to the supra- and infraglenoid tubercles respectively, but for the flexors the degree of rotation of the forearm is also an important factor. As far as rotation is concerned, the power of the flexors is greater when the forearm is pronated, because biceps is stretched, thereby increasing the efficiency of its action. The flexor efficiency ratio of biceps for pronation and supination is of the order of 5:3.

Biceps works most efficiently between 80° and 90°, brachialis between 90° and 100°, and brachioradialis between 100° and 110°. Consequently, as a whole, the flexor muscles work at their best advantage when the elbow is flexed at 90°. In this position the muscles are at their optimal lengths and their direction of pull (line of action), particularly for biceps and brachialis, is almost at 90° to the forearm. Most of the force of contraction is thus directed towards moving the forearm and not to the maintenance of the integrity of the elbow joint. The reverse is true when the elbow is extended, because in this situation the direction of pull is nearly parallel to the forearm, rather than perpendicular to it. Because the attachment of the flexor muscles, between the fulcrum (elbow joint) and the resistance (weight of the forearm and hand together with any applied loads), conforms to that of a third class lever, it follows that the flexors favour range and speed of movement at the expense of power.

Triceps is most efficient when the elbow is flexed to between 20° and 30°. As flexion increases, its tendon becomes wound around the superior surface of the olecranon, which thus acts as a pulley. At the same time the muscle fibres become passively stretched. Both of these features help to compensate for the loss of efficiency of the muscle in flexion. Triceps is more powerful when the shoulder is flexed, and also when both the elbow and shoulder are being extended simultaneously from a flexed position, for example when executing a karate chop. On the other hand triceps is weakest when the elbow is extended at the same time as the shoulder is being flexed.

There are, therefore, preferential positions of the upper limb in which the muscle groups achieve maximum efficiency. For extension this is with the arm and forearm hanging by the side with an angle of 20°–30° between them. For flexion it is with the arm and forearm stretched above the shoulder. Thus the muscles of the upper limb have retained their adaptation for climbing, developed in the dawn of human evolution.

Elbow joint replacement

Total elbow arthroplasty is now a well established procedure for any condition that leads to elbow dysfunction with pain, stiffness and instability following traumatic or arthritic conditions being the main indications. A common pathology leading to replacement is rheumatoid arthritis; however osteoarthritis and fracture of the distal humerus may also lead to total joint replacement. Initial designs were one of two types, linked and unlinked; however, a new generation of modular implants are being introduced (Fig. 2.89). The advantage of linked prostheses is their inherent stability due to the linkage between the components; consequently they are generally used where there is traumatic soft tissue pathology. Linked prostheses are also used in revision procedures because osseous and ligamentous integrity is often compromised. In unlinked prostheses there is no physical connection between the components; however they require precise insertion to ensure the correct alignment and tracking of the components. Semiconstrained implants consist of two components which are closely congruent although not physically linked, often replicating the normal anatomy. Modular implant systems offer the advantage of interchangeability between linked and unlinked designs; they allow a more precise restoration of the anatomical flexion-extension axis and a more careful soft tissue balance enabling the reproduction of normal elbow kinematics that is likely to decrease mechanical failure.

Section summary	
Elbow joint	
Type	Synovial hinge joint
Articular surfaces	Capitulum and trochlea of humerus, trochlear notch of ulna and head of radius
Capsule	Complete fibrous capsule surrounds the joint including the superior radioulnar joint, radial, olecranon and coronoid fossae
Ligaments	Ulnar collateral
	Radial collateral
Stability	Provided by shape of articular surfaces, collateral ligaments and muscles crossing joint
Movements	Flexion and extension

Radioulnar articulations

Introduction

As well as articulating independently with the humerus at the elbow joint, the radius and ulna also articulate with each other at their proximal and distal ends by synovial joints of the pivot type (Fig. 2.90), and by an interosseous membrane along their shafts (Fig. 2.91) in the manner of a syndesmosis. The predominant movement between the two bones is rotation of the radius around the ulna, so that the two bones cross in space, producing pronation; the reverse movement brings the bones into parallel alignment producing supination. In pronation and supination the hand is carried with the rotation of the forearm, thereby giving a further axis of movement of the hand at the wrist. In functional terms the combination of pronation and supination of the forearm and the movements possible at the

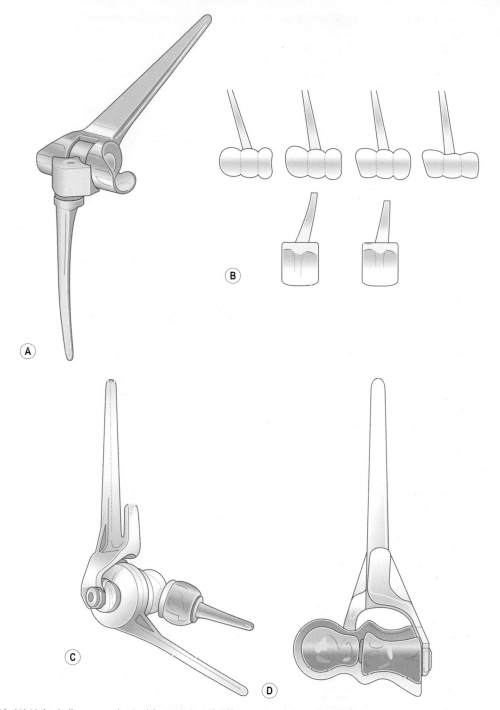

Figure 2.89 (A) Linked elbow prosthesis, (B) examples of different size unlinked capitellocondylar implants, (C) modular elbow prosthesis, (D) distal humeral hemi-arthroplasty implant.

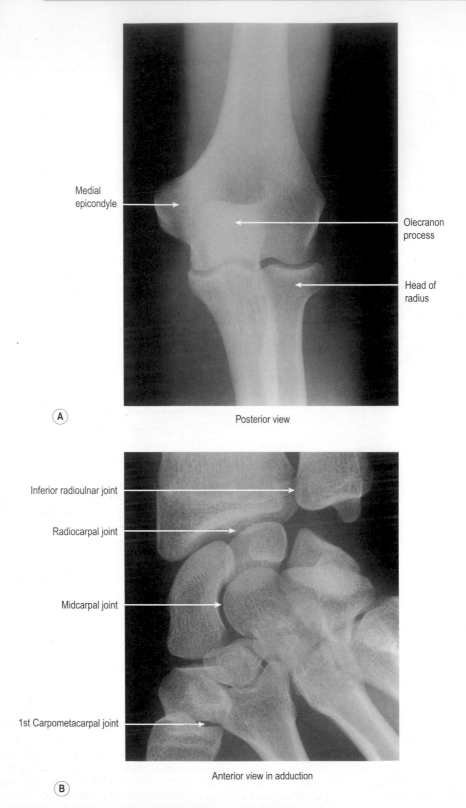

Medial epicondyle

Olecranon process

Head of radius

(A)

Posterior view

Inferior radioulnar joint

Radiocarpal joint

Midcarpal joint

1st Carpometacarpal joint

Anterior view in adduction

(B)

Figure 2.90 Radiographs of (A) the superior and (B) the inferior radioulnar joints.

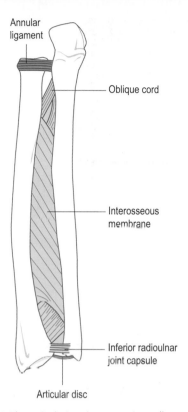

Annular
ligament

Oblique cord

Interosseous
membrane

Inferior radioulnar
joint capsule

Articular disc

Figure 2.91 The articulations between the radius and the ulna.

wrist means that the hand is united to the arm by a universal joint. The fact that the various axes about which movement occurs do not all pass through a common point, and that several joints are involved, gives stability to the hand when performing delicate tasks.

Without freely movable joints between the radius and ulna, perhaps the evolutionary development of the hand as a manipulative tool would not have been so successful. Nor perhaps would there have been so much development and enlargement of the brain, particularly of the cerebral cortex.

The superior radioulnar joint

Articular surfaces

The articulation is between the head of the radius rotating within the fibro-osseous ring formed by the radial notch of the ulna and the annular ligament.

Head of the radius

The bevelled circumference of the head of the radius is covered by hyaline cartilage continuous with that on its upper concave surface, forming a smooth surface for articulation with the ulna and annular ligament (Fig. 2.92). The anterior, medial and posterior parts of the circumference tend to be wider than the lateral part, for direct articulation with the ulna. The head of the radius is slightly oval, with the major axis lying obliquely anteroposteriorly. The major and minor axes have a length ratio of approximately 7:6.

Radial notch

The hyaline-covered radial notch is continuous with the trochlear notch of the ulna on its lateral side, being separated from it by a blunt ridge (Fig. 2.81A,C). It forms approximately one-fifth of the articular fibro-osseous ring; it is concave anteroposteriorly but almost flat vertically.

Annular ligament

A strong flexible well defined band attached to the anterior and posterior margins of the radial notch of the ulna completing the remaining four-fifths of the articular surface which encircles the head and neck of the radius (Fig. 2.92). Its flexibility enables the oval head of the radius to rotate freely during pronation and supination. Posteriorly the ligament widens where it attaches to adjacent areas of the ulna above and below the posterior margin of the radial notch. The diameter between its lower borders is narrower than that above (Fig. 2.92C), so it cups under the head of the radius and acts as a restraining ligament preventing downward displacement of the head through the ring.

Superiorly the annular ligament is supported by fusion of the radial collateral ligament and blending of the lateral part of the fibrous capsule of the elbow joint in front and behind (Fig. 2.83B). Inferiorly a few loose fibres attach the ligament to the neck of the radius beyond the epiphyseal line. These fibres are too loose to interfere with movements at the joint, but give some support to a dependent fold of synovial membrane. The upper part of the ligament is lined with fibrocartilage continuous with the hyaline cartilage of the radial notch. The lower part of the ligament is lined by the synovial membrane.

Joint capsule and synovial membrane

The superior radioulnar joint is continuous with the elbow joint and consequently shares the same joint capsule (Fig. 2.93). For details of the capsular attachments see page 131. The synovial membrane associated with the elbow part of the joint space attaches to the upper margin of the fibrocartilage lining of the annular ligament. From the lower border of the fibrocartilage, and lining the lower part of the annular ligament, the synovial membrane extends below the lower border of the ligament hanging as a redundant fold which has a loose attachment to the neck of the radius (Fig. 2.93). The membrane lies on the upper surface of the quadrate ligament, which limits and supports it, and passes medially from the radius to attach to the lower border of the radial notch of the ulna. The redundancy of the synovial membrane below the annular

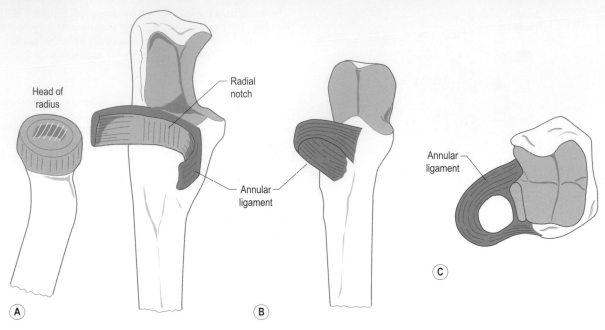

Figure 2.92 Articular surfaces of the superior radioulnar joint: (A) exposed, (B) from the front, (C) from above.

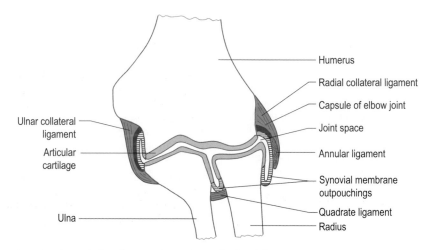

Figure 2.93 Coronal section through the elbow and superior radioulnar joints.

ligament accommodates twisting of the membrane that accompanies rotation of the radius.

Ligaments

Although the annular ligament provides an important support for the head of the radius, it is not sufficient by itself to provide the only support to the superior radioulnar joint, because of its need to change shape with rotation of the radius. Indeed, this constant need to accommodate to the changing orientation of the head of the radius may lead

to stretching of the ligament. Consequently, additional structures provide support to the joint, the quadrate ligament and the interosseous membrane. (For details of the interosseous membrane see p. 146.)

Quadrate ligament

This stretches from the lower border of the radial notch of the ulna to the adjacent medial surface of the neck of the radius proximal to the radial tuberosity (Fig. 2.94). Its fibres run in a criss-cross manner between the radius and ulna, so that, irrespective of their relationship, some fibres

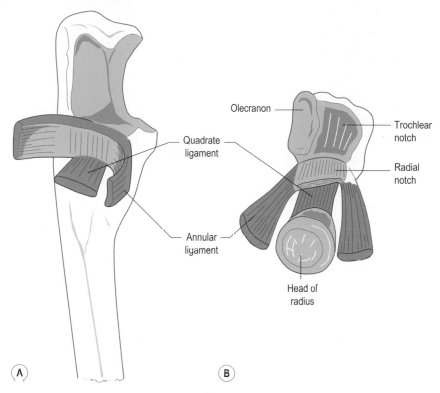

Figure 2.94 The quadrate ligament: (A) attachment to the ulna, (B) seen from above

are always under tension. The overall tension within the ligament therefore remains relatively constant in all positions of pronation and supination. Its two borders are strengthened by fibres from the lower border of the annular ligament.

Blood and nerve supply

The arterial supply to the superior radioulnar joint is by branches from vessels supplying the lateral part of the elbow joint, namely the middle and radial collateral branches of the profunda brachii, and the radial and interosseous recurrent branches from the radial and common interosseous arteries respectively. Venous drainage is by similarly named vessels eventually draining to the brachial vein. Lymphatic drainage is by vessels travelling with the arteries to small nodes associated with the main arteries and then to the lateral group of axillary nodes.

The nerve supply to the joint is by twigs from the posterior interosseous branch of the radial nerve, the musculocutaneous and median nerves, with a root value of C5, 6 and 7.

Surface marking and palpation

The line of the superior radioulnar joint can be palpated posteriorly. Having identified the head of the radius in the depression on the posterolateral aspect of the elbow, a vertical groove between the radius and ulna can be felt medially. This is the position of the joint line. During pronation and supination, the head of the radius can be felt rotating against the ulna.

Relations

Anteriorly the joint is crossed by the tendon of biceps passing to its attachment on the radial tuberosity; posteriorly is the fleshy belly of anconeus. Medial to the tendon of biceps is the brachial and then the radial artery from above downwards.

Stability

The joint has a reasonable degree of inherent stability. However, in children, the head of the radius may be pulled from the confines of the annular ligament in traction dislocation. Tears of the annular ligament will also result in dislocation at the joint. For further details see page 134.

Movement

The main movement that occurs at the superior radioulnar joint is rotation of the head of the radius within the fibro-osseous ring of the annular ligament and radial notch of

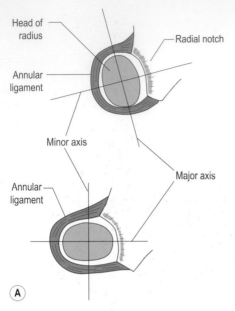

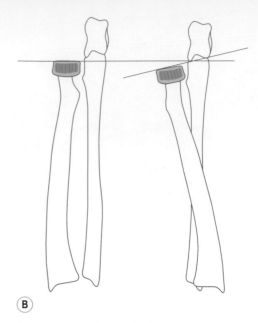

Figure 2.95 Superior radioulnar joint showing axes of movement during pronation and supination (A) from above and (B) anteriorly.

the ulna (Fig. 2.95A). The movement is probably limited by tension developed in the quadrate ligament.

In addition to this principal movement, there are four other related movements, these are:

1. rotation of the superior concave surface of the radial head in relation to the capitulum of the humerus
2. gliding of the bevelled ridge of the radial head against the capitulotrochlear groove of the humerus
3. lateral displacement of the head of the radius as its major axis comes to lie transversely (Fig. 2.95A)
4. lateral and inferior tilting of the plane of the radial head during pronation due to the radius moving obliquely around the ulna (Fig. 2.95B).

Accessory movements

With the head of the radius gripped between the thumb and index finger, it can be moved anteroposteriorly with respect to both the ulna and the capitulum.

The Inferior radioulnar joint

Articular surfaces

The articulation is between the head of the ulna and the ulnar notch on the lower end of the radius. The joint is closed inferiorly by an articular disc between the radius and ulna, thereby separating the inferior radioulnar joint from the radiocarpal joint of the wrist.

Head of the ulna

This is the slightly expanded distal end of the ulna (Fig. 2.96). The crescent-shaped articular surface is situated on its anterior and lateral aspects and is covered with hyaline cartilage continuous with that on the distal end of the ulna over a rounded border. The distal end of the head of the ulna articulates with an intra-articular disc.

Ulnar notch of the radius

Situated between the two edges of its interosseous border the ulnar notch of the radius faces medially (Fig. 2.96). It is concave anteroposteriorly and plane or slightly concave vertically. The notch is lined by hyaline cartilage.

Articular disc

A triangular, fibrocartilaginous articular disc is the principal structure uniting the radius and ulna. It attaches by its apex to the lateral side of the root of the ulna styloid process, and by its base to the sharp inferior edge of the ulnar notch between the ulnar and carpal surfaces of the radius (Fig. 2.96B). The disc is thicker peripherally than centrally; it is rarely perforated.

It is an essential part of the total bearing surface of the inferior radioulnar joint by its articulation with the distal surface of the head of the ulna. Inferiorly it participates in the radiocarpal joint. Perforation of its central part would therefore lead to a communication between the inferior radioulnar and radiocarpal joints.

Joint capsule and synovial membrane

The relatively weak and loose fibrous capsule is formed by transverse bands of fibres attaching to the anterior and posterior margins of the ulnar notch of the radius and

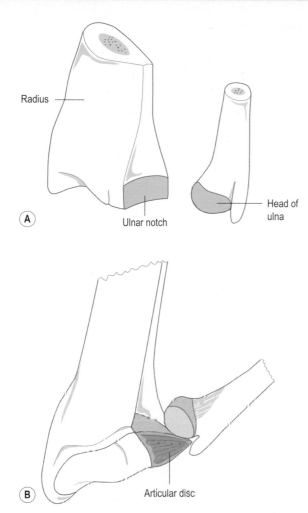

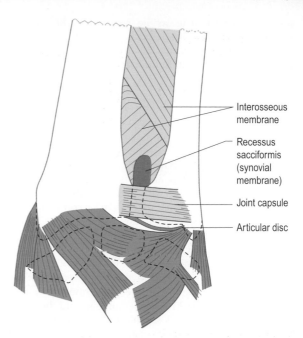

Figure 2.97 Anterior view of the inferior radioulnar joint.

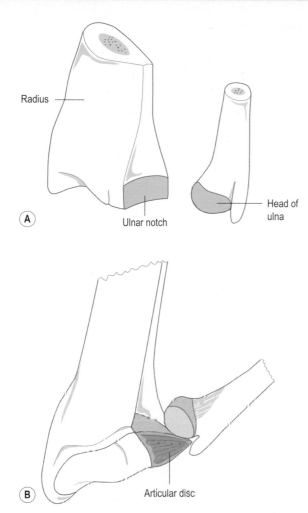

(A)

Radius

Ulnar notch

Head of ulna

(B)

Articular disc

Figure 2.96 Articular surfaces of the inferior radioulnar joint: (A) without the articular disc, (B) with the articular disc.

Interosseous membrane

Recessus sacciformis (synovial membrane)

Joint capsule

Articular disc

corresponding regions on the head of the ulna (Fig. 2.97). The inferior margins of these bands blend with the anterior and posterior edges of the articular disc; however, superiorly they remain separated.

The synovial membrane is large in relation to the size of the joint, extending upwards above the margins of the joint capsule between the radius and ulna in front of the interosseous membrane forming the recessus sacciformis (Fig. 2.97).

Blood and nerve supply

The arterial supply to the joint is by branches from the anterior and posterior interosseous arteries, and the dorsal and palmar carpal networks, which receive branches from the radial and ulnar arteries. Venous drainage is by similarly named vessels into the deep system of veins. The lymphatic drainage of the joint is by vessels accompanying the deeper blood vessels, some of which pass to nodes in the cubital fossa, but most go directly to the lateral group of axillary nodes.

The nerve supply to the joint is by twigs from the anterior and posterior interosseous nerves, with a root value of C7 and 8.

Relations

Passing directly behind the inferior radioulnar joint is the tendon of extensor digiti minimi, enclosed within its synovial sheath, on its way to the little finger (see Fig. 2.110). Anteriorly lies the lateral part of flexor digitorum profundus enclosed within the common flexor sheath. Proximal to the joint, pronator quadratus passes between the radius and ulna holding them together, thereby protecting the joint.

Stability

Although the joint capsule is loose permitting movement between the radius and ulna, the inferior radioulnar joint is extremely stable and is rarely dislocated. Joint stability is due primarily to the articular disc, but also to the interosseous membrane and pronator quadratus.

A fall on the outstretched hand with the wrist extended frequently results in a transverse fracture in the lower 2 or 3 cm of the radius (Colles' fracture), with the fragment being displaced posteriorly. The ulna is usually not

145

involved except that its styloid process may be torn off. In a Colles' fracture the hand is displaced laterally and dorsally. Alternatively, the fall may result in dislocation at the radiocarpal joint, but not at the inferior radioulnar joint.

Movements

The main movement at the inferior radioulnar joint is a rotation of the distal end of the radius around the head of the ulna during pronation and supination. However, because during everyday activities the axis of pronation and supination coincides with the axis of the hand along the middle finger, radial rotation is accompanied by movement of the head of the ulna. As the radius rotates about the ulna, the ulna is also displaced with respect to the radius (Fig. 2.98). The ulna displacement observed during rotation is the result of two elementary movements: slight extension and medial displacement of the ulna at the elbow. The slight side-to-side movement possible between the trochlea of the humerus and the trochlear notch of the ulna is mechanically amplified at the lower end of the ulna to become a movement of appreciable magnitude. Both the extension and lateral displacement of the ulna are brought about by the action of anconeus, and therefore occur simultaneously during pronation. The arc of the movement described by the head of the ulna does not involve rotation; it remains parallel to itself throughout, that is, the ulnar styloid process remains posteromedial (Fig. 2.98).

Accessory movements

With the distal ends of the radius and ulna gripped firmly, the head of the ulna can be moved anteroposteriorly with respect to the radius.

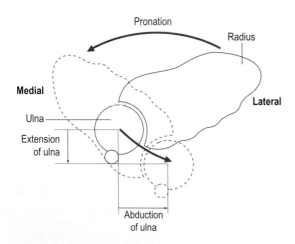

Figure 2.98 Movements of the radius and ulna at the inferior radioulnar joint.

Palpation

The line of the inferior radioulnar joint can be palpated on the posterior aspect of the wrist, running vertically between the two bones.

Interosseous membrane

A strong, fibrous sheet stretching between the interosseous borders of the radius and ulna; the fibres pass predominantly obliquely downward and medially (Fig. 2.99). Deficient superiorly, the free oblique border is attached 2–3 cm below the radial tuberosity passing to a slightly lower level on the ulna. Inferiorly the membrane is continuous with the fascia on the posterior surface of pronator quadratus, attaching to the posterior of the two lines into which the radial interosseous border divides. An opening in the lower part of the membrane enables the anterior interosseous vessels to pass into the posterior compartment of the forearm. On the posterior part of the membrane there are a small number of fibrous bands which pass obliquely downwards and laterally (Fig. 2.99). During pronation and supination, tension in the membrane varies, being greatest in the midprone position.

The oblique direction of its fibres serves to transmit forces from the radius to the ulna. Through its articulation at the radiocarpal joint the radius receives impacts and

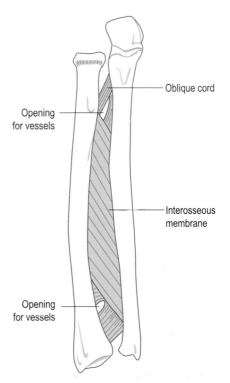

Figure 2.99 The interosseous membrane.

forces from the scaphoid and lunate. At the elbow, however, it has a rather ineffective articulation with the humerus, whereas the ulna has a large and firm articulation. The membrane serves to transmit forces carried from the hand through the radius to the ulna and thence to the elbow joint and humerus.

As well as providing a firm connection between the radius and ulna the interosseous membrane separates and increases the area of attachment of the deep muscles of the anterior and posterior compartments of the forearm.

Above the upper free border of the interosseous membrane the oblique cord passes upwards and medially from the radius to the ulna (Fig. 2.99). It is a slender, flattened fibrous band, said to represent a degenerated part of flexor pollicis longus or supinator, attached just below the radial tuberosity and to the lateral border of the ulnar tuberosity. In the gap between the oblique cord and the interosseous membrane, the posterior interosseous vessels pass to and from the posterior compartment of the forearm.

Pronation and supination

In the supine position, the bones of the forearm lie parallel to one another (Fig. 2.100A); in the anatomical position the palm of the hand therefore faces forwards. In the prone position, the radius and ulna cross one another (Fig. 2.100), with the radius lying anterior to the ulna; with reference to the anatomical position the palm faces backwards. Pronation is the movement which causes the radius to cross the ulna, while supination is the movement causing them to lie parallel to each other. Movements between the radius and ulna occur at the superior and inferior radioulnar joints (see pp. 143, 146 for details of the movements at each joint).

The muscles producing pronation are the pronators teres and quadratus, with pronator teres being the more powerful. Flexor carpi radialis, because of its oblique course, can and does assist in pronation. Supination is produced by supinator and biceps brachii, of which biceps is by far the stronger. However, when the elbow is fully extended biceps is unable to act as a supinator as its tendon runs almost parallel to the shaft of the radius, and so cannot produce radial rotation. Of the two movements supination is the more powerful. Because the majority of the population are right-handed, screws have a right-hand thread. If you are trying to remove a particularly stubborn screw from a cabinet or door frame, ask a left-handed friend to do it for you! Both pronation and supination are most powerful when the elbow is flexed to 90°.

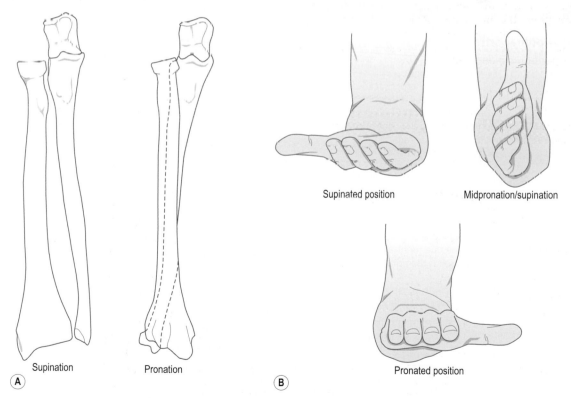

Figure 2.100 (A) Relation of the radius and ulna in pronation and supination, (B) movement of the hand in pronation and supination, anterior view with elbow at 90°.

The axis of pronation and supination varies depending on which finger the movement is occurring about. It always passes through the centre of the head of the radius, but at the level of the wrist it can pass through any point between the ulnar and radial styloid processes. Nevertheless, it will tend to lie in the medial half of this region in most instances. Therefore, to state that the axis runs between the centre of the radial head and the base of the ulnar styloid process is not strictly correct. When rotation occurs about a more laterally placed centre at the wrist, ulnar movement at the trochlea is insufficient. Consequently, with the elbow flexed the movement is supplemented by rotation of the humerus.

The forearm can be pronated through almost 180°, without medial rotation of the humerus (Fig. 2.100B). The constraint to further movement comes predominantly from passive resistance of the opposing muscles, and not from ligamentous ties. However, if the humerus is allowed to rotate then it becomes possible to turn the hand through almost 360°.

Pronation and supination are frequently used movements in many activities. Consequently loss of the ability to pronate and supinate can be a marked disability. When these movements are lost it is less disabling if the forearm is fixed in a mid position, so that the palm faces medially.

Section summary

Radioulnar articulations

Superior radioulnar joint

Type	Synovial pivot joint
Articular surfaces	Rounded head of radius, fibro-osseous ring formed by radial notch of ulna and annular ligament
Capsule	Continuous with that of elbow joint
Ligaments	Annular and quadrate
Stability	Mainly due to ligaments

Inferior radioulnar joint

Type	Synovial pivot joint
Articular surfaces	Head of ulna and ulnar notch of radius and intra-articular disc
Capsule	Weak and loose
Stability	Very stable due to intra-articular disc, interosseous membrane and pronator quadratus
Movements	Pronation and supination, which occur at the superior and inferior radioulnar joints

---

The wrist

Introduction

The wrist joint is not a single joint but comprises the articulations between the carpal bones (intercarpal and midcarpal joints) and the articulation with the forearm (radiocarpal joint) (Fig. 2.101). Functionally, however, the eight carpal bones are arranged and move as two rows of bones: a proximal row, comprising from lateral to medial, scaphoid, lunate, triquetral and pisiform; and a distal row, again from lateral to medial, formed by the trapezium, trapezoid, capitate and hamate. The two rows articulate with each other at the midcarpal joint, a sinuous articular area convex laterally and concave medially (Fig. 2.101). The distal surface of the distal row of carpal bones articulates with the bases of the metacarpals.

Because of the functional interdependence of the wrist and hand, all movements of the hand are accompanied by movements at the radiocarpal and intercarpal joints. The wrist complex is capable of movement in two directions. However, when combined with pronation and supination of the forearm the hand appears to be connected to the forearm by a ball and socket joint, having great intrinsic stability because of the separation of the three axes about which movement occurs.

The radiocarpal joint

Formed between the distal surfaces of the radius and articular disc, and the scaphoid, lunate and triquetral of the proximal row of carpal bones. It is a synovial joint of the ellipsoid type allowing movement in two planes.

Articular surfaces

Distal surface of the radius and articular disc

The radius and articular disc form a continuous, concave ellipsoid surface, shallower about its transverse long axis than about its shorter anteroposterior axis (Fig. 2.102). The articular cartilage on the radius is divided by a low ridge into a lateral triangular and medial quadrangular area.

Proximal carpal row

The proximal row of carpal bones presents an almost continuous convex articular surface (Fig. 2.102B). The three carpal bones are closely united by interosseous ligaments continuous with the cartilage on the proximal surfaces of the bones. In the anatomical position, the scaphoid lies opposite the lateral area on the radius, the lunate opposite the medial radial area and the articular disc, while the triquetral is in contact with the medial part of the joint capsule (Fig. 2.102B).

Joint capsule and synovial membrane

A fibrous capsule completely encloses the joint (Fig. 2.103). It is attached to the distal edges of the radius and ulna anteriorly and posteriorly, and to the radial and ulnar styloid processes laterally and medially, respectively. Distally the capsule is firmly attached anteriorly and

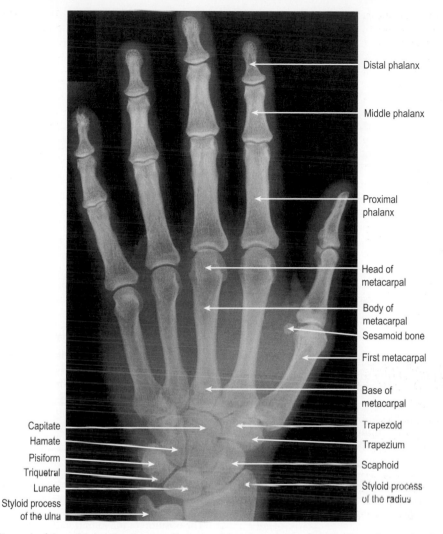

Distal phalanx

Middle phalanx

Proximal
phalanx

Head of
metacarpal

Body of
metacarpal

Sesamoid bone

First metacarpal

Base of
metacarpal

Trapezoid

Trapezium

Scaphoid

Styloid process
of the radius

Capitate

Hamate

Pisiform

Triquetral

Lunate

Styloid process
of the ulna

Figure 2.101 Radiograph of the wrist showing the carpal bones and their arrangement forming the radiocarpal and midcarpal joints.

posteriorly to the margins of the articular surfaces of the proximal row of carpal bones. Medially it passes to the medial side of the triquetral, and laterally to the lateral side of the scaphoid. Both the anterior and posterior parts of the capsule are thickened and hence strengthened, while at the sides it blends with the collateral carpal ligaments.

Capsular ligaments

These are distinct bands of fibres passing between specific bones. As well as strengthening the capsule, their arrangement determines that the hand follows the radius in its movements and displacements.

Dorsal radiocarpal ligament: This extends from the posterior edge of the lower end of the radius to the posterior surface of the scaphoid, lunate and triquetral (Fig. 2.103A). Its fibres run downwards and medially, principally to the triquetral, and are continuous with the dorsal intercarpal ligaments.

Palmar radiocarpal ligament: A broad band of fibres passing downwards and slightly medially from the anterior edge of the lower end of the radius and its styloid process, to the anterior surfaces of the proximal row of carpal bones (Fig. 2.103B). Some fibres are prolonged and extend to attach to the capitate.

Palmar ulnocarpal ligament: This is formed by fibres extending downwards and laterally from the anterior edge of the articular disc and the base of the ulnar styloid process to the anterior surfaces of the proximal carpal bones (Fig. 2.103B).

149

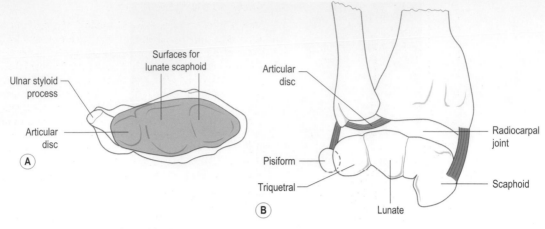

Figure 2.102 The articular surfaces of (A) the radius and articular disc and (B) the proximal row of carpal bones.

The anterior and posterior capsular ligaments become taut in extension and flexion, respectively, of the radiocarpal joint.

Synovial membrane

A relatively lax synovial membrane lines the deep surface of the joint capsule attaching to the margins of all articular surfaces. It presents numerous folds, particularly posteriorly. Because of the presence of the articular disc of the inferior radioulnar joint and the completeness of the interosseous ligaments uniting the proximal surfaces of the proximal carpal row, the synovial cavity is limited to the radiocarpal space. Only occasionally does it communicate with the inferior radioulnar joint via a perforation in the articular disc, or with the intercarpal joint when one of the interosseous ligaments is incomplete.

Ligaments

At the sides of the radiocarpal joint, collateral ligaments reinforce and strengthen the joint capsule. They are active in limiting abduction and adduction at the joint. In adduction, the radial ligament becomes taut while the ulnar relaxes; in abduction the reverse occurs.

Radial collateral carpal ligament

This passes from the tip of the radial styloid process to the lateral side of the scaphoid, immediately adjacent to its proximal articular surface and lateral side of the trapezium (Fig. 2.103).

Ulnar collateral carpal ligament

A rounded cord attached to the ulnar styloid process above and at the base of the pisiform and medial and posterior non-articular surfaces of the triquetral below (Fig. 2.103).

By its attachment to the pisiform it also blends with the medial part of the flexor retinaculum.

Blood and nerve supply

The arterial supply to the joint is by branches from the dorsal and palmar carpal arches, with venous drainage going to the deep veins of the forearm. Lymphatic drainage of the joint follows the deep vessels.

The nerve supply to the joint is by twigs from the anterior interosseous branch of the median nerve, the posterior interosseous branch of the radial nerve, and the dorsal and deep branches of the ulnar nerve, with root value C7 and 8.

Surface marking

The position of the joint is indicated by a line, slightly convex proximally, between the radial styloid process and the head of the ulna, so that the concavity of the radius and articular disc face distally, medially and slightly anteriorly.

Movements

Flexion and extension, and adduction and abduction are possible at the radiocarpal joint. However, each is also contributed to by movements between the proximal and distal row of carpal bones at the midcarpal joint.

Flexion and extension

Flexion and extension occur about a transverse axis more or less in the sagittal plane so that the hand moves towards the front and back of the forearm, respectively. Flexion is freer than extension and has a maximum range of 50°, whereas extension has a maximum range of 35° (Fig. 2.104). The movements are checked by the margins of the radius; as

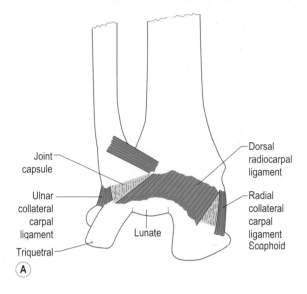

Joint capsule

Ulnar collateral carpal ligament

Triquetral

Dorsal radiocarpal ligament

Radial collateral carpal ligament

Lunate

Scaphoid

(A)

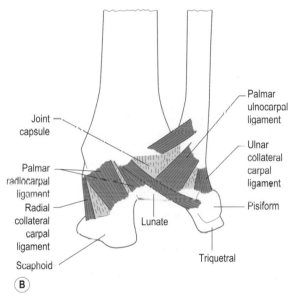

Joint capsule

Palmar radiocarpal ligament

Radial collateral carpal ligament

Scaphoid

Palmar ulnocarpal ligament

Ulnar collateral carpal ligament

Pisiform

Lunate

Triquetral

(B)

Figure 2.103 The capsular and collateral ligaments of the radiocarpal joint: (A) posterior, (B) anterior.

the posterior margin extends further distally than the anterior, extension is checked earlier than flexion.

During flexion the scaphoid and lunate move within the concave distal surface of the radius so their proximal surfaces face posterosuperiorly. In addition, the scaphoid twists about its long axis so that its tubercle becomes less prominent in full flexion. During extension the twisting of the scaphoid about its long axis makes the tubercle more prominent in full extension.

Abduction and adduction

Also referred to as radial and ulnar deviation, abduction and adduction are lateral or medial movements respectively of the proximal row of carpal bones in relation to the distal end of the radius (Fig. 2.104). The radial styloid process extends further distally than the ulnar styloid process; consequently abduction is more limited at the radiocarpal joint having a range of only 7°, whereas adduction has a range of 30°. In adduction the scaphoid rotates so that its tubercle moves away from the radial styloid process, enabling the lunate to move laterally so that it comes to lie entirely distal to the radius. The triquetral lies distal to the articular disc. In abduction the triquetral moves medially and distally to be clear of the radius; the lunate follows so that its centre lies distal to the inferior radioulnar joint. The movement is limited by the impact of the scaphoid tubercle against the radial styloid process.

Accessory movements

An anteroposterior gliding of the proximal row of carpal bones against the radius and articular disc can be produced by firmly gripping the distal end of the radius and ulna with one hand, and the proximal row of carpal bones with the other. Alternate anterior and posterior pressure elicits a palpable gliding movement at the radiocarpal joint. With the same grip, a longitudinally applied force along the line of the forearm pulls the carpal bones away from the radius and articular disc.

The intercarpal joints

The carpal bones are arranged as two transverse rows between which is the important midcarpal joint (Fig. 2.105). For the majority, the joints between the adjacent individual carpal bones are of the plane synovial type, permitting only slight movement between the bones involved. The only bone which moves appreciably is the capitate.

Joints of the proximal row

Plane synovial joints exist between the distal parts of the adjacent surfaces of the scaphoid, lunate and triquetral (Fig. 2.105). However, because the bones are bound together by interosseous, dorsal and palmar intercarpal ligaments there is minimal movement between them.

The interosseous intercarpal ligaments are short bands attaching to the margins of the joint surfaces involved in the radiocarpal articulation (Fig. 2.105), uniting the bones along their whole anteroposterior length. The palmar and dorsal intercarpal ligaments are transverse bands passing from scaphoid to lunate and from lunate to triquetral on the anterior and posterior aspects of the bones respectively.

The pisiform rests on the palmar surface of the triquetral and has a separate synovial joint with it, which is completely enclosed by a thin but strong fibrous capsule.

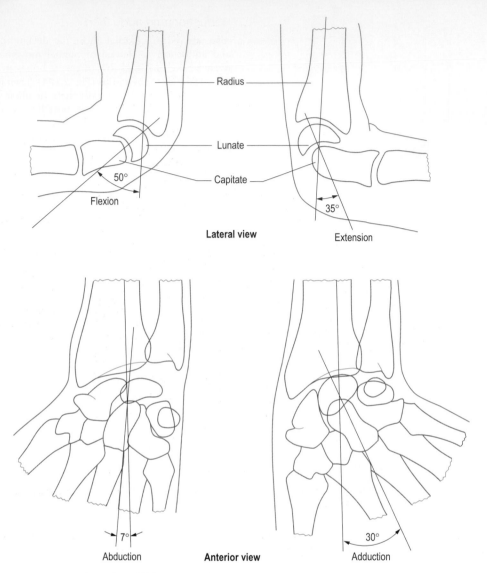

Figure 2.104 Movements at the radiocarpal joint.

The pisiform is also anchored to the hook of the hamate by the pisohamate ligament, as well as to the base of the fifth metacarpal by the pisometacarpal ligament. These two ligaments resist the pull of flexor carpi ulnaris, and transfer its action to the hamate and base of the fifth metacarpal. In this way the ligaments form extensions of the muscle.

Joints of the distal row

As in the proximal row, the four bones of the distal row are united by interosseous, palmar and dorsal intercarpal ligaments, with the joints between the individual bones being of the plane synovial type. Because of these ligaments, movement between adjacent bones is minimal.

The interosseous ligaments are not as extensive as in the proximal row, leaving clefts between the bones which communicate with the midcarpal joint proximally, and with the common carpometacarpal joint distally (Fig. 2.105). Occasionally, the midcarpal and common carpometacarpal joints communicate with each other between the bones of the distal row. This occurs when one of the interosseous ligaments is incomplete with communication being around the borders of the ligament, or when one ligament is absent (most commonly that between the trapezium and trapezoid). Dorsal and palmar ligaments generally run

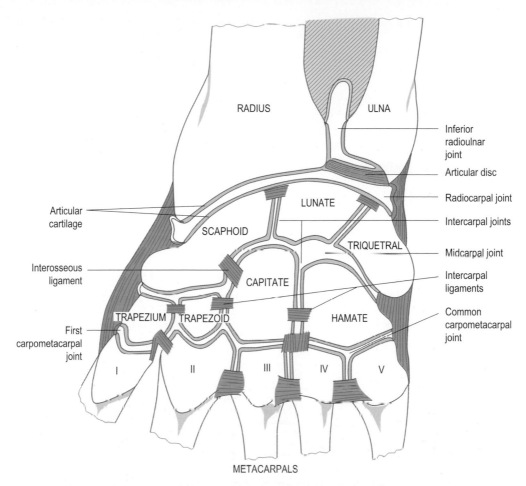

RADIUS

ULNA

Inferior radioulnar joint

Articular disc

Radiocarpal joint

Intercarpal joints

Midcarpal joint

Intercarpal ligaments

Common carpometacarpal joint

LUNATE

SCAPHOID

TRIQUETRAL

Articular cartilage

Interosseous ligament

First carpometacarpal joint

CAPITATE

TRAPEZIUM   TRAPEZOID

HAMATE

I   II   III   IV   V

METACARPALS

Figure 2.105 Coronal section through the wrist showing the relationship between the radiocarpal, midcarpal, intercarpal and the first and common carpometacarpal joints.

transversely across the appropriate surfaces of the bones, uniting trapezium to trapezoid, trapezoid to capitate, and capitate to hamate.

The midcarpal joint

The articulation between the proximal and distal rows of carpal bones, each considered to act as a single functional unit (Fig. 2.106). The lateral part of the joint consists of two plane surfaces arranged to form a slight convexity directed distally. The larger medial part of the joint is concave distally in all directions.

Articular surfaces

Laterally, plane joint surfaces on the trapezium and trapezoid articulate with the slightly rounded distal surface of the scaphoid. The head of the capitate articulates with the scaphoid and lunate in the central part of the joint. The apex of the hamate also articulates with the lunate,

while its ulnar surface articulates with the triquetral (Figs. 2.105 and 2.106).

Joint capsule

The midcarpal joint is surrounded by a fibrous capsule, composed primarily of irregular bands of fibres running between the two rows of bones. Anteriorly and posteriorly these bands constitute the palmar and dorsal intercarpal ligaments. At the sides of the midcarpal joint the capsule is strengthened by collateral ligaments.

Intercarpal synovial cavity

The joint cavity is large and complex (Fig. 2.105) extending from side to side between the two rows of carpal bones. However, this may be partially or completely interrupted by an interosseous ligament between the scaphoid and capitate. Extensions of the cavity pass proximally between the scaphoid, lunate and triquetral as far as the interosseous

153

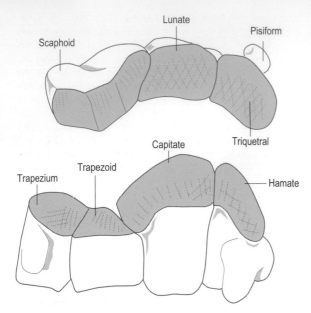

Figure 2.106 Articular surfaces of the midcarpal joint.

ligaments between them (Fig. 2.105). Rarely is there communication with the radiocarpal joint cavity. Further extensions of the cavity pass distally between the trapezium, trapezoid, capitate and hamate. If the interosseous ligaments connecting these bones do not extend the full depth of the articulation, or one is missing (usually that between trapezium and trapezoid) then the intercarpal joint cavity communicates with the carpometacarpal joint and is prolonged between the bases of the medial four metacarpals (Fig. 2.105). The intercarpal cavity does not, however, communicate with the first carpometacarpal or the pisiform–triquetral joint spaces.

The synovial membrane lines the capsule and all non-articular surfaces, attaching to the margins of all joint surfaces.

Ligaments

Palmar intercarpal ligament

From the bones of the proximal row predominantly to the head of the capitate (Fig. 2.107). It is sometimes referred to as the *radiate capitate ligament*.

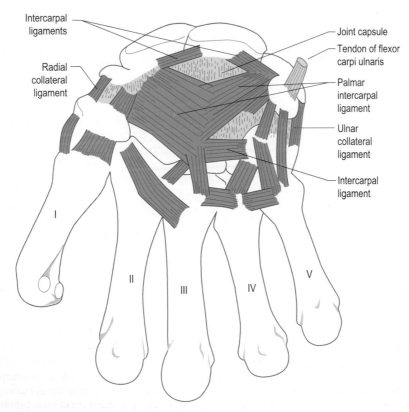

Figure 2.107 Ligaments associated with the midcarpal joint, palmar aspect.

Dorsal intercarpal ligament

Passes from the bones of one row to those of the other.

Radial collateral ligament

A strong distinct band passing from the scaphoid to the trapezium (Fig. 2.107). It is a continuation of the radial collateral carpal ligament of the radiocarpal joint.

Ulnar collateral ligament

Connects the triquetral and hamate, and is a continuation of the ulnar collateral carpal ligament of the radiocarpal joint (Fig. 2.107).

Interosseous ligament

An occasional slender interosseous ligament passes from the lateral side of the capitate to the scaphoid near its trapezoid articular surface (Fig. 2.107).

Blood and nerve supply

The arterial supply to all the intercarpal joints is by branches from the palmar and dorsal carpal arches.

The nerve supply to the joints is by twigs from the anterior and posterior interosseous nerves, and the deep and dorsal branches of the ulnar nerve; root value C7 and 8.

Movements

Movements at the intercarpal joints, except the midcarpal joint, are small, accompanying and facilitating movements at the radiocarpal and midcarpal joints. Movements possible at the midcarpal joint are flexion and extension, and abduction and adduction, which occur about transverse and anteroposterior axes passing through the head of the capitate.

Flexion and extension

In flexion the hand moves towards the front of the forearm, while in extension it moves towards the back of the forearm (Fig. 2.108). Extension is freer than flexion, having a range of 50°; flexion has a range of only 35°. In these movements the head of the capitate rotates within the concavity formed by the scaphoid and lunate, while the hamate rotates against the triquetral. Accompanying these movements is a compensatory swing of the scaphoid on the lunate in order to receive the head of the capitate.

Abduction and adduction

During adduction the capitate rotates so that its distal part moves medially; the hamate approaches the lunate and separates from the triquetral (Fig. 2.108). In abduction the capitate comes close to the triquetral, separating the hamate from the lunate. Accompanying abduction and adduction is a complex movement of torsion between the two rows of carpal bones. During abduction the distal row undergoes a 'rotation' in the direction of supination and extension, while the proximal row 'rotates' in the direction of pronation and flexion. The twisting of the scaphoid delays its impact on the radial styloid process by bringing its tubercle forwards; it also makes the tubercle more easily palpable. In adduction a reverse twisting motion occurs so that the proximal row 'rotates' in the direction of supination and extension, while the distal row moves in the direction of pronation and flexion. It must be emphasized that these movements are of extremely small magnitude. It is debatable whether they contribute much to the normal functioning of the wrist.

The range of abduction and adduction are 8° and 15° respectively. The principal limit to abduction is a closing of the lateral part of the joint space between the scaphoid and the trapezium.

Accessory movements

Anteroposterior gliding movements of any two adjacent carpal bones can be produced if one is stabilized while the other is moved. This can be achieved by gripping each bone between the thumb and index finger.

Anteroposterior movement at the midcarpal joint can be elicited using a similar technique to that described for the radiocarpal joint (p. 151). A firm circular grip is applied around each carpal row. While the proximal row is stabilized, the distal row can be moved anteroposteriorly. Applying the same grip, a longitudinally applied force separates the two joint surfaces.

Relations

All of the structures entering or leaving the hand have to cross the region of the wrist. Some of these lie directly against the carpal bones; others are separated by intervening soft tissues. The nature of the arrangement of the carpal bones, in which they form part of a fibro-osseous canal, makes the anterior aspect of the wrist an extremely important region.

The carpal bones of each row form a transverse arch with a palmar concavity (Fig. 2.109A). The principal structure maintaining the bones in this position is the flexor retinaculum. Consequently, it is considered by some to be an accessory ligament to the intercarpal joints. The flexor retinaculum attaches medially to the pisiform and hook of the hamate, and laterally to the scaphoid tubercle and to both lips of the groove on the trapezium, so forming a small lateral compartment separate from the rest of the canal. Passing through this lateral compartment is the tendon of flexor carpi radialis enclosed within its own synovial sheath (Fig. 2.109B). Through the larger main part of the canal pass:

1. the tendon of flexor pollicis longus most laterally, deep in the concavity of the carpal bones
2. the four tendons of flexor digitorum profundus lying side by side directly over the capitate

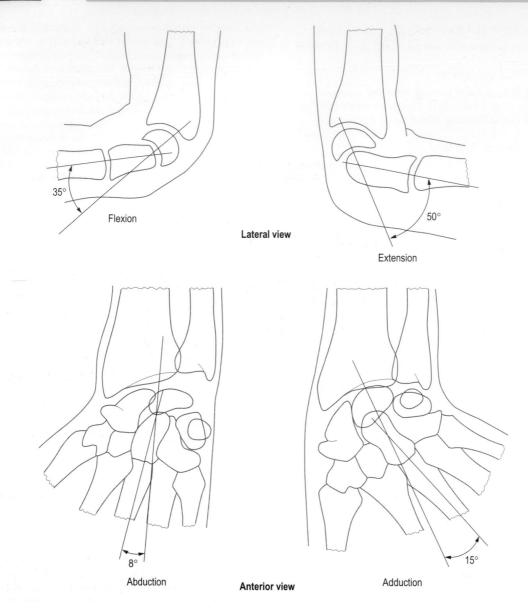

35°

Flexion

Lateral view

50°

Extension

8°

Abduction

Anterior view

15°

Adduction

Figure 2.108 Movements at the midcarpal joint.

3. the four tendons of flexor digitorum superficialis overlying those of profundus, with the tendons to the third and fourth digits anterior to those of the second and fifth: the tendons of flexors digitorum superficialis and profundus are enclosed within the same synovial sheath

4. the median nerve lying lateral to the superficialis tendons (Fig. 2.109B).

Inflammation of the synovial sheaths within the so-called *carpal tunnel* may lead to compression of the median nerve giving rise to *carpal tunnel* (or *median nerve*) *syndrome*. This leads to paraesthesia and diminution of sensory acuity in the region of the median nerve's sensory distribution, loss of power and limitation of some thumb movements, together with some wasting of the thenar eminence. Passing anterior to and blending with the flexor retinaculum is the tendon of palmaris longus. Also passing superficial to the flexor retinaculum medially is the ulnar nerve with the ulnar artery lateral to it. In addition, the palmar cutaneous branches of the median and ulnar nerves, and the superficial palmar branch of the radial artery enter the hand by crossing the retinaculum.

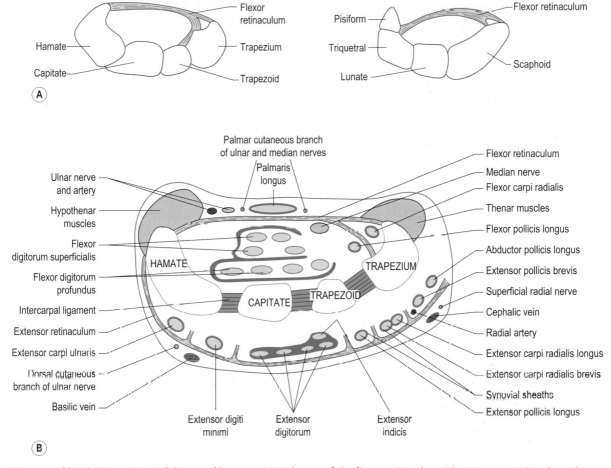

Figure 2.109 (A) Arrangement of the carpal bones and attachment of the flexor retinaculum, (B) transverse section through the wrist region showing the relationships of the various structures which pass into the hand.

On the posterior aspect of the carpal bones, the extensor tendons pass into the hand. They are separated by the fibrous septa passing from the deep surface of the extensor retinaculum to ridges on the radius, ulna and capsular tissues of the joint (Fig. 2.109B). The six longitudinal compartments formed transmit the tendons of the nine muscles of the extensor compartment. Most laterally, over the lateral surface of the radial styloid process and continuing over the scaphoid and trapezium, pass the tendons of abductor pollicis longus and extensor pollicis brevis within the same synovial sheath (Fig. 2.110A) In the adjacent compartment over the radius lateral to the dorsal tubercle, and then over the scaphoid and the most medial part of trapezium and the trapezoid, run the tendons of the extensors carpi radialis longus and brevis (Fig. 2.110A). In the third compartment, in a groove on the medial side of the dorsal tubercle, is the tendon of extensor pollicis longus (Fig. 2.110A). However, because this tendon uses the dorsal tubercle as a pulley, it

deviates laterally towards the thumb and so crosses the scaphoid and trapezium between the tendons of the two previous compartments. Running over the most medial part of the dorsum of the radius, and then over the adjacent parts of the scaphoid and lunate and the capitate are the four tendons of extensor digitorum, with the tendon of extensor indicis deep to them (Fig. 2.110B). All five tendons are enclosed with a common synovial sheath. Crossing the posterior surface of the inferior radioulnar joint, the lunate and the adjacent surfaces of the capitate and hamate is the tendon of extensor digiti minimi (Fig. 2.110B). Finally, the tendon of extensor carpi ulnaris passes in a groove on the back of the ulna and onto the triquetral before attaching to the base of the fifth metacarpal (Fig. 2.110B). Crossing the extensor retinaculum to enter the dorsum of the hand on its medial and lateral sides are the dorsal branch of the ulnar nerve and the terminal branches of the superficial radial nerve, respectively. The other major structure to enter the

157

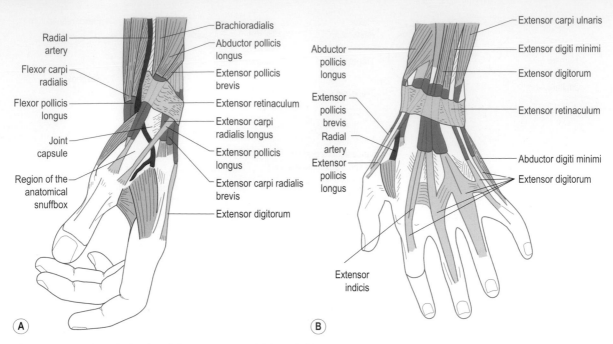

Figure 2.110 Principal relations of the wrist region: (A) lateral, (B) posterior.

hand is the radial artery, and it does so by a convoluted route (Fig. 2.110A). In the forearm proximal to the flexor retinaculum, the radial artery can be palpated lateral to the tendon of flexor carpi radialis. It then turns laterally to cross the radial collateral carpal ligament, the scaphoid and trapezium, being crossed by the tendons of abductor pollicis longus and extensors pollicis brevis and longus before passing into the palm of the hand between the two heads of the first dorsal interosseous. The hollowed region between the tendons of abductor pollicis longus and extensor pollicis brevis laterally, and extensor pollicis longus medially when the thumb is extended is known as the 'anatomical snuffbox' (Fig. 2.110A). The radial artery crosses the floor of this hollow, formed from proximal to distal by the radial styloid process, scaphoid, trapezium and base of the first metacarpal; its pulsations can be felt readily by applying firm pressure between the tendons.

Stability

Because of the attachment of the flexor retinaculum and the many tendons crossing the joint both anteriorly and posteriorly, the wrist is a relatively stable region (Figs. 2.109 and 2.110). Nevertheless, abnormal stresses applied to this region may result in dislocation or fracture.

A fall on the outstretched hand may result in a dislocation at the radiocarpal and/or midcarpal joints, involving anterior dislocation of the lunate. Usually this can be reduced by manipulation. Care must be taken, however, in not confusing a dislocation with a Colles' fracture. A fall on the hand is more likely to result in the force being transmitted through the trapezium and trapezoid to the scaphoid, which tends to fracture across its waist. Persistent pain on applying pressure in the anatomical snuffbox is characteristic of scaphoid fracture. Care should be taken when setting the fracture that the two fragments are aligned and in contact, otherwise non-union and/or avascular necrosis may result if viable blood vessels reach only one of the fragments. (The blood supply to the scaphoid in the majority of individuals is from distal to proximal.)

Movements

The movements which occur at the radiocarpal and midcarpal joints take place at the same time. The total range of flexion and extension is therefore 85° in each direction (Fig. 2.111). Flexion is limited by tension in the extensor tendons and is greatly reduced if the fingers are fully flexed. The main muscles producing flexion are flexors carpi radialis and ulnaris. The main muscles producing extension are extensors carpi ulnaris and radialis longus and brevis. Radiographic film shows that flexion and extension movements at the wrist occur about a single transverse axis through the head of the capitate (Fig. 2.111).

The total range of abduction and adduction possible at the wrist is also the sum of the ranges possible at the radiocarpal and midcarpal joints. Consequently, abduction has a range of 15° and adduction a range of 45° (Fig. 2.111).

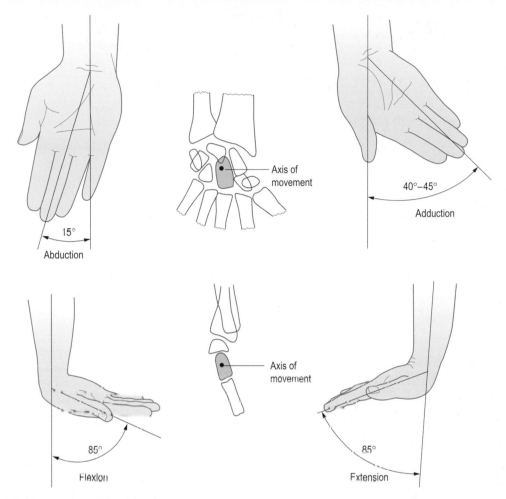

Figure 2.111 Movements at the wrist region.

The movements occur about a single anteroposterior axis passing through the head of the capitate slightly more distal to the axis for flexion and extension (Fig. 2.111). Abduction is more limited than adduction primarily because the radial styloid process projects further distally than the ulnar styloid process. Abduction is produced by the flexor carpi radialis and extensors carpi radialis longus and brevis, while adduction is produced by flexor and extensor carpi ulnaris.

Biomechanics

The lines of action of the muscles of the wrist are always oblique with respect to the axes of movement (Fig. 2.112). By using only one muscle the movement produced is not a pure movement. For example, contraction of flexor carpi radialis produces flexion and abduction at the wrist. To produce pure flexion the unwanted abduction has to be cancelled; this is achieved by contracting

flexor carpi ulnaris. Thus by combining various forces acting in different directions any desired movement can be produced within the complete range of motion of the joint.

The carpal flexors and extensors fix the wrist during extension or flexion of the fingers, thereby preventing the digital muscles from losing power and efficiency, which would occur if they also acted on the radiocarpal and midcarpal joints. When powerful movements of the fingers are required, both the flexors and extensors of the wrist contract simultaneously. The importance of such actions is obvious when attempts are made to grip strongly with the finger flexors when the wrist is already flexed. Extending the wrist stretches these muscles so that they can now exert considerable power. Slight extension of the wrist is the position naturally adopted when the hand is used for gripping: look at your own wrist when writing or picking up a mug. If the wrist is likely to become fixed through disease, it should be secured in a position of slight extension so that a powerful and precise grip can still be achieved.

159

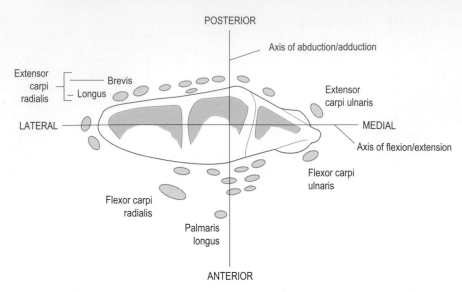

Figure 2.112 Relationship of the tendons of muscles producing movement at the wrist to the axes of flexion/extension and abduction/adduction.

The extrinsic finger flexors are the major force-producing muscles during exertions of the hand. Deviation of the wrist causes these tendons to move against the adjacent walls of the carpal tunnel. When the wrist is flexed the tendons are supported by the flexor retinaculum, and when extended by the carpal bones. The force between the tendons and the retinaculum may compress the median nerve and be an important factor in carpal tunnel syndrome; such compression has been confirmed by direct pressure measurement. As well as the median nerve being compressed, the synovial sheaths surrounding the flexor tendons are also compressed, both in flexion and extension. This may lead to their inflammation and subsequent swelling, leading to further compression of the median nerve. Taking into account wrist size, the loading of the flexor retinaculum in flexion is 14% greater in females than in males. This may be one of the reasons why carpal tunnel syndrome is between 2 and 10 times more prevalent in women than men.

In some cases of limitation or absence of movement at the wrist, often associated with persistent pain, total wrist arthroplasty can relieve pain and improve mobility.

Capsule	Complete fibrous capsule reinforced by ligaments
Ligaments	Dorsal radiocarpal
	Palmar radiocarpal
	Palmar ulnocarpal
	Radial collateral carpal
	Ulnar collateral carpal
Movements	Flexion and extension
	Abduction (radial deviation) and adduction (ulnar deviation)

Midcarpal joint

Type	Complex synovial joint
Articular surfaces	Distal surfaces of scaphoid, lunate and triquetral with proximal surfaces of trapezium, trapezoid, capitate and hamate
Capsule:	Complete fibrous capsule reinforced by ligaments
Ligaments:	Palmar intercarpal
	Dorsal intercarpal
	Radial collateral
	Ulnar collateral
Movements:	Flexion and extension
	Abduction (radial deviation) and adduction (ulnar deviation)

- Movements at the radiocarpal and midcarpal joints combine to give a greater range of movement at the wrist.
- Stability at the wrist is mainly due to ligaments and tendons crossing the joints.

Section summary

The wrist joints

Radiocarpal joint

Type	Synovial ellipsoid joint
Articular surfaces	Distal surface of radius and articular disc, proximal surfaces of scaphoid, lunate and triquetral of the proximal row of carpal bones

Articulations within the hand

Introduction

Just as the foot has evolved as an organ of support and locomotion so the hand has developed into an instrument of manipulation endowed with fine sensory discrimination. It is often hard to accept that the hand and wrist and the foot and ankle have similar building blocks in terms of bony and muscular constituents, patterns of innervation and blood supply. Undoubtedly, the refinements that have occurred in the hand followed its release from the burden of supporting and propelling the body. The extent to which the hand is used indicates its importance in everyday life. We use hands to grip and manipulate; they enable us to dress, eat, play instruments and games. The hand has to be capable of applying large gripping forces between the fingers and thumb while also performing precision movements. However, its sensory functions must not be overlooked; it relays information regarding texture and surface contour, warns against extremes of hot and cold, and prevents collisions, especially when sight cannot be used. All of these motor and sensory functions require considerable representation in the motor and sensory cortices of the brain. As the hand developed, so the cerebral cortex enlarged to obtain maximum benefit for the new freely mobile, sensitive structure.

Much of the motor functioning of the hand is due to its ability to grip objects. Yet prehension can be observed in many animals from the pincers of the crab to the hand of the great ape. It is the concomitant development of hand and brain forming an interacting functional pair that has led to human dominance in the animal kingdom. Because of the uses to which the hand is put, it is particularly disabling when part or all of it is injured. It is especially vulnerable because it is usually unprotected. Lesions of the peripheral and central nervous systems and infections, accidental amputations, burns, lacerations and penetrating wounds, as well as diseases of the joints, all serve to disable the hand.

In many respects the arrangement of the bones and their intervening joints are simpler in the hand than the foot, principally because the carpus is limited to the wrist (Fig. 2.113), whereas the tarsus forms the hindfoot. The metacarpals articulate with the wrist region via the carpometacarpal joints, of which the first is separate and different from the remainder, and with each other via the intermetacarpal joints. The head of each metacarpal articulates with the base of the proximal phalanx at the metacarpophalangeal joint (Fig. 2.113A). Adjacent phalanges articulate via interphalangeal joints (Fig. 2.113A). However, because the thumb has only two phalanges it has only one interphalangeal joint, whereas the fingers have three phalanges and two interphalangeal joints.

Care has to be taken when using the terms *fingers* and *digits* as confusion can often arise. There are four fingers and a thumb, or five digits. If finger and thumb is the preferred terminology then, to avoid confusion, use of an appropriate prefix is advised, that is index, middle, ring and little.

The axis of the hand runs along the middle finger (third digit), and is in line with the long axis of the forearm (Fig. 2.113B). Certain movements of the digits are made with reference to this axis. In describing movements of the thumb, remember that it is rotated through 90° with respect to the remaining digits.

However fine the movements produced in the hand, they must be controlled from a stable base, so that the origins of intrinsic muscles of the hand remain fixed by the musculature of the forearm which is brought into play. The origins of the muscles in the forearm in turn require fixation at the elbow by muscles of the arm, and these in turn require fixation at their origin at the shoulder and pectoral girdle. Even writing therefore involves use of the shoulder muscles as well as those of the fingers and thumb.

The common carpometacarpal joint

The carpometacarpal joints are the sites of articulation between the carpal and metacarpal bones. The bases of the medial four metacarpals and the medial three carpal bones of the distal row form the common carpometacarpal joint, which has an irregular joint line. The joints are, on the whole, plane synovial joints, the only exception being the slightly bevelled joint surfaces between the hamate and the base of the fifth metacarpal.

Articular surfaces

The base of the second metacarpal fits into a recess formed by the medial side of the trapezium, the distal surface of the trapezoid and the anterolateral corner of the capitate. The third metacarpal base articulates only with the distal surface of the capitate. The base of the fourth metacarpal articulates mainly with the anterolateral distal surface of the hamate, but also just catches the anteromedial corner of the capitate. Finally, the base of the fifth metacarpal abuts against the anteromedial part of the distal surface of the hamate (Figs. 2.114 and 2.116).

Joint capsule and synovial membrane

A fibrous capsule surrounds the common carpometacarpal joint; various capsular thickenings can be identified. Synovial membrane lines the capsule and all non-articular surfaces, attaching to the articular margins. The joint cavity extends proximally between the carpal bones, and usually communicates with the midcarpal joint (p. 153). Distally the joint space extends between the bones of the medial four metacarpals.

Ligaments

The dorsal and palmar carpometacarpal ligaments are little more than thickenings of the joint capsule.

Dorsal carpometacarpal ligaments

These present as a series of bands of fibres which pass from the distal row of carpal bones to the bases of the metacarpals; each metacarpal generally receives two bands. Those to the second metacarpal come from the trapezium and trapezoid; those to the third from the trapezoid and capitate, and those to the fourth from the capitate and hamate. The base of the fifth metacarpal receives only a single band arising from the hamate.

Palmar carpometacarpal ligaments

The arrangement of the fibrous bands constituting the palmar carpometacarpal ligaments is similar to those for the dorsal ligaments, except that the base of the third

metacarpal receives three bands arising from the trapezoid, capitate and hamate (Fig. 2.115).

Interosseous ligament

A short interosseous ligament is usually present passing from the adjacent inferior angles of the capitate and hamate to the base of the third or fourth metacarpal or both (Fig. 2.116). Occasionally, the ligament divides the joint space into medial and lateral compartments.

Blood and nerve supply

The arterial supply to the joint is from the palmar and dorsal carpal networks, while the nerve supply is by twigs from

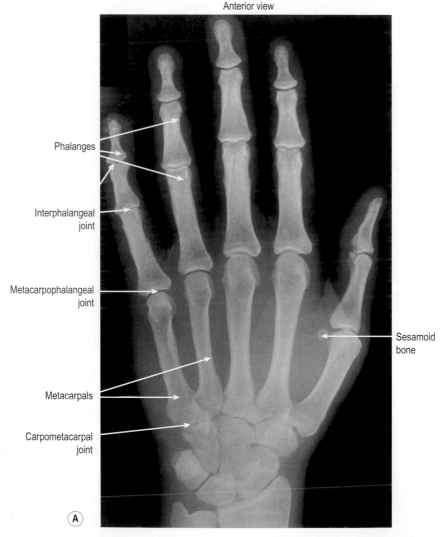

Anterior view

Phalanges

Interphalangeal joint

Metacarpophalangeal joint

Sesamoid bone

Metacarpals

Carpometacarpal joint

(A)

Figure 2.113 (A) Radiograph of the hand,

Continued

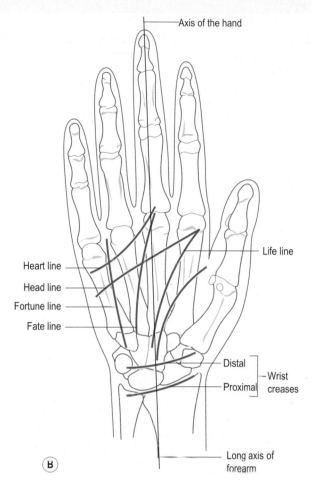

Figure 2.113, Cont'd (B) relation of bony elements to surface features.

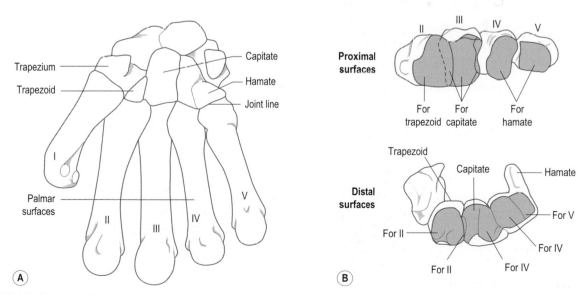

Figure 2.114 (A) Relationship of the metacarpals (I–V) with the distal row of carpal bones, (B) articular surfaces of the common carpometacarpal joint.

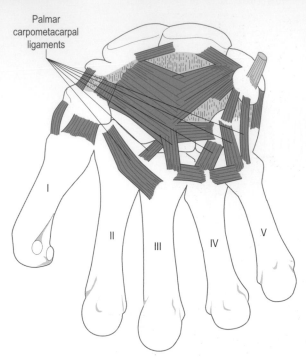

Figure 2.115 Ligaments of the common carpometacarpal joint, palmar view.

the anterior and posterior interosseous nerves and the deep and dorsal branches of the ulnar nerve, root value C7 and 8.

Relations

The carpometacarpal joints of the fingers lie deep to the tendons of flexors digitorum superficialis and profundus. Most laterally on the palmar surface, the tendon of flexor carpi radialis crosses the joint to insert into the base of the second metacarpal, while the tendon of flexor carpi ulnaris passes most medially. Also overlying the joint medially are the muscles of the hypothenar eminence.

On the posterior aspect of the joint are the tendons of the extensor muscles as they pass into the hand. From lateral to medial these are: extensors carpi radialis longus and brevis, extensor pollicis longus, extensor indicis, extensors digitorum and digiti minimi, and extensor carpi ulnaris.

Stability

The joint is extremely stable, providing a firm base between the joints of the wrist and the hand.

Movements

There is little movement at the carpometacarpal joints of the fingers. The second and third metacarpals are essentially immobile, while a slight gliding may occur between

the fourth metacarpal and the hamate. Only the fifth metacarpal shows any appreciable movement as it glides on the hamate; this is because of the bevelled joint surfaces. The movement that occurs is flexion, and is seen during a tight grasp and in opposition of the thumb to the little finger. In addition, there is also a slight rotation during opposition due to the action of opponens digiti minimi.

Accessory movements

A slight degree of anteroposterior gliding can be produced between the base of the metacarpal and the adjacent carpal bone if the appropriate pressure is applied.

The intermetacarpal joints

These are plane synovial joints between the adjacent sides of the bases of the second and third, third and fourth, and fourth and fifth metacarpals. The joints are closed anteriorly, posteriorly and distally by palmar and dorsal metacarpal ligaments, and interosseous ligaments respectively, that pass transversely between the adjacent bones (Fig. 2.116). The joint spaces are continuous with the common carpometacarpal joint distally. The blood and nerve supply to these joints is similar to that for the common carpometacarpal joint. Movements at these joints accompany movements of the metacarpals against the distal row of carpal bones. In accessory movements, a small amount of anteroposterior gliding can be produced between any two metacarpal bases by appropriately applied pressure.

The joints of the thumb

The thumb is an extremely mobile and specialized digit, both of which are important prerequisites for the movement of opposition, and for the normal prehensile functioning of the hand.

The carpometacarpal joint

Although extremely mobile, the carpometacarpal joint of the thumb nevertheless provides a stable base from which it can work effectively and efficiently. It plays a vital role in movements of the thumb allowing movement in three dimensions.

Articular surfaces

The joint is between the trapezium and the base of the first metacarpal (Fig. 2.117A). It provides the best example in the body of a saddle type of synovial joint, with the two surfaces being reciprocally concavoconvex and covered with hyaline cartilage.

The articular surface of the trapezium is concave more or less in an anteroposterior direction, and convex perpendicular to this. The base of the first metacarpal has reciprocal curvatures. The concavities and convexities of the surfaces

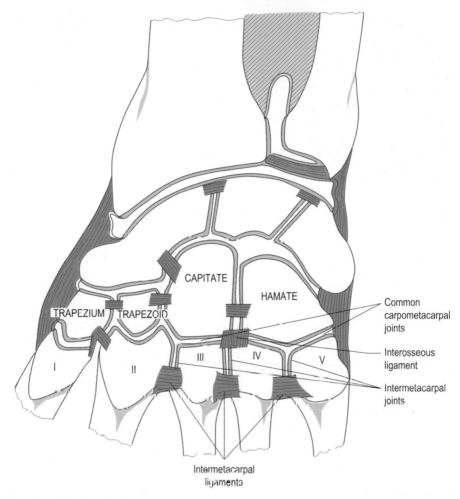

Figure 2.116 Coronal section through the carpus.

do not lie strictly within the transverse and anteroposterior planes, as will become evident when the axes about which movement occurs are considered.

Joint capsule and synovial membrane

A loose but strong fibrous capsule completely encloses the joint, attaching to the articular margins of both bones (Fig. 2.117B). It is lined with synovial membrane. The capsule is thickened laterally by the radial carpometacarpal ligament, and anteriorly and posteriorly by anterior and posterior oblique ligaments.

Ligaments

Radial carpometacarpal ligament

This passes between the adjacent lateral surfaces of the trapezium and first metacarpal (Fig. 2.117B).

Anterior and posterior oblique ligaments

These pass from their respective surfaces of the trapezium to the medial side of the first metacarpal, converging as they do so (Fig. 2.117B). The posterior oblique ligament becomes taut in flexion of the thumb, and the anterior during extension.

Blood and nerve supply

The arterial supply to the joint is by branches from the palmar and dorsal carpal networks, while its nerve supply is by twigs from the anterior and posterior interosseous nerves, root value C7 and 8.

Relations

The carpometacarpal joint of the thumb lies deep to the thenar muscles, which cover its anterior aspect. The tendon of flexor pollicis longus lies medial, while those of extensors pollicis longus and brevis lie laterally.

165

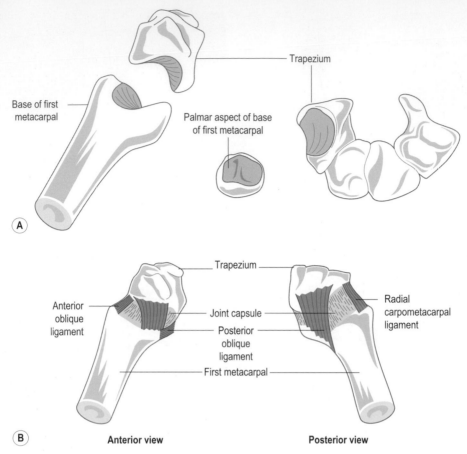

Figure 2.117 (A) Articular surfaces of the carpometacarpal joint of the thumb, (B) ligaments of the carpometacarpal joint of the left thumb.

Stability

The joint is principally stabilized and the surfaces kept in opposition by the tone of the muscles whose tendons cross the joint. The shape of the articular surfaces and the looseness of the fibrous capsule allow the joint its large degree of mobility, and hence play only a minor role in its stability.

Movements

Because the thumb is rotated approximately 90° with respect to the plane of the hand, the terminology used to describe its movements appears at first to be confusing. The terms flexion, extension, abduction and adduction are used as if the thumb were in line with the fingers. That this is not so can be clearly seen when observing your own hand; the thumbnail faces almost laterally while the finger nails face posteriorly. Thus flexion and extension occur in a coronal plane, as does abduction and adduction of the remaining four digits; and abduction and adduction occur in a sagittal plane, as do flexion and extension of the other digits. Due

to the nature of the joint surfaces and the looseness of the capsule, a certain amount of rotation is also possible at the joint. Because movements at the joint are brought about mainly by muscles whose tendons lie parallel to the metacarpal, compression across the two opposing surfaces always accompanies the movements. Consequently, the joint surfaces tend to grind against each other, instead of there being simple rolling or gliding.

Flexion and extension

This occurs in the plane of the palm of the hand, so that in flexion the thumb moves medially and in extension it moves laterally (Fig. 2.118). The axis about which the movement occurs passes through the base of the metacarpal at the centres of curvature of the concave trapezium and the convex metacarpal. It does not lie exactly in an anteroposterior plane but is set slightly obliquely from posterior, lateral and proximal to anterior, medial and distal.

The total range of flexion and extension is between 40° and 50°. Towards the end of full flexion, tension developed

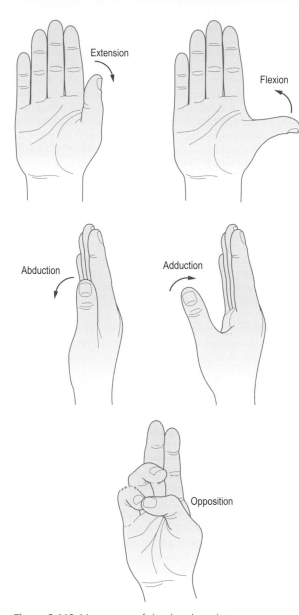

Figure 2.118 Movements of the thumb at the carpometacarpal joint.

in the posterior oblique ligament results in a medial rotation of the metacarpal causing the palmar surface of the thumb to face posteriorly. Conversely, towards the end of full extension, lateral rotation of the metacarpal occurs because of tension developed in the anterior oblique ligament. Flexion at the joint occurs as a secondary movement of flexors pollicis longus and brevis acting on the interphalangeal and metacarpophalangeal joints respectively. Similarly extension is due to the actions of extensors pollicis longus and brevis as they act on the interphalangeal and metacarpophalangeal joints respectively.

Abduction and adduction

Both occur at right angles to the palm so that in abduction the thumb is carried forwards away from the palm and in adduction it is moved back towards the palm (Fig. 2.118). The axis about which movement occurs is perpendicular to that for flexion and extension. It passes through the trapezium at the centres of curvature of the concave metacarpal and convex trapezium, and runs slightly obliquely from medial, posterior and distal to lateral, anterior and proximal.

The range of abduction and adduction is about 80°, with adduction being brought about by adductor pollicis pulling on the proximal phalanx. Abduction is brought about by the direct action of abductor pollicis longus on the joint, and by the secondary action of abductor pollicis brevis acting on the metacarpophalangeal joint.

Opposition

This is described as a movement in which the distal pad of the thumb is brought against the distal pad of any of the remaining four digits (Fig. 2.118). It is an essential movement of the hand; its loss markedly reduces the functional capacity of the hand. The movement is complex involving flexion, abduction and rotation followed by adduction at the carpometacarpal joint, as well as movements at other joints of the thumb.

Essentially, opposition consists of three elementary movements. Initially, flexion and abduction of the thumb occur simultaneously at the carpometacarpal joint due to the action of flexors pollicis longus and brevis and abductor pollicis longus. This produces a certain amount of passive axial rotation of the metacarpal, possible because of the looseness of the joint capsule, with the rotation directed medially due to the posterior oblique ligament becoming taut. At some point during this movement, opponens pollicis contracts to produce active rotation of the metacarpal: this is the second elementary movement. Finally, adduction occurs at the carpometacarpal joint, produced by adductor pollicis, to bring the metacarpal back towards the plane of the palm of the hand.

Movements of the thumb at the metacarpophalangeal joint contribute significantly to the overall movement of opposition. At the same time as the carpometacarpal joint is being flexed and abducted, so is the metacarpophalangeal joint. Again the simultaneous movements of flexion and abduction of the proximal phalanx bring about a degree of axial rotation at this joint. Consequently, the pad of the thumb comes to face posteromedially.

Returning the thumb to the anatomical position has no specific name. It is brought about mainly by the contraction of the extensor muscles of the thumb. Perhaps it should therefore be referred to as exposition.

The metacarpophalangeal joint

This is of similar design to those of the fingers, being a synovial condyloid joint between the head of the first metacarpal and the base of the proximal phalanx.

Articular surfaces

The articulation is between the rounded head of the metacarpal and the shallow, oval concavity of the base of the proximal phalanx (Fig. 2.119). Both surfaces are covered

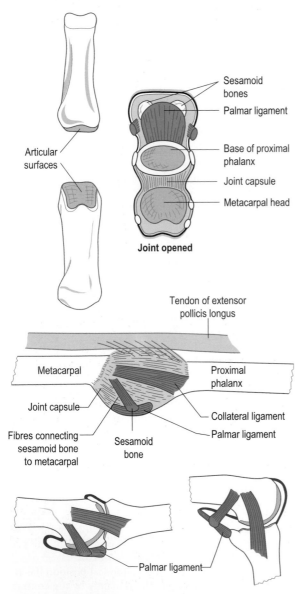

Figure 2.119 Articular surfaces and ligaments of the metacarpophalangeal joint of the thumb.

with hyaline cartilage. The biconvex metacarpal head is wider anteriorly than posteriorly; indeed the articular surface does not extend very far on its posterior surface. In addition, the curvature of the metacarpal head is greater in the transverse than in the anteroposterior plane. The base of the proximal phalanx has a much smaller articular area than the metacarpal head. It is increased anteriorly, however, by the presence of a fibrocartilaginous plate known as the palmar ligament (Fig. 2.119), which is attached to the anterior surface of the base of the phalanx by a small fibrous band which functions like a hinge.

Joint capsule and synovial membrane

A loose fibrous capsule surrounds the joint, being attached closer to the articular margins posteriorly than anteriorly (Fig. 2.119). It is strengthened at the sides by collateral ligaments. Anteriorly the capsule is mainly replaced by the palmar ligament, which has a weak attachment to the neck of the metacarpal. Posteriorly the capsule is strengthened or entirely replaced by the expansion of extensor pollicis longus.

Synovial membrane lines all non-articular surfaces of the joint and presents anterior and posterior recesses when the joint is extended.

Ligaments

Palmar ligament

A dense fibrocartilaginous pad that increases the phalangeal articular surface anteriorly. It is firmly attached to the anterior surface of the base of the proximal phalanx, and loosely attached to the anterior aspect of the neck of the metacarpal (Fig. 2.119). The collateral ligaments of the joint blend with the sides of the palmar ligament. The palmar ligament contains two small sesamoid bones which are attached to the phalanx and the metacarpal by straight and cruciate fibres. It is grooved on its anterior aspect by the tendon of the flexor pollicis longus.

Collateral ligaments

These strong ligaments on either side of the joint are attached proximally to the tubercle and adjacent depression on the side of the metacarpal head, and pass to the palmar aspect of the side of the base of the proximal phalanx (Fig. 2.119). Although cord-like in appearance, they fan out slightly from proximal to distal, gaining attachment to the margins of the palmar ligament. The collateral ligaments are relatively lax during extension, becoming increasingly taut with flexion of the joint.

Blood and nerve supply

The arterial supply to the joint is by branches from the princeps pollicis artery, while its nerve supply is by twigs from the median nerve, root value C7.

Stability

The metacarpophalangeal joint of the thumb is stabilized by the collateral ligaments, as well as the tendons of flexor and extensor pollicis longus as they pass in front of and behind the joint to their insertion on the distal phalanx. Flexor and extensor pollicis brevis and abductor pollicis brevis also cross the joint to insert into the base of the proximal phalanx.

Movements

Being a condyloid joint, the metacarpophalangeal joint has, according to its shape, two degrees of freedom of movement, these being flexion and extension, and abduction and adduction. However, as in the carpometacarpal joint, there is a third movement of axial rotation which occurs passively due to the small degree of elasticity of the associated ligaments.

Flexion and extension

At the metacarpophalangeal joint this occurs about a single, fixed axis which passes transversely through the metacarpal at approximately nine-tenths of its midline length from its base. In passing from flexion to extension, the area of contact shifts from the palmar aspect of the phalangeal base to its distal end.

Flexion has a range of 45°, while extension is zero under normal circumstances, both actively and passively (Fig. 2.120). Only in full extension does the anterior part of the metacarpal head articulate with the palmar ligament. As flexion proceeds, the palmar ligament gradually loses contact with the metacarpal head. At the same time the synovial recesses progressively become unfolded (Fig. 2.119).

Flexion at the joint is brought about primarily by flexor pollicis brevis, aided by flexor pollicis longus. Similarly, extension from the flexed position is due primarily to the extensor pollicis brevis, with some help from extensor pollicis longus.

Abduction and adduction

These are limited due to the width of the metacarpal head. The 15° of abduction and negligible adduction occurs about an anteroposterior axis through the head of the metacarpal. As well as the bony limitations to the movements, the collateral ligaments also become taut, adding to the restriction.

Abduction is caused by contraction of abductor pollicis brevis. Although adductor pollicis attaches to the base of the proximal phalanx, because of the severe limitation of adduction at the metacarpophalangeal joint, its action is seen principally at the carpometacarpal joint.

Some degree of *axial rotation* is possible at the metacarpophalangeal joint of the thumb, which is of importance during opposition (Fig. 2.120). The movement can be produced actively by the co-contraction of flexor and abductor pollicis brevis, or passively as when pressing the thumb

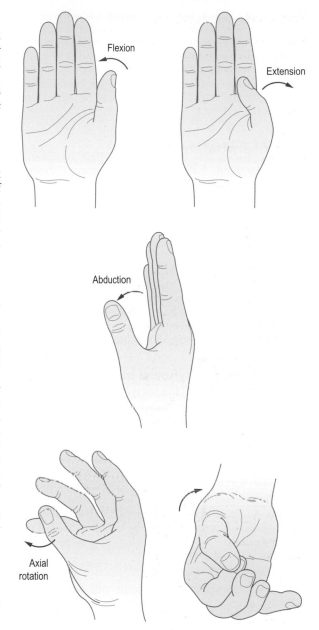

Figure 2.120 Movements at the metacarpophalangeal joint of the thumb.

against the index finger. The actively produced rotation is always medially directed, while that produced passively can be in either direction depending on which side of the thumb comes into contact with the finger.

Movements at the joint during opposition

These include a secondary flexion at the metacarpophalangeal joint following that at the carpometacarpal joint

169

during opposition. At the same time there is abduction at the joint, which continues after the metacarpal becomes adducted. The degree of abduction is greatest when contact is made with the pad of the little finger. The flexion and abduction movements at the metacarpophalangeal joint initially cause an active axial rotation at the joint. Following contact, the degree of rotation may be augmented passively.

The interphalangeal joint

Because the thumb only contains two phalanges, there is only one interphalangeal joint. Like those of the fingers, however, it is a synovial hinge joint permitting movement in one direction only.

Articular surfaces

The articulation is between the pulley-shaped head of the proximal phalanx and the base of the distal phalanx, which has a median ridge separating two shallow facets (Fig. 2.121). As in the metacarpophalangeal joint, a fibrocartilaginous plate (palmar ligament) is attached to the anterior margin of the base of the distal phalanx.

Joint capsule and synovial membrane

A fibrous capsule completely surrounds the joint, being replaced by the palmar ligament anteriorly and strengthened at the sides by the collateral ligaments (Fig. 2.121).

Ligaments

Collateral ligaments

These are attached to the sides of the head of the proximal phalanx, and pass to the palmar aspect of the base of the distal phalanx. They blend with the lateral margins of the palmar ligament.

Palmar ligament

As in the metacarpophalangeal joint, this ligament is a fibrocartilaginous plate attached to the anterior margin of the base of the distal phalanx. It is loosely attached to the front of the neck of the proximal phalanx via the joint capsule.

Blood and nerve supply

The blood and nerve supply is the same as that for the metacarpophalangeal joint (p. 168).

Movements

Being a hinge joint supported by strong collateral ligaments, movement is allowed only in one plane (Fig. 2.121B).

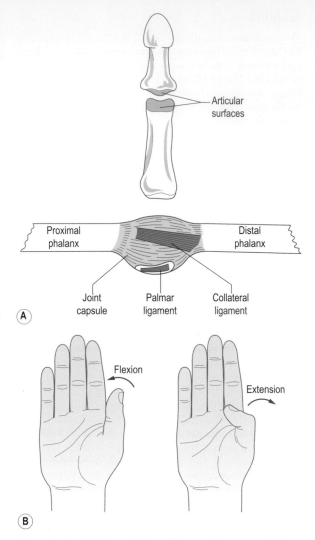

Figure 2.121 (A) The articular surfaces and ligaments of the interphalangeal joint of the thumb, (B) movements of the thumb.

Flexion and extension

These occur about a transverse axis passing approximately through the neck of the proximal phalanx. The range of flexion is in excess of 90°, while extension is normally no more than 10°. However, passive hyperextension may be marked in some individuals who apply large forces using the thumb, for example butchers and physiotherapy manipulators.

Accessory movements of joints of the thumb

Carpometacarpal joint

Gripping the trapezium between the thumb and index finger of one hand, and the base of the first metacarpal with the other, the metacarpal base can be moved in both anteroposterior and mediolateral directions. With the same grip longitudinal gapping and rotation can be achieved.

Metacarpophalangeal and interphalangeal joints

If the principle of stabilizing the proximal bone and moving the distal one is employed, then anteroposterior gliding movements can be demonstrated at both the metacarpophalangeal and interphalangeal joints. Once again, best results are obtained when each bone is held firmly between the thumb and index finger. A good range of rotation, as well as longitudinal gapping, is also possible at the metacarpophalangeal joint.

Joints of the fingers

In general, the fingers act in one plane to close around an object and so form a pincer action with the opposed thumb. The size of the object being grasped determines whether, and to what extent, two-dimensional movement at the metacarpophalangeal joints occurs.

The articulation of the second to fifth metacarpals with the distal row of carpal bones has already been considered (p. 161); consequently, this section considers the metacarpophalangeal joints and the two interphalangeal joints of each finger.

The metacarpophalangeal joint

The metacarpophalangeal joint of the fingers is structurally and functionally similar to that of the thumb.

Articular surfaces

The articular surface of the metacarpal head is biconvex, with unequal curvatures transversely and anteroposteriorly. It is broader anteriorly than posteriorly, with the hyaline cartilage extending further proximally on its anterior aspect (Fig. 2.122).

The base of the proximal phalanx is biconcave, but has a smaller articular surface than the metacarpal head (Fig. 2.122A). The surface area is increased by the presence of the palmar ligament attached to the anterior margin of the articular surface.

Joint capsule and synovial membrane

The fibrous capsule surrounding the joint is loose, and is attached closer to the articular margins on the posterior aspects of the bone than anteriorly. The capsule is strengthened on each side by collateral ligaments and replaced anteriorly by the palmar ligament. Posteriorly the extensor hood of the long extensor tendon replaces the capsule, blending at the sides with the collateral ligaments. The posterior part of the capsule also receives fibres from the distal slips of the palmar aponeurosis (Fig. 2.122).

The capsule is lined by the synovial membrane, which also covers all non-articular surfaces. Synovial-lined anterior and posterior recesses of the capsule permit freedom of movement, particularly during flexion.

Ligaments

In addition to the collateral and palmar ligaments associated with each metacarpophalangeal joint, the heads of the second to fifth metacarpals are united by the deep transverse metacarpal ligaments.

Collateral ligaments

These pass from the tubercle and adjacent depression on the side of the head of the metacarpal to the palmar aspect of the side of the base of the proximal phalanx (Fig. 2.122A). They are strong and tend to fan out in passing from metacarpal to phalanx. Anteriorly they blend with the palmar ligament, while posteriorly the extensor expansion joins them.

Palmar ligament

This is a dense fibrocartilaginous plate firmly attached to the anterior margin of the base of the proximal phalanx (Fig. 2.122A). Proximally it is loosely attached to the neck of the metacarpal by the joint capsule. On each side it receives fibres from the collateral ligaments. The palmar ligament acts as a mobile articular surface faciliating flexion at the joint.

Deep transverse metacarpal ligaments

These are a series of short ligaments connecting the palmar ligaments of the four metacarpophalangeal joints of the fingers (Fig. 2.122B). They are continuous with the palmar interosseous fascia and blend with the fibrous flexor sheaths. Consequently, they act to bind the heads of the four medial metacarpals together, and so limit their movement apart. (There is no ligament between the first and second metacarpal, hence the independence and freedom of movement of the thumb.) The deep transverse metacarpal ligaments also receive fibres from the distal slips of the palmar aponeurosis, as well as part of the extensor expansion as it passes forwards on each side of the head of the metacarpal.

Passing posterior to the deep transverse metacarpal ligaments are the tendons of both the dorsal and palmar interossei, while the tendons of the lumbricals pass anteriorly.

Blood and nerve supply

The arterial supply to the joints is by branches from the adjacent digital arteries, while the nerve supply is by twigs from the median, and possibly the radial nerve for the

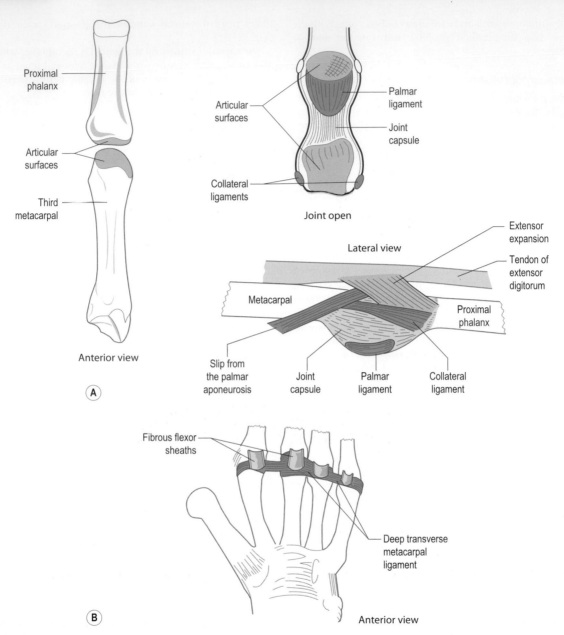

Figure 2.122 (A) Articular surfaces and ligaments of the metacarpophalangeal joints of the fingers, (B) deep transverse metacarpal ligament.

index and middle finger, and the ulnar nerve for the ring and little fingers. Root value of the nerve supply is C7.

Relations

On the posterior aspect of the joint is the expansion of the long extensor tendon (Fig. 2.123A), part of which passes around the sides of the metacarpal head blending with

the deep transverse metacarpal ligament. The tendon of the lumbrical muscle passes lateral to the joint (Fig. 2.123A,C), anterior to the deep transverse metacarpal ligament, before attaching to the base of the proximal phalanx and the dorsal digital expansion. Exactly which interosseous tendons pass medial and lateral to each metacarpophalangeal joint depends on which finger is being considered:

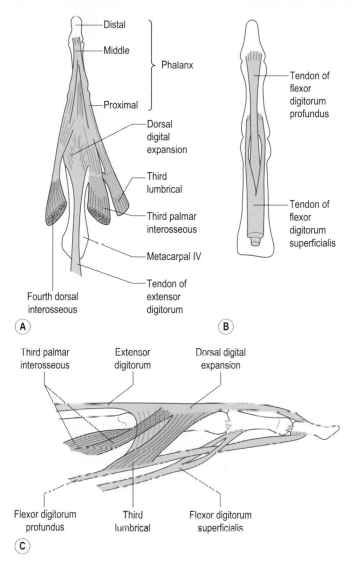

Figure 2.123 Relations of the metacarpophalangeal joint of the left ring finger: (A) posterior, (B) anterior, (C) lateral.

1. for the *index finger*, the first dorsal interosseous is lateral and the first palmar interosseous is medial
2. for the *middle finger*, the second and third dorsal interossei are lateral and medial respectively
3. for the *ring finger*, the second palmar interosseous is lateral and the fourth dorsal interosseous is medial (Fig. 2.123A)
4. for the *little finger*, the third palmar interosseous is lateral and the tendon of the abductor digiti minimi is medial.

Immediately anterior to the joint is the tendon of flexor digitorum profundus, and anterior to that, the tendon of flexor digitorum superficialis, which at the level of the joint splits into two (Fig. 2.123B). Flexor digiti minimi brevis is situated on the anterolateral aspect of the joint of the little finger. Digital branches from the dorsal and palmar metacarpal arteries, together with digital branches from the median, ulnar and radial nerves, depending on the finger, pass either side of the metacarpophalangeal joint.

Stability

The metacarpophalangeal joint is stabilized primarily by the long flexor and extensor tendons crossing the joint, as well as by the lumbricals and interossei. Dislocations of the joint do occur; however, they can often be reduced by manipulation.

Movements

Active movement at the metacarpophalangeal joint takes place about two axes, each of which is located in the metacarpal head approximately nine-tenths of the midline length of the metacarpal from its base. These are flexion and extension, and abduction and adduction; passive axial rotation can be added to these movements.

Flexion and extension

These occur about a transverse axis through the head of the metacarpal (Fig. 2.124A). The geometry of the articular surfaces dictates that the intersection of the longitudinal axes of the proximal phalanx and the metacarpal moves distally during flexion. In extension the anterior surface of the metacarpal head articulates with the palmar ligament (Fig. 2.124A). During flexion the ligament moves past the metacarpal head, turning upon itself to glide along the palmar surface of the shaft (Fig. 2.124A). As this is occurring, the capsule and its synovial lining unfold so as not to limit movement prematurely. The range of flexion is slightly less than 90° for the index finger, but progressively increases towards the little finger. Flexion of one joint in isolation is limited by tension developed in the deep transverse metacarpal ligaments; flexion is ultimately resisted by tension in the collateral ligaments. The range of active extension is variable between subjects but may reach 50°. Passive extension may reach as much as 90° in individuals with lax ligaments. A study by Youm et al. (1978)[2] of over 400 hands gave the following total ranges of active/passive flexion and extension for each of the fingers: index, 148°/155°; middle, 145°/151°; ring, 149°/159° and little, 152°/172°.

Flexion at the metacarpophalangeal joint is brought about primarily by the lumbrical muscles, aided by the tendons of flexors digitorum profundus and superficialis, as well as the interossei. In the little finger, flexor and abductor digiti minimi also contribute to the movement. Extension is achieved at all of the metacarpophalangeal joints by extensor digitorum, with the addition of extensor indicis in the index finger and extensor digiti minimi in the little finger.

Abduction and adduction

These occur about an anteroposterior axis through the metacarpal head; the movement occurs away from or towards the middle finger respectively (Fig. 2.124A). The movement is easier and has a greater range when the finger is extended, being as much as 30° in each direction (Fig. 2.124B). Tension developed in the collateral ligaments in flexion of the joint severely limits the side-to-side movement, so much so that at 90° flexion the total range may be no more than 10°. The total range of active/passive abduction and adduction movements, with the fingers in a neutral position, has been reported by Youm et al. (1978)[2] as follows: index, 50°/62°; middle, 40°/53°; ring, 38°/55° and little, 57°/68°.

Abduction at the joint is brought about by the dorsal interossei for the index, middle and ring fingers, and by abductor digiti minimi for the little finger. At the index and middle fingers, the movement may be assisted by the first and second lumbricals respectively via their attachment to the extensor hood. If the joint is hyperextended, then extensor digitorum will also aid abduction. Adduction towards the middle finger is achieved by the palmar interossei and can be assisted by the third and fourth lumbricals for the ring and little fingers. If the joint is being flexed simultaneously, then adduction is assisted by flexors digitorum superficialis and profundus.

Active rotation

This is not possible except in the little finger; however, because of the shape of the joint surfaces and the relative laxity of the associated ligaments, some degree of passive rotation can occur (Fig. 2.124A) having a maximum range of 60°. In the index finger, the range of medial rotation is of the order of 45°, while lateral rotation is negligible. The range of medial and lateral rotation in the remaining fingers is approximately equal.

The interphalangeal joints

Because each finger consists of three phalanges it contains two interphalangeal joints: a proximal joint between the head of the proximal and base of the middle phalanx, and a distal joint between the head of the middle and base of the distal phalanx. All of the joints are hinge joints permitting flexion and extension only, with the articular surfaces covered by hyaline cartilage.

Articular surfaces

The articulation is between the pulley-shaped head of the phalanx and two shallow facets separated by a ridge on the base of the immediately distal phalanx (Fig. 2.125). The groove and ridge on the head and base respectively do not lie exactly in an anteroposterior direction, except for the joints of the index finger. In all other joints they run slightly obliquely from posterolateral to anteromedial, with the obliquity increasing from the middle to the little finger.

The articular surface of the phalangeal head is greater than that on the adjacent base, extending further distally on its anterior aspect. The head is also wider anteriorly than posteriorly. A fibrocartilaginous plate, the palmar ligament, similar to that associated with the metacarpophalangeal joint, acts as a mobile articular surface.

[2]Youm Y, Gillespie TE, Flatt AE, Sprague BL (1978) Kinematic investigation of normal MCP joint. *Journal of Biomechanics*, 11, 109–118.

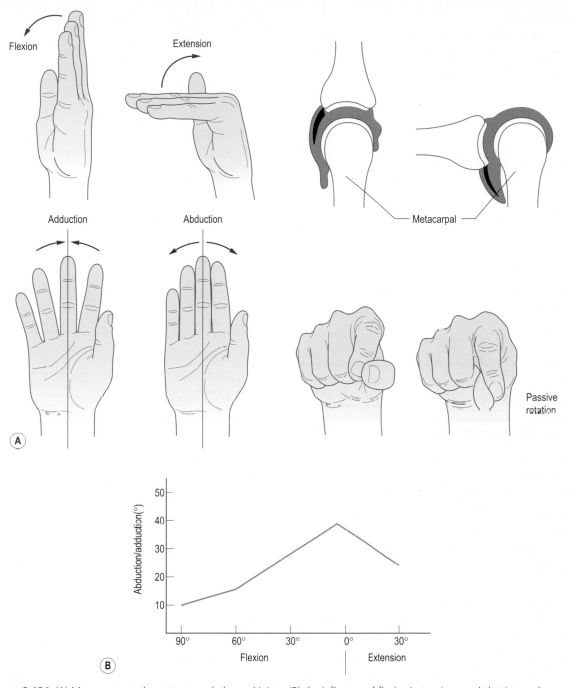

Figure 2.124 (A) Movements at the metacarpophalangeal joints, (B) the influence of flexion/extension on abduction and adduction at the joint. *(Adapted from Youm Y, Gillespie TE, FLatt AE, Sprague BL (1978) Kinematic investigation of normal MCP joint. Journal of Biomechanics, 11, 109–118.)*

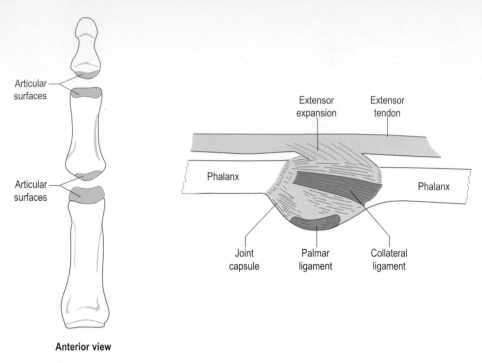

Anterior view

Figure 2.125 Articular surfaces and ligaments of the interphalangeal joints.

Joint capsule and synovial membrane

A loose fibrous capsule surrounds the joint, being strengthened at the sides by collateral ligaments, and partly replaced anteriorly and posteriorly by the palmar ligament and extensor expansion respectively. Synovial membrane lines all non-articular surfaces, including the anterior and posterior recesses of the capsule.

Ligaments

Collateral ligaments

These attach to the sides of the head of the most proximal phalanx and the sides of the base of the adjacent more distal phalanx, blending with the margins of the palmar ligament (Fig. 2.125). They tend not to be as obliquely orientated as the collateral ligaments of the metacarpophalangeal joints. They become increasingly tense with flexion at the joint.

Palmar ligament

This is a mobile fibrocartilaginous plate attached to the anterior margin of the base of the adjacent phalanx. It is loosely attached by the joint capsule to the front of the neck of the immediately preceding phalanx (Fig. 2.125). Also attached to the palmar ligament is the fibrous flexor sheath of the digit.

Blood and nerve supply

The arterial supply to the joints is by branches from the digital arteries running along the sides of each finger. Anteriorly the digital arteries arise from the palmar metacarpal arteries, while posteriorly they come from the dorsal metacarpal arteries. Venous drainage is by similarly named vessels, eventually draining into the venae comitantes associated with the radial and ulnar arteries. Posteriorly some venous drainage passes to the dorsal venous plexus on the dorsum of the hand, and thence into the basilic and cephalic veins. Lymphatic drainage from the joints is by vessels which follow the arteries, with the majority of lymph draining to the lateral group of axillary nodes, although some may pass to cubital or brachial nodes.

The nerve supply to each joint is by twigs from adjacent digital nerves, and has a root value of C7. For the index, middle and lateral side of the ring fingers, the digital nerves are branches of the median nerve anteriorly and the radial nerve posteriorly. For the medial side of the ring and the little fingers, the digital nerves all arise from the ulnar nerve.

Relations

On the anterior aspect of the proximal interphalangeal joint, enclosed within the fibrous flexor sheath, are the tendons of flexors digitorum superficialis and profundus (Fig. 2.126B(ii)); only the tendon of profundus lies in front of the distal interphalangeal joint (Fig. 2.126B(i)).

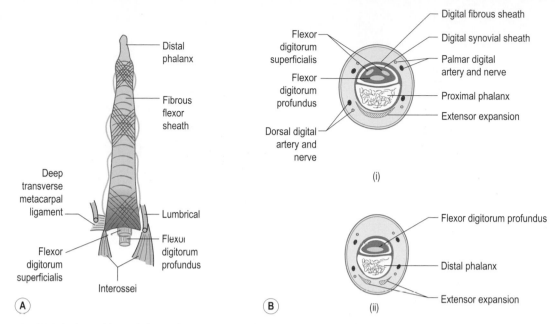

Figure 2.126 Relations of interphalangeal joints viewed (A) anteriorly and (B) in transverse sections.

The fibrous flexor sheaths are relatively thin and loose over the interphalangeal joints, with a cruciate arrangement of fibres as they pass from the side of one phalanx to the opposite side of the preceding phalanx (Fig. 2.126A). Immediately beyond the distal interphalangeal joint, the flexor sheath attaches to the palmar surface of the distal phalanx.

Posterior to the proximal interphalangeal joint is the central slip of the dorsal digital expansion. On the back of the middle phalanx the two collateral slips of the expansion come together, so that a single tendon crosses the posterior aspect of the distal interphalangeal joint.

Stability

The interphalangeal joints are fairly stable because of the presence of the long flexor and extensor tendons. Nevertheless dislocations can and do occur; they can often be reduced by manipulation.

Movements

Because of the nature of the joint surfaces, the only active movements possible at the interphalangeal joints are flexion and extension (Fig. 2.127A). However, a small degree of passive side-to-side movement is possible, particularly at the distal interphalangeal joint.

Flexion and extension

These take place about a transverse axis, which for the middle, ring and little fingers runs with increasing obliquity from lateral and distal to medial and proximal (Fig. 2.127B). The axis is approximately perpendicular to the groove on the phalangeal head, so that when the medial fingers are flexed at the interphalangeal joints the movement does not occur in a sagittal plane, but enables these fingers to oppose the thumb more easily. Flexion of the index finger, however, occurs in a sagittal plane.

The range of flexion at the proximal interphalangeal joint is greater than 90° for all fingers, and gradually increases towards the little finger so that this is capable of flexing 135°. At the distal interphalangeal joint, the range of flexion for the little finger is 90°, and gradually decreases towards the index finger. Active extension at the interphalangeal joints is minimal, being no more than 5° at the distal and only 1° or 2° at the proximal interphalangeal joints. Passive extension may be considerably greater.

Flexion at the proximal interphalangeal joint is primarily due to the action of flexor digitorum superficialis, assisted by the flexor digitorum profundus. Only profundus flexes the distal interphalangeal joint. Extension of the interphalangeal joints is produced by contraction of the lumbricals and interossei via their attachments to the dorsal digital expansion. They are assisted in each finger by extensor digitorum, and in the index and little fingers by extensors indicis and digiti minimi respectively.

Simultaneous flexion at one interphalangeal joint and extension at the other is produced by a controlled balance between the activity of the flexor and extensor muscles.

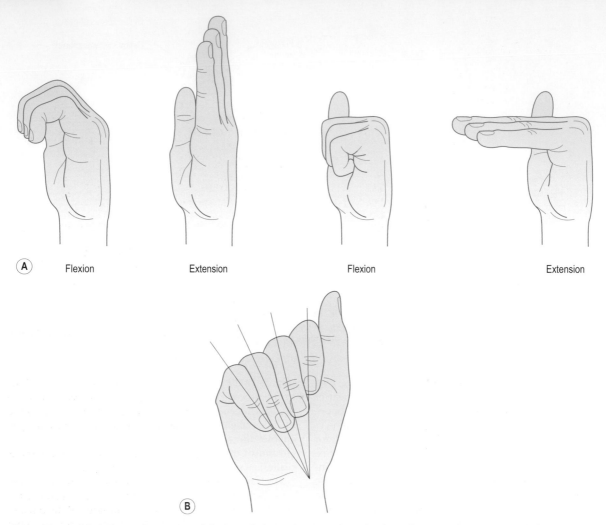

Figure 2.127 (A) Flexion and extension of the interphalangeal joints with and without flexion at the metacarpophalangeal joints, (B) the obliquity of the joint axes causes the fingers to converge toward the lateral side of the wrist in flexion.

Flexion of the wrist facilitates extension of the fingers and opening the fist. The functional position of the wrist (i.e. in extension) puts the finger flexors beyond their natural length and so enables greater tension to be developed in them, facilitating a powerful grip. Similarly, flexing the interphalangeal joints places the extensors of the wrist under increased tension. In stabilizing the wrist, a certain amount of flexor strength and extensor power is sacrificed. Only 70% of the strength of the finger flexors is available from flexion of the interphalangeal joints. Weakness of the wrist extensors, by failing to maintain the position of function, greatly interferes with the strength of the finger flexors and the ability to carry out a forceful closure of the fist.

Accessory movements of the joints of the fingers

Similar accessory movements of anteroposterior gliding and rotation are possible at each of the metacarpophalangeal and interphalangeal joints of all four fingers. With the proximal component stabilized between the thumb and index finger, the more distal segment can be made to execute the accessory movements.

Biomechanics

The arrangement of bones, tendons and ligaments within the hand is such that in the so-called position of rest the palm is hollowed, the fingers are flexed and the thumb is

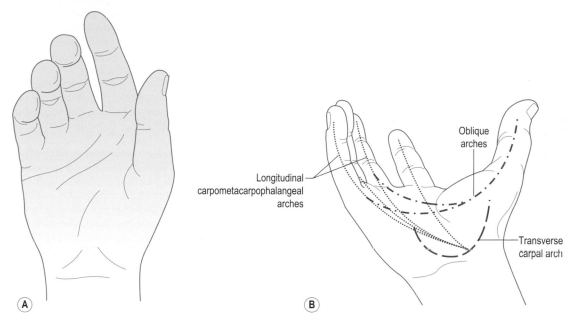

Figure 2.128 (A) The resting position of the hand, (B) the arch arrangements of the various bony elements of the wrist and hand.

slightly opposed (Fig. 2.128A). The flexion of the fingers increases progressively from the index to the little finger. Underlying the hollowing of the palm and facilitating gripping movements, the bony skeleton forms a series of arches running in three different directions (Fig. 2.128B). The transverse carpal arch is maintained by the flexor retinaculum. It continues distally as far as the heads of the metacarpals as the metacarpal arch. The long axis of this 'gutter' crosses the lunate, capitate and third metacarpal. The concavity of this arch at the level of the metacarpal heads is much shallower, and more widespread than at the level of the carpal bones. The shallowness of the arch at the metacarpal heads can be readily seen and appreciated in the relaxed hand. Running longitudinally are the carpometacarpophalangeal arches, which open out from the wrist and are formed for each digit by the corresponding metacarpal and phalanges. These arches are concave on their palmar surface, with the keystone of each being at the level of the metacarpophalangeal joint. Consequently, muscular imbalance at this point interferes with the concavity of the arch. Of these longitudinal arches, the two most important are those of the middle and index fingers. During opposition of the thumb to the fingers, oblique arches are formed running from the thumb into the finger being opposed. The most important oblique arch is that linking the thumb and index finger because of its use in holding objects, such as a pen.

When the palm is hollowed an oblique gutter, the palmar gutter, runs across the oblique arches formed with the thumb from the base of the hypothenar eminence, where the pisiform can be palpated, to the head of the second metacarpal. It corresponds approximately to the palmar crease ('life line') (see Fig. 2.113B), and is the direction taken by the handle of a tool when fully grasped by the hand.

In use, the hand does not always utilize the various arches. When carrying large and heavy flat objects the hand spreads out and becomes flattened, so that contact with the object is made at the thenar and hypothenar eminences, the metacarpal heads and the anterior surfaces of the phalanges. This provides a large area of contact and support, with movement of the object limited by friction between it and the skin.

The digits may be used individually as circumstances dictate. Being functionally separate from the remaining digits, the thumb can be moved and used independently. Although not completely separate, the index finger has a considerable degree of freedom, for example it can be used independently in pointing and gesturing. The relative freedom of the index finger is important in grasping. The remaining fingers cannot be used independently throughout their full range of movement principally because of the linkage between the tendons of extensor digitorum on their dorsal surfaces.

Prehension (grip)

The way in which the hand is used depends upon several factors, not least of which are the size, shape and weight of the object as well as the use to which it is being put.

In general terms, the grip can be classified as being either a 'precision grip' or a 'power grip'. The thumb and the fingers combine in various ways to produce the former, while the hand becomes involved in the latter.

Precision grips: In precision grips the object is usually small, and sometimes fragile. It is seized between the pads of the digits, which spread around the object conforming to its shape. The action involves rotation at the carpometacarpal joint of the thumb, and at the metacarpophalangeal joints of the thumb and finger(s) involved. The muscles involved in precision grips are all the small muscles of the hand, as well as flexors digitorum profundus and superficialis, and pollicis longus. Within this class of grip, several types can be identified:

Terminal opposition (pincer) grip (Fig. 2.129A) is where the tips of the pads or sometimes the edges of the nails are used to pick up fine objects, such as a pin: this is the finest and most precise of all precision grips, and consequently is the most easily upset in traumas of the hand.

Subterminal opposition in which the palmar surfaces of the thumb, index and possibly other fingers come into contact. The most common example is the tripod grip (Fig. 2.129B) in which the thumb, index and middle finger come into contact as when holding a pen. The *pinch grip* (Fig. 2.129C) involves the pads of the thumb and index finger when they oppose as in gripping a clothes peg or tweezers. The efficiency of the latter can be tested by attempting to pull a sheet of paper from between the thumb and index finger.

Subtermino-lateral opposition (key) grip (Fig. 2.129D) is where the pad of the thumb presses against the side of the index finger to hold an object such as a key or plate between them. It is less fine but strong and can replace other precision grips when the distal phalanx of the index finger has been lost.

Power grips: In power grips, where considerable force may be required, the hand is used in addition to the fingers. The flexors and extensors of the wrist work strongly to fix its position, while the long finger flexors and intrinsic muscles of the hand work to grip the object.

Oblique palmar grip (Fig. 2.125E) is the most powerful in which the whole hand and fingers wrap around the object, whose long axis lies obliquely across the palm. The thumb acts as a buttress and the fingers close around the object, the size of which determines the strength of the grip. The shapes of tool handles are sculpted to facilitate this grip, which is maximal when the thumb can still touch the index finger as in using a screwdriver.

Ball grip (Fig. 2.129F) In this grip the object, which is ball shaped, is enclosed by the palm, fingers and thumb. The fingers and thumb are increasingly abducted as the diameter of the object increases.

If the object is circular but flat, as the lid of a canister or jar, the palm may not make contact which will increase the power needed in the fingers and thumb in order to maintain grip whilst opening. This may be called a *span grip* (Fig. 2.129G).

Cylinder grip (Fig. 2.129H) is a grip where the object is held transversely across the palm, fingers and thumb. The cylindrical shape of the hand thus formed allows for variation in the size of the object and the power applied to it.

Hook grip (Fig. 2.129I) is where the object, such as a suitcase handle, is held across the palmar surfaces of the flexed fingers. The thumb plays no part in a true hook grip, but should the weight of the object increase the grip will change to a palmar grip. A hook grip is relatively secure but in one direction only, towards the fingers.

One activity in which both types of grip can be seen, albeit in different hands, is when hammering in a nail. The nail is held precisely between the thumb and index finger, while the hammer is firmly held using a palmar grip.

Joint forces

Because of the size and weight of many objects that are carried or manipulated, the forces transmitted across various joints within the hand and digits can reach considerable magnitudes. An indication of the magnitudes of these forces has been required in order to refine finger prostheses. Joint reactions at the carpometacarpal, metacarpophalangeal, proximal and distal interphalangeal joints are on average twice, three times, ten times and six times the applied load respectively. The magnitude and direction of action of these forces at various joints have led to predictions of clinical deformities and joint damage.

Pathology

Due to the delicate balance between the soft tissues of the hand, it is highly susceptible to trauma and disability. Well over half of all injuries leading to disability involve the hand and fingers. Various bones may be fractured, or joints dislocated, but the most common injuries are accidental amputation of all or part of the hand and finger(s), burns, tendon lacerations, penetrating wounds and nerve injuries.

The most common deformity in rheumatoid arthritis is induced by synovitis of the metacarpophalangeal joint. This allows a narrowing of the articular cartilage and attenuation of the collateral ligaments. Under the influence of these two effects, there is a palmar subluxation of the proximal phalanx on the metacarpal head with laxity of the flexor complex. Under the pull of the flexor tendons there may be further subluxation of the proximal phalanx associated with an ulnar deviation, particularly in the index and middle fingers.

Joint replacement

As with all joint replacements the primary object is the relief of pain and restoration of as full a range of movement as possible. However, when considering finger joint replacements the main problem appears to be the maintenance of the cortical bone. Stresses on the endosteal surface of the bone have a tendency to produce gradual deformity, and sometimes erosion so that the prosthesis protrudes through the shaft of the bone. The inserted foreign material may be a contributory factor in this erosion through disruption of the endosteal arterial supply.

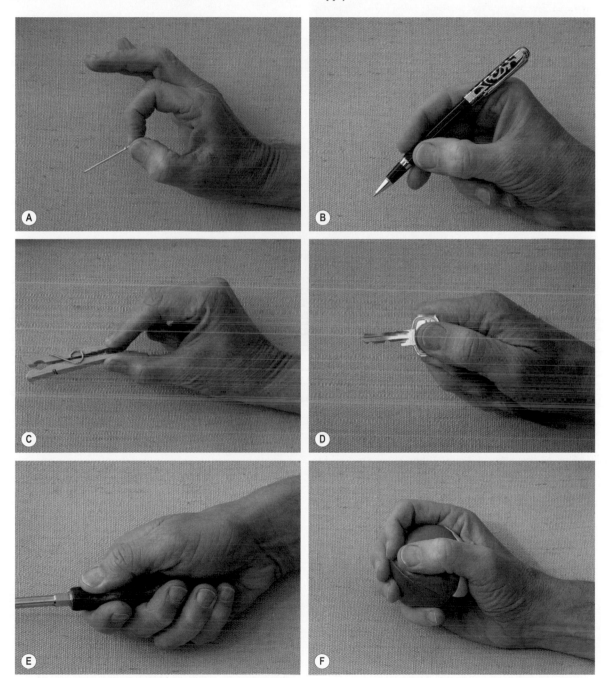

Figure 2.129 Types of grip: (A) pincer, (B) tripod, (C) pinch, (D) key, (E) oblique palmar, (F) ball,

Continued

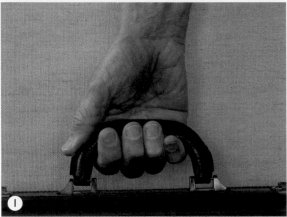

Figure 2.129, Cont'd   (G) span, (H) cylinder, (I) hook.

In normal bone, the tensile stresses applied to the outer surface of the bone by ligamentous and periosteal attachments make a significant contribution to maintaining the integrity of the cortical bone. When these stresses are replaced by prostheses that depend on transmission of stress by the endosteal surface, the bone is clearly reacting and remodelling to an entirely different pattern of stresses.

Section summary	
Joints of the hand	
Common carpometacarpal joint	
Type	Synovial plane
Articular surfaces	Distal surfaces of trapezium, trapezoid, capitate and hamate with bases of second, third, fourth and fifth metacarpals
Capsule	Complete and surrounds joint
Ligaments	Dorsal carpometacarpal
	Palmar metacarpal interosseous

Movements	Some flexion and rotation of the fifth metacarpal only
Carpometacarpal joint of thumb	
Type	Synovial saddle joint
Articular surfaces	Distal surface of trapezium with base of first metacarpal
Capsule	Loose but strong
Ligaments	Radial carpometacarpal anterior and posterior oblique
Stability	Good due to tendons crossing joint
Movements	Flexion and extension
	Abduction and adduction
	Opposition
Metacarpophalangeal joints	
Type	Synovial condyloid
Articular surfaces	Head of metacarpal with base of proximal phalanx
Capsule	Loose but reinforced by ligaments

Ligaments	Palmar
	Collateral
	Deep transverse metacarpal (index to little finger)
Stability	Mainly by tendons crossing joint
Movements	Flexion/extension
	Abduction/adduction

Interphalangeal joints

Type	Synovial hinge joints
Articular surfaces	Head of proximal phalanx with base of middle phalanx and head of middle phalanx with base of distal phalanx
Capsule	Encloses joint, replaced by palmar ligament anteriorly
Ligaments	Palmar
	Collateral
Stability	Mainly by tendons and ligaments
Movements	Flexion and extension

NERVE SUPPLY

The brachial plexus

Introduction

Knowledge of the distribution of the major peripheral nerves of the upper limb is necessary in clinical practice for the diagnosis and assessment of peripheral nerve injuries and other neurological disorders. The course and distribution of the various peripheral nerves are described in detail later in this section. Their cutaneous distribution follows and complements the segmental dermatome arrangement. A prerequisite for learning the muscular distribution of the major peripheral nerves is to appreciate the groupings of muscles into compartments. The muscles of the upper limb are grouped into four compartments. In the arm, there is an anterior compartment containing coracobrachialis, biceps brachii and brachialis supplied by the musculocutaneous nerve, and a posterior compartment containing triceps and anconeus supplied by the radial nerve. In the forearm, the anterior compartment contains all the flexors of the wrist, the long flexors of the digits, pronators teres and quadratus and is supplied by the median and ulnar nerves; while the posterior compartment contains the extensors of the wrist and digits, abductor pollicis longus, supinator and brachioradialis and is supplied by the radial nerve. The intrinsic muscles of the hand are supplied by the median and ulnar nerves.

The nerves supplying the structures in the arm are all derived from the brachial plexus, a complex of intermingling nerves originating in the neck (Fig. 2.130).

The brachial plexus is formed by the *ventral rami* of the *lower four cervical nerves* and the *first thoracic nerve* giving it a root value of C5, 6, 7, 8 and T1. Occasionally, there may be a contribution from C4 or T2 or both.

The ventral (anterior) rami are the anterior divisions of the spinal nerves, formed just outside the intervertebral foramen, lying between scalenus anterior and medius. They are collectively termed the *roots* of the plexus. Each spinal nerve receives an autonomic contribution, C5 and 6 receive grey rami communicantes from the middle cervical ganglion, while C7, 8 and T1 receive them from the inferior or cervicothoracic ganglion.

Generally, the upper two roots (C5 and 6) unite to form the *upper trunk*, the C7 root continues as the *middle trunk* and the lower two roots (C8 and T1) unite to form the *lower trunk*. These three trunks are found between the scalene muscles and the upper border of the clavicle in the posterior triangle of the neck. The lower trunk may groove the upper surface of the first rib behind the subclavian artery; the T1 root is always in contact with the rib.

Just above the clavicle, each trunk divides into *anterior* and *posterior divisions* which supply the flexor and extensor compartments of the arm respectively. The three posterior divisions unite to form the *posterior cord*, while the anterior divisions of the upper and middle trunks unite to form the *lateral cord*, and the anterior division of the lower trunk continues as the *medial cord*. The three cords pass downwards into the axilla, running first posterolateral to the axillary artery, and then in their named positions with respect to the second part of the axillary artery posterior to pectoralis minor. The cords and axillary artery are bound together in an extension of the prevertebral fascia layer of the cervical fascia, the axillary sheath, which protrudes into the axilla.

Applied anatomy

The brachial plexus is subject to direct injury; consequently an understanding of its formation is helpful in determining exactly which parts, and at what levels, the damage has occurred. Traction injuries occur when the roots of the plexus are torn from the spinal cord, or when the constituent parts are partially or completely torn. In extreme cases, the whole plexus may be disrupted to produce a completely denervated arm. If the upper roots are completely torn, *Erb's paralysis* affecting the musculature of the arm is produced, whereas if the lower roots are completely torn, *Klumpke's paralysis* affecting the hand and forearm will result.

Nerves arising from the brachial plexus and their distribution

The simplest way in which to describe the nerves of the brachial plexus is to put them into terminological order relating to the part of the plexus from which they originate (and to indicate their root value).

183

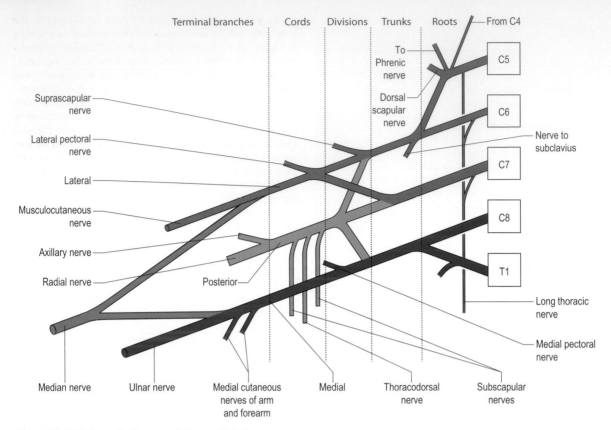

Figure 2.130 Schematic diagram of the brachial plexus.

Branches from the roots

1. Nerves to the scalene and longus colli muscles (C5, 6, 7, 8).
2. A branch to the phrenic nerve (C5).
3. Dorsal scapular nerve (C5).
4. Long thoracic nerve (C5, 6, 7).

Branches from the trunks

1. Nerve to subclavius (C(4), 5, 6).
2. Suprascapular nerve (C(4), 5, 6).

There are no nerves arising from the divisions.

Branches from the cords of the plexus

Medial cord

1. Medial pectoral nerve (C8, T1).
2. Medial cutaneous nerve of the forearm (C8, T1).
3. Medial cutaneous nerve of the arm (T1).
4. Ulnar nerve (C(7), 8, T1).
5. Medial part of the median nerve (C8, T1).

Posterior cord

1. Upper subscapular nerve (C(4), 5, 6, (7)).
2. Thoracodorsal nerve (C(6), 7, 8).
3. Lower subscapular nerve (C5, 6).
4. Axillary nerve (C5, 6).
5. Radial nerve (C5, 6, 7, 8, (T1)).

Lateral cord

1. Lateral pectoral nerve (C5, 6, 7).
2. Musculocutaneous nerve (C5, 6, 7).
3. Lateral part of the median nerve (C(5), 6, 7).

Branches from the roots

The muscular supply to the scalene and longus colli muscles arises by twigs from the upper surface of the anterior rami as they emerge from the intervertebral foramina, directly entering the muscles. The C5 contribution to the phrenic nerve arises at the lateral border of scalenus anterior.

The **dorsal scapular nerve** (C5) (Fig. 2.130) passes through scalenus medius to the deep surface of levator scapulae, from where it runs onto the anterior surface of the rhomboids. It supplies rhomboid major, rhomboid minor and levator scapulae.

The C5 and 6 roots of the **long thoracic nerve** (see Fig. 2.130) unite after piercing scalenus medius and are joined by the C7 root on the anterior surface of the muscle. The nerve passes behind the trunks of the plexus between the first rib and the axillary artery to the outer (axillary) surface of serratus anterior which it supplies. The upper two digitations of the muscle are supplied by C5, the next two by C6, and the remaining four by C7.

The long thoracic nerve may be damaged by direct pressure on it from above the shoulder. The resulting paralysis of serratus anterior causes the characteristic 'winged' scapula, with the inability to perform activities, such as abduction of the arm, where the scapula is stabilized or laterally rotated.

Branches from the trunks

The **nerve to the subclavius** (C(4), 5, 6) is a small branch from the upper trunk. It descends anterior to subclavian artery to supply subclavius. It may communicate with the phrenic nerve.

The **suprascapular nerve** (C(4), 5, 6) (Fig. 2.130) is a large branch from the upper trunk. It passes inferolaterally above and parallel to the trunks through the suprascapular notch deep to trapezius to enter the supraspinous fossa of the scapula. It then runs deep to supraspinatus and enters the infraspinous fossa via the spinoglenoid notch. The suprascapular nerve supplies both supraspinatus and infraspinatus, and also gives articular filaments to the shoulder and acromioclavicular joints.

All of the above branches arise from the plexus above the clavicle, whereas those considered below all arise below the level of the clavicle in the axilla.

Branches from the cords

The **lateral pectoral nerve** (Fig. 2.130) arises from the lateral cord of the plexus with a root value of C5, 6 and 7. It crosses anteromedially in front of the axillary artery, giving a branch to the medial pectoral nerve before piercing the clavipectoral fascia to the deep surface of pectoralis major which it supplies.

The **medial pectoral nerve** (Fig. 2.130) arises from the medial cord with a root value C8 and T1. It receives a contribution from the lateral pectoral nerve and passes between the axillary artery and vein to the deep surface of pectoralis minor which it supplies. It then the pierces pectoralis minor to end in pectoralis major which it also supplies.

The **upper and lower subscapular nerves** (Fig. 2.130) both arise from the posterior cord with root values C(4), 5, 6, (7) and C5, 6 respectively. From behind the axillary artery they descend towards the subscapular fossa where they both supply subscapularis. In addition, the lower subscapular nerve enters and supplies teres major.

The **thoracodorsal nerve** (Fig. 2.130) arises from the posterior cord (root value C(6), 7, 8) between the two subscapular nerves. It passes inferomedially along the posterior wall of the axilla and the anterolateral surface of latissimus dorsi before entering its deep surface to supply it.

The medial cutaneous nerve of the arm is a small nerve arising from the medial cord (root value T1). It descends through the axilla on the medial side of the axillary vein, and then along the medial side of the brachial artery. It pierces the deep fascia to supply skin and fascia on the medial side of the proximal half of the arm, extending onto both anterior and posterior surfaces (Fig. 2.131). It may be partly or entirely replaced by the intercostobrachial nerve (root value T2 and 3).

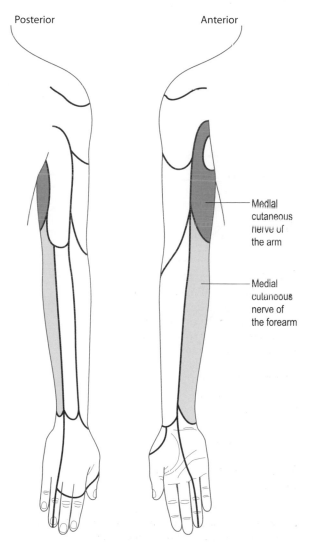

Posterior   Anterior

Medial cutaneous nerve of the arm

Medial cutaneous nerve of the forearm

Figure 2.131 The cutaneous distribution of the medial cutaneous nerves of the arm and forearm.

The **medial cutaneous nerve of the forearm** arises directly from the medial cord with a root value of C8 and T1. It descends on the medial side of the axillary and brachial arteries. It pierces the deep fascia, together with the basilic vein, in the middle of the arm, and descends with it to the elbow, where it divides into anterior and ulnar branches. The medial cutaneous nerve of the forearm supplies the skin over the lower part of biceps, the medial side of the forearm as far as the wrist, and part of the medial side of the posterior surface of the forearm (Fig. 2.131).

Axillary nerve

The axillary nerve (Fig. 2.132) arises from the posterior cord of the brachial plexus and has a root value of C5 and 6. In the axilla it descends behind the axillary artery and in front of subscapularis, at the lower border of which it passes backwards close to the inferior part of the shoulder joint in company with the posterior circumflex humeral vessels. It then passes through the *quadrilateral space* (Fig. 2.71) where it

supplies the shoulder joint and divides into anterior and posterior branches. The anterior branch winds around the surgical neck of the humerus, deep to and as far as the anterior part of deltoid, which it supplies. The posterior branch supplies teres minor and the posterior part of deltoid. It then passes around deltoid as the *upper lateral cutaneous nerve of the arm*, which pierces the deep fascia to supply the skin over the lower part of deltoid and the lateral head of triceps as far as the middle part of the arm (Fig. 2.132).

Section summary	
Axillary nerve	
From	Posterior cord of brachial plexus
Root value	C5, 6
Muscles supplied	Deltoid
	Teres minor
Cutaneous branch	Upper lateral cutaneous nerve of arm

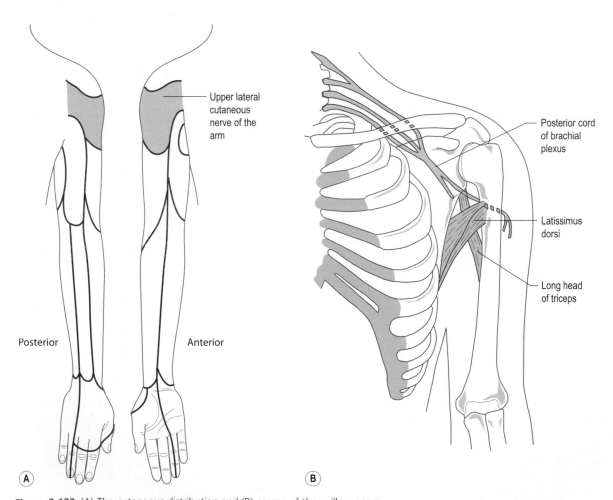

Figure 2.132 (A) The cutaneous distribution and (B) course of the axillary nerve.

The axillary nerve is frequently injured when the shoulder is dislocated because of its close proximity to the joint. Paralysis of deltoid and teres minor results, and consequently there is an inability to abduct the arm beyond that possible by the action of supraspinatus. This, in conjunction with an area of anaesthesia over the posterior part of deltoid and lateral head of triceps, allows a clinical diagnosis of nerve injury to be made.

Musculocutaneous nerve

The musculocutaneous nerve (Fig. 2.133) arises from the lateral cord of the brachial plexus and has a root value C5, 6 and 7. Initially it lies lateral to the axillary artery and then descends between the artery and coracobrachialis which it supplies and pierces before running distally between biceps and brachialis to reach the lateral side of the arm. At the elbow, the musculocutaneous nerve pierces the deep fascia between biceps and brachioradialis as the *lateral cutaneous nerve of the forearm*.

In the arm the musculocutaneous nerve supplies both heads of the biceps brachii and two-thirds of brachialis as well as coracobrachialis.

Section summary	
Musculocutaneous nerve	
From	Lateral cord of brachial plexus
Root value	C5, 6, 7
Muscles supplied	Coracobrachialis
	Biceps brachii
	Brachialis (medial two-thirds)
Cutaneous branch	Lateral cutaneous nerve of forearm

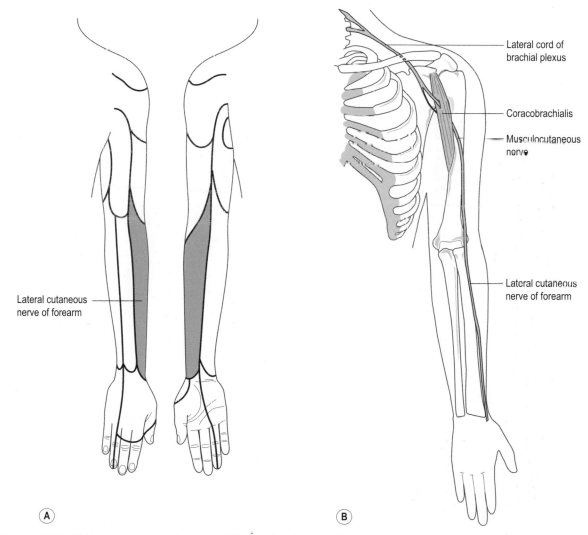

Figure 2.133 (A) The cutaneous distribution and (B) course of the musculocutaneous nerve.

The *lateral cutaneous nerve of the forearm* (Fig. 2.133) divides into anterior and posterior branches. The anterior branch supplies the skin on the lateral half of the forearm as far as the ball of the thumb, while the posterior branch supplies a variable area over the extensor muscles of the forearm, wrist and occasionally the first metacarpal.

Ulnar nerve

The ulnar nerve (Fig. 2.134) is one of the terminal branches of the medial cord of the brachial plexus, having a root value C8 and T1, but frequently contains fibres from C7.

It descends on the medial side of the axillary artery behind the medial cutaneous nerve of the forearm, and continues downwards medial to the brachial artery, anterior to triceps. In the distal half of the arm, the ulnar nerve passes posteriorly and pierces the medial intermuscular septum to enter the posterior compartment of the arm, where it lies on the front of the medial head of triceps. Continuing its descent in the posterior compartment of the arm, the ulnar nerve passes between the medial epicondyle of the humerus and the olecranon of the ulna, lying in the ulnar groove behind the medial epicondyle. The ulnar nerve then enters the anterior compartment of the forearm by passing between the two heads of flexor carpi ulnaris,

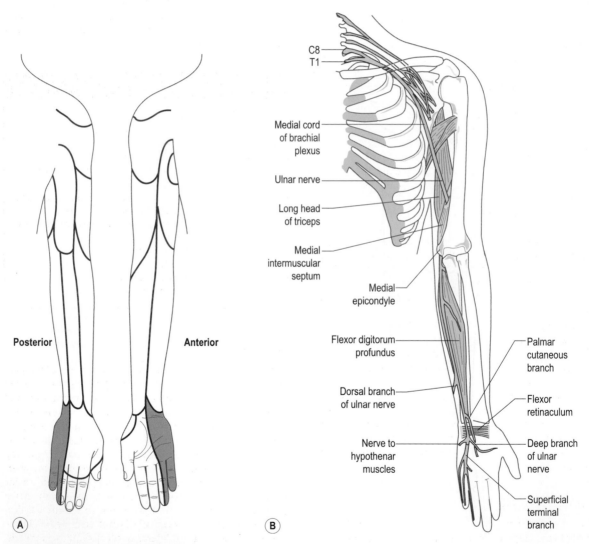

Figure 2.134 (A) The cutaneous distribution and (B) course of the ulnar nerve.

initially in contact with the ulnar collateral ligament of the elbow. As it descends on the medial side of the forearm, the ulnar nerve lies on flexor digitorum profundus, lateral to the ulnar artery, covered in its upper part by the belly of flexor carpi ulnaris, but in the lower half, only by its tendon. Proximal to the flexor retinaculum it pierces the deep fascia to lie lateral to flexor carpi ulnaris, and passes anterior to the flexor retinaculum lateral to the pisiform where it divides into superficial and deep branches.

During its course, the ulnar nerve gives an articular branch to the elbow joint, and supplies flexor carpi ulnaris and the medial half of flexor digitorum profundus.

A *palmar cutaneous branch* arises from the ulnar nerve piercing the deep fascia in the distal third of the forearm and descends to supply the skin over the medial part of the palm (Fig. 2.134).

The *dorsal branch* of the ulnar nerve also arises in the distal third of the forearm, passes backwards deep to flexor carpi ulnaris to pierce the deep fascia on the medial side to become superficial. On the medial side of the wrist, it crosses the triquetral, against which it can be palpated, and gives branches to the dorsal surface of the wrist and hand. Here it divides into two or three dorsal digital nerves which supply the skin on the dorsum of the hand and the dorsal surfaces of the medial one and a half or two and a half fingers, excluding the skin over the distal phalanx. The dorsum of the distal phalanges is supplied by branches from the median or ulnar nerves derived from the palm.

The *superficial branch of the ulnar nerve* lies deep to palmaris brevis on the medial side of the hand where it can be compressed against the hook of the hamate. It supplies palmaris brevis, the skin on the medial side of the palm of the hand, and the skin on the palmar surface of the little and adjacent half of the ring fingers, extending onto the dorsal surface supplying the skin and nail bed of the distal phalanx.

The *deep branch of the ulnar nerve* eventually runs with the deep branch of the ulnar artery and thus loops across the palm from medial to lateral deep to the flexor tendons. It passes initially between abductor digiti minimi and flexor digiti minimi and pierces opponens digiti minimi, supplying all three muscles. As it passes across the deep part of the palm, it supplies the medial two lumbricals, all of the dorsal and palmar interossei and adductor pollicis. Rarely, the ulnar nerve also supplies the thenar muscles. The deep branch gives articular filaments to the wrist joint.

Applied anatomy

The ulnar nerve may be damaged in the groove behind the medial epicondyle either by trauma or by entrapment leading to a partial or complete loss of muscular and sensory

innervation. At the wrist, the nerve can easily be cut or lacerated due to its superficial position. The clinical picture can be complicated if the lesion occurs below the level where the dorsal and palmar cutaneous branches are given off, as a considerable portion of the skin on the ulnar side of the hand still has a sensory supply. The result of an ulnar nerve lesion often gives the typical 'ulnar claw-hand' deformity. This is due to loss of power in the intrinsic muscles of the hand and the unopposed actions of antagonistic muscle groups which extend the interphalangeal joints of the ring and little fingers (Fig. 2.135A). It is possible to correct this deformity using a splint to counteract the pull of the unopposed muscles which mas allow more functional use of the hand (Fig. 2.135B). There is 'guttering' between the metacarpals, an inability to abduct the fingers or adduct the thumb, with marked wasting of adductor pollicis (Fig. 2.135A). The area of sensory loss usually follows the outline of the sensory map (Fig. 2.134A).

Section summary	
Ulnar nerve	
From	Medial cord of brachial plexus
Root value	(C7) C8, T1
Muscles supplied	Flexor carpi ulnaris
	Flexor digitorum profundus (medial half)
	Palmaris brevis
	From deep branch: abductor, flexor and opponens digiti minimi medial 2 lumbricals palmar and dorsal interossei adductor pollicis
Cutaneous branches	Palmar cutaneous branch
	Dorsal branch
	Superficial branch
	Digital branches

Radial nerve

The radial nerve (Fig. 2.136) is the major nerve of the posterior cord, root value C5, 6, 7, 8 (T1), being one of its terminal branches. In the axilla, it lies behind the axillary and upper part of the brachial arteries, passing anterior to the tendons of subscapularis, latissimus dorsi and teres major. The radial nerve, together with the profunda brachii artery, enters the posterior compartment of the arm by passing through the triangular space, formed by the humerus laterally, the long head of triceps medially and teres major above (Fig. 2.136). In passing through this space, the nerve enters the spiral (or radial) groove of the humerus, descending obliquely between the lateral and medial heads of triceps, reaching the lateral border of the humerus in the distal third of the arm. The nerve pierces the lateral

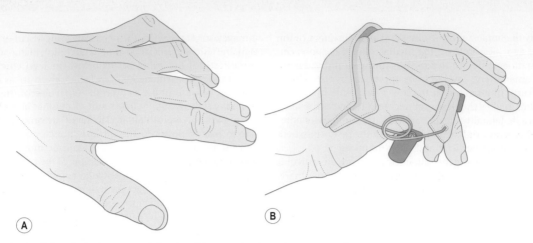

Figure 2.135 (A) Typical ulnar nerve deformity, (B) example of an ulnar nerve splint.

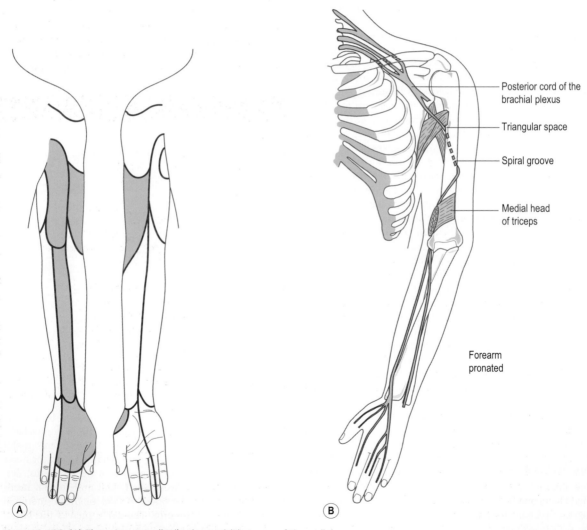

Posterior cord of the brachial plexus

Triangular space

Spiral groove

Medial head of triceps

Forearm pronated

Figure 2.136 (A) The cutaneous distribution and (B) course of the radial nerve.

intermuscular septum to enter the anterior compartment where it lies in a muscular groove between brachialis and brachioradialis. In front of the lateral epicondyle of the humerus, the radial nerve divides into its terminal *superficial* and *deep branches* (Fig. 2.136).

In the arm, the radial nerve supplies all three heads of triceps, anconeus, the lateral part of brachialis, brachioradialis and extensor carpi radialis longus. The branches to triceps all arise before the radial nerve enters the spiral groove; anconeus is supplied by one of the branches to the medial head of triceps.

The radial nerve also gives articular branches to the elbow joint and has three cutaneous branches which supply the skin on the back of the arm and forearm (Fig. 2.136A).

The *posterior cutaneous nerve of the arm* arises in the axilla, piercing the deep fascia near the posterior axillary fold. It supplies the skin on the posterior surface of the proximal third of the arm.

The *lower lateral cutaneous nerve of the arm* arises before the radial nerve, pierces the lateral intermuscular septum, and becomes cutaneous just below deltoid. It supplies the skin over the lower lateral part of the arm and a small area on the forearm.

The *posterior cutaneous nerve of the forearm* arises just below the lower lateral cutaneous nerve of the arm, and supplies a variable area of the skin on the dorsum of the forearm as far as the wrist, or occasionally beyond.

The *superficial branch* is the direct continuation of the radial nerve, beginning in front of the lateral epicondyle and descending along the anterolateral side of the forearm. It is entirely sensory. It lies on supinator, pronator teres, flexor digitorum superficialis and flexor pollicis longus covered by brachioradialis with the radial artery medial to it. In the distal third of the forearm, the nerve emerges posteriorly from below the tendon of brachioradialis and pierces the deep fascia to become superficial. It supplies the skin on the dorsum of the wrist, the lateral dorsal surface of the hand and dorsum of the thumb, and then divides into four or five digital nerves. The digital nerves supply the skin on the dorsum of the thumb, index, middle and adjacent half of the ring finger as far as the distal interphalangeal joint. The digital branches also give articular branches to the metacarpophalangeal and proximal interphalangeal joints of all five digits.

The *deep branch*, more often called the *posterior interosseous nerve*, is entirely muscular and articular. It begins in front of the lateral epicondyle of the humerus and enters the posterior compartment of the forearm by passing between the two heads of supinator, thereby curving around the lateral and posterior surfaces of the radius. During its course, the nerve supplies both extensor carpi radialis brevis and supinator. It then descends between the deep and superficial groups of extensor muscles, accompanied by the posterior interosseous artery, supplying all the muscles in the extensor compartment of the forearm: extensor digitorum, extensor digiti minimi, extensor carpi ulnaris, extensor pollicis longus, extensor indicis, abductor pollicis longus and extensor pollicis brevis.

In the lower part of the forearm, the posterior interosseous nerve lies on the interosseous membrane and ends in a flattened expansion, which gives articular branches to the intercarpal joints.

Applied anatomy

The radial nerve is often injured in its course close to the humerus, either as the result of a fracture or by pressure from a direct blow or incorrect use of a crutch. Triceps usually escapes paralysis as it derives its innervation from branches given off high in the arm, but a total paralysis of the extensors of the wrist and digits leads to the deformity of a 'dropped wrist' (Fig. 2.137A). As a result, any attempt to grip or make a fist leads to increased flexion of the wrist and an inability to carry out effective movement. This is due to the loss of the synergic action of the wrist extensors which usually prevent the unwanted flexion of the wrist produced by the continued action of the finger flexors.

Section summary	
Radial nerve	
From	Posterior cord of brachial plexus
Root value	C5, 6, 7, 8 (T1)
Muscles supplied	Triceps (all 3 heads)
	Anconeus
	Brachialis (lateral third)
	Brachioradialis
	Extensor carpi radialis longus
	From posterior interosseous nerve (deep branch):
	Supinator
	Extensor carpi radialis brevis
	Extensor digitorum
	Extensor digiti minimi
	Extensor carpi ulnaris
	Extensors pollicis longus and brevis
	Extensor indicis
	Abductor pollicis longus
Cutaneous branches	Posterior cutaneous nerve of the arm
	Lower lateral cutaneous nerve of the arm
	Posterior cutaneous nerve of the forearm
	Superficial radial branch

The interphalangeal joints of the fingers can be extended by the lumbricals and interossei which have an attachment to the dorsal digital expansion, but proper use of the hand requires an effective form of dynamic splint which compensates for the paralysed muscles. The

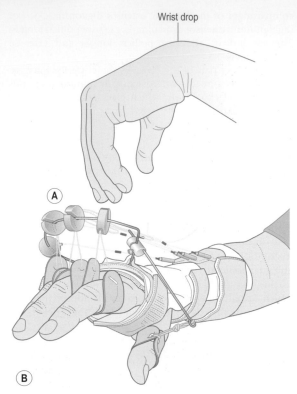

Wrist drop

(A)

(B)

Figure 2.137 (A) Typical radial nerve lesion, (B) example of a dynamic radial nerve splint.

dynamic splint shown in Fig. 2.137B holds the wrist in extension and recoils to pull the fingers and thumb from flexion into extension, allowing the grip to release. Even though the sensory distribution of the radial nerve on the dorsum of the hand appears extensive, overlap by adjacent cutaneous nerves means that the area of exclusive radial nerve supply is a small patch on the dorsum of the thumb web.

Median nerve

The median nerve (Fig. 2.138) arises partly from the lateral cord (C5, 6, 7) and partly from the medial cord (C8, T1) of the brachial plexus. These two contributing heads unite by embracing the third part of the axillary artery. Once formed, the nerve descends under cover of biceps brachii initially laterally to the brachial artery and then medially, having crossed it anteriorly. In the lower part of the arm the median nerve lies on brachialis, and in the cubital fossa is protected by the bicipital aponeurosis which crosses it.

The median nerve enters the forearm by passing between the two heads of pronator teres, and then runs below the tendinous arch connecting the heads of flexor digitorum superficialis to access its deep surface. Closely bound to the deep surface of flexor digitorum superficialis, it descends on flexor digitorum profundus until just above the wrist where it becomes superficial by passing between the tendons of flexors digitorum superficialis and carpi radialis, deep to palmaris longus. The median nerve enters the hand deep to the flexor retinaculum, passing anteriorly to the long flexor tendons. It is, therefore one of the structures found within the carpal tunnel.

During its course the median nerve gives articular branches to the elbow joint and supplies pronator teres, flexor carpi radialis, palmaris longus and flexor digitorum superficialis.

The *palmar cutaneous nerve* arises in the distal third of the forearm. It pierces the deep fascia and enters the palm by passing superficial to the flexor retinaculum. It supplies a small area of skin on the lateral side of the palm and thenar eminence.

In the cubital fossa, the *anterior interosseous nerve* arises from the median nerve and descends, with the anterior interosseous artery, on the anterior surface of the interosseous membrane between flexor pollicis longus and flexor digitorum profundus. It then runs deep to pronator quadratus to end at the wrist by giving articular branches to the radiocarpal and intercarpal joints. The anterior interosseous nerve supplies flexor pollicis longus, the lateral half of flexor digitorum profundus and pronator quadratus.

Once the median nerve has passed through the carpal tunnel to enter the hand, it divides into lateral and medial terminal branches. The *lateral branch* passes laterally and proximally to enter the thenar eminence and supply abductor pollicis brevis, flexor pollicis brevis, opponens pollicis and the first lumbrical. It gives sensory branches to the adjacent sides of the thumb and index finger.

The *medial branch* of the median nerve divides into a variable number of branches, the palmar digital nerves, the most lateral of which supplies the second lumbrical. These nerves are sensory to the palmar surface of adjacent sides of the index and middle, and middle and ring fingers (Fig. 2.138A). Each digital nerve produces a dorsal branch which passes posteriorly to supply the dorsal aspect of the distal phalanx and nail bed, and a variable amount of the middle phalanx of the same digits.

The *digital nerves* lie deep to the palmar aponeurosis and superficial palmar arch, but superficial to the long flexor tendons. As well as the sensory innervation, they also give articular branches to the interphalangeal and metacarpophalangeal joints.

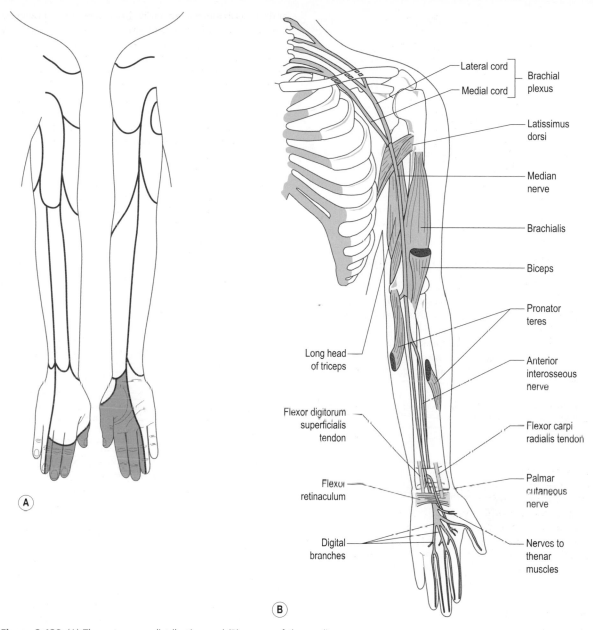

Figure 2.138 (A) The cutaneous distribution and (B) course of the median nerve.

Applied anatomy

The median nerve can be injured in the forearm by deep cuts with a resultant loss of flexion at all interphalangeal joints, except the distal ones in the ring and little fingers (Fig. 2.139). The metacarpophalangeal joints of these same fingers can still be flexed by the lumbricals and interossei but the movement of pronation is severely restricted.

In the hand, the thumb is held in extension and adduction, thus losing its ability to oppose and abduct. This, combined with the sensory loss, proves a major disability. More commonly the nerve is damaged just proximal to the flexor retinaculum by laceration, or deep to it in the carpal tunnel where compression gives rise to carpal tunnel syndrome. In this instance only the thenar muscles, lateral

and do contribute to an individual dermatome; however, all will have the same single root value.

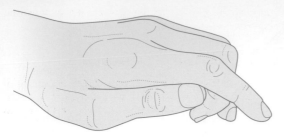

Figure 2.139 Typical median nerve deformity.

two lumbricals and sensation in the hand will be affected. Static splinting following median nerve lesions often involves holding the thumb in abduction and some opposition to prevent loss of the thumb webspace.

Dermatomes of the upper limb

Figures 2.131–2.140A show the distribution of the cutaneous branches of the major nerves arising from the brachial plexus. During development cells from the dermomyotome spread out to form the skin (see Figs. 1.20 and 1.26). The dermatome represents an area of skin innervated by a single nerve root; Fig. 2.140B shows the dermatomes of the upper limb. Obviously branches from more than one nerve can

Section summary	
Median nerve	
From	Medial and lateral cords of the brachial plexus
Root value	C5, 6, 7 (lateral cord) and C8, T1 (medial cord)
Muscles supplied	Pronator teres
	Flexor carpi radialis
	Palmaris longus
	Flexor digitorum superficialis
	Abductor pollicis brevis
	Flexor pollicis brevis
	Opponens pollicis
	Lateral 2 lumbricals
	From anterior interosseous nerve:
	Flexor pollicis longus
	Lateral half flexor digitorum profundus
	Pronator quadratus
Cutaneous branches	Palmar cutaneous branch
	Lateral and medial branches
	Digital branches

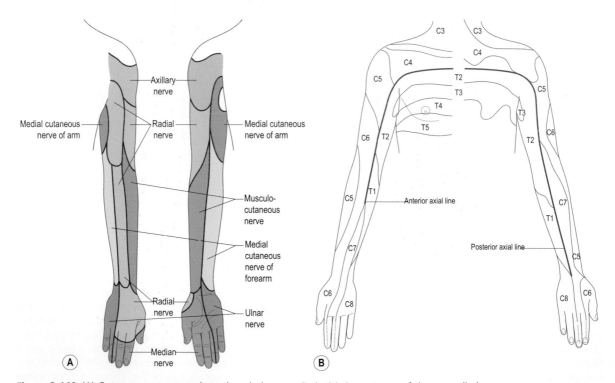

Figure 2.140 (A) Cutaneous nerve supply to the whole upper limb, (B) dermatomes of the upper limb.

BLOOD SUPPLY

The arteries and pulses

The main arterial stem of the upper limb passes through the root of the neck, the axilla and the arm before dividing into two in the forearm (Fig. 2.141). It changes its name in each of the regions as it crosses particular bony or muscular landmarks.

Subclavian artery

The right subclavian artery lies entirely within the root of the neck, having arisen from the brachiocephalic trunk. The left subclavian artery arises from the aortic arch in

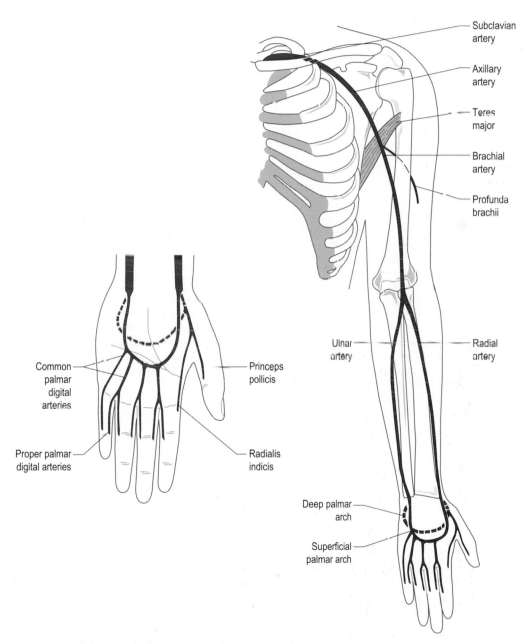

Figure 2.141 Arteries of the upper limb (anterior view) and arteries of the hand (palmar view).

the superior mediastinum to enter the root of the neck. Both arteries pass laterally over the first rib towards the axilla, and end by becoming the axillary artery at its lateral border. The subclavian artery is conveniently divided into three parts by scalenus anterior, which crosses it anteriorly.

In the neck, the artery runs from the upper border of the sternoclavicular joint to the middle of the clavicle. The course is convex upwards. The artery can be compressed against the first rib by a downward and backward pressure applied behind the clavicle, lateral to the posterior border of sternomastoid.

Branches from the subclavian artery supply structures within the neck, and the anterior chest wall, and give an important supply to the brain via the vertebral artery.

Axillary artery

The axillary artery is the continuation of the subclavian artery at the lateral border of the first rib and ends by becoming the brachial artery at the lower border of teres major, at the level of the lateral extremity of the posterior axillary fold. For descriptive purposes, it is divided into three parts by pectoralis minor; however, the length of each part depends on the position of the arm. The cords of the brachial plexus are named according to their position with respect to the second part of the axillary artery.

The course of the artery is represented by a line drawn from the midpoint of the clavicle passing immediately below the coracoid process to the medial lip of the intertubercular groove behind coracobrachialis. It thus describes a curve with the concavity facing downwards and medially.

The pulse of the axillary artery is readily palpated in the lateral wall of the axilla in the groove behind coracobrachialis. This is a useful pressure point to control distal bleeding, although paraesthesia may result from the inevitable pressure on the median, ulnar and radial nerves which are in close relation to the artery at this point.

Branches from the axillary artery supply the shoulder and pectoral regions, as well as the lateral chest wall. They anastomose with branches from the subclavian artery and the posterior intercostal arteries from the descending thoracic aorta. These anastomoses and their vessels may be enlarged in conditions where there is a narrowing of the aorta beyond the origin of the left subclavian artery (coarctation of the aorta), and serve as a collateral system bypassing the restriction.

Brachial artery

The continuation of the axillary artery beyond the lower border of teres major ending in the cubital fossa opposite the neck of the radius. It lies successively on the long and medial heads of triceps, the insertion of coracobrachialis and brachialis. Anteriorly it is covered by the medial border of biceps brachii, and is crossed about half way down the arm from lateral to medial by the median nerve. In the cubital fossa it lies beneath the bicipital aponeurosis, which separates it from the median cubital vein, with the median nerve lying medial to the artery and the biceps tendon lateral. The brachial artery divides into the radial and ulnar arteries.

High division of the brachial artery sometimes occurs proximal to the cubital fossa. Indeed, it may divide at any point between the axilla and the cubital fossa, in which case the two arteries descend side by side following the normal course of the brachial artery.

The *brachial pulse* may be felt along the whole course of the artery by compressing it against the humerus, directing lateral pressure proximally and dorsolateral pressure distally. It is best felt just medial to the bicipital aponeurosis at the level of the medial epicondyle of the humerus, and it is at this point that one listens for *Korotkoff's sounds* when measuring blood pressure.

A major branch of the brachial artery is the *profunda brachii*, which runs with the radial nerve in the spiral groove between the lateral and medial heads of triceps to pass into the posterior compartment of the arm. Branches from both the brachial artery and profunda brachii supply the muscles of the arm and contribute to the anastomosis around the elbow joint.

Radial artery

This begins in the cubital fossa opposite the neck of the radius and ends by completing the deep palmar arch in the hand. It is usually thought of and described as having three parts. The first part is in the forearm, the second curves laterally around the wrist as far as the first interosseous space, and the third passes through the interosseous space into the palm.

If the arm is placed in a midpronated position and brachioradialis tensed, then the course of the first part of the radial artery may be indicated by a slightly convex line beginning at the biceps tendon and running down the medial side of brachioradialis to a point just medial to the radial styloid process on the anterior aspect. As it curves around the wrist, the radial artery is within the 'anatomical snuffbox' and lies on the radiocarpal ligament, the scaphoid and trapezium. It is crossed by the tendons of abductor pollicis longus and extensors pollicis brevis and longus from lateral to medial.

The third part of the artery passes between the two heads of the first dorsal interosseous and adductor pollicis before completing the deep palmar arch.

The *radial pulse* may be felt against the distal border of the radius lateral to flexor carpi radialis, and in the 'anatomical snuffbox' against the scaphoid.

Branches from the first part of the artery are involved in the elbow anastomosis and supply the muscles on the radial side of the forearm. From the second part arise branches supplying the wrist and dorsum of the hand and thumb. Before completing the deep palmar arch, the third part of the artery gives off the *princeps pollicis* and *radialis indicis* branches to the thumb and index fingers respectively.

Ulnar artery

This begins in the cubital fossa as a terminal branch of the brachial artery, and ends at the pisiform by dividing into deep and superficial palmar arteries. It may be represented by a line passing, medially convex, from the tendon of biceps to the pisiform and from there to the hook of the hamate. In its course, it lies on brachialis, flexor digitorum profundus and the flexor retinaculum, and is crossed anteriorly, from above downwards by: pronator teres, the median nerve, flexor carpi radialis, palmaris longus and flexor digitorum superficialis, being overlapped lower down by flexor carpi ulnaris. Just below the radial tuberosity, the common interosseous artery is given off, and this divides into anterior and posterior interosseous arteries which run down either side of the interosseous membrane, supplying the deep muscles of both the flexor and extensor compartments. Branches from the proximal and distal ends of the artery are involved in the supply to the elbow and wrist joints respectively.

Superficial palmar arch

Formed mainly by the ulnar artery, with a contribution from the superficial palmar branch of the radial artery, it lies deep to the palmar aponeurosis. The distal convexity of the arch lies level with the flexor surface of the extended thumb.

Four *common palmar digital arteries* arise from the superficial arch with the most medial running along the medial side of the little finger. The other three divide into two proper digital arteries each, which supply adjacent sides of the little, ring, middle and index fingers.

Deep palmar arch

This is formed mainly by the radial artery with a contribution from the deep branch of the ulnar artery. It lies deep to the long flexor tendons and their synovial sheaths on the bases of the metacarpals and gives rise to the *palmar metacarpal arteries*. Its distal convexity is 2 cm distal to the distal crease of the wrist.

The veins

The veins (Fig. 2.142) of the upper limb are divided into a superficial group which lie in the superficial fascia, and a deep group which accompany the arteries. Both groups of veins have valves which allow proximal drainage (a fact which should be borne in mind when using massage) and drain into the axillary vein.

Deep veins

Apart from the axillary artery, which is accompanied by a single vein, all of the arteries are accompanied by two *venae comitantes.*

The *axillary vein* is the continuation of the basilic vein at the lower border of teres major. It ends by becoming the subclavian vein at the lateral border of the first rib. Its course is identical to that of the axillary artery, which lies lateral to it.

Superficial veins

These are arranged in irregular networks in the superficial fascia. They are connected with the deep veins by inconstant perforating veins which pierce the deep fascia. The blood is drained from the superficial system principally by the basilic and cephalic veins.

The main superficial channels are as follows:

Dorsal venous arch

This would be better named the dorsal venous plexus, for its arch-like nature is seldom apparent. It lies on the back of the hand, its position and pattern being highly variable.

Basilic vein

This arises from the ulnar side of the arch and ascends along the ulnar side of the distal half of the forearm before inclining forwards to pass in front of the medial epicondyle of the humerus entering the medial bicipital furrow. Opposite the insertion of coracobrachialis, it pierces the deep fascia to ascend along the medial side of the brachial vessels to become the axillary vein at the lower border of teres major. In the forearm it can usually be clearly seen, particularly in males. It is joined by tributaries from the forearm and by the median cubital vein in front of the elbow.

Cephalic vein

This arises from the radial end of the dorsal venous arch and receives the dorsal veins of the thumb. Inclining forwards, it ascends on the anterolateral part of the forearm as far as the elbow, and then along the lateral side of the biceps tendon to reach the groove in front of the shoulder between deltoid and pectoralis major (the deltopectoral groove). It ascends in this groove to the level of the coracoid process, where it turns medially between pectoralis major and pectoralis minor. It pierces the clavipectoral fascia and ends in the axillary vein at a point just below the middle of the clavicle. It receives several tributaries in the forearm, and at the elbow is connected to the basilic vein by the median cubital vein.

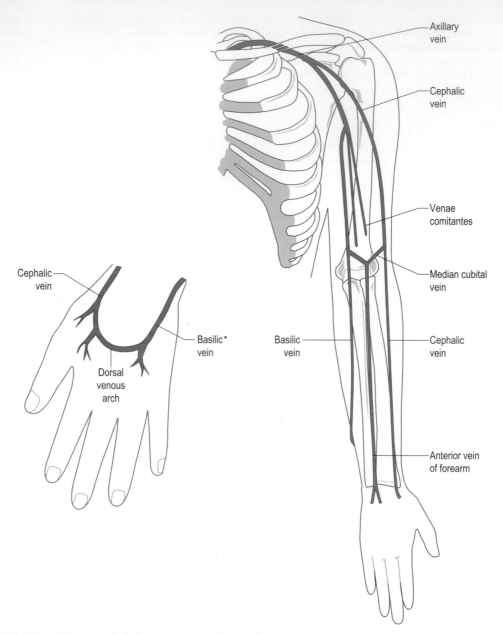

Figure 2.142 Veins of the upper limb (anterior view) and veins of the hand (dorsal view).

Median cubital vein

A short, wide vein, useful for venepuncture it usually runs upwards and medially across the bicipital aponeurosis. The latter separates it from the underlying brachial artery. It joins the basilic vein just above the medial epicondyle.

Anterior median vein of the forearm

When present, this vein runs up the middle of the anterior aspect of the forearm and may join the basilic or cephalic vein, or it may divide at the cubital fossa into the median cephalic and median basilic veins.

LYMPHATICS

The lymphatic drainage of the upper limb is by a superficial intermeshing network of vessels just below the skin, and by deep lymphatic channels below the deep fascia. The larger lymph vessels contain numerous valves which allow lymph to move only in a proximal direction. Both groups of vessels drain proximally and end by passing through some of the 25–30 lymph nodes in the axilla. This mass of lymph nodes serves to filter the lymph contained in the system and acts as an important defence mechanism in preventing the spread of infection.

The axillary lymph nodes are distributed in the axillary fat throughout the axilla but can be divided into five groups, four of which lie below pectoralis minor and one (the apical group) above pectoralis minor. Ultimately, all lymph from the upper limb passes through the apical group of nodes, from where the efferent lymph channels condense to form the subclavian trunk. On the left hand side, the subclavian trunk joins the thoracic duct, while on the right it drains into the subclavian vein directly or via the right lymphatic duct.

Superficial nodes and lymph vessels

The superficial lymph vessels are found in the skin and drain lymph from the superficial tissues (Fig. 2.143). In the hand a fine meshwork of vessels exists, which drain into progressively larger channels as they pass up the arm. The only superficial lymph vessels that have any consistent course are the larger ones which follow the major superficial veins. These end by passing into the axilla.

In the cubital fossa one or two lymph nodes lie medial to the basilic vein, receiving lymph from the medial fingers and ulnar half of the hand and forearm. There are also one or two lymph nodes in the infraclavicular fossa associated with the cephalic vein. These receive vessels from the shoulder and breast. A single node may be found in the deltopectoral groove.

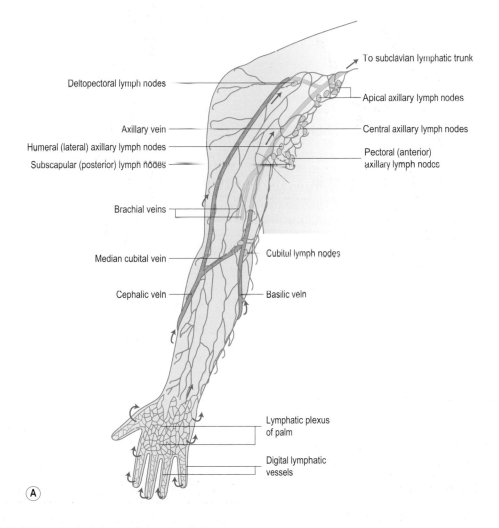

Figure 2.143 (A) Lymphatic drainage of the upper limb,

Continued

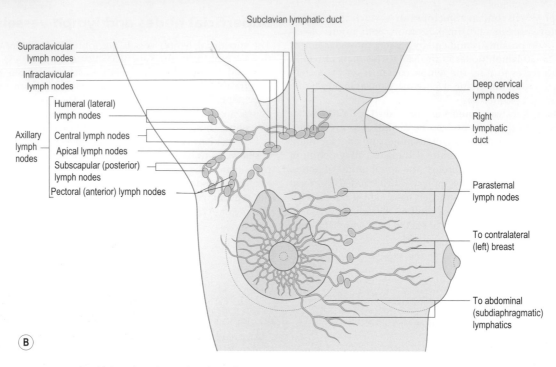

Subclavian lymphatic duct

Supraclavicular
lymph nodes

Infraclavicular
lymph nodes

Humeral (lateral)
lymph nodes

Axillary
lymph
nodes

Central lymph nodes

Apical lymph nodes

Subscapular (posterior)
lymph nodes

Pectoral (anterior) lymph nodes

Deep cervical
lymph nodes

Right
lymphatic
duct

Parasternal
lymph nodes

To contralateral
(left) breast

To abdominal
(subdiaphragmatic)
lymphatics

B

Figure 2.143, Cont'd (B) lymph nodes within the axilla.

Deep nodes and lymph vessels

The deep lymph vessels of the upper limb are less numerous than the superficial vessels with which they have many connections. Lying below the deep fascia, they accompany the major arteries in the arm. Most pass directly to the lateral group of axillary nodes. Small nodes may occur along both the radial and ulnar arteries and deep within the cubital fossa. Efferents from all of these nodes pass to the lateral group of axillary nodes. These latter nodes lie alongside the axillary vein, and from the pectoral and subscapular nodes pass along the lateral thoracic and subscapular arteries respectively. A central group of nodes is formed above the floor of the axilla. Although the group receives lymph from all areas, the main drainage of the limb is to the lateral group. Drainage of the breast and anterior chest wall is to the pectoral nodes; that of the scapular region and upper back is to the subscapular nodes. Efferents from these groups pass to the central and then the apical group of nodes, the latter also receiving efferents from the superficial infraclavicular nodes (Fig. 2.143B).

Application

The fact that the larger lymph vessels contain valves is important during massage techniques aimed at reducing oedema. The massage strokes are applied from distal to proximal, ending at the axilla, with sufficient depth to compress the lymph vessels and encourage drainage.

Active muscle contraction will also cause compression of the lymph vessels and encourage drainage proximally. This effect can be further enhanced by placing an elastic compressive support or bandage on the upper limb and encouraging active rhythmical contraction of the arm muscles. Elevation of the arm above the level of the axilla will allow gravity to assist the lymphatic drainage.

Pneumatic splints which apply a rhythmically alternating compressive force to the arm, using a small electric compressor pump, can also achieve the effect of increasing lymphatic flow proximally, utilizing the same principles as massage.

Part | 3 |

The lower limb

CONTENTS

INTRODUCTION

The human lower limb is adapted for weight-bearing, locomotion and maintaining the unique, upright, bipedal posture. For all of these functions a greater degree of strength and stability are required than in the upper limb. The bones of the lower limb are larger and more robust than their upper limb counterparts, and vary in their characteristics in relation to muscular development and body build. Many bones, particularly the innominate and, to a lesser extent the femur, show sexual differences; variations in the female pelvis for example being an adaptation to childbearing.

The form and structure of individual bones are adapted to the function of supporting and resisting mechanical stresses. Their internal architecture is arranged to resist all such stresses and forces; this is particularly marked in the articular regions. During growth and throughout life continuous modifications are being made to maintain the functions of support and resistance to stress as the stresses change. The attainment of an habitual, upright posture and bipedal gait has resulted in a change in both the mechanical and functional requirements of all bones of the lower limb. Consequently, during evolution, the lower limb has been the subject of major change.

The pelvic girdle, formed by the right and left innominate articulating anteriorly at the symphysis pubis and posteriorly by the sacrum via the sacroiliac joint, connects the lower limb to the vertebral column. The sacroiliac joint provides great strength for weight transference from the trunk to the lower limb at the sacrifice of almost all mobility. The human ilium has developed so that it is no longer blade-like but is shortened and tightly curved backwards and outwards (Fig. 3.1), changing the actions of the gluteal muscles. These changes in the pelvis have resulted in a shift from it lying in an essentially horizontal to an essentially vertical position. This has enabled the trunk to be held vertically, but has necessitated a change in the orientation of the sacrum with respect to the ilium: the result is that the axis of the pelvic canal lies almost at right angles to the vertebral column. During evolution there has been a relative approximation of the sacral articular surface to the acetabulum which makes for greater stability in the transmission of the weight of the trunk to the hip joint. This increase in the magnitude has resulted in an increase in the contact area between sacrum and ilium relative to the area of the ilium as a whole. For the same reason the acetabulum and femoral head have also increased in relative size during evolution. The shortening of the ischium

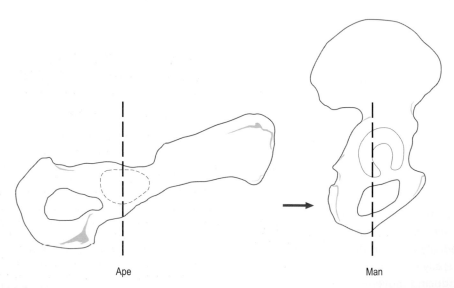

Ape   Man

Figure 3.1 The evolutionary changes that have occurred in the human pelvis as part of the adaptation to the erect posture and bipedalism.

that has occurred is an adaptation for speed and rapid movements, which is of great importance in bipeds. Thus power of action has been sacrificed for speed.

Changes have also occurred at the knee, with the femoral condyles being more parallel in humans compared with other primates. The major change, however has been in bringing the knees inwards towards the midline, which appears to be part of the overall pattern of centring the body mass, thus reinforcing skeletal rather than muscular equilibrium.

In humans, the tibia and fibula are held tightly together, with the tibia being the weight-bearing component while the fibula is mainly for muscle attachments. There has been a loss of rotation of the fibula with respect to the tibia.

It is the foot, however, which has undergone the greatest change during evolution (Fig. 3.2), reflecting not so much the evolution of a new function as a reduction in the original primate functions, with the foot changing from a grasping, tactile organ to a locomotor prop. Although some non-locomotor function is still possible, the foot has evolved from a generalized to a specialized organ. The joints of the human foot permit much less internal mobility – an adaptation to ground walking. In locomotion, the foot acts as a lever adding propulsive force to that of the leg, with the point of pivot being the subtalar joint. The forefoot has been shortened relative to the hindfoot, where the main thrust in walking is developed – the power capabilities are thus accentuated.

The arches of the foot are formed by the shape of the bony elements, and are supported by ligaments and tendons. They convert the foot into a complex spring under tension and allow it to transmit the stresses involved in walking, both when body momentum is checked at heel-strike and when the foot is used in propelling the body forward. The lateral arch helps to steady the foot on the ground, while the medial arch transmits the main force of thrust in propulsion. The arched foot is important in providing one of the major determinants of gait, i.e. helping to minimize energy expenditure and thus increase the efficiency of walking.

An important consequence of the upright, bipedal posture is that the centre of gravity of the body has been brought towards the vertebral column, so that in humans it lies slightly behind and at about the same level as the hip joint, thus reducing the tendency of gravity to pull the trunk forward. The centre of gravity projection then passes anterior to the knee and ankle joints (see Fig. 3.4); at the knee the line of weight transmission passes towards the outside. Because of the angulation of the femur, during walking the foot, tibia and knee joint of each leg stay close to the line followed by the centre of gravity, and thus energy expenditure is minimal in maintaining the centre of gravity above the supporting limb. Balance is thus improved and there is more time for the free leg to swing forward promoting an increase in stride length. The alteration in the line of weight transmission is carried into the foot, where it passes to the inner side. However, it must be remembered that weight is also transmitted through the outside of the foot, bringing the entire foot into use as a stabilizing element.

In order to reduce the possibility of collapse or dislocation, due to the forces to which they are subjected, the joints of the lower limb are structurally more stable than those of the upper limb. This increased stability is due to either the shape of the articular surfaces, the number and strength of the ligaments, or the size of the muscles related to the joint. Generally, each of these factors contributes to a varying extent.

DEVELOPMENT OF THE MUSCULOSKELETAL SYSTEM

Mesodermal somites

By the end of the third week after fertilization the paraxial mesoderm begins to divide into *mesodermal somites* which are easily recognizable during the fourth and fifth weeks (Fig. 3.3B). Eventually some 44 pairs of somites develop, although not all are present at the same time; however, the paraxial mesoderm at the cranial end of the embryo remains unsegmented. There are 4 occipital somites, followed by 8 cervical, 12 thoracic, 5 lumbar, 5 sacral and 8–10 coccygeal somites. The growth and migration of these somitic cells are responsible for the thickening of the body wall, as well as the development of bone and muscle. The deeper layers of the skin are also of somitic origin. Somite-derived tissue spreads medially to form the

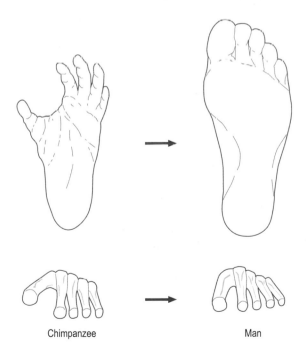

Chimpanzee   Man

Figure 3.2 The evolutionary changes that have occurred within the foot during its adaptation to bipedalism.

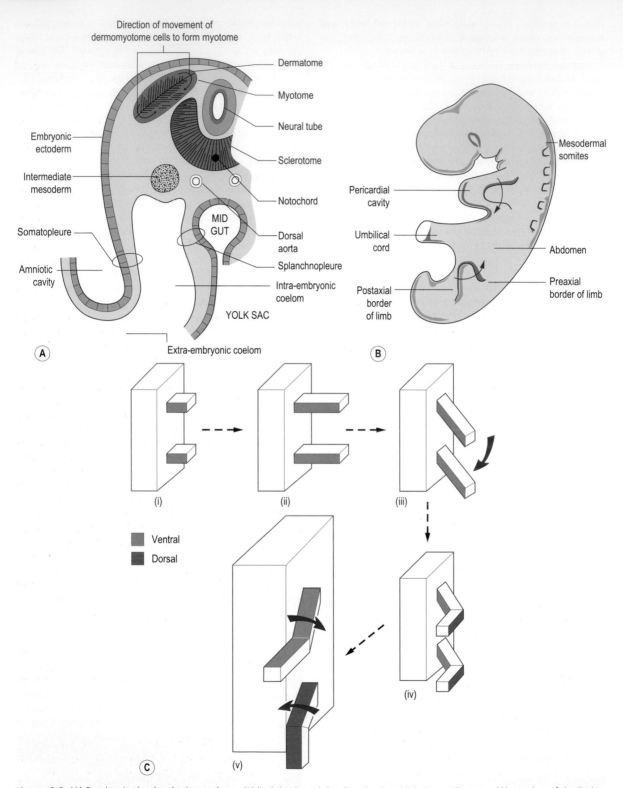

Figure 3.3 (A) Somites in the developing embryo, (B) limb buds and the direction in which they will rotate, (C) rotation of the limbs.

vertebrae, dorsally to form the musculature of the back, and ventrally into the body wall to form the ribs, and the intercostal and abdominal muscles.

Soon after its formation each somite becomes differentiated into three parts. The ventromedial part forms the *sclerotome*, which migrates medially towards the notochord and neural tube to take part in the formation of the vertebrae and ribs (Fig. 3.3A). The remainder of the somite is known as the *dermomyotome*. The cells of the dorsal and ventral edges proliferate and move medially to form the *myotome*, whose cells migrate widely and become differentiated into myoblasts (primitive muscle cells). The remaining thin layer of cells forms the *dermatome*, which spreads out to form the dermis of the skin.

The myotome of each somite receives a single spinal nerve which innervates all the muscle derived from that myotome, no matter how far it eventually migrates. The dorsal aortae lie adjacent to the somites and give off a series of intersegmental arteries which lie between them.

Development of the limbs

The limbs initially appear as flipper-like projections (the limb buds), with the forelimbs appearing first, between 24 and 26 days, each bud consisting of a mass of mesenchyme covered by ectoderm with a thickened ectodermal ridge at the tip; the ectodermal ridge controls normal development of the limb. Consequently, damage to it will result in some trauma to the limb. At the beginning of the second month the elbow and knee prominences project laterally and backwards. At about the same time, the hand and foot plates appear as flattened expansions at the end of the limb bud. Between 36 and 38 days, five radiating thickenings forming the fingers and toes can be distinguished; the webs between the thickenings disappear freeing the digits. Appropriate spinal nerves grow into the limbs in association with migration of the myotomes: C5, 6, 7, 8 and T1 for the upper limb, and L4, 5, S1, 2 and 3 for the lower limb. The limb bones differentiate from the mesenchyme of the bud. The limbs grow in such a way that they rotate in opposite directions, the upper limb laterally and the lower limb medially (Fig. 3.3C). Consequently, the thumb becomes the lateral digit of the hand while the great toe is the medial digit of the foot.

During development the lower limb bud appears as a swelling from the body wall (Fig. 3.3C(i)); the lower limb follows the upper limb in appearance and development. At first, each limb bud projects at right angles to the surface of the body, having ventral and dorsal surfaces, and cephalic (preaxial) and caudal (postaxial) borders (Fig. 3.3C(ii)). As the limb increases in length it becomes differentiated, during which time it is folded ventrally so that the ventral surface becomes medial (Fig. 3.3C(iii)), with the convexities of the elbow and knee directed laterally (Fig. 3.3C(iv)). At a later stage, the upper and lower limbs rotate in opposite directions such that the convexity of the knee is directed towards the cranial end of the body

(Fig. 3.3C(v)). The ventral rami entering a limb bud pass anterior to the myotomes and eventually divide into dorsal and ventral divisions, which unite with the corresponding branches of the adjacent ventral rami to form the nerves of the dorsal and ventral aspects of the limb bud. These nerves will supply respectively the sheets of muscle and overlying skin of the dorsal (extensor) and ventral (flexor) surfaces of the limbs prior to their rotation into the adult pattern. During the folding and rotation of the limb and the migration of muscle masses, these nerve–muscle connections are maintained.

The rotation of the lower limb in the course of development results in its extensor and flexor surfaces coming to lie anteriorly and posteriorly respectively. This rotation is reflected in the arrangement of the innervation. The muscles on the front of the thigh and leg are supplied by nerves coming from the posterior part of the lumbar and lumbosacral plexuses, the femoral and common fibular (peroneal) nerves, while those at the back of the thigh and leg and in the sole of the foot are from the anterior aspect of the lumbosacral plexus, the tibial nerve.

As in the upper limb, many muscles cross several joints and exert their actions on each. It is unusual for one joint of the lower limb to be moved in isolation. In standing and walking the joints and muscles form a coordinated mechanism.

The uppermost limit of the lower limb is a line joining the iliac crest, inguinal ligament, symphysis pubis, ischiopubic ramus, ischial tuberosity, sacrotuberous ligament, and the dorsum of the sacrum and coccyx. The bulge of tissue running between the innominate and the upper part of the femur forms the buttock, but the region is usually named gluteal after the underlying muscles. The lower limb is divided into the thigh between the hip and knee, the leg between the knee and ankle, and the foot distal to the ankle joint. The foot is divided into the foot proper and the toes, and has a superior surface (dorsum) and an inferior (plantar) surface, which is the sole of the foot (Fig. 3.4).

The bones of the lower limb are those of the pelvic girdle (*innominate* and *sacrum*), the *femur* in the thigh, the medial *tibia* and lateral *fibula* in the leg, the seven *tarsal* bones and five *metatarsals* in the foot, and the *phalanges* of the toes – two in the big toe and three in each remaining toe (Fig. 3.4).

BONES

Pelvic girdle

Introduction

The pelvic girdle (Fig. 3.5) comprises the two *innominates* and *sacrum*; the ring of bone formed uniting the trunk and the lower limbs. A large, irregular bone, the innominate consists of two expanded triangular blades twisted 90° to each other in the region of the *acetabulum*; each blade

REGIONS

BONES

JOINTS

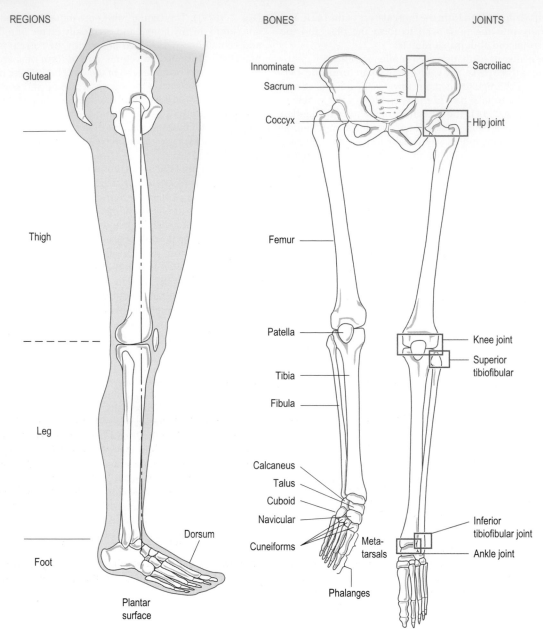

Gluteal

Thigh

Leg

Foot

Dorsum

Plantar
surface

Innominate

Sacrum

Coccyx

Femur

Patella

Tibia

Fibula

Calcaneus

Talus

Cuboid

Navicular

Cuneiforms

Meta-
tarsals

Phalanges

Sacroiliac

Hip joint

Knee joint

Superior
tibiofibular

Inferior
tibiofibular joint

Ankle joint

Figure 3.4 The regions, bones and joints of the lower limb.

is also twisted within itself. The innominate is formed from three separate bones, the *ilium*, *ischium* and *pubis*, which come together and fuse in the region of the acetabulum such that in the adult it appears as a single bone. The sacrum consists of five fused vertebrae and is roughly triangular. The *coccyx*, which is the remnant of the tail, consists of four fused coccygeal vertebrae. Each innominate articulates with the sacrum posteriorly by joints which are synovial anteriorly and fibrous posteriorly. Each innominate also articulates with its counterpart anteriorly at the pubic symphysis, by a secondary cartilaginous joint.

The pelvic girdle has several functions:

1. It supports and protects the pelvic viscera.
2. It supports body weight transmitted through the vertebrae, thence through the sacrum, across the sacroiliac joints to the innominate and then to the femora in the standing position, or to the ischial tuberosities when sitting.

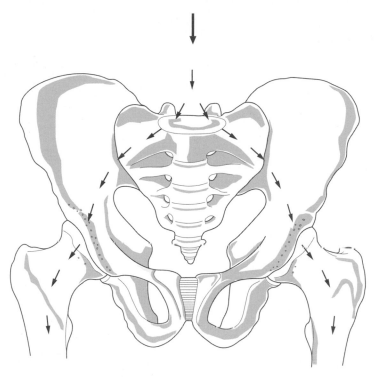

Figure 3.5 Transfer of weight from the vertebral column, through the pelvis to the femur.

3. During walking, the pelvis swings from side to side by a rotatory movement at the hip joint which occurs together with small movements at the lumbar intervertebral joints; when the hip joints are fused pelvic rotation is taken up by the thoracic spine enabling a patient to walk reasonably well.

4. As with all but a few small bones in the hand and foot, the pelvis provides attachments for muscle.

5. In females it provides bony support for the birth canal.

The pelvic girdle transmits the weight from the vertebral column to the lower limbs (Fig. 3.5), its bony and ligamentous components reflecting this function. There are some differences in the pelvis between females and males. In the female, these are due to adaptation for childbearing and the transmission of the relatively large fetal head during childbirth. The pelvis is essentially a basin, the upper part being known as the greater or false pelvis containing abdominal viscera. That part below the pelvic brim or pelvic inlet is the lesser or true pelvis. In the anatomical position, the pelvic inlet forms an angle of about 60° with the horizontal (Fig. 3.6). The acetabulum is directed laterally and inferiorly with the acetabular notch directed inferiorly. The *anterior superior iliac spines* and the *pubic tubercles* lie in the same vertical frontal plane. The lowest part of the sacrum lies above the level of the *symphysis pubis*.

The anterior superior iliac spines can be easily palpated in the living subject, particularly in females, where they tend to be further apart than in males. The iliac crest can also be palpated about 10 cm above the greater trochanter of the femur. In the sitting position, the ischial tuberosities can be felt, with the weight of the body resting on them. The body of the right and left pubic bones can also be palpated, they separate the anterior abdominal wall from the genitalia.

The innominate (hip) bone

An irregularly shaped bone (Figs. 3.7 and 3.8) consisting of three bones fused together, the *ilium*, the *pubis* and the *ischium*.

The ilium

The upper broad blade for the attachment of ligaments and large muscles. It forms the pelvic brim between the hip joint and the articulation with the sacrum. The anterior two-thirds forms the *iliac fossa* medially, which is part of the lateral and posterior abdominal wall, and the gluteal surface laterally, for attachment of the gluteal muscles. The posterior one-third of the medial surface, which is thicker, carries the *auricular surface* for the articulation with the sacrum; behind this is a prolonged rough part, the *iliac tuberosity*, for the attachment of the strong sacroiliac ligaments which bear the body weight. The upper border

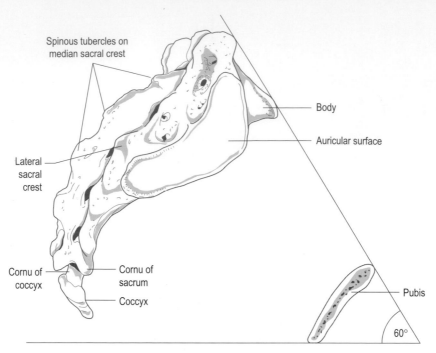

Spinous tubercles on
median sacral crest

Lateral
sacral
crest

Cornu of
coccyx

Cornu of
sacrum

Coccyx

Body

Auricular surface

Pubis

60°

Figure 3.6 Relative positions of the sacrum and pubis in the anatomical position, lateral view.

is the *iliac crest*, which is convex superiorly as well as anteroposteriorly with the anterior part curved outwards. The iliac crest ends as the *anterior superior iliac spine* anteriorly and the *posterior superior iliac spine* posteriorly. Both spines and the whole of the crest can be palpated (p. 211). Behind the anterior spine on the outer border is the prominent *tubercle of the crest*. Below the anterior superior spine the *anterior inferior iliac spine* is separated by a shallow notch. Similarly, below the posterior superior spine the *posterior inferior iliac spine* is separated by a shallow notch; the two posterior spines are closer together than the two anterior spines. Below the posterior inferior spine is the deep *greater sciatic notch*. The iliac fossa is separated from the sacropelvic surface of the ilium by the *arcuate line*, which forms part of the pelvic brim. The anterior part of this line has an elevation, the *iliopubic eminence*, marking the junction of the ilium with the pubis. The part which participates in the formation of the acetabulum is the body of the ilium.

The outer gluteal surface follows the curvature of the iliac crest and has three curved gluteal lines, which demarcate the attachments of the gluteal muscles. The most obvious is the *posterior gluteal line*, which passes down from the iliac crest to the front of the posterior inferior spine. The *anterior gluteal line* is a series of low tubercles from the iliac crest curving upwards and backwards below the iliac tubercle and then down towards the greater sciatic notch. The *inferior gluteal line* is less prominent, curving from below the

anterior superior iliac spine towards the apex of the greater sciatic notch. Below the inferior gluteal line is an area of multiple vascular foramina. Fusion of the ilium and the ischium is marked by a rounded elevation between the acetabulum and the greater sciatic notch; above this the ilium forms the major part of the notch. The gluteal surface is succeeded inferiorly by the acetabular part of the ilium.

The iliac fossa is the smooth internal concavity of the ala of the ilium. It narrows inferiorly, ending at the roughened iliopubic eminence, the line of junction between the ilium and the pubis. Its deepest part, high in the fossa, is composed of paper-thin translucent bone. The pelvic brim, marked by the arcuate line of the ilium, is the posteroinferior limit of the iliac fossa. Behind and below the iliac fossa and the arcuate line is the sacropelvic surface of the ilium. Posterior to the iliac fossa, this region has the auricular surface for articulation with the first two segments of the sacrum and, behind and above it, the tuberosity. The roughened tuberosity provides attachment for the short posterior sacroiliac ligaments and fibres of erector spinae and multifidus. The auricular area extends from the pelvic brim to the posterior inferior iliac spine. Its surface is gently undulating, being convex above to concave below, and roughened by numerous tubercles and depressions. The surface is covered with hyaline cartilage forming a synovial joint, which is immobile, with the ala of the sacrum. In later years, fibrous bands usually join the articular surfaces within the joint space.

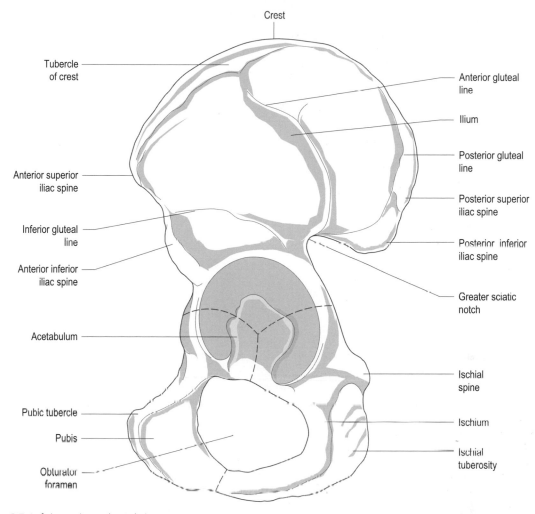

Crest

Tubercle
of crest

Anterior gluteal
line

Ilium

Posterior gluteal
line

Anterior superior
iliac spine

Posterior superior
iliac spine

Inferior gluteal
line

Posterior inferior
iliac spine

Anterior inferior
iliac spine

Greater sciatic
notch

Acetabulum

Ischial
spine

Pubic tubercle

Ischium

Pubis

Ischial
tuberosity

Obturator
foramen

Figure 3.7 Left innominate, lateral view.

The pubis

The quadrilateral body has a medially directed oval symphyseal surface crossed by several transverse ridges to which the fibrocartilage of the symphysis pubis is attached; the surface is coated with hyaline cartilage for the secondary cartilaginous joint that constitutes the symphysis pubis. The upper border of the body is the pubic crest, marked laterally by the *pubic tubercle* from which two ridges diverge laterally into the superior ramus. The upper ridge is the pectineal line, which is continuous with the arcuate line of the ilium, and forms part of the pelvic brim. The lower, rounded ridge is the obturator crest, passing downwards into the anterior margin of the acetabular notch. Between the two ridges is the iliopubic eminence. Below the obturator crest on the superior pubic ramus is the deep obturator groove. The superior ramus continues laterally to join the ilium and ischium at the acetabulum; the pubis forms

one-fifth of the acetabulum. A thin and flattened inferior ramus extends inferiorly and posterolaterally from the body to fuse with the ischium below the *obturator foramen*.

The ischium

The angulated posterior inferior part of the innominate; the angulation is in the same plane as the pubis. The blunt rounded apex of the angulation is the *ischial tuberosity*, divided transversely by a low ridge. A smooth oval above this ridge is further subdivided by a vertical ridge into lateral and medial areas. In the sitting position, the weight of the body rests on the two ischial tuberosities. Anteriorly the tuberosity passes upwards as the ischial ramus, being continuous with the inferior pubic ramus to form the ischiopubic ramus. The body of the ischium forms two-fifths of the acetabulum. The posterior border of the body is continuous above with the ilium, forming the greater

209

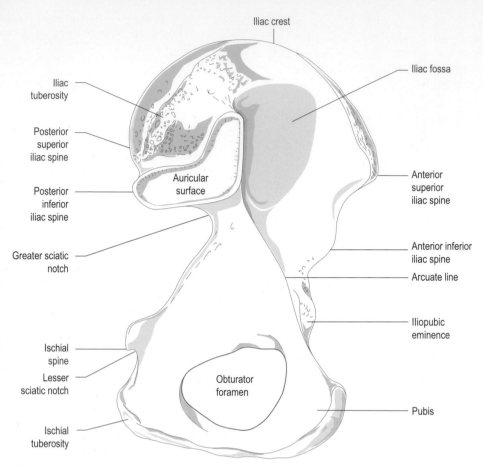

Iliac crest

Iliac fossa

Iliac
tuberosity

Posterior
superior
iliac spine

Auricular
surface

Posterior
inferior
iliac spine

Anterior
superior
iliac spine

Greater sciatic
notch

Anterior inferior
iliac spine

Arcuate line

Iliopubic
eminence

Ischial
spine

Lesser
sciatic notch

Obturator
foramen

Ischial
tuberosity

Pubis

Figure 3.8 Left innominate, medial view.

sciatic notch; inferiorly it ends as the blunt medially pro-jecting *ischial spine*, below which is the groove forming the *lesser sciatic notch*. The pelvic surface of the body is con-tinuous with the pelvic surface of the ilium, forming part of the lateral wall of the pelvis.

Acetabulum

The acetabulum is formed by the fusion of the three com-ponents of the innominate; the ilium, ischium and pubis meet at a Y-shaped cartilage which forms their epiphyseal junction. The anterior one-fifth is formed by the pubis, the posterosuperior two-fifths by the body of the ilium, and the posteroinferior two-fifths by the body of the is-chium. It is a hemispherical hollow on the outer surface of the innominate, facing downwards, forwards and later-ally. The prominent rim of the acetabulum is deficient in-feriorly, forming the acetabular notch. The rim gives attachment to the acetabular labrum of the hip joint; its uneven internal edge provides an attachment for the

synovial membrane of the joint. The acetabular labrum continues across the acetabular notch to produce the trans-verse ligament. The transverse ligament and margins of the notch give attachment to the ligament of the head of the femur. The heavy wall of the acetabulum consists of a semi-lunar articular portion covered with hyaline cartilage, open below, and a deep, central non-articular portion, the ace-tabular fossa. The acetabular fossa is formed mainly from the ischium, and its wall is frequently thin.

Obturator foramen

A large aperture ringed by the sharp margins of the pubis and ischium, those of the pubis overlapping each other in a spiral forming the obturator groove, which runs obliquely forwards and downwards from the pelvis into the thigh, being converted into a canal by a specialization of the obturator fascia. The obturator membrane is at-tached to the margins of the foramen, except superiorly at the obturator groove.

Ossification

Each innominate bone ossifies from eight centres – three primary centres, one each for the ilium, ischium and pubis; and five secondary centres, one each for the iliac crest, the anterior inferior iliac spine, the ischial tuberosity, the pubic symphysis and the triradiate cartilage at the centre of the acetabulum. The sequence of ossification has functional significance because of the support given to the pelvic organs and its role in weight transmission. The primary ossification centres appear during the third, fourth and fifth months of development in the ilium, ischium and pubis respectively. At birth the ilium, ischium and pubis are quite separate and the secondary ossification centres have not yet appeared. By the age of 13 or 14 years, the major parts of the ilium, ischium and pubis are completely bony but are still separated by the Y-shaped triradiate cartilage in the acetabulum. At the age of 8 or 9 years, three major centres of ossification appear in the acetabular cartilage. The largest appears in the anterior wall of the acetabulum and fuses with the pubis. Further centres appear in the iliac acetabular cartilage superiorly, which fuses with the ilium, and in the ischial acetabular cartilage posteriorly, which fuses with the ischium. Fusion of the three bones in the acetabulum occurs between 16 and 18 years. The other secondary ossification centres appear at about puberty and unite with the major bones between 20 and 22 years.

Palpation

The anterior superior iliac spines can easily be palpated in the living, particularly in females where they tend to be further apart. They are situated at the anterior end of the iliac crest, being found in the upper part of the pocket area. Tracing backwards from these spines, the iliac crest is easily palpable, having a large tuberosity about 5 cm from its anterior end. Following the crest as far back as possible, the smaller posterior superior iliac spines can be felt. These are situated in dimples in females, while in males each appears as a small raised tubercle.

About 10 cm below the centre of the iliac crest, the greater trochanter of the femur can be clearly felt. In sitting, the body rests on the ischial tuberosity of each innominate bone; if the hands are placed under this area the tuberosities can be easily felt. This part of the tuberosity is covered with a bursa which often becomes painful and swollen when sitting for too long on a hard surface; this is termed a bursitis.

If the hands are drawn down the front of the abdominal wall, about 5 cm above the genitalia, a bony ring can be felt. This has a central depression where the pubic symphysis is situated with each pubic tubercle about 1 cm above and lateral on either side.

The sacrum

A triangular bone with the apex inferior, it consists of five fused vertebrae broadened by the incorporation of large costal elements and transverse processes into heavy *lateral masses*, which lie lateral to the transverse tubercles on the back of the sacrum extending between the *anterior sacral foramina* onto the front of the bone; the *auricular surface* lies entirely on the lateral mass. It is wedged between the posterior parts of the two innominates with which it articulates at the sacroiliac joints. The pelvic (anterior) surface (Fig. 3.9A) is concave and relatively smooth, being marked by four *transverse ridges* separating the original bodies of the five sacral vertebrae. Lateral to each ridge is the anterior sacral foramen, representing the anterior part of the intervertebral foramen; the foramina are directed laterally and anteriorly.

The dorsal surface (Fig. 3.9B) is convex and highly irregular with *posterior sacral foramina*, medial to which the vertebral canal is closed over by the fused laminae. However, the spinous processes and laminae of the fourth and fifth sacral vertebrae are usually absent, leaving the vertebral canal open. This is the *sacral hiatus*, an inferior entrance to the vertebral canal, which may be used, for example during labour, to introduce an anaesthetic agent to block the sacral nerves. Posteriorly, in the midline, the reduced spinous processes form the *median sacral crest*. Lateral to the posterior sacral foramina are the prominent *lateral sacral crests*, representing the transverse processes, which provide attachment for the dorsal sacroiliac ligaments, and inferiorly for the sacrotuberous and sacrospinous ligaments. Medial to the posterior sacral foramina are the indistinct *intermediate sacral crests*, representing the fused articular processes. The superior articular processes of the first sacral vertebra are large and oval, being supported by short heavy pedicles. Their facets, for articulation with the inferior articular surfaces of the fifth lumbar vertebra, are concave from side to side and face posteromedially. The tubercles of the inferior articular processes of the fifth sacral vertebra form the *sacral cornua* and are connected to the cornua of the coccyx.

The lateral surface (Fig. 3.10A) is triangular, being narrower below. The upper part is divided into an anterior smoother pitted *auricular surface*, covered in cartilage, for articulation with a similar area on the ilium. The rougher posterior area has three deep impressions for attachment of the powerful posterior sacroiliac ligaments. The superior surface (Fig. 3.10B) faces anterosuperiorly and has a central oval area which is the upper surface of the first sacral vertebra; it is separated from the fifth lumbar vertebra by a thick intervertebral disc. Its anterior projecting border is the *sacral promontory*. On each side of the body of the sacrum is the *ala*, formed by the fusion of the costal and transverse processes of the first sacral vertebra. When articulated with the innominate, the ala of the sacrum is continuous with the ala of the ilium.

Ossification

Primary centres appear in the sacrum between the third and eighth month *in utero*; one for each vertebral body, one for each half of each vertebral arch, and one for each costal

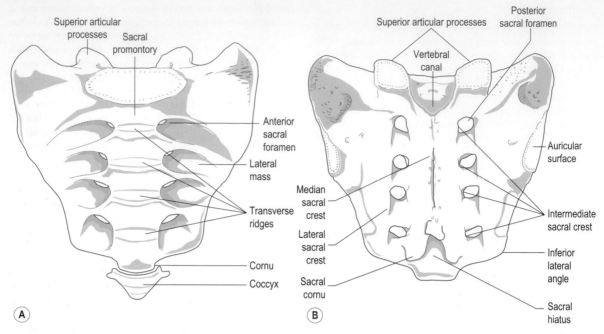

Figure 3.9 Sacrum: (A) anterior view, (B) posterior view.

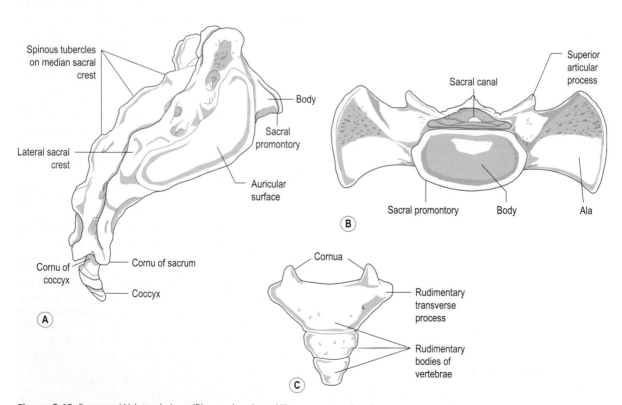

Figure 3.10 Sacrum: (A) lateral view, (B) superior view, (C) coccyx, anterior view.

element in the upper four vertebrae. The costal elements fuse with the arches by the age of 5 years, the arches with the body slightly later, with the two parts of each arch uniting between the ages of 7 and 10. The segments of the lateral masses fuse together during puberty, with secondary centres appearing for the vertebral bodies at about the same time. Bodies and epiphyses fuse between 18 and 25 years. Several secondary centres appear at the ends of the costal and transverse processes from which two epiphyses are formed, one of which covers the auricular surface while the other completes the lower margin of the sacrum.

The coccyx

It usually consists of four fused vertebrae forming a single piece or two pieces of bone (Fig. 3.10C), which mainly represent the bodies. The pelvic surface is concave and relatively smooth, while on the posterior surface the rudimentary articular processes are present as a row of tubercles. Superiorly the larger pair form the coccygeal *cornua* which articulate with the sacral cornua, and enclose the fifth sacral intervertebral foramen. The posterior wall of the vertebral canal of the coccyx is absent so that the sacral hiatus continues downwards over the back of the coccyx.

Section summary

Bones of the pelvic girdle

Innominate
- Consists of three bones (ilium, ischium, pubis) fused together at the acetabulum.
- Orientated so that pelvis faces superoanteriorly.
- The two innominates and sacrum form the pelvic girdle.
- Articulates with: sacrum by the sacroiliac joints; pubis of other side by the symphysis pubis; head of femur by the hip joint.

Sacrum
- Curved bone consisting of five fused vertebrae (concave anteriorly).
- With innominates forms pelvic girdle.
- Articulates with: innominate by the sacroiliac joints; fifth lumbar vertebra by the lumbosacral joint; and the coccyx by sacrococcygeal joint.

The femur

The longest and strongest bone in the body (Fig. 3.11), it transmits body weight from the ilium to the upper end of the tibia. It has a shaft and two extremities.

The upper end of the femur consists of a *head*, *neck*, and *greater* and *lesser trochanters*. The head is slightly more than half a sphere, is entirely smooth and covered with articular cartilage except for a small hollow just below its centre, the *fovea capitis*, which provides attachment for the ligament of the head of the femur. Connecting the head to the *shaft* is the neck, which is approximately 5 cm long and forms an angle of 125° with the shaft. The angle varies a little with age and sex. The neck is flattened anteroposteriorly, giving upper and lower rounded borders; the upper border is concave along its long axis and the lower is straight. The anterior surface of the neck joins the shaft at the *intertrochanteric line* and the posterior at the *intertrochanteric crest*, marked at its centre by the large *quadrate tubercle*.

The large quadrilateral greater trochanter is situated on the lateral aspect of the upper part of the shaft lateral to the neck. It has an upper border marked by a tubercle, an anterior border marked by a depression and posterior and inferior borders both roughened for muscular attachment. Its lateral surface is crossed by a diagonal roughened line running downwards and forwards having above it a smooth area covered by a bursa. The medial surface, above the neck, is small and has a deep *trochanteric fossa* at its centre.

The conical lesser trochanter is situated medially, behind and below the neck, and is smaller than the greater trochanter. Its tip is drawn forwards and presents a roughened ridge running downwards and forwards.

The shaft is strong, and except for a prominent posterior border, almost cylindrical in cross-section. It is gently convex anteriorly, being narrowest at its centre becoming stouter as it approaches the upper and lower extremities. Its posterior border, the *linea aspera*, is rough for muscle attachments, has medial and lateral lips with a central flattened area between. In the upper and lower quarters of the shaft the two lips diverge, producing a posterior surface. The upper surface is marked medially by the narrow vertical *pectineal line*, whereas the lateral truncated border is continuous upwards with the posterior border of the greater trochanter to form the *gluteal tuberosity*. The lower surface, between the *supracondylar lines* above and the *condyles* below, forms the *popliteal surface* of the femur. The rest of the shaft is slightly flattened on its anterior, posteromedial and posterolateral aspects.

The lower end of the femur consists of two large condyles, each projecting backwards beyond the posterior surface of the shaft, with the lateral being stouter than the medial. The inferior, posterior and posterosuperior surfaces of the condyles are smooth and continuous anteriorly with the triangular shaped *patellar surface*, which is grooved vertically, giving larger lateral and smaller medial regions. The two condyles are separated posteriorly and inferiorly by the *intercondylar notch*, marked on its lateral wall posteriorly by the attachment of the anterior cruciate ligament (ACL) and on its medial wall anteriorly by the posterior cruciate ligament (PCL). The separating lips of the linea aspera continue downwards onto the upper aspect of the medial and lateral condyles as the supracondylar lines, the medial presenting at its lower end as the *adductor tubercle*.

The lateral surface of the lateral condyle is roughened, marked just below its centre by the *lateral epicondyle* below

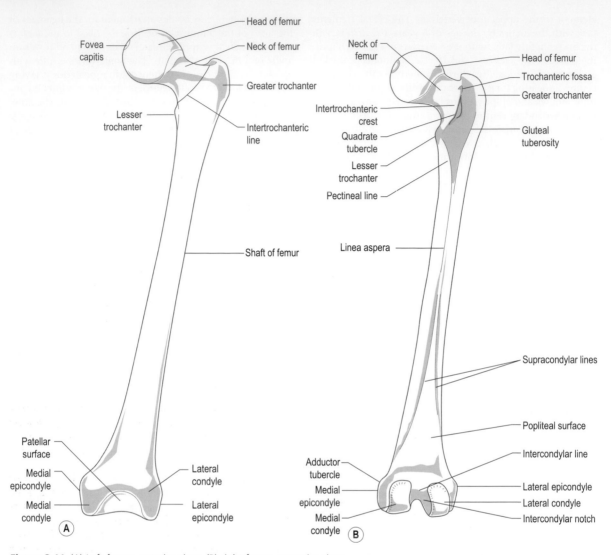

Figure 3.11 (A) Left femur, anterior view, (B) right femur, posterior view.

which is a smooth groove for the popliteus tendon. The medial surface of the medial condyle is also roughened and again marked just below its centre by the *medial epicondyle*.

Ossification

The primary ossification centre for the shaft appears at 7 weeks *in utero*. At birth, growth plates separate the bony shaft from the upper and lower cartilaginous epiphyses (Fig. 3.12). A secondary ossification centre appears in the lower epiphysis shortly before birth. Secondary ossification centres appear in the upper epiphysis for the head at 1 year and in the greater trochanter at 4 years (Fig. 3.12). The last secondary ossification centre appears in the cartilaginous lesser trochanter at 12 years. The upper epiphysis fuses with the shaft at about the 18th year, the last to do so being the

head. The lower epiphysis fuses with the shaft at about 20 years. The neck of the femur is ossified as part of the body (shaft) and not from the upper epiphysis.

Palpation

The femur is almost completely surrounded by muscles and is only palpable in limited areas. At its upper end, the greater trochanter is an obvious landmark, projecting just a little more laterally than the iliac crest, being easily located by running the hands down from the middle of the crest some 7–10 cm. The greater trochanter is perhaps easier to feel if the fingers are brought forward from the hollows on the sides of the buttocks in the region of the back pocket. Its posterior border can be palpated for about 5 cm, running down towards the shaft, while its upper

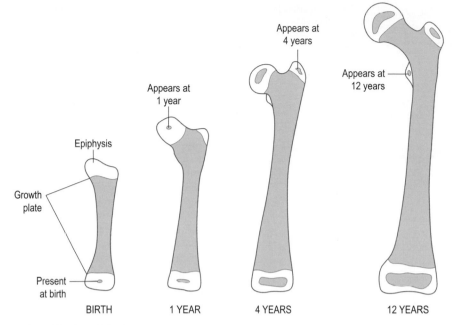

Figure 3.12 Stages in ossification of the femur.

border is an important landmark to locate the level of the hip joint.

At its lower end, the femur is well covered with muscle until just above the knee joint. As the fingers pass down the medial side of the thigh the medial condyle can be palpated. This is marked just behind its centre by the medial epicondyle, above which the adductor tubercle can be palpated with the tendinous part of adductor magnus attaching to it. On the lateral side of the knee, the lateral condyle can be palpated with the lateral epicondyle projecting from its outer surface. At the lower edge of each of these condyles, the knee joint line can be palpated particularly as it passes forwards.

If the knee is fully flexed, the patella is seen to move downwards revealing on the front of the knee the two femoral condyles, covered by the lower part of quadriceps femoris and its retinacula.

Section summary

The femur

- Long bone of the thigh having a proximal head, neck, greater and lesser trochanters; shaft with linea aspera posteriorly; distal end with medial and lateral condyles.
- Head articulates with acetabulum of innominate bone forming the hip joint.
- Condyles articulate with tibia and patella forming the tibiofemoral and patellofemoral joints respectively; both are part of the knee joint.

The patella

A triangular sesamoid bone (Fig. 3.13) formed in the tendon of quadriceps femoris, with its *apex* pointing inferiorly and its base uppermost. It is flattened from front to back, having anterior and posterior surfaces and superior, lateral and medial borders.

The anterior surface of the patella is marked by a series of roughened vertical ridges produced by the fibres of quadriceps which pass over it. It is slightly convex forwards and its shape varies according to the pull of the muscle.

The posterior surface has a large, smooth oval facet covered with hyaline cartilage for articulation with the patellar surface of the femur. It is divided by a broad vertical ridge into a smaller *medial* and a larger *lateral facet*. The cartilage on each of these facets is marked by two horizontal lines dividing each surface into upper, middle and lower sections. Below, there is a roughened area on the posterior aspect of the apex for the upper attachment of the ligamentum patellae.

The base of the patella is roughened for the attachment of rectus femoris and vastus intermedius. The medial and lateral borders are rounded but also roughened, receiving attachments of vastus medialis and lateralis.

Ossification

At birth the patella is cartilaginous, ossifying from a single centre or several centres between 3 years and puberty. Occasionally the patella may be absent.

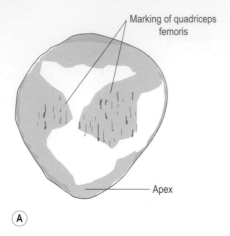

Marking of quadriceps femoris

Apex

(A)

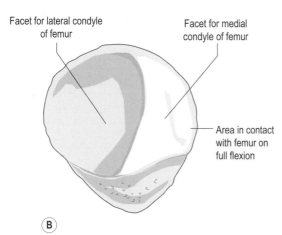

Facet for lateral condyle of femur

Facet for medial condyle of femur

Area in contact with femur on full flexion

(B)

Figure 3.13 The patella: (A) anterior surface, (B) posterior surface.

Palpation

As the patella lies subcutaneously, the whole margin as well as its anterior surface can be palpated.

The tibia

A long bone (Fig. 3.14), which transmits body weight from the medial and lateral condyles of the femur to the foot. It is the larger of the two bones of the leg, being situated medial to the fibula. It consists of a shaft and two extremities, the upper extremity being much larger than the lower.

The upper end is expanded in all directions, but particularly posteriorly where it projects beyond the shaft. It consists of two *condyles* having between them anteriorly a large, truncated area elongated in its vertical axis, roughened in its upper and smooth in its lower parts. This is the *tibial tuberosity*, the roughened area giving attachment to the ligamentum patellae. The lateral condyle projects

further laterally than the shaft and has a round articular facet on its posterolateral part for articulation with the head of the fibula. Posteriorly the space between the condyles is smooth. On the superior surfaces of the two condyles are areas for articulation with the femoral condyles. These are divided by two raised tubercles, the medial and lateral intercondylar tubercles, which are close together and termed the *intercondylar eminence*. In front of and behind the eminence is an uneven non-articular area which is narrower close to the eminence, becoming wider as it passes anteriorly and posteriorly. This area gives attachment to some important structures of the knee joint. Anterior to the intercondylar eminence three structures are attached; most anteriorly is the anterior horn of the medial meniscus, whilst closest to the eminence is the anterior horn of the lateral meniscus, and between them the ACL. The area behind the intercondylar eminence also gives attachment to three structures; most posteriorly is the PCL, whilst closest to the eminence is the posterior horn of the lateral meniscus; between them is the posterior horn of the medial meniscus.

The *shaft* is triangular in cross-section, tapering slightly from the condyles for about two-thirds of its length and widening again at its lower end. It has an *anterior border* which runs from the lower part of the tibial tuberosity downwards to the anterior part of the *medial malleolus*. The *medial border* begins just below the posterior aspect of the medial condyle, and although not always easy to see, can be traced to the *posterior part* of the medial malleolus. The *interosseous border* begins just below the articular facet on the lateral condyle and runs in a curved line with its concavity forwards, down to the roughened triangular area on the lateral side at the lower end of the bone.

The shaft therefore has three surfaces – medial, posterior and lateral. The smooth *medial surface*, sloping posteriorly from the anterior border, is subcutaneous for the whole of its length, from the medial condyle above to the medial malleolus below and is commonly called the shin. The *lateral surface* between the anterior and interosseous borders is slightly concave, particularly in its upper two-thirds, and gives attachment to tibialis anterior. Inferiorly it becomes continuous with the anterior surface of the lower end of the bone. The posterior surface between the interosseous and medial borders is crossed by two raised lines, one running obliquely from just below the lateral condyle downwards and medially to join the posterior border about halfway down; this is the *soleal line*. The area above it is roughened for the attachment of popliteus. Below the soleal line is a *vertical line* to which the fascia covering tibialis posterior is attached. It divides the lower part of the posterior surface into two roughened areas for the attachment of muscle, laterally tibialis posterior and medially flexor digitorum longus.

The lower end is expanded, but to a lesser extent than the upper. It has a prominent medial malleolus which is

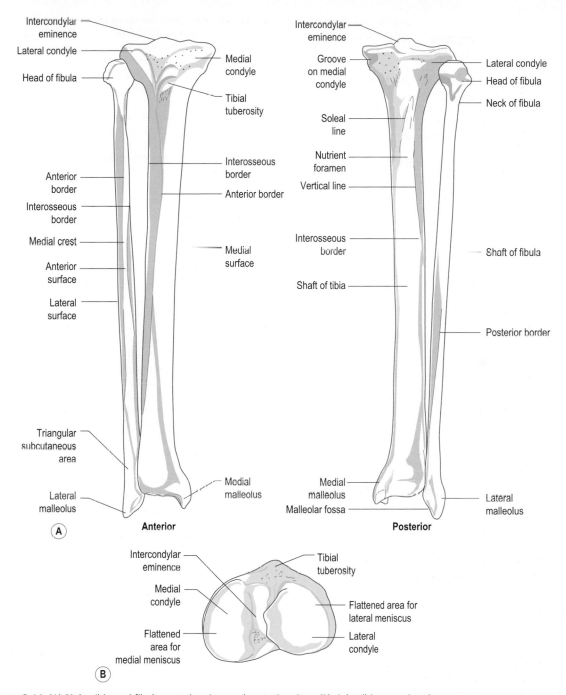

Figure 3.14 (A) Right tibia and fibula, anterior view and posterior view, (B) right tibia, superior view.

continuous with the medial surface of the shaft projecting downwards from its medial side. The inferior surface is smooth for articulation with the superior surface of the body of the talus. Medially it is continuous with the malleolar articular surface. It usually turns upwards on the lateral surface

where it becomes concave anteriorly for articulation with the fibula. It continues superiorly as a rough triangular area for the attachment of the interosseous ligament. The posterior surface is coarse and grooved by tendons passing into the foot. The anterior surface is smooth and slightly convex.

Ossification

The primary ossification centre for the tibia appears in the shaft during the seventh week *in utero* and spreads so that at birth only the ends are cartilaginous. The secondary centre for the proximal end, which includes the tibial tuberosity, appears at birth, spreading down to the tuberosity after the tenth year. An independent centre for the tuberosity may appear; if it does so, this appears at 11 years. The secondary centre for the distal end appears during the second year. Fusion of the proximal epiphysis with the shaft occurs between 19 and 21 years, and of the distal epiphysis with the shaft a few years earlier, between 17 and 19.

Palpation

The tibial tuberosity is easily recognizable at the upper end of the anterior border of the tibia (the shin), with the ligamentum patellae attaching to its upper part. The medial and lateral condyles can be palpated about 2 cm higher, as they are subcutaneous as far as the hamstring muscles on either side. The upper edge of the condyles indicates the knee joint line.

Just below and behind the midpoint on the lateral side, the head of the fibula stands out clearly. Running down the whole length of the tibia from the medial surface of the medial condyle is the medial surface of the shaft, being subcutaneous as far as the medial malleolus. Both anterior and posterior borders are palpable at its edges. The medial malleolus is subcutaneous and its medial surface, borders and tip are easily palpable.

Applied anatomy

Due to the fact that the medial surface of the tibia is subcutaneous, the risks of damage to and fracture of this bone are increased. In addition, the likelihood of infection, and delayed/non-union of the bone is very high and a common complication in this region. The most common area of damage is at the junction between the upper two-thirds and the lower third of the shaft, this being its thinnest part, and unfortunately the area with the poorest blood supply.

The fibula

A long slender bone (Fig. 3.14A) being expanded both at its upper and lower ends.

The upper end, or *head*, is expanded in all directions, having on its superomedial side a facet for articulation with the lateral condyle of the tibia. Lateral to the facet is the apex of the head, which projects upwards. The rest of the upper end is roughened for the attachment of biceps femoris. Just below the head is the *neck*, around which runs the common fibular (peroneal) nerve.

The *shaft* varies considerably in individuals and its features are often difficult to recognize; it has three borders and three surfaces. The *anterior border* is more prominent inferiorly where it widens into a smooth, *triangular,*

subcutaneous area continuous with the lateral surface of the malleolus. It runs from below the anterior aspect of the head passing vertically down to the triangular subcutaneous area. Medial to the anterior border is the *interosseous border*, again often poorly marked. Extending from the neck, it lies close to the anterior border in its upper third, but then passes posteriorly and medially to join the apex of the roughened triangular area superior to the malleolar articular surface. The *posterior border* begins below the lateral aspect of the head and neck, and passes down to the medial margin of the posterior surface of the *lateral malleolus*; this border is rounded and more difficult to trace. The *lateral surface* of the fibula is concave and posterolateral to the anterior border. It becomes convex as it winds round posterior to the triangular subcutaneous area to the posterior surface of the malleolus. It is roughened for the attachment of the fibularis (peroneal) muscles. The *anterior surface* is a very narrow strip between the anterior and interosseous borders at its upper end, expanding as it continues downwards. The posterior surface is more expanded than the anterior and lateral surfaces, being divided by a vertical ridge, the *crest*, into a medial and lateral part similar to the tibia. The region between the crest and the interosseous border is concave and usually divided by an oblique line, whereas the region between the crest and posterior border is flat and roughened in its upper part by the attachment of soleus.

The lower end can be recognized readily, being flattened on its medial and lateral sides and having posteriorly a deep *malleolar fossa*. On its medial side, just above the fossa, is a triangular area which is smooth for articulation with the lateral surface of the body of the talus. Just above this medial articular area is an elongated roughened area for attachment of the interosseous ligament of the inferior tibiofibular joint, just below which is the malleolar fossa. The fibula varies in shape according to the muscles that are attached to it, and it can be seen that it carries no weight, but contributes to the lateral stability of the ankle joint.

Ossification

The primary centre appears in the shaft during the seventh week *in utero*, again spreading so that at birth only the ends are cartilaginous. The secondary centre for the distal end appears during the second year and fuses with the body between 17 and 19 years. The secondary centre for the proximal end appears slightly later, during the third or fourth years, and also fuses with the body later, between 19 and 21 years.

At birth, the fibula is relatively thick, being about half as thick as the tibia in the third postnatal month. As development and growth continue, the thickness of the fibula and tibia progressively changes to approach adult proportions. The distal end of the fibula does not reach below the medial malleolus until after its ossification has begun, that is after the second year. It is only after this time that the adult relations of the malleoli can be seen.

Palpation

The head of the fibula can be readily palpated on the posterolateral side below the knee joint. If the hand is placed on the lateral side of the calf and moved up towards the knees, the bony head can be felt projecting laterally. The head can also be easily palpated if the fingers are placed in the hollow on the lateral side of the knee when it is flexed to 90°. Little of the shaft can be palpated as it is surrounded by muscles; however, in its lower third an elongated triangular area can be palpated, the lateral aspect of which can be traced down to the lateral malleolus. The lateral malleolus is easily palpated on the lateral side of the ankle projecting down to a point 2.5 cm below the level of the ankle joint.

Section summary

Bones of the leg

Tibia

- Medial bone of leg having proximal expanded tibial condyles; shaft with tibial tuberosity anteriorly and sharp, lateral facing interosseous border; slightly expanded distal end with medial malleolus.
- Proximally the medial and lateral tibial condyles articulate with the medial and lateral femoral condyles forming the knee joint; the facet below the lateral condyle articulates with the head of the fibula forming the superior tibiofibular joint.
- Distally articulates with fibula forming the inferior tibiofibular joint; the talus forming the ankle.

Fibula

- Lateral bone of leg having proximal head; irregular shaft with sharp medial facing interosseous border; expanded distal end with lateral malleolus projecting inferiorly.
- Articulates with the tibia superiorly and inferiorly forming the superior and inferior tibiofibular joints; the talus inferiorly forming part of the ankle joint.

The bony structure of the foot

The foot (Figs. 3.15 and 3.16) consists of many small bones; posteriorly are the tarsus and anteriorly the *metatarsals* and *phalanges*. The tarsus and metatarsals comprise the foot proper, whereas the phalanges comprise the toes. The largest bone in the foot is the *calcaneus*, while the largest metatarsal is the most medial, having anterior to it the two phalanges of the big toe or hallux. The other metatarsals of the foot each have three phalanges distal to them. Along the medial longitudinal arch from posterior to anterior are the calcaneus, the *talus* (situated more on top of the calcaneus), the *navicular*, the three *cuneiform bones*, and the first, second and third metatarsals with their phalanges.

Along the lateral longitudinal arch of the foot, again from posterior to anterior, are the calcaneus, the *cuboid*, the fourth and fifth metatarsals with their phalanges.

The tarsal bones

The calcaneus

Lies inferior to the talus and projects backwards to form the prominence of the heel (see also Figs. 3.140 and 3.141); it is strongly bound to all the tarsal bones by ligaments. It is the largest bone in the foot, being oblong in shape and having six surfaces. The anterior surface faces forwards for articulation with the cuboid; it is slightly convex from top to bottom and more or less flat from side to side. The medial part of this surface extends onto the medial side of the calcaneus to accommodate a backward projection of the cuboid. The posterior surface is rounded and has three areas – the upper part is smooth where a bursa lies between it and the tendocalcaneus; the middle is smooth and convex, except at its lower margin where it ends as a jagged rough edge, receives the attachment of the tendocalcaneus; the lowest subcutaneous part is roughened and covered by the strong fibrous tissue and fat of the heel pad. This lowest part of the posterior surface transmits body weight from the heel to the ground during heel-strike in walking, and curves forwards onto the inferior surface.

Here are found the larger medial and smaller lateral tubercles projecting forwards. The inferior surface continues forwards as a rough area terminating as the anterior tubercle; the long plantar ligament attaches to the rough area. The lateral surface is slightly roughened and nearly flat and has two tubercles, one for the attachment of part of the lateral ligament of the ankle joint, and the other, which is slightly lower and more anterior, provides attachments for the inferior fibular (peroneal) retinaculum. The latter fibular (peroneal) tubercle is elongated, with grooves above and below. The medial surface is smooth and hollowed, being overhung anteriorly by the sustentaculum tali, below which is the groove for the tendon of flexor hallucis longus. On the superior surface of the sustentaculum tali is the middle articular surface for the head of the talus. Behind the middle articular surface is a deep groove, the sulcus calcanei, which continues across the superior surface of the calcaneus in a posteromedial direction. In front of the sinus calcanei is a roughened area for the attachment of muscles and ligaments, while behind it is the posterior articular surface, convex from front to back and flat from side to side, for articulation with the undersurface of the body of the talus. Behind this articular surface is a further roughened area, concave upwards from front to back and convex from side to side.

The trabeculae of the calcaneus (see Fig. 3.157) have a particular arrangement due to the weight-bearing nature of the bone. From the posterior articular surface, the supporting trabeculae pass downwards and backwards to the

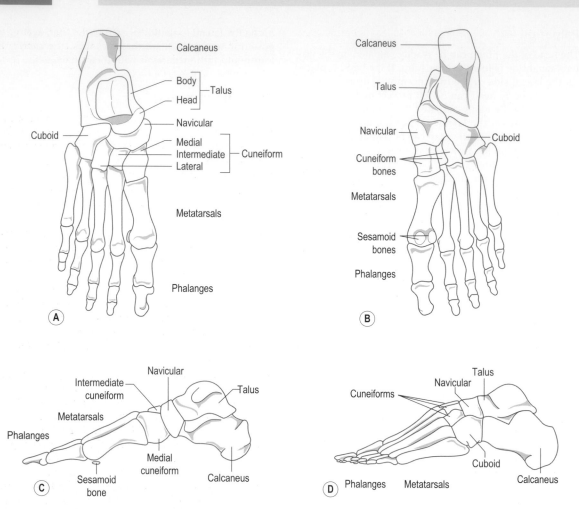

Figure 3.15 (A) Right foot, superior view, (B) right foot, inferior (plantar) surface, (C) right foot, medial view, (D) left foot, lateral view.

heel and downwards and forwards to the articular area for the cuboid. Running from the heel to the anterior surface are superior and inferior arcuate systems serving to tie the bone together. Between all of these systems there is an area of less dense and therefore weaker bone.

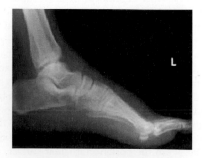

Figure 3.16 Radiograph of the left foot, medial view.

The talus

Situated above the calcaneus, the talus (see also Fig. 3.141) has a head and neck directed forwards and medially. It transmits the body weight from the tibia to the calcaneus and navicular. The body is wedge-shaped from front to back and lies between the malleoli of the tibia and fibula, it is wider anteriorly. Its upper surface is convex from front to back and slightly concave from side to side, being pulley-shaped, and articulates with the trochlear surface of the tibia. The lateral surface is triangular in shape with its apex pointing downwards; it articulates with the medial surface of the lateral malleolus. The medial surface is partly articular, with its upper articular part being comma-shaped. The medial and lateral articular surfaces are continuous with the upper surface of the talus. Below the medial articular surface is a depressed, roughened area for attachment of the deep part of the deltoid ligament. The inferior surface of the body is also articular, being concave from front to back, and articulates with the posterior facet on the upper surface of the calcaneus.

At the posterior aspect of the talus is a groove running downwards and medially for the tendon of flexor hallucis longus. Lateral and medial to this groove are the lateral and medial tubercles of the talus.

From the anteromedial aspect of the body the neck projects forwards and medially. Its upper, medial and lateral surfaces are roughened, whilst its inferior surface presents an area for articulation with the calcaneus on the upper surface of the sustentaculum tali. Just behind this articular surface is a deep groove termed the sulcus tali, which lies immediately over the sulcus calcanei and forms with it the sinus tarsi.

The head of the talus is slightly flattened anteriorly and articulates with the posterior surface of the navicular. Below this main articulation are two smaller articular areas, one for the upper surface of the 'spring' ligament and the other, which continues onto the inferior surface of the neck, for the anterior articular area of the calcaneus.

The navicular

Lies anterior to the head of the talus. On its inferomedial side is a large tuberosity. The posterior surface is concave for articulation with the head of the talus. The anterior surface is subdivided into three triangular areas by two faint ridges for articulation with the three cuneiform bones. The inferior surface is narrow and roughened for the attachment of ligaments and muscles. The small lateral and subcutaneous upper surfaces are rough near their edges for the attachment of interosseous ligaments, but together they form a curved surface.

The cuboid

Situated lateral to the navicular, anterior to the calcaneus and posterior to the fourth and fifth metatarsals the cuboid has six surfaces, but in reality is a cube that has been flattened from above downwards. The posterior surface is slightly concave from top to bottom but flat from side to side, articulating with the anterior surface of the calcaneus. The medial surface is smooth on its anterior two-thirds for articulation with the lateral cuneiform and occasionally the navicular, while the posterior third is usually roughened for the attachment of ligaments. Anteriorly it is nearly flat, being divided by a slight ridge into two facets for articulation with the bases of the fourth and fifth metatarsals.

The lateral surface is the smallest due to the convergence of the anterior and posterior surfaces as they pass laterally. Nearly the entire surface is taken up by a deep groove passing downwards and forwards through which the tendon of peroneus longus passes. The groove is continued on the under surface of the bone and crosses medially and anteriorly towards the medial cuneiform. The groove is very close to the anterior border of the cuboid and is limited by a prominent ridge posterior to it. The remainder of the undersurface is rough for the attachment of the long and short plantar ligaments. The dorsal surface is roughened, and as in the case of the dorsal surfaces of the cuneiforms and navicular, is subcutaneous.

The cuneiforms

There are three cuneiform bones: medial, intermediate and lateral. As the name implies they are wedge-shaped, triangular at their anterior and posterior ends with three rectangular surfaces along their length.

Medial cuneiform This has the apex projecting superiorly and its base inferiorly. The anterior and posterior surfaces are smooth for articulation with the first metatarsal and anterior surface of the navicular respectively. The smooth lateral surface articulates with the intermediate cuneiform in its posterior two-thirds, and the base of the second metatarsal on its anterior third. The superior, medial and inferior surfaces form a continuous surface on the medial side of the foot, which is roughened by ligament attachments; it has a smooth impression at the anteroinferior part of its medial aspect over which the tendon of tibialis anterior runs. This is the largest of the three cuneiforms.

Intermediate cuneiform This has its base superiorly and its apex inferiorly. It is shorter than the other two cuneiforms and is only non-articular on its dorsal surface. It articulates medially with the medial cuneiform, laterally with the lateral cuneiform, anteriorly with the second metatarsal and posteriorly with the navicular. Part of the medial surface is roughened for the attachment of ligaments.

Lateral cuneiform The apex projects inferiorly and base superiorly. The medial surface articulates mainly with the middle cuneiform, having a small facet anteriorly for articulation with the second metatarsal. The lateral surface articulates with the medial surface of the cuboid, the posterior surface with the navicular and the anterior surface with the third metatarsal. The nonarticular parts of the medial and lateral surfaces are roughened for the attachment of ligaments.

The fact that the medial cuneiform has its base projecting inferiorly, whilst the other two have their bases superior, contributes to the arch shape across the foot from medial to lateral. With the addition of the cuboid on their lateral side, the cuneiforms make up part of the transverse tarsal arch.

The metatarsals

There are five metatarsal bones in each foot, the most medial of which is the stoutest, although it is also the shortest. The second metatarsal is the longest, whilst the fifth can be recognized by the large tubercle projecting posterolaterally from its base. All five metatarsals have certain features in common – a shaft, with a head distally, and a base proximally. The bases articulate with the tarsal bones while the heads articulate with the proximal phalanx of each toe (Fig. 3.15A,B).

The base of the first metatarsal is concave from side to side and flat from above downwards, articulating with the anterior surface of the medial cuneiform. Its lateral surface has a facet for articulation with the base of the second metatarsal, whilst its inferior surface projects downwards ending as a tuberosity. The base of the second metatarsal articulates with the intermediate cuneiform posteriorly. Medially it articulates with the medial cuneiform and first metatarsal, and laterally with the lateral cuneiform and third metatarsal. The base of the third metatarsal is flat and articulates with the lateral cuneiform, and on either side with the adjacent metatarsals; it is roughened on its upper and lower surfaces. The fourth and fifth metatarsal bases articulate with the anterior surface of the cuboid. The fourth has a small facet on either side for articulation with the adjacent metatarsals, whereas the base of the fifth is more expanded, having a large tubercle on its lateral side. The upper and lower surfaces of each are roughened.

All the shafts are more or less cylindrical, the first being the thickest and the second usually the thinnest. All become narrower as they pass forward towards their heads.

The heads are smooth, convex from above downwards as well as from side to side. On either side, just behind the head, is a tubercle in front of which is a small depression for the attachment of ligaments. The superior non-articular surface is roughened, while the inferior surface is marked by a groove passing forwards, which gives passage to the long and short flexor tendons. The head of the first metatarsal is large and wide, forming the ball of the great toe. It articulates with the base of the first phalanx and the two sesamoid bones. The plantar surface of this bone is grooved, on each side of a prominent central ridge, by the sesamoid bones in the tendons of the short muscles passing inferior to it.

The phalanges

There are two phalanges in the great toe and three in each of the other toes. They are miniature long bones having a shaft and two extremities and with certain features in common. Each of the bases of the proximal phalanges has a smooth concave proximal surface for articulation with the head of its metatarsal. The remaining phalanges have a proximal surface divided into two by a vertical ridge. Each bone is flattened on its plantar surface and rounded on its dorsum. The head of each bone, except the terminal phalanges, is divided into two condyles by a vertical groove giving it a pulley shape. The articular surface tends to be more extensive on the plantar surface of the head where it joins the flattened surface of the shaft. The sides of the heads are roughened, being marked by a small tubercle at the centre.

The head of each distal phalanx is flattened on its dorsum and has no articular area. This surface is the nail bed.

Ossification of the bones of the foot

The tarsus

Each tarsal bone ossifies from a primary centre which appears in the cartilaginous precursor. The calcaneus is the only tarsal bone to have a secondary centre. The primary centres for the calcaneus and talus appear before birth in the sixth and eighth months *in utero* respectively. That for the cuboid appears at 9 months *in utero* and may therefore be present at birth; if not, it appears soon afterwards. The centres of ossification for the remaining bones appear as follows: at the end of the first year for the lateral cuneiform, during the third year for the medial cuneiform and navicular, and during the fourth year for the intermediate cuneiform. Ossification of these bones is completed shortly after puberty.

The secondary centre for the calcaneus appears at about 9 years in its posterior end and extends to include the medial and lateral tubercles. Fusion occurs between 15 and 20 years. Occasionally, the lateral tubercle may ossify separately.

Because the ossification centres for the calcaneus, talus and cuboid are usually present before birth, they can be used to assess the skeletal maturity of a newborn child. They may be used in conjunction with the secondary centres in the distal end of the femur and in the proximal end of the tibia.

The metatarsals

A primary centre appears in the body of each metatarsal at 9 weeks *in utero*, so that at birth they are well ossified. Secondary centres appear in the base of the first metatarsal and in the heads of the remaining metatarsals during the second and third years, with the medial ones appearing earlier. Fusion of the epiphyses with the bodies occurs between 15 and 18 years. In the lateral metatarsals, the epiphyses may occasionally be found in the bases instead of the heads.

The phalanges

The primary centres for the distal and proximal phalanges appear during the fourth month *in utero*, with the distal ones appearing first. The primary centre for the middle phalanx appears between 6 months and birth. Secondary centres for the bases of all phalanges appear during the second and third years and fuse with the bodies between 15 and 20 years.

It is interesting to note that the first metatarsal has an ossification pattern similar to that of the phalanges. It could be argued that instead of the middle phalanx being missing in the great toe it is the metatarsal that is missing, so that what we now refer to as the first metatarsal is in fact an enlarged proximal phalanx.

Palpation of the bones of the foot

Posteriorly the calcaneus can clearly be identified being subcutaneous on its lateral, posterior and medial aspects.

The inferior surface is covered with thick plantar fascia, but the medial and lateral tubercles are identifiable on deep palpation posteriorly. Medially, 1 cm below the tip of the medial malleolus, the sustentaculum tali appears as a horizontal ridge, whilst on the lateral aspect, the fibular (peroneal) tubercle lies approximately 2 cm below and anterior to the tip of the lateral malleolus, with the lateral tubercle of the calcaneus (for the attachment of the calcaneofibular ligament) being slightly posterior.

The head and neck of the talus can be gripped between the finger and thumb in the two hollows just anteroinferior to the medial malleolus, the tubercle of the navicular forming a clear landmark anterior to the medial hollow. Midway along the lateral border of the foot, the base of the fifth metatarsal, with its tubercle directed posteriorly, is prominent. The bases of the fourth to the first metatarsals can be identified across the dorsum of the foot, the base of the first being 1 cm anterior to the tubercle of the navicular, the medial cuneiform being interposed.

The shafts and heads of the metatarsals can be readily palpated on the dorsum of the foot with the bases proximal and heads towards the toes. If the toes are extended, the heads of the metatarsals, especially the first, become palpable under the forefoot. The heads are a little less obvious on the dorsum of the foot when the metatarsophalangeal joints are flexed.

The proximal phalanx of each toe is easily recognized, being the longest of the three; the rest are hidden to a certain extent by the pulp of the toe.

Section summary

Bones of the foot

Tarsus

- Seven irregular bones located in the hindfoot and midfoot.
- *Talus* articulates with: the tibia and fibula forming the ankle joint; the calcaneus forming the subtalar joint; the navicular as part of the talocalcaneonavicular joint.
- *Calcaneus* articulates with: the talus forming the subtalar joint; the cuboid forming the calcaneocuboid joint.
- *Navicular* articulates with: the talus as part of the talocalcaneonavicular joint; the cuneiforms forming the cuneonavicular joints.
- *Cuneiforms* (medial, intermediate, lateral) articulate with: the navicular forming the cuneonavicular joints; the bases of medial three metatarsals forming the tarsometatarsal joints.
- *Cuboid* articulates with: the calcaneus forming the calcaneocuboid joint; the bases of the lateral two metatarsals forming tarsometatarsal joints.

Metatarsals

- Five long bones (*metatarsals*) in foot each with quadrilateral base proximally, shaft, rounded head distally.
- Bases articulate with the cuneiforms, cuboid and adjacent metatarsals.
- Heads articulate with base of corresponding proximal phalanx forming metatarsophalangeal joints.

Phalanges

- Fourteen individual bones with two in big toe (proximal, distal) and three in each of other toes (proximal, middle, distal) each having: base; shaft; head.
- Bases articulate with heads of corresponding metatarsal.
- Heads of middle phalanges articulate with base of middle or distal phalanges forming interphalangeal joints.

MUSCLES

Muscles around the hip joint

The hip joint is situated deep within the gluteal region. It is a ball and socket joint capable of movement in many directions.

To produce these movements there is a complex arrangement of muscles around the joint which either act on the thigh with respect to the pelvis or on the pelvis with respect to the thigh. It must be remembered that during many of these movements the hip joint is weight-bearing, transmitting the weight of the body above it, via the lower limbs, to the ground.

Thus muscles surrounding the joint have a dual role; they must be capable of immediate controlled power when needed for sudden powerful activities such as running uphill or upstairs, yet retain the ability to maintain a set position for long periods of time as in standing, leaning forwards and sitting.

The hip joint is completely surrounded by muscles, which are thicker and stronger posteriorly and laterally; consequently the joint appears to be situated towards the front of the region.

Muscles anterior to the joint tend to be flexors, those posterior tend to be extensors, those medial tend to be adductors and those lateral abductors. Both medial and lateral rotation occurs at the joint because of the obliquity of some of the muscle fibres. This is explained more fully under the individual muscles.

Some of the muscles in this region have their effect on more than one joint. When this is the case, reference is made to the muscle in both regions, but details are only found in one section.

Muscles extending the hip joint

Gluteus maximus
Hamstrings
    Semitendinosus
    Semimembranosus
    Biceps femoris

Gluteus maximus

As its name implies, this is the largest of the gluteal muscles. It is very powerful and is situated on the posterior aspect of the hip joint. In lower primates gluteus maximus is an adductor of the hip; this was also the case in early, primitive humans. However, with the changes that have occurred in the human pelvis associated with the erect posture, gluteus maximus has become mainly an extensor of the hip. It is the muscle mainly responsible for the erect position, therefore freeing the forelimbs (upper limbs) from a weight-bearing role, enabling them to become the precision implements that they are today.

Gluteus maximus (Fig. 3.17) is quadrilateral in shape consisting of bundles of muscle fibres laid down in the line of pull of the muscle, giving its surface a coarse appearance.

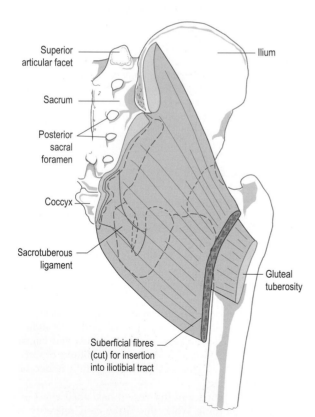

Figure 3.17 Attachments of the right gluteus maximus (posterior view) with direction of fibres in diagrammatic form.

It is a thick muscle formed in two layers as it passes down to its lower attachment. Above, it attaches to the *gluteal surface* of the *ilium* behind the posterior gluteal line, the *posterior border* of the *ilium* and the adjacent part of the *iliac crest*. It also arises from the *side* of the *coccyx* and the *posterior aspect* of the *sacrum*, including the *upper part* of the *sacrotuberous ligament*. Its upper fibres attach to the *aponeurosis* of the *sacrospinalis* while its deep anterior fibres come from the *fascia* covering gluteus medius.

The fibres pass downwards and forwards towards the upper end of the femur. The most superficial fibres, about three-quarters, form a separate lamina, which narrows down and attaches between the two layers of the fascia lata, helping to form the *iliotibial tract*. The remaining deeper one-quarter of the fibres form a broad aponeurosis which attaches to the *gluteal tuberosity* of the *femur*.

Nerve supply

By the *inferior gluteal nerve* (root value L5, S1 and 2). The skin covering the muscle, however, is mainly supplied by branches from L2 and S3.

Action

When acting from above, the muscle pulls the shaft of the femur backwards, producing extension of the flexed hip joint. As its lower attachment is nearer to the lateral side of the thigh, it also tends to rotate the thigh laterally during extension. The lower fibres can adduct the thigh, while the upper fibres may help in abduction.

Those fibres attaching to the iliotibial tract can produce extension of the knee joint because the lower end of the tract attaches to the lateral tibial condyle anterior to the axis of movement. Through the iliotibial tract, gluteus maximus provides powerful support on the lateral side of the knee. If the femur is fixed, contraction of gluteus maximus pulls the ilium and pelvis backwards around the hip joint, but this time the pelvis and trunk are the moving parts, and a lifting of the trunk from a flexed position occurs.

Functional activity

Being a powerful extensor of the thigh, especially when the hip has been flexed, means that gluteus maximus is ideally suited for fulfilling its role in such powerful movements as stepping up onto a stool, climbing and running. However, it is not used greatly as an extensor in ordinary walking.

With the hamstrings, it participates in raising the trunk from a flexed position, as in standing upright from a bent forward position. Indeed, gluteus maximus and the hamstrings provide the main control in forward bending movement of the body, as the movement primarily occurs at the hip joint.

It plays an important role in balancing the pelvis on the femoral heads, thus helping to maintain the upright posture; its ability to aid lateral rotation of the femur when

The image labels (Figure 3.17):
Superior articular facet
Sacrum
Posterior sacral foramen
Coccyx
Sacrotuberous ligament
Superficial fibres (cut) for insertion into iliotibial tract
Ilium
Gluteal tuberosity

standing assists in raising the medial longitudinal arch of the foot.

The role of gluteus maximus during sitting should not be dismissed. Although the ischial tuberosities support the majority of the weight of the trunk when sitting, pressure is regularly relieved from these bony points by a static or sometimes dynamic contraction of the muscle which raises the tuberosities of the ischium from the supporting surfaces. The muscle is then relaxed and the weight is then lowered. Sometimes the weight is shifted from side to side with the alternate use of gluteus maximus of each side.

Paralysis of gluteus maximus leads to flattening of the buttock, and an inability to climb stairs and run. However, it must be kept in mind that there are other muscles which can be brought into action to extend the hip, although it is then a much weaker movement. Gluteus maximus can be developed to produce a functional extension of the knee in patients where quadriceps femoris is either very weak or paralysed. This is not a powerful movement, but may be sufficient to enable the patient to extend the knee and enable the lower limb to become weight-bearing during walking or standing.

Palpation

First locate the iliac crest approximately at the belt level; on moving the hand backwards along the crest a small bony process can be felt; this is the posterior superior iliac spine. With the fingers running inferiorly and medially place the centre of the palm over this point. The hand will now just about cover the upper attachment of gluteus maximus; the palm is over the posterior part of the ilium, the sacrum and the back of the sacroiliac joint, while the tips of the fingers are on the edge of the coccyx and the upper end of the sacrotuberous ligament. The bulk of the muscle is now under the palm; follow this path to the greater trochanter of the femur. Now try the following:

1. Extend the lower limb whilst in the standing position, keeping the hand on the muscle; it goes hard and produces a much clearer shape.
2. Place the foot onto a stool and put the hand in the same position as before and step up. Again the muscle will be felt coming into action very strongly.
3. Take up the standing position and place the hands on each gluteus maximus as if they were in the back pocket. Raise the medial borders of the feet as if to shorten the medial longitudinal arch of the foot. As the arch is raised gluteus maximus will be felt working quite strongly, with the femur tending to rotate laterally.
4. Finally, when sitting place a hand under each buttock so that the ischial tuberosity now rests on the hand. Now move the weight from side to side as if getting tired of sitting. Gluteus maximus now contracts alternately, taking the weight off the tuberosity and then lowering it down again.

Hamstrings

Semitendinosus, semimembranosus and biceps femoris are collectively known as the hamstrings.

Semitendinosus

Its upper attachment is from the *lower medial facet* of the *lateral section* of the *ischial tuberosity* (Fig. 3.18). Its tendon of

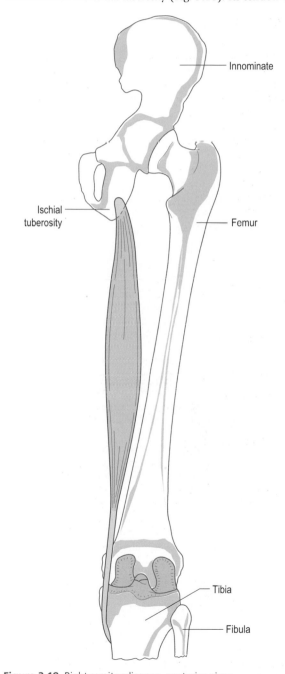

Figure 3.18 Right semitendinosus, posterior view.

attachment is combined with that of the long head of biceps femoris and the two muscles run together for a short distance. It then forms a fusiform muscle belly which quickly gives way to a long tendon, hence its name. The tendon passes downwards and medially behind the medial condyle of the femur, being separated from the medial collateral ligament by a small bursa, to attach to a *vertical line* on the *medial surface* of the *medial condyle* of the *tibia* just behind the insertion of sartorius and behind and below the attachment of gracilis. Near its insertion it is separated from gracilis by a bursa, and with gracilis is separated from sartorius by another bursa.

Nerve supply

By the *tibial division* of the *sciatic nerve* (root value L5, S1 and 2). The skin covering the muscle is supplied mainly by S2.

Action

The semitendinosus, when working from below, helps to extend the hip joint when the trunk is bent forward. When working from above, it aids in flexion of the knee joint; if the knee is semiflexed it produces medial rotation of the knee. If the foot is fixed, semitendinosus acts as a lateral rotator of the femur and pelvis on the tibia.

Semimembranosus

Situated on the posteromedial side of the thigh in its lower part, deep to semitendinosus, semimenbranosus attaches by a strong membranous tendon to the *upper lateral facet* on the rough part of the *ischial tuberosity* (Fig. 3.19) and passes downwards and medially. It becomes fleshy on the medial side of the tendon, being deep to semitendinosus and biceps femoris. From the lower part of the muscle a second aponeurotic tendon arises narrowing down towards its lower attachment to a *horizontal groove* on the *posteromedial surface* of the *medial tibial condyle*. From here its fibres spread in all directions, but particularly upwards and laterally forming the *oblique popliteal ligament*. Bursae separate the muscle from the medial head of the gastrocnemius and from the tibia near its attachment.

Nerve supply

By the *tibial division* of the *sciatic nerve* (root value L5, S1 and 2). The nerve supply to the skin covering the muscle is the same as that for semitendinosus, being mainly from S2.

Action

As for semitendinosus.

Biceps femoris

Situated on the posterolateral aspect of the thigh, biceps femoris (Fig. 3.20) arises by two heads which are separated by a considerable distance.

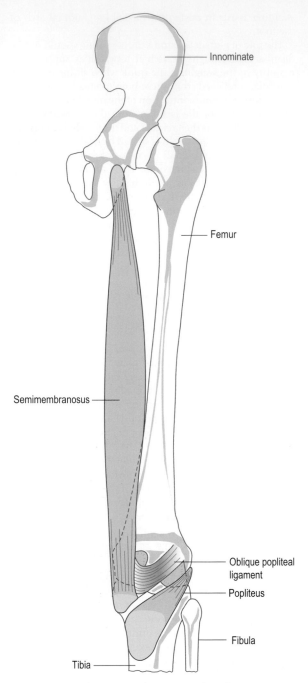

Figure 3.19 Right semimembranosus and popliteus, posterior view.

The long head attaches to the *lower medial facet* on the *ischial tuberosity* with the tendon of semitendinosus, spreading onto the *sacrotuberous ligament*. These two tendons descend together for a short distance then separate into the two individual muscles, the long head of biceps far up as the attachment of gluteus maximus and running down onto the *upper half* of the *lateral supracondylar line* of the *femur*; some fibres arise from the lateral intermuscular septum. The fibres of the short head gradually blend with the narrowing tendon of the long head which lies superficial to it.

On approaching the knee, the tendon can be felt crossing its posterolateral aspect running towards the head of the fibula.

Prior to its attachment to the *head* of the *fibula* the tendon of biceps femoris is split in two by the fibular (lateral) collateral ligament. Some fibres of the tendon join the ligament, while a few others attach to the lateral tibial condyle and some to the posterior aspect of the lateral intermuscular septum which lies just in front of it. A bursa separates the tendon from the lateral collateral ligament.

Nerve supply

The long head is supplied by the *tibial division* of the *sciatic nerve*, while the short head is supplied by the *common (fibular) peroneal division* (the root value of both is L5, S1 and 2). The skin covering the muscle is supplied mainly by S2.

Action

Biceps femoris helps the other two hamstrings to extend the hip joint, particularly when the trunk is bent forwards and is to be raised to the erect position. All three hamstrings will control forward flexion of the trunk; however in this case they are working eccentrically. Biceps femoris aids semimembranosus and semitendinosus in flexing the knee joint. With the knee in a semiflexed position biceps femoris rotates the leg laterally on the thigh or if the foot is fixed, rotates the thigh and pelvis medially on the leg.

Functional activity of the hamstrings

The hamstrings make up the large mass of muscle which can be palpated on the posterior aspect of the thigh. All three muscles cross the posterior aspect of both the hip and knee joints. Flexion of the knee and their stabilizing effect is a very important function of these muscles, although for this action a much smaller muscle bulk would be sufficient. Extension of the hip joint when the thigh is the moving part would also require a smaller group of muscles, especially when it is remembered that gluteus maximus is better situated to do this. Raising the trunk from a flexed position on the other hand requires a great deal more power as the muscles are working with a very short lever arm – the ischium and its ramus. The weight of the trunk acting on the other side of the hip joint is considerable.

The mode of action of this group of muscles may well be the reason it is injured so frequently during sporting activities. The most common cause of sports injury appears to be in the running, being more common in the first 10–20 m of a sprint. This is often blamed on inadequate preparation

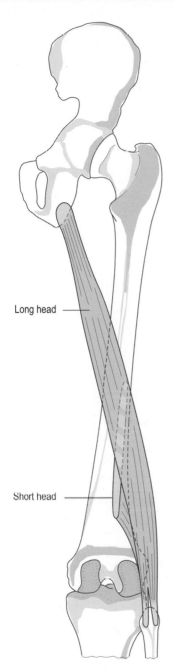

Figure 3.20 Right biceps femoris, posterior view.

Long head

Short head

forming a fusiform muscle running downwards and laterally across the posterior aspect of the thigh superficial to the sciatic nerve. In the lower third of the thigh the long head narrows and is joined on its deep aspect by the short head of biceps.

The short head has its upper attachment from the *lower half of* the *lateral lip* of the *linea aspera* reaching almost as and warm-up before the start, and to some extent this may be true. It must, however, be remembered that at this stage in a race the hamstrings are contracting strongly and are acting over two joints.

At the start of a race the athlete is in a forward-lean position in order to gain as much forward motion as possible. Starting blocks serve to increase the degree of forward leaning. The hamstrings are therefore working to their maximum, either to raise the trunk to an upright position, or to hold the trunk in such a position that forward collapse of the body as a whole is imminent. At the same time the lower limb is being thrust forward to gain as much ground as possible, with flexion of the knee to prevent the foot touching the ground. The hamstrings are under immense strain in this position and it is not surprising that the muscle may tear.

The hamstrings play an important part in the fine balance of the pelvis when standing, particularly when the upper trunk is being moved from vertical. Working in conjunction with the abdominal muscles anterosuperiorly and gluteus maximus posteroinferiorly, the anteroposterior tilt of the pelvis can be altered. This will have an effect on the lumbar lordosis.

Finally, the hamstrings have a role in decelerating the forward motion of the tibia when the free swinging leg is extended during walking, and so prevents the knee snapping into extension.

Muscles abducting the hip joint

Gluteus maximus (p. 224)
Gluteus medius
Gluteus minimus
Tensor fascia lata (p. 244)

Gluteus medius

A fan-shaped muscle situated on the lateral and upper part of the buttock, just below the iliac crest, gluteus medius (Fig. 3.21) is broader above narrowing to its tendon below. Filling the space between the iliac crest and greater trochanter of the femur, it is overlapped posteriorly by gluteus maximus.

Its upper attachment is to the *gluteal*, or *lateral*, *surface* of the *ilium* between the posterior and anterior gluteal lines (Fig. 3.7). This area is quite extensive, reaching to the iliac crest above and almost as far as the sciatic notch below. The muscle is covered with a strong layer of fascia from the deep surface of which it has a firm attachment. It shares the posterior part of this fascia with gluteus maximus.

The posterior fibres pass downwards and forwards, the middle fibres pass straight downwards and the anterior fibres pass downwards and backwards. The fibres come together and form a flattened tendon which attaches to a roughened area, which runs downwards and forwards, on the *superolateral side* of the *greater trochanter* of the *femur*.

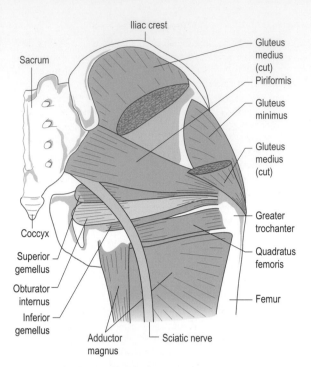

Figure 3.21 Right gluteal region with gluteus maximus removed, posterior view.

The tendon is separated from the trochanter by a bursa, whose position is given by a smooth area in front of the tendon's attachment.

Nerve supply

By the *superior gluteal nerve* (root value L4, 5 and S1). The skin covering the muscle is mainly supplied from L1 and 2.

Action

With the pelvis fixed, gluteus medius pulls the greater trochanter of the femur upwards. However, as the fulcrum of the movement is at the hip joint, this causes the femoral shaft to move laterally producing abduction.

If the lower attachment of the muscle is fixed it pulls the wing of the ilium down, producing a downward tilting of the pelvis to the same side and raising of the pelvis on the opposite side. In addition, acting from a fixed pelvis the anterior fibres of gluteus medius helps with medial rotation of the femur. Acting with the femur fixed, these fibres rotate the opposite side of the pelvis forward.

Functional activity

Gluteus medius plays a vital role in walking, running and single limb weight-bearing. When the opposite limb is taken off the ground the pelvis on that side would tend to drop through loss of support from below. Gluteus medius on the supporting side works hard to maintain, or even raise, the opposite side of the pelvis, allowing the raised limb to be brought forward for the next step. If the muscle is paralysed the pelvis drops on the opposite side during this manoeuvre.

In walking or running, not only is gluteus medius important for support, but with the help of other muscles, such as the gluteus minimus and tensor fascia lata, it produces a rotation of the hip joint. This time with the femur the more fixed point, it controls pelvic rotation on the same side. If the muscle is unable to work efficiently due to paralysis or poor mechanics of the hip joint, the pelvis will drop on the opposite side. This is referred to as a *Trendelenburg sign.* Walking in this case is awkward and difficult, and running virtually impossible.

Palpation

Find the middle of the iliac crest, which is directly above the greater trochanter of the femur. About two fingers' breadth below this region is the bulk of the muscle. Now stand alternately on one limb and then the other; the muscle becomes hard as the weight is borne on the same limb. Place the fingers of the other hand on the opposite side; walk slowly down the room. The two muscles can be felt coming into action alternately.

A patient with a Trendelenburg gait, either on one or both sides, compensates for the lack of support of the swing limb by throwing the trunk over the supporting limb so that the weight is balanced over the hip, thus giving time to swing the limb through.

Gluteus minimus

Although the smallest of the gluteal muscles it has the largest attachment from the gluteal surface of the ilium. It is triangular in shape, being wide at the top and narrowing to a tendon below (Fig. 3.21).

Its upper attachment is from the *gluteal surface* of the *ilium* in front of the anterior and above the inferior gluteal lines (Fig. 3.7), reaching as far forward as the anterior border of the ilium in front and almost to the sciatic notch behind. Its fibres pass downwards, backwards and slightly laterally forming a tendon which attaches to a small depression on the *anterosuperior aspect* of the *greater trochanter* of the *femur.*

Nerve supply

By the *superior gluteal nerve* (root value L4, 5 and S1). The skin overlying the muscle is mainly supplied by L1.

Action

If the upper attachment of the muscle is fixed, contraction of its anterior fibres medially rotates the femur. This is because the femoral attachment lies lateral to the fulcrum of the movement, the hip joint. If the lower attachment is fixed, the muscle raises the opposite side of the pelvis in a way similar to gluteus medius. It will also, by pulling the front of ilium outwards, swing the opposite side of the pelvis forwards.

Functional activity

Gluteus minimus appears to play its most important role in the support and control of pelvic movements. It is a well-developed and powerful muscle, using its power to a maximum in walking and running when the opposite limb is off the ground. As the limb is swung forward, the pelvis on the same side is also swung forward. This uses the hip of the weight-bearing limb as the fulcrum of the movement, with the gluteus medius and minimus both supporting the pelvis and swinging it forward on the opposite side.

Palpation

Find the anterior superior iliac spine at the front of the iliac crest. Allow the pads of your fingers to slip downwards and backwards towards the greater trochanter of the femur. Within two fingers' breadth you will be on the muscle bulk. Now rotate the lower limb medially and the muscle belly can be felt contracting hard. Do the same on the opposite side of the body and then begin to walk forward. The muscles can be felt contracting alternately, as each limb becomes weight-bearing.

Muscles adducting the hip joint

Adductor magnus
Adductor longus
Adductor brevis
Gracilis (p. 239)
Pectineus (p. 234)

These muscles are situated on the medial side of the hip joint running down the medial side of the thigh.

Adductor magnus

The largest and most posterior of the adductor muscles, adductor magnus (Fig. 3.22) lies posterior to adductors brevis and longus (Fig. 3.23), and anterior to semimembranosus and semitendinosus. It is really composed of two parts, an adductor part and a hamstring part, forming a large triangular sheet of muscle with a thickened medial margin.

Its upper attachment is from the *femoral surface* of the *ischiopubic ramus* running down to the *lateral part* of the *inferior surface* of the *ischial tuberosity* (see Fig. 3.26). The part of the muscle which attaches anteriorly to the ischiopubic ramus represents a sheet of muscle which twists before attaching to the femur while the posterior fibres of adductor magnus, from the ischial tuberosity, pass vertically downwards as a thickened cord.

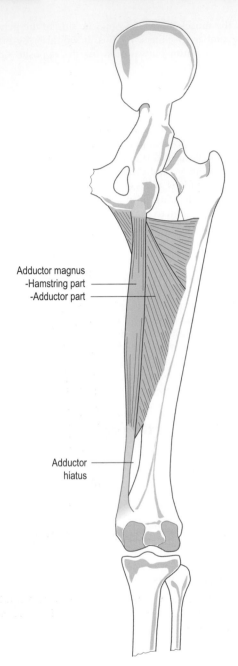

Adductor magnus
-Hamstring part
-Adductor part

Adductor
hiatus

Figure 3.22 Right adductor magnus, posterior view.

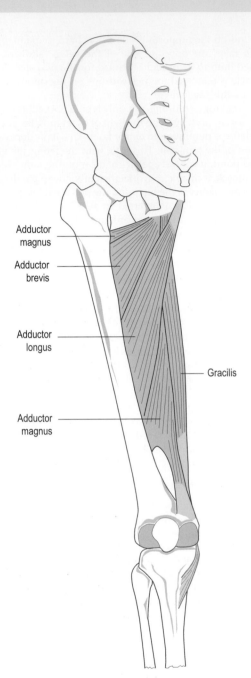

Adductor
magnus

Adductor
brevis

Adductor
longus

Gracilis

Adductor
magnus

Figure 3.23 Right adductor muscles, anterior view.

The ischiopubic fibres fan out and form a large triangular muscular sheet. The most anterior of these fibres pass laterally and slightly backwards to attach to the *upper part* of the *linea aspera* continuing upwards as far as the *greater trochanter* medial to the attachment of gluteus maximus. These upper fibres may be fused with quadratus femoris. Fibres from the posterior part of the ischiopubic ramus attach to the *whole length* of the *linea aspera* and *medial supracondylar ridge*. This attachment to the femur is not continuous as there are small fibrous arches close to the bone which allow vessels and nerves to pass from the medial (adductor) to the posterior compartment of the thigh. The posterior ischial fibres pass downwards and attach mainly to the *adductor tubercle* situated on top of the medial condyle of the femur at the lower end of the medial supracondylar ridge. Some of these fibres continue downwards to fuse with the medial collateral ligament of the knee.

Nerve supply

Because of its two parts, adductor magnus has a dual nerve supply. The adductor part from the ischiopubic ramus is by the *posterior division* of the *obturator nerve* (root value L2 and 3), while the hamstring part from the ischial tuberosity is by the *tibial division* of the *sciatic nerve* (root value L4), the skin covering the inner side of the thigh being mainly from L3.

Action

Working as a whole, the muscle is an adductor of the hip joint, although the posterior portion aids in extension of the hip. Some believe that this muscle, together with adductor longus, medially rotates the hip joint, although it was believed in the past that they also acted as lateral rotators. Whether the muscle acts as a medial or lateral rotator will depend on the position of the thigh, and the line of action of the muscle with respect to the mechanical axis of the femur (p. 295). All the adductor muscles are important in preventing lateral overbalancing during the support phase of walking.

It is worth noting that the medial collateral ligament of the knee joint appears to be a downwards continuation of the tendon of adductor magnus and as such the muscle may at some time have crossed the knee joint and therefore have been a flexor of the knee in a similar fashion to gracilis.

Palpation

Adductor magnus is a deep muscle and is therefore difficult to palpate, nevertheless, if the fingers are pushed in just above the medial condyle of the femur, the adductor tubercle can be identified (p. 214). If the inside of the same foot is now pressed against a stationary obstacle, the vertical part of the muscle can be felt contracting. The muscle can be traced about one-third of the way up the thigh until it becomes hidden by other muscles.

Adductor longus

A long, slender, triangular muscle, adductor longus (Fig. 3.24) is situated on the medial aspect of the thigh, overlying the middle part of adductor magnus. Its upper, narrower attachment comes from a small roughened area just below the *medial end* of the *obturator crest* on the *anterior aspect* of the *body of the pubis* (see Fig. 3.26).

Its fibres pass downwards and laterally, spreading out as they go to attach to the *middle two-quarters* of the *linea aspera*, anterior to adductor magnus below and adductor brevis above and posterior to vastus medialis.

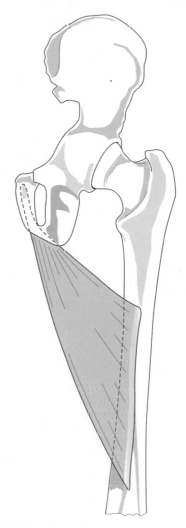

Figure 3.24 Right adductor longus, posterior view.

Nerve supply

By the *anterior division* of the *obturator nerve* (root value L2, 3 and 4). The skin covering the area of the adductor longus is supplied by L3.

Action

Adductor longus is an adductor of the thigh, but as a rotator of the thigh there is some doubt (see also adductor magnus, p. 229). Adductor longus can also flex the extended thigh, and extend the flexed thigh.

Adductor brevis

Adductor brevis is a triangular muscle situated on the medial aspect of the thigh (Fig. 3.25).

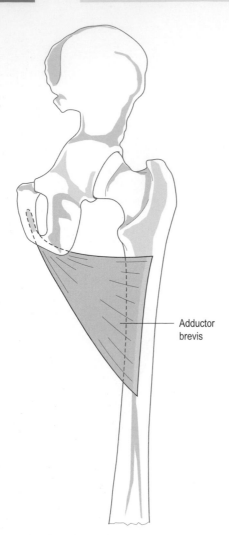

Figure 3.25 Right adductor brevi, posterior view.

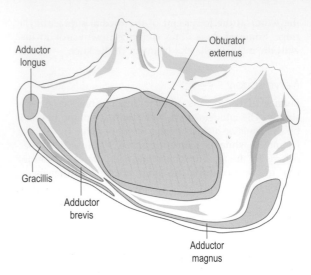

Figure 3.26 Muscle attachments to the outer surface of the pubis, ischium and obturator membrane.

Its upper attachment is from the *lateral part* of the *front* of the *body* and *inferior ramus* of the *pubis* (Fig. 3.26). Its fibres pass downwards, laterally and backwards to attach to the *upper half* of the *linea aspera* anterior to adductor magnus. Its upper part is posterior to pectineus and its lower part posterior to adductor longus.

Nerve supply

By the *anterior division* of the *obturator nerve* (root value L2, 3 and 4). The skin covering the area of the adductor brevis is supplied by L2.

Action

Adductor brevis is an adductor of the thigh.

Palpation

If the fingers are placed high up on the inside of the thigh and the lower limb is adducted against resistance a mass of muscle can be palpated running down towards the thigh. These are the adductors; however, it is difficult to distinguish between the different muscle masses.

The adductors

Gracilis (see Fig. 3.30) is the most medial muscle of the adductor group. It is an adductor of the thigh as well as a flexor of the knee (p. 239).

Functional activity

Although it is clear that these muscles adduct the thigh, they appear to work most strongly when the hip joint is in the neutral position, which is the anatomical position. They certainly work strongly, synergically, when the knee and hip are being flexed and extended when weight-bearing. There is still some confusion over whether these muscles are involved in either medial or lateral rotation of the thigh. They work strongly during walking, as they pull the supporting leg into adduction, thereby moving the line of gravity over the supporting foot. They also contribute to the delicate balancing of the pelvis on the hip joint. The adductors, as a group, are used very strongly when an object is being held between the knees in the sitting position, for example, when sitting on a horse, particularly when the horse is moving.

Muscles flexing the hip joint

Psoas major
Iliacus
Pectineus
Rectus femoris (p. 241)
Sartorius (p. 240)

Psoas major

A large, thick powerful muscle situated mainly in the abdominal cavity (Fig. 3.27). Within its substance is the lumbar plexus. Psoas major has important relations – at its upper end the diaphragm and medial arcuate ligament lie anterior, whilst lower down, the kidney, the psoas minor (when present), the renal vessels and ureter are anterior relations. On the right side, the psoas major is overlapped by the inferior vena cava and the ileum. The ascending and descending colon lie lateral to the right and left psoas respectively. Medially is the lumbar part of the vertebral column, whilst directly posterior are the transverse processes of the lumbar vertebrae. The segmental lumbar nerves emerging from the intervertebral foramina are directly behind the muscle and pass forward into its substance.

The upper attachment of psoas major is to the *adjacent margins* of the *bodies of the vertebrae* and intervening *intervertebral discs*. The uppermost attachment is to the *lower margin* of the *body of T12*, whilst the lowermost attachment is to the *upper margin* of the *body of L5*. Psoas major also has an attachment to the *anterior medial part* of each *transverse process*, and from *tendinous arches* over the constricted part of the lumbar vertebral bodies.

The fibres of the muscle pass downwards and forwards towards the pelvic brim, the individual digitations from the vertebral column joining together to form a thick muscle which gradually narrows as it passes over the pelvic brim under the inguinal ligament. At this point the tendon changes direction, becoming more vertical; it then passes downwards, backwards and laterally. The muscle is separated from the pubis and hip joint capsule by a large bursa. Before passing over the pelvic brim psoas is joined, on its lateral side, by fibres from iliacus, which continues to blend with psoas even after it becomes tendinous, until it attaches to the *tip* and *posterior aspect* of the *lesser trochanter* of the *femur*. Some fibres of iliacus attach to the femur on a line running downwards and forwards from the lesser trochanter.

Action

Psoas major is a flexor of the hip joint. Because of its attachment to the lumbar spine, and using the lower attachment as the fixed point, it will also flex the lumbar spine. There has been much discussion of the role of psoas in rotation at the hip joint. It was thought at first that as it attached to the posterior and medial aspect of the femur, it must be a lateral rotator of the thigh. However, when rotation is around an axis drawn through the head of the femur and the lateral condyle of the tibia, it would be expected to be a medial rotator. Electromyographically, however, it shows little activity during either medial or lateral rotation. The question is still not answered. If psoas major of one side acts on its own, with reversed origin and insertion, it produces lateral flexion of the lumbar spine to that side.

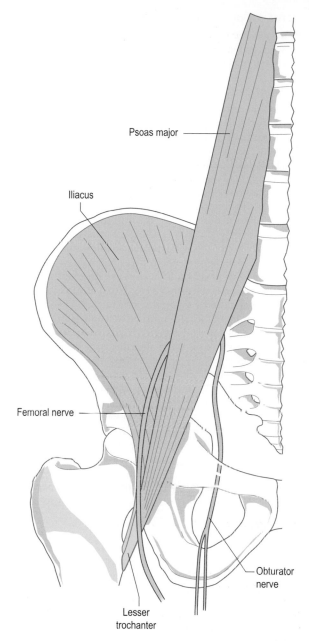

Psoas major

Iliacus

Femoral nerve

Obturator nerve

Lesser trochanter

Figure 3.27 Right psoas major and iliacus, anterior view.

Nerve supply

By the *anterior rami* of L1, 2, 3 and sometimes 4. The muscle only appears near the surface in the area of the groin and this small area of skin is supplied by L1.

Functional activity

The action and functional activity of psoas are included with those of iliacus as far as flexion of the hip is concerned. However, psoas major does act independently on the lumbar spine when its lower end is fixed. In sitting up from a lying position, both psoas muscles help to pull the considerable weight of the trunk up, and this is at a time when the abdominal muscles are working hard to flex the trunk. It is in fact very important that the abdominal muscles are brought into action early as this will prevent the lumbar spine being drawn forwards before the trunk begins to rise. Pulling the head up first will prevent this unwanted and potentially damaging movement occurring.

Raising both lower limbs at the same time whilst in the supine position is the cause of much back trouble and, unfortunately, it can be a popular exercise with lay teachers. The mechanics of this area must be well understood before exercising is begun as it is better to prevent back problems rather than try to treat them after they have occurred.

Each lower limb is approximately 15% of the body weight; thus when both legs are raised off the floor the hip flexors will be lifting approximately 30% of body weight. This initial lift will involve psoas major, and for about the first 30°, the lumbar spine is pulled forwards due to the muscle working with reversed origin and insertion. This dragging forward of the lumbar spine can cause considerable damage to the area, particularly if some degenerative changes have already occurred. It is believed, erroneously, that this is a good abdominal exercise and will reduce the waistline because the abdominal muscles are working hard.

Palpation

It is almost impossible to palpate psoas major as most of its bulk lies within the abdominal cavity. It appears near the surface in the groin, but it is still quite difficult to feel as it is covered by other structures.

Iliacus

A large fleshy triangular muscle situated mainly in the pelvis (Fig. 3.27). Its larger upper attachment comes mainly from the *upper* and *posterior two-thirds* of the *iliac fossa* with some fibres arising from the *ala* of the *sacrum* and *anterior sacroiliac ligament* (see Fig. 3.54). Its fibres pass downwards, forwards and medially, blending with the lateral side of psoas major. This blending of the muscles continues over the pelvic brim where they change direction to pass downwards, backwards and slightly laterally to insert into the *lesser trochanter* of the *femur*, blending with the insertion of psoas major from the tip of the lesser trochanter. A few fibres are attached to the hip joint capsule.

Nerve supply

By the *femoral nerve* (root value L2 and 3). The skin covering the area, where the tendon passes over the brim of the pelvis, is supplied by L1.

Action

Its effect on the hip is similar to that of psoas major. If its upper attachment is the fixed point it pulls the thigh forwards as in flexion of the hip. If the lower attachment is the fixed point it draws the pelvis forwards, thus tilting it forwards, being again flexion of the hip but this time with the trunk doing the moving.

Functional activity

This muscle is used with psoas major in all activities of pulling the lower limb up in front of the trunk as in drawing the lower limb forward in walking, running and jumping. It also helps to draw the trunk forward from a lying supine position to a sitting position. There is the same controversy over its role in rotation of the femur as there is for psoas major.

Palpation

This muscle is almost impossible to palpate (see also psoas major).

Pectineus

A quadrilateral muscle situated at the upper and medial part of the thigh, deep in the groin (Fig. 3.28). It appears to consist of two layers, superficial and deep, which are generally supplied by different nerves.

Its upper attachment is to the *superior ramus (pecten)* of the *pubis*, the *iliopubic eminence* and the *pubic tubercle*. It also attaches to the *fascia* covering it. The fibres pass downwards, backwards and laterally between psoas major and adductor longus to attach to a line running from the *lesser trochanter* of the *femur* to the top of the *linea aspera*, anterior to the upper part of the adductor brevis. This is often called the *pectineal line*.

Nerve supply

By the *femoral nerve* (root value L2 and 3) and occasionally the *obturator* or the *accessory obturator nerve* (root value L3). The skin covering this area of the groin is supplied by L1.

Action

Pectineus flexes and adducts the hip joint. Some authorities also believe the muscle to be a medial rotator of the hip.

is at an angle of 45° to the horizontal – as if the legs were going to be crossed. The muscle fibres now pass forwards and upwards, passing well behind the axis of the rotating thigh. The action of the muscle in this position will now be adduction as before, but also extension and lateral rotation; in fact a movement very similar to that of crossing the legs except that the thigh is being pulled down onto the opposite thigh. This movement is comparable with the initial stages of rising from a very low chair, or from the squat position, especially if the movement is being carried out at some speed and under load.

If the argument is taken one stage further, the muscle is obviously a flexor and adductor in the upright position with perhaps some medial rotation. The muscle is an extensor and lateral rotator in the fully flexed position, but still performs adduction. Therefore, as in the case of many muscles in the body, pectineus can perform different actions according to its starting position and the relative position of its origin and insertion. It is not surprising, therefore, that it is supplied by nerves from both the flexor and adductor compartments of the thigh.

Remembering the dual nerve supply, the dual action, the closeness of the muscle to the hip joint and its important relations, it is surprising that pectineus only merits a few lines in most anatomy texts. It must have played a vital role in locomotion with a flexed hip, either in climbing or when all four limbs were on the ground. Has its role diminished that much or are we overlooking the true action and worth of pectineus?

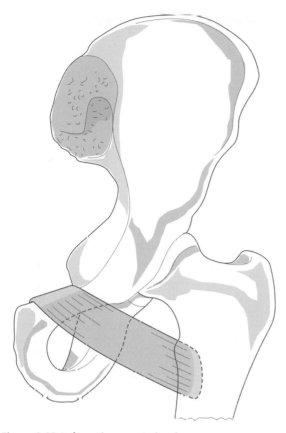

Figure 3.28 Left pectineus, anterior view.

Functional activity

It is easy to see how pectineus acts as a flexor and adductor of the hip by considering the direction of its fibres, which pass downwards, backwards and laterally. Contraction of the muscle therefore draws the thigh inwards and forwards. Most authorities dismiss the rotation element as there is certainly disagreement, although some feel that there must be more to the rotation than has yet been deduced.

There is no doubt that the insertion of pectineus is lateral and in front of the mechanical axis of the femur (the line around which rotation occurs in the anatomical position). Thus the muscle in this case would produce medial rotation (p. 298). However, when the foot is off the ground, as in the swing phase in walking, the axis would still pass through the hip joint, but now would vary considerably according to the position of the thigh and also that of the pelvis. In fact it would depend very much on the swing of the lower limb.

So far, the functional activity of this muscle has only been considered in the upright, standing or walking position. Much of the time is spent sitting with the hip joint flexed at a right angle. The relationship between the origin and insertion of this muscle is now reversed. To make the situation clearer, the thigh can be raised off the seat until it

Muscles medially rotating the hip joint

Muscles laterally rotating the hip joint

Piriformis

Located posterior to the hip joint in the same plane as gluteus medius piriformis (Fig. 3.29) is a triangular muscle with its base in the pelvis and its apex in the gluteal region.

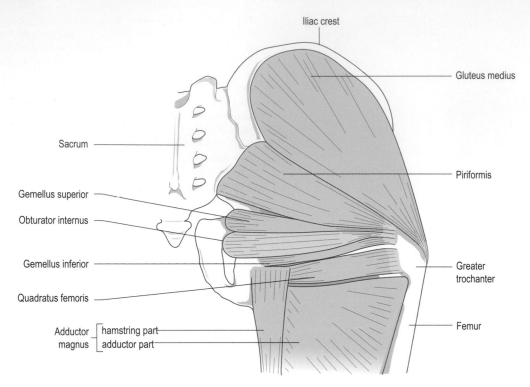

Figure 3.29 Muscles responsible for laterally rotating the hip joint: right side, posterior view.

Its upper attachment is to the *front* of the *second to fourth sacral segments* between and lateral to the anterior sacral foramina. It has an additional attachment from the *gluteal surface* of the *ilium* and the *pelvic surface* of the *sacrotuberous ligament* as it passes out of the pelvis through the greater sciatic foramen into the gluteal region. Its fibres continue to pass downwards, laterally and forwards, narrowing into a tendon which attaches to the *upper border* and *medial side* of the *greater trochanter* of the *femur*. The fibres run in a straight line from the origin to the insertion through the greater sciatic foramen.

Nerve supply

By the *anterior rami of the sacral plexus* (L5, S1 and 2; mainly S1). The skin covering this area is supplied by the same nerve roots but it must be remembered that this is a deep muscle and gluteus maximus intervenes between the two.

Action

In the anatomical position piriformis is a lateral rotator of the thigh, even though it is situated a little high. However, when seated it is concerned in the important action of abduction, being well positioned for this action. It is an important muscle in holding the head of the femur in the acetabulum.

Functional activity

It must be remembered that many times muscle action is only considered in the anatomical position. In the sitting position, the pull of muscles changes and the movement produced may bear little or no relation to the previous action. Piriformis is particularly important in the following activities:

1. abduction when sitting, for example in moving from one chair to another without standing up
2. moving the legs to the outside of a car in preparation for standing up
3. stabilizing the pelvis when the trunk is rotated
4. controlling the balance of the pelvis when standing on a moving bus.

It is not, therefore, surprising that strain of this muscle is quite common, but unfortunately easily overlooked.

Palpation

This is not easy as piriformis is situated deep within the gluteal region. Dig your fingers into the buttock just lateral to the sacrum and then push the outside of the thigh up against the leg of a table or some such resistance; the muscle can then be felt to contract, although it must be remembered that a large part of the muscle is in the pelvis.

Obturator internus

A triangular-shaped muscle, obturator internus (Fig. 3.29) is situated partly in the pelvis and partly in the gluteal region posterior to the hip joint. It arises from the *internal surface* of the *obturator membrane* and *surrounding bony margin*, except at the obturator canal. The bony attachment extends backwards as far as the *pelvic surface* of the *ilium*. The muscle fibres pass laterally, but mainly backwards towards the lesser sciatic foramen, through which they pass, narrowing and becoming tendinous.

As the tendon passes through the lesser sciatic foramen deep to the sacrotuberous ligament it changes direction to pass forwards and laterally to insert into the *medial surface* of the *greater trochanter* of the *femur* in front of and above the trochanteric fossa. Before attaching to the femur, the tendon is commonly joined by the tendons of gemellus superior above, and gemellus inferior below. Occasionally the two gemelli tendons insert into the greater trochanter above and below the tendon of obturator internus.

The inner (pelvic) surface of obturator internus is covered by the obturator fascia, from which arises part of levator ani. As the tendon of obturator internus turns around the lesser sciatic notch the surface of the bone in this region is grooved and covered with cartilage. A bursa intervenes between tendon and cartilage.

Nerve supply

By the *nerve to obturator internus* (root value L5, S1 and 2). The skin covering this area is mainly supplied by S3.

Action

In the anatomical position the obturator internus is a lateral rotator of the thigh pulling the greater trochanter backwards using the hip joint as the fulcrum. However, when the hip is flexed to a right angle, it pulls the upper end of the femur medially, and therefore the lower end moves laterally as in abduction.

Functional activity

As with the piriformis (p. 236), obturator internus is used when moving sideways in the seated position, in swinging the lower limb sideways, as in placing the limb outside a car, and in balancing and controlling the stability of the trunk when the seated person is being rocked from side to side. For the same reasons, moving around on the floor or on a platform, either sitting or crawling requires considerable activity in this muscle.

Gemellus superior

As obturator internus passes out of the pelvis around the lesser sciatic notch it is joined by gemellus superior and inferior.

Gemellus superior (Fig. 3.29) arises from the *gluteal surface* of the *ischial spine*. It runs laterally and slightly downwards to blend with the *superior aspect* of the *tendon* of the *obturator internus*. Sometimes its fibres are prolonged onto the medial surface of the greater trochanter of the femur.

Nerve supply

By the *nerve to the obturator internus* (root value L5, S1 and 2).

Gemellus inferior

The gemellus inferior (Fig. 3.29) arises from the *upper part* of the *ischial tuberosity*. It runs laterally and slightly upwards to blend with the *inferior aspect* of the *tendon* of *obturator internus*.

Nerve supply

By the *nerve to the quadratus femoris* (root value L4, 5 and S1).

Action

The gemelli aid obturator internus in its action. As obturator internus turns around the lesser sciatic notch it loses some of its power; this is compensated for by the action of the gemelli.

Quadratus femoris

A flat quadrilateral muscle, quadratus femoris is situated below gemellus inferior and above the upper margin of adductor magnus (Fig. 3.29). It is separated from the hip joint by obturator externus.

It attaches to the *ischial tuberosity* just below the lower rim of the acetabulum. The fibres pass laterally to attach to the *quadrate tubercle* situated *halfway down* the *intertrochanteric crest* of the *femur* and the area of bone surrounding it.

Nerve supply

By the *nerve to the quadratus femoris* (root value L4, 5 and S1).

Action

In the anatomical position quadratus femoris is a lateral rotator of the hip joint, but with the hip flexed, it acts as an abductor of the hip.

Obturator externus

A triangular muscle with its muscular base attached to the *outer surface* of the *obturator membrane* and the *surrounding margins* of the *pubis* and *ischium*, excluding the area superiorly around the obturator canal (Fig. 3.26). The muscle fibres converge onto a tendon which runs in a groove below the acetabulum across the back of the neck of the femur, which it grooves, to insert into the *trochanteric fossa* of the *femur*. The muscle lies deep to quadratus femoris.

Nerve supply

By the *posterior branch* of the *obturator nerve* (root value L3 and 4).

Action

In the anatomical position obturator externus laterally rotates the femur. However, when the hip is flexed it pulls the upper part of the femur medially with the lower part passing laterally, as in abduction.

Functional activity

The functional activities of these muscles must be considered together. Piriformis, the gemelli, obturator internus and externus, and quadratus femoris are always considered in the anatomical position. In this position they perform an important role in controlling the pelvis, particularly when only one foot is on the ground and even more so in walking. They are responsible, together with gluteus maximus and the posterior part of gluteus medius, for producing lateral rotation of the lower limb in the forward swing-through phase of gait. However, in sitting, crawling and turning over when lying down they have a completely different role, producing abduction of the hip and thereby controlling the movements of the pelvis on the flexed thigh.

Palpation

These muscles are situated deep to the thick gluteus maximus and it is almost impossible to distinguish their contraction through the overlying muscle tissue, especially as gluteus maximus is usually contracting at the same time. However, overactivity or strain of these muscles may result in acute tenderness deep to the back of the hip joint; the production of this pain with the relevant movement is then obvious.

It is difficult to precisely determine the actions of these lateral rotators primarily because of their depth in the gluteal region, but also because much of their action is concerned with controlling the movements of the hip and pelvis. Thus they may be in a state of contraction even when the opposite movement to their primary action is occurring.

To see these muscles, gluteus maximus must be removed – the muscles resembling the rungs of a ladder, consequently, they are often referred to as the *ladder of muscles*.

Section summary

Movements at hip joint

The hip joint consists of the head of the femur and the acetabulum of the innominate (hip) bone. It is capable of a wide range of movement produced by the following muscles:

Movement	Muscles
Extension	Gluteus maximus
	Hamstrings
	Semitendinosus
	Semimembranosus
	Biceps femoris
Abduction	Gluteus maximus
	Gluteus medius
	Gluteus minimus
	Tensor fascia lata
Adduction	Adductors magnus, longus and brevis
	Gracilis
	Pectineus
Flexion	Psoas major
	Iliacus
	Pectineus
	Rectus femoris
	Sartorius
Lateral rotation	Gluteus maximus
	Piriformis
	Obturator internus and externus
	Gemellus superior and inferior
	Quadratus femoris
Medial rotation	Gluteus medius and minimus
	Tensor fascia lata

- All the above muscles contribute to the stability of the hip joint.
- Long muscles acting across two joints (e.g. the hamstrings, rectus femoris) can only work efficiently across one joint at a time. For example, the hamstrings can only extend the hip strongly with the knee in extension.
- The abductors and adductors of the hip have an important function during gait when they work with their distal (femoral) attachments fixed to stabilize the pelvis.
- Some muscles attaching to the pelvis and lumbar spine are important in producing movements of the trunk, e.g. gluteus maximus, psoas major.

Muscles producing movement of the knee joint

The main movement at the knee joint is flexion and extension. With the knee in a semiflexed position and the foot off the ground, there is, in addition, medial and lateral rotation of the tibia with respect to the femur. If the feet are on the ground when the knees are flexed, the rotation at the knee joint is taken up by the femur, which rotates about a vertical axis, running approximately through the intercondylar eminence. This allows the upper part of the femur to move from side to side as in moving along sideways from one seat to another.

The movement between the patella and the patellar surface of the femur must also be taken into account when considering the movements of the knee (pp. 306 and 327).

Muscles flexing the knee joint

Hamstrings (semitendinosus, semimembranosus and
    biceps femoris) (pp. 225–228)
Gastrocnemius (p. 248)
Gracilis
Sartorius
Popliteus

The hamstrings

Functional activity

It must be emphasized that, with the exception of the short head of biceps femoris, the three hamstring muscles pass over, and act upon both the hip and knee joints. Their action is therefore extremely complex. Although details of their attachments and much of their action are covered on pages 225–228, the functional activities of these muscles with respect to the knee joint have yet to be considered.

Rotation of the knee joint by the hamstrings is usually considered to take place when the foot is off the ground; however this is not exactly true. It is certainly easier to describe the rotation that occurs when the foot is off the ground, but in practice the rotation occurs when the feet are firmly on the ground. For example, consider moving sideways from seat to seat. The feet are fixed and body weight is taken onto them; however, the person stays in a sitting position, just allowing the buttock to come clear of the seat. The trunk is then moved to one side by a swivelling of the femur on the upper surface of the tibia. This is achieved by the combined action of the medial rotators of one knee and the lateral rotators of the other.

Finally, the simultaneous action of the hamstrings on both the hip and knee joints must be considered. Such a situation arises in athletes accelerating towards a bend. Here the hamstrings are functioning to lift the trunk into a more upright position as well as to flex the knee of the leg that is being swung-through ready for the next stride. As the body is being forced around the bend, the hamstrings also have to produce a rotation of the knee in order to produce this turning force. It is thought that the hamstrings act as a tie between the back of the pelvis and the tibia, and can therefore adjust the relationship between the two bones. This is particularly important when the body is changing posture during active movement as there

are additional forces due to the acceleration of body segments. This concept would certainly go some way to explain why there are so many injuries to the hamstrings in athletes.

Gastrocnemius

The muscle (see Fig. 3.38) is mainly a strong plantarflexor of the ankle joint and is dealt with in that section. Nevertheless, it is also a strong flexor of the knee.

The medial and lateral heads of gastrocnemius cross the knee joint on their respective sides. It appears to come into action when the foot is fixed and the body is being pulled forwards on the feet. This is best seen when pulling forwards on the slide of a rowing boat seat, or manoeuvring the fully reclined body. The turning and pulling down of the body when in bed or on a plinth is very important considering that we spend one-third of our lives lying down.

Gracilis

A long, thin muscle, gracilis, as its name implies, is situated on the medial side of the thigh (Fig. 3.30). Gracilis is the most superficial of the adductor group. Its upper attachment is to the *front* of the *body* of the *pubis* and its *inferior ramus*, just encroaching onto the *ramus of* the *ischium* (Fig. 3.26). As it descends between semimembranosus posteriorly and sartorius anteriorly, gracilis develops a fusiform-shaped belly at about its middle. It becomes tendinous above the knee and crosses the joint before expanding to attach to a *short vertical line* on the *upper part* of the *medial surface* of the *shaft* of the *tibia*. This attachment is above that of semitendinosus and behind and blends with that of sartorius. Bursae separate the tendon of gracilis from those of sartorius and semitendinosus.

Nerve supply

By the *anterior division* of the *obturator nerve* (root value L2 and 3). The skin covering this area is innervated by roots L2 and 3; the upper part by the obturator nerve and the lower part by the femoral nerve.

Action

Although gracilis is situated with the adductor group of muscles, its action of adduction at the hip joint is not so important as its action on the knee. It is mainly a flexor of the knee, but with the knee in a semiflexed position it aids medial rotation of the leg on the thigh.

Functional activity

As a flexor of the knee, gracilis helps the hamstrings in simple flexion activities, such as the beginning of the swing phase in walking when the knee needs to be flexed. It also

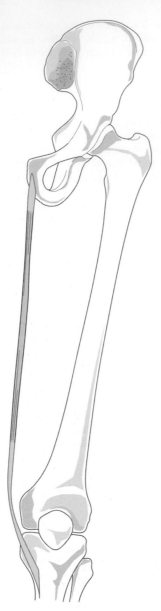

Figure 3.30 Left gracilis, anterior view.

helps when strong flexion is required, as when pulling the body forward on the sliding seat of a rowing boat. In horse riding, gracilis is used in all its actions. When the rider is gripping the horse, gracilis helps the adductor muscles, whilst at the same time helping to control the flexed knee.

Palpation

In the sitting position with the medial aspect of the foot against a solid object, such as the leg of a table, or when the toes are pointing towards the midline, the tendon

can be felt on the posteromedial aspect of the knee joint, being the upper of the two obvious tendons. If traced upwards, the muscle belly can be palpated and traced to its attachment on the front of the pubic body.

Sartorius

A long strap-like muscle with flattened tendons at each end, sartorius is the most superficial muscle in the anterior compartment of the thigh (Figs. 3.31 and 3.36). Its lower part is mainly on the medial side anterior to gracilis. It is renowned as the longest muscle in the body, getting its name from its action, which is to produce most of the actions needed in the lower limb to produce cross-legged sitting, the position that tailors used to use when making clothes.

Its upper attachment is to the *anterior superior iliac spine* and the area just below. From here the muscle passes medially and inferiorly to attach to a *vertical line* on the *medial side* of the *shaft* of the *tibia*, in front of both semitendinosus and gracilis, partly blending with the latter. A few fibres from the lower tendon go to the medial collateral ligament of the knee joint and to the fascia of the leg. A bursa separates sartorius from gracilis at its lower end. The medial border of the upper third forms the lateral boundary of the femoral triangle whilst the middle third forms the roof of the adductor canal.

Nerve supply

By the *femoral nerve* (root value L2 and 3). The area of skin covering this muscle is also supplied by L2 and 3.

Action

Sartorius produces many of the movements which are combined to produce cross-legged sitting, i.e. flexion of the hip and knee, lateral rotation and abduction of the thigh, and medial rotation of the tibia on the femur. These actions can be summarized by saying that it places the heel on the medial side of the opposite knee.

Functional activity

Going into the cross-legged or the tailor sitting position is a functional activity; however, sartorius will help to produce any activity which involves flexion of the knee and hip together, combined with lateral rotation of the thigh as in drawing up the lower limbs when using the breast stroke in swimming.

Palpation

Sartorius is most easily palpated at its proximal end just below the anterior superior iliac spine. Here its strap-like shape can be easily palpated, particularly when the leg, with the knee slightly flexed, is raised some 15 cm from the floor when lying supine.

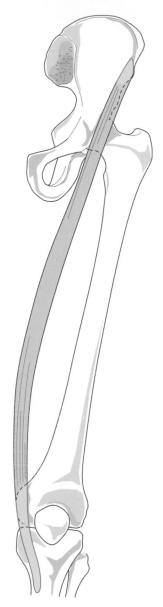

Figure 3.31 Left sartorius, anterior view.

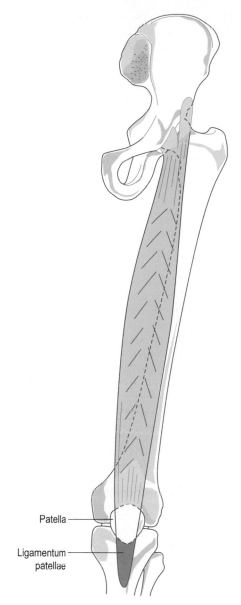

Patella

Ligamentum
patellae

Figure 3.32 Left rectus femoris, anterior view.

Muscles extending the knee joint

Quadriceps femoris:

    Rectus femoris
    Vastus lateralis
    Vastus medialis
    Vastus intermedius
    Tensor fascia lata

Quadriceps femoris is the large muscle bulk on the anterior surface of the thigh. As its name implies, it is composed of four main parts. One part, rectus femoris, has its origin above the hip joint, whilst the other three parts take origin from the shaft of the femur. All four join together around the patella to form a thick strong tendon called the ligamentum patellae, which inserts into the tibial tuberosity.

Rectus femoris

A spindle-shaped bipennate muscle, rectus femoris (Figs. 3.32 and 3.36) is seen to stand out on the front of the thigh. Its upper attachment is by two heads which are continuous with each other, one to the *anterior inferior iliac spine* (the straight head) and the other to a *rough area immediately above the acetabulum* (the reflected head). It is thought that the straight head is a human acquisition associated with the evolution of an upright posture. From this continuous origin, a single tendon descends from which arises the fleshy muscle. About two-thirds of the way down the thigh, the muscle begins to narrow to a thick tendon attaching to the *upper border* of the *patella*. From here, some fibres pass around the patella, helping to form the ligamentum patellae. The deep surface of the muscle is tendinous and smooth, allowing free movement over a similar surface on vastus intermedius, thus permitting its independent action on the hip joint.

Vastus lateralis

Situated on the anterolateral aspect of the thigh lateral to rectus femoris (Fig. 3.33). It has an extensive linear attachment from the *upper lateral part* of the *intertrochanteric line*, the *lower border* of the *greater trochanter*, the *lateral side* of the *gluteal tuberosity* and the *upper half* of the *lateral lip* of the *linea aspera*. It also attaches to the fascia lata and lateral intermuscular septum. From this origin the muscle fibres run downwards and forwards with those at the top passing almost vertically downwards.

The muscle bulk is mainly situated in the upper half of the lateral side of the thigh, from which a broad tendon arises which narrows down as it approaches the lateral side of the patella. This tendon inserts into the *tendon of rectus femoris* and the *base* and *lateral border* of the *patella*. Some fibres pass to the front of the lateral condyle of the tibia, blending with the iliotibial tract helping to form the expansion that finally attaches to the line running towards the tibial tuberosity. To a large extent this part of its attachment replaces the knee joint capsule in this region.

Vastus medialis

Situated on the anteromedial aspect of the thigh medial to rectus femoris (Fig. 3.33), with most of its bulk showing at the lower third just above the patella. It has an extensive linear origin from a line beginning at the *lower medial end* of the *intertrochanteric line*, running downwards around the *medial aspect* of the *upper end* of the *shaft* on the *spiral line*, the *medial lip* of the *linea aspera*, continuing on to the *upper two-thirds* of the *medial supracondylar line*, the *medial intermuscular septum* and the tendon of adductor magnus.

Its upper fibres pass mainly downwards, whilst its lower fibres tend to pass almost horizontally laterally. These two sets of fibres which make up vastus medialis are considered by some to be anatomically and functionally distinct, with the oblique fibres being called vastus medialis obliquus. The muscle attaches to the *tendon of rectus femoris*, the *medial border* of the *patella*, and the *front* of the *medial condyle* of

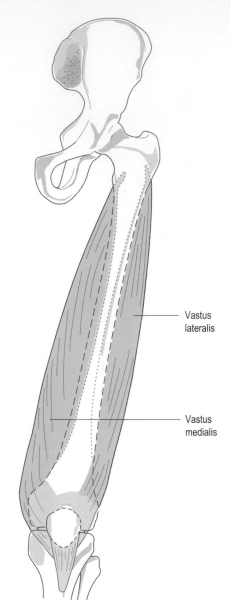

Figure 3.33 Left vastus lateralis and medialis, anterior view.

the *tibia*. The expansions which pass across the knee joint to attach to the tibia replace the joint capsule in this region and become fused with the deep fascia. This attachment also runs to the *tibial tuberosity* (p. 308).

Vastus intermedius

The deepest part of quadriceps femoris lying between vastus lateralis and medialis, and deep to rectus femoris (Figs. 3.34 and 3.36).

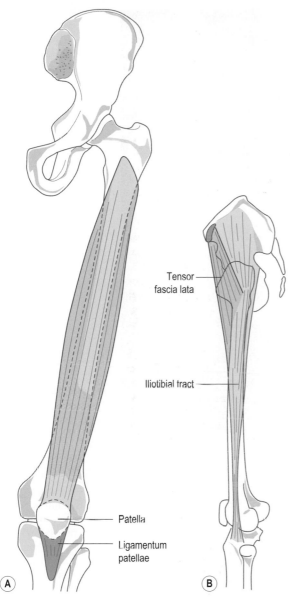

Tensor
fascia lata

Iliotibial tract

Patella

Ligamentum
patellae

(A)   (B)

Figure 3.34 (A) Left vastus intermedius, anterior view, (B) left tensor fascia lata, lateral view.

It arises by fleshy fibres from the *upper two-thirds* of the *anterior* and *lateral surfaces* of the *femur;* its fibres pass downwards to form a broad tendon on its more superficial aspect. This attaches to the *deep surface* of the *tendon* of *rectus femoris* and the other vastus muscles, and to the *base* of the *patella.* In the middle of the thigh, vastus intermedius is difficult to separate from vastus lateralis, whilst lower down it is impossible to separate from vastus medialis.

Articularis genus

Some of the deep fibres of vastus intermedius, arising from a small area on the *inferior third* of the *anterior surface* of the *femur*, pass downwards to attach to the *upper part* of the *suprapatellar bursa* of the *knee joint* which lies deep to vastus intermedius. These fibres are the articularis genus. Its main function is to prevent the synovial membrane becoming trapped and interfering with the normal movements of the knee joint.

Ligamentum patellae

All four of the quadriceps tendons contribute to the formation of the ligamentum patellae. It runs from the *apex* of the *patella* to the *upper part* of the *tibial tuberosity* acting as the tendon of insertion of quadriceps femoris. The patella is really a sesamoid bone in the tendon of rectus femoris and vastus intermedius helping to relay the pull of the quadriceps over the front of the femur.

Nerve supply

By the *femoral nerve* (root value L2, 3 and 4), including articularis genu. The skin covering the quadriceps is supplied by L2 and 3.

Quadriceps femoris

Action

Although rectus femoris is part of the quadriceps femoris, by crossing anterior to the hip joint it also flexes the thigh.

Quadriceps femoris is the main extensor of the knee joint. Rectus femoris crosses in front of the hip joint and therefore is also a flexor of that joint. Each muscle appears to have its particular role in extension of the knee and often comes in at different ranges of the movement. For example, vastus medialis is more obviously active in the final stage of extension and is believed to resist the tendency of the patella to move laterally caused by the angulation of the femur. Specific exercises designed to strengthen the oblique fibres of vastus medialis are advocated by some practitioners in order to affect tracking of the patella, resisting dislocation and possibly helping reduce anterior knee pain in certain circumstances.

Rectus femoris works particularly strongly in straight leg raising or in the combined movement of flexion of the hip and extension of the knee.

Functional activity

Quadriceps femoris is used strongly in stepping activities, for example stair climbing and squats. Rectus femoris will perform its function particularly in the swing phase of walking when the lower limb is being carried forward and the knee is being extended. Vastus medialis, in the final stages of extension of the knee, helps in the locking mechanism of the joint when the femur is allowed to rotate medially.

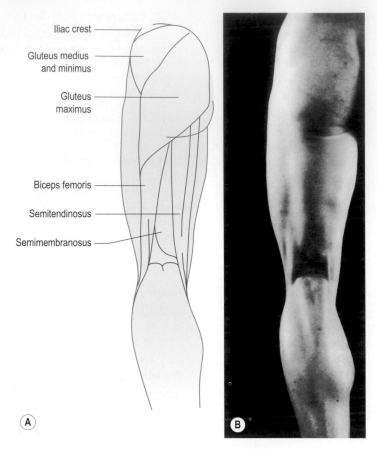

Figure 3.35 Left thigh: (A, B) posterior view;

Surprisingly, in the standing position very little or no action is recorded in the quadriceps as in this position the knees are in the close-packed position. It is at these times that if the knees are knocked from behind, forward collapse will almost certainly occur. However, when standing on a moving vehicle the quadriceps will be active. When standing on one leg, all the muscles around the knee work statically to provide stability at the joint.

Quadriceps femoris is a powerful and important muscle. It must work strongly throughout its full range. It will lose strength and bulk rapidly if there is any injury to it or to the knee joint. It may take months to regain power, but only days to lose it.

Palpation

When sitting on a chair and straightening the knee joint, particularly against resistance, the separate parts of quadriceps femoris, except vastus intermedius, can easily be palpated; the medialis on the lower medial aspect, lateralis in the upper half of the lateral side and rectus femoris running down the centre. Stand with the knees semi-flexed, place your hands on the front of each thigh; the three parts of the muscle (as above) can be readily palpated.

In straight leg raising, the muscle should be able to extend the knee into a few degrees of hyperextension, this being the extra range required for the knee to be able to lock. Patients not able to do this often complain of their knee giving way during walking.

Tensor fascia lata

Situated anterolateral to the hip joint, tensor fascia lata (Fig. 3.34B) is superficial to gluteus minimus. It attaches above to the *anterior part of* the *outer lip* of the *iliac crest*, between and including the *iliac tubercle* and the *anterior superior iliac spine*, the area of the *gluteal surface* just below it, the fascia between it and gluteus minimus and that covering its superficial surface. Inferiorly it attaches between the two layers of the *iliotibial tract*, below the level of the greater trochanter.

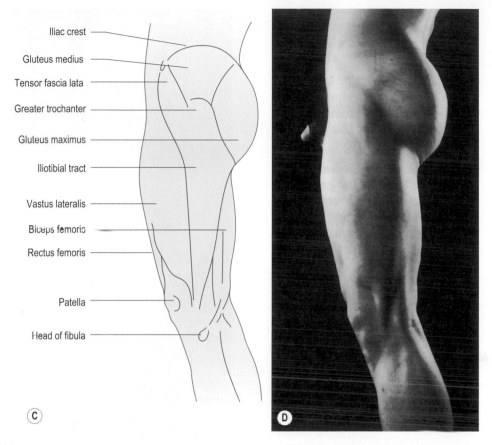

Iliac crest
Gluteus medius
Tensor fascia lata
Greater trochanter
Gluteus maximus
Iliotibial tract
Vastus lateralis
Biceps femoris
Rectus femoris
Patella
Head of fibula

(C)

(D)

Figure 3.35, Cont'd   (C, D) lateral view. *(B, D, reproduced with permission from Keogh B, Ebbs S (1984) Normal Surface Anatomy, William Heinemann Medical Books Ltd.)*

Nerve supply

By the *superior gluteal nerve* (root value L4 and 5), with the skin overlying the muscle supplied by L1.

Action

Tensor fascia lata overlies gluteus minimus and helps in flexion, abduction and medial rotation of the hip joint. It also straightens out the backward pull of gluteus maximus on the iliotibial tract.

Acting with the superficial fibres of gluteus maximus it tightens the iliotibial tract, and through its attachment to the lateral condyle of the tibia, extends the knee joint. Acting with gluteus minimus it medially rotates the hip joint and its posterior fibres may help in abduction of the thigh.

Functional activity

Due to the fact that tensor fascia lata, together with gluteus maximus, links the pelvis with the tibia it helps to steady and control the movements of the pelvis and femur on the tibia when the limb is weight-bearing. It produces strong medial rotation when the hip is in extension and the lower limb, pelvis and trunk are prepared to take the thrust relayed through the lower limb by the calf muscles during the 'toe-off' phase of walking.

When quadriceps femoris is paralysed, tensor fascia lata can be developed to produce sufficient extension of the knee to enable the patient to walk, but its action is only weak and limited in range.

Palpation

Place the fingers halfway between the anterior superior iliac spine and the greater trochanter of the femur. When the lower limb is medially rotated, the muscle can be felt to contract powerfully. If the weight is taken on the limb and the pelvis is rotated to the same side, a similar contraction of the muscle will be observed.

Muscles laterally rotating the tibia at the knee joint

Biceps femoris (p. 226)

245

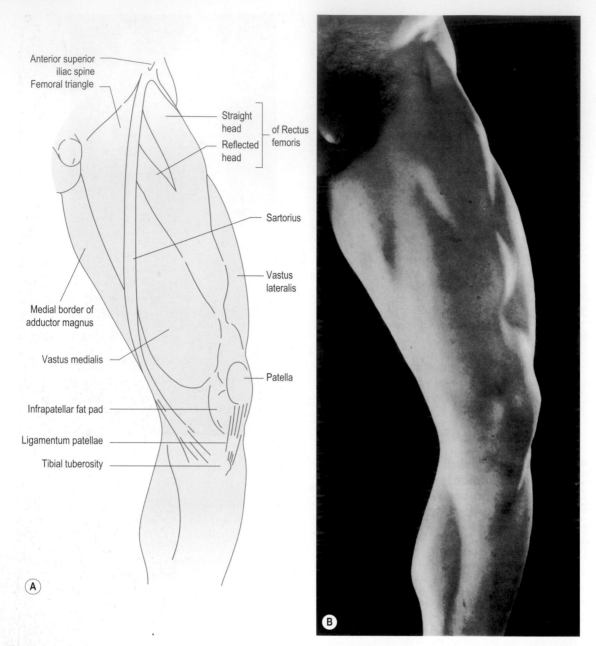

Figure 3.36 (A, B) Left thigh, anteromedial aspect. *(B, reproduced with permission from Keogh B, Ebbs S (1984) Normal Surface Anatomy, William Heinemann Medical Books Ltd.)*

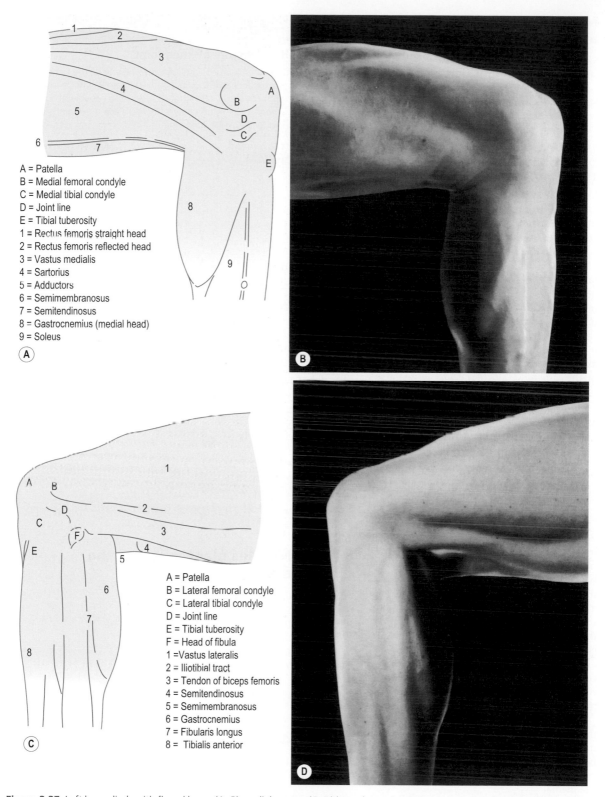

A = Patella
B = Medial femoral condyle
C = Medial tibial condyle
D = Joint line
E = Tibial tuberosity
1 = Rectus femoris straight head
2 = Rectus femoris reflected head
3 = Vastus medialis
4 = Sartorius
5 = Adductors
6 = Semimembranosus
7 = Semitendinosus
8 = Gastrocnemius (medial head)
9 = Soleus

A = Patella
B = Lateral femoral condyle
C = Lateral tibial condyle
D = Joint line
E = Tibial tuberosity
F = Head of fibula
1 = Vastus lateralis
2 = Iliotibial tract
3 = Tendon of biceps femoris
4 = Semitendinosus
5 = Semimembranosus
6 = Gastrocnemius
7 = Fibularis longus
8 = Tibialis anterior

Figure 3.37 Left lower limb with flexed knee: (A, B) medial aspect, (C, D) lateral aspect. *(B, D, reproduced with permission from Keogh B, Ebbs S (1984) Normal Surface Anatomy, William Heinemann Medical Books Ltd.)*

Muscles medially rotating the tibia at the knee joint

Popliteus

A triangular-shaped muscle situated deep in the popliteal fossa, below and lateral to the knee joint (Fig. 3.19). It arises within the joint capsule from a tendinous attachment from the *anterior aspect* of the groove on the *outer surface* of the *lateral condyle* of the *femur*, below the lateral epicondyle and the attachment of the lateral collateral ligament. The tendon passes backwards, downwards and medially, crossing the line of the joint over the outer border of the lateral meniscus to which it is attached. This upper part, within the capsule of the knee joint, is enveloped in a double layer of synovial membrane until it leaves the capsule under the arcuate popliteal ligament, from which it has a fleshy origin. Continuing downwards and medially, popliteus attaches by fleshy fibres to a *triangular area* on the *posterior surface* of the *tibia* above the soleal line, and the fascia covering the muscle.

Nerve supply

By a branch from the *tibial division* of the *sciatic nerve* (root value L5) which enters the muscle on its anterior surface after winding around its inferolateral border. The skin covering the area is supplied mainly by S2.

Action

Popliteus laterally rotates the femur on the tibia when the foot is on the ground, thus releasing the knee from its close-packed or locked position allowing it to flex. By exerting a backward pull on the lateral surface of the lateral condyle of the femur, the condyle is rotated laterally about a vertical axis running through it just medial to its centre. This allows the medial condyle of the femur to glide forward, releasing the ligaments and muscles involved in maintaining its close-packed position.

When strong flexion of the knee is required, popliteus comes into action, drawing the tibia backwards on the femoral condyles, and if the foot is off the ground, it also aids the medial hamstrings in medial rotation of the tibia.

Through its attachments to the lateral meniscus, it pulls the meniscus backwards during lateral rotation of the femur, preventing it from being trapped between the moving bones. This is believed by some to be the reason for the lateral meniscus being damaged much less frequently than the medial.

Muscles plantarflexing the ankle joint

Gastrocnemius

The shape of the calf is mainly due to the two fleshy bellies of gastrocnemius (Fig. 3.38A), being situated on the back of the leg with its muscle bulk mainly in the upper half.

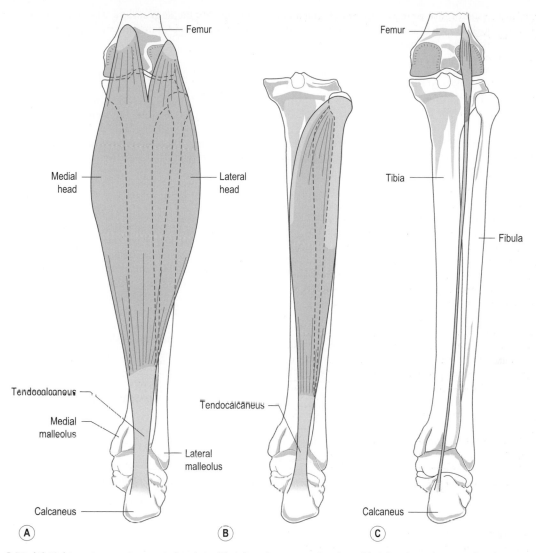

Figure 3.38 (A) Right gastrocnemius, posterior view, (B) right soleus, posterior view, (C) right plantaris, posterior view.

Together with soleus and plantaris, the gastrocnemius forms a composite muscle referred to as the triceps surae. The two heads of gastrocnemius form the lower boundaries of the popliteal fossa, which can only really be seen when the knee is flexed. The two heads arise from the *medial* and *lateral condyles* of the *femur*: the medial head, from behind the *medial supracondylar ridge* and the *adductor tubercle* on the *popliteal surface* of the femur, the lateral head from the *outer surface* of the *lateral condyle* of the *femur* just above and behind the lateral epicondyle. Each head has an additional attachment from the *capsule* of the *knee joint* and from the *oblique popliteal ligament*, below which each head is separated from the capsule by a bursa. The bursa associated with the medial head often communicates with the knee joint; that under the lateral head rarely does. There

is often a sesamoid bone, the fabella, in the lateral head as it crosses the lateral condyle of the femur. Less commonly there may be one associated with the medial head.

From each head a fleshy bulk of muscle fibres arise which gradually come together, although not actually blending with each other, to insert into the posterior surface of a broad membranous tendon which fuses with the tendon of soleus to form the upper part of the tendocalcaneus. This broad tendon gradually narrows, becoming more rounded until it reaches about three fingers' breadth above the calcaneus, where it begins to expand again and continues to do so, until its insertion into the *middle part* of the *posterior surface* of the *calcaneus*. The fibres of the tendon spiral as they pass from the myotendinous junction to their insertion, so that the most medial fibres superiorly become posterior at the site of insertion. This helical arrangement results in less buckling when the tendon is lax and less deformity of the individual strands when they are put under tension. A bursa lies between the tendon and the upper part of the calcaneus while a pad of fat lies between the tendon and the posterior aspect of the ankle joint. Inferior to the insertion is the fat pad of the heel.

Nerve supply

Each head by a branch from the *tibial nerve* (root value S1 and 2). The area of skin covering the muscle is supplied by roots L4, 5 and S2.

Action

Gastrocnemius, together with the soleus, is the main plantarflexor of the ankle joint. It provides the propelling force for locomotion. As it crosses the knee joint, gastrocnemius is also a powerful flexor of that joint. However, it is not able to exert its full power on both joints simultaneously. For example, if the knee is flexed, gastrocnemius cannot exert maximum power at the ankle joint and vice versa.

Functional activity

In running, walking and jumping gastrocnemius provides a considerable amount of the propulsive force. When one considers the power needed to launch the body into the air, the triceps surae must be one of the most powerful muscle groups in the body.

The habitual wearing of shoes with a high heel can cause considerable shortening of the fibres of gastrocnemius, as the two attachments of the muscle are brought closer together. If shortening has occurred, difficulty in walking in flat shoes or bare feet may be experienced due to limited dorsiflexion at the ankle joint.

Soleus

Situated deep to gastrocnemius, soleus is a broad flat muscle wider in its middle section and narrower below (Fig. 3.38B). It arises from the *soleal line* on the *posterior surface* of the *tibia*, the *posterior surface* of the *upper third* of the *fibula* (including the head) and a *fibrous arch* between these bony attachments. The fibres pass downwards, forming a belly about halfway down the calf to the deep surface of a membranous tendon, which faces posteriorly. This tendon glides over a similar one on the deep surface of gastrocnemius, thereby enabling independent movement of the two muscles. Inferiorly the two tendons fuse to form the upper part of the tendocalcaneus, which passes behind the ankle joint to insert into the *middle part* of the *posterior surface* of the *calcaneus*.

Nerve supply

By two branches from the *tibial nerve* (root value S1 and 2), one of which arises in the popliteal fossa and enters the superficial surface of the muscle, while the other arises in the calf entering the deep surface. The skin over the region of the muscle is predominantly supplied by root S2.

Action

Soleus is one of the two main plantarflexors of the ankle joint. It is so placed to prevent the body falling forwards at the ankle joint during standing, and as such is an important postural muscle. Intermittent contraction of the muscle during standing aids venous return (the soleal pump) due to the communicating vessels joining the deep and superficial venous systems which pass through its substance.

Plantaris

A long, slender muscle which is variable in its composition (Fig. 3.38C). It may have one muscle belly high up in the calf, or two smaller bellies separated by a tendon. It arises from the *lowest part* of the *lateral supracondylar ridge*, the adjacent part of the *popliteal surface* of the *femur* and the *knee joint capsule*. The tendon passes obliquely downwards between gastrocnemius and soleus to emerge on the medial side of the tendocalcaneus. It may insert into the *tendocalcaneus* or into the *medial side* of the *posterior surface* of the *calcaneus*.

Nerve supply

By the *tibial nerve* (root value S1 and 2).

Action

Plantaris is a weak flexor of the knee and plantarflexor of the ankle joint.

The tendocalcaneus (Achilles tendon)

This is considered to be the thickest and strongest tendon in the body, being the tendon by which the calf muscles exert their force on the posterior part of the foot during the propulsive phase of many activities, for example, walking, running and jumping. It has been suggested that the tendocalcaneus is able to withstand strains of up to 10 tonnes. As its fibres pass downwards they spiral through some 90°, with the medial fibres passing posteriorly. This unusual arrangement is thought to explain the apparent elastic qualities of the tendon. For example, when jumping the body will land in an upright position with the foot held in plantarflexion by the active triceps surae. The strain is then taken by the tendocalcaneus, which produces a recoil effect.

Functional activity of the calf muscles

The action of the calf muscles is to plantarflex the foot at the ankle joint. Gastrocnemius acts as the propelling force, working mainly on the ankle but also producing flexion of the knee if working strongly enough. Soleus, on the other hand, is better situated to act more as a postural muscle. This is because its lower attachment is the fixed point and prevents the leg from moving forwards under the influence of body weight, because the vertical projection from the centre of gravity of the body falls in front of the ankle joint.

Gastrocnemius is composed of muscle fibres which give it a pale appearance; consequently it is often referred to as 'white' muscle, whereas soleus has fibres which give it a red appearance and is therefore termed a 'red' muscle.

Plantaris takes very little part in plantarflexion of the ankle and, in fact, sometimes causes pain and disability when it is torn. This condition, referred to as 'tennis leg', occurs during a game of tennis, when the player believes that they have been struck on the back of the calf by a tennis ball. The tendon is often completely ruptured and may have to be surgically removed.

Palpation of the calf muscles

When standing, draw your hand down the back of the knee. The two large muscular bellies of gastrocnemius can be felt on either side of the upper part of the calf. The medial head projects slightly higher and lower than the lateral. Both can be felt joining a broad flattened tendon just over halfway down the calf. The junction between the muscle fibres and the tendon is very clear and it is along this line that many injuries of the calf occur.

Soleus is not quite so easy to palpate being deep to gastrocnemius, its lateral boundary appearing as a flattened elevation below and lateral to the lateral head of gastrocnemius when the ankle is plantarflexed. When standing on tiptoe, soleus can be seen and felt to bulge either side of gastrocnemius. Passing the hand further down the calf it will encounter the flattened tendocalcaneus, which is felt to narrow and become rounded at the level of the ankle joint. It then expands slightly to its insertion into the middle section of the posterior surface of the calcaneus.

Muscles dorsiflexing the ankle joint

Tibialis anterior
Extensor digitorum longus (p. 258)
Extensor hallucis longus (p. 257)
Fibularis tertius (p. 256)

Tibialis anterior

A long fusiform muscle situated on the front of the leg lateral to the anterior border of the tibia (Fig. 3.39). It is covered by strong fascia and gains its upper attachment from the deep surface of this fascia, the *upper two-thirds* of the *lateral surface* of the *tibia* and the adjoining part of the *interosseous membrane*. The muscle becomes tendinous in its lower third, passing downwards and medially over the distal end of the tibia. The tendon continues through both

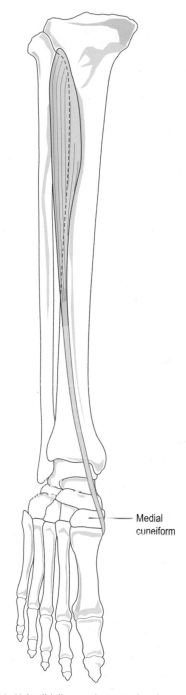

Medial cuneiform

Figure 3.39 Right tibialis anterior, anterior view.

the superior and inferior extensor retinacula to insert into the *medial side* of the *medial cuneiform* and *base* of the *first metatarsal*, the insertion reaching the undersurface of both bones to blend with that of peroneus longus.

251

Nerve supply

By the *deep fibular (peroneal)* (Fig. 3.51C) *nerve* (root value L4 and 5). The skin covering the muscle is also supplied by roots L4 and 5.

Action

Tibialis anterior is a dorsiflexor of the foot at the ankle joint. When working with tibialis posterior it acts to invert the foot, in which the sole of the foot is turned to face medially.

Functional activity

As with other muscles in the leg, tibialis anterior is concerned with balancing the body on the foot. It works with the surrounding muscles to maintain body balance during activities of the upper part of the body which change the distribution of weight.

Not only is tibialis anterior responsible for dorsiflexing the foot as the lower limb is carried forward during the swing-through phase of walking, so preventing the toes catching the ground, it also controls placement of the foot on the ground following initial ground contact by the heel. On close observation, especially in slow motion, it can be seen that the heel does not strike the ground and remain immobile at the initiation of the stance phase, but glides onto the surface and acts as the first braking force of the lower limb's forward movement. Overactivity of tibialis anterior accounts for the wear patterns seen on the posterolateral aspect of the heel, due to the frictional forces between the shoe and the ground. The rest of the foot is then gradually lowered to the ground in a controlled manner taking up the undulations of the surface concerned. The landing of the foot on the ground is similar to the landing of an aeroplane; the main wheels touch down first applying the initial braking force followed by a controlled lowering of the front of the aircraft as the speed decreases.

Tibialis anterior in association with the other dorsiflexors, therefore, plays an important part in lowering the forefoot to the ground in walking or running and is put under stress in extended activities, particularly over rough terrain. The anterior calf muscles are enclosed in a particularly tight fascia which allows very little expansion of the tissues. The result is a compression of the muscle during activity and a dragging on the attachments of the surrounding fascia, particularly where it attaches to bone. This leads to a painful condition of this area commonly called 'shin splints'.

Paralysis of tibialis anterior causes footdrop because the remaining dorsiflexors are not strong enough to raise the toes and so prevent them dragging along the ground. The patient may overcome this by flexing the knee more than normal during walking, or a 'toe-raise' orthosis may be fitted to patients or their shoe.

Palpation

Both the muscle belly and tendon can be seen and felt when the foot is dorsiflexed against resistance, the tendon being the most medial at the ankle joint.

Muscles inverting the foot

Tibialis posterior
Tibialis anterior (p. 251)

Tibialis posterior

The deepest muscle in the posterior compartment of the leg (Fig. 3.40). It arises from the *upper half of* the *lateral aspect* of the *posterior surface* of the *tibia* below the soleal line, the *interosseous membrane*, the *posterior surface* of the *fibula* between the medial crest and interosseous border, and the fascia covering it posteriorly. The tendon, enclosed in its own synovial sheath, passes behind the medial malleolus

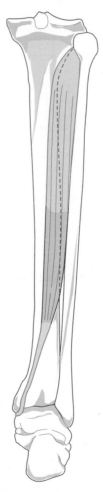

Figure 3.40 Right tibialis posterior, posterior view.

grooving it, being medial to flexor hallucis longus and flexor digitorum longus. It lies superficial to the deltoid ligament. Lying inferior to the plantar calcaneonavicular ligament, the tendon passes downward to attach principally to the *tubercle* on the *medial side* of the *navicular* and the *plantar surface* of the *medial cuneiform* (Fig. 3.41). Tendinous expansions pass to the *plantar surfaces* of *all* the *tarsal bones* except the talus, although a strip passes back to the tip of the *sustentaculum tali*, and the *bases* of the *middle three metatarsals*.

Nerve supply

By a branch of the *tibial nerve* (root value L4 and 5). The skin over the area on the back of the calf is supplied by root S2.

Action

Tibialis posterior is the main invertor of the foot, acting with tibialis anterior. By its attachment to the tubercle of the navicular, it pulls upwards and inwards and therefore rotates the forefoot so that the plantar surface faces medially. It must be noted that inversion and eversion of the foot involve movement at the midtarsal joint, whereby the navicular and cuboid move against the head of the talus and the calcaneus respectively.

The muscle is also a plantarflexor of the foot at the ankle joint, but its contribution is small; gastrocnemius and soleus are better situated and have a more direct line of action. Nevertheless, if the tendocalcaneus is ruptured, tibialis posterior can produce plantarflexion. Because of its attachments to both the tibia and fibula, contraction of tibialis posterior tends to bring the two bones closer together. Consequently, during plantarflexion, the malleoli are approximated to maintain their firm grip on the narrower posterior part of the trochlear surface of the talus.

Functional activity

Tibialis posterior helps to maintain the balance of the tibia on the foot, particularly when the body weight is tending to move laterally. Being a strong invertor, it controls the forefoot in walking and running by positioning the foot so that the medial arch is not completely flattened. Its many tendinous expansions help to maintain all the various arches of the foot.

Palpation

It is not possible to palpate the belly of the muscle due to other muscles covering it. It is, however, quite easy to feel the tendon as it passes behind the medial malleolus and particularly as it attaches to the tubercle of the navicular. When lying supine, the tendon can be felt and seen behind the medial malleolus when inversion of the plantarflexed foot against resistance is attempted. From just above the flexor retinaculum to its insertion, it is surrounded by a synovial sheath and it is in this area that the tendon can

become painful if the muscle has been overactive. The pain is sharp and knifelike and is termed tenosynovitis.

Muscles everting the foot

Fibularis (peroneus) longus
Fibularis (peroneus) brevis
Fibularis (peroneus) tertius

Fibularis (peroneus) longus

A long, thin fusiform muscle (Figs. 3.41–3.43) situated on the lateral side of the leg, with a long belly and an even longer tendon. It is also unique in that the tendon changes direction three times on its way to its insertion on the medial side of the sole of the foot.

It arises from a small area on the *lateral condyle* of the *tibia* (in conjunction with extensor digitorum longus) and the *upper two-thirds* of the *lateral surface* of the *fibula*, its lower

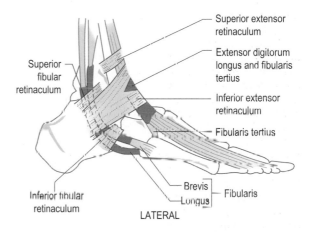

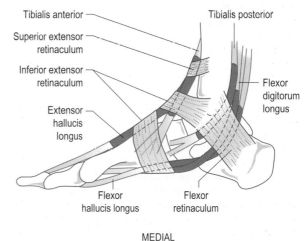

Figure 3.41 Lateral and medial relations of the ankle joint.

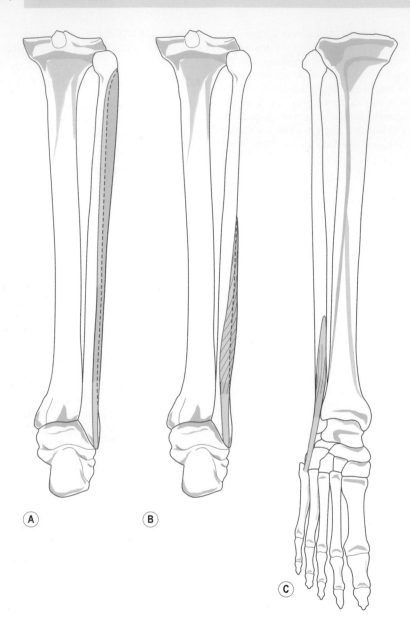

Figure 3.42 (A) Right fibularis (peroneus) longus, posterior view, (B) right fibularis (peroneus) brevis, posterior view, (C) right fibularis (peroneus) tertius, anterior view.

half lying behind the upper part of the origin of peroneus brevis. It also takes origin from the *lateral side* of the *head* of the *fibula*, leaving a small area around the neck for the anterior passage of the common fibular (peroneal) nerve. In front and behind, it attaches to the intermuscular septa and to the fascia surrounding the muscle.

The tendon forms about a hand's breadth above the lateral malleolus and lies superficial to that of fibularis (peroneus) brevis, sharing the same synovial sheath. It runs in a shallow groove behind the lateral malleolus passing deep to the superior fibular (peroneal) retinaculum. From here, the tendon passes downwards and slightly forwards to pass below the fibular (peroneal) tubercle on the calcaneus, being held in position by the inferior band of the inferior fibular (peroneal) retinaculum. At this point the tendon is enclosed in a separate synovial sheath. As it

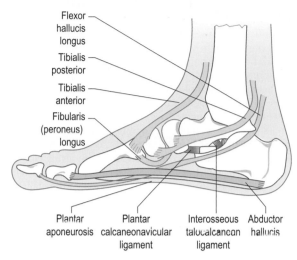

Flexor hallucis longus
Tibialis posterior
Tibialis anterior
Fibularis (peroneus) longus

Plantar aponeurosis
Plantar calcaneonavicular ligament
Interosseous talocalcanean ligament
Abductor hallucis

Figure 3.43 The medial longitudinal arch of the foot and muscles associated with its maintenance.

reaches the inferolateral side of the cuboid, which it grooves, the tendon turns to enter the groove on the inferior aspect of the cuboid. This groove is converted into a tunnel by fibres from the long plantar ligament the tibialis posterior tendon; while in the tunnel the tendon is still surrounded by a synovial sheath. The tunnel conveys the tendon forwards and medially across the foot to its attachment to the *plantar* and *lateral surfaces* of the *medial cuneiform* and *base* of the *first metatarsal*.

Nerve supply

By the *superficial fibular (peroneal) nerve* (root value L5 and S1). The skin covering the muscle is supplied by roots L5 and S1.

Action

Fibularis (peroneus) longus is an evertor of the foot because of the fact that it arises from the lateral side of the leg and passes around the lateral side of the foot. In passing from behind the lateral malleolus to the medial cuneiform and first metatarsal, it produces plantarflexion of the foot, with the medial side of the foot being drawn downwards, as in pronation.

It is worth noting that its insertion is to the same two bones as tibialis anterior, although the latter muscle approaches its insertion from the medial side of the foot. This is believed to provide a stirrup for the arches of the foot and help control their height during activity. The attachment of both muscles to the medial cuneiform and base of the first metatarsal emphasizes the importance of the control of the medial side of the foot during activity, particularly on uneven terrain.

Functional activity

In standing, fibularis (peroneus) longus, in company with other surrounding muscles, helps to maintain the erect position. It controls mediolateral sway by pressing the medial side of the foot onto the ground. This function is better seen and appreciated when standing on one leg when fibularis (peroneus) longus works hard to maintain the leg over the foot and prevent the body falling to the opposite side. Its main functional activity, however, is during powerful action of the foot as in running, particularly over rough ground. Here, its control, together with that of tibialis anterior, over the medial side of the foot and the first metatarsal (carrying the great toe), is vital.

Palpation

When sitting, place the fingers on the lateral side of the knee joint and locate the head of fibula just below the joint level. The tendon of biceps femoris can be identified coming from the back of the thigh. Run the fingers downwards, keeping the tip of the index finger on the head of the fibula spreading the rest of the finger tips down the lateral side of the fibula. Keeping the fingers in this position, lift up the outer side of the foot. The long vertical belly can be felt contracting. If the fingers are now taken down to the lateral malleolus and placed below and behind it and the same manoeuvre is performed, the tendons of fibularis (peroneus) longus fibularis (peroneus) brevis can be palpated and traced to the fibular (peroneal) tubercle where they part, longus passing below and brevis above.

Fibularis (peroneus) brevis

Situated on the lateral side of the leg enclosed in the same osseofascial compartment as fibularis (peroneus) longus (Fig. 3.42B). It arises from the *lower two-thirds* of the *lateral surface* of the *fibula*, the upper half being anterior to fibularis (peroneus) longus. It also attaches to the intermuscular septa at its sides.

The muscle belly is fusiform and short, soon becoming a tendon which accompanies that of fibularis (peroneus) longus to pass behind the lateral malleolus in a common synovial sheath. The tendon then passes forwards and downwards into a groove above the fibular (peroneal) tubercle on the calcaneus, which is converted into a tunnel by the superior band of the inferior fibular (peroneal) retinaculum. It then passes forward to its insertion into the *tubercle* on the *lateral side* of the *base* of the *fifth metatarsal*. Above the tubercle, the tendon is surrounded by a synovial sheath, which is separated from that of fibularis (peroneus) longus. A slip from the tendon usually joins the long extensor tendon to the little toe. Other separate slips may join fibularis (peroneus) longus, or pass to the calcaneus or cuboid.

Nerve supply

By the *superficial fibular (peroneal) nerve* (root value L5 and S1). The skin covering the muscle is innervated by roots L5, S1 and 2.

Action

Fibularis (peroneus) brevis is an evertor of the foot. Because of its course and attachments, the pull of its tendon is in such a direction as to produce plantarflexion of the ankle at the same time.

Functional activity

This muscle is also well positioned to prevent mediolateral sway when standing. When standing on one leg, it helps to prevent the body falling to the opposite side, thus working with a reversed origin and insertion. In walking or running, especially over rough ground, it plays an important part in controlling the position of the foot and should prevent the foot from becoming too inverted. In many cases, however, this mechanism does not always appear to work correctly, and the foot over-inverts, causing the weight to come down on the lateral side of the foot forcing the foot into further inversion. This can severely damage or even snap the tendon of the muscle and often the anterior talofibular ligament of the ankle joint.

Palpation

With the fingers placed on the belly of fibularis (peroneus) longus moving downwards to the lower half of the fibula (but in the same vertical line), the belly of fibularis (peroneus) brevis can be palpated when the foot is everted and plantarflexed. Its tendon can easily be traced to the groove just above the fibular (peroneal) tubercle and then forwards to its insertion into the tubercle of the fifth metatarsal.

Fibularis (peroneus) tertius

Situated on the lower lateral aspect of the leg, fibularis (peroneus) tertius (Fig. 3.42C) appears to have been part of extensor digitorum longus. It arises from the *front* of the *lower quarter* of the *fibula* in continuation with the attachment of extensor digitorum longus (with no gap between them), and from the intermuscular septum and adjoining fascia. Its fibres pass downwards and laterally into a tendon which passes deep to the superior and through the inferior extensor retinacula to insert into the *medial* and *dorsal aspect* of the *base* of the *fifth metatarsal*.

Nerve supply

By the *deep fibular (peroneal) nerve* (root value L5 and S1). The area of skin covering the muscle is also supplied by roots L5 and S1.

Action

The muscle is a weak evertor and dorsiflexor of the foot at the ankle joint.

Functional activity

It is difficult to assess the importance of this small muscle as its actions appear to be covered by other muscles which have a much better mechanical leverage. Indeed in some subjects it is absent. It does, however, pass over the anterior talofibular ligament of the ankle joint, and it is well known that this is very often damaged in inversion injuries. It is, therefore, well placed to help prevent excessive inversion during sporting activities, for example, and may be responsible for reducing the number of injuries. Unfortunately, the muscle is often torn and may be completely ruptured during violent inversion, which is the cause of considerable pain and swelling. It is possible that with the attainment of bipedalism, fibularis (peroneus) tertius is assuming a more important role because eversion of the foot is a peculiarly human characteristic.

Palpation

Fibularis (peroneus) tertius is very difficult to palpate. However, it can be felt by drawing the fingers downwards from the anterior part of the lateral malleolus into the small hollow found there. The tendon can be felt crossing the lateral part of the hollow to its insertion into the medial side of the base of the fifth metatarsal. Take care not to confuse the tendon of fibularis (peroneus) tertius with that of fibularis (peroneus) brevis, which lies lateral to this point as it passes forwards to insert into the tubercle on the lateral side of the fifth metatarsal.

Section summary

Movements at joints of ankle and foot

The ankle joint is only capable of dorsiflexion and plantarflexion with inversion and eversion occurring at the subtalar and transverse (mid) tarsal joints. In addition to the major muscles working on these joints, other muscles crossing the ankle and foot to reach more distal insertions on the toes can contribute to these movements.

Movement	Muscles
Dorsiflexion	Tibialis anterior
	Fibularis tertius
	Extensor digitorum longus
	Extensor hallucis longus
Plantarflexion	Gastrocnemius
	Soleus
	Plantaris
	Fibularis longus
	Fibularis brevis
	Tibialis posterior
	Flexor digitorum longus
	Flexor hallucis longus
Inversion	Tibialis posterior
	Tibialis anterior
Eversion	Fibularis longus, brevis and tertius

- Muscles in italics have their primary function in flexing or extending the toes, and once this has been achieved they can aid movement at the ankle in continued action.
- All of the muscles crossing the ankle have an important function in balance.
- Movements of the forefoot are described as pronation and supination.

Muscles extending the toes

Extensor hallucis longus
Extensor digitorum longus
Extensor digitorum brevis
Lumbricals

Extensor hallucis longus

A unipennate muscle deep to and between the tibialis anterior and extensor digitorum longus on the front of the leg (Fig. 3.44). Arising from the *middle half* of the *anterior surface* of the *fibula* and the adjacent *interosseous membrane*, the muscle fibres pass downwards and medially to the tendon which forms on its anterior surface. The tendon passes under the superior extensor retinaculum, through the upper part of the inferior extensor retinaculum in a separate compartment enclosed in its own synovial sheath, and then deep to the lower band of the inferior extensor retinaculum on its way towards the base of the great toe. Generally, the tendon does not form a fully developed extensor hood but passes to attach to the *base* of the *distal phalanx* on its *dorsal surface*. Tendinous slips may be given off to the dorsal aspect of the base of the proximal phalanx and the first metatarsal.

Nerve supply

By the *deep fibular (peroneal) nerve* (root value L5 and S1). The skin covering this area is supplied by roots L4 and 5.

Action

As its name implies, extensor hallucis longus extends all of the joints of the great toe, but mainly the metatarsophalangeal joint. It is also a powerful dorsiflexor of the foot at the ankle joint.

Functional activity

In running, the great toe is the last part of the foot to leave the ground and therefore the final thrust comes from the long flexors of the toes. After this, the toe must be brought back into the extended position at the same time as the foot is dorsiflexed and slightly inverted, ready for the heel to be placed on the ground for the next weight-bearing phase. By extending the great toe and dorsiflexing the foot,

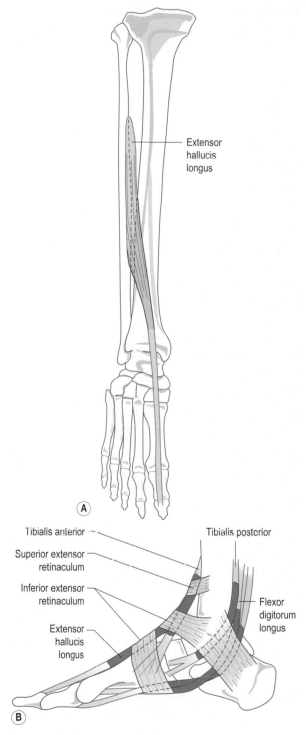

Extensor hallucis longus

(A)

Tibialis anterior
Superior extensor retinaculum
Inferior extensor retinaculum
Extensor hallucis longus

Tibialis posterior
Flexor digitorum longus

(B)

Figure 3.44 (A) Extensor hallucis longus, anterior view, (B) medial relations of the ankle joint.

clearance of the surface is also achieved. It should be noted that the great toe does not have a lumbrical or interossei associated with it. Consequently, extension of the interphalangeal joint depends entirely on extensor hallucis longus. Paralysis of the muscle results in flexion of the joint and buckling of the toe during the last phase of gait, due to the unopposed action of the flexor muscles.

Palpation

If the great toe is extended, the tendon of the muscle is clearly visible as it crosses the first metatarsophalangeal joint to its insertion into the base of the distal phalanx. Trace the fingers up the tendon; it can be felt and seen crossing the anterior aspect of the ankle joint lateral to the tendon of tibialis anterior. From here the tendon can be felt passing upwards and laterally before passing deep to the surrounding muscles. Continue to move the fingers upwards for another 12 cm and allow them to pass a little laterally; when the great toe is rhythmically extended and flexed, the muscle can be felt contracting under the fingers.

Extensor digitorum longus

A unipennate muscle situated on the anterior aspect of the leg, being lateral to tibialis anterior, and overlying extensor hallucis longus (Fig. 3.45). It has a linear origin from the *upper two-thirds* of the *anterior surface* of the *fibula*, the deep fascia and the *upper part* of the *interosseous membrane* with its upper fibres reaching across to the *lateral condyle* of the *tibia* in conjunction with those of fibularis (peroneus) longus. Its tendon appears on the medial side, with the muscle fibres passing downwards and medially to reach it. The tendon passes over the front of the ankle joint deep to the superior extensor retinaculum and then through the inferior extensor retinaculum accompanied by fibularis (peroneus) tertius. At the level of the inferior extensor retinaculum or immediately distal, it gives rise to four tendons which run to the lateral four toes. The four separate tendons are enclosed in a common synovial sheath at the level of the inferior extensor retinaculum. On the dorsal surface of the proximal phalanx, each tendon forms a triangular membranous expansion, the *extensor hood* (dorsal digital expansion). Each hood is joined on its medial side by the tendon of the lumbrical and on the lateral side for the second to fourth toes by the tendon of the extensor digitorum brevis. The interossei of the foot do not have an attachment to the extensor hood.

As the extensor hood passes forwards over the proximal phalanx it divides into three parts before reaching the dorsum of the proximal interphalangeal joint. The central portion attaches to the *base* of the *middle phalanx*, while the two outer portions unite before inserting onto the *base* of the *distal phalanx*. An attachment of extensor hood to the dorsal aspect of the proximal phalanx has also been described.

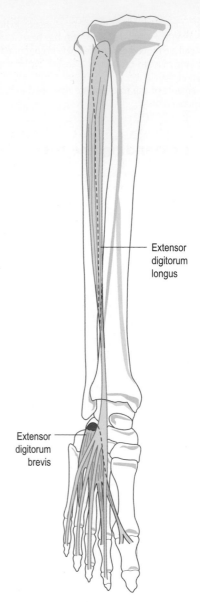

Figure 3.45 Right extensor digitorum longus and brevis, anterior view.

Nerve supply

By the *deep fibular (peroneal) nerve* (root value L5 and S1). The skin covering the muscle is supplied by root L5.

Action

As its name implies, extensor digitorum longus is an extensor of the lateral four toes at the metatarsophalangeal joints, and also assists in extension at the interphalangeal joints. However, it is unable to perform the latter action unaided, which is primarily performed by the lumbricals. If the lumbricals are paralysed, extensor digitorum longus produces hyperextension of the metatarsophalangeal joint, while the interphalangeal joints become flexed. As the muscle passes across the front of the ankle joint, it also aids in dorsiflexion of the foot.

Functional activity

During walking and running, extensor digitorum longus pulls the toes upwards after they have been flexed prior to toe-off, and keeps them clear of the ground until the heel and foot make contact with the ground again. Unfortunately, the lateral four toes in most individuals tend to be flexed at the proximal interphalangeal joint and extended at the distal interphalangeal joint. Consequently, extensor digitorum longus will lift the toes in this adapted position.

Palpation

The muscle belly is easily palpated on the anterolateral aspect of the leg. From the head of the fibula on the lateral side of the leg, just below the knee joint, run the fingers downwards and medially for about 2 cm. When raising the toes off the floor, the muscle can be felt contracting. Now place the fingers over the front of the ankle joint; the tendon can be identified standing out clearly, being lateral to those of tibialis anterior and extensor hallucis longus. From here the tendon can now either be traced upwards, under the superior part of the extensor retinaculum to join the muscle belly, or downwards where it breaks up into four individual tendons running towards each of the lateral four toes. Each tendon stands clear of the metatarsophalangeal joint as it passes towards the dorsum of the toe.

Extensor digitorum brevis

A thin muscle on the dorsum of the foot beyond the inferior part of the extensor retinaculum (Fig. 3.45), lateral to and partly covered by the tendons of fibularis (peroneus) tertius and extensor digitorum longus. It arises from the *anterior roughened part* of the *upper surface* of the *calcaneus* and the deep fascia covering the muscle, including the stem of the inferior extensor retinaculum. From the small belly, short tendons pass forwards and medially, the most medial of which crosses the dorsalis pedis artery to insert separately onto the *dorsal aspect* of the *base* of the *proximal phalanx* of the *great toe*. The remaining three tendons join the lateral side of the *extensor hood* of the *second*, *third* and *fourth toes*. The most medial part of the muscle may develop a separate belly, sometimes referred to as the extensor hallucis brevis.

Nerve supply

By the *deep fibular (peroneal) nerve* (root value L5 and S1). The skin covering the muscle is supplied by roots L5 and S1.

Action

The medial part of the muscle aids extensor hallucis longus in extending the great toe at the metatarsophalangeal joint, while the other three tendons aid extensor digitorum longus. As with the long extensor tendons, extensor digitorum brevis helps the lumbricals to extend the interphalangeal joints; however, it is unable to do this independently.

Functional activity

Extensor digitorum brevis helps extensor digitorum longus and extensor hallucis longus to raise the toes clear of the ground in running and walking.

Palpation

Place the fingers on the tendon of extensor digitorum longus as it splits into its four parts. When the toes are extended extensor digitorum brevis can be felt just lateral and deep to the tendon. The tendons are difficult to trace distally as they become inseparable from those of the extensor digitorum longus.

The lumbricals

Four small muscles associated with the tendons of flexor digitorum longus; they pass from the flexor to the extensor compartment of the foot (see Fig. 3.47B). The most medial lumbrical arises from the *medial side* of the *tendon* to the *second toe*, adjacent to the attachment of flexor accessorius (quadratus plantae) to the main longus tendon. The remaining lumbricals arise by two heads from *adjacent sides of two tendons*, that is, the second from the tendons to the second and third toes, the third from the tendons to the third and fourth toes, and the fourth from the tendons to the fourth and fifth toes. Each muscle then passes forwards superficial to the deep transverse metatarsal ligament on the medial side of the toe, winding obliquely upwards to attach to the *medial side* of the *extensor hood* and *base* of the *proximal phalanx*.

Nerve supply

The first and most medial lumbrical is supplied by the *medial plantar nerve* (root value S1 and 2) and the lateral three by the *lateral plantar nerve* (root value S2 and 3), both being terminal branches of the tibial nerve. The skin on the dorsum of the foot at the point of attachment is supplied by roots L5 and S1. The skin of the plantar aspect of the foot overlying the muscles is supplied by the medial and lateral plantar nerves, which have the same root values as the supply to the muscles. It should be noted, however, that only the most lateral of the lumbricals has skin over its plantar aspect.

Action

There has been much discussion over the role of these small, almost insignificant muscles. They have a long muscle belly compared with their tendon and they link the flexors of the toes with the extensors. By their attachment to the proximal phalanx, contraction of the lumbricals produces flexion of the toes at the metatarsophalangeal joint. However, because they also insert into the extensor hood, the lumbricals extend the interphalangeal joints. Indeed, this latter action is primarily due to the lumbricals and not the long and short extensor tendons.

Functional activity

The action of the lumbricals prevents clawing of the toes during the propulsive phase of gait. Paralysis of these muscles results in the extensor muscles pulling the toes into hyperextension at the metatarsophalangeal joints. Even at rest the toes become clawed.

The nerves supplying the muscles appear to have many more fibres than would be necessary for such a small muscle and a great number of these are sensory. This leads to the conclusion that they may have an important role in providing information related to the tension developed between the long flexor and extensor muscles. This sort of information is of great importance in locomotion, especially as the point of attachment of the lumbricals is a long way from the muscle bellies of the extensors and flexors.

Palpation

It is not possible to palpate these muscles as they lie deep in the sole of the foot covered by many of the small muscles of the sole and the long flexor tendons.

Muscles flexing the toes

Flexor digitorum longus
Flexor accessorius (quadratus plantae)
Flexor digitorum brevis
Flexor hallucis longus
Flexor hallucis brevis
Flexor digiti minimi brevis
Interossei (pp. 266–267)
Lumbricals (p. 259)

Flexor digitorum longus

Situated on the back of the calf for most of its course (Fig. 3.46) it arises from the *medial part* of the *posterior surface* of the *tibia* below the soleal line and the deep transverse fascia surrounding it. The tendon forms about three fingers' breadth above the medial malleolus, lying next to that of tibialis posterior, which has crossed anterior to it to come to lie on its medial side, and medial to the tendon of extensor hallucis longus. Passing deep to the flexor retinaculum the tendon lies in its own synovial sheath along the medial

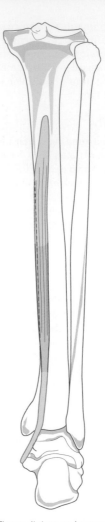

Figure 3.46 Right flexor digitorum longus, posterior view.

aspect of the sustentaculum tali, sometimes grooving it, to enter the sole of the foot deep to abductor hallucis. Passing forwards and laterally, it crosses the tendon of flexor hallucis longus (on its plantar aspect), usually receiving a slip from it which passes into the medial two of its four digitations. About halfway along the sole, on its lateral side, the tendon is joined by flexor accessorius (quadratus plantae) (Fig. 3.47B) and at this point divides into its four individual tendons; one for each of the lateral four toes. Just distal to the attachment of flexor accessorius (quadratus plantae) the lumbrical muscles arise.

Just distal to the metatarsophalangeal joint, the tendons enter their respective fibrous sheaths, together with the appropriate tendon of flexor digitorum brevis which lies superficial to it. The tendon of brevis then splits to enable that of longus to pass through and reach the *plantar surface* of the *base* of the *distal phalanx* where it inserts. Both tendons share a common synovial sheath.

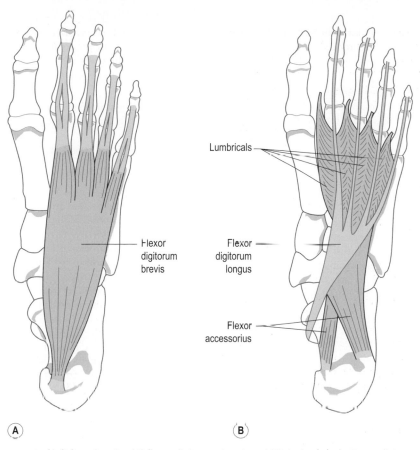

Figure 3.47 Plantar aspect of left foot showing (A) flexor digitorum brevis and (B) the lumbricals, flexor digitorum longus and flexor accessorius.

Nerve supply

By the *tibial nerve* (root value L5, S1 and 2). The skin covering this area on the medial and posterior aspect of the calf and sole is supplied by roots L4, 5 and S1.

Action

Flexor digitorum longus flexes the lateral four toes, flexing the distal interphalangeal joints first, then the proximal interphalangeal joints and finally the metatarsophalangeal joints. Its course behind the medial malleolus means that flexor digitorum longus also helps to plantarflex the foot at the ankle joint. With the ankle plantarflexed, its flexing action on the toes is diminished.

Functional activity

In the propulsive phase of running, jumping or walking, flexor digitorum longus pulls the toes firmly downwards towards the ground to get the maximum grip and thrust during the toe-off phase. When the body is in the standing position, the toes tend to grip the ground to improve balance.

Palpation

Flexor digitorum longus is very difficult to distinguish as its origin is deep to soleus in the calf, while its tendons in the foot, with the lumbricals, lie deeply. However, with care, the tendon can just be identified as it passes alongside the sustentaculum tali.

Flexor accessorius (quadratus plantae)

Lying deep to flexor digitorum brevis, flexor accessorius (Fig. 3.47B) arises by two heads from the *medial* and *lateral tubercles* of the *calcaneus* and the adjacent long plantar ligament. A flattened muscular band is formed by the merging of the two heads which inserts into the *tendon of flexor digitorum longus* in the midpoint of the sole, proximal to the origin of the lumbricals.

Nerve supply

By the *lateral plantar nerve* (root value S2 and 3). The skin over the region is supplied by root S1.

Action

Flexor accessorius helps the long flexor tendons flex all the joints of the lateral four toes. By pulling on the lateral side of the tendon of flexor digitorum longus, it changes the direction of pull so that the toes flex towards the heel and not towards the medial malleolus.

Functional activity

It has an important role to play in gait when flexor digitorum longus is already shortened due to plantarflexion of the ankle joint. The muscle exerts its action on the long flexor tendons so that the toes can be flexed to grip the ground giving support and thrust during the propulsive phase. This action essentially means that flexor digitorum longus can be considered to act powerfully across two joints at the same time – an unusual phenomenon.

Palpation

Lying deep in the sole of the foot, flexor accessorius cannot be palpated.

Flexor digitorum brevis

Situated in the sole of the foot just deep to the central part of the plantar aponeurosis, flexor digitorum brevis (Fig. 3.47A) lies between abductor hallucis medially and abductor digiti minimi laterally. Arising from the *medial tubercle* of the *calcaneus*, the deep surface of the central portion of the plantar aponeurosis and the muscular septa on either side, the fibres pass forwards in the middle of the sole, and separate into four tendons, which pass to the lateral four toes. Just distal to the metatarsophalangeal joint, within their respective fibrous flexor sheaths, each tendon splits into two for the passage of flexor digitorum longus tendon, which passes from deep to superficial. After rotating through almost 180°, the outer margins of the slips of each tendon rejoin, leaving a shallow groove along which the tendon of flexor digitorum longus slides. After passing over the proximal interphalangeal joint, the tendon again splits to insert into the *sides* of the *base* of the *middle phalanx*.

Nerve supply

By the *medial plantar nerve* (root value S2 and 3). The skin covering this area is supplied by roots L5 and S1.

Action

Flexor digitorum brevis primarily flexes the proximal interphalangeal joint of the lateral four toes, followed by flexion of the metatarsophalangeal joints.

Functional activity

Flexor digitorum brevis is obviously concerned, as is flexor digitorum longus, in producing the thrust from the toes when the demand arises.

Palpation

It is almost impossible to palpate as it is covered with some of the thickest fascia in the body and its tendons lie deep within the foot.

Flexor hallucis longus

A powerful unipennate muscle (Fig. 3.48) situated deep to the triceps surae below the deep fascia of the calf. It arises from the *lower two-thirds* of the *posterior surface* of the *fibula* and the adjacent fascia.

The muscle fibres pass to a central tendon which lies on its superficial surface, with those on the lateral side extending lower. The tendon passes downwards, deep to the flexor retinaculum in its own synovial sheath, to cross the posterior aspect of the ankle joint lateral to flexor digitorum longus. During its course, it grooves the lower end of the tibia, the back of the talus (between the medial and posterior tubercles) and the inferior surface of the sustentaculum tali, where it is held in position by a synovial-lined fibrous sheath forming a tunnel for it to run through.

In the sole of the foot, the tendon lies superficial to the plantar calcaneonavicular ligament lying lateral to the tendon of flexor digitorum longus. As it passes forwards, the tendon of flexor hallucis longus crosses deep to that of flexor digitorum longus, and in doing so usually gives a slip to its medial two tendons. It then enters the fibrous digital sheath of the great toe, passing between the two sesamoid bones situated on either side of the base of the proximal phalanx, to insert into the *plantar surface* of the *base* of the *distal phalanx*.

Nerve supply

By a branch of the *tibial nerve* (root value S1 and 2). The skin covering this area is supplied by root S2.

Action

Flexor hallucis longus flexes all the joints of the great toe. It first acts on the interphalangeal joint and then the metatarsophalangeal joint. As it crosses the ankle joint, it helps to produce plantarflexion of the foot.

Functional activity

Flexor hallucis longus is of great importance as it produces much of the final thrust from the foot during walking. At this point in the gait cycle, the calf muscles have already produced their maximum power and the flexors of the lateral four toes are just completing their maximum contraction. Flexion of the great toe is thus the final act before

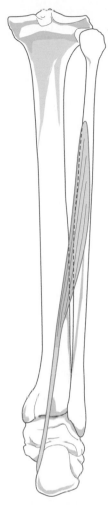

Figure 3.48 Right flexor hallucis longus, posterior view.

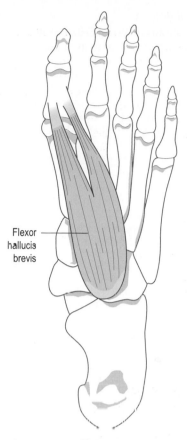

Flexor
hallucis
brevis

Figure 3.49 Plantar aspect of left foot showing flexor hallucis brevis.

the foot is lifted from the ground ready for the next step. It must also be remembered that the muscle plays an important role in maintaining the medial longitudinal arch.

Palpation

Flexor hallucis longus is almost impossible to palpate as it lies deep to the calf muscles, flexor retinaculum, plantar aponeurosis and the muscles in the foot. Its tendon is set deep within both the calf and plantar aspect of the foot.

Flexor hallucis brevis

Flexor hallucis brevis is a short muscle (Fig. 3.49) situated deep in the sole of the foot between abductor hallucis medially and flexor digitorum brevis laterally. It arises from the *medial side* of the *plantar surface* of the *cuboid*, behind the groove for fibularis (peroneus) longus, and the *adjacent*

surface of the *lateral cuneiform* and from the tendon of tibialis posterior.

The muscle fibres run forwards and medially towards the great toe, separating into two fleshy bellies which lie either side deep to the tendon of flexor hallucis longus. The tendon from each belly inserts onto the appropriate side of the *base* of the *proximal phalanx*. The medial tendon joins with that of abductor hallucis, while the lateral tendon joins with that of adductor hallucis, thereby giving common insertions. Small sesamoid bones, which run in shallow grooves on the head of the first metatarsal, develop in each tendon.

Nerve supply

By the *medial plantar nerve* (root value S1 and 2). The skin covering the area is supplied by root L5.

Action

The action of flexor hallucis brevis is to flex the metatarsophalangeal joint of the great toe.

Functional activity

Flexor hallucis brevis also aids flexor hallucis longus in the final push-off from the ground during activity. Being accompanied at its insertion by abductor and adductor hallucis suggests that steadying of the great toe during propulsion is of great importance, probably to ensure the generation of maximum force. When the great toe is deformed as in hallux valgus, where the tip of the toe points laterally and the base medially, this thrust is lost and the patient finds it difficult to run or sometimes walk, even at slow speeds.

It is interesting to note that it is not uncommon for injuries to occur to the sesamoid bones, particularly in individuals who put considerable strain on the great toe. Such injuries produce an inflamed region where the sesamoid bone slides against the metatarsal. This can cause considerable pain and altered function.

Palpation

This muscle is set so deep within the plantar surface of the foot that it is not possible to palpate. Only the sesamoid bones found within its tendons can be felt, and then only with considerable practice.

Flexor digiti minimi brevis

A small muscle (Fig. 3.50) situated on the lateral side of the plantar surface of the foot. It arises from the *plantar aspect* of the *base* of the *fifth metatarsal* and the sheath of the tendon of fibularis (peroneus) longus. It inserts into the *lateral side* of the *plantar surface* of the *proximal phalanx* of the *little toe*, in conjunction with abductor digiti minimi.

Nerve supply

By the *lateral plantar nerve* (root value S2 and 3). The skin covering the muscle is supplied by root S1.

Action and functional activity

Flexor digiti minimi brevis flexes the metatarsophalangeal joint; it also helps to support the lateral longitudinal arch of the foot.

Palpation

Flexor digiti minimi brevis can be felt contracting, by applying deep pressure, on the middle part of the lateral plantar aspect of the foot when the little toe is flexed.

Abduction and adduction of the toes

In the hand, it is the middle finger which is regarded as the central digit when considering abduction and adduction. In the foot, however, the central digit when considering these movements is the second toe. Therefore, if the great toe is drawn medially it is said to abduct, whereas if all

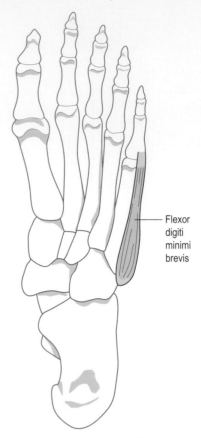

Flexor digiti minimi brevis

Figure 3.50 Plantar aspect of left foot showing flexor digiti minimi brevis.

the other toes are drawn laterally, i.e. away from the second toe, this is also termed abduction. If all the toes are drawn towards the second toe they are said to adduct.

Muscles abducting the toes

Abductor hallucis
Abductor digiti minimi
Dorsal interossei

Abductor hallucis

A powerful and important muscle (Fig. 3.51A) found superficially on the medial side of the plantar aspect of the foot, lying deep to the medial part of the plantar aponeurosis. It arises, in part, from the plantar aponeurosis, the *plantar aspect* of the *medial tubercle* of the *calcaneus*, the flexor retinaculum and the intermuscular septum separating it from flexor digitorum brevis.

The fibres pass forwards forming a tendon which passes over the medial side of the metatarsophalangeal joint of the

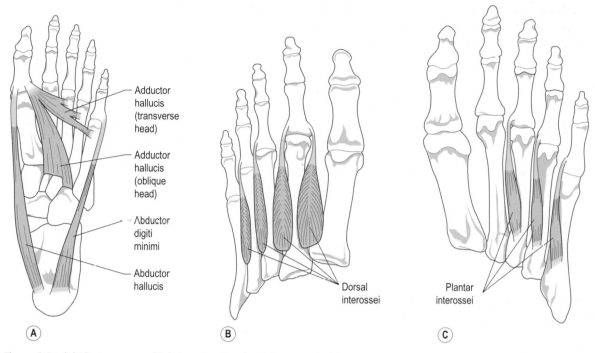

Figure 3.51 (A) Plantar aspect of left foot showing the abductor and adductor muscles, (B) dorsal aspect of left foot showing the dorsal interossei, (C) plantar aspect of left foot showing the plantar interossei.

great toe, to insert into the *medial side* of the *base* of the *proximal phalanx* in conjunction with the tendon of flexor hallucis brevis.

Nerve supply

By the *medial plantar nerve* (root value S1 and 2), with the skin covering the muscle supplied by root L5.

Action

As its name implies, the muscle abducts the great toe at the metatarsophalangeal joint and also helps to flex it at this joint.

Functional activity

Abduction of the great toe is not of importance as such, except perhaps as a party trick and then very few people are able to perform the action easily! However, it is strong and bulky, and must therefore be assumed to have an important role to play in specific activities.

Due to its position along the medial side of the foot, together with the fact that it is attached to either end of the medial longitudinal arch, it can act as a bowstring to the arch when the foot is being used for propelling the body forwards. Its attachment to the medial side of the great

toe also helps in controlling the central position of this toe when it is being flexed.

It should be noted that when the muscle contracts hard, the great toe does indeed move medially, but more importantly, the foot is positioned laterally, thus improving the relationship between the great toe and medial side of the foot. Indeed, if this alignment of the foot and toes was encouraged from an early age many deformities of the toes might be prevented.

Palpation

Place the fingers on the medial plantar aspect of the foot, under the medial longitudinal arch. On flexing the toes the belly of the muscle can be easily palpated towards the heel. Tracing forwards from the heel, the tendon of the muscle can be felt.

Abductor digiti minimi

Situated on the lateral side of the plantar aspect of the foot (Fig. 3.51A), lying deep to the plantar aponeurosis from which it gains part of its attachment. It also arises from the *medial* and *lateral tubercles* of the *calcaneus* and intervening area, as well as the intermuscular septum separating it from flexor digitorum brevis.

The fibres pass forwards forming a tendon which inserts into the *lateral side* of the *base* of the *proximal phalanx of the fifth toe*.

Nerve supply

By the *lateral plantar nerve* (root value S2 and 3), with the skin covering the muscle being supplied by root S1.

Action

On contraction abductor digiti minimi abducts the fifth toe at the metatarsophalangeal joint and also helps to flex it at this joint.

Functional activity

Because the muscle runs from the posterior to the anterior parts of the lateral longitudinal arch, it acts as a bowstring to this arch in a way similar to abductor hallucis on the medial side of the foot, except of course that the lateral arch can hardly be called a true arch. Nevertheless, the muscle certainly comes into action in running and jumping activities to ensure that this arch is maintained under stress.

Palpation

Unless a subject can abduct the fifth toe easily, the muscle is difficult to palpate.

Dorsal interossei

Four small bipennate muscles (Fig. 3.51B) situated between the metatarsals. Each arises from the *proximal half of the sides of adjacent metatarsals*, forming a central tendon which passes forwards, deep to the deep transverse metatarsal ligament. It passes between the metatarsal heads to attach to the *side* of the *proximal phalanx* and capsule of the metatarsophalangeal joint. The tendons do not attach to the extensor hood.

The first, or most medial, arises from the adjacent sides of the first and second metatarsals and attaches to the medial side of the base of the proximal phalanx of the second toe. The second arises from the adjacent sides of the second and third metatarsals and attaches to the proximal phalanx of the second toe but to the lateral side. The third and fourth dorsal interossei attach to the lateral side of the proximal phalanx of the third and fourth toes respectively.

Nerve supply

By the *lateral plantar nerve* (root value S2 and 3), with those in the fourth interosseous space being from the superficial branch and the remainder from the deep branch. The skin covering this area on the dorsum of the foot is supplied by root L5 medially and S1 laterally.

Action

The dorsal interossei abduct the toes at the metatarsophalangeal joint; however, this action, as such, is of little importance in the foot. Acting with the plantar interossei, they produce flexion of the metatarsophalangeal joint.

Functional activity

The dorsal interossei are powerful muscles and their activity in combination with the plantar interossei controls the direction of the toes during violent activity, thus enabling the long and short flexors to perform their appropriate actions.

These muscles, because of their relationship to the metatarsophalangeal joint, can flex these joints and so raise the heads of the second, third and fourth metatarsals, thus helping to maintain the anterior metatarsal arch. They also help, to a limited extent, with the maintenance of the medial and lateral longitudinal arches of the foot.

Palpation

Place the finger tips between the proximal parts of the metatarsals on the dorsum of the foot; when the toes are abducted the muscles can be felt to contract.

Muscles adducting the toes

Adductor hallucis
Plantar interossei

Adductor hallucis

Situated deep within the plantar aspect of the foot (Fig. 3.51A) it arises by two heads, oblique and transverse. The oblique head comes from the *plantar surface* of the *bases* of the *second*, *third* and *fourth metatarsals* and the tendon sheath of fibularis (peroneus) longus. The transverse head comes from the *plantar surface* of the *lateral three metatarsophalangeal joints* and the deep transverse metatarsal ligament.

The muscle fibres of the oblique head pass forwards and medially while those of the transverse head pass medially. The two heads unite and blend with the medial part of flexor hallucis brevis to insert into the *lateral side* of the *base* of the *proximal phalanx* of the *great toe*.

Nerve supply

By the *lateral plantar nerve* (root value S2 and 3); the skin covering this area is supplied by root S1.

Action

It adducts the great toe towards the second toe, and flexes the first metatarsophalangeal joint.

Functional activity

Working with abductor hallucis, adductor hallucis helps to control the position of the great toe so that active flexion can be produced and thereby provide the final thrust needed in walking, running or jumping. Due to its transverse position across the forefoot it also helps to maintain the anterior metatarsal arch of the foot.

The pull of adductor hallucis is almost at right angles to the phalanx and therefore has a better mechanical advantage than abductor hallucis. If the medial longitudinal arch is allowed to fall, allowing the foot to drift medially and the toes laterally, the pull of adductor hallucis overcomes that of abductor hallucis, thus adding to the deformity often seen in the great toe.

Palpation

This muscle is too deep to be palpated.

Plantar interossei

Smaller than their dorsal counterparts and fusiform in shape (Fig. 3.51C), the plantar interossei are found in the lateral three interosseous spaces. Each arises from the *plantar* and *medial aspect* of the *base* and *proximal end* of the *shaft* of the *metatarsal*. The tendon formed passes forwards deep to the *deep transverse metatarsal ligament* to insert into the *medial side of the base* of the *proximal phalanx of the same toe*.

Nerve supply

By the *lateral plantar nerve* (root value S2 and 3), with that in the fourth interosseous space being supplied by the superficial branch of the nerve. The skin covering the area is supplied on the lateral side by root S1 and medially by root L5.

Action

The plantar interossei adduct the third, fourth and fifth toes towards the second. In conjunction with the dorsal interossei they flex the metatarsophalangeal joints of the lateral three toes.

Functional activity

With the help of the dorsal interossei and abductor digiti minimi, the plantar interossei help to control the position of the third, fourth and fifth toes during the push-off phase of walking and running. They also help to prevent splaying of the toes when weight is suddenly applied to the forefoot.

Palpation

These muscles are too deep to be palpated.

Section summary

Movements at the toe joints

The great (big) toe (hallux) has two joints: interphalangeal (IP) and metatarsophalangeal (MTP). Each of the remaining four toes has a distal interphalangeal (DIP), proximal interphalangeal (PIP) and metatarsophalangeal (MTP) joint. As in the hand, the muscles are capable of moving one or more of these joints. The joints that the muscles act on are given in brackets.

Movement	Muscles
Toe extension	Extensor hallucis longus (IP, MTP)
	Extensor digitorum longus (DIP, PIP, MTP)
	Extensor digitorum brevis (DIP, PIP, MTP)
	Lumbricals (DIP, PIP)
Toe flexion	Flexor digitorum longus (DIP, PIP, MTP)
	Flexor digitorum brevis (PIP, MTP)
	Flexor hallucis longus (IP, MTP)
	Flexor hallucis brevis (MTP)
	Flexor accessorius (DIP, PIP, MTP)
	Flexor digiti minimi brevis (MTP)
	Interossei (MTP)
	Lumbricals (MTP)
Toe abduction	Abductor hallucis (MTP)
	Abductor digiti minimi (MTP)
	Dorsal interossei (MTP)
Toe adduction	Adductor hallucis (MTP)
	Plantar interossei (MTP)

- Interaction between the extrinsic and intrinsic muscles allows the various arches of the foot to be formed and maintained.
- Variation in the shape and size of the arches allows the foot to function both as a rigid lever and as a pliable platform.

Fasciae of the lower limb

Fascial projections from the abdomen into the thigh

The fascial lining of the abdomen and pelvis is a continuous membranous bag outside the peritoneum. The transversalis fascia is that part lying on the deep surface of transversus abdominis; inferiorly it is thick and strong where it attaches along the inguinal ligament and iliac crest. From the anterior superior to the posterior superior iliac spines it becomes known as the iliac fascia, being reflected over the surface of iliacus. As the femoral vessels pass from the abdomen, where they are inside the fascia, they drag with them a covering of this fascia which is the femoral sheath. The anterior part of the sheath is derived from the transversalis part of the fascia, while posteriorly it is derived from the iliac part. It is divided into three compartments by septa which pass from its anterior to posterior walls. About three fingers' breadth below the inguinal ligament, the femoral sheath blends with the adventitia of the femoral vessels.

As the transversalis fascia passes posteriorly, it becomes continuous with the anterior layer of the thoracolumbar fascia covering quadratus lumborum and then with the fascia over the anterior surface of psoas major. At the upper margin of quadratus lumborum and psoas major the fascia is thickened forming the lateral and medial arcuate ligaments respectively (see p. 434). The fascia surrounding psoas major forms a sheath completely enclosing it from its upper attachments in the abdomen to the lesser trochanter of the femur.

Fasciae of the lower limb

Functionally there are two types of fascia in the lower limb. One merges with and acts as a base for the skin; this is the *superficial fascia* and enables the skin to move freely over the underlying tissue. The other is composed of dense, tough, fibrous tissue and is known as the investing layer of the *deep fascia*. As this investing layer passes over bony projections it usually becomes attached to them. From its deep surface, a sheet of similar tissue passes between groups of muscles forming the intermuscular septa. These septa are usually attached to bone and serve to maintain the shape of the limb and to exert a compression force on the contents of the osseofascial compartments formed.

Superficial fascia of the lower limb

The superficial fascia is continuous with that of the abdominal wall, the perineum and the back. The two layers present in the lower part of the abdominal wall and perineum continue into the upper part of the anterior thigh. The deeper membranous layer crosses superficial to the inguinal ligament to enter the thigh, fusing with the deep fascia along a line approximately one finger's breadth below and parallel to the inguinal ligament. This attachment limits the spread of fluid into the thigh from the perineum or deep to the superficial abdominal fascia.

The superficial fascia is thick and fatty in the gluteal region, the fat contributing to the shape of the buttock, and forms the gluteal fold. There is usually a deposit of fat over the lateral part of the female thigh – a secondary sexual characteristic. Over the ischial tuberosity the fascia has many dense strands of tissue enclosing fat. Such an arrangement helps to distribute high pressures due to the weight of the seated body, and so prevent tissue damage. Similarly on the sole the superficial fascia is characterized by its thickness and the presence of pads of fat under the heel, and balls and pads of the toes. Again they serve to protect underlying structures from high pressures. The heel pad may be as much as 2 cm thick. The fascia covering the rest of the leg and foot shows no particular features.

Deep fascia of the lower limb

The deep fascia is composed of much stronger fibres which tend to be laid down in the same direction as the applied stresses. It covers the limb in a way similar to the superficial fascia, but attaches to the most prominent bony points and all round the groin and buttocks.

Proximally it attaches to the outer lip of the iliac crest from the anterior superior to the posterior superior iliac spines; the posterior aspect of the ilium and sacrum; the sacrotuberous ligament; the ischial tuberosity; the anterior surface of the ischiopubic ramus; the anterior surface of the body of the pubis and pubic tubercle, and finally onto the inguinal ligament. It can be seen that this forms a complete ring of attachment around the upper end of the thigh.

Below this attachment the deep fascia forms a strong cylinder around the thigh; on the medial side it is thin, but on the lateral side is extremely thick and tough, being composed of two distinct layers called the iliotibial tract. The iliotibial tract is attached above to the tubercle of the iliac crest and below to the lateral side of the lateral tibial condyle (Fig. 3.34B). The major part of gluteus maximus and all of tensor fascia lata insert between these two layers about one-third of the way down.

In the upper part of the front of the thigh there is an opening in the fascia for the long (great) saphenous vein as it passes to drain into the femoral vein. This saphenous opening (see Fig. 3.170) is about three fingers' breadth inferior and lateral to the pubic tubercle. Fascia from the inguinal ligament passes downwards and laterally, forming the falciform margin of the saphenous opening, it then passes deep to the great saphenous vein and wraps around the femoral vein to pass superiorly to attach to the superior pubic ramus. Medially the opening has a smooth margin formed from the fascia covering pectineus. The two margins are joined by the cribriform fascia, which is a thin, perforated layer of fibrous and fatty tissue.

In the lower part of the thigh, intermuscular septa pass from the deep surface of the fascia to the femur. The lateral intermuscular septum separates the quadriceps muscles anteriorly from the hamstrings posteriorly, while the medial septum passes between the adductors anteriorly and the hamstrings posteriorly. Each septum is prolonged downwards onto the medial and lateral supracondylar ridges as far as the medial or lateral femoral condyles. Higher up in the thigh, a thickening of the fascial septum deep to sartorius forms the roof of the adductor canal.

Around the knee the deep fascia is continuous with that of the leg, and is attached to the medial and lateral condyles of the tibia, the head of the fibula and in front to the patella. The patella is held to the tibial condyle by thickened bands of the deep fascia, the medial and lateral patellar retinacula. Behind the knee, over the popliteal fossa, the fascia is reinforced by transverse fibres.

Below the knee, the deep fascia encloses the leg, attaching mainly to the anterior and medial borders of the tibia and the medial and lateral malleoli. Where the tibia and fibula are subcutaneous, the deep fascia blends with the periosteum of the bone. Intermuscular septa pass to the fibula separating the fibularis (peroneal) muscles from the extensors anteriorly and the flexors posteriorly. A further septum passes across the back of the leg separating the superficial and deep flexor muscles. Around the ankle, the fascia is thickened by numerous transverse fibres forming retaining bands for the tendons of the muscles which pass across the ankle.

The fascia of the foot is continuous with that of the leg. On the dorsum it is thin, wrapping around either side of the foot to become continuous with the plantar aponeurosis, while anteriorly it splits to cover the dorsum of the toes.

Plantar aponeurosis

This comprises some of the thickest fascia in the body, being up to 80 layers thick. It is continuous with the fascia over the heel, and on the sides of the foot with the dorsal fascia. It is triangular in shape with the apex at the heel, attaching to the inferior aspect of the calcaneus just behind the medial tubercle, and spreads out anteriorly into five slips which pass forwards and become continuous with the fibrous flexor sheaths of the toes. Most of its fibres run longitudinally, except anteriorly where it splits with transverse fibres binding the five slips together. As each slip approaches the head of the metatarsal it splits into superficial and deep layers. The former attaches to the superficial fascia and produces the deep cleft in the tissue under the base of the toe. The deeper layer splits again into medial and lateral parts which attach either side of the base of the proximal phalanx of each toe and the deep transverse metatarsal ligament. This is the beginning of the fibro-osseous tunnel which houses the flexor tendons of the toes.

The central part of the aponeurosis gives partial attachment to the muscles lying deep to it. At its edges strong septa pass upwards separating abductor hallucis, flexor digitorum brevis and abductor digiti minimi. Between the five slips, close to the base of the toes, there is a space which gives access to the nerves and vessels supplying the toes.

The plantar aponeurosis is an extremely important structure in the maintenance of the longitudinal arches of the foot. With the toes dorsiflexed, the proximal phalanx winds its slip of the aponeurosis around the metatarsal head, a 'windlass' effect, which tightens the aponeurosis and raises the longitudinal arches.

Fibro-osseous tunnels

Where the anterior attachment of the plantar aponeurosis splits to attach to either side of the proximal phalanx it forms the beginning of a fibro-osseous tunnel, which runs under the toe and accommodates the flexor tendons. It is composed of arching fibres passing over the tendons and attaching to the flat plantar surface of each phalanx.

At the interphalangeal joints the fibres criss-cross from the medial side of the head of one phalanx to the lateral side of the base of the adjacent phalanx, and vice versa. Each tunnel is lined with a double layer of synovial membrane which facilitates movement of the tendons in the tunnel.

Retinacula

Around the ankle the deep fascia is thickened by transversely orientated bands forming retinacula. These serve to hold the tendons passing across the ankle joint in position and prevent bowstringing. They are named according to the tendons they serve, that is flexor, extensor and fibular (peroneal).

The flexor retinaculum passes between the back of the medial malleolus and the medial tubercle of the calcaneus. From its deep surface the septa pass to the tibia and ankle joint capsule forming four tunnels. From medial to lateral these tunnels transmit the tendons of tibialis posterior, flexor digitorum longus, the posterior tibial vessels and tibial nerve, and finally the tendon of flexor hallucis longus. Each tendon is enclosed in its own synovial sheath.

The extensor retinacula are the most extensive. The superior extensor retinaculum (Fig. 3.52) runs horizontally across all the extensor tendons between the tibia and fibula just above the ankle joint. The inferior extensor retinaculum is a Y-shaped thickening on the dorsum of the foot (Fig. 3.52B). The stem of the Y attaches to the upper surface of the calcaneus anteriorly and floor of the sinus tarsi. The upper part passes to attach to the medial malleolus, while the lower part blends with the deep fascia on the medial side of the foot. This lower band splits into superficial and deep layers, with a septum passing between the two, so forming two tunnels. In the most medial of these tunnels runs the tendon of tibialis anterior, with that of extensor hallucis longus in the other compartment. Each tendon has its own synovial sheath. Passing under the stem laterally are the tendons of extensor digitorum longus and fibularis (peroneus) tertius in a common synovial sheath. Both the tendons of tibialis anterior and extensor hallucis usually pass deep to the upper band of the extensor retinaculum, each in its own synovial sheath, although tibialis anterior may pierce it. The anterior tibial vessels and the deep fibular (peroneal) nerve pass deep to both extensor retinacula.

The fibular (peroneal) retinacula are also in two parts (Fig. 3.52B). The superior fibular (peroneal) retinaculum attaches to the lateral side of the calcaneus and passes over the tendons of fibularis (peroneus) longus and brevis to the posterior border of the lateral malleolus. The two tendons are contained within a single synovial sheath. The inferior fibular (peroneal) retinaculum binds the two fibular (peroneal) tendons to the lateral side of the calcaneus. A septum passes from its deep surface to the fibular (peroneal) tubercle to form two separate tunnels, one above the tubercle for the tendon of fibularis (peroneus)

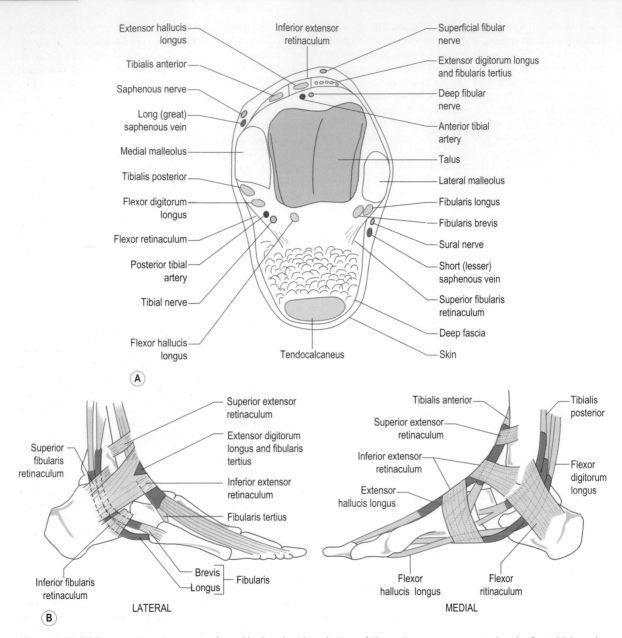

Figure 3.52 (A) Transverse section across the ankle showing the relations of the various structures entering the foot, (B) lateral and medial relations of the ankle joint.

brevis and one below for fibularis (peroneus) longus. At this point each tendon is enclosed in a separate synovial sheath.

Synovial sheaths

As the various tendons pass under the appropriate retinacula around the ankle joint, they are invaginated from the lateral side into a double layer of synovial membrane. This lines the compartments so formed, protruding approximately 1 cm above each of the retinacula. The sheath of tibialis posterior continues down to its insertion, while those of the tendons of flexors digitorum longus and hallucis longus continue as far as the base of the metatarsals.

Laterally the tendons of fibularis (peroneus) longus and brevis initially share the same sheath, but after passing below the lateral malleolus they occupy separate compartments each lined by a separate synovial sheath which continues as far as the lateral side of the cuboid. Fibularis (peroneus) longus is surrounded by another synovial sheath as it passes through the fibrous tunnel formed on the plantar surface of the cuboid.

Anteriorly both tibialis anterior and extensor hallucis longus are surrounded by separate synovial sheaths. The former reaches down almost to the insertion of the tendon, whereas the latter only passes as far as the mid-metatarsal region. The tendons of extensor digitorum longus and fibularis (peroneus) tertius share the same sheath, which is only present as the tendons pass deep to the extensor retinacula (Fig. 3.52B).

The sheaths facilitate the sliding of the tendons under the retinacula and normally are not evident. However, when the sheaths have been damaged in some way, for example by trauma, infection or overactivity, they become inflamed and swollen. They then present a 'sausage-like' swelling along the length of the sheath and around the tendon. This often proves to be acutely painful and incapacitating. The condition is referred to as tenosynovitis and can occur around any tendon where it is surrounded by a synovial sheath. It is one of the conditions in a group commonly labelled repetitive strain injury (RSI) and may occur anywhere in the body where tendons pass through sheaths in confined spaces.

Simple activities of the lower limb

The activities considered here are:

walking
standing up from sitting
climbing steps
cycling
squats
rowing.

It is only possible to give a basic outline of the activities listed above, as it must be remembered that each differs considerably according to the height, weight and build of the individual. There is an additional problem in that everyone has their own characteristic pattern, which may be so clearly individual that they can be recognized by these movements. Many activities can be influenced by the clothing worn, for example high heels, tight jeans, trousers or skirts. The muscular work involved and joint activity also varies depending on the speed at which the activity is performed and the surrounding environment. Nevertheless, there are certain factors which tend to be common to most individuals and these are considered here.

Walking

Walking involves the whole of the body, consequently it must be remembered that a change in the pattern of movement of the upper body will affect the walking pattern. This section only considers the lower limbs and lower trunk in any detail, with the upper body referred to in outline only.

Each lower limb performs a cycle of events which is similar, but performed half a cycle out of phase with each other. Thus, the left limb will be weight-bearing while the right is off the ground; as the toes are pushing off in one limb, the heel of the other limb is contacting the ground, and as one limb is being taken forward the other is being drawn backwards.

By examining the cycle of events associated with one limb, an understanding of the composite movement of both limbs can be obtained. Movements of the right lower limb are presented in the following account; any reference to the other limb is clearly distinguished. When considering walking, it is often easier to break up the pattern observed into different phases; however, this should not detract from the fact that this is a continuous cycle of events performed smoothly, precisely and efficiently.

Toe-off phase

This account begins where the foot is being powerfully plantarflexed to push the body forward. At this point the trunk becomes flexed, abducted and rotated at the supporting hip. This rotation is equivalent to medial rotation of the femur. At the hip joint the femur is extended, adducted and medially rotated, the knee extended and the ankle dorsiflexed. Although the toes are in a neutral position, they are gripping the supporting surface.

Powerful ankle plantarflexion is achieved by the calf muscles, *gastrocnemius* and *soleus*. As the movement progresses, the toes are forced into extension, but after receiving this initial stretch, the *toe flexors* also work powerfully, the lateral four toes first, closely followed by the great toe. Because *quadriceps femoris* is holding the knee almost fully extended, the powerful thrust from the foot is transmitted to the hip, pelvis and trunk, which, because of their forward inclination, are pushed forwards and upwards.

Carry-through phase

As the great toe leaves the ground the first part of the carry-through phase begins. There is extension of the toes, brought about by *extensors hallucis longus* and *digitorum longus*, dorsiflexion of the ankle by *tibialis anterior* and *extensor digitorum longus*, flexion of the knee by the *hamstrings*, flexion of the hip by *psoas major*, *iliacus*, *rectus femoris*, *sartorius* and *pectineus*. The hip is laterally rotated by *piriformis*, *obturators internus* and *externus*, *quadratus femoris* and the two *gemelli*. This continues until the foot passes a point immediately below the hip joint. During this phase the unsupported side of the pelvis is also moving forward, firstly because of the thrust it receives at toe-off, and secondly because it is pivoting around the opposite hip joint. The upper trunk, however, tends to rotate in the opposite direction so that the same shoulder, carrying the arm with it, moves backwards. To keep the head facing forwards, the neck rotates towards the opposite side.

After the foot passes below the hip joint the limb begins to extend again. The *dorsiflexors* allow the ankle and toes to assume a neutral position, or even a slightly flexed position, with the foot being slightly inverted by *tibialis anterior* and *posterior*. Contraction of *quadriceps femoris* extends the knee to just short of full extension. The hip is still being flexed by the same group of muscles and the pelvis is still pivoting forwards under the action of *gluteus medius* and *minimus*, and *tensor fascia lata* of the supporting limb. The carry-through phase terminates when the foot makes contact with the ground again, and the support phase begins once more.

Heel-strike phase

Because the foot is slightly inverted the heel comes into contact with the ground on its lateral side. The frictional forces generated have a dragging effect on the heel which slows the foot allowing it to land and make full contact with the supporting surface. As the foot takes the full weight of the body, the heel is compressed and the intrinsic musculature contracts to support the arches of the foot. The weight is then relayed, via the lateral side of the foot, forwards to the forefoot. The *intrinsic muscles* of the foot convert it into a semirigid lever, which enables it to absorb the stresses associated with foot contact and yet prepare it for the next propulsive phase preceding toe-off.

Support phase

As the weight is taken onto the foot the body is moving forward, the ankle is dorsiflexed, the knee undergoes a small flexion wave and the hip is being extended by the momentum of the body aided, particularly in walking fast, by *gluteus maximus* and the *hamstrings*. The pelvis is maintained in a more or less level position by the action of the *abductors* of the right hip, that is, *gluteus medius* and *minimus*, allowing the opposite foot to be raised from the ground.

After the foot passes behind the line of the hip joint the *calf muscles* again contract strongly, thus completing the cycle mentioned earlier. From the point where the weight is borne on the heel to that where the toes push the body forward and upwards, the other limb is being lifted from the ground to perform the same carry-through phase as described above.

Standing up from the sitting position

There are many different ways in which to stand up from a sitting position. As well as the factors listed earlier, the height and type of chair being used may have a profound effect on the pattern of movement. Nevertheless, the fundamentals of rising from a seated position will be outlined, bearing in mind that there are many variations.

Sitting

The starting position is, in this case, sitting upright on a wooden chair which has a firm seat but no arm rests. The feet are both on the floor lying parallel to one another about 10 cm apart. The ankles are at right angles, as are the knees and hips. The arms are by the sides, and in this case will give no assistance to the movement except as a mechanism to aid balance. The line of gravity through the trunk and upper limbs falls through the seat between, but just anterior to, the ischial tuberosities. During movement the base is changed from the seat to the feet; consequently some readjustment of the body weight and foot position must be carried out.

Preparation for standing

The first phase in preparation for standing is to move the trunk forward and the feet backwards so that the centre of gravity of the upper body is brought as far forward as possible. The feet are drawn backwards by the *hamstrings* of both limbs; flexion of the knees and dorsiflexion of the ankles are both increased. However, full contact is still maintained between the feet and the ground. At the same time the trunk is flexed forward by the *abdominal muscles*, the pectoral girdle is protracted by *serratus anterior* and *pectoralis minor*, the neck is flexed by the *prevertebral muscles*, while the head is extended by *rectus capitis posterior major* and *minor*. (Extension of the head is not, however, a natural movement when standing from the seated position.) If leaning forwards does not bring the centre of gravity of the upper body sufficiently over the feet to allow the transfer of weight, then the whole trunk must be shifted forwards.

The next phase is to transfer the weight of the body over the new base, that is, the feet. This is brought about by additional contraction of the *abdominal muscles* to bring the upper part of the body forward. The weight is now taken by the feet and the *intrinsic foot muscles* contract to maintain the various arches.

Standing

There is now plantarflexion of the ankle to bring it back to just short of a right angle. Plantarflexion is brought about mainly by *soleus* while the knee is extended by *quadriceps femoris*. Extension at the hip is brought about by a combination of the *hamstrings* (which are acting as a tie-mechanism), being drawn down by the tibia which is moving forwards in relation to the femur and *gluteus maximus*, which, in addition to its effect on the hip, pulls on the iliotibial tract to aid in extension of the knee. At the same time as the hip and knee joints are being extended, the back is extended by the long back muscles, that is, *sacrospinalis*. The pectoral girdle is retracted by the *rhomboids* and *middle fibres* of *trapezius*, while the neck is extended by the *upper fibres* of *trapezius* and *splenius cervicis*. The head is brought to the neutral position by contraction of *longus capitis*.

All of these movements occur simultaneously until the erect standing position is achieved.

Climbing steps

Variations in this activity are again so wide that it makes description difficult. There are many combinations of step height and tread which can only but modify the pattern of movement. The example given here is with a step height of 25 cm, the movement being performed fairly slowly.

Toe-off phase

The starting position is with the left foot already up one step and the right about to push off. Each limb again performs a complete cycle of movement, with the limbs being out of phase, such that as the left limb is weight bearing the right is being carried through. The right foot pushes off with a strong plantarflexion of the ankle brought about by the *calf muscles* producing a passive extension of the toes, thus stretching the flexors. The latter respond immediately by flexing the toes, thereby producing the final thrust from the right foot. It is at this point that the whole body is inclined forward; however, the knee and hip are maintained in an extended position while the trunk flexes to bring bodyweight forwards. At the same time the left *quadriceps femoris* is working maximally to extend the left knee and raise the body to the next step.

Carry-through phase

The right limb now begins its carry-through phase. This involves extension of the toes by *extensors digitorum longus* and *hallucis longus*; dorsiflexion of the ankle by *tibialis anterior* and the *long toe extensors*; flexion at the knee, by the *hamstrings*; flexion of the hip by *psoas major* and *iliacus*, and a raising and forward rotation of the pelvis on the same side by the action of *gluteus medius* and *minimus* of the weight-bearing side. This movement continues until the foot has passed the left leg and lies just above the next step.

Foot down and step-up phase

The foot is then lowered onto the step by the eccentric contraction of the *hip flexors* and weight is transferred to it, thereby beginning the next weight-bearing phase. The *intrinsic muscles* of the *right foot* now contract to stabilize the arches while the *long flexors* pull the toes down towards the supporting surface. The ankle, which was in slight dorsiflexion because of the forward inclination of the tibia, now comes into a neutral position brought about partly by *soleus* and partly by extension of the knee. The movements at both ankle and knee contribute to the extension force at the hip which, aided by *gluteus maximus* and the *hamstrings*, produces a backward tilting of the pelvis. The back muscles use this firm base to extend the trunk into an upright position.

The carry-through phase of the limb is augmented by forward rotation of the pelvis at the hip of the supporting side. This is brought about by the action of *gluteus medius* and *minimus* of the weight-bearing limb.

When the hip and knee are fully extended the other foot is placed onto the step above. The ankle of the supporting leg is plantarflexed to throw the body forward onto the left limb and so complete the cycle.

Cycling

Cycling is a very popular activity either as a pleasurable pastime, a means of transport or a form of keeping fit. It is important, therefore, to understand the movements of the joints that are involved and the muscles producing these movements. The activity varies according to both the type and size of machine used and the stature and ability of the individual. Although only the activity of the lower limbs is considered, cycling is an activity which exercises the body as a whole; muscles of the trunk and upper limbs contribute to the stability, control, balance and counter pressure needed for efficient and effective power to be applied by the lower limbs.

The following analysis considers an individual using a mountain bike with straight handle-bars and a fairly low saddle. The trunk will therefore be slightly inclined forwards with some of the body weight being transferred through the arms to the handle-bars.

Each lower limb performs a similar action, being 180° out of phase with each other. As one lower limb is pushing hard against the pedal the other is passing through a recovery phase. Consequently only one limb needs to be considered.

The thrust phase

The thrust phase begins immediately after the pedal has reached its highest point. Although the power generated is usually the same throughout the downward thrust, maximum work is produced when the crank shaft of the pedal is at right angles to the calf. Experienced cyclists learn to apply the maximum pressure to the pedal approximately 45° on either side of this point, easing off near the top and bottom of this thrust phase.

At the beginning of the thrust phase the hip is flexed to almost 90° and the ankle is fully dorsiflexed. From this position to the mid-thrust position, the hip extends some 70° being brought about by the concentric and isotonic action of *gluteus maximus* and *hamstrings*. Slight abduction of the weight-bearing hip is brought about by powerful concentric contraction of *gluteus medius* assisted by the *anterior fibres* of *gluteus maximus* and the *posterior fibres* of *gluteus minimus*.

A small degree of medial rotation of the working hip occurs at the lower part of the thrust phase as the pelvis rotates forward in preparation for the equivalent phase in the opposite limb. This medial rotation is brought about by *gluteus minimus* and *tensor fascia lata*, aided by the *anterior fibres* of *gluteus medius*, all working concentrically.

Powerful extension at the knee is brought about by the concentric action of *quadriceps femoris*, the knee joint moving through approximately 90° to become just short of full extension. There is also powerful plantarflexion at the ankle joint due to the concentric action of the *calf (triceps surae)* and *posterior tibial muscles*.

The toes, particularly the great toe, are flexed at their metatarsophalangeal and interphalangeal joints by *flexor digitorum longus*, with *quadratus plantae* and *flexor hallucis longus*, being aided by *flexor digitorum brevis*, *flexor hallucis brevis* and the *interossei*.

The height of the saddle will considerably influence the range of joint movement as well as the degree of muscle activity, which in turn will critically affect the efficiency of the downward thrust. For professional cyclists the saddle height can dramatically affect their performance.

The recovery phase

The recovery phase produces elevation of the opposite side of the pelvis. In this phase the upward-moving pedal pushes the foot upwards as it rises. Weight is reduced on the pedal maintaining contact so the foot is positioned ready for the next thrust phase.

There is extension of the toes and dorsiflexion at the ankle, both through almost their full range brought about by the upward movement of the pedal and controlled by eccentric work of the *flexors of the toes* and *plantarflexors of the ankle* respectively. At the same time there is flexion at both the knee and hip, due to the same upthrust of the pedal, being controlled by *quadriceps femoris* and by *gluteus maximus* and the *hamstrings* respectively, all working minimally but eccentrically. In addition the *hip abductors* are working to maintain the position of the lower limb, although the pelvis is raised by the *flexors* of the same side.

As the foot passes the highest point of the cycle, the thrust phase begins again.

If the pedal is fitted with a toe clip the recovery phase becomes a dynamic 'pulling-up' phase involving active extension of the toes, dorsiflexion at the ankle, and flexion at the knee and hip, with the appropriate muscles all working concentrically to augment the thrust phase of the opposite limb. Stationary (exercise) cycles have become extremely popular. However, the joint movement and muscle analysis is essentially the same as that described above. The power of the thrust phase, however, is normally controlled by friction being applied to a fly wheel driven by the pedals. Cycling is a good way of improving and maintaining cardiovascular fitness.

Squats

Explanation of movement

Squats are a common form of physical activity, being particularly advantageous for building up the calf muscles, quadriceps femoris, the glutei and the extensor muscles of the back. They are convenient, require no apparatus and can be performed in a limited space. Performance can be assessed by numerical progression; they are frequently used to enhance stamina. There is full range activity at the ankle and knee joints and, except for the last few degrees of extension, at the hip and the vertebral column. As in other activities in this section only the lower limbs will be dealt with in detail. Small variations in the performance of squats can easily be included: the subject can rise onto the toes or leave the feet flat on the floor; the trunk can be bent over the knees or remain vertical when reaching the fully squat position. In the following description the heels are raised and the trunk is allowed to bend forward over the knees.

Starting position

The activity begins from the standing position. The ankles are slightly dorsiflexed; the knees are fully extended with *quadriceps femoris* relaxed; the hips are in a neutral position, maintained by slight contraction of *gluteus maximus* and the *hamstrings* and slightly laterally rotated; the vertebral column is erect.

Heel raise phase

There is plantarflexion of the ankle joints brought about by concentric contraction of *gastrocnemius*, *soleus* and *plantaris*. The longitudinal and transverse arches of the feet are raised by the *extrinsic* and *intrinsic muscles* of the *foot*.

The weight of the body is then transferred to the metatarsal heads, particularly the first and the fifth, and to all of the toes, the latter being passively forced into extension by raising the heel. The long (*flexors digitorum longus* and *hallucis longus*) and short (*flexors hallucis brevis*, *digitorum brevis*, *accessorius*) flexors of the toes initially contract concentrically and then statically to maintain the balance of the whole body.

The knees remain in their close-packed position with *quadriceps femoris* contracting strongly to maintain this position. The hips become slightly extended with the *glutei* acting statically to maintain balance. The *flexor*, *extensor* and *lateral flexor muscles* of the *trunk* all contract statically to maintain the position of the trunk and balance of the body.

Knees bend phase

The knees are 'unlocked' by the action of *popliteus* pulling on the lateral side of the lateral femoral condyle, thus rotating the femur laterally and sliding the medial condyle slightly forwards. The knee then flexes under the action of body weight, controlled by the eccentric action of *quadriceps femoris*. The power needed to control this movement increases as the knee flexes. Movement is arrested as the buttocks and posterior thigh make contact with the heels and calf, respectively.

The ankles become dorsiflexed under both the action of body weight and the change in position of the tibia; however, the *posterior calf muscles* act powerfully eccentrically to control this movement. The feet maintain their high arches as the weight is still borne on the metatarsal heads and the pads of the toes. All of the muscles crossing the ankle joint and of the feet are interacting to maintain balance.

The hips are flexed because the trunk bends forward to maintain equilibrium, the movement being controlled by powerful eccentric activity of *gluteus maximus* and the *hamstrings*. There is interplay between the abductors (*gluteus medius* and *minimus*) and adductors (*longus*, *brevis* and *magnus*) to maintain balance.

The rising phase

The knees are extended by powerful concentric contraction of *quadriceps femoris*, with maximum force being applied when the knees are fully flexed, decreasing as they become more extended. At full extension the knees move into a 'close-packed' position as the femur rotates medially with respect to the tibia, and the medial femoral condyle slides backwards on the tibial plateau. As the knees extend, the ankles plantarflex and the hip and vertebral column both extend. The ankles are plantarflexed by the concentric action of the *posterior calf muscles*; the hips are extended by the concentric contraction of *gluteus maximus* and the *hamstrings*, while the trunk is extended by the concentric contraction of the *postvertebral muscles*.

Rowing

The joint movement and muscle work of the lower limbs in rowing is intimately related with the timing of the movements of the upper limb (p. 99). It is this timing, power and balance combined with many other factors which determines the efficiency of the rower and the speed of movement through the water.

Outline of activity

The description begins from the fully forward position, as in the upper limb (p. 99). The seat is fully forward on the runners. The whole body is then pushed backwards by extension of the lower limbs, the trunk extended and the oars drawn backwards by the upper limbs.

Starting position

The feet are usually strapped to the stretcher, the ankles being fully dorsiflexed and held at this point by the ankle dorsiflexors (*tibialis anterior*, *extensors hallucis longus* and *digitorum longus*), which also produce extension of the toes. The knees are fully flexed and held by the *hamstrings* working statically. The hips are fully flexed, held by the hip flexors (*psoas major*, *pectineus*, *rectus femoris*). The trunk is also flexed (p. 99).

Sequence of movement

Stroke phase At the beginning of this phase, the lower limbs are forcefully extended. There is plantarflexion of the ankles to the neutral position partly brought about by concentric contraction of the calf muscles (*gastrocnemius* and *soleus*), and partly due to the change in position of the knees.

The toes are thrust against the stretcher mainly by static work of *flexors digitorum longus* and *brevis* and *hallucis longus*. There is vigorous and full extension of the knee joints brought about by strong concentric contraction of *quadriceps femoris* and there is extension of the hip joints from the flexed position to just short of neutral, brought about by powerful concentric contraction of *gluteus maximus* and the *hamstrings*. This is one of the occasions when the hamstrings act as a tie between the knee and hip. As the knees are being extended the lower attachment of the hamstrings is moving downwards while its fibres are contracting concentrically to extend the hip joints.

The trunk and neck are extended while the head is stabilized in a neutral position (see upper limb and trunk action, p. 99).

Recovery phase At the end of the stroke phase, when the hands and oar reach the abdomen, the wrists are extended, the oars are removed from the water and begin their movement backwards, parallel with the water. The rower, on the seat, now begins to move forwards. The ankles are dorsiflexed by concentric action of *tibialis anterior*, *extensor digitorum longus*, *extensor hallucis longus* and *fibularis (peroneus) tertius*.

The knees are flexed by the concentric action of the hamstrings (*semitendinosus*, *semimembranosus* and *biceps femoris*). The hips are flexed by the concentric action of *psoas major*, *iliacus*, *pectineus* and *rectus femoris*. The trunk is flexed by the *abdominal muscles* working concentrically (p. 99).

When the forward position is reached the wrists are flexed to the neutral position, the oar is again placed in the water (p. 99), and the full cycle is ready to begin again.

JOINTS

Joints of the pelvis

Introduction

The pelvic girdle is a ring of bone providing articulation for the lower limbs with the trunk, and consists of the two innominate bones and the sacrum (Fig. 3.53A). Each innominate articulates with the sacrum posteriorly by a synovial joint, and with each other anteriorly at the symphysis pubis by a secondary cartilaginous joint. The articulation of the pelvis with the lower limb is in the region of the acetabulum (p. 210), while the articulation with the trunk is via

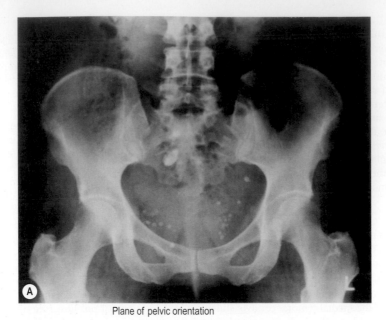

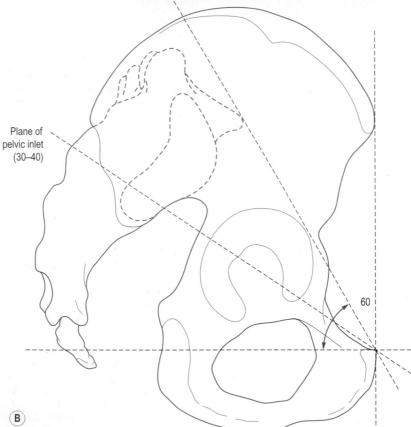

Figure 3.53 (A) Anteroposterior radiograph of the pelvis showing the sacroiliac articulations, (B) anatomical position of the pelvis.

the sacrum. Superiorly the sacrum articulates with the fifth lumbar vertebra at the lumbosacral joint, and inferiorly with the coccyx at the sacrococcygeal joint. Being part of the vertebral column, these latter two joints are both secondary cartilaginous joints.

The skeletal pelvis is arranged in such a way as to provide great strength for the transference of weight from the trunk to the lower limbs when standing, or to the ischial tuberosities when sitting. This major function of stability has been achieved with loss of mobility at both the sacroiliac joint and the symphysis pubis. However, under certain conditions, namely childbirth, a greater degree of movement is possible at these two joints.

As well as enabling weight transfer from the trunk to the lower limbs, the pelvis also supports the pelvic viscera, provides extensive attachments for muscles of the trunk and lower limb, and in the female gives bony support for the birth canal.

The evolutionary development of the pelvis and of its articulation with the vertebral column via the lumbosacral joint is considered by many to lag behind the adaptations of the remaining skeleton. The anatomical position of the pelvis is neither vertical, as in quadrupeds, nor horizontal (Fig. 3.53B). Consequently, special provision has to be made to prevent the sacrum being pushed downwards and forwards under the superincumbent body weight. The lumbosacral junction represents the transition between the mobile and immobile portions of the vertebral column. It has consequently become the least stable part of the vertebral column, being more exposed to static stresses and less well-equipped to meet them adequately than any other part of the column.

The sacroiliac joint

Articular surfaces

A sacroiliac joint is a synovial joint between the auricular surface of the ilium and that of the ala of the sacrum (Fig. 3.54A). Since the region behind the synovial joint is united by powerful interosseous ligaments, some authorities consider the joint to be both synovial (anterior) and fibrous (posterior). The auricular surfaces are approximately L-shaped, broader above and narrower below, and show marked reciprocal irregularities (Fig. 3.54C). The central part of the sacral auricular surface is concave with raised crests on either side; conversely, the ilial auricular surface has a central crest lying between two furrows. The lower parts of the auricular surfaces are shaped so that the widest part of the sacral surface is on its pelvic side. On the sacrum, the auricular surface occupies the upper two vertebral elements in females, and usually extends onto the third element in males. However, the shape and degree of irregularity of the surfaces vary considerably between individuals, and often between the two sides within the same

individual. The auricular surface of the sacrum is covered with hyaline cartilage, while the cartilage on the corresponding surface on the ilium is usually a form of fibrocartilage. With increasing age, particularly in males, the joint cavity becomes partially, or occasionally completely, obliterated by fibrous bands or fibrocartilaginous adhesions between the articular surfaces. In very old individuals the joint may show partial bony fusion.

Joint capsule and synovial membrane

A fibrous capsule completely surrounds the joint attaching to the articular margins on both bones. Synovial membrane lines the non-articular surfaces of the joint.

Ligaments

Because of the nature and position of the joint it is richly endowed with ligaments. Extremely strong posterior and slightly weaker anterior ligaments surround the joint capsule; while accessory ligaments situated some distance from the joint provide additional stability against unwanted movement.

Anterior sacroiliac ligament

Broad and flat, the anterior sacroiliac ligament consists of numerous thin bands on the pelvic side of the joint (Fig. 3.54B and C). It stretches from the ala and pelvic surface of the sacrum, above and below the pelvic brim, to the adjoining margin of the auricular surface of the ilium. The ligament is stronger in females, indenting a pre-auricular groove on the ilium just below the pelvic brim.

Posterior sacroiliac ligaments

The ligaments lying behind and above the joint are much thicker and stronger than those that are anterior. Several distinct bands can be identified as they fill the space between the sacrum and the tuberosity of the ilium.

1. *Interosseous sacroiliac ligament.* The deepest of all the posterior ligaments, being short, thick and extremely strong. It fills the narrow cleft between the rough areas on the bones immediately behind and above the auricular surfaces (Fig. 3.54C). Small accessory joint cavities, usually no more than one or two, may sometimes be found within the ligament between facets near the posterior superior iliac spine and the transverse tubercles of the sacrum.
2. *Long and short posterior sacroiliac ligaments.* Superficial to the interosseous ligament, the posterior ligament consists of numerous bands passing between the two bones. In general the longer fibres of the posterior ligament run obliquely downwards and medially. However, within this arrangement two sets of fibres can usually be identified. The short posterior sacroiliac

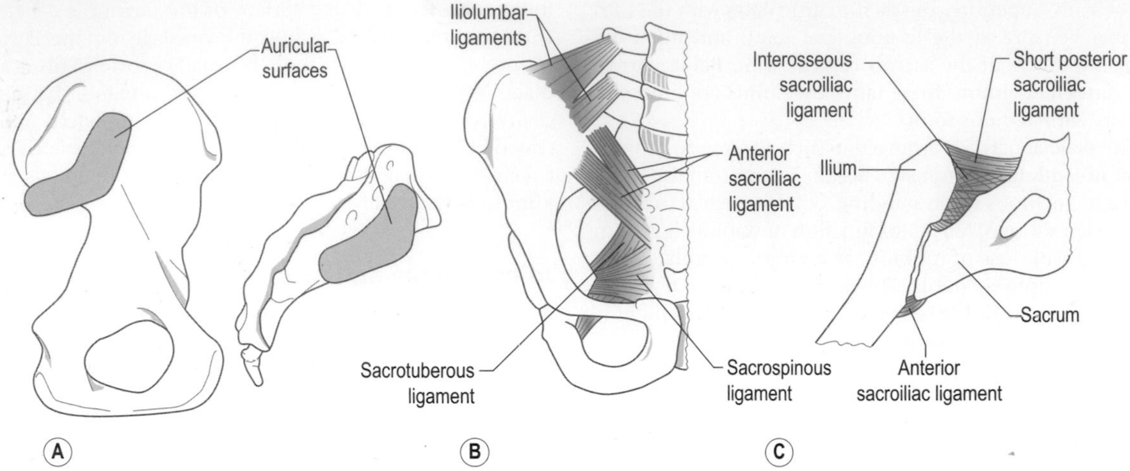

Figure 3.54 (A) The articular surfaces of the ilium and sacrum, (B) anterior sacroiliac ligaments, (C) posterior sacroiliac ligaments, horizontal section.

ligament is found in the upper part of the cleft between the two bones passing horizontally between the first and second transverse tubercles of the sacrum and the iliac tuberosity (Fig. 3.55A). They are arranged to resist forward movement of the sacral promontory. The long posterior sacroiliac ligament has the longest and most superficial fibres of the posterior complex. It runs almost vertically downwards from the posterior superior iliac spine to the third and fourth transverse tubercles of the sacrum (Fig. 3.55A). Its fibres are arranged to resist downward movement of the sacrum with respect to the ilium.

Accessory ligaments

In addition to the above ligaments, accessory ligaments confer added stability to the joint. The most important accessory ligaments are the sacrotuberous and sacrospinous ligaments which help to stabilize the sacrum on the innominate by preventing forward tilting of the sacral promontory. They also convert the greater and lesser sciatic notches of the pelvis into the greater and lesser sciatic foramina (Fig. 3.55B), through which several important structures leave the pelvis.

As well as the sacrotuberous and sacrospinous ligaments, the iliolumbar ligament assists in strengthening the bond between the ilium and sacrum (for details of this ligament see p. 283).

1. *Sacrotuberous ligament.* A flat, triangular band of great strength (Fig. 3.55B) attaching superiorly to the posterior border of the ilium between the posterior superior and posterior inferior iliac spines, to the back and side of the sacrum below the auricular surface, and to the side of the upper part of the coccyx. From this extensive attachment the fibres pass downwards and laterally towards the ischial tuberosity, converging as they do so. However, before attaching to the medial surface of the ischial tuberosity, the fibres twist upon themselves and diverge again so that the attachment is prolonged along the lower margin of the ischial ramus. This prolongation is known as the falciform process and lies just below the pudendal canal. The ligament as a whole is narrower in its middle part than at each end.

The most superficial fibres of the sacrotuberous ligament attaching to the ischial tuberosity are closely associated with the long head of biceps femoris. Consequently, the ligament is considered to be derived from biceps femoris, being the degenerated tendon of the origin of the long head. The posterior surface of the ligament gives attachment to gluteus maximus.

2. *Sacrospinous ligament.* Lying deep to the sacrotuberous ligament (Fig. 3.55B) its broad base is attached to the edge of the lower sacral and upper coccygeal segments in front of the sacrotuberous ligament. As the ligament passes laterally it narrows; the apex attaches to the ischial spine. On the pelvic surface of the sacrospinous ligament and closely blended with it is coccygeus; in fact the ligament can be considered to be a fibrous part of the muscle.

Blood and nerve supply

The arterial supply to the joint is by branches of the iliolumbar artery anteriorly and the superior gluteal artery posteriorly (Fig. 3.56B). This is reinforced by branches from the lateral sacral arteries both anteriorly and posteriorly. Venous drainage is to correspondingly named veins which eventually drain into the internal iliac vein. The lymphatic drainage of the joint follows the arteries to the internal iliac group of nodes.

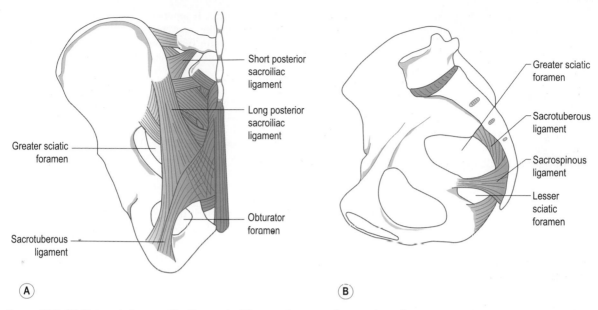

Figure 3.55 (A) The posterior sacroiliac ligaments, (B) sacrotuberous and sacrospinous ligaments.

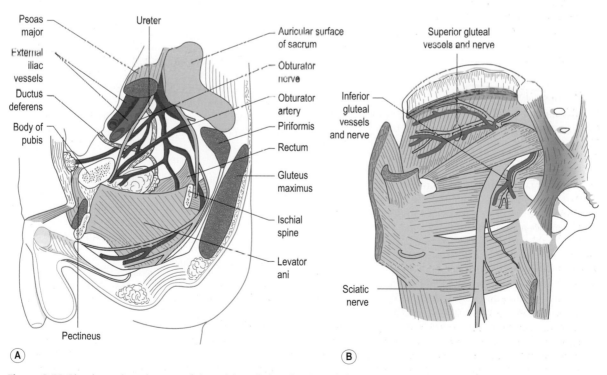

Figure 3.56 Blood vessels and nerves of the pelvic and gluteal regions: (A) sagittal section, (B) posterior view.

The nerve supply to the joint is by twigs directly from the sacral plexus and dorsal rami of the first and second sacral nerves. In addition it also receives branches from the superior gluteal and obturator nerves as they pass close to the joint. The joint is therefore supplied by roots L4 to S2.

Relations

Anterior to the joint, at the level of the lumbosacral intervertebral disc, the common iliac artery divides into its terminal branches: the internal and external iliac arteries (Fig. 3.56A). Crossing anterior to this bifurcation, directly in contact with the artery, the ureter enters the lesser pelvis on its way towards the bladder. Posterior to the joint is the erector spinae muscle mass. Indeed, deep to its lateral limb, tendinous fibres of the muscle blend with the posterior sacroiliac, sacrotuberous and sacrococcygeal ligaments.

Medial to the joint running over the ala of the sacrum are from lateral to medial: the obturator nerve, the iliolumbar artery and the lumbosacral trunk, while laterally in the iliac fossa is iliacus and its covering fascia.

Surface marking

The line of the sacroiliac joint is far too deep for it to be palpable; however, its surface projection can be estimated as follows. Identify the posterior superior iliac spine; an oblique line at approximately 25°, passing from superolateral to inferomedial and extending 2 cm in each direction represents the joint line.

Stability

The resilient support of the weight of the trunk, head and upper limbs given by curvatures of the vertebral column imposes additional stresses on the sacroiliac joints, because the line of weight passes anterior to them. There is, therefore a tendency for the sacral promontory to move downwards into the pelvis, and for the lower part of the sacrum and coccyx to tilt upwards. These tendencies are resisted by a number of factors, all of which are entirely dependent on various ligaments associated with the joint.

Because the line of weight passes anterior to the joint, the bony surfaces are not weight-bearing *per se*. Body weight is suspended by the sacroiliac ligaments, which sling the sacrum below the iliac bones. Providing the sacroiliac joints are intact, then the slight wedging of the auricular surfaces, together with their reciprocal irregularities, are sufficient to help resist rotation and gliding movements of the sacrum with respect to the innominate bones. The strong interosseous and posterior sacroiliac ligaments usually maintain such opposition. Rotation of the sacrum and coccyx is also resisted by the strong sacrotuberous and sacrospinous ligaments, which hold the lower sacral segments forwards, thereby preventing them rotating backwards.

The iliolumbar ligaments, as well as helping to oppose any simple gliding movements of the joint surfaces, also help to prevent the fifth lumbar vertebra from slipping forwards on the surface of the first sacral segment.

Movements

The arrangement of the joint surfaces and the ligamentous support given to the joint allow very little movement. There is a slight gliding and rotatory movement between the two bones. Investigations have shown that when standing, compared with lying supine, the sacrum moves downwards some 2 mm, and undergoes forward rotation of some 5°. Obviously any appreciable movement would lead to instability in the erect posture.

During childbirth there is a complex movement of the sacrum which has been likened to nodding of the head. The movement is possible because of the slight softening of the sacroiliac and associated ligaments which occurs during the latter part of pregnancy. The result is that the diameters of the pelvic inlet and outlet increase to facilitate passage of the fetal head. First, the sacral promontory moves superiorly and posteriorly, increasing the anteroposterior diameter of the pelvic inlet by between 3 and 13 mm (Fig. 3.57A). After the fetal head has entered the pelvic canal, the sacral promontory then moves inferiorly and anteriorly. This increases the anteroposterior diameter of the pelvic outlet by some 15–18 mm (Fig. 3.57B). Although the ligaments have become softened, the tension developed in them still limits the degree of movement possible. The extent of the softening, together with the extent of the irregularity of the opposing joint surfaces, accounts for the range of variation reported in pelvic diameter changes.

The sacroiliac joints are important clinically, as a sudden bending forward can tear the posterior ligaments and possibly even dislocate the adjacent joint surfaces. Either condition is extremely painful in flexion of the trunk and may be disabling. Treatment is in many cases difficult. Manipulation often produces a successful result.

Accessory movements

The ligaments of the sacroiliac joint are so arranged as to allow very little accessory movement. With the subject lying prone, so that the pelvis is supported by the two anterior iliac spines and the pubic region, place the heel of the hand on the apex of the sacrum and apply a downward pressure. A small rotation of the sacrum with respect to the pelvis can be elicited.

Biomechanics

Trabecular systems

Two trabecular systems arise in the region of the auricular surface of the innominate, being continuous with those converging towards the auricular surface of the sacrum. Both systems become continuous with trabeculae in the

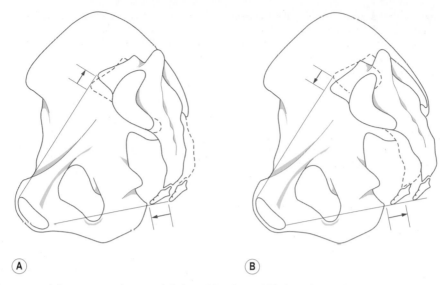

Figure 3.57 Movement of the sacrum to increase (A) the pelvic inlet and (B) the pelvic outlet.

head and neck of the femur as well as uniting with other systems within the pelvis (Fig. 3.58). From the upper part of the auricular surface, trabeculae converge on the posterior border of the greater sciatic notch. Some of the trabeculae fan out laterally towards the inferior aspect of the acetabulum, while the remainder pass downwards into the ischium, intersecting the trabeculae of the acetabular

rim as they do so. These trabeculae are compressive, bearing the weight of the body in the sitting position.

Trabeculae arising from the lower part of the auricular surface converge at the level of the pelvic brim. From here some pass laterally towards the superior aspect of the acetabulum, while the remainder pass into the superior pubic ramus towards the body of the pubis, so completing the pelvic ring of the trabeculae.

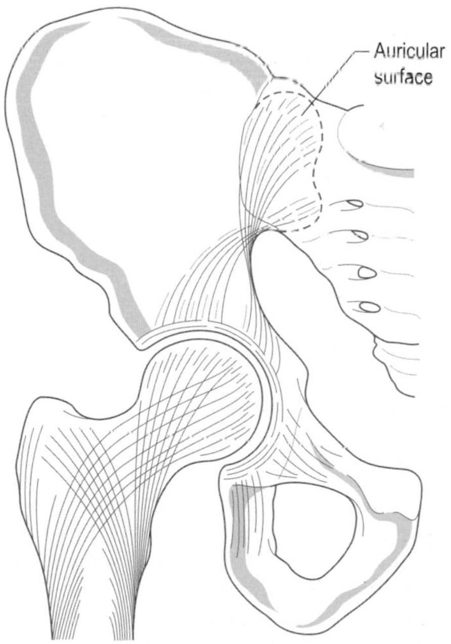

The symphysis pubis

Articular surfaces

A secondary cartilaginous joint between the medial surfaces of the bodies of each pubic bone. Each oval articular surface is irregularly ridged and grooved, with the irregularities fitting snugly together (Fig. 3.59A). The whole articular surface of each bone is covered with a thin layer of hyaline cartilage, which is joined to the cartilage of the opposite side by a fibrocartilaginous interpubic disc, thicker in females than in males (Fig. 3.59B).

In the upper posterior part of this disc a small fluid-filled cavity appears in early life, but it is never lined with synovial membrane. In females this cavity may eventually extend throughout the greater part of the disc.

Ligaments

Above, below and in front of the joint, thickenings of fibrous tissue form ligaments.

Superior pubic ligament

Attached to the pubic crests and tubercles of each side, strengthening the anterosuperior aspect of the joint (Fig. 3.59B).

Figure 3.58 Trabecular systems in the region of the sacroiliac joint.

Auricular surface

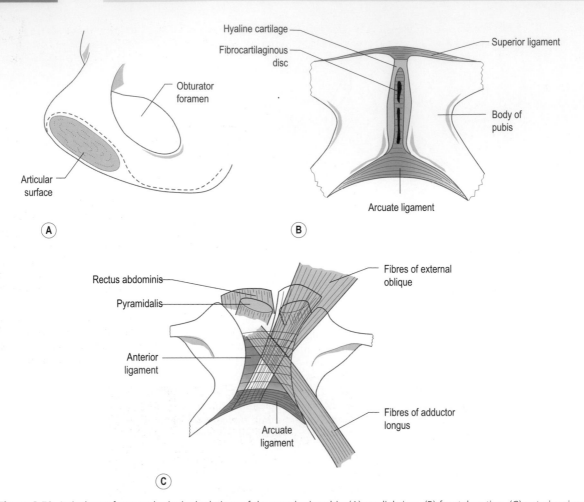

Figure 3.59 Articular surfaces and principal relations of the symphysis pubis: (A) medial view, (B) frontal section, (C) anterior view.

Arcuate pubic ligament

Arches across between the inferior pubic rami, rounding off the subpubic angle and strengthening the joint inferiorly (Fig. 3.59B). Between this rather thick, ligamentous arch and the transverse perineal ligament of the urogenital diaphragm is a small gap through which passes the dorsal vein of the penis or clitoris to gain entry to the pelvis.

Relations

Overlying the interpubic disc anteriorly are the decussating tendinous fibres of rectus abdominis, external oblique and adductor longus (Fig. 3.59C). These fibres serve to strengthen the joint and provide additional anterior stability. Indeed, some authorities consider this dense feltwork of fibres to constitute a thick anterior pubic ligament. Behind the joint lies the bladder, being separated from it for the most part by the retropubic fat pad.

Palpation

The line of the symphysis pubis can be palpated anteriorly as a groove between the bodies of the two pubic bones. The alignment of the joint can be checked by placing the hand on the lower abdomen with the finger pointing towards the subject's feet. The index and ring fingers are then placed on the upper surface of the pubis, with the middle finger placed on the intervening disc.

Movements

The nature of the joint is such that there is normally no movement between the bones involved. However, during pregnancy the ligaments associated with the joint, as well as those of other joints, soften and allow a small degree of movement so that there is some separation at the symphysis pubis. The separation is small, being of the order of 2 mm. Nevertheless, it increases the circumference of the pelvic inlet, which probably makes it easier for the fetal head to pass through the pelvic cavity. Occasionally the bone adjacent to the joint is absorbed facilitating separation at the symphysis.

Pathology

Occasionally slipping of one pubic body with respect to the other occurs at the symphysis pubis. This is known as *osteitis pubis*, and seems to affect some females following childbirth, and surprisingly some professional footballers. The aetiology is essentially unknown although it is thought to be related to abnormal stresses across the symphysis. The unevenness of the pubic arch can clearly be seen on X-ray. Pain associated with the condition is usually referred to the hip joint.

The lumbosacral joint

Between the last lumbar vertebra, usually the fifth, and the first sacral segment (Fig. 3.60). The superior surface of the sacrum is inclined approximately 30° to the horizontal, while the lumbosacral angle, formed between the axis of L5 and the sacral axis averages 140°. Being part of the vertebral column, the two bones are joined, like all typical vertebrae, by an intervertebral disc, anterior and posterior

ligaments, ligamenta flava, interspinous and supraspinous ligaments, and by synovial joints between their adjacent articular processes. Further details of vertebral articulations can be found on page 445.

Ligaments

In addition to the ligaments mentioned above, the iliolumbar and lateral lumbosacral ligaments help to stabilize the lumbosacral joint.

Iliolumbar ligament

A strong ligament passing inferiorly and laterally from the tip of the transverse process of the fifth lumbar vertebra to the posterior part of the inner lip of the iliac crest (Fig. 3.61). In reality it is the thickened lower border of the anterior and middle layers of the thoracolumbar fascia. Occasionally an additional smaller ligamentous band passes from the tip of the transverse process of the fourth lumbar vertebra to the iliac crest behind the main iliolumbar ligament (Fig. 3.61). There are usually some fibrous strands passing between the transverse process of L4 and the iliac crest; only when these strands become condensed can they be considered to constitute a true ligament.

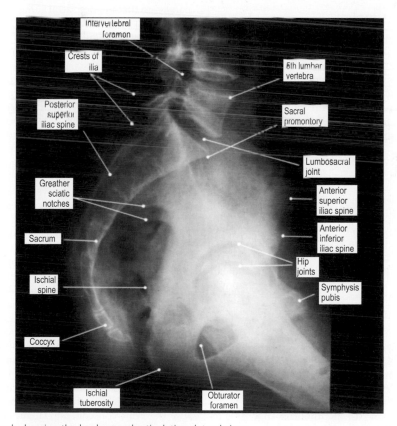

Figure 3.60 Radiograph showing the lumbosacral articulation, lateral view.

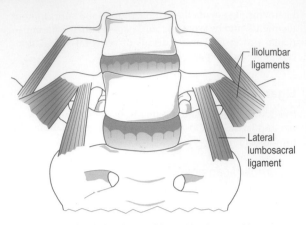

Figure 3.61 The iliolumbar and lateral lumbosacral ligaments.

Lateral lumbosacral ligament

Partially continuous with the lower border of the iliolumbar ligament and passes obliquely downwards from the lower border of the transverse process of the fifth lumbar vertebra to the ala of the sacrum, intermingling with the anterior sacroiliac ligament (Fig. 3.61). It consists of bundles of fibres of varying strength.

Blood and nerve supply

The blood supply to the joint is by small branches from the median sacral and iliolumbar arteries. However, it must be remembered that the major component of the joint, the intervertebral disc, is essentially an avascular structure, obtaining its nutrients by diffusion from the adjacent vertebral bodies. The nerve supply to the joint and associated ligaments is by twigs from both the anterior and posterior rami of L5 and S1.

Stability

Because of the inclination of the superior surface of the sacrum, there is a tendency for the fifth lumbar vertebra to slide inferiorly and anteriorly. Consequently, the lumbosacral junction is the weak link in the vertebral column. The tendency for slippage to occur is prevented by the overlapping of the articular processes of the vertebrae involved. This bony arrangement, together with the spinous and iliolumbar ligaments, is sufficient to prevent any abnormal movements occurring.

Movements

The degree of flexion, extension, and lateral flexion possible at the lumbosacral joint varies from individual to individual and with age. There is no axial rotation at this joint.

Flexion and extension

Between the ages of 2 and 13 the lumbosacral joint is responsible for as much as 75% of the total range of flexion

and extension of the lumbar spine. The average range of flexion and extension possible at this joint is of the order of 18°. However, this is greatly reduced from age 35 onwards. The iliolumbar ligaments tend to limit flexion (superior band when present) and extension (inferior band) of the joint.

Lateral flexion

The range of lateral flexion at the lumbosacral joint is minimal as it drops from 7° in the child to only 1° in the adult, and zero in the elderly. Again the iliolumbar ligament plays an important role in limiting movement, with the contralateral ligament becoming taut and the ipsilateral ligament slackening. The importance of the presence of the superior band is that it will restrict lateral flexion of L4 with respect to the sacrum.

Accessory movements

These movements, possible at the lumbosacral joint, are similar to those between any two lumbar vertebrae.

Pathology

Normally, under the component of body weight acting parallel to the superior sacral surface, the inferior articular process of L5 fits tightly into the superior sacral surface of S1, binding the lumbar and sacral processes tightly together (Fig. 3.62A). The associated forces act through the pars interarticularis, that part of the vertebral arch between the superior and inferior articular processes. If the pars interarticularis becomes fractured or destroyed then the condition of *spondylolysis* exists. Consequently, the buttressing of L5 on S1 no longer occurs, and the body of the fifth lumbar vertebra slips forwards and downwards giving rise to *spondylolisthesis* (Fig. 3.62B). The only structure now providing support to the joint and preventing further slippage is the lumbosacral intervertebral disc, which is put under tension, and the paravertebral muscles, which go into spasm and account for the pain associated with spondylolisthesis. The extent of the anterior movement of L5 relative to S1 can be assessed on oblique radiographs.

It has become apparent recently that spondylolisthesis can arise as a slowly developing fracture of the pars interarticularis, being most common in adolescents who participate in contact sports and gymnastics. The mechanism of the injury is probably through impact loading while the L5–S1 interspace is repetitively flexed and extended. The repetitive nature of the loading is important because unless there is some congenital anomaly, the pars interarticularis can withstand the stress induced by a single normal impact. However, with repetitive frequent loading the bending stresses eventually produce a small crack on the tensile side of the pars, which slowly extends across the bone with loading. The bone consequently fatigues and eventually gives way. If the fracture is incomplete when diagnosed, then

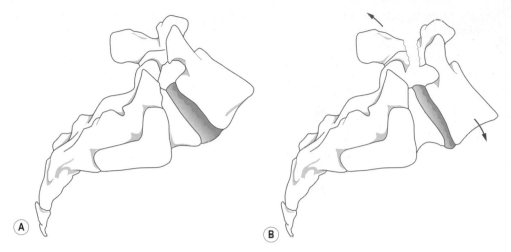

Figure 3.62 (A) Normal anatomy of the lumbosacral junction, (B) fracture of the vertebral arch giving rise to spondylolisthesis.

by avoiding repetitive bending stresses it may heal. However, once the pars interarticularis has been ruptured the joint becomes relatively unstable. Unfortunately, conservative measures that bring about repair in a stable fracture do not appear to work, and spinal fusion may have to be performed.

The sacrococcygeal joint

The articulation between the last sacral and the first coccygeal segments (Fig. 3.63A) via an intervening interosseous ligament similar to an intervertebral disc. The articular surfaces are elliptical with their long axes lying transversely,

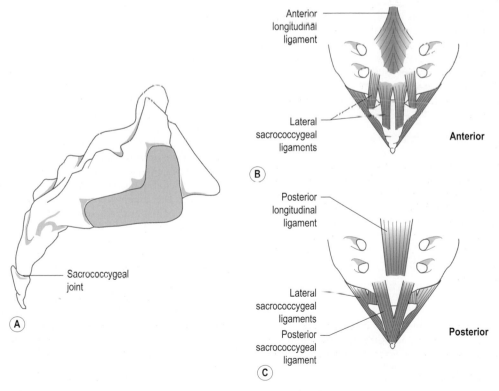

Figure 3.63 The sacrococcygeal joint: (A) lateral view, (B) anterior view of the associated ligaments, (C) posterior view of the associated ligaments.

that on the sacrum being convex and that on the coccyx concave. The joint is completely surrounded and reinforced by longitudinal fibrous strands, collectively called the *sacro-coccygeal ligaments* (Fig. 3.63B and C). The lateral part of these strands forms the lateral boundary of the foramen transmitting the anterior ramus of the fifth sacral nerve.

The sacrococcygeal joint frequently becomes partially or completely obliterated in old age.

Flexion and extension are the only movements possible at the joint, and these are essentially passive, occurring during defecation and labour. The increase in the antero-posterior diameter of the pelvic outlet following movement of the sacrum can be further increased by extension of the coccyx.

Palpation

The line of the sacrococcygeal joint can be felt as a horizontal groove between the apex of the sacrum and the coccyx deep within the natal cleft. By moving the palpating finger downwards so that it lies against the back of the coccyx, an applied forward pressure produces a degree of rotation of the coccyx against the sacrum.

Section summary

Joints of the pelvis

Sacroiliac joint

Type	Synovial plane anteriorly; fibrous posteriorly
Articular surfaces	The auricular surfaces of the ilium and sacrum
Capsule	Complete fibrous capsule surrounds the joint attaching to the articular margins
Ligaments	Anterior and posterior (interosseous, long and short) sacroiliac ligaments accessory sacrotuberous and sacrospinous ligaments
Stability	Provided by the wedge-shape of the sacrum, interlocking articular surfaces and ligaments
Movements	Slight gliding and rotation between the surfaces; during childbirth the pelvic inlet/outlet diameters increase

Symphysis pubis

Type	Secondary cartilaginous
Articular surfaces	Medial aspects of the pubic body with a fibrocartilaginous disc
Ligaments	Superior and arcuate pubic ligaments
Movements	Normally very little movement possible; during childbirth the joint surfaces separate slightly

Lumbosacral joint

Type	Secondary cartilaginous
Articular surfaces	Inferior surface of the body of L5 with the superior surface of the sacrum, separated by an intervertebral disc
Ligaments	Iliolumbar and lateral lumbosacral and all ligaments associated with joints between vertebrae

Sacrococcygeal joint

Articular surfaces	Inferior surface sacrum and coccyx
Ligaments	Sacrococcygeal
Movements	Passive flexion and extension

The hip joint

Introduction

The articulation between the head of the femur and the acetabulum of the innominate (Figs. 3.64 and 3.65). It is a synovial ball and socket joint, and as such permits a wide range of movements compatible with a wide range of locomotor activities. The hip joint connects the lower limb to the trunk, and therefore is involved in the transmission of weight. Indeed, the mechanical requirements of the joint are severe. It must be capable not merely of supporting the entire weight of the body, as in standing on one leg, but of stable transference of the weight, particularly during movement of the trunk on the femur, as occurs during walking and running. The joint must therefore possess great strength and stability, even at the expense of limitation of range of movement. The stability of the joint is determined by the shape of the articular surfaces (a deep socket securely holding the femoral head), the strength of the joint capsule and associated ligaments and the insertion of muscles crossing the joint, which tend to be at some distance from the centre of movement.

Although the hip is a ball and socket joint, when standing erect the femoral head is not completely covered by the acetabulum, the anterosuperior aspect being exposed. Coincidence of the articular surfaces can be achieved by flexing the hip to 90°, abducting by 5° and laterally rotating by 10°. This position of the thigh corresponds to the quadrupedal position. Thus one of the consequences of attaining the erect position and adopting bipedalism as a means of locomotion is the loss of coincidence of the articular surfaces. This loss of coincidence puts the hip in a potentially vulnerable position regarding its stability.

The acetabulum of the pelvis faces laterally, anteriorly and inferiorly to articulate with the head of the femur, which due to the anteversion of the femoral neck faces medially, anteriorly and superiorly. Consequently, there is an angle of 30° to 40° between the axes of the acetabulum and the femoral neck (Fig. 3.65), so that the anterior part of the femoral head articulates with the joint capsule. In addition, the lateral

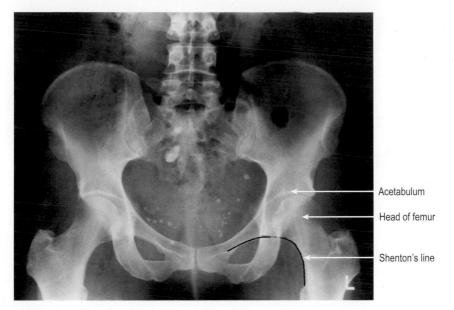

Figure 3.64 Radiograph of the adult hip.

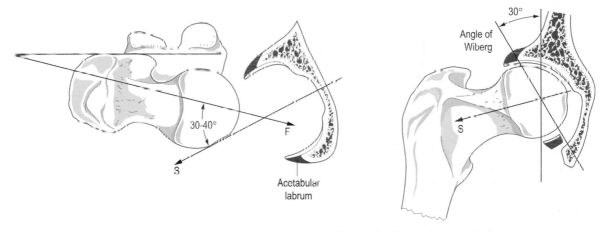

Figure 3.65 Relationship of the femoral head and acetabulum: F, axis of femoral neck; S, axis of acetabulum.

inferior inclination of the acetabulum forms an angle of 30° to 40° with the horizontal (Fig. 3.65), so that the superior part of the acetabulum overhangs the femoral head laterally. An angle of 30° is thus formed between a vertical line through the centre of the femoral head and a line from this centre to the bony margin of the acetabulum. This is the angle of Wiberg and can be easily measured on radiographs. Decreases in this angle have implications for stability of the joint.

In a normal individual, a line drawn along the upper margin of the obturator foramen and the inferior margin of the femoral neck to the medial side of the shaft describes a smooth curve: this is *Shenton's line*. The curve is not influenced by small changes in position; however, in fractures and dislocations of the femur it may become greatly distorted.

Articular surfaces

Acetabulum

A hemispherical hollow on the outer surface of the innominate, formed by the fusion of its three component parts, the ilium, ischium and pubis, which meet at a Y-shaped cartilage forming their epiphyseal junction. The anterior one-fifth of the acetabulum is formed by the pubis, the superior posterior two-fifths by the body of the ilium and the inferior posterior two-fifths by the ischium (Fig. 3.66A). The prominent rim of the acetabulum is deficient inferiorly as the *acetabular notch*. The heavy wall of the acetabulum consists of a semilunar articular part, covered with hyaline cartilage, which is open below, and a deep

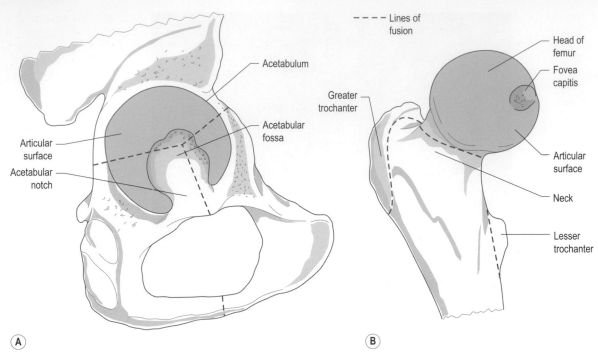

Figure 3.66 The articular surfaces of the hip joint: (A) acetabulum, (B) head of femur.

central non-articular part, the *acetabular fossa*. The acetabular fossa is formed mainly from the ischium and its wall is frequently thin.

Head of the femur

Approximately two-thirds of a sphere, being slightly compressed in an anteroposterior direction; however, the difference between the two principal axes is small. Nevertheless, it is best thought of as being an ellipsoid. The head is covered in articular (hyaline) cartilage, except for a small area superolaterally adjacent to the *neck* and at the *fovea capitis*, a pit on the posteromedial part of the head. Anteriorly the cartilage extends onto the femoral neck for a short distance (Fig. 3.66B). This is thought to be a reaction to the pressure from the iliopsoas tendon crossing the joint in this region. Both the femoral head and the acetabulum are composed of cancellous bone covered by a thin layer of compact bone.

Congruence

Although the articular surfaces are reciprocally curved, the hip joint consists of two incongruent shapes. Incongruity implies limited contact between the two surfaces under low loading conditions, with a gradual increase in the area of contact with increasing load. Because of this it is usual to think of the incongruity as a means of distributing load and protecting the underlying cartilage from excessive stress. An important factor in the functioning of such a mechanism is

the compressibility of the cartilage. The incongruity of the hip joint is determined by an arched acetabulum and a rounded femoral head.

Because of the relationship between the femur and the pelvis, the superior surface of the head of the femur and that of the acetabulum generally sustain the greatest pressures. Consequently, the articular cartilage is thicker in these regions than elsewhere. In the unloaded acetabulum, the cartilage surface is very nearly spherical. The small deviations seen on the cartilage surface (less than 150 μm) reflect the much larger deviations at the cartilage–bone interface (frequently greater than 500 μm). Ultrasonic measurement of the acetabular region has shown that the cartilage and calcified interface surfaces are not concentric, confirming the clinical finding of thinner cartilage in the anteromedial aspect. It is only when weight is taken with the hip joints in extreme flexion, as when squatting, that this anteromedial region of the acetabulum, articulating with the inferior part of the femoral head, becomes involved in weight-bearing.

Joint capsule

Attachments

The fibrous capsule of the hip is very strong, being thicker anteriorly, where the head of the femur articulates, and superiorly, enhancing the stability of the joint. Proximally the capsule surrounds the acetabulum, attaching directly to the bone outside the labrum above and behind, and to the bone and outer edge of the labrum in front and below

(Fig. 3.67A). Opposite the acetabular notch the capsule attaches to the transverse ligament. On the femur the capsule attaches anteriorly to the intertrochanteric line and to the junction of the neck with the trochanters (Fig. 3.67B). Posteriorly the capsule has an arched free border and covers the medial two-thirds of the neck only (Fig. 3.67C), approximately as far laterally as the groove formed by the tendon of obturator externus. Consequently, part of the femoral neck is intracapsular and part extracapsular. It is worth noting that the epiphyseal line for the head of the femur is intracapsular,

while the trochanteric epiphyseal lines are extracapsular (Fig. 3.66B). The capsule may be likened to a cylindrical sleeve enclosing the joint and the majority of the femoral neck. Anteromedially the capsule is strengthened by the deep fibres of the reflected head of rectus femoris, and laterally by deep fibres from gluteus minimus.

Direction of fibres

The majority of the fibres of the capsule run from the pelvis to the femur with several distinct sets being identified

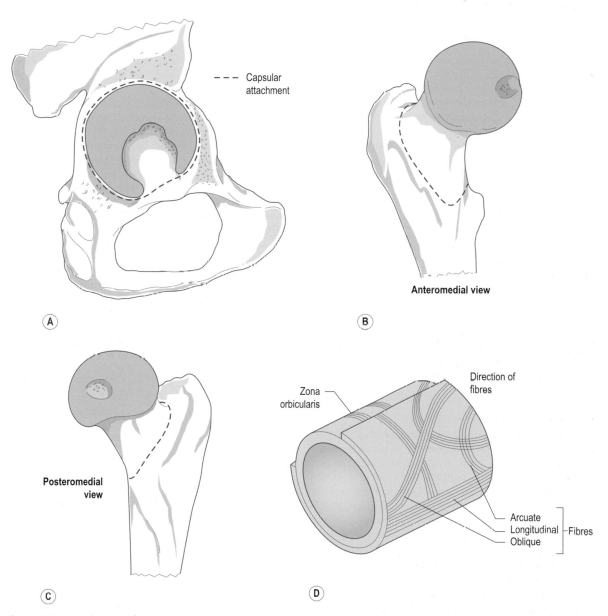

Figure 3.67 Attachments of the hip joint capsule to (A) the innominate and (B, C) femur, (D) arrangement of fibres within the joint capsule.

(Fig. 3.67D). *Longitudinal fibres*, running parallel to the axis of the cylinder, pass from the acetabular to the femoral attachment of the capsule. They, together with *oblique fibres* which spiral around the cylinder between their attachments, unite the articular surfaces. A series of *arcuate fibres* run in an arching manner from one part of the acetabular rim to another, helping to keep the femoral head within the acetabulum. Deeper fibres run circularly around the capsule and have no bony attachments. They appear to be most marked on the posterior of the capsule and constitute the *zona orbicularis*, where they are reinforced by the deep part of the ischiofemoral ligament. They can be clearly seen on the deep surface of the capsule.

Some of the deeper longitudinal fibres, on reaching the femoral neck, turn upwards towards the articular margin forming the retinacular fibres. These bundles are most marked on the upper and lower surfaces of the neck and convey blood vessels to the head and neck regions. The main longitudinal capsular fibres form named thickened bands, which resist the tensile stresses to which the capsule is subjected. The bands are named after their regional attachment around the acetabulum (Fig. 3.67D). They may not always be readily identifiable.

Capsular ligaments

Iliofemoral ligament

A very strong triangular ligament of considerable thickness situated anterior to the joint (Fig. 3.68A). The apex attaches to the lower part of the anterior inferior iliac spine and adjacent part of the acetabular rim, and the base to the intertrochanteric line. However, because the central part is thinner it is often referred to as being Y-shaped, with the stem corresponding to the apex and the two limbs to the base. The outer bands of the ligament attaching to the upper and lower parts of the intertrochanteric line are the strongest parts, with the central area being thinner and weaker.

Pubofemoral ligament

Strengthens the inferior and anterior aspects of the joint capsule (Fig. 3.68B) and runs from the iliopubic eminence and superior pubic ramus to the lower part of the intertrochanteric line, blending with the inferior band of the iliofemoral ligament. Between the iliofemoral and pubofemoral ligaments the capsule is at its thinnest. However, it is crossed here by the tendon of iliopsoas. Between the tendon and the capsule is *psoas bursa*, which usually communicates with the joint cavity through a perforation in this part of the capsule.

Ischiofemoral ligament

Less well-defined than either the iliofemoral or pubofemoral ligaments, the ischiofemoral ligament spirals laterally and upwards around the capsule. It arises from the body of the ischium behind and below the acetabulum attaching to the superior part of the neck and root of the greater trochanter (Fig. 3.68C). Some of the deeper fibres are continuous with the zona orbicularis.

Role of the ligaments

The three capsular ligaments have important roles in limiting and controlling the various movements of which the hip is capable. When standing erect, all three ligaments are under moderate tension. On flexing the hip, they all become relaxed. However, on extending the hip they all become taut, with the inferior band of the iliofemoral ligament being under the greatest tension as it runs almost vertically and is thus responsible for checking the posterior tilt of the pelvis. The concerted action of all three ligaments seen in flexion and extension of the hip is not seen in abduction/adduction or medial/lateral rotation. The attachments of the ligaments demand that during each of these latter movements some will become taut while others relax. During adduction, the superior band of the iliofemoral ligament becomes taut, the inferior band only slightly, while the pubofemoral and ischiofemoral ligaments slacken. The opposite occurs in abduction of the hip. In lateral rotation, both anterior ligaments become taut, that is the iliofemoral and pubofemoral, while the ischiofemoral ligament slackens. In medial rotation of the hip it is the ischiofemoral ligament which becomes taut, the others slackening.

Intracapsular structures

Transverse ligament of the acetabulum

The inferior deficiency in the acetabular rim is completed by the transverse ligament of the acetabulum (Fig. 3.69D), creating a foramen with the acetabular notch through which vessels and nerves may enter the joint. The superficial edge of the ligament is flush with the acetabular rim; it consists of strong bands of fibrous tissue.

Acetabular labrum

The acetabulum is deepened all round by the fibrocartilaginous acetabular labrum attached to the bony rim and to the transverse ligament (Fig. 3.69B). The labrum is triangular in cross-section with the apex forming the thin free edge (Fig. 3.69C). The diameter of this free edge is smaller than its fixed edge, and is also somewhat less than the maximum diameter of the femoral head. Extending further laterally than the equatorial region of the femoral head, the labrum cups around the head, holding it firmly in the acetabular socket (Fig. 3.69D).

Ligamentum teres

Within the hip joint there is a weak, flattened band of connective tissue, the ligamentum teres (ligament of the head of the femur). It is attached to the adjacent margins of the acetabular notch and the lower border of the transverse ligament (Fig. 3.69B), narrowing as it passes to insert into

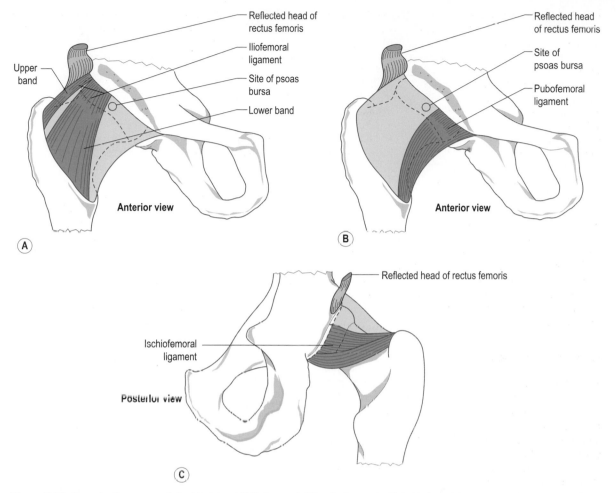

Figure 3.68 Capsular ligaments of the hip joint: (A) iliofemoral, (B) pubofemoral, (C) ischiofemoral.

the *fovea capitis* on the head of the femur (Fig. 3.69A). Between its two attachments the ligament is enclosed in a sleeve of synovial membrane, so that although it is intracapsular it is extrasynovial. It lies in the acetabular fossa below the femoral head. The ligament appears to be of little importance in strengthening the hip joint, being variable in size and occasionally absent. Its function in the adult is uncertain. It is stretched when the flexed thigh is adducted or laterally rotated. However, in many cases it is too weak to have any definite ligamentous action. No apparent disability arises from its rupture or absence. In early life a small artery is always found within its substance which becomes obliterated in late childhood.

Acetabular pad of fat

This lies within the acetabular fossa; it is really a fibroelastic fat pad. It is said to contain numerous proprioceptive nerve endings, so that when compressed and/or partially extruded from the acetabular fossa beneath the transverse ligament, additional proprioceptive information regarding hip joint movements is provided.

Synovial membrane

Lines the internal surface of the fibrous capsule and covers the acetabular labrum. At the acetabular notch it is attached to the medial margin of the transverse ligament, almost completely covering the fatty tissue in the acetabular fossa. It extends like a sleeve around the ligament of the head of the femur, attaching to the margins of the fovea capitis. At the femoral attachment of the joint capsule, the synovial membrane is reflected back towards the head as far as the articular margin. The retinacular fibres raise this reflected part into prominent folds within which blood vessels run towards the head. An extension of the synovial membrane beyond the free margin of the capsule at the back of the femoral neck serves as a bursa for the tendon of obturator externus.

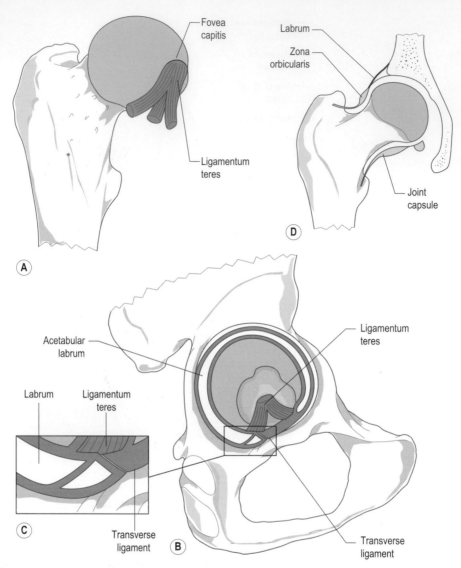

Figure 3.69 The transverse acetabular ligament, acetabular labrum and ligamentum teres.

The communicating psoas bursa breaches the joint capsule between the inferior limb of the iliofemoral ligament and the pubofemoral ligament.

Blood and nerve supply

Blood supply

The hip joint receives its blood supply from the *medial* and *lateral circumflex femoral arteries*, the *obturator artery* and the *superior* and *inferior gluteal arteries*, which together form a *periarticular anastomosis* around the joint (Fig. 3.70A, B). Within the ligamentum teres there is always a small artery derived from the obturator artery, which gains access to the joint by passing below the transverse ligament bridging the acetabular notch. However, this may only be of significance prior to the fusion of the epiphysis of the head with the femoral neck (Fig. 3.70C). The greatest volume of blood reaches the joint via the periarticular anastomosis, the branches piercing the joint capsule at its femoral attachment and passing in the retinacula, between the capsule and synovial membrane, to reach and supply the femoral neck and head.

The adequacy of the periarticular arterial anastomosis is of critical importance for the nutrition of the bone, particularly for the proximal femoral epiphysis until ossification is completed, which is usually between 16 and 20 years. Similarly, following a fracture, healing of the bone requires a good blood supply. Of particular concern is fracture of the

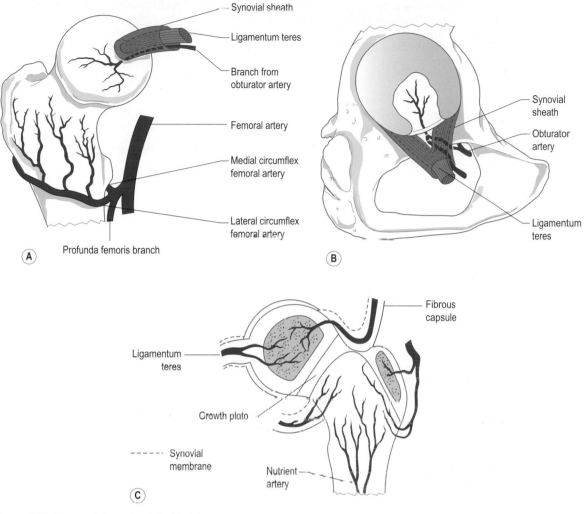

Figure 3.70 The arterial supply of the hip joint.

femoral neck, which frequently occurs in older females. Such fractures tend to be intracapsular and realigning the head and neck fragments can be a problem. Furthermore, the blood vessels serving the head and neck may be ruptured, resulting in ischaemic degeneration of the femoral head; the artery of the head of the femur frequently provides an inadequate alternative blood supply. For this reason early prosthetic replacement of the femoral component or of the joint as a whole is often performed in subcapital fractures. In the same way dislocation of the hip and slipping of the femoral epiphysis are both potentially damaging to the retinacular blood vessels.

Lymphatic drainage

By vessels accompanying the arteries supplying the joint. The lymph vessels arise primarily from the synovial membrane, and drain to the deep inguinal and iliac lymph nodes, and thence to the common iliac nodes.

Nerve supply

From the lumbar plexus, the nerve supply is by twigs from the femoral and obturator nerves, and the sacral plexus, by twigs from the superior gluteal nerve and the nerve to quadratus femoris, with a root value of L2 to S1. This is a typical example of articular innervation, in that the nerve supply to the joint is derived from the same nerves which supply the muscles crossing it. The articular supply consists of sensory nerve fibres, transmitting proprioceptive information, and vasomotor fibres.

In posterior dislocation of the hip, the sciatic and other nerves of the posterior hip region may be stretched over the head of the femur or otherwise damaged. Injury to the superior gluteal nerve results in collapse of the pelvis on the opposite side as the leg is raised, due to paralysis of gluteus medius and minimus (positive Trendelenburg sign). Primary disease of the hip joint frequently manifests itself in pain referred to the knee because the same nerves provide branches to that joint.

Relations

Anteriorly from below upwards, the hip joint is related to pectineus, the tendon of psoas major, iliacus and rectus femoris (Fig. 3.71A and C). Indeed, the deep part of the reflected head of rectus femoris is so intimately related to the joint that it strengthens the superior and anterior aspects of the joint capsule. Between the tendons of iliacus and psoas, and separated from the joint capsule by these structures, lies the femoral nerve, while the femoral artery and vein lie on the psoas tendon and pectineus.

Posteriorly, from above downwards, are piriformis, the tendon of obturator internus with the two gemelli and

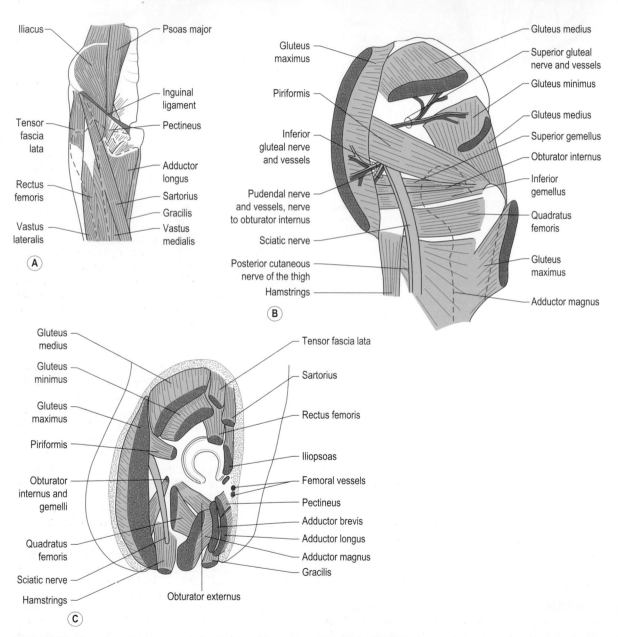

Figure 3.71 Muscular relations around the hip joint: (A) anterior, (B) posterior, (C) lateral.

quadratus femoris (Fig. 3.71B and C). All these lie close to the joint capsule. The nerve to quadratus femoris is deep to obturator internus and therefore lies directly on the joint capsule. The sciatic nerve is separated from the capsule by obturator internus and the gemelli.

In addition to the reflected head of rectus femoris superiorly, gluteus minimus is found more laterally, part of which may blend with the joint capsule. Inferiorly obturator externus winds backwards below the capsule lying between it and quadratus femoris.

The deep fascia of the thigh, of which the fascia lata is the uppermost subdivision, invests the soft tissue of the lower limb. The fascia lata is a strong membranous fascia, thicker where it is reinforced by tendinous contributions and thinner in the gluteal region. It has a continuous bony and ligamentous attachment superiorly. Between the iliac crest and the superior border of gluteus maximus the fascia is thickened by vertical tendinous fibres, the gluteal aponeurosis. The remaining gluteal portion and adductor part of the fascia lata are thin except for a lateral band, the iliotibial tract, which is especially strong.

An opening in the front of the fascia lata, the saphenous opening, provides passage for the long (great) saphenous vein to its termination in the femoral vein. The efferent vessels from the superficial inguinal nodes also pass through this opening, mainly to travel within the femoral sheath to the external iliac nodes.

Outside the fascia lata, the subcutaneous connective tissue contains a considerable amount of fat over the hip and thigh, although it varies in different regions. In the gluteal region the fat is deposited in a thick layer which contributes to the contour of the buttock and the formation of the gluteal fold. At the fold of the groin the subcutaneous tissue is separable into a superficial fatty layer and a deeper membranous layer, between which are found the subcutaneous blood vessels and nerves.

Palpation

Because it is completely surrounded by muscles, the hip joint cannot be directly palpated. However, the approximate position of the joint centre, in relation to the anterior surface, can be determined in the living body. The joint centre lies in a horizontal plane passing through the top of the greater trochanters; these can be palpated. This plane passes 1 cm below the middle third of the inguinal ligament. This is the position of the joint centre. Passive movement of the thigh on the pelvis and palpation in this area will improve the estimation of the centre of movement for the hip joint.

Stability

The stability of the hip is determined by the shape of the bones, the strong reinforced capsule, the acetabular labrum and the muscles crossing the joint. Although the articular surfaces of the femur and pelvis fit well together and provide a good degree of support for the joint, the periarticular muscles are essential for the continued stability of the joint, particularly those muscles that cross the joint transversely.

Muscular factors

Muscles whose fibres run parallel to the femoral neck will tend to keep the femoral head in firm contact with the acetabulum. Muscles in this category are psoas, iliacus and pectineus anteriorly, gluteus minimus superiorly, and gluteus medius, obturator internus and externus, the gemelli, quadratus femoris and piriformis posteriorly. On the other hand, muscles whose fibres run parallel to the femoral shaft, that is longitudinally, may have a tendency to cause dislocation of the joint superiorly, particularly if the roof of the acetabulum is everted. The adductor muscles run longitudinally, and in some positions of the joint, for example attempting adduction of the extended and laterally rotated thigh, their action is such that without the synergistic activity of other muscles around the hip, there would be a tendency for anteromedial dislocation of the joint. It should not be inferred from this that the hip joint is continually liable to dislocation; in the normal joint it is not. However, if there is some malformation of the acetabular socket, such as eversion of the roof, then the probability of dislocation occurring is increased. Eversion of the roof is an acetabular malformation present in congenital dislocation of the hip (CDH). In the presence of this malformation the adductors can cause dislocation, especially when the limb is adducted. However, when the limb is abducted the dislocating tendency of the adductors decreases until in full abduction the adductors eventually favour apposition of the joint surfaces.

Bony factors

The direction of the femoral neck in both the frontal and horizontal planes is of considerable importance in the stability of the hip (Fig. 3.72). During embryonic development, and even after birth, the femoral shaft becomes adducted and medially rotated, as a result of which the head and neck become angulated against the shaft in both the frontal and horizontal planes. In the frontal plane the angle between the neck and shaft indicates the adaptation of the femur to the parallel position of the legs assumed in bipedalism. This is the *angle of inclination* (Fig. 3.72A), which in the newborn is about 150° decreasing in the early years to the adult value of 125°. In the horizontal plane the angle between the neck and shaft is also due to the assumption of an upright stance. The neck and head are outwardly rotated against the shaft. This *angle of anteversion* is about 10° in the adult (Fig. 3.72C), having decreased from about 25° in infants and young children.

In some pathological conditions, such as CDH, the angle of inclination can be as much as 140° producing a *coxa valga*, and the angle of anteversion may be as large as 40°, thereby increasing the risk of anterior dislocation of

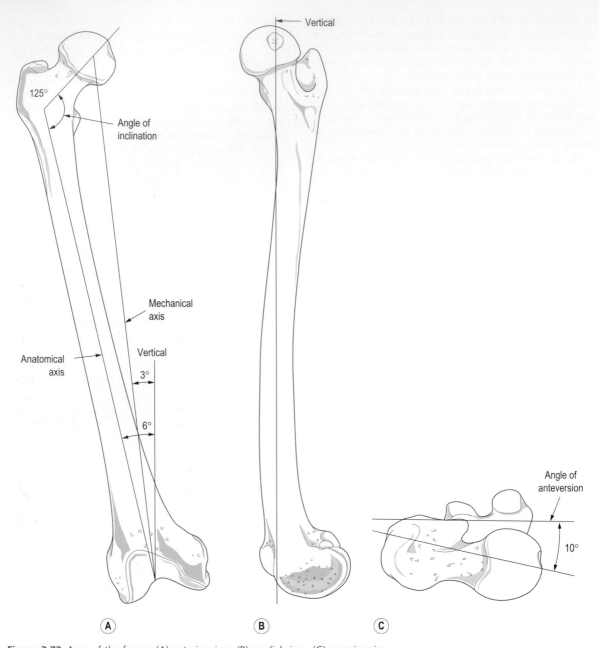

Figure 3.72 Axes of the femur: (A) anterior view, (B) medial view, (C) superior view.

the joint. In other conditions the angle of inclination may be reduced, for example in acquired dislocation of the hip (see Fig. 3.76B), and there may be a reduced or even reversed angle of anteversion, that is the head and neck are retroverted with respect to the femoral shaft.

The values for the angles of inclination and anteversion given above are mean values. Within a normal population there will be considerable racial and individual variation in

these angles, which will have important consequences as far as joint stability is concerned. If the angle of inclination and/or the angle of anteversion are greater than 130° and 15° respectively, the coincidence and thus the stability of the two joint surfaces is decreased.

Because of the length of the femoral neck and its angulation with respect to the shaft, in both the horizontal and frontal planes, a line joining the centre of the femoral head to the middle of the intercondylar notch passes mainly outside the bone on its medial side. This is the *mechanical axis* of the femur about which medial and lateral rotation occurs. The *anatomical axis* of the femur deviates by some 3° in the frontal plane from the mechanical axis. In turn, the mechanical axis deviates some 3° from the vertical, so that its upper end is lateral to its lower end.

Dislocations and fractures

Traumatic dislocation of the hip is not common except as a consequence of a car accident. Posterior dislocation is more common, being favoured by the usual direction of the dislocating force in such accidents. The joint capsule is ruptured, the head of the femur is in the posterior iliac fossa and the limb is shortened, adducted and medially rotated; the knee on the affected side overlies the normal knee. The position is almost diagnostic, being contrasted with the position in fracture of the femoral neck, in which the limb is shortened but laterally rotated.

Fracture of the upper third of the femoral shaft exemplifies the effects of muscle pull. The proximal fragment has the attachment of iliopsoas to the lesser trochanter, and the gluteal and other muscles of the posterior femoral group to the greater trochanter. This fragment is therefore strongly flexed, abducted and laterally rotated, while the distal fragment is displaced upwards and medially by the adductors and hamstrings. Realignment requires some form of traction with surgical intervention often necessary in the adult.

Movement

The movements possible are those of a typical ball and socket joint, being flexion and extension around a transverse axis, adduction and abduction around an anteroposterior axis, and medial and lateral rotation around a vertical axis. Circumduction is also allowed. The three axes intersect at the centre of the femoral head. Since the head is at an angle to the shaft, all movements involve conjoint rotation of the head.

Flexion and extension

Flexion of the hip joint is free, being limited by contact of the thigh with the anterior abdominal wall when the knee is flexed (Fig. 3.73A). With the knee extended, hip flexion is limited by tension in the hamstring muscles. Extension of the hip beyond the vertical is limited partly by tension in the associated ligaments and partly by the shape of the articular surfaces. Extension beyond 30° is not normally possible (Fig. 3.73A). In going from a position of flexion to one of full extension, the capsular ligaments become increasingly tense pulling the femoral head more tightly against the acetabulum. Because ligaments are slightly extensible and articular cartilage deforms slightly under compressive loads, the point at which no further extension is possible will vary slightly with the magnitude of the extending force. During extension of the hip joint, part of the acetabular fat pad is extruded from the acetabular fossa below the transverse ligament. Flexing the joint draws the fat pad back in again to fill the potential space created by the lateral movement of the femoral head, which occurs with flexion.

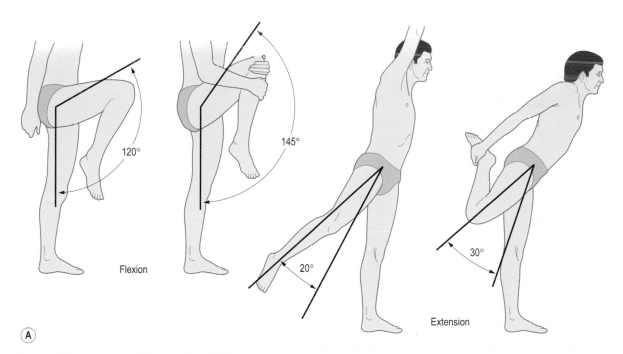

(A)

Figure 3.73 Movements at the hip joint: (A) flexion and extension,

Continued

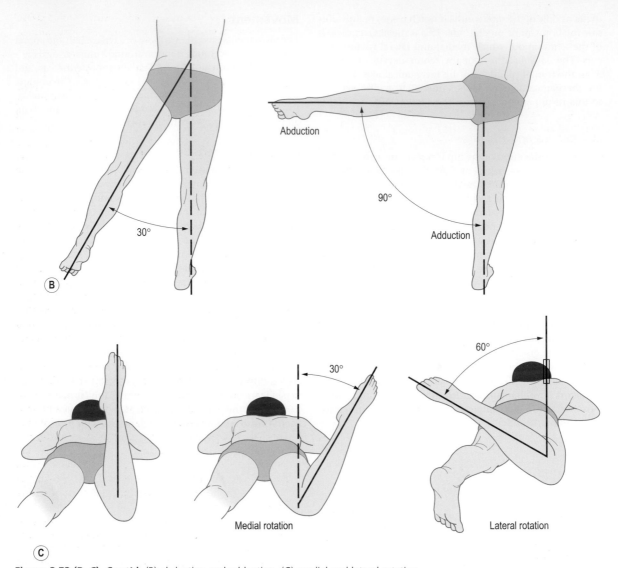

Figure 3.73 (B, C), Cont'd (B) abduction and adduction, (C) medial and lateral rotation.

Abduction and adduction

These movements, some 45° each (Fig. 3.73B), are free in all positions of the lower limb, except of course for adduction in the anatomical position. Abduction is greatest when the hip is also partly flexed. It is limited by tension in the adductor muscles and the pubofemoral ligament. Adduction is easier with the hip flexed than when extended; it is limited by the opposite limb, tension in the abductor muscles and the lateral part of the iliofemoral ligament.

Rotation

Rotation occurs about the mechanical axis of the femur and *not* about the long axis of the shaft of the femur.

Consequently, in medial rotation the femoral shaft moves anteriorly around the mechanical axis, carrying with it the calf and foot so that the toes point towards the midline. In lateral rotation, the femoral shaft moves posteriorly, with the result that the toes point away from the midline. Rotation in both directions is freer when combined with flexion rather than extension of the hip. Of the two, lateral rotation is freer than medial rotation, and is also the more powerful movement (Fig. 3.73C). Lateral rotation is limited by tension in the medial rotators of the thigh and the iliofemoral and pubofemoral ligaments; medial rotation is limited by tension in the lateral rotators and the ischiofemoral ligament. A small amount of rotation of the hip joint occurs automatically in association with terminal extension and the beginning of flexion at the knee, particularly with the foot fixed. The total range of medial and lateral rotation is some 90°.

Because the axis about which rotation occurs is mainly outside the femur, there has been some confusion as to the role of certain muscles in producing medial or lateral rotation at the hip joint. In general terms, any muscle whose line of action passes anterior to the mechanical axis of the femur will produce medial rotation, while those whose line of action passes posterior to the axis will produce lateral rotation. However, it should be remembered that the position of the line of action of a muscle with respect to the mechanical axis can change depending on the degree of flexion or extension of the joint. Consequently, in some positions of the joint a muscle may be a medial rotator, while in others it may be a lateral rotator. Similarly, for large muscles crossing the hip joint it may be more convenient to consider more than one line of action for the muscle, in which case one part of the muscle may medially rotate and another laterally rotate the hip joint. The line of axis through the hip joint also varies according to whether the foot is on the ground or not (p. 235).

When assessing the range of movement at the hip joint it is important to determine that there is no movement of the pelvis or vertebral column. An apparent increase in flexion or extension may be due to flexion or extension of the vertebral column, especially in the lumbar region. Similarly, an apparent increase in abduction or adduction may be produced by a lateral bending of the trunk to the opposite or same side respectively.

Hip joint axis

The greater trochanters can normally be palpated, and in the erect standing position a horizontal line through the tips of the greater trochanters represents the common hip joint axis in the frontal plane. The position of the greater trochanter with respect to the hip joint varies little with flexion and extension. However, with respect to the centre of the hip joint, it moves upwards in abduction and downwards in adduction.

Limits to mobility

In conditions other than those involving the hip there may be some loss of hip mobility. For example, in *osteitis pubis* there may be some loss of hip mobility, particularly medial rotation and in some cases lateral rotation, a finding typically associated with *upper femoral epiphysiolysis*. In situations requiring free medial rotation of the hip joint in both flexion and extension, when movement is restricted stress will be applied across the hip joint to the pelvis. This will be specifically a shearing stress causing anteroposterior movement of one half of the pelvis with respect to the other in extension, or proximodistal movement in flexion. Such forces are perhaps less liable to be applied in other instances of pathological restriction of hip joint movement,

as in *degenerative joint disease*, where soft tissue tension rather than abnormal joint shape is the restricting factor.

Developmental features of the hip joint

For normal acetabular development to occur as the pelvis enlarges, a delicate balance must be maintained between the growth of the acetabular and triradiate cartilages and that of the adjacent bone. As all the secondary centres of ossification for the acetabulum appear (at 8–9 years) more bone is formed at the periphery than at the inner part of the acetabulum, thus increasing depth and contributing to the cup-like shape of the cavity.

The acetabular cartilage complex is composed of epiphyseal growth-plate cartilage adjacent to the bones, of articular cartilage around the acetabular cavity, and, for the most part, of hyaline cartilage. Interstitial growth within the triradiate part of the cartilage complex causes the acetabular socket to expand during growth. The concavity of the acetabulum develops in response to the presence of the femoral head (Fig. 3.74A). The depth of the acetabulum increases during development as a result of interstitial growth in the acetabular cartilage, of appositional growth at the periphery of this cartilage, and of periosteal new-bone formation at the acetabular margin.

CDH is thought to be due to faulty development of the acetabulum, especially of its upper rim, in addition to which the femoral head may also be poorly developed. In such hips, a cartilaginous ridge is present in the acetabulum dividing the socket into two sections. This ridge may be formed exclusively by a bulge of acetabular cartilage or by a bulge of acetabular cartilage covered by an inverted acetabular labrum. The acetabular cartilage usually shows signs of degeneration, whereas the triradiate cartilage is normal. The condition is much more common in young females.

From radiographs of the developing hip, a growth quotient, the *acetabular index*, can be determined. The index shows periods of limited intensive growth of the hip. These periods occur in both sexes between the ages of 5 and 8 and again between the ages of 9 and 12 in females and 11 and 14 in males. In pathological hip joints, the triradiate cartilage closes earlier. The development of the acetabular roof and the closure of the triradiate cartilage are closely related.

During growth, ossification of the femoral head occurs at a faster rate in breadth than in height (Fig. 3.74B). The growth rate in breadth also exceeds that of the proximal metaphysis. The radiographic appearance of these two regions is therefore characteristic of the individual's age. Fortunately, the osseous acetabulum also grows at a faster rate in breadth than in depth (Fig. 3.74B). Up to the age of 15 the width of the articular space gradually decreases and approaches adult values. However, in the region of the acetabular fossa the width of the articular space does not decrease in a parallel manner; the result is the appearance of the double-arched pattern of the acetabulum.

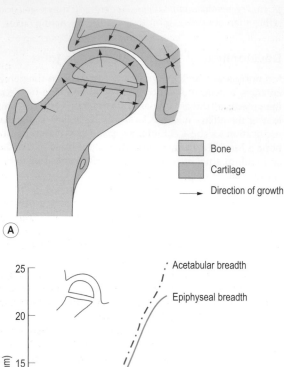

(A)

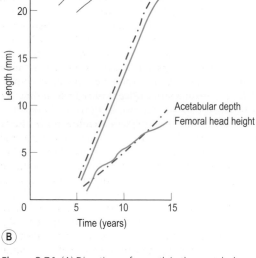

(B)

Figure 3.74 (A) Directions of growth in the acetabulum and the proximal end of the femur, (B) growth in height of the osseous femoral head and depth of the acetabulum, and breadth of the epiphysis and acetabulum. *(Adapted from Meszaros T, Kery L (1980) Quantitative analysis of growth of the hip. A radiological study.* Acta Orthopaedica Scandinavian, *51, 275–284.)*

Biomechanics

Joint forces

The estimation of hip joint forces during simple activities, such as walking, has a long history. Such estimates are continually being refined with the use of increasingly

sophisticated measuring techniques and the advent of powerful microcomputers. The use of such estimates has been invaluable in the design of replacement joints. The majority of studies, however, are still limited to determining joint forces during walking. In one-legged stance, hip joint forces of between 1.8 and 3 times body weight have been reported. In the stance phase of walking, hip joint forces increase to between 3.3 and 5.5 times body weight. In running, higher forces would be expected. A typical pattern of these forces during walking is shown in Fig. 3.75 from which it can be seen that the resultant joint forces reach a maximum shortly after heel-strike and just prior to toe-off. The major component of this resultant is the vertical joint force. However, the anteroposterior force (Y in Fig. 3.75) exceeds body weight at both heel-strike and toe-off, indicating that there is considerable stress on the joint in this direction at these times. Even during straight leg raises, when the heels are lifted 5 cm off the surface, the resultant hip joint force is of the order of two times body weight.

With estimates of hip joint forces in normal subjects ranging up to nearly six times body weight, the trabeculae within the acetabulum and the head and neck of the femur must be arranged to minimize bending and shearing stresses within the bone.

Trabecular systems

The trabecular systems which arise in the pelvis and run towards the acetabulum become continuous with those

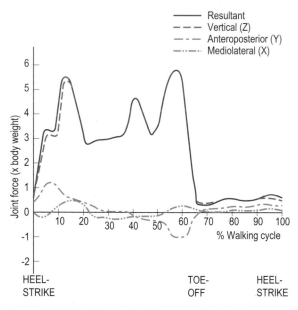

Figure 3.75 Hip joint reaction forces; resultant and components.

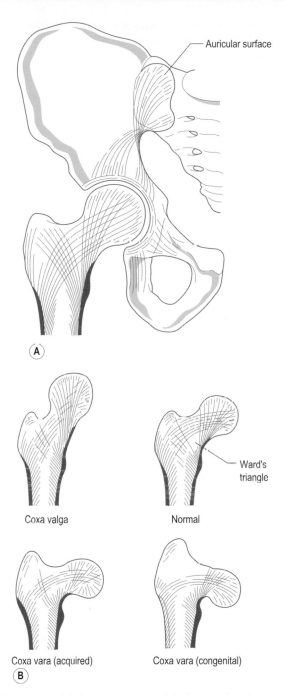

Auricular surface

Ward's triangle

(A)

Coxa valga

Normal

Coxa vara (acquired)

Coxa vara (congenital)

(B)

Figure 3.76 (A) The arrangement of the bony trabeculae in the innominate and upper part of the femur, (B) changes in the angle of inclination in coxa valga and coxa vara showing the internal remodelling that has occurred as a result.

in the femoral head and neck (Fig. 3.76A). Two trabecular systems arise in the region of the auricular surface of the innominate. From the upper part of this surface the trabeculae converge onto the posterior surface of the greater sciatic notch; from here they are reflected towards the inferior aspect of the acetabulum. This system lines up with one in the femur that arises from the cortical layer of the lateral aspect of the shaft to the inferior aspect of the cortical layer of the neck and inferior part of the femoral head. A second system arising from the lower part of the pelvic auricular surface converges at the level of the superior gluteal line; from here it is reflected laterally towards the upper aspect of the acetabulum. This system lines up with one that arises from the cortical layer of the medial aspect of the femoral shaft running to the superior aspect of the femoral head. This latter system develops in response to the compressive forces transmitted across the joint. If the pattern of stresses applied to the hip changes, as for example in acquired dislocation of the hip, the two trabecular systems remodel to realign with the new stress patterns. The outward appearance is a change in the relation of the head and neck of the femur with respect to the shaft (Fig. 3.76B).

There are within the upper end of the femur two accessory trabecular systems. The first is a trochanteric bundle arising from the cortical layer of the medial aspect of the shaft; this system develops in response to the tension forces applied to the greater trochanter with muscular contraction. The second trabecular system is entirely within the greater trochanter.

The hip joint is the pivot upon which the human body is balanced, particularly in gait. The acetabulum and femoral head, being cancellous bone, provide some elasticity to the joint, having the ability to deform without sustaining structural damage. The presence of large quantities of relatively deformable bone suggests spreading under load and, indeed, spreading occurs and is essential if the stress (force per unit area) on the articular cartilage is to be kept within tolerable limits. Since the bony portions of the hip deform under load, maximum contact area and congruence must occur in the deformed position. This congruity should occur only under full load. A congruous fit under no load would lead to an incongruous fit under load when the femoral head tends to flatten.

Deformation of the bone under load has a significant effect on protecting the overlying articular cartilage from impulsive loads. The true sparing effect of trabecular bone on the overlying cartilage involves an increase in the available potential contact area. However, excessive deformation of trabecular bone, particularly when sustained repetitively, can lead to microfracture with subsequent remodelling and stiffening of the underlying trabecular network. It has been suggested that stiffening and loss of congruence can lead to deterioration of the articular surfaces and to osteoarthrosis.

Contact area

The area of contact between the femoral head and acetabulum increases with increasing load. Under light loads there are two distinct areas on the anterior and posterior aspects of the head, which merge superiorly as the load increases (Fig. 3.77A). With loads applied across the hip joint ranging from 150 to 3200 N, the contact area increases from 2470 to 2830 mm². Dividing the applied load by contact area gives the mean contact stress. It appears that during walking the mean contact stress in the hip is in the range 2–3 MN/m². It is interesting to note that the measured contact areas coincide closely with regions of stiff cartilage. The stiffest cartilage is located in bands of variable size extending over the superior surface of the femoral head round to the anterior and posterior facets; there is a very soft area of cartilage near the margin of the fovea capitis (Fig. 3.77B). These observations suggest a direct link between the load or normal stress distribution and the pattern of cartilage stiffness.

In situations where the load-bearing area of the femoral head is diminished, perhaps because of deformity, the stress on the remaining load-bearing cartilage is concentrated and degenerative changes frequently follow.

Both abnormally large joint incongruities and abnormally low cartilage compliance cause the load to shift away from the superior weight-bearing area of the hip toward the periphery of the contact area. As a consequence transverse compressive stresses, which may be of appreciable magnitude but make no contribution to weight-bearing, are built up through much of the superior and central portions of the femoral head. Most small changes in the overall cartilage thickness or in its thickness distribution, when considered in isolation from hip compliance changes, have only minor effects on the internal stress distribution. However, an important exception is cartilage thinning at the superior margin, which can result in abrupt longitudinal compressive stress concentrations. Such aberrations of the normal patterns of stress transmission may contribute to sclerosis or the formation of osteophytes or cysts in the osteoarthritic hip.

Joint space

There is still a joint space in the loaded hip although it is variable in size between individuals, and changes in shape with different joint positions in the same individual under comparable loads. The intra-articular joint space in the hip is to allow synovial fluid to have access to the joint for lubrication and nutrition, with the acetabular fossa acting as a fluid reservoir. Under high loading of the joint there is movement of synovial fluid from the joint space to the reservoir; when the pressure in the joint space falls below that in the reservoir the flow is reversed, thus lubricating and nourishing the joint.

Pathology

If the blood supply to the femoral head is severely disrupted infarct of part of the femoral head may result. The intact viable bone surrounding a weakened infarct preferentially takes up additional load, with the degree of stress relief being dependent upon both the degree of necrotic stiffness deficit and the geometry of the infarct. Infarcted bone in the superior weight-bearing area must continue to carry high stresses even though load is progressively transferred out through the periphery of the lesion. Vulnerability to mechanical overload would appear to be more pronounced superiorly than inferiorly, and may partially explain the clinically observed collapse sequence which usually begins just beneath the subchondral bone. The elastic instabilities found to occur at the peripheral aspects of

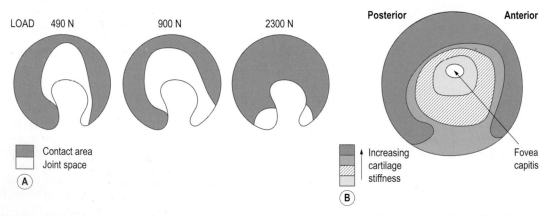

Figure 3.77 (A) The contact area of the acetabulum, (B) the distribution of cartilage stiffness over the femoral head.

the viable, weight-bearing subchondral region suggest that fracture of the subchondral bone is caused by excessive compressive hoop stress. The increased functional demand placed upon viable bone bordering on an infarct tends to enhance the sclerosing effects of hypervascularity.

Femoroacetabular impingement is a cause of early osteoarthritis of the hip occurring as a result of abnormal anatomy of the hip leading to abutment of the femur against the acetabulum. Cam impingment is the result of an aspherical femoral head-neck junction, while pincer impingement is due to acetabular overcoverage of the femoral head. Contact between the femur and acetabulum damages both the articular cartilage and the acetabular labrum, producing the early signs of degenerative joint diseases. Patients typically present with pain around the hip and groin and may have a history of developmental hip disorders. Provocation tests can be used to elicit symptoms. In the impingement test the patient lies supine and the hip is passively internally rotated and flexed to 90° and then adducted; if any pathology is present pain in the groin is elicited. The Fitzgerald test is in two parts, one for anterior and one for posterior labral involvement. For anterior labral involvement the hip is initially flexed, externally rotated and abducted before being brought into extension, internal rotation and adduction; a positive result produces pain with or without an audible click (a strong sign of mechanical injury). To test for posterior acetabular impingement the hip begins in extension, external rotation and abduction, and is moved through flexion, internal rotation and adduction. Treatment of femoroacetabular impingement is predominantly surgical, aiming to improve joint clearance by resecting bony prominences on the femur and/or acetabulum; it can be carried out with open or arthroscopic surgery.

Male/female considerations

Although the radii of the right and left femoral heads are practically identical for any individual, the femoral head is significantly smaller in females than in males in relation to pelvic dimensions (Fig. 3.78). This may result in an increased stress level in the hip joints of females. Because of the greater breadth of the female lesser pelvis, due to adaptation for childbearing, the weight of the individual acting through the body's centre of gravity lies at a greater distance from the centre of the hip joint (Fig. 3.79). This longer resistance arm, coupled with the shorter lever arm of gluteus medius, reduces the mechanical advantage of the abductor muscles. Consequently, greater abductor force is required to control pelvic tilt. Moreover, the greater force requirement acts over a smaller femoral head, resulting in greater femoral head pressures in females. It would appear, therefore, that the adaptation to childbearing, and its effects on the female pelvis, is a major contributing factor to mechanical dimorphism of the human hip.

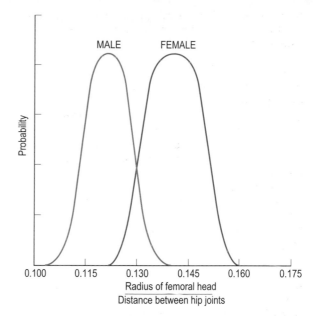

Figure 3.78 Male/female distribution of the ratio of femoral head radius to distance between the hip joints. *(Adapted from Brinckman P, Hoefert H and Jongen HT (1981) Sex differences in the skeletal geometry of the human pelvis and hip joint.* Journal of Biomechanics, *14, 427–430.)*

Section summary

Hip joint

Type	Synovial ball and socket
Articular surfaces	Head of the femur with the acetabulum of the innominate and labrum
Capsule	Strong fibrous capsule attaching to the margins of the acetabulum and intertrochanteric line (anterior) and neck of the femur (posterior). Has longitudinal, oblique, arcuate and zona orbicularis fibres
Ligaments	Iliofemoral Pubofemoral Ischiofemoral Ligamentum teres Transverse ligament of acetabulum
Stability	Shape of articular surfaces; associated muscles and ligaments
Movements	Flexion and extension Abduction and adduction Medial and lateral rotation

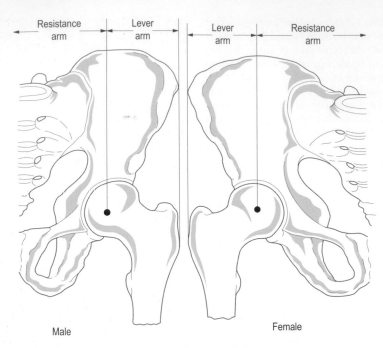

Resistance arm | Lever arm | Lever arm | Resistance arm

Male   Female

Figure 3.79 The lever and resistance arms acting about the hip joint in males and females.

The knee joint

Introduction

The largest and one of the most complex joints of the body, the knee joint is a synovial bicondylar hinge joint between the condyles of the femur and those of the tibia, and with the patella anteriorly (Fig. 3.80). Three separate articulations may be identified – the two femorotibial joints and the femoropatellar articulation. The arrangement of the synovial membrane suggests that these three articulations were completely separate at some stage in human evolution. However, in *Homo sapiens* the three joint cavities are not separate, being connected by restricted openings, thus forming a large single joint cavity.

The knee joint satisfies the requirements of a weight-bearing joint by allowing free movement in one plane only combined with considerable stability, particularly in extension. Usually stability and mobility are incompatible functions of a joint, with the majority of joints sacrificing one for the other. However, at the knee both functions are secured by the interaction of ligaments and muscles, and complex gliding and rolling movements at the articular surfaces. Nevertheless, the relatively poor degree of interlocking of the articular surfaces, which is essential for great mobility, renders it liable to strains and dislocations. Although functionally the knee joint is a hinge joint, allowing flexion and extension in the sagittal plane, it also permits a small amount of rotation of the leg, particularly when the knee is flexed and the foot is off the ground.

The support of the weight of the body on the vertically opposed ends of the two largest bones in the body is obviously an unstable arrangement. However, security at the knee is ensured by a number of compensating mechanisms. Among these are an expansion of the weight-bearing surfaces of both the femur and the tibia, the presence of strong collateral and intracapsular ligaments, a strong capsule and the reinforcing effects of aponeuroses and tendons.

The knee joint plays an important role in locomotion, being the shortener and lengthener of the lower limb. It can also be considered to work by axial compression under the action of gravity. Being endowed with powerful muscles it acts with the ankle joint as a strong forward propeller of the body. It receives and absorbs vigorous stresses which the lateral movements of the body in the frontal plane and axial rotations in the transverse plane impart to it.

Because the femoral neck overhangs the shaft, the anatomical axes of the femur and tibia do not coincide, but form an outward opening angle between 170° and 175° (*the femorotibial angle*). However, the joint centres of the hip, knee and ankle all lie on a straight line, which is the mechanical axis of the lower limb (Fig. 3.81). In the leg, this axis coincides with the anatomical axis of the tibia, while in the thigh it forms an angle of some 6° with the axis of the femoral shaft. Because the hip joints are further apart than the ankles, the mechanical axis of the lower limb runs obliquely inferiorly and medially, and forms an angle of 3° with the vertical. The wider the pelvis the larger the latter angle, as in females. In some pathological conditions, the

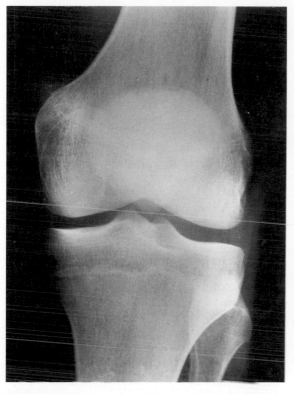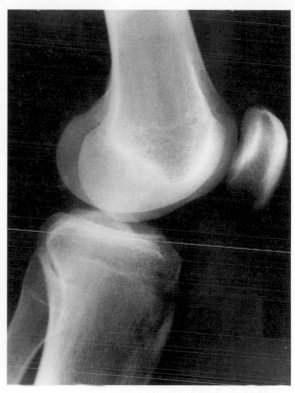

Figure 3.80 Anteroposterior and lateral radiographs of the knee.

femorotibial angle may be increased or decreased, giving the appearance of 'bowlegs' (genu varus) or 'knock-knees' (genu valgus). Genu valgus is not uncommon in toddlers, generally disappearing with growth.

The transverse axis of the knee joint is horizontal, with the common axis of both knee joints lying in a frontal plane. The axis does not bisect the femorotibial angle, so that the angle between it and the tibia is larger (about 93°) than that with the femur (about 84°) (Fig. 3.81). Comparing the axes of the hip, knee and ankle joints in upright standing, the knee maintains a position of inwards rotation with respect to both the head and neck of the femur, and the lower end of the tibia.

Articular surfaces

Femur

The articular surfaces of the femur are the two condylar areas which rest on the tibia, and the patellar surface which unites the condyles in front and is opposed to the deep surface of the patella (Fig. 3.82A). At the junction of the condylar and patellar surfaces are two faint grooves. Laterally this groove is almost transverse, being emphasized at each end, while on the medial condyle the medial end of the groove begins further forwards, passing obliquely backwards, and disappears before reaching the lateral edge.

In this region there is a narrow crescentic area marked off where the patella articulates in acute flexion. The medial condylar area can be divided into two parts, a posterior part parallel to the lateral condyle and equal in extent, and an anterior triangular extension which passes obliquely and laterally. The patellar surface is divided by a well-marked groove into a smaller medial part and a larger and more prominent lateral part.

When viewed from below the femoral condyles form two prominences convex in both planes, being longer anteroposteriorly than transversely. They are not identical since the medial condyle juts out more than the lateral and is also narrower. In addition, their long axes are not parallel but diverge posteriorly (Fig. 3.82B). The intercondylar notch, however, continues the line of the groove of the patellar surface. In the transverse plane the convexity of the femoral condyles corresponds more or less to the concavity of the tibial condyles, whereas in the sagittal plane the radius of curvature of the condyles is not uniform but varies as a spiral. The manner in which the condyles become increasingly flatter from before backwards is particularly significant in the mechanics of the joint.

The spirals of the femoral condyle are not simple spirals, even though their radii of curvature increase regularly, because they do not have a single centre of rotation but a series of centres which themselves lie on a spiral. Thus the

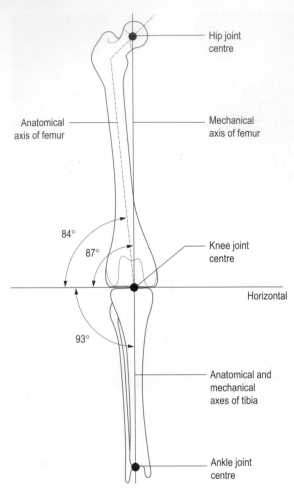

Figure 3.81 The anatomical and mechanical axes of the femur and tibia.

surface is larger, oval and slightly concave. The lateral articular surface is smaller, rounded and concave from side to side, but concavoconvex from front to back (Fig. 3.83B). Posteriorly the lateral surface extends downwards over the condyle in relation to the tendon of popliteus. The fossae of the articular surfaces of the tibia are deepened by the menisci, intervening between the femur and tibia. The menisci rest on a flattened strip at the periphery of each tibial condyle. Again the internal architecture reflects the shape of the upper end of the tibia and the stresses to which it is subjected (see Fig. 3.109).

The basic shapes of the articular surfaces of the tibia and femur essentially only allow movement in one plane, flexion and extension. Axial rotation involves the twisting of the femur against the tibia, or vice versa, during which the intercondylar eminence of the tibia, lodging in the intercondylar notch of the femur, acts as a pivot. The pivot consists of the intercondylar tubercles which form the lateral border of the medial condyle and the medial border of the lateral condyle. Through this latter point runs the vertical axis about which axial rotation occurs.

Patella

The articular surface of the patella is oval and can be divided into a larger lateral and a smaller medial area by a vertical ridge, for corresponding areas on the femur. Two faint transverse ridges separate three facets on each side, and a further faint vertical ridge separates a medial perpendicular facet from the main medial area (Fig. 3.83C). In acute flexion, this medial facet articulates with the crescentic facet of the medial femoral condyle. The remaining facets articulate in succession from above downwards with the patellar surface of the femur as the joint moves from flexion to full extension. Because of the stresses to which the patella is subjected, particularly during locomotion, the cartilage on its deep surface is extremely thick, and may be the thickest of anywhere in the body.

Joint capsule

The knee joint is surrounded by a thick ligamentous sheath composed mainly of muscle tendons and their expansions (Fig. 3.84). There is no complete, independent fibrous capsule uniting the two bones; only occasionally are there true capsular fibres running between them. The capsular attachment to the femur is deficient anteriorly, where it blends with the fused tendons of the quadriceps muscles. Its attachment to the tibia is more complete, being deficient only in the region of the tibial tuberosity, which gives attachment to the ligamentum patellae. In spite of its composite nature, it is nevertheless convenient to think of the joint capsule as a cylindrical sleeve passing between the femur and tibia, with a deficiency anteriorly lodging the patella.

Posteriorly the true capsular fibres arise from the femoral condyles, just above the articular surfaces, and the

curve of the condyles represents a spiral of a spiral (Fig. 3.82C). On each condyle can be identified the groove separating the condylar and patellar surface areas. Both anterior and posterior to this point the radius of curvature decreases. However, it must be remembered that the two condyles do not show the same curvature, thus their radii of curvature will differ. The internal architecture of the femoral condyles will reflect their geometry as well as the stresses to which they are subjected (see Fig. 3.109).

Tibia

The articular surfaces of the tibia are cartilage-covered areas on the upper surface of each tibial condyle, being separated from each other by the intercondylar eminence and triangular intercondylar areas in front and behind (Fig. 3.83A). The articular areas are comparatively flat following the slight (3–5°) posteroinferior inclination of the tibial condyles with respect to the horizontal. The medial articular

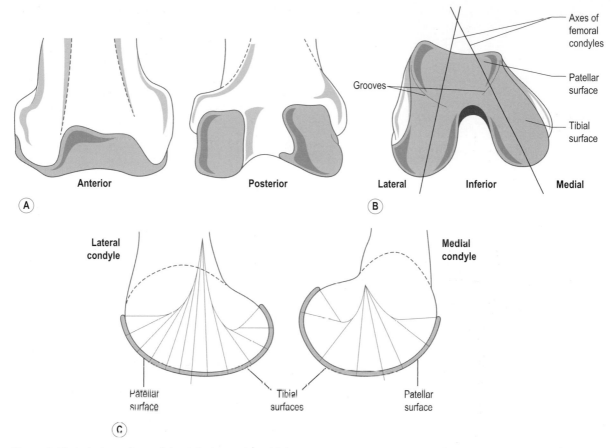

Figure 3.82 Articular surfaces of the right femur viewed (A) anteriorly and posteriorly and (B) from below, (C) the spiral profiles of the femoral articular condyles (paramedian sections).

intercondylar line, and pass vertically downwards to attach to the posterior border of the upper end of the tibia. At the side of the joint, capsular fibres descend from the femoral to the tibial condyles. They blend posteriorly with a ligamentous network, and anteriorly with the various tendinous expansions of the quadriceps muscles.

Capsular strengthening

The majority of what is seen and taken for as the knee joint capsule is in fact the ligamentous feltwork associated with the joint. This feltwork is extremely important as it provides the capsule with its strength, as well as providing the necessary control and restriction to movement that is required at the knee.

Oblique popliteal ligament

The central region of the posterior part of the capsule is strengthened by the oblique popliteal ligament (Fig. 3.84A), an expansion of the semimembranosus tendon. It passes upwards and laterally to attach to the intercondylar line of the femur. It has large foramina for the vessels and nerves which perforate it.

Arcuate popliteal ligament

The lower lateral part of the capsule is strengthened by the arcuate popliteal ligament (Fig. 3.84A), as it passes from the back of the head of the fibula arching upwards and medially over the popliteus tendon to spread out over the posterior surface. The most medial part of the arcuate ligament arches downwards onto the posterior part of the intercondylar area of the tibia, while the most lateral fibres appear as a separate band that runs to the back of the lateral femoral condyle.

On the lateral side of the joint the true capsular fibres are short and weak, only bridging the interval between the femoral and tibial condyles. Indeed, some of them are separated from the lateral meniscus by the passage of the popliteal tendon.

307

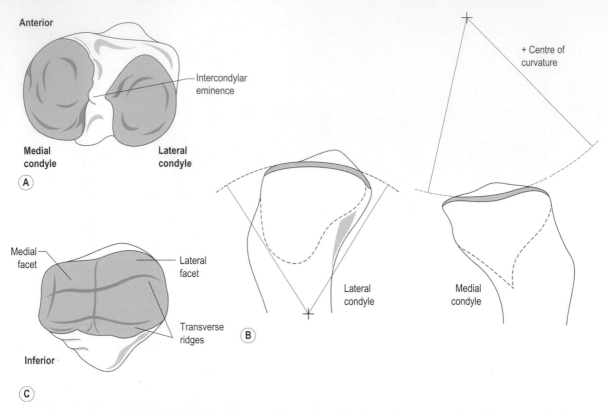

Figure 3.83 (A) Articular surfaces of the right tibia viewed from above, (B) schematic representation of the anteroposterior condylar curvatures, (C) articular surface of the right patella.

Because the points of attachment of both collateral ligaments lie behind the vertical axis of the bones, they are most tightly stretched in extension and so prevent hyperextension. They also prevent any abduction or adduction of the femur or tibia. In extension, the lateral collateral ligament is directed downwards, while the medial collateral ligament runs downwards and forwards. This relationship prevents rotation of the femur medially or the tibia laterally. With the knee flexed the ligaments are relaxed and rotation can occur relatively easily.

At the front of the joint the ligamentous sheath is mainly composed of the fused tendons of the various quadriceps muscles (Fig. 3.84B). The fibres of these tendons descend from above and each side to attach to the margins of the patella as far as the attachment of the ligamentum patellae. Superficial tendinous fibres pass downwards over the patella into the ligamentum patellae. The aponeurotic tendons of vastus medialis and vastus lateralis expand over the sides of the joint capsule as the medial and lateral patellar retinacula respectively, to insert into the front of the tibial condyles and oblique lines of the condyles as far to the sides as the collateral ligaments. Medially the retinaculum blends with the periosteum of the tibial shaft; while laterally it blends with the overlying iliotibial tract. Some tendinous fibres of the vastus muscles pass obliquely downwards across the patella into the opposite retinaculum. Other deeper fibres from each side of the patella pass across to the front of each femoral epicondyle.

Ligamentum patellae

The continuation of the tendon of quadriceps femoris (Fig. 3.84B). It is a strong, flat band attaching around the apex of the patella continuous over its front with fibres of the quadriceps tendon. It extends to the tibial tuberosity, ending somewhat obliquely as it is prolonged further downwards laterally than medially. Between the ligament and the bone, immediately above its insertion, is the deep infrapatellar bursa, while in the subcutaneous tissue over the ligament is the large subcutaneous infrapatellar bursa.

Superficial to the fibrous bands associated with the quadriceps complex are strong expansions of the fascia lata covering the front and sides of the joint. As it descends to attach to the tibial tuberosity and oblique lines of the condyles, it overlies and blends with the patellar retinacula.

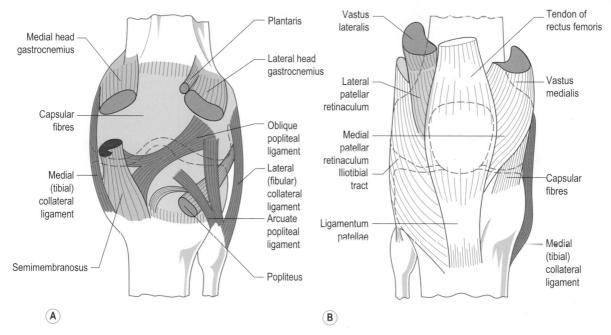

Figure 3.84 The knee joint capsule from the front and back showing the contributions from the various muscles and ligaments crossing the joint: (A) posterior, (B) anterior.

The thickened and strong iliotibial tract descends across the anterolateral part of the joint to insert into the lateral tibial condyle after blending with the joint capsule (Fig. 3.84B). A strong band passes forwards from the tract to attach to the upper part of the lateral edge of the patella; this is the superior patellar retinaculum. On the thinner medial side of the patella, the fascia lata sends a few fibres inferiorly to blend with the expansion of sartorius.

The whole of the anterior covering of the knee joint, together with the patella, is kept tense by the tone of the extensor muscles, being tightly braced when they are active in extension. This appears to contradict the philosophy of hinge joints in that the extensor portion of joint capsules should be loose in the extended position in order to permit free flexion to occur. At the knee, the extensor part of the capsule is in reality continued into the extensor muscles above rather than being attached directly to bone. Consequently, it may be kept relatively taut in all positions of the joint.

Ligaments

As is characteristic of all hinge joints, collateral ligaments are found at the sides of the knee joint, although the form of the medial and lateral ligaments differs greatly.

Medial (tibial) collateral ligament

A strong flat band extending from the medial epicondyle of the femur passing downwards and slightly forwards to attach to the medial condyle of the tibia and the medial side of the shaft (Fig. 3.85A and B). A few fibres at the femoral attachment can usually be traced upwards into adductor magnus. Consequently, the medial collateral ligament has been regarded as having been formed, in part at least, from an original tibial insertion of adductor magnus. The most superficial fibres of the ligament descend below the level of the tibial tuberosity; deeper fibres have a shorter course from femur to tibia, with the deepest fibres spreading triangularly to attach to the medial meniscus (Fig. 3.85A and B). The medial collateral ligament is 8–9 cm long, being well-defined anteriorly where it blends with the medial patellar retinaculum; however, bursae may partially separate them. Under cover of the ligament's posterior border a downward expansion from semimembranosus reaches the tibial shaft, considerably strengthening this aspect of the joint capsule. The inferior medial genicular vessels and nerve pass between the semimembranosus expansion and the ligament.

Lateral (fibular) collateral ligament

A rounded cord some 5 cm long standing clear of the thin lateral part of the fibrous capsule (Fig. 3.85A and C). It is attached to the lateral epicondyle of the femur above and behind the groove for popliteus, and passes down to attach to the lateral surface of the head of the fibula in front of the apex, splitting the tendon of biceps femoris as it does so. Occasionally the lateral collateral ligament continues into the upper part of peroneus longus, and can be thought of as a femoral origin of the muscle.

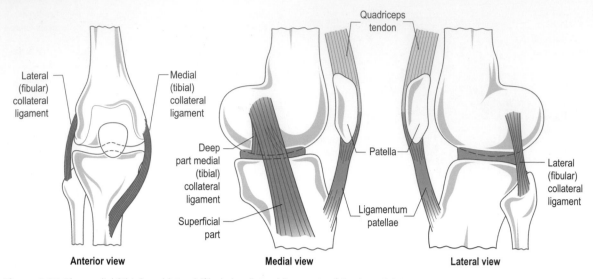

Figure 3.85 The medial (tibial) and lateral (fibular) collateral ligaments of the knee joint.

Passing deep to the ligament is the tendon of popliteus, together with the inferior lateral genicular vessels and nerve as they pass forwards.

Synovial membrane and bursae

The joint cavity of the knee is the largest within the body, having an irregular shape with identifiable regions in free communication with each other. Synovial membrane lines the joint capsule and is reflected onto the bone as far as the margins of the articular cartilage. The central part of the joint space lies between the deep surface of the patella anteriorly and the patellar surface of the femur and the cruciate ligaments posteriorly. It extends outwards to pass between the femoral and tibial condyles as well as above and below the menisci. From the articular margins of the patella, the synovial membrane passes in all directions onto the deep surface of the anterior joint capsule. Below the patella the synovial membrane is pushed backwards into the joint space by the infrapatellar fat pad, lying on the deep surface of the ligamentum patellae and extending towards the intercondylar notch. The synovial reflections over this fat pad are thrown into a series of folds (Fig. 3.86A). A vertical crescentic fold, the infrapatellar fold, passes in the medial plane towards the cruciate ligaments, attaching to the intercondylar fossa in front of the ACL and lateral to the PCL. The infrapatellar fat pad is the remains of a septum which divided the embryonic knee into two compartments; the attachment of the synovium to the intercondylar fossa is called the infrapatellar plica. From the edges of the patella horizontal double folds of synovial membrane, the alar folds, project into the interior of the joint; they also cover collections of fat. There are two further reflections

(plicae) of synovium in the knee joint which have been consistently observed in arthroscopy. The infrapatellar plica extends backwards from the infrapatellar fat pad; the suprapatellar plica is a horizontal fold level with the superior border of the patella; while the mediopatellar plica forms a shelf-like fold partway between the femur and patella. The significance of these plicae is that if they become trapped between the joint surfaces they can become inflamed, causing pain on certain movements or in certain positions of the knee. Arthroscopic removal of an inflamed plica usually resolves this problem.

Several recesses extend from the central part of the joint cavity. The suprapatellar bursa extends approximately 6 cm above the patella between the femoral shaft and quadriceps femoris. Initially it develops as a separate bursa, but soon communicates freely with the joint space. To the upper part of the suprapatellar bursa is attached articularis genu (Fig. 3.87A), bundles of muscle fibres from the deep surface of vastus intermedius, which serve to maintain the bursa during extension of the knee. There are also recesses behind the posterior part of each femoral condyle. In this region the two heads of gastrocnemius overlie the joint capsule; there is usually a bursa between each muscle head and the capsule (Fig. 3.87). The bursa under the medial head of gastrocnemius occasionally communicates with the joint cavity; it may also enlarge and present as a cyst behind the knee joint. The recesses behind the femoral condyles are separated by the interposition of the cruciate ligaments, which are intracapsular but extrasynovial. A vertical fold of synovial membrane covers the cruciate ligaments anteriorly and on their sides, but not posteriorly where the tibial attachment of the PCL blends with the joint capsule (Fig. 3.86B).

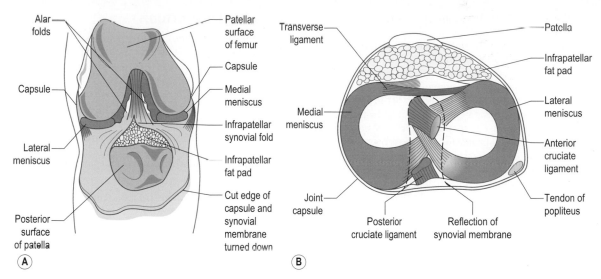

Figure 3.86 (A) The right knee joint opened from the front showing the infrapatellar fat pad and synovial folds, (B) intracapsular reflections of the synovial membrane on the tibia.

The remaining recess of the joint cavity is associated with the tendon of popliteus (Fig. 3.87B). As the tendon passes from its intracapsular femoral attachment, the synovial membrane invests its medial side thereby separating it from the lateral meniscus, except where the tendon attaches to the meniscus. As the popliteal tendon emerges through the back of the joint capsule, it takes with it a synovial extension associated with its deep surface which lies between the tendon and the superior tibiofibular joint. Occasionally, the superior tibiofibular joint cavity communicates

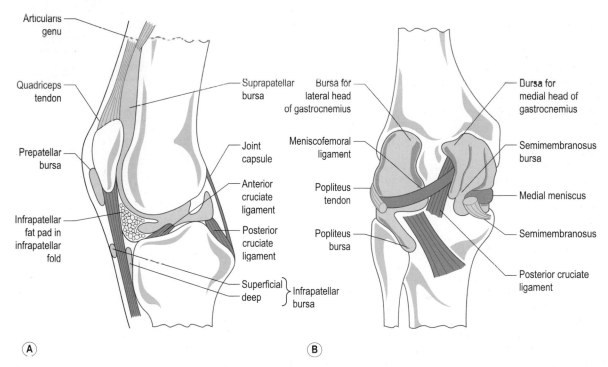

Figure 3.87 Synovial membrane and bursae of the knee: (A) paramedian section, (B) posterior view.

with the subpopliteal bursa and therefore with the knee joint itself.

Since all of the tendons and muscles which cross the knee run parallel to the bones involved, and thus pull lengthwise across the joint, there are numerous other bursae associated with it – the subcutaneous prepatellar bursa (Fig. 3.87A) lies between the skin and the lower part of the patella while the subcutaneous infrapatellar bursa overlies the patellar tendon; the subfascial prepatellar bursa lies between the tendinous and fascial expansions that pass over the patella, while the subtendinous prepatellar bursa separates the deep and superficial tendinous fibres; finally the deep infra-patellar bursa lies between the patellar tendon and the up-per part of the tibia, being separated from the knee joint by the infrapatellar fat pad (Fig. 3.87A). On the lateral side of the joint is the subtendinous bursa of biceps femoris be-tween the lateral collateral ligament and the tendon. As well as the subpopliteal bursa an additional bursa may in-tervene between the popliteus tendon and the lateral collat-eral ligament. Medially the bursa anserina separates the tendons of sartorius, gracilis and semitendinosus from the medial collateral ligament. There may, in addition, be bursae intervening between the individual tendons. The semimembranosus bursa (Fig. 3.87) lies between the muscle and the tibia and medial collateral ligament.

Intra-articular structures

Within the capsule of the knee joint are two sets of structures which play an important part in knee function, anatomically and mechanically. These are the cruciate ligaments, anterior and posterior, and the menisci, medial and lateral. Of these, the cruciate ligaments are extrasynovial.

Cruciate ligaments

The cruciate ligaments are so called because of the way they cross each other between their attachments, the anterior and posterior ligaments being named according to their tibial attachments.

Anterior cruciate ligament (ACL) is attached to the tibia immediately anterolateral to the anterior tibial spine (Figs. 3.86B and 3.88). It passes beneath the transverse ligament, blending with the anterior horn of the lateral meniscus, and runs posteriorly, laterally and proximally to attach to the posterior part of the medial surface of the lateral femoral condyle (Fig. 3.88). The femoral attachment is not as strong as the tibial and takes the form of a segment of a circle. During its passage from tibia to femur the ACL undergoes a medial spiral of some 110°. It can be divided anatomically into two parts – an anteromedial band attaching to the anteromedial region

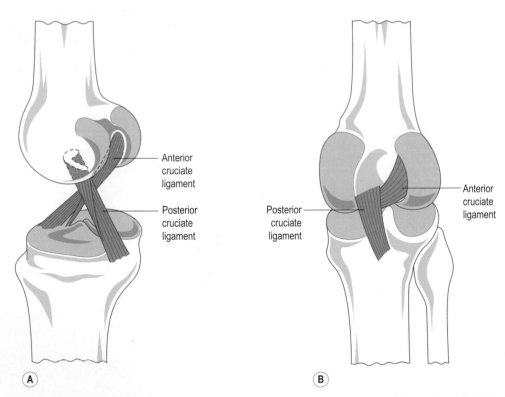

Figure 3.88 The anterior and posterior cruciate ligaments: (A) oblique view showing twisting of fibres, (B) posterior view showing the ligaments crossing in space.

of the tibial attachment, and the posterolateral band which constitutes the remainder of the ligament. The posterolateral band of the ligament is said to be taut in extension, with the anteromedial band lax (and vice versa in flexion). Functionally the ligament should be considered as a continuum with part of it being taut throughout the whole range of knee movement, thereby having a restraining influence in all positions of the joint.

Posterior cruciate ligament (PCL) attaches to a depression in the posterior intercondylar area of the tibia (Figs. 3.88 and 3.86B). It runs anteriorly, medially and proximally, passing medially to the ACL to attach to the anterior part of the lateral surface of the medial femoral condyle (Fig. 3.88). The PCL is shorter and less oblique in its course, as well as being almost twice as strong in tension, than the ACL. It is closely aligned to the centre of rotation of the knee joint and, as such, may be its principal stabilizer. Like the ACL, the PCL can be divided into two parts, an anterolateral band and a posteromedial band. The more superficial posteromedial fibres also have an attachment to the posterior horn of the lateral meniscus, and constitute the meniscofemoral ligament.
Functionally, however, it is best considered as a continuum.

The crossing and twisting of the cruciate ligaments as they pass from their tibial to their femoral attachments arises because on the tibia the attachments are more or less in a sagittal plane, while on the femur they are almost in a frontal plane. The cruciate ligaments are said to have a constant length ratio of 5:3 (ACL:PCL); however this varies between individuals. Nevertheless, this ratio together with their sites of attachment is a primary factor in controlling the type of movement seen at the knee joint (Fig. 3.89). They provide, almost exclusively, the resistance to anterior and posterior displacements of the tibia with respect to the femur. The ACL provides approximately 86% of the restraint to anterior displacement, and the PCL about 94% of the restraint to posterior displacement of the tibia on the femur. Rupture of the ACL results in very little increase in the anterior draw (anterior tibial displacement at 90° flexion), while rupture of the PCL results in a posterior draw of up to 25 mm. The latter is probably due to lack of collateral resistance to posterior displacement and a lax capsule posteriorly. In addition to their role in an anteroposterior direction, the cruciate ligaments also provide some mediolateral stability. The PCL provides 36% of the resistance to lateral displacement and the ACL 30% of the resistance to medial displacement of the tibia.

Microscopically, the cruciate ligaments are composed mainly of collagen fibres, with a small proportion of elastic fibres (10%), giving the ligaments high tensile strength. These fibres are arranged to form fasciculi a few millimetres in diameter. In the ACL two types of fasciculi can be identified – those that run directly between the tibial and femoral attachments, and those that spiral around the longitudinal axis of the ligament, with a helical axis of 25°. This helical arrangement of many of the fibres has important consequences. When loaded lightly only a few fibres are under tension, but as the load increases the ligament unwinds, bringing more fibres into play and effectively increasing its strength. Two phases of loading of the cruciate ligaments can be identified; prior to the yield point the

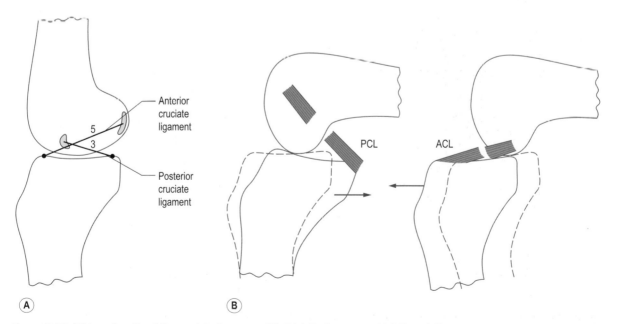

Figure 3.89 (A) Length ratio of the cruciate ligaments, (B) tibial displacement with PCL or ACL rupture.

deformation is elastic due to the stretching and unwinding of the ligament. Further loading, however, causes disruption of the cross-links between fibres and permanent deformation occurs. Although gross damage may not be apparent until this later phase, microfailure can occur at stresses below the yield point. The strength of the ligaments is influenced by several factors:

1. Tensile strength, but not stiffness, decreases significantly with age.
2. Cyclical loading, as in gait, softens the ligaments causing a decrease in yield point. However, recovery occurs within a few hours.
3. Immobilization may cause a decrease in tensile strength of up to 60%, even in otherwise healthy ligaments. This is probably due to disuse atrophy; recovery does occur but it may take many months.
4. Internal rotation reduces tensile strength by as much as 6%; consequently torsional forces are potentially much more damaging.
5. Exercise may increase strength slightly.

The bony attachments of the ligaments show a complex interdigitation of the collagen fibres from the bone and the ligament, between which is interposed a transitional zone of fibrocartilage. This allows a gradual change in stiffness and prevents stress concentrations in this region.

The cruciate ligaments possess a fairly good blood supply, derived mainly from the middle genicular artery, with a small contribution from the inferior lateral genicular artery. The blood vessels form a periligamentous sheath around the ligaments from which small penetrating vessels arise.

Mechanoreceptors, resembling Golgi tendon organs, are situated near the femoral attachments of the ligaments, located around the periphery where maximum bending occurs, and running parallel to the long axis of the ligament. They probably convey information regarding angular acceleration, and may be involved in reflexes to protect the knee from potential injury. The nerve supply enters and leaves via the femoral attachment of each ligament.

The menisci

So called because of their 'half-moon' or 'meniscal' configuration; they may also be referred to as semilunar cartilages. Intra-articular discs are found where three conditions exist: (i) in joints with a large degree of rotation about an axis perpendicular to the articular surface; (ii) in joints with flat articular surfaces; and (iii) where the forces across a joint tend to bring the two surfaces together in a rotatory movement. The menisci in the knee are slightly inclined to the tibial surface, thus enabling a viscous lubricating film to be set up in the region of the loadline of the joint. Joints without menisci (surgically removed) show an increased mechanical curvature. The function of the menisci is fivefold:

1. increase the congruence between the articular surfaces of the femur and tibia
2. participate in weight-bearing across the joint
3. act as shock absorbers
4. aid lubrication
5. participate in the locking mechanism.

The menisci are first recognizable as a region of closely packed cells with a long axis transverse to that of the limb at about 8 weeks *in utero*. By week 19, both menisci are well developed with well-defined collagen arrangements. During the next 4 weeks, extensive collagenous rearrangement occurs so that at week 23 the collagen bundles lie parallel to one another and show the adult meniscal morphology even though there is no fibrocartilage present. However, the fetal meniscus has a high cellular content and is well vascularized. Postnatal changes in the menisci are:

1. a gradual decrease in cellularity
2. a decrease in vascularity from centrally outwards
3. growth to match the femoral and tibial enlargements
4. a configurational change which accommodates the changing contact areas, with growth of the menisci being linearly related to that of the tibial condyles
5. an increase in collagen content.

The higher vascularity of the young meniscus results in an increased efficiency of repair and a reduced tendency to injury. As age increases, the ratio of collagenous to non-collagenous proteins increases, leading to a lower tensile strength. This changing biochemical balance is an important determinant of collagen fibre arrangement. The histological arrangement is accounted for by the adaptation of the menisci to weight-bearing, with the collagen fibres becoming more circumferentially arranged. This latter refinement occurs at about 3 years, following the attainment of an upright posture and gait in the infant. The change in fibrillar arrangement is an important factor confirming the role of the menisci in weight-bearing.

The menisci are crescent-shaped structures triangular in cross-section interposed between the femoral and tibial condyles (Fig. 3.90); they are normally composed of fibrocartilage. Their thick peripheral border is convex and attached to the deep surface of the joint capsule. These capsular fibres attach the menisci to the tibial condyles and constitute the medial and lateral coronary ligaments. Through its capsular attachment, the medial meniscus is anchored to the medial collateral ligament, whereas the lateral meniscus is attached only to the weak fibres of the lateral side of the capsule and not to the lateral collateral ligament. In addition, where the lateral meniscus is crossed by the popliteal tendon, the periphery of the meniscus is not attached to the joint capsule. Consequently, the lateral meniscus is much more mobile than the medial. The inner border of each meniscus is thin and concave forming the free inner edge. Superiorly the meniscal surface is smooth and concave articulating with the femoral condyles. The horns of the menisci are the sites of attachment to the tibia,

Anterior

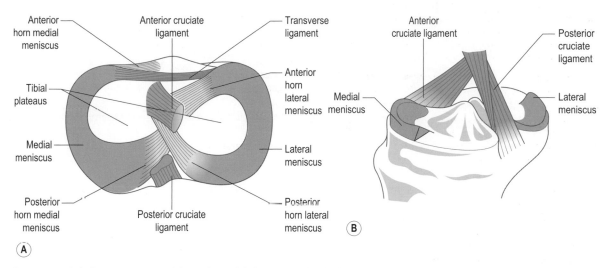

Figure 3.90 (A) The menisci viewed from above, (B) the menisci (cut), oblique view showing their triangular cross-section.

being regions where fibrocartilage gives way to bands of fibrous tissue.

Medial meniscus The larger of the two menisci, the medial meniscus is semicircular in shape, with its posterior part being broader than that anteriorly (Fig. 3.90). The anterior horn is attached to the anterior part of the intercondylar area on the tibia immediately in front of the ACL. The most posterior of these fibres are continuous with the transverse ligament of the knee. The posterior horn attaches to the posterior intercondylar area between the PCL posteriorly and the posterior horn of the lateral meniscus anteriorly. Its entire periphery attaches to the joint capsule.

Lateral meniscus It forms about four-fifths of a circle and is of uniform breadth throughout (Fig. 3.90). The two horns of the lateral meniscus are attached close together, the anterior horn attaching in front of the intercondylar eminence posterolateral to the ACL with which it partially blends. In this region it is twisted upwards and backwards as it rests on the sloping bone of the tibial condyle. The posterior horn attaches behind the intercondylar eminence anterior to the posterior horn of the medial meniscus. Posterolaterally the lateral meniscus is grooved by tendon of popliteus, from which it receives a few fibres.

Meniscal attachments As well as their capsular attachments, the two menisci are attached anteriorly by a fibrous band of variable thickness, the transverse ligament of the knee. A posterior transverse ligament is also present in 20% of knees. The posterior part of the lateral meniscus usually contributes a ligamentous slip to the PCL, which splits on the lateral side of the ligament to run both in front of and behind the PCL. The part which runs anteriorly is the anterior meniscofemoral ligament, while that running posteriorly is the posterior meniscofemoral ligament; of the two, the posterior is the more constant. However, when present, both end on the medial femoral condyle in association with the PCL. A few fibres from the anterior horn of the medial meniscus may run with the ACL (Fig. 3.91).

The connection between the lateral meniscus and the PCL is in keeping with the typical lower mammalian situation, in which the lateral meniscus is attached posteriorly to the medial femoral condyle behind the PCL rather than to the tibia.

Posteriorly the lateral meniscus gives rise to some fibres of popliteus, while the medial meniscus is attached via the joint capsule to the oblique popliteal ligament, the expansion of semimembranosus.

Structure Histologically, the fibrocartilaginous menisci are described as being between dense fibrous cartilage and hyaline cartilage, containing large collagen bundles embedded in a matrix. The fibrocartilage does, however, retain the ability to become either hyaline or fibrous cartilage, as seen in regenerating menisci where the hypertrophic cells form a purely fibrous structure.

Microscopically, the collagen fibres arrange themselves into circumferential (passing between the tibial attachments) and radial patterns, which are determined by meniscal age and functional requirements (Fig. 3.92). The radially arranged fibres are primarily found associated with the

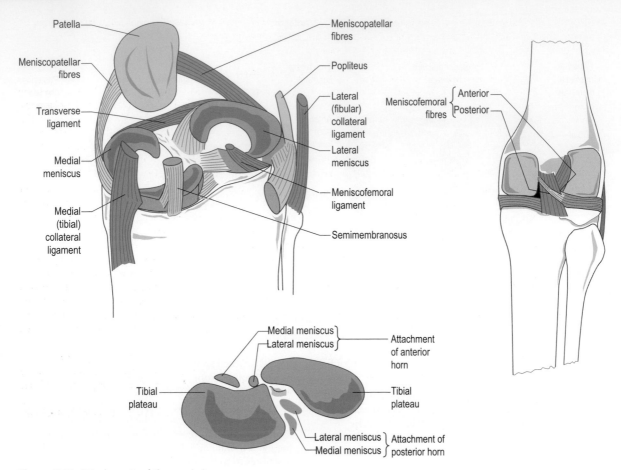

Figure 3.91 Attachments of the menisci.

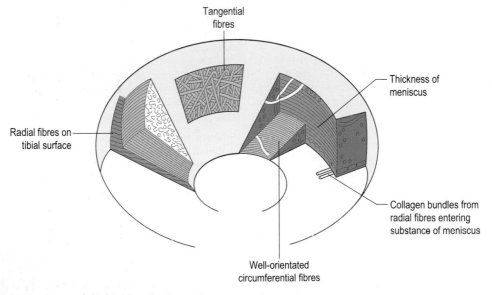

Figure 3.92 Arrangement of the bundles of collagen fibres within the meniscus.

femoral and tibial surfaces, many curving to travel perpendicular to the meniscal surface. The orientation of the collagen fibres is closely related to the direction of stresses during weight-bearing. Under compression, the major force is outwards and is resisted by the circumferential fibres. The radial fibres probably act as ties to prevent longitudinal splitting.

Severe strain on a meniscus, usually involving rotation, can result in longitudinal and sometimes transverse splitting of the fibrocartilage (Fig. 3.93). Because of its attachment to the medial collateral ligament, the medial meniscus is more frequently injured, with the inner thinner portion separating from the thicker outer part (buckethandle rupture). The detached part may move into the centre of the joint, thereby preventing full extension of the knee.

The menisci receive a blood supply from the middle, medial and lateral inferior genicular arteries. There is a peripheral perimeniscal plexus which gives rise to small penetrating branches which reach the menisci via the coronary ligaments. In the fetus and young child, the entire meniscus is vascularized, but with increasing age the central regions become avascular. By age 11 the adult pattern is seen, whereby only the peripheral 20% and the horns are vascular. The vessels themselves are located mainly within the deep part of the meniscus, with the various surfaces receiving only a limited blood supply. The main source of nutrition of the menisci is via diffusion from the synovial fluid. The outer third of the menisci may be innervated by a few myelinated and non-myelinated fibres. However, no specialized nerve endings have been found so their precise role

is not understood, although some will undoubtedly be vasomotor.

Blood supply

At the knee there is an important genicular anastomosis consisting of a superficial plexus above and below the patella, and a deep plexus on the joint capsule and adjacent condylar surfaces of the femur and tibia. Ten vessels are involved in the anastomosis – two descending from above (the *descending branch* of the *lateral circumflex femoral artery* and the *descending genicular branch* of the femoral artery), three ascending from below (the *circumflex fibular artery* from the posterior tibial artery, and the *anterior* and *posterior tibial recurrent branches* from the anterior tibial artery), and the remainder being branches of the *popliteal artery* (Fig. 3.94). The popliteal artery lies deep within the popliteal fossa, lying on (from above downwards) the fat covering the popliteal surface of the femur, the posterior knee joint capsule, and the fascia covering popliteus. The five branches of the popliteal artery involved in the anastomosis are the *lateral superior* and *inferior*, the *medial superior* and *inferior*, and the *middle genicular arteries* (Fig. 3.94).

Venous drainage of the joint is by corresponding veins accompanying the arteries. However, venostasis is a major contributory factor in degenerative changes of the knee joint.

The lymphatics of the joint drain to the popliteal and inguinal lymph nodes.

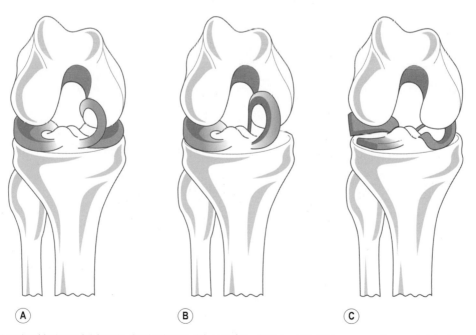

Figure 3.93 Meniscal lesions: (A) longitudinal splitting, (B) complete detachment, (C) transverse tear and detachment of anterior horn.

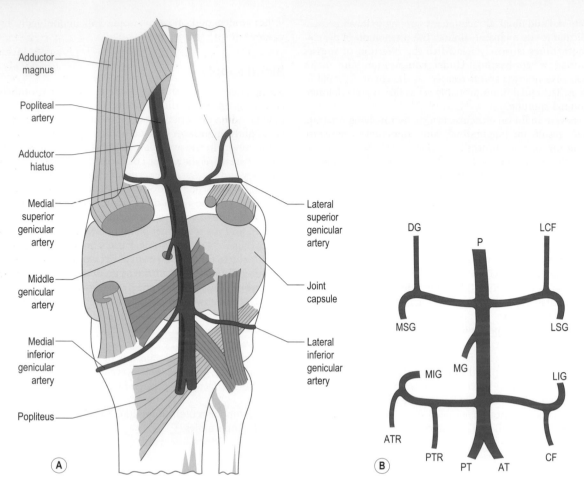

Figure 3.94 (A) Blood supply to the knee joint, together with (B) a schematic representation of the vessels involved in the genicular anastomosis, posterior view. AT, anterior tibial; ATR, anterior tibial recurrent; CF, circumflex fibular; DG, descending genicular; LCF, lateral circumflex femoral; LIG, lateral inferior genicular; LSG, lateral superior genicular; MG, middle genicular; MIG, medial inferior genicular; MSG, medial superior genicular; P, popliteal; PT, posterior tibial; PTR, posterior tibial recurrent.

Nerve supply

The nerve supply to the joint is from many sources. However, it must be remembered that the articular cartilage has no direct nerve supply. Proprioceptive information is via nerve endings located in the bone, periosteum and cruciate ligaments, while pain and pressure sensitivity comes from endings in the collateral ligaments and joint capsule. The root values of the branches supplying the knee joint are L2 to S3. From the femoral nerve, articular branches reach the joint via the nerves to vastus medialis and through the saphenous nerve. The posterior division of the obturator nerve, after supplying adductor magnus, ends in the knee joint. Both the tibial and common fibular (peroneal) nerves send articular branches to the joint. Those from the tibial nerve follow the two medial genicular arteries and the middle genicular artery, while those from the common fibular (peroneal) nerve follow the lateral genicular arteries and the anterior tibial recurrent artery.

Palpation

Several parts of the knee joint can be readily palpated, particularly at the sides and front, thus confirming the superficial position of the joint within the lower limb. Although many musculotendinous structures cross the joint, they tend to do so posteriorly, together with the major blood vessels and nerves. The following parts of the joint can be easily palpated:

1. whole of the circumference of the patella, particularly its medial border
2. articular margin of each femoral condyle
3. anterior articular margin of each tibial condyle

4. joint line medially, anteriorly and laterally
5. tibial tuberosity together with the ligamentum patellae attaching to it
6. adductor tubercle, and the medial and lateral epicondyles of the femur.

No part of the joint can be identified posteriorly. However, the tendons of semitendinosus and semimembranosus medially, and that of biceps femoris laterally, stand out and can be easily identified in the knee flexed against resistance (Fig. 3.95).

The medial collateral ligament can be palpated on the medial side of the knee, and can be tested for laxity by applying a laterally directed force to the medial side of the ankle with the knee in extension. The lateral collateral ligament can be palpated as a rounded cord above the head of the fibula on the lateral side of the knee, and can be tested for laxity by applying a medially directed force to the lateral side of the ankle. The ligamentum patellae can be palpated running between the apex of the patella and tibial tuberosity.

The cruciate ligaments are placed deep within the joint, and so cannot be palpated. However, they can be tested for laxity by placing the knee of the supine subject at 90° with the foot flat on the bed, and applying alternative anterior and posterior pressure on the upper end of the tibia. Pressure directed backwards tests the PCL, while forward pressure tests the ACL.

When sitting with the knees flexed at 90°, place the fingers of both hands on the anterior surface of the right patella. The skin and fascia can be moved from side to side against the prepatellar bursa revealing the vertical ridges on the bone produced by the fibres of quadriceps femoris. Moving to either side of the patella, trace down to the apex and locate the ligamentum patellae. This broad, strong tendon of quadriceps femoris, approximately 4 cm long and 2 cm wide, joins the patella to the upper part of the tibial tuberosity; its lower half is prominent, being covered by the superficial infrapatellar bursa. The surface marking of the knee joint is represented by a horizontal line bisecting this tendon.

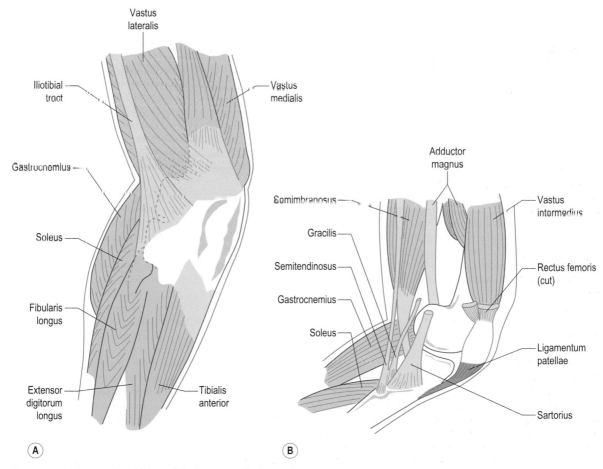

Figure 3.95 (A) Lateral and (B) medial relations of the knee joint.

On each side of the ligamentum patellae is a triangular depression bounded above by the appropriate femoral condyle, below by the tibia and by the margin of the ligamentum patellae. Deep within each hollow can be palpated the broad, horizontal anterior margin of the knee joint. On medial tibial rotation the medial meniscus protrudes forward, while on lateral rotation the lateral meniscus can be felt, but to a lesser extent. Movement of the tibia beneath the femur, however, is quite clear.

On the medial side the joint line becomes difficult to palpate just posterior to the midpoint, where the broad medial collateral ligament crosses the joint. Similarly, on the lateral side the joint line becomes difficult to palpate where the thickened articular capsule and the lateral collateral ligament cross the joint. The lateral collateral ligament can be traced down to the head of the fibula, which lies just below and posterior to the joint line with the tendon of biceps femoris attaching to its upper posterior part.

On the medial side both the tibial and femoral condyles are easily palpable with the tendons of gracilis and semitendinosus approximately 5 cm posteroinferior to the medial collateral ligament. These become distinct if the knee is flexed against resistance; gracilis is the more medial of the two tendons.

Posteriorly lies the popliteal fossa, containing several structures and a variable amount of fat, and thereby not permitting palpation of the knee joint.

From the anterior surface of the patella, draw your fingers up to its base. On either side is a narrow groove between the patella and the femoral condyles. On extending the knee the patella can be felt gliding upwards against the femoral condyles, with the upper part of the patella coming clear of the patellar surface of the femur. The tendon of quadriceps femoris lies between your hands, while the belly of vastus medialis lies under the left or right hand depending on which knee is being palpated.

On flexion of the knee the patella can be felt gliding down the patellar surface of the femur, coming to lie on the lower part of the femoral condyles on full flexion. The whole of the patellar surface of the femur can now be palpated, even though it is covered by the tendon of quadriceps femoris.

Relations

On the posterior aspect of the joint is a region known as the popliteal fossa, which contains from superficial to deep the major nerves, veins and arteries passing between the thigh and leg. The fossa is diamond-shaped, bound below by the medial and lateral heads of gastrocnemius and above by the diverging hamstring tendons. Laterally the tendon of biceps femoris crosses the lateral head of gastrocnemius, while medially the superimposed tendons of semitendinosus and semimembranosus cross the medial head of gastrocnemius. The floor of the popliteal fossa is formed from above downwards by the popliteal surface of the femur, the posterior part of the knee joint capsule, reinforced by the oblique popliteal ligament, and popliteus with its covering fascia.

The whole of the fossa is covered by dense popliteal fascia, which is continuous with both the fascia of the thigh and that of the leg, being strengthened in this region by transverse fibres as well as by expansions from the tendons of sartorius, gracilis, semitendinosus and biceps femoris (Fig. 3.96). Directly under the fascia are found the tibial and common fibular (peroneal) divisions of the sciatic nerve, together with the posterior cutaneous nerve of the thigh. Within the fossa, the tibial nerve emerges from under cover of semimembranosus to pass obliquely across the back of the popliteal vessels to lie medial to them on popliteus, deep to gastrocnemius and plantaris. It gives muscular branches to gastrocnemius, plantaris and popliteus, genicular branches to the knee joint, and the cutaneous sural nerve which eventually pierces the deep fascia in the middle third of the posterior surface of the leg. The common fibular (peroneal) nerve passes under cover of biceps femoris and its tendon, coursing downwards and along the lateral margin of the fossa to reach the back of the head of the fibula. It passes subfascially into the leg by curving around the neck of the fibula where it can be readily palpated. Within the popliteal fossa the common fibular (peroneal) nerve gives genicular branches to the knee joint, and two cutaneous branches – the lateral cutaneous nerve of the leg, and the fibular (peroneal) communicating branch which joins the sural nerve.

Just above the popliteal fossa, the femoral vessels lie opposite the posterior border of the femur and the lower end of the adductor canal, against the tendon of adductor magnus. Continuing their backward inclination, they pass through the adductor hiatus into the popliteal fossa, and so become the popliteal vessels. The popliteal artery is the deepest of the major structures within the popliteal fossa, lying next to the bone, where it is well protected from external trauma but is vulnerable to supracondylar fractures of the femur. It ends by dividing into anterior and posterior tibial arteries at the distal border of popliteus, level with the distal part of the tibial tuberosity. With the knee flexed the popliteal pulse can be felt. Within the fossa the artery gives off the five genicular branches supplying the knee joint and participating in the genicular anastomosis.

The popliteal vein is formed by the anterior and posterior tibial veins at the lower border of popliteus. Ascending through the fossa it crosses from medial to lateral posterior to the artery, being bound to it by a dense fascial sheath. As well as receiving the genicular veins, the popliteal vein also receives the small (lesser) saphenous vein, which pierces the fascial roof of the fossa.

Both the popliteal artery and vein are embedded in a considerable amount of fat and areolar connective tissue, which also lodges the popliteal lymph nodes.

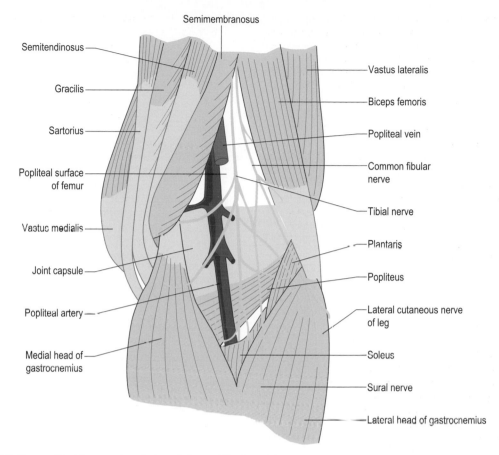

Semimembranosus

Semitendinosus

Gracilis

Sartorius

Popliteal surface
of femur

Vastus medialis

Joint capsule

Popliteal artery

Medial head of
gastrocnemius

Vastus lateralis

Biceps femoris

Popliteal vein

Common fibular
nerve

Tibial nerve

Plantaris

Popliteus

Lateral cutaneous nerve
of leg

Soleus

Sural nerve

Lateral head of gastrocnemius

Figure 3.96 The popliteal fossa and posterior relations of the knee joint.

Stability

The stability of the knee joint is primarily maintained by the collateral and cruciate ligaments reinforced by the musculotendinous ties that cross the joint. In fact it is the interaction between the two sets of ligaments which provides the all-round stability of the joint.

Since the anatomical axis of the femur runs inferiorly and medially, the force applied to the tibia (F) can be resolved into vertical (V) and horizontal (T) components (Fig. 3.97A). The horizontal component tends to tilt the joint widening the medial joint interspace; this potential disruption is normally resisted by the medial collateral ligament. The greater the degree of the genu valgus, the greater the demand made upon the medial ligament.

During walking and running, the knee is continually subjected to side-to-side stresses. There is no danger of tearing the collateral ligaments during such activities unless accompanied by the application of a violent transverse force. However, when the knee is severely sprained, abnormal side-to-side movements can be demonstrated about an anteroposterior axis. Such movements are best seen with

the knee fully extended or hyperextended. In such a position, if the leg can be displaced laterally this indicates disruption of the medial collateral ligament (Fig. 3.97D). Similarly, medial displacement of the leg indicates disruption of the lateral collateral ligament (Fig. 3.97E).

As well as the collateral and cruciate ligaments, musculotendinous ties are also important in securing the stability of the joint. Laterally the lateral collateral ligament is aided by the iliotibial tract, which can be tightened by tensor fascia lata and/or gluteus maximus. Medially the medial collateral ligament is assisted by sartorius, semitendinosus and gracilis. In addition the collateral ligaments are aided by the retinacular fibres of quadriceps femoris. Those fibres which do not cross the midline prevent opening out of the joint space on the same side, while those which cross the midline prevent it opening out on the opposite side. Consequently, each vastus muscle influences the stability of the joint on both its medial and lateral aspects. It is therefore not surprising that atrophy of the quadriceps muscles may weaken the stability of the knee so that it tends to 'give way'.

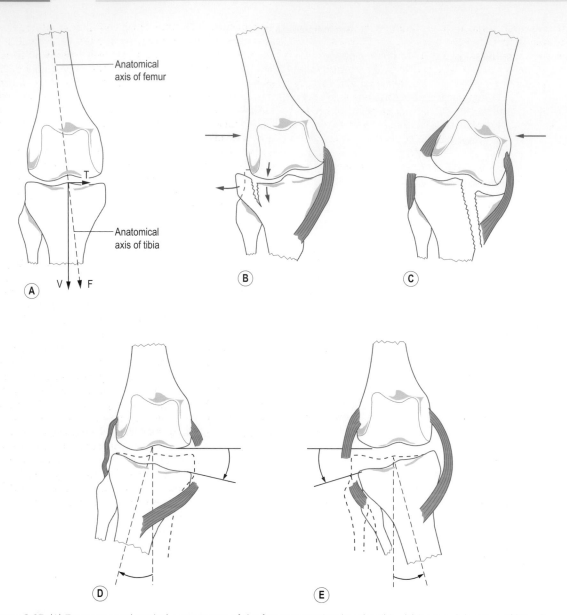

Figure 3.97 (A) Transverse and vertical components of the forces transmitted to the tibia, (B) impact-dislocation of lateral tibial condyle, (C) fracture-dislocation of medial tibial condyle, (D) rupture of the medial collateral ligaments, (E) rupture of the lateral collateral ligament.

The structures involved in maintaining the anteroposterior stability of the knee are position dependent. If the knee is flexed, even slightly, the projection of the line of gravity falls behind the knee joint axis and so acts to cause further flexion. This is resisted by quadriceps femoris. If, on the other hand, the knee is hyperextended, the centre of gravity projection falls in front of the joint axis; there is then a natural tendency for the hyperextension to increase. In this latter position of the knee the cruciate ligaments have both become taut, which, together with the posteriorly placed

arcuate and oblique popliteal ligaments as well as the collateral ligaments, act to limit further hyperextension. Although the ligaments are sufficiently strong to limit hyperextension, their action is reinforced by the flexor muscles crossing the joint. Without this active component the ligaments would become lax through stretching, eventually leading to decreased stability at the joint.

In patients with paralysis of quadriceps femoris, an exaggerated hyperextension of the knee (genu recurvatum) is often seen. This allows the patient to stand erect and even to walk. However, it has to be remembered that when the knee is hyperextended the mechanical axis of the femur runs obliquely inferiorly and posteriorly. The resultant force vector can be resolved into vertical and horizontal components; the posteriorly directed horizontal component acts to accentuate hyperextension. Thus if the genu recurvatum is too severe, the cruciate, collateral and posterior ligaments of the joint become stretched and a vicious circle is set up which results in further accentuation of hyperextension.

Rotational stability at the knee is also provided by the collateral and cruciate ligaments. Axial rotation can only occur with the knee flexed; in the extended knee it is prevented by tension in both the cruciate and collateral ligaments. In the flexed knee lateral rotation of the tibia with respect to the femur causes the cruciate ligaments to become more vertical and separate, so that they become relaxed. However, at the same time the collateral ligaments become more oblique with a corresponding increase in their tension. Lateral rotation of the tibia with respect to the femur is therefore resisted by the collateral ligaments, especially the lateral (Fig. 3.98A).

When the tibia medially rotates with respect to the femur the collateral ligaments become more vertical and relaxed, particularly the lateral collateral ligament, while the cruciate ligaments become coiled around each other and are effectively shortened and tightened. The most effective element in resisting tibial rotation is the PCL, as it wraps around the anterior, and the medial collateral ligament (Fig. 3.98B).

It is clear then that the cruciate and collateral ligaments work in concert to provide three-dimensional stability of the knee joint. However, it must also be remembered that the menisci, by their movements which accompany those of the femur and tibia, interact with the ligaments in providing stability at the joint. Disruption of the ACL leads to decreased stability, and subsequent increased

movement, in all actions in which it is put under tension, as well as in those movements limited by the PCL as it wraps around the ACL. Experiments have shown that isolated section of the ACL significantly increases anterior displacement of the tibia (Fig. 3.99), with a loss of the coupled tibial rotation accompanying this displacement. By itself, medial meniscectomy has no influence on anterior displacement at the joint, but when associated with ACL section, the anterior displacement observed is significantly greater than that in isolated ACL section, thus highlighting the fact that while it is possible to assign specific roles to individual stabilizing structures in the knee joint, there is considerable interaction between them which may not always be fully appreciated.

Movements

With movements of the knee, two separate articulations have to be considered – that between the femur and tibia, which is the most important as it controls the lengthening and shortening of the lower limb, and that between the patella and the femur, with the patella acting as a pulley for the quadriceps tendon, changing its line of action.

The main movements that occur at the knee joint are flexion and extension, together with a limited amount of active rotation when the joint is flexed. Consequently, it is termed a modified hinge joint. It differs from a typical hinge joint, for example the interphalangeal joints, not only because there is rotation but also because the axis about which movement occurs, together with the contact area between the articular surfaces, moves forwards during extension and backwards during flexion. The change in the position of the axis of rotation is due to the constantly changing radius of curvature of the femoral condyles. In addition, there is an accompanying passive rotation of the joint towards the end of extension.

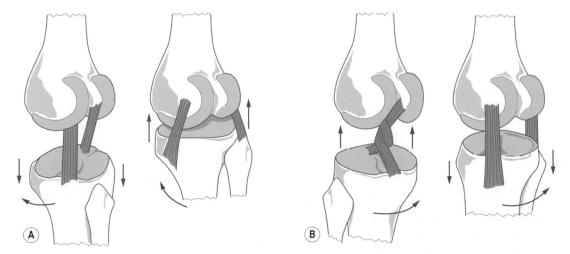

Figure 3.98 Action of the cruciate and collateral ligaments in limiting lateral (A) and medial (B) axial rotation of the knee.

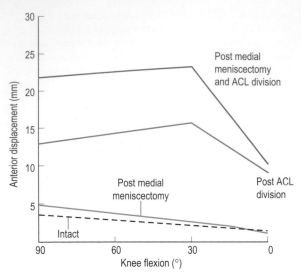

Figure 3.99 The influence of anterior cruciate ligament (ACL) division and medial meniscectomy on the anterior displacement of the tibia. *(Adapted from Levy IM, Torzilli PA and Warren RF (1982) The effect of medial meniscectomy on anterior–posterior motion of the knee. Journal of Bone and Joint Surgery, 64A, 883–887.)*

Flexion and extension

Flexion is movement of the leg so that the posterior aspect of the calf moves towards the posterior aspect of the thigh, extension being the opposite movement. Some movement, usually passive, of the tibia beyond the position of alignment of the long axis of the leg and thigh may be possible. This movement is usually referred to as hyperextension.

The range of flexion achieved is dependent on the position of the hip, and also whether the movement is performed actively or passively. Maximum flexion is achieved when the movement is carried out passively.

Active flexion of the knee can reach 140° if the hip is already flexed (Fig. 3.100A), but only 120° if the hip is extended (Fig. 3.100A) because the hamstrings lose some of their efficiency with hip extension. If the hamstrings contract abruptly and powerfully, however, some measure of passive flexion follows the active movement so that the range may be increased. Passive movement of the knee joint allows the heel to touch the buttock, giving a range of flexion of 160° (Fig. 3.100A). Normally knee flexion is only limited by contact of the thigh and calf muscles. If, however, the movement is arrested before this occurs it may be due to contraction of the quadriceps muscles or by shortening of the capsular ligaments. In such circumstances, the effectiveness of intervention procedures can be assessed in terms of the distance between the heel and buttock.

Because the hamstrings are both extensors of the hip and flexors of the knee, their action on the knee depends on the position of the hip. As the hip is flexed, the distance between the hamstring attachments progressively increases as they wrap around the ischial tuberosity (Fig. 3.100B). Thus the more the hip is flexed, the greater the degree of relative shortening of the hamstrings and the more stretched they become. With hip flexion greater than 90° it becomes increasingly difficult to keep the knee fully extended because of the stretching of the hamstrings (Fig. 3.100B(iii)). However, when the hamstrings are stretched by hip flexion their efficiency as knee flexors increases. Similarly, extension of the knee will also increase the efficiency of the hamstrings as extensors of the hip.

From the foregoing, it might seem that the efficiency of knee flexion is subject to modification according to the position of the hip. This is not necessarily the case as both the short head of biceps and popliteus, crossing only the knee joint, have the same efficiency irrespective of hip position.

Movement of the femoral condyles is achieved by a combination of rolling and gliding, with the ratio of rolling to gliding changing during flexion and extension (Fig. 3.101). Beginning with full extension, the femoral condyles begin to roll without gliding, and then the gliding movement becomes progressively more important so that towards the end of flexion the condyles glide without rolling. There are, however, differences between the condyles in the extent of their rolling action. For the medial condyle, pure rolling occurs only during the initial 10–15° of flexion, while for the lateral condyle it continues until 20° flexion. This initial 15–20° of pure rolling corresponds to the normal range of movement at the knee during the support phase of gait, when stability is the prime requirement.

The change from rolling to gliding is very significant for the functioning of the knee joint in which both stability and mobility are required. Beyond 20° flexion the knee becomes looser as the radius of the femoral condyles becomes smaller. The tibia moves closer to the axis of movement in the femur, consequently the ligaments passing between the two bones become relaxed. This loosening up prepares the joint for a wider range of axial rotation.

Not only do the relative positions of the femur and tibia change during flexion and extension of the knee joint, with the contact area moving anteriorly and also increasing with extension thereby promoting greater stability of the joint, but the two menisci interposed between them also move (Fig. 3.102). If this were not so then the range of movements possible at the joint would be severely impaired. However, as well as following the movement of the femur with respect to the tibia, the menisci also undergo considerable distortion during their movement. In going from extension to flexion both menisci move posteriorly, with the lateral receding twice as far as the medial, approximately 12 and 6 mm respectively. During this movement the lateral meniscus undergoes greater distortion than the medial, primarily because its anterior and posterior horns are situated closer together.

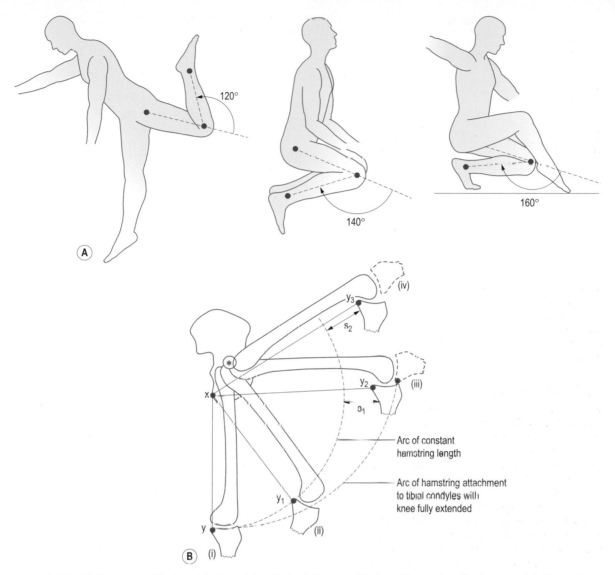

Figure 3.100 (A) The range of flexion of the knee joint, (B) the influence of limb position on the effectiveness of the hamstrings as flexors of the knee. s, degree of lengthening; x, y, hamstring attachments to ischium and tibia.

Only one passive element is involved in displacement of the menisci; the femoral condyles push them anteriorly during extension. In contrast, there are many active elements involved in meniscal movements. During extension both menisci are pulled forwards by the meniscopatellar fibres, which are stretched and these pull the transverse ligament forwards. In addition, the posterior horn of the lateral meniscus is pulled anteriorly by the meniscofemoral ligament as the PCL becomes taut.

During flexion, the lateral meniscus is pulled posteriorly by the attachment of popliteus to it. At the same time the medial meniscus is drawn posteriorly by the semimembranosus expansion attached to its posterior edge; the anterior horn is pulled posterosuperiorly by fibres of the ACL attached to it.

While these various movements of the femur, tibia and menisci are occurring, it is clear that the patella is also changing its relationship to the femur. This must be so given the fact that the meniscopatellar fibres attaching to the margins of the patella become stretched and pull the menisci forwards during extension. Movement of the patella during flexion occurs along the groove of the patellar surface of the femur as far as the intercondylar notch. This almost vertical displacement, over a distance of twice its

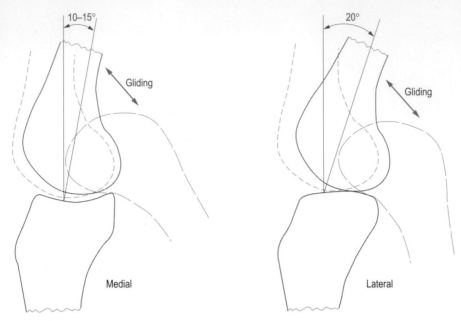

Figure 3.101 The extent of the rolling and gliding action of the femoral and tibial condyles during flexion and extension. – – – – fully extended; —— —— fully flexed; ———— limit of pure rolling action.

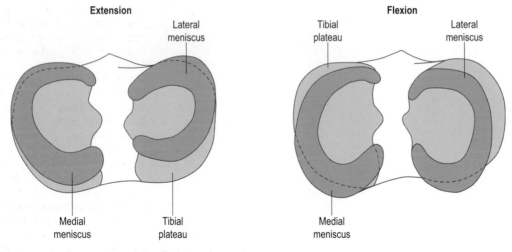

Figure 3.102 Meniscal movements during flexion and extension.

length, is accompanied by a turning movement of the patella about a transverse axis (Fig. 3.103A). Considering the tibia as the moving bone, and going from extension to flexion, the deep surface of the patella initially faces posteriorly then superiorly. However, if the tibia is considered fixed so that it is the femur which moves, the patella tilts on itself some 35° so that its deep surface which initially faced posteriorly faces posteroinferiorly in full flexion (Fig. 3.103B).

It is the capsular recesses in relation to the patella which facilitate these movements, that is the suprapatellar bursa superiorly and the parapatellar recesses on either side. Inflammatory adhesions developed in these recesses obliterating their cavities, resulting in the patella being held firmly against the femur so that it cannot move down to the intercondylar notch. This is one of the causes of post-traumatic or post-infective 'stiff knee'.

Normally there is no transverse movement of the patella, being held against the femur by the quadriceps tendon, increasingly more firmly as flexion increases. In full extension the appositional force is diminished; there may even be

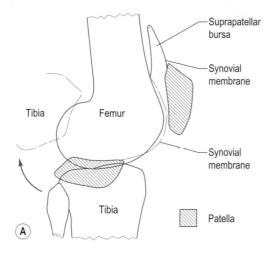

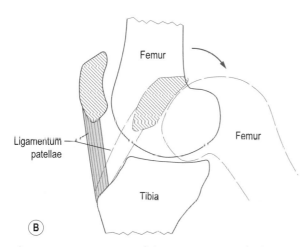

Figure 3.103 Movements of the patella: (A) with the femur fixed, (B) with the tibia fixed.

separation of the femur and patella in hyperextension. If this occurs there is a tendency for the patella to move laterally due to the pull of quadriceps femoris. Lateral displacement is prevented by the lateral lip of the patellar surface of the femur, which projects further forwards than the medial lip, and the horizontally placed fibres of vastus medialis attaching to the medial border of the patella. If, however, the lateral lip is underdeveloped or there is weakness of vastus medialis, the patella is no longer held firmly in place and may dislocate laterally in extension. A severely underdeveloped lateral lip may lead to lateral patellar dislocation during flexion movements. This is the mechanism underlying recurrent dislocation of the patella.

Lateral displacement or misalignment of the patella can cause knee pain as the articular surfaces are subjected to increased compressive forces over a smaller area. Lateral movement of the patella may be due to abnormal tracking of the patella associated with a tight patellofemoral retinaculum, which is often combined with a tight iliotibial tract. The resulting pain has led to a number of surgical procedures including lateral release techniques and transposition of the tibial tuberosity.

However, conservative treatment of the patellar tracking mechanism is being increasingly used to reduce the pain associated with patellar misalignment. The treatment involves taping techniques combined with specific isometric and eccentric reeducation of the oblique fibres of vastus medialis. By selectively training this part of quadriceps, lateral tracking of the patella may be minimized.

It has been suggested that the above technique also corrects soft tissue abnormalities around the knee, and thus decreases the effect of abnormal foot mechanics.

Axial rotation

Rotation of the leg about its long axis is only possible with the knee flexed. Medial and lateral rotation brings the toes to face medially or laterally respectively (Fig. 3.104). The ranges of rotation are influenced slightly by the degree of knee flexion, and hence the efficiency of the appropriate part of the hamstrings. With the knee flexed to 90° active medial and lateral rotation are 30° and 40° respectively; this may be increased to 35° and 50° if the movement is performed passively.

In lateral rotation of the tibia on the femur, the lateral femoral condyle moves forwards over the lateral tibial condyle while the medial femoral condyle moves backwards over the medial tibial condyle. In medial rotation the reverse occurs. The extent of this anteroposterior movement differs for the medial and lateral condyles. The medial condyle moves hardly at all, while the lateral moves an appreciable amount and in so doing comes to lie slightly higher than the medial condyle. Although small, this height difference is nevertheless real. The shape difference between the two tibial condyles is reflected in the configuration of the intercondylar tubercles. The medial tubercle acts as a shoulder against which the medial femoral condyle rests, while the lateral femoral condyle moves easily past its respective tubercle. Consequently, the axis of rotation passes through the medial tubercle and not between the two tubercles.

As might be expected, the two menisci follow the displacements of the femoral condyles during axial rotation (Fig. 3.105). During lateral femoral rotation, the medial meniscus is pulled forwards over the tibia while the lateral meniscus is drawn posteriorly; the reverse occurs in medial femoral rotation. Again, during their movements both menisci undergo distortion. The meniscal displacements during axial rotation are mainly passive, being due to movement of the femoral condyles. However, the

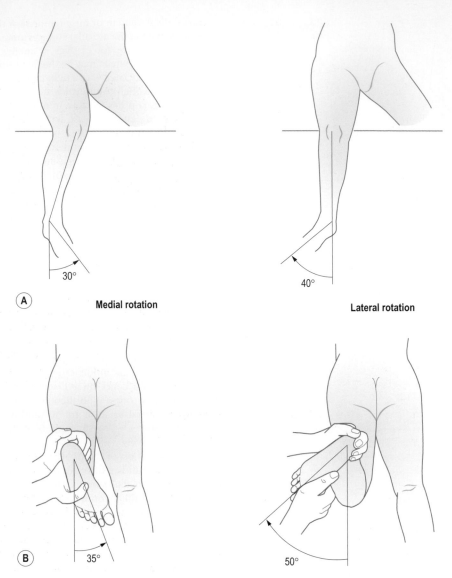

Figure 3.104 The ranges of axial rotation of the knee joint determined: (A) actively and (B) passively.

meniscopatellar fibres become taut as a result of movement of the patella with respect to the tibia; this tension draws one of the menisci anteriorly. Failure of the menisci to follow movements of the femoral condyles can lead to transverse tears, longitudinal splitting, detachment of the anterior horn, or complete detachment of the meniscus from the capsule. As soon as a meniscus is damaged, the injured part fails to follow the normal movements and may become wedged between the femoral and tibial condyles. Consequently, the knee 'locks' in a position of flexion, with the locking being more marked the more posterior the damage; full extension becomes impossible.

During axial rotation the patella moves in a frontal plane with respect to the tibia. In medial rotation of the tibia with respect to the femur, the patella is dragged laterally, so that the ligamentum patellae runs obliquely inferiorly and medially. The opposite occurs in lateral tibial rotation (Fig. 3.106).

There is also an automatic, involuntary rotation of the knee associated with the terminal part of extension and the beginning of flexion. This can be described with respect to either the femur or the tibia, depending on which of the two limb segments, thigh or leg, is fixed. With the tibia fixed, the femur undergoes a medial rotation in the

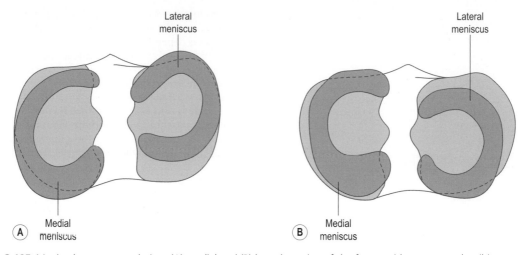

Figure 3.105 Meniscal movements during: (A) medial and (B) lateral rotation of the femur with respect to the tibia.

terminal stages of extension. Consequently, during the initial part of flexion the femur is laterally rotated by popliteus. When the knee is in the fully extended rotated position it is said to be close-packed, denoting the maximum contact between the femoral and tibial articular surfaces, with all the ligaments under tension. The joint is then at its most stable. The rotation that occurs is due to three independent mechanisms:

1. the unequal length of the articular profiles of the femoral condyles, with the lateral condyle rolling over a greater distance than the medial
2. the shape of the tibial condyles whereby the medial femoral condyle is more or less contained within the concave medial tibial condyle, while the lateral femoral condyle glides more freely over the convex lateral tibial condyle
3. the direction of the collateral ligaments, which means that the medial collateral ligament becomes stretched less rapidly than the lateral.

In addition to the above mechanisms, there are two force couples acting to produce rotation. At the beginning of flexion, lateral femoral rotation with respect to the tibia is brought about by the action of flexors gracilis, sartorius and semitendinosus, and by popliteus. Towards the end of extension, it is the tension developed in the ACL which produces medial femoral rotation, again with respect to the tibia.

Accessory movements

The amount of accessory movement available at the knee joint is controlled to a great extent by its position, and hence the degree of tension in the ligaments. In the close-packed position no accessory movement is possible. However, if the knee is flexed to 25° a number of accessory movements can be demonstrated. The tibia can be moved anteriorly and posteriorly on the femur by applying forces in the appropriate directions. Also in this position rotation of the tibia on the femur can be taken further than the available range by applying a firm rotatory force to the tibia. In this position the tibia can also be rocked medially and laterally. Lastly, it is possible to distract the tibia away from the femur if a longitudinal pulling force is applied.

Biomechanics

Joint forces

When considering knee joint forces, those acting on both compartments of the joint, that is the femorotibial and femoropatellar, must be taken into account. During level walking the force across the femorotibial joint can reach five times body weight, although for most of the gait cycle it is usually between two and four times body weight. The force across the femoropatellar joint in similar circumstances is approximately half body weight. The precise level of loading depends on internal factors, such as alignment between the femur and tibia and any residual deformity, and external factors, such as the speed of walking and environmental conditions. Ascending or descending ramps and stairs, for example, appear to have little influence on femorotibial forces. In contrast, femoropatellar forces increase significantly to between one-and-a-half and two times body weight when ascending, and to between two-and-a-half and three times body weight when descending ramps and stairs. However, the greatest forces across the femoropatellar joint, in everyday as opposed to vigorous activities, are found when getting out of a chair without using the arms. In such an activity, femoropatellar forces reach three-and-one-half times body weight, while the corresponding femorotibial forces are four times body weight. Activities such as running and jumping will significantly increase the magnitude of all of these forces.

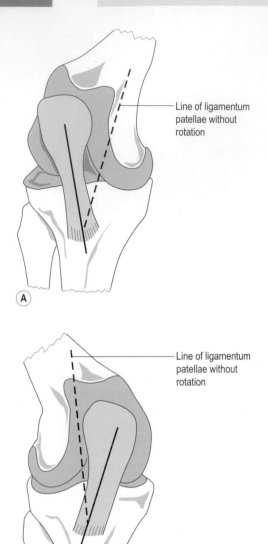

Line of ligamentum
patellae without
rotation

Ⓐ

Line of ligamentum
patellae without
rotation

Ⓑ

Figure 3.106 Patella movements during axial rotation of the tibia with respect to the femur: (A) medially and (B) laterally.

In jumping, for example, femorotibial joint forces may reach 24 times body weight, and femoropatellar joint forces 20 times body weight. The articular cartilage of the two joints is therefore subjected to extremely high stresses. It is not surprising, therefore that gymnasts present with articular cartilage degeneration of the knee, accompanied by significant bone buttressing of the femoral and tibial condyles. The pattern and magnitude of the forces in level walking is shown in Fig. 3.107A. Joint force increases

rapidly following heel-strike, reaching its peak value as the foot makes full contact with the ground. The magnitude of this peak is dependent on two factors, one being the acceptance of full body weight onto the supporting limb and the other being the result of muscle activity across the joint to prevent collapse of the knee. The second joint force peak is associated with the propulsive part of the stance phase. Its magnitude is primarily determined by the level of muscle activity across the joint, quadriceps femoris acting to extend the knee and gastrocnemius to plantarflex the foot, thus propelling the body forwards. When walking up and down stairs, the increased activity of quadriceps femoris is responsible for the three- to sixfold increase in femoropatellar joint forces.

In contrast to the high vertical forces across the joint, mediolateral joint forces are low, being approximately one-quarter body weight.

Because of the angulation between the femur and tibia, the two parts of the femorotibial joint, that is medial and lateral, are not equally loaded. Indeed, the degree of angulation, varus or valgus, has a profound influence on the loading pattern. During the gait of a normal subject, the centre of force is located just medial to the midline of the joint for the majority of the stance phase (Fig. 3.107B). This does not necessarily mean that the contact stresses are greater on the medial side. Indeed, because of the shape of the femoral and tibial condyles it is likely that the contact stresses on the lateral side of the joint are greater. In varus deformities the medial contact forces will increase dramatically. In 2.5° varus, for example, these medial forces increase by 70% and in 5° varus by 95%. Consequently, the centre of force location throughout the stance phase lies medial to the midline of the joint (Fig. 3.108, line *a*). Similarly, in valgus deformities there is an increase in the contact forces on the lateral side of the joint. In 2.5° valgus the lateral forces increase by 50% and in 5° valgus by 75%. In such conditions, the centre of force location is now on the lateral side of the joint midline (Fig. 3.108, line *b*). However, when the valgus deformity is of the order of 15° the centre of force location fluctuates about the midline of the joint (Fig. 3.108, line *c*). This is achieved by the patient compensating for the valgus load in order to reduce the lateral compartment loading. One compensatory mechanism the patient could adopt would be to use the hip abductors during the stance phase to produce a large lateral horizontal force at the foot. This would necessitate an abduction movement on the tibia, thus loading the knee medially. Alternatively, and perhaps more commonly, the hip may be abducted during the swing phase, placing the foot medially at heel-strike. This would also require a similarly large lateral force at the foot during stance in order to maintain equilibrium, resulting in increased loading of the medial compartment of the knee.

It would appear then that individuals can modify force transmission by adopting compensatory mechanisms that

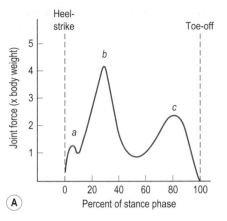

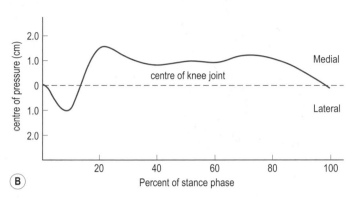

Figure 3.107 (A) Loading profiles in normal knees; peaks *a*, *b* and *c* correspond to hamstring, quadriceps and gastrocnemius contraction, (B) the centre of joint pressure during walking for normal individuals.

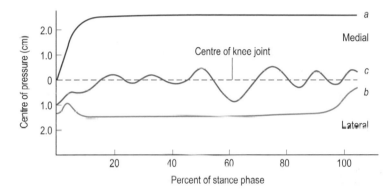

Figure 3.108 The centre of joint pressure during level walking. *a*, with a varus deformity; *b*, *c*, with a valgus deformity.

result in unloading a particular compartment of the knee. Studies suggest that it is easier to compensate for a valgus than a varus deformity. Whether femoropatellar joint forces increase with a valgus or varus deformity is not known; however, it is likely that the pattern of loading at this joint changes.

Trabecular arrangement

As well as being subject to high contact forces during activity, the knee is also subjected to considerable side-to-side stresses. This is reflected in the internal architecture of the articulating bones, where the trabeculae are arranged along the lines of mechanical stress, either tension or compression.

The distal end of the femur shows two main sets of trabeculae arising from the cortex of the shaft, and a transverse set uniting the two condyles (Fig. 3.109). From the cortical region of the shaft medially and laterally, trabeculae fan out into both condyles. The system running to the ipsilateral condyle resists compression forces, while that which runs

to the contralateral condyle resists traction forces. The proximal end of the tibia has similarly arranged trabecular systems (Fig. 3.109). From the cortex of the shaft, trabeculae radiate into both tibial condyles. Ipsilaterally they again resist compression forces while contralaterally they resist traction forces. A transverse system unites the two tibial condyles. As in the femur, the trabecular systems arising from the medial and lateral sides of the tibial shaft cross each other approximately at right angles. An important set of trabeculae run into the tibial tuberosity from the anterior aspect of the shaft. This system has developed in response to the tensile stresses to which the tuberosity is subject by the pull of the ligamentum patellae.

Contact areas

Femorotibial joint Because the radius of curvature of the femoral condyles increases anteriorly, the greatest area of contact between the femur and tibia is in the fully extended position. However, not only is the contact area position-dependent, it is also load-dependent. At low

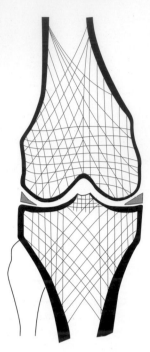

loading levels, up to 500 N, the joint, including the menisci, is not congruous, but with loads greater than 1500 N it becomes markedly congruous. The menisci play a very important role in load transmission across the joint, as can be seen from Fig. 3.110A, which shows that the total contact area is doubled with the menisci present. With the knee fully extended and with an applied load of 1000 N the contact area is approximately 11.5×10^2 mm^2 with the menisci present and only 5.2×10^2 mm^2 with the menisci removed. This indicates that with the menisci removed the average stress across the joint doubles compared with that of the intact knee. Peak pressure at 1000 N loading on the intact knee is approximately 3 MPa, while with the menisci removed it is closer to 6 MPa. The highest pressure areas are on the lateral meniscus as well as the uncovered part of the articular cartilage of the lateral compartment.

The menisci actually account for about 70% of the total contact area at 1000 N loading. At lower loads they account for a greater proportion of the total contact area (Fig. 3.110B). It can also be seen from Fig. 3.110B that the two menisci do not contribute the same percentage contact areas in the medial and lateral compartments. This is because of the different shapes of the medial and lateral menisci, as well as the relative positions of their sites of attachment to the intercondylar eminence. Rather surprisingly, reported contact areas

Figure 3.109 The trabecular arrangement in the distal femur and proximal tibia.

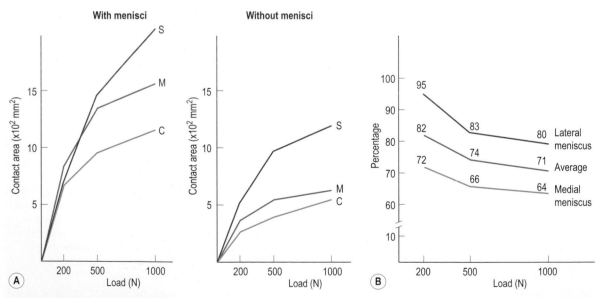

Figure 3.110 (A) Contact area of the femorotibial joint with and without menisci. (B) Percent contact area of the menisci compared with total contact area. C, control; M, mild osteoarthritis; S, severe osteoarthritis. *(Adapted from Fukubayashi T and Kurosawa H (1980) The contact area and pressure distribution pattern of the knee. A study of normal and osteoarthritic knee joints.* Acta Orthopaedica Scandinavian, *51, 871–880.)*

in osteoarthritic knees were significantly larger than in normal knees, both with and without the menisci present (Fig. 3.110A), suggesting that the menisci play a less significant role in osteoarthritic knees.

From the above, it is clear that the menisci give elastic stability to the joint, having both load-bearing and load-spreading functions. They provide surface compliance and serve to transmit stresses across a wider area to the periphery of the joint. The menisci, therefore, help to avoid stress concentrations both in the articular cartilage and in the subchondral bone.

Femoropatellar joint This has two complex mechanisms for ameliorating the forces transmitted across it. With increasing flexion, the extensor lever arm is lengthened by the fact that the axis of rotation of the knee joint moves posteriorly, particularly in the range 30–70° flexion. In addition, there is an increasing area of contact between the patella and the femur. Within the 30–70° range of flexion the patella is solely responsible for transmitting the quadriceps force to the femur. The deep surface of the patella is covered by the thickest cartilage in the body, and therefore not surprisingly is the most frequent site of cartilage degeneration. Stresses applied to the cartilage of between 2 and 4 MN/m^2 have been calculated for level walking and ascending/descending steps.

Between 30° and 90° of flexion, the contact area almost triples (Fig. 3.111A). This is achieved by an increasingly larger proportion of the patella coming into contact with the femur (Fig. 3.111B). However, by 135° flexion, this contact area has decreased (Fig. 3.111C), being limited to the upper part of the lateral and the odd facets.

The Q-angle and valgus vector explain the predominance of pathological lesions on the lateral side of the joint as well as the associated dislocations, subluxations, lateral pressure syndromes and patellofemoral arthrosis. The Q-angle is the angle between the anatomical axis of the femur and that of the tibia (Fig. 3.81), being determined on the lateral side of the leg. A number of anatomical factors affect tracking of the patella as it moves in the groove on the femur. Physiotherapy techniques which involve specific exercises for parts of vastus medialis together with carefully applied skin taping are used in certain cases to improve tracking and help reduce anterior knee pain.

Mechanical role of the menisci

The average pressure in the articular cartilage of the femorotibial joints is much less with load-bearing menisci than without. With loads distributed over a larger contact area, not only is the average pressure lower, but the pressure gradients are also lower. Consequently, cartilage deformation is small. Following meniscectomy the pressure distribution

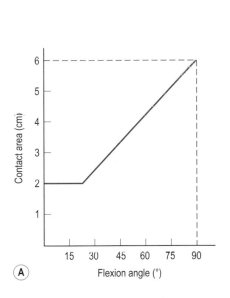

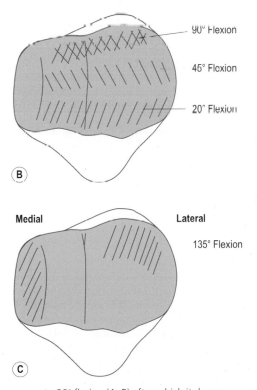

Figure 3.111 The area of contact between the patella and femur increases up to 90° flexion (A, B) after which it decreases again (C).

becomes non-uniform with a high average pressure, higher stress gradients and large cartilage deformations. As articular cartilage contains 65–85% water, the less uniform the pressure distribution the more rapid will be the loss of fluid from high pressure zones, and therefore the greater the rate of increase in cartilage deformation. With several thousand stress reversals within the course of a single day cartilage degeneration often ensues.

Therefore, failure or removal of a meniscus usually has detrimental effects on the adjacent articular cartilage. The area of cartilage most regularly affected after meniscectomy is that on the tibial condyle (originally enclosed by the meniscus) because it is unavoidably subjected to increased pressure every time the joint is loaded. There may also be a modification in the shapes of the femoral and tibial condyles, a response of the bone to the change in pressure distribution. The articular cartilage on the tibial condyle within the embrace of the meniscus constitutes an area of non-habitual contact, and is said to be functionally different from the cartilage in other areas of the joint. Age-dependent surface degeneration is commonly encountered at these sites. Consequently, it is generally accepted that the menisci should be regarded as an important and integral part of the articular surfaces of the tibia, with their function being to distribute load over a large area of the articular surfaces at pressures which the cartilage can tolerate.

The importance of the menisci in lubricating, and hence providing nutrients for, the articular surfaces must not be underestimated. A particular pattern of articular degeneration can be seen in many knees, being in the form of a triangle of erosion or osteophyte formation on the medial femoral condyle, sometimes with associated strips on the lateral condyle (Fig. 3.112). The base of the triangle on the medial condyle is where rotation occurs with full extension. The regions of degeneration are those which are normally in contact with the anterior horns of the menisci in full extension, and are only found in association with a flexion contracture. In other words, the lesion arises in those areas which have no opposing articular surface because of the inability to extend the joint fully.

As if the above functions were not sufficient, the menisci also play an important role in absorbing shock transmitted across the knee joint. Meniscectomy reduces the knee's shock-absorbing capacity by some 20%.

Patella pathology

Stress analysis of the normal patella suggests that areas of high tensile stress on its medial aspect may exist during knee flexion. The high tensile stress is a function of Q-angle, reduction of which dramatically reduces the values of combined stress within the cartilage, thereby reducing the possibility of fatigue failure of the cartilage collagen.

Vertical tensile stresses can occur over the medial side of the lateral facet. If this tensile stress is combined with high contact stresses applied perpendicular to the surface, a

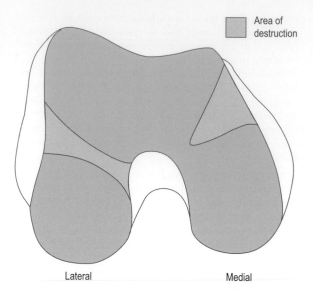

Figure 3.112 Commonly observed pattern of articular cartilage destruction on the femoral condyles.

situation exists of very high shear stresses which may fatigue the deeper layers of articular cartilage (Fig. 3.113). If this combination of stresses persists it may lead to a closed type of chondromalacic lesion centred on the medial region and commonly extending onto the lateral surface.

Chondrosclerotic lesions, which are rare, are caused by an extreme compression phenomenon, which is the result of high bending stresses combined with high contact stresses over the lateral facet.

To some extent the high stresses within the patella articular cartilage may be relieved by a forward or forward-and-medial displacement of the tibial tuberosity. More recently this problem has been approached by using osteochondral grafts to replace the damaged area, followed by rehabilitation aimed at correcting the patellar glide.

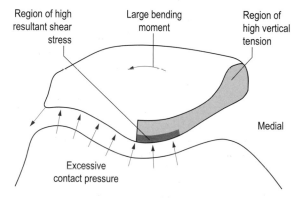

Figure 3.113 The pattern of patellar stresses which may lead to the formation of a closed chondromalacic lesion of the patella.

Cruciate ligament replacement

Injuries to the knee joint are rarely isolated; other damage is frequently obscured at the time of injury due to swelling and only becomes apparent later. Secondary injury may develop as a result of abnormal forces acting across the joint. The cruciate ligaments, because of their blood supply, are capable of a certain amount of repair; however, severe ruptures have to be dealt with surgically. The primary aim of cruciate ligament repair is restoration of static and dynamic stability in the flexed knee since this is the usual position of function. It is vital that any reparative measures restore normal biomechanical functioning of the joint. If this fails to happen then the altered pattern of forces transmitted across the joint may lead to secondary changes both at the knee and at other joints in a manner similar to that which would occur in the absence of surgical intervention.

The complexity of the cruciate ligaments, with their spiralling fibres and intracapsular position, makes reconstruction difficult. Surgical repairs with sutures have in general not been very successful. Replacements of the ligaments are now becoming more common and successful as new surgical techniques are being developed. There are two broad categories of cruciate ligament replacement – single and double strand autologous tendon grafts.

Synthetic prostheses are still used in some centres but have two major disadvantages – they cannot withstand the forces generated in the joint and often break, and they can induce foreign body changes, such as fibrosis, within the joint. The latter can lead to compromised joint function.

One way of avoiding rejection of the implant is to use the patient's own tendons as the graft. By preserving the graft's original blood supply, the success of the procedure is enhanced, since the tendon is capable of repair and remodelling. Nevertheless, the strength of tendon grafts is still less than that of the original cruciate ligaments, so rupture may occasionally occur. Tendon grafts are essentially of two types – static or dynamic stabilizers. Static stabilizers attempt to mimic the course of the original ligament, with the ends of the graft secured to the tibia and femur, often in specially prepared tunnels.

Tendons commonly used in autologous grafts are those of semitendinosus, gracilis and part of the patellar tendon. Although initially quite effective, the grafts often stretch, with a recurrence of the original problems. To counteract this, the graft is often plaited using three or four strands to increase its strength. In addition, the fixation sites are prone to rarefication and may become loose.

In contrast to static tendon grafts, dynamic tendon grafts preserve the muscular attachment of the tendon. In such situations, the insertion of the tendon is detached and passed through the joint capsule and a tunnel in one bone and tied to the other by staples or a bone plug. The course of the tendon is such that it mimics that of the original ligament. The ACL may be replaced by the iliotibial tract, semitendinosus, gracilis, sartorius or by a middle patellar tendon graft. Dynamic tendon grafts give little passive resistance to movement, and the 'drawer sign' usually persists, although it may be slightly decreased. During use, however, subluxing forces are resisted by the active contraction of the muscle. Such grafts may also provide protection to excessive movements by initiating muscle stretch reflexes to oppose the movement.

However, advances in cruciate ligament repair have been such that dynamic tendon grafts are only used when other procedures have failed. Very often the type of 'salvage' operation gives good results, allowing moderate functional activity but precluding high impact sports. Although extremes of range are sometimes protected using knee braces, early mobilization, particularly weight-bearing, has been found to improve the effectiveness of ligament reconstruction.

Replacement of the menisci

The treatment of meniscal injuries includes complete or partial meniscectomy, repair and prosthetic replacement. Complete meniscectomy can cause instability and osteoarthritic changes at the knee and other joints. Loss of the shock-absorbing capacity of the meniscus may lead to back pain and headaches. Partial meniscectomy, with a peripheral rim being preserved, has two functional consequences. It maintains some stability and prevents high stress concentrations, and because the periphery is vascular it allows meniscal regeneration. The new meniscus is formed from the synovial membrane and resembles the original except that its attachments are usually thicker and it projects less into the centre of the joint cavity. The success of regeneration is age-dependent, being more successful in younger patients.

Meniscal repair is an important consideration for the competitive athlete, who wants to return to training as soon as possible. Repair of the meniscus by suturing is considered if the tear is in a peripheral vascular part of the meniscus. While repair of the meniscus is taking place rehabilitation progresses slowly in the early stages. However, if this procedure is successful, the long-term results are claimed to be better than partial or complete meniscectomy.

Knee joint replacement

Because of the complex nature of the knee, it presents many problems in designing a suitable prosthesis. The purpose of total knee replacement is to restore normal function and range of movement by relieving pain and disability and restoring normal limb alignment. The common indications for total knee replacement include severe and unremitting degenerative bone disease, osteoarthritis and inflammatory arthritis which conservative medical treatment has failed to manage. Extreme deformity and structural damage are also frequent reasons for replacement, particularly in rheumatoid arthritis and other inflammatory joint conditions.

Early designs were simple hinge devices, which had a high rate of loosening because they did not allow the normal rotatory movements of the knee to occur. Nevertheless, hinge replacements are still used in patients with badly deformed or unstable knees. In reality they are only suitable for sedentary patients who will not make great demands on the prosthesis. There are now available a large range of prostheses, which all attempt to mimic, to a greater or lesser extent, the movement of the knee. The particular prosthesis implanted will depend on several factors, including the adequacy of the collateral and cruciate ligaments and the amount of bone stock available. Many devices provide a series of options for implant depending on the prevailing anatomy, for example as shown in Fig. 3.114. All modern devices allow at least 110° of movement, being the range required for normal functional activities. Of interest and current debate is whether the PCL should be retained, sacrificed or substituted as part of the replacement. Anatomical designs (Fig. 3.115) aim to retain as much normal joint anatomy as possible, especially the posterior and occasionally the anterior cruciate ligaments. These designs therefore rely on the remaining soft tissues of the joint to provide stability and adequate support; however PCL substituting models are available (Fig. 3.115B). Functional designs often ignore knee joint anatomy and therefore do rely on any remaining soft tissues to operate efficiently. Non-anatomical joint surfaces have been created to improve congruence and reduce complications, such as polyethylene wear. Some of these prostheses utilise mobile joint surface components that can be manipulated to preserve the PCL, but more commonly the PCL is removed entirely and replaced with a cruciate substituting mechanism.

Irrespective of the prosthetic design or whether the posterior cruciate ligament has been retained, sacrificed or substituted, correct alignment of the implant is of crucial importance in the transmission of forces across the joint and the possibility of loosening. Loosening is a fairly common problem, particularly of the tibial component. As the knee goes into extension the anterior part of the tibial implant is pushed downwards, which tends to raise the

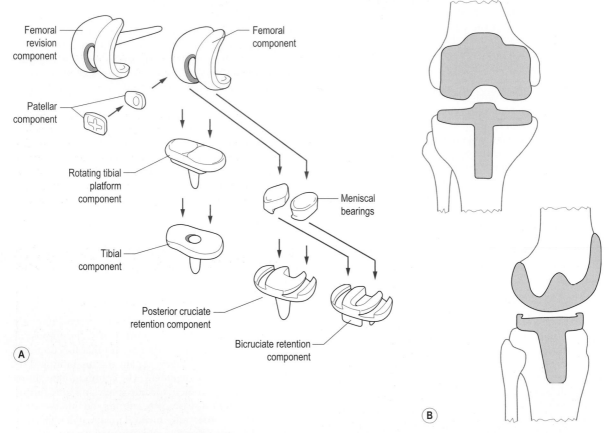

Figure 3.114 An example of one type of prosthetic knee joint replacement: (A) component parts, (B) *in situ*.

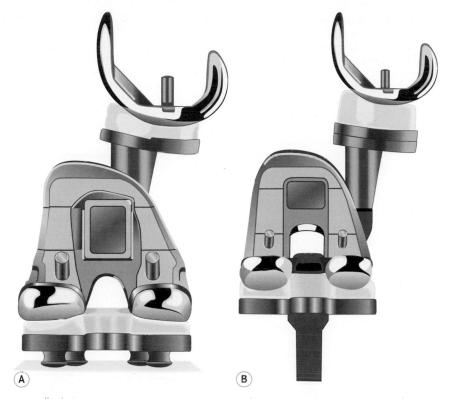

Figure 3.115 Anatomically designed prostheses. (A) this design has a femoral condyle spacing and tibial 'cut-out' for retaining the posterior cruciate ligament, (B) this design has an intercondylar cam and tibial post posterior to it which substitutes for the posterior cruciate ligament.

posterior part. The bone resection required at operation greatly reduces the amount of bony buttress at the back of the tibia, and this coupled with the cement's inability to resist tensile stresses promotes prosthesis loosening.

To give the implant some shock-absorbing features, and so reduce the transmission of high frequency, high magnitude forces across the joint, the tibial component is usually polyethylene. A few prostheses have attempted to insert artificial menisci between the two replacement joint surfaces in an attempt to (i) improve stability, (ii) increase the area of contact and so spread the load more evenly, and (iii) improve the shock-absorbing capacity of the joint.

Even if the replacement has been successful clinically, in that there has been a reduction or complete loss of pain on movement, and a more or less full range of movement has been achieved, there may still be abnormalities of gait and difficulty in negotiating stairs. Total knee arthroplasty is not advised for young active individuals unless the deterioration in joint function is so severe that it limits normal functional activities. If only one knee is involved every effort should be made through medical treatment to restore function. In some osteoarthroses only the medial, or less commonly, the lateral half of the joint is involved. In such cases it is often possible to avoid joint replacement and provide

adequate relief by realigning the axis of the knee so that most of the forces are redirected through the less affected half of the joint. This procedure, tibial osteotomy, is useful in young patients with high levels of activity and a long life expectancy. As an alternative to the tibial osteotomy it is now possible to insert a unicondylar replacement, thus preserving more of the undamaged parts of the knee joint.

Section summary	
Knee joint	
Type	Synovial bicondylar hinge joint
Articular surfaces	Condyles of femur with condyles of tibia; posterior surface of patella with patellar surface of femoral condyles
Capsule	Thick ligamentous sheath mainly composed of tendinous expansions; attached to articular margins of femoral condyle (except anterosuperiorly) and medial and lateral margins of tibial condyles

Ligaments	Oblique and arcuate popliteal ligaments
	Ligamentum patellae
	Tibial and fibular collateral ligaments
Intra-articular structures	Anterior and posterior cruciate ligaments
	Medial and lateral meniscus
	Transverse ligament
	Coronary ligaments
Stability	Provided by collateral and cruciate ligaments (most stable in full extension – 'close-packed'); muscles crossing the joint
Movements	Flexion and extension
	Medial and lateral rotation

Tibiofibular articulations

Introduction

Altough the tibia and fibula articulate together there is no active movement between them, there is, however some slight movement between the two bones which appears to be mechanically linked to movement at the ankle joint. The two bones are united by a synovial joint at their proximal ends, a fibrous joint at their distal ends and an interosseous membrane connecting their shafts.

The superior tibiofibular joint

A plane synovial joint between the circular or oval facet on the *head* of the *fibula* and a similar *facet* on the *posterolateral aspect* of the *undersurface* of the *lateral tibial condyle* (Fig. 3.116). The fibular articular facet faces anteriorly, superiorly and medially, while that on the tibia faces posteriorly, inferiorly and laterally. A fibrous capsule attaches to the margins of the facets on both tibia and fibula, which is strengthened anteriorly and posteriorly by accessory ligaments.

The fibrous bands of the short, thick *anterior ligament of the head of the fibula* pass obliquely upwards and medially from the front of the head of the fibula to the front of the lateral tibial condyle. Posteriorly a single band, the *posterior ligament of the head of the fibula*, runs in a similar direction between the head of the fibula and the back of the lateral tibial condyle. The tendon of popliteus is intimately related to the posterosuperior aspect of the joint as it crosses the posterior ligament. The popliteal bursa, prolonged under the tendon from the knee joint, occasionally communicates with the synovial cavity of the tibiofibular joint through an opening in the upper part of the capsule.

The blood supply to the joint is from the *lateral inferior genicular* and *anterior tibial recurrent arteries*; lymphatic drainage is to the popliteal nodes. The nerve supply to the joint is by twigs from the recurrent branch of the common fibular (peroneal) nerve and the branch to popliteus from the tibial nerve. The root value of this supply is L5.

The joint is overhung by the apex of the head of the fibula, to which attaches part of the tendon of biceps femoris, the remainder going to the lateral aspect of the head below the apex. The lateral collateral ligament attaches between the biceps insertion and the joint.

There is slight movement at the superior tibiofibular joint, which gives a small degree of flexibility to the relationship between the tibia and fibula during movements of the ankle joint, and also in response to the pull of the muscles attached to the fibula.

The inferior tibiofibular joint

A fibrous joint (syndesmosis) between the rough, *triangular convex surface* of the *medial aspect* of the *lower end* of the *fibula* above the articular facet, and a corresponding rough, *triangular concave surface*, the fibular notch, on the *lateral side* of the *tibia* (Fig. 3.117). Uniting the two bones is the strong *interosseous ligament*, continuous with the interosseous membrane above, and consisting of short fibrous bands. It forms the principal connection between the two bones, and is supplemented by anterior, posterior and transverse tibiofibular ligaments.

The *anterior* and *posterior tibiofibular ligaments* are longer and more superficial bands stretching from the borders of the fibular notch of the tibia to the anterior and posterior surfaces of the lateral malleolus of the fibula. Of the two, the posterior ligament is thicker and broader. The inferior edges of both ligaments cover the lateral ridge of the trochlear surface of the talus, the anterior during flexion of the ankle joint, and the posterior during extension. Both ligaments run downwards and laterally. Under cover of the posterior ligament is the *transverse tibiofibular ligament*, which attaches along the whole length of the inferior border of the posterior surface of the tibia and the upper part of the malleolar fossa of the fibula. This strong, thick ligament projects below the margin of the bones, so closing the posterior angle between the tibia and fibula and in doing so forms part of the articular surface for the posterior part of the trochlear surface of the talus.

A synovial-lined recess of the ankle joint cavity usually extends upwards between the tibia and fibula (approximately 1 cm), being blocked above by the distal end of the interosseous membrane. Occasionally the articular cartilage on the lower ends of the tibia and fibula extends upwards for a short distance on the walls of the recess.

The blood supply to the joint is from the *fibular (peroneal)* and *anterior tibial arteries*. The nerve supply is by twigs from the deep fibular (peroneal) and tibial nerves, with roots L4 to S2.

Anterior to the inferior tibiofibular joint runs the tendon of fibularis (peroneus) tertius, with that of extensor digitorum longus lying medially together with the

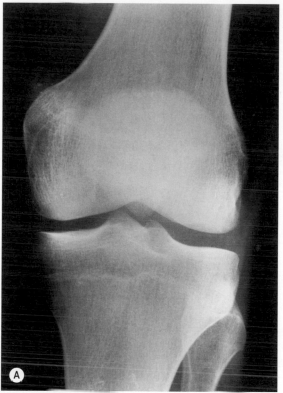

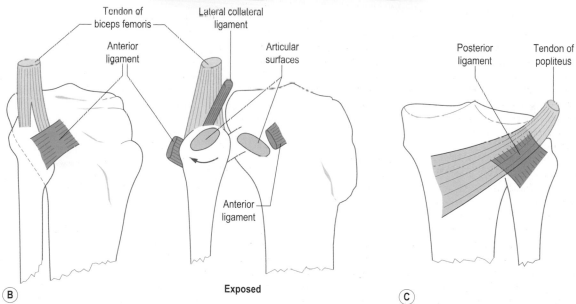

Figure 3.116 (A) Radiograph of the left superior tibiofibular joint, (B) the right superior tibiofibular joint: anterior and (C) posterior view.

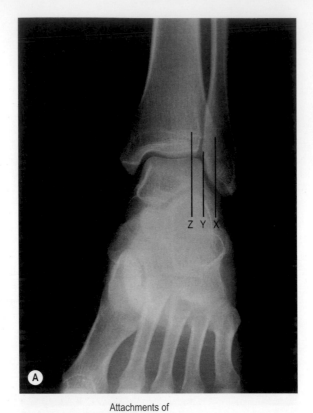

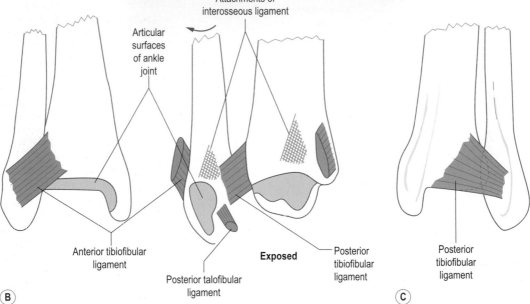

Figure 3.117 (A) Radiograph of the left inferior tibiofibular joint, (B) the right inferior tibiofibular joint: anterior and (C) posterior view.

anterior tibial artery and deep fibular (peroneal) nerve. Posteriorly the tendons of fibularis (peroneus) longus and brevis lie lateral to the joint, with brevis perhaps overlying it slightly.

The inferior tibiofibular joint provides a firm union between the tibia and fibula, and makes a significant contribution to the integrity of the malleolar mortise of the ankle joint. The fibrous tissue of the joint allows a slight yielding of the bones for the accommodation of the talus in movements of the ankle joint. In radiographs centred on the ankle, the fibula shadow encroaches upon the anterior border of the fibular notch. If the distance YZ is greater than XY then diastasis of the ankle joint is said to be present (Fig. 3.117A).

Interosseous membrane

This is often regarded as a form of fibrous joint uniting the tibia and fibula. It is tightly stretched between the interosseous borders of the two bones, and consists predominantly of fibres passing laterally and downwards from the tibia to the fibula (Fig. 3.118). The upper margin of the membrane does not reach as far as the superior tibiofibular joint, to allow the anterior tibial vessels to gain

access to the anterior compartment of the leg. The membrane is, however, continuous below with the interosseous ligament of the inferior tibiofibular joint. Distally there is a small opening for the passage of a branch from the fibular (peroneal) artery.

The interosseous membrane separates the muscles of the anterior and posterior compartments of the leg, as well as giving attachment to some of the muscles of each group (Fig. 3.119).

Movement of the fibula

Plantarflexion and dorsiflexion at the ankle joint automatically cause passive movements at both tibiofibular joints. Although the magnitude of these movements is small, they are nevertheless real.

In dorsiflexion the broader, anterior part of the trochlear surface of the talus is forced into the narrower, posterior part of the tibiofibular socket. This causes a separation of the tibia and fibula, with increased tension in the interosseous and transverse tibiofibular ligaments. Consequently, the talus is held securely between the two malleoli. In plantarflexion the narrower part of the talus moves into the broader part of the socket. The malleoli come together again to retain their grip on the talus; however, in full plantarflexion some side-to-side movement can be demonstrated. The grip of the malleoli upon the talus is a function of the inferior tibiofibular joint, the strong ligaments of which provide the spring mechanism involved.

This moving apart and coming together of the two malleoli during ankle movements imparts a complex movement to the fibula, which is partly guided by the shape and orientation of the lateral surface of the talus, and partly by the tension developed in various ligaments associated with the inferior tibiofibular joint. When the malleoli are separated, as in ankle dorsiflexion, both the anterior and posterior tibiofibular ligaments are put under tension. As the fibres in each of these ligaments run downwards and laterally, there is a tendency for the fibula to be lifted superiorly in an attempt to maintain the integrity of the ligaments. There are no bony constraints to this movement as the lateral articular surface of the talus is convex anteroposteriorly and concave superoinferiorly. At the same time as the lateral malleolus is moving laterally and superiorly, the fibula itself is undergoing a small degree of axial rotation (Fig. 3.120A). The direction of rotation depends upon the shape of the lateral talar surface, which if convex anteroposteriorly, results in a slight medial rotation. If, however, the lateral talar surface is plane then a slight lateral rotation of the fibula occurs.

When plantarflexion at the ankle joint occurs the two malleoli come together again, partly because of the tension developed in the anterior, posterior and interosseous tibiofibular ligaments but also under the action of tibialis posterior, particularly in full plantarflexion. As the lateral malleolus moves medially, it also moves inferiorly with

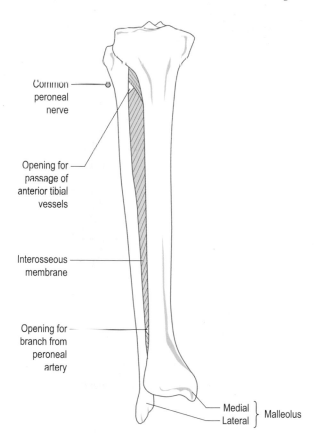

Common peroneal nerve

Opening for passage of anterior tibial vessels

Interosseous membrane

Opening for branch from peroneal artery

Medial } Malleolus
Lateral }

Figure 3.118 The right interosseous membrane, anterior view.

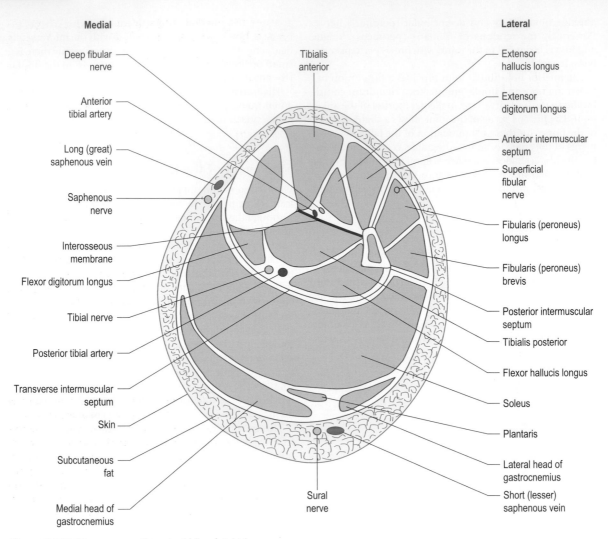

Figure 3.119 Transverse section at middle of right leg.

the fibula undergoing axial rotation in the opposite direction to that experienced in dorsiflexion (Fig. 3.120B).

These movements of the fibula at the inferior tibiofibular joint are transmitted to the synovial superior tibiofibular joint, which has plane articular surfaces and offers no resistance to movement.

Accessory movements

Superior tibiofibular joint Accessory movements of the superior tibiofibular joint are an anteroposterior glide of the fibula on the tibia. The movement can be produced by pressure in the appropriate direction when the head of the fibula is gripped between the thumb and index finger.

Inferior tibiofibular joint Accessory movements are difficult to produce in the inferior tibiofibular joint,

but a slight anteroposterior movement can be felt if the lateral malleolus is gripped between thumb and index finger, and pressed anteriorly then posteriorly.

Palpation

The head of the fibula can be palpated below the posterior part of the lateral condyle of the tibia, on approximately the same horizontal level as the tibial tuberosity. The tendon of biceps femoris can be felt and seen passing to its insertion on the head. Immediately below the head is the narrower neck, around which passes, on the lateral side, the common fibular (peroneal) nerve (Fig. 3.118).

In the region of the ankle both malleoli can be palpated, the lateral malleolus extending further distally, and lying

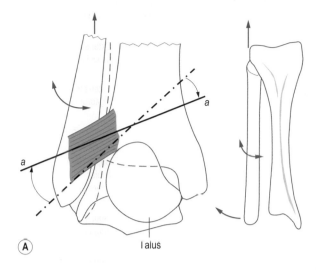

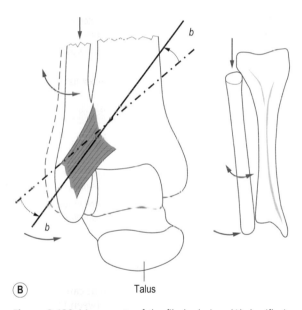

Figure 3.120 Movements of the fibula during: (A) dorsiflexion and (B) plantarflexion of the ankle joint. Red arrows indicate movement of the fibula. – – – – direction of fibres in the tibiofibular and interosseous ligaments in the neutral position; —— new direction of these same fibres in either full dorsiflexion (*aa*) or full plantarflexion (*bb*).

more posteriorly than the medial malleolus. Various tendons pass behind both malleoli on their way to attach in the foot. Passing in front of the ankle joint are the extensor tendons to the dorsum of the foot. All of the tendons are bound down by thickenings of the deep fascia – the retinacula.

Section summary

Tibiofibular articulations

Superior tibiofibular joint

Type	Synovial plane joint
Articular surfaces	Head of fibula with undersurface lateral tibial condyle
Capsule	Attaches to articular margins
Ligaments	Anterior and posterior ligaments

Interosseous membrane

Attaches to the interosseous borders of tibia and fibula

Inferior tibiofibular joint

Type	Syndesmosis
Articular surfaces	Distal end of fibula and tibia
Ligaments	Interosseous Anterior and posterior tibiofibular Transverse tibiofibular
Movements	Slight accessory rotation accompanying movement at ankle joint

The ankle joint

Introduction

The ankle joint is a synovial hinge joint with one degree of freedom of movement allowing only plantarflexion and dorsiflexion (flexion and extension respectively). The forward and backward fluctuation of the line of gravity, which in normal standing falls in front of the joint, is regulated at the ankle so that it is kept within the limits of the supporting surface. The joint has the appearance of a mortise and tenon, with the boxlike mortise being formed by the distal ends of the tibia and fibula, and the body of the talus forming the tenon (Fig. 3.121).

In the anatomical position the axis of the ankle joint is horizontal but set obliquely to the frontal plane some 20–25°, so that as it passes laterally it also runs posteriorly (Fig. 3.122). Although both the knee and the ankle joint axes lie in a horizontal plane, simultaneous movements at both joints can only be achieved if movement is allowed at other joints to compensate for the obliquity of the ankle joint axis. This compensation is essentially provided by the subtalar joint. Indeed, movement of the foot at the ankle joint is rarely performed alone; it is invariably combined with subtalar and midtarsal joint motion, so that plantarflexion is associated with adduction and supination of the foot, and dorsiflexion with abduction and pronation of the foot. These combined movements at the subtalar and midtarsal joints are usually referred to as inversion and eversion of the foot respectively.

The ankle and foot are the structures which provide restraint and propulsion at each step so that equilibrium can be maintained while the body is in motion. To

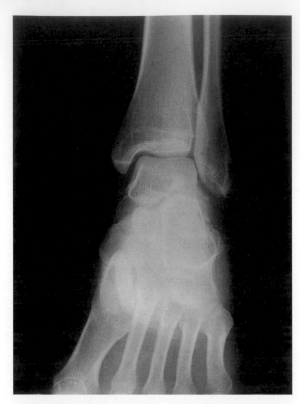

Figure 3.121 Radiograph of the left ankle joint.

accomplish this there have been changes in some of the tarsal bones, and the establishment of interrelated articulations. A single joint, the ankle joint, has been established between the leg and the foot which controls the foot in the sagittal plane. It is responsible for adjusting the line of gravity during standing and in providing the propulsion and restraint required during gait. A second articulation has been established between the talus and the calcaneus. This is the subtalar joint (p. 360). Lastly a joint has been established which interrupts the structure of the foot in the middle of the tarsus; this is the midtarsal joint (p. 367).

This series of joints, assisted by axial rotation of the knee, is equivalent to a single joint with three degrees of freedom of movement; they allow the foot to take up any position in space and to adapt to any irregularities in the ground during walking. However, unlike a single joint, these three separate articulations provide a high degree of stability without the usual sacrifice to mobility.

Articular surfaces

The distal ends of the tibia and fibula proximally and the body of the talus distally. The weight-bearing surfaces are the trochlear surfaces of the tibia and talus. Stabilizing surfaces are those of the medial and lateral malleoli, which grip the body of the talus.

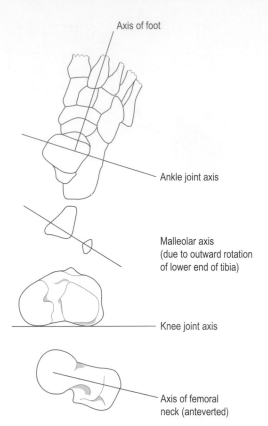

Figure 3.122 Relationship of the right ankle joint axis to the knee joint and femoral neck axes.

Tibia

The distal end of the tibia provides a continuous articular surface which receives the trochlear surface and medial edge of the body of the talus (Fig. 3.123). The trochlear surface is concave anteroposteriorly and slightly convex transversely with a blunt sagittal ridge which fits into a corresponding groove on the talus, being slightly wider in front than behind. On either side of the ridge are medial and lateral gutters which receive the corresponding lips of the talar trochlear surface. The posterior part of this surface projects slightly downwards; it is sometimes known as the posterior malleolus.

The cartilage on the trochlear surface of the tibia is continuous medially with that on the lateral surface of the medial malleolus, the junction being a rounded angle.

Fibula

The medial surface of the lateral malleolus of the fibula forms the lateral surface of the mortise of the joint. The articular surface is approximately triangular with the inferior apex being slightly convex (Fig. 3.123). The base of the triangle extends superiorly as far as the tibial articular surface, with the joint cavity usually extending between the

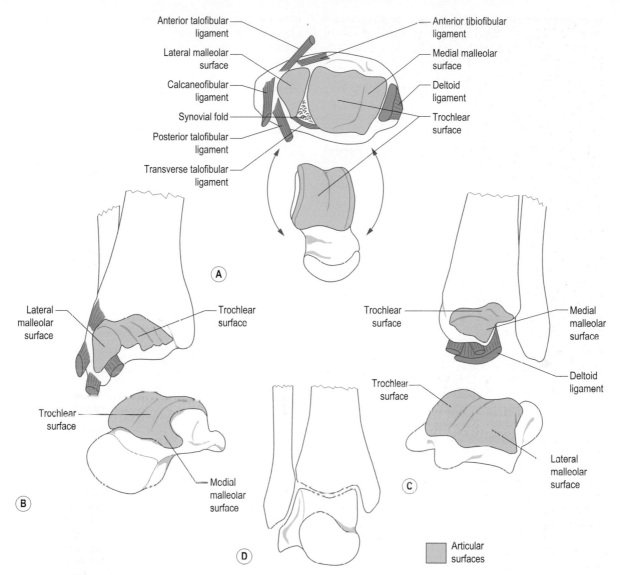

Figure 3.123 The articular surfaces of the ankle joint: (A) trochlear surfaces of tibia and talus, (B) medial anterior oblique view, (C) lateral anterior oblique view, (D) coronal section.

two bones. In the angle between the lateral malleolus and the trochlear surface of the tibia is a narrow cleft distal to the interosseous ligament. This cleft disappears posteriorly as it is filled by the transverse tibiofibular ligament.

Talus

The body of the talus forms the whole of the distal surface of the ankle joint, articulating superiorly and medially with the tibia and laterally with the fibula (Fig. 3.123). The trochlear surface is convex anteroposteriorly with a central

longitudinal groove bound by medial and lateral lips; it is slightly broader in front than behind. The groove and lips make the trochlear surface slightly concave transversely. The cartilage covering the trochlear surface is continuous with that on the sides of the body of the talus.

The medial surface is nearly plane, except anteriorly where it is inclined medially, and can be likened to a comma placed on its side with the tail pointing backwards. It lies in a sagittal plane and articulates with the lateral surface of the medial malleolus. The lateral articular surface runs obliquely anteriorly and laterally, being concave superoinferiorly as well as anteroposteriorly. The surface is triangular and much larger than that medially, curving inferiorly to a laterally projecting apex. It articulates with the medial surface of the lateral malleolus, and with the deep surface of the transverse tibiofibular ligament posterosuperiorly at the angle between the trochlear and lateral surfaces. The interval between the tibia, fibula and the transverse tibiofibular ligament posteriorly is padded by a synovial fold.

Joint capsule

A fibrous capsule completely surrounds the joint, attaching above to the articular margins of the tibia and fibula, and below to just outside the edges of the corresponding articular areas of the talus, except anteriorly where it attaches to the neck of the talus. The capsule is thin and weak in front and behind to accommodate plantarflexion and dorsiflexion of the joint, but is strengthened laterally and medially by collateral ligaments. Posteriorly the capsule is attached to the posterior tibiofibular ligament.

Synovial membrane

The synovial membrane of the ankle joint is loose and capacious. It lines the joint capsule and is reflected anteriorly onto the neck of the talus before attaching to the articular margins. It covers well-marked fatty pads that lie in relation to its anterior and posterior parts. The synovial membrane extends upwards between the tibia and fibula as far as the interosseous ligament of the inferior tibiofibular joint and may be covered in part by an extension of the articular cartilage on the tibia and fibula.

Ligaments

As with all hinge joints, there is an extremely strong set of collateral ligaments associated with the ankle joint. Medially is the deltoid ligament while laterally there are three separate ligaments. Each set of ligaments radiate downwards from the respective malleolus and both have a middle band to the calcaneus and anterior and posterior bands to the talus. The deltoid and anterior and posterior lateral ligaments blend with the joint capsule.

Deltoid ligament

A strong, roughly triangular ligament composed of several bands of fibres fused together, the various bands only being differentiated by their distal attachments (Fig. 3.124A). It can be considered to have deep and superficial parts, attaching by its apex to the anterior and posterior borders and to the fossa at the tip of the medial malleolus. The thick base forms a continuous attachment from the navicular anteriorly to the body of the talus posteriorly.

The deeper parts of the ligament are the *anterior* and *posterior tibiotalar* bands. The anterior tibiotalar band is the

most anterior and runs obliquely forwards and downwards to attach to the medial part of the neck of the talus. The posterior tibiotalar band is the most posterior and thickest part of the deltoid ligament. Its fibres run laterally and backwards to the medial side of the talus under the tail of the comma-shaped articular facet and to the medial tubercle of its posterior process.

The more superficial parts of the deltoid ligament partly overlie the anterior tibiotalar band, and have a continuous attachment from the navicular to the sustentaculum tali of

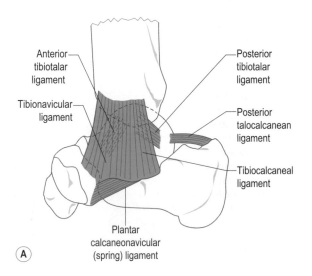

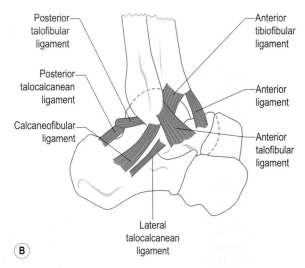

Figure 3.124 (A) The deltoid and (B) lateral collateral ligaments of the ankle.

the calcaneus. The *tibionavicular* band runs forwards and downwards towards the tuberosity on the navicular attaching to its upper and medial parts. Continued backwards this band blends below with the upper medial border of the plantar calcaneonavicular (spring) ligament. It is succeeded by the *tibiocalcaneal* band whose fibres descend almost vertically to attach to the whole of the length of the sustentaculum tali.

Lateral ligaments

The lateral collateral ligament is composed of three separate parts, the *anterior* and *posterior talofibular* and the *calcaneofibular* ligaments (Fig. 3.124B). These separate ligaments are not as strong a ligamentous structure as the deltoid ligament, as evidenced by the fact that most ankle sprains involve the lateral ligaments. The anterior talofibular ligament is a flat band stretching between the anterior border and tip of the lateral malleolus to the neck of the talus. Its fibres run anteromedially. The posterior talofibular ligament is a strong, thick ligament running almost horizontally. It arises from the bottom of the malleolar fossa of the lateral malleolus and passes posteromedially to the lateral tubercle of the posterior process of the talus. Above it lies the posterior tibiofibular ligament. In plantarflexion these two ligaments lie edge-to-edge, while in dorsiflexion they diverge medially.

Between the two talofibular ligaments is the calcaneofibular ligament, a narrow rounded cord which is free from the fibrous capsule, but has the two talofibular ligaments fused with its upper part. It arises from the front and tip of the lateral malleolus and passes downwards and slightly backwards to attach above and behind the fibular (peroneal) tubercle on the middle of the lateral surface of the calcaneus.

Anterior and posterior ligaments

These are localized thickenings of the joint capsule (Fig. 3.125). The anterior ligament runs obliquely from the anterior margin of the lower end of the tibia to the upper surface of the anterior part of the neck of the talus. The posterior ligament has fibres arising from both the tibia and fibula, which converge to attach to the medial tubercle of the posterior surface of the talus.

Role of the collateral ligaments

Laxity of the ankle joint is dependent on its position, full dorsiflexion being the position of least laxity, reflecting talar geometry and the inferior tibiofibular syndesmosis. The ligaments of the joint are primarily responsible for maintaining its stability and controlling the movements of which it is capable. Damage to some or all of the collateral ligaments will seriously impair the integrity of the joint (Fig. 3.126).

Sectioning the lateral ligaments is associated with an increase in the range of dorsiflexion, but not plantarflexion,

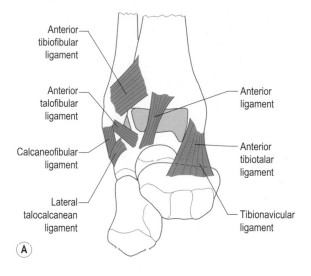

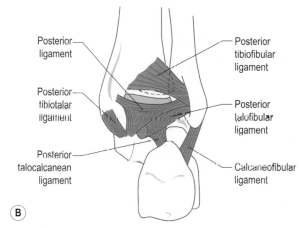

Figure 3.125 Capsular thickenings of the ankle joint (capsule removed): (A) anterior, (B) posterior.

and an increase in internal rotation of the talus, particularly when the ankle is plantarflexed. Only when all the collateral ligaments have been severed is an increase in external talar rotation observed, which is more marked in dorsiflexion. Talar tilt is also increased with gradually increasing injury. Prior to complete severing of all collateral ligaments talar tilt is maximum in plantarflexion, but after total disruption of the ligaments talar tilt is most marked in dorsiflexion.

The role of the various parts of the collateral ligaments in maintaining stability can be summarized as follows. The tibiocalcaneal and tibionavicular bands control abduction of the talus, while adduction is controlled by the

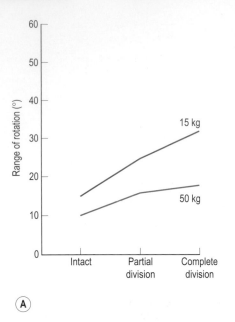

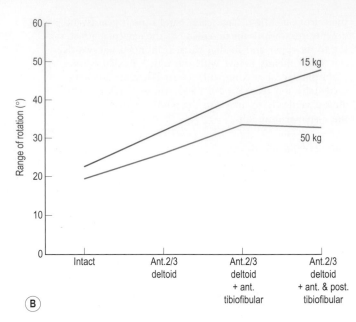

Figure 3.126 The role of the (A) lateral and (B) medial collateral ligaments in rotational stability of the ankle joint. *(Adapted from McCullough CJ, Burge PD (1980) Rotatory stability of the load-bearing ankle – an experimental study.* Journal of Bone and Joint Surgery, 62B, 460–464.)

calcaneofibular ligament. The anterior tibiotalar band and the anterior talofibular ligament control plantarflexion, while dorsiflexion is resisted by the posterior tibiotalar band and the posterior talofibular ligament. In combination, both the anterior tibiotalar and tibionavicular bands control external rotation, and, together with the anterior talofibular ligament, internal rotation of the talus (Fig. 3.126). In isolation, neither the anterior nor the posterior tibiotalar ligament appears to play any major role in ankle stability. In contrast, the anterior talofibular ligament appears to be an extremely important structure, and can probably be considered the primary stabilizer of the ankle. It provides significant resistance to varus tilt of the talus in all positions of flexion.

Apart from the pain involved in a sprain of the ankle, the accompanying instability is probably due to involvement of the anterior talofibular ligament, as most sprains are due to a turning of the ankle outwards (foot inwards) and an associated straining of the lateral ligaments (inversion stress).

Blood and nerve supply

The blood supply of the joint is from the *malleolar branches* of the *anterior tibial, fibular (peroneal)* and *posterior tibial arteries,* which form an anastomosis around the malleoli. Venous drainage is by the corresponding venae comitantes accompanying the arteries. The lymphatics drain into the deep system of vessels, which again accompany the arteries.

The nerve supply to the joint is from roots L4 to S2 by articular branches from the tibial nerve and lateral branch of the deep fibular (peroneal) nerve.

Relations

All of the muscles, vessels and nerves entering the foot cross the ankle joint. Anteriorly, from medial to lateral, are the tendons of tibialis anterior, extensor hallucis longus, extensor digitorum longus and fibularis (peroneus) tertius as they pass through the inferior extensor retinaculum (Fig. 3.127A). The first two tendons lie within separate compartments surrounded by their own synovial sheaths, while the last two share a compartment and synovial sheath. Deep to the extensor retinaculum, behind the middle compartment, the anterior tibial artery passes; this becomes the dorsalis pedis artery at the inferior border of the retinaculum, and the deep fibular (peroneal) nerve. Anterior to the extensor retinaculum is the superficial fibular (peroneal) nerve, and in front of the medial malleolus the long (great) saphenous vein draining from the dorsal venous plexus, and the saphenous branch of the femoral nerve (Fig. 3.127A).

Behind the medial malleolus and bound down by the flexor retinaculum are, from anteromedial to posterolateral, the tendons of tibialis posterior, flexor digitorum longus, the posterior tibial artery, the tibial nerve and the tendon of flexor hallucis longus (Fig. 3.127A). The latter tendon passes in a groove on the posterior part of the talus between the medial and lateral tubercles, and enters the

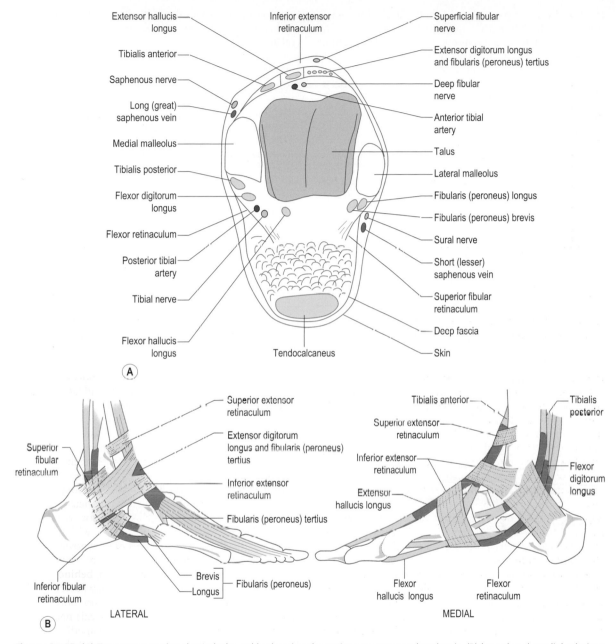

Figure 3.127 (A) Transverse section through the ankle showing the various structures related to it, (B) lateral and medial relations of the ankle joint.

foot by passing below the sustentaculum tali. As they pass behind the medial malleolus, both the posterior tibial artery and the tibial nerve divide into their terminal medial and lateral plantar branches. All of the tendons are surrounded by individual synovial sheaths.

The tendons of fibularis (peroneus) longus (anterolateral) and fibularis (peroneus) brevis (posteromedial) pass in a common synovial sheath behind the lateral malleolus bound down by the superior fibular (peroneal) retinaculum (Fig. 3.127B). Below the level of the joint these two tendons diverge to pass either side of the fibular (peroneal) tubercle of the calcaneus – longus below, brevis above. At this point each tendon has its own synovial sheath. Behind these two tendons as they pass posterolateral to the ankle joint are the sural nerve and short (lesser) saphenous vein.

Posterior to the joint, but separated from it by an extensive fat pad, is the tendocalcaneus (Achilles tendon) passing to its attachment on the posterior tubercle of the calcaneus.

Palpation

Both the medial and lateral malleoli can be readily palpated on their anterior and posterior borders, as well as at their tips. The two malleoli are, however, basically different. The lateral malleolus is larger than the medial, extending further distally and lying more posteriorly. The fibularis (peroneal) tendons can be palpated behind the lateral malleolus (Fig. 3.127B), and those of tibialis posterior and flexor digitorum longus behind the medial malleolus (Fig. 3.127B). All of the tendons crossing anterior to the joint can be identified and palpated.

The pulsations of the anterior tibial artery can be felt between the tendons of extensor hallucis longus and extensor digitorum, while those of posterior tibial artery can be felt behind the medial malleolus posterior to the tendon of flexor digitorum longus.

The ankle joint, which lies on a horizontal line 1 cm above the tip of the medial malleolus and 2 cm above the tip of the lateral malleolus, can be palpated on the dorsal surface. Starting medially, the joint can be identified by applying firm pressure along the inner border of the medial malleolus. If the extensor tendons are moved aside the lower end of the tibia can be palpated, as well as the medial edge of the lateral malleolus.

Stability

The anteroposterior stability of the ankle joint and the coaptation of its articular surfaces depend upon the effect of gravity to keep the tibia pressed against the superior surface of the talus. In addition, the anterior and posterior margins of the tibial surface form bony spurs, due to its concave shape, which help prevent the talus from escaping posteriorly or anteriorly respectively. The collateral ligaments are passively responsible for the coaptation of the articular surfaces and are assisted by the muscles crossing the joint, provided that the joint is intact. In a subluxed or dislocated joint the ligaments and muscles may act to cause further joint distraction.

The ankle joint, by virtue of its structure, should not be able to exhibit movements other than plantarflexion and dorsiflexion. Transverse stability depends on the interlocking of its articular surfaces (Fig. 3.128A). Provided that the distance between the malleoli remains relatively unchanged, they grip the talus on each side. This can only be accomplished when the malleoli and the ligaments of the inferior tibiofibular joint are intact. Furthermore, the collateral ligaments prevent any rolling movements of the talus about its long axis.

When the foot is forcibly moved laterally, as in a violent movement of abduction, the lateral surface of the talus knocks against the lateral malleolus. The sequence of events which follow depends on the severity of the movement, the state of the bone and the integrity of the various ligaments of the joint. Following rupture of the inferior tibiofibular ligaments the grip of the malleoli on the talus is disrupted, leading to a widening of the mortise (diastasis of the ankle). The talus is no longer held tightly and can move from side to side (rattling of the talus) (Fig. 3.128B). It is also able to rotate about its long axis (tilting of the talus), which is made all the more easy if the deltoid ligament is sprained (Fig. 3.128C), and rotate about its vertical axis so that the posterior part of the body of the talus abuts against the posterior margin of the tibia, possibly fracturing it. If the abduction movement continues both medial and lateral malleoli may become fractured, the lateral above the inferior tibiofibular joint (Fig. 3.128E). This is one form of Pott's fracture. Occasionally the lateral malleolus does not fracture and the fibular fracture occurs at the level of the neck.

Both inferior tibiofibular ligaments do not always rupture; very often the anterior tibiofibular ligament resists tearing. There is still fracture of both malleoli, but with the lateral malleolus fracturing through the inferior tibiofibular joint (Fig. 3.128F). This is another form of Pott's fracture. Instead of the medial malleolus fracturing, the deltoid ligament may be ruptured, but again the lateral malleolus fractures through the inferior tibiofibular joint (Fig. 3.128G).

In all of these fractures, a chip of bone is often broken off the posterior margin of the tibia; this can present as a separate fragment of bone or form a single unit with the malleolar fragment.

In violent adduction movements of the foot, the talus is forced to rotate about its vertical axis fracturing both malleoli, the lateral below the inferior tibiofibular joint. In these bimalleolar adduction fractures, the inferior tibiofibular and both collateral ligaments remain intact (Fig. 3.128H).

Obviously these lesions require proper treatment if full structural and functional integrity of the joint is to be restored. If the bony and soft tissue damage is excessive then a fusion of the joint may be the only practical way of restoring stability.

Movement

This takes place about a transverse axis level with the tip of the lateral malleolus and slightly below the level of the medial malleolus. Strictly speaking the axis is not horizontal, but slopes slightly downwards and laterally, passing through the lateral surface of the talus just below the apex

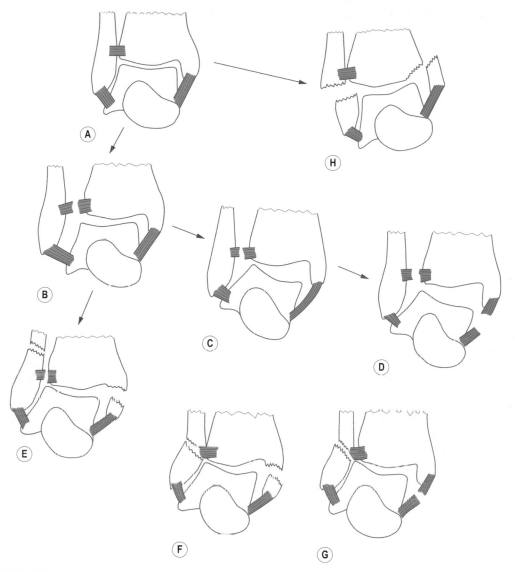

Figure 3.128 Disruption of the transverse stability of the ankle joint.

of the articular triangle, and through the medial surface at a higher level, just below the concavity of the comma-shaped articular area. The axis also changes slightly during movement because the upper surface of the talus is elliptical rather than being an arc of a circle; however this change is of no practical importance. Because of the obliquity of the joint axis there is a slight movement resembling inversion on full plantarflexion, and eversion on full dorsiflexion. These 'inversion/eversion' movements are not true inversion and eversion.

The movements possible at the ankle joint are dorsiflexion and plantarflexion of the foot through a maximum range approaching 90°. In the normal standing position

the foot makes a right angle with the leg; this is the neutral position of the joint (Fig. 3.129A). In dorsiflexion the foot is drawn upwards towards the leg; plantarflexion is movement in the opposite direction from the neutral position. The ranges of dorsiflexion and plantarflexion are essentially determined by the profiles of the articular surfaces; dorsiflexion having a range of 30° and plantarflexion of 50°. There is, however, considerable individual variation in the extent of these movements (Fig. 3.129B).

In dorsiflexion the broader, anterior part of the trochlear surface of the talus is forced between the narrower posterior part of the tibiofibular mortise, causing a slight separation of the tibia and fibula and increased tension in the

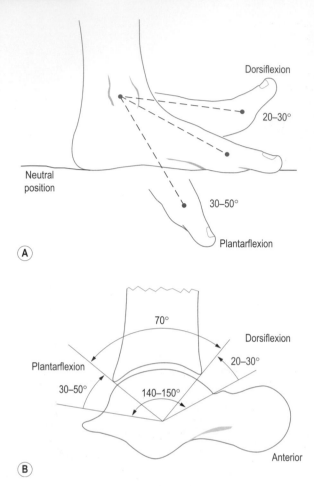

Figure 3.129 (A) Movement of the foot in dorsiflexion and plantarflexion, (B) the range of dorsiflexion and plantarflexion at the ankle joint is determined by the profiles of the articular surfaces.

interosseous and transverse tibiofibular ligaments. It is in this close-packed position that the stability of the ankle joint is greatest. Movement of the fibula away from the tibia during dorsiflexion causes it to rotate about its long axis and to move upwards (see tibiofibular joints, p. 338). Dorsiflexion is produced by tibialis anterior, extensor hallucis longus, extensor digitorum and fibularis (peroneus) tertius crossing the joint anteriorly. It is limited by tension in gastrocnemius and soleus, the posterior part of the deltoid ligament, the calcaneofibular ligament and the posterior joint capsule, and wedging of the talus between the malleoli. Should dorsiflexion continue, the anterior margin of the tibia can come into contact with the upper surface of the neck of the talus. If sufficient force is behind the movement, the neck of the talus or the anterior tibial margin or both can be fractured. The anterior part of the joint

capsule is prevented from being nipped between the tibia and talus by being pulled up by the extensor muscles, whose sheaths are attached to the capsule (Fig. 3.130A).

Shortening of gastrocnemius and soleus may check dorsiflexion prematurely. If the shortening is severe the ankle may be permanently fixed in plantarflexion (talipes equinus); lengthening of the tendocalcaneus by surgical intervention will be required to restore full function.

In plantarflexion the narrower, posterior part of the trochlea of the talus moves forward into the broader part of the tibiofibular mortise. The malleoli tend to come together again to retain their grip upon the talus. In full plantarflexion some rotation, abduction and adduction, and side-to-side movement of the talus is possible; this is the position of least stability of the ankle joint. Approximation of the medial and lateral malleoli during plantarflexion is partly passive (recoil of the stretched interosseous and transverse tibiofibular ligaments) and partly active under the action of tibialis posterior, which attaches to both the tibia and fibula. The accessory movements of the fibula are the reverse of those seen during dorsiflexion (p. 342).

Plantarflexion is brought about mainly by soleus and gastrocnemius. However, all of the muscles which enter the foot behind the malleoli produce plantarflexion at the ankle, i.e. tibialis posterior, flexor digitorum longus, flexor hallucis longus, fibularis (peroneus) longus and fibularis (peroneus) brevis. The movement is checked by tension in the anterior muscles, the anterior part of the deltoid ligament, the anterior talofibular ligament and the anterior joint capsule. If these structures fail to arrest a forceful plantarflexion movement, then the posterior margin of the tibia may come into contact with the tubercles on the posterior surface of the talus, particularly the lateral. Rarely does this latter tubercle become fractured. The posterior joint capsule avoids becoming trapped between the bones by the presence of the posterior talofibular ligament and by its attachment to the sheath of flexor hallucis longus medially and the fibularii (peronei) laterally (Fig. 3.130B).

The wedge-shaped form of the joint surfaces helps to prevent backward displacement of the foot on the leg when coming to a sudden stop in jumping or running. The maintenance of this positive grip upon the talus is the function of the inferior tibiofibular joint.

When the body is erect, muscular effort, mainly by soleus and gastrocnemius, is necessary to prevent forward collapse. This destabilizing effect of gravity can be, and usually is, minimized by turning the feet out a little laterally, so that the inclination between the two ankle joint axes is increased.

Accessory movements

Two accessory movements are possible at the ankle joint. The first is a longitudinal distraction in which the talus is pulled away from the tibia and malleoli. It is performed with the subject lying supine. The calcaneus at the heel and the head of the talus on the dorsum of the foot are

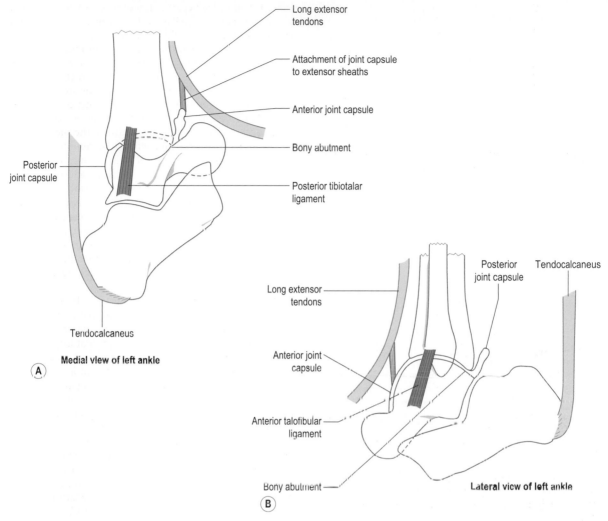

Figure 3.130 Factors involved in limiting: (A) dorsiflexion and (B) plantarflexion of the ankle joint.

gripped and a longitudinal pull is exerted along the line of the tibia.

The second movement is in an anteroposterior direction. Again it is performed with the subject lying supine. The knee is flexed to 90° keeping the sole of the foot firmly pressed against the table. In this position the ankle joint is partly plantarflexed. Gripping the lower ends of the tibia and fibula, they can be moved anteriorly and posteriorly by applying pressure in the appropriate direction.

Biomechanics

Although the total range of ankle joint motion approaches 90°, a much smaller range than this is used during gait; however with increasing speed of walking ankle joint motion decreases, being mainly a decrease in plantarflexion (Fig. 3.131).

At heel-strike the ankle is slightly plantarflexed, immediately after which plantarflexion increases and then begins to decrease when full foot contact is made. As the body moves over the supporting foot, the ankle goes into dorsiflexion, reaching a maximum just prior to the heel leaving the ground, after which it decreases. Just prior to toe-off the ankle once again becomes plantarflexed. During the swing phase there is a second wave of dorsiflexion, which together with a second wave of knee flexion ensures clearance between the toes and ground. Towards the end of the swing phase, the ankle becomes plantarflexed prior to heel-strike. Not only does an increase in cadence decrease the range of motion seen during the stance phase, it also brings forward

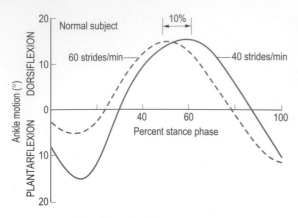

Figure 3.131 The effect of walking speed on ankle joint motion. *(Adapted from Stauffer RN, Chao EYS and Brewster RC (1977) Force and motion analysis of the normal, diseased, and prosthetic ankle joint.* Clinical Orthopaedics and Related Research, *127, 189–196.)*

the peak dorsiflexion wave before the heel leaves the ground (Fig. 3.131).

Joint forces

During the gait cycle there are both tangential shear forces and compressive forces acting across the ankle joint. The tangential shear forces are in an anteroposterior direction and are the result of a combination of internal musculotendinous forces and external forces as the body moves over the foot. The force is biphasic and mainly directed posteriorly during the stance phase, reaching 80% of body weight as the heel leaves the ground (Fig. 3.132). During the last 15% of the stance phase, as the ankle goes into plantarflexion again, the shear force is directed anteriorly reaching 20% of body weight prior to toe-off.

Compressive forces across the ankle joint rise to three times body weight between heel-strike and full foot contact. Following heel-lift these forces rise to five times body weight and then decrease to toe-off (Fig. 3.133). With increasing speed, the relatively smooth pattern of compressive forces becomes markedly biphasic. Although the absolute magnitudes of these forces following heel-strike and prior to toe-off are not much greater than at lower speeds, the relative unloading of the cartilage between the two peaks effectively means that the articular cartilage undergoes two loading cycles per gait cycle. This rapid and alternating pattern of loading of the articular cartilage during fast walking may over a long period of time be responsible for some cartilage degeneration.

It would appear that patients with ankle joint disease have a poor tolerance to both compressive and shear forces across the joint (see the dashed line in Figs. 3.132 and 3.133). Consequently, they modify their gait patterns in an attempt to reduce the magnitude of these forces.

Contact area and contact stresses

The occurrence of osteoarthrosis of the ankle joint is small in comparison with the other joints of the lower limb. If degenerative disease is the result of a fatigue process, it becomes important to know the magnitude of the contact stresses as well as the loading cycle. To calculate contact stresses the contact area of the joint needs to be known. Studies have suggested that the mean contact area at the ankle joint is between 1000 and 1500 mm^2. Bearing in mind that compressive forces across the joint are between three and five times body weight for the majority of the stance phase, this gives compressive contact stresses of between 1.4 MN/m^2 and 3.5 MN/m^2. These values are very similar to those calculated for the hip joint, which shows a much higher incidence of osteoarthrosis. This difference in the incidence of pathology under seemingly similar magnitudes of contact stresses can probably be accounted for by the much more complex loading patterns at the hip.

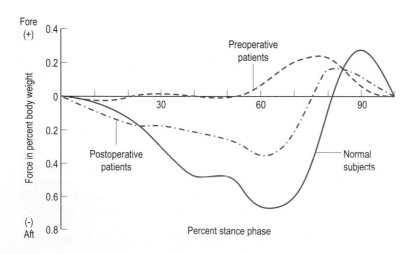

Figure 3.132 The pattern of tangential ankle force during the stance phase of gait. *(Adapted from Stauffer RN, Chao EYS and Brewster RC (1977) Force and motion analysis of the normal, diseased, and prosthetic ankle joint.* Clinical Orthopaedics and Related Research, *127, 189–196.)*

Figure 3.133 The pattern of compressive ankle forces during the stance phase of gait. *(Adapted from Stauffer RN, Chao EYS and Brewster RC (1977) Force and motion analysis of the normal, diseased, and prosthetic ankle joint.* Clinical Orthopaedics and Related Research, *127, 189–196.)*

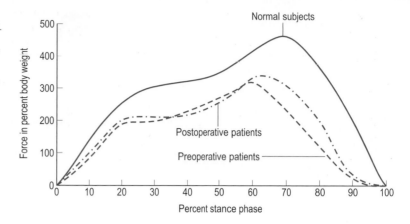

Trabecular arrangement

The transmission of mechanical forces across the joint is reflected in the arrangement and direction of the trabeculae within the associated bones. Two sets of trabeculae can be seen crossing the joint. One arises from the cortex on the anterior surface of the tibia and passes posteroinferiorly through the talus to the calcaneus. The other begins in the cortex on the posterior surface of the tibia and passes forwards and inferiorly through the talus. Medially, this system can be traced to the head of the first metatarsal, while laterally it can be traced to the head of the fifth metatarsal. The trabecular systems within the bones of the foot are considered in the section dealing with biomechanics of the foot (p. 381).

Joint replacement/fusion

With total replacement of the ankle joint the range of motion during gait is generally within normal limits. However, the patterns are often abnormal. Normally in gait, initial foot contact is with the heel, with the ankle slightly plantarflexed. In contrast, in patients with total replacement, initial foot–floor contact is with the entire foot with the ankle in maximum available passive plantarflexion. At the end of the stance phase the normal change from dorsiflexion to plantarflexion is not seen. Such alterations in the pattern of motion do not appear to be related to ligament instability, stiffness or pain in the ankle or foot. Consequently, ankle fusion remains the procedure of choice for many painful ankle conditions, particularly if the subtalar and midtarsal joints are not involved and if the patient is going to subject the extremity to high levels of activity. Arthrodesis should be performed in a neutral position. An equinus position is unfavourable as it prevents an effective heel-strike occurring at the beginning of the stance phase of gait.

Total joint replacement has been, and continues to be, limited to a select group of patients, in whom only a few degrees of motion may allow a degree of independence that would otherwise not be achieved. The use of ankle prostheses in single joint involvement, such as after severe trauma or joint degeneration or after osteochrondritic damage to the articular surface, will result in early breakdown of the prosthesis and later conversion to an arthrodesis. In general ankle prostheses do not tolerate a normal degree of activity even in an individual leading a relatively sedentary life.

First generation design ankle prostheses were not very successful, with many patients experiencing loosening; consequently many were removed and the ankle fused. However, total joint replacement is becoming more common with the advent of the second generation of designs, which have paid attention to factors such as reproducing normal ankle anatomy, joint kinematics, ligament stability and mechanical alignment. Two- and three-component designs are currently available (Fig. 3.134), with the latest designs improving fixation by allowing increased bone ingrowth as well as screws. With improvements in surgical experience, instrumentation, implant design and patient selection, total ankle arthroplasty is becoming more successful.

Section summary	
Ankle joint	
Type	Synovial hinge joint
Articular surfaces	Distal end of tibia and inner surfaces of medial and lateral malleoli with trochlear surface and sides of talus
Capsule	Thin, loose capsule attaching to articular margins except for attachment to neck of talus anteriorly
Ligaments	Deltoid (medial collateral) Ligament lateral collateral ligament – Anterior and posterior talofibular – Calcaneofibular – Capsular anterior and posterior ligaments
Stability	Provided by coaptation of articular surfaces, collateral ligaments and muscles
Movements	Dorsiflexion (extension) and plantarflexion (flexion)

Figure 3.134 Examples of current ankle prostheses. (A) the Agility two-component design, (B) the Buechel–Pappas three-component design, (C) the HINTEGRA three-component design.

Joints of the foot

Introduction

It is widely believed that the human foot has evolved from the mobile, prehensile organ seen in many primates to the specialized, supporting structure necessary for bipedal locomotion. In contrast to the anthropoid foot the human foot is characterized by a reduction in the ability to oppose the great toe (Fig. 3.135A). This ability is not completely lost, however, as the requisite musculature is still present. Under certain circumstances, for example congenital absence of the upper limbs, opposition of the great toe

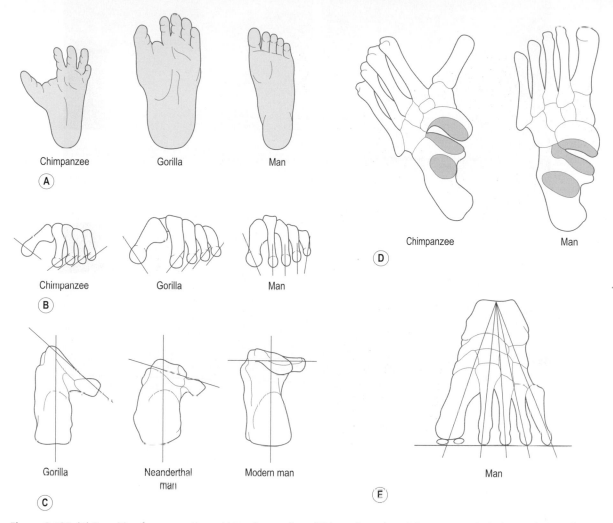

Figure 3.135 (A) Transition from an anthropoid to a human foot, (B) loss of rotation of the metatarsals during evolution of the human foot, (C) changes in the obliquity of the sustentaculum tali, (D) reduction in the angulation of the great toe, (E) anterior support of the foot, showing that all of the metatarsals participate.

can become remarkably well developed, to the extent of holding a pen or fork. However, with this general reduction in opposability the metatarsal heads are no longer rotated towards each other as the gripping function of the toe requires, but are directed anteroposteriorly (Fig. 3.136B). With this, the axis of leverage of the foot has shifted to come to lie from between the second and third to between the first and second metatarsals. Correspondingly the axis of abduction/adduction of the toes is through the second digit, compared with the third in the hand (see Fig. 2.124).

With the assumption of bipedal locomotion important changes have occurred within the calcaneus. The sustentaculum tali has become more massive, as has the calcaneus as a whole, and has assumed a more horizontal position

(Fig. 3.135C), in order to be able to support the body of the talus and the superincumbent body weight. The calcaneus as a whole has changed its relative position within the foot from a downward to an upward one.

In primates adduction of the great toe gives the forefoot a medial direction. In the human foot this angulation disappears so that the axis of the foot follows a straight line (Fig. 3.135D). Consequently, the first and second metatarsals run more parallel to each other. With the medial shift of the axis of leverage, the medial border of the foot becomes flattened and depressed, and a transverse arch develops so that the supporting ball occupies the entire width of the foot (Fig. 3.135E).

The developmental changes which have occurred within the foot are essentially based upon the requirements to

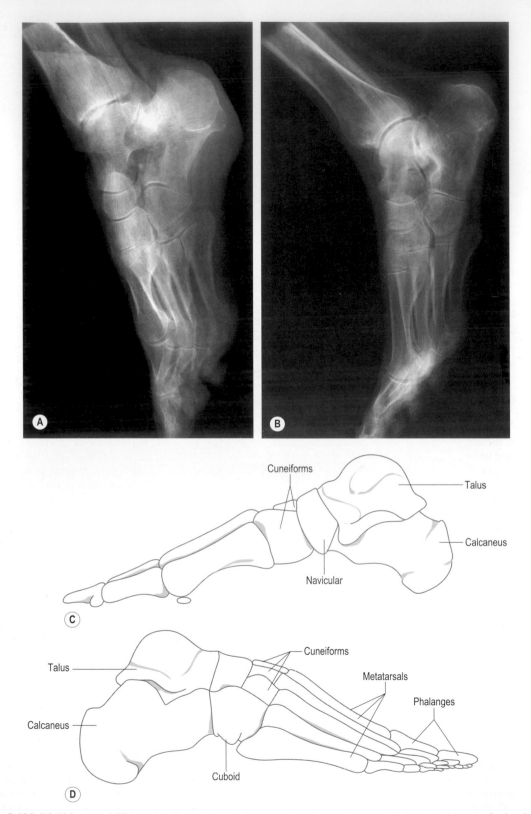

Figure 3.136 (A) Oblique and (B) lateral radiograph of the foot showing the tarsal bones, (C) the medial longitudinal arch, (D) the lateral longitudinal arch.

adjust the centre and line of gravity to a small supporting surface area. Having achieved this and adjusted the gravitational stresses to the area of support for a bipedal gait, further changes enabling a bipedal gait to be adopted had to be undertaken. An alternating bipedal gait has been the stimulus for the development of certain articulations to provide for static balance during standing and dynamic propulsion during walking and running.

The ankle and foot are the structures which provide propulsion and restraint at each step. A single joint, the ankle joint, has been established between the leg and the foot, which controls the foot in the sagittal plane (p. 343). A second articulation, which allows side-to-side adjustment of the line of gravity in standing, as well as participating in restraint and propulsion, has been established between the talus and the calcaneus. This is the subtalar joint. Lastly a composite, functional joint has been established which interrupts the structure of the foot in the middle of the tarsus, the transverse (mid) tarsal joint. This latter joint plays an important role in putting the spring into the propulsive phase of gait, by allowing the anterior part of the foot to adjust itself against the posterior. By so doing the anterior footplate is able to maintain full contact with the supporting surface independent of the posterior part of the foot.

Besides forcing the foot into a right-angled posture with respect to the leg, bipedal gait has also produced some changes in the arrangement of the musculature around the ankle and within the foot. In humans, tibialis anterior has lost its insertion into the great toe. Both extensor digitorum longus and extensor hallucis longus split off from the primitive extensor plate as separate units, and a new unit, fibularis (peroneus) tertius, is formed. Fibularis (peroneus) tertius is an important muscle which aids in pronation of the formerly supinated foot. Fibularis (peroneus) longus and brevis run behind the human ankle joint axis and are consequently now plantarflexors, as opposed to their previous role as dorsiflexors. In addition, because of the increased stress put on the anterior part of the foot in bipedal locomotion, the insertion of fibularis (peroneus) longus has migrated across the sole of the foot. Consequently, it now acts as a tie helping to maintain the arches of the foot against depression. Finally, tibialis posterior has developed a fan-shaped insertion to all of the tarsal bones except the talus, thereby providing one of the principal supports of the longitudinal arches of the foot.

The human foot is strong to support the weight of the body, but also flexible and resilient to absorb the shocks transmitted to it and to provide spring and lift during activity. These properties are achieved by the presence of a series of arches, convex above, composed of a number of bones and their interconnecting joints. The joints and ligaments, together with muscle action, provide for spring for they yield when weight is applied and recoil when the weight is removed.

The bones of the foot are arranged in longitudinal and transverse arches (Fig. 3.136). The longitudinal arch, sometimes regarded as having two parts, lateral and medial, is supported posteriorly on the tuberosity of the calcaneus, and anteriorly on the metatarsal heads. The talus is at the summit of this arch, being primarily related to the navicular, the three cuneiforms and the medial three metatarsals; this is the medial longitudinal arch. The calcaneus is more directly related to the cuboid and lateral two metatarsals; this is the lateral longitudinal arch. These differences appear in the function of the foot, for the medial longitudinal arch has a greater curvature and is more elastic than the lateral. The flatter, more rigid, lateral arch makes contact with the ground and provides a firm base for support. The transverse arch results from (i) the shape of the tarsal bones in the distal row, and (ii) the bases of the metatarsals. Being broader dorsally, the bones articulate in a domed curve, thus forming a transverse arch.

The maintenance of these arches depends on the integrity of the tarsal, tarsometatarsal and intermetatarsal joints, because it is here that the bones are held in their proper relationships as segments of the arches. Some ligaments are consequently more important than others, being extremely strong to resist undue yielding of the joints and collapse of the arches. They are found on the plantar aspect of the joints and are themselves supported by the plantar aponeurosis and the intervening musculature.

As well as the movements of plantarflexion and dorsiflexion of the foot, both of which occur at the ankle, the foot can also be adducted or abducted about the long axis of the leg (Fig. 3.137A), and pronated or supinated about its own longitudinal axis (Fig. 3.137B). The movements of adduction (toes pointing towards the midline) and abduction take place in a transverse plane, and are only possible when the knee is flexed, when axial rotation of the tibia at the knee is possible. The adduction/abduction which occurs when the knee is extended is the result of medial and lateral rotation of the femur at the hip joint. The total range of abduction and adduction when they occur exclusively in the foot is 35–45°. However, contributions from the leg (with the knee flexed) or whole limb (at the hip joint) can increase this range to 90° in each direction, as seen in ballerinas. Movement of the foot about its own long axis causes the sole to face medially (supination) or laterally (pronation). The range of supination is about 50°, while that of pronation is only 25–30°.

Because of the arrangement of the joints of the foot, neither adduction and abduction nor supination and pronation can occur as pure movements. Adduction of the foot is always accompanied by supination, while abduction is always accompanied by pronation. The above combinations of movements are known as inversion and eversion respectively; adding plantarflexion and dorsiflexion respectively increases the range of these movements. An apparently pure movement of supination can be achieved by laterally rotating the leg at the knee to

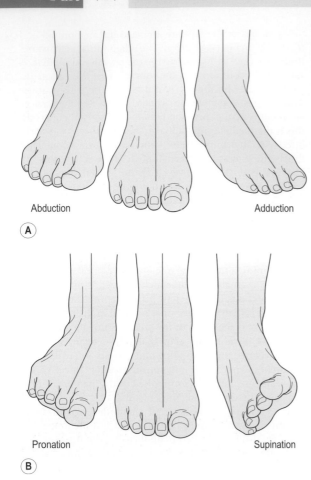

Abduction   Adduction

(A)

Pronation   Supination

(B)

Figure 3.137 (A) Abduction and adduction of the foot about the longitudinal axis of the leg, (B) pronation and supination about the longitudinal axis of the foot.

compensate for the accompanying adduction. Similarly, medial rotation of the leg at the knee can compensate for the linked abduction of the foot to produce an apparently pure movement of pronation. When balancing on one leg, a lateral rotation of the leg (i.e. relative adduction of the foot) is accompanied by pronation of the forefoot in an attempt to maintain full foot contact with the supporting surface. Supination of the forefoot accompanies medial rotation of the leg under similar circumstances.

The joints of the foot can be divided into four groups:

1. intertarsal
2. tarsometatarsal and intermetatarsal
3. metatarsophalangeal
4. interphalangeal.

Intertarsal joints

These are the subtalar, talocalcaneonavicular, calcaneocuboid, transverse (mid) tarsal, cuneonavicular, intercuneiform

and cuneocuboid joints. The transverse tarsal joint is a functional description and comprises the separate talocalcaneonavicular joint medially and the calcaneocuboid joint laterally. The most important of these joints are those between the talus, calcaneus and navicular, and between the calcaneus and cuboid. All of the joints are characterized by interosseous, dorsal and plantar ligaments, of which the plantar ligaments are much stronger than the dorsal. The bones and ligaments receive their blood supply from branches of the dorsalis pedis, medial plantar and lateral plantar arteries. They are supplied on the dorsal aspect by the deep fibular (peroneal) nerve and on their plantar aspect by the medial and lateral plantar nerves.

The subtalar joint

Articular surfaces

This is a synovial joint formed between the concave facet on the undersurface of the body of the talus and the convex posterior facet on the upper surface of the calcaneus (Fig. 3.138). The articular facet on the calcaneus is roughly oval, with the long axis running anterolaterally; it is about this axis that the facet is convex, being plane or concave about the other axis. Consequently, the joint surface can be considered to be cylindrical, with the axis of the cylinder running obliquely from anterior, lateral and superior to posterior, medial and inferior.

The corresponding surface of the talus also has this cylindrical shape with a similar radius and a similar axis.

Joint capsule

A thin, loose, fibrous capsule surrounds the joint attaching close to the margins of the articular surfaces. It is thickened medially, posteriorly and laterally forming the medial, posterior and lateral talocalcanean ligaments. The capsule is lined with synovial membrane; the joint cavity does not communicate with that of any other joint.

The anterior part of the capsule is thin and attaches to the floor and roof of the sinus tarsi. (The sinus tarsi is a narrow tunnel running obliquely forwards and laterally between the talus and calcaneus, in front of the subtalar joint. Its anterolateral end opens onto the dorsum of the foot.) Also attaching within the sinus tarsi is the posterior part of the talocalcaneonavicular joint capsule. Where the two capsules are adjacent to each other they are thickened and form the interosseous (talocalcanean) ligament.

Ligaments

Interosseous (talocalcanean) ligament

This is a strong band composed of several laminae of fibres with fatty tissue between them. The interosseous ligament can best be thought of as two thick quadrilateral bands,

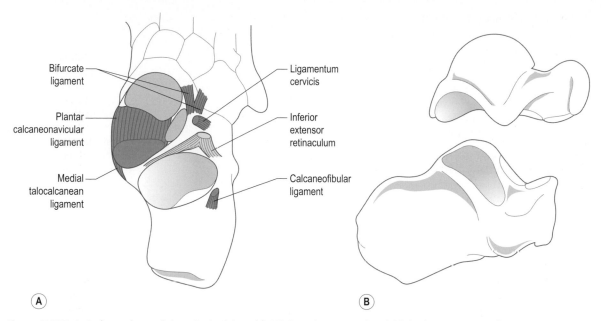

Figure 3.138 Articular surfaces of the subtalar joint with (A) the talus removed and (B) the bones separated.

anterior and posterior (Fig. 3.139). The dense fibres of the anterior band pass obliquely superiorly, anteriorly and medially from the floor of the sinus tarsi to the inferior surface of the neck of the talus just behind the articular surface of the head. The thick fibres of the posterior band run from the floor of the sinus tarsi obliquely superiorly, posteriorly and medially to just in front of the posterior articular facet of the talus.

Between the two bands of the interosseous ligament lies the deep extension of the lateral limb of the inferior extensor retinaculum, which attaches to the floor of the sinus tarsi. (Fig. 3.138).

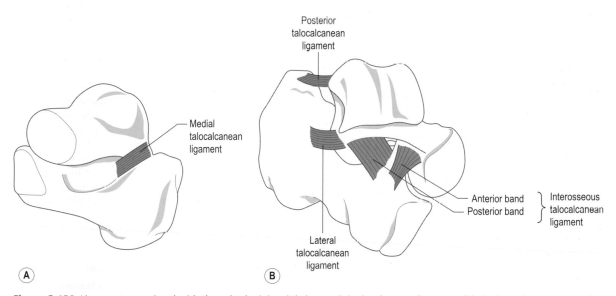

Figure 3.139 Ligaments associated with the subtalar joint: (A) the medial talocalcanean ligament, (B) the lateral, posterior and interosseous talocalcanean ligaments.

Medial talocalcanean ligament

This runs from the medial tubercle of the posterior process of the talus to the posterior border of the sustentaculum tali (Fig. 3.138A).

Posterior talocalcanean ligament

This is a short band whose fibres radiate out from a narrow attachment on the lateral tubercle of the talus to the upper and medial surfaces of the calcaneus (Fig. 3.139B).

Lateral talocalcanean ligament

This lies parallel and deep to the calcaneofibular ligament. It runs obliquely inferiorly and posteriorly from the lateral tubercle of the talus to the lateral surface of the calcaneus (Fig. 3.139B).

Ligamentum cervicis

At the lateral end of the sinus tarsi is the ligamentum cervicis (Fig. 3.138), a strong discrete band attaching to the neck of the talus above and to the calcaneus below. It forms a strong ligamentous connection between the two bones, becoming taut in inversion.

The calcaneofibular ligament and talocalcaneal part of the deltoid ligament of the ankle joint act as accessory ligaments for the subtalar joint and provide additional support.

Stability

The interosseous talocalcanean ligament plays an essential part in maintaining the stability of the subtalar joint, both at rest and during activity. It occupies a central position between the subtalar and talocalcaneonavicular joints, lying directly below the long axis of the leg (Fig. 3.140). Acting as the fulcrum around which movements of the leg and foot occur, the interosseous talocalcanean ligament is continually subjected to twisting and stretching.

The calcaneal parts of the medial and lateral ligaments of the ankle joint confer a considerable degree of stability at the subtalar joint by holding the talus between the leg and the calcaneus. In addition, the fibularis (peroneal) muscles laterally and flexor hallucis longus medially reinforce the ligamentous support. Indeed, without active muscle support the ligaments around the joint, as well as the capsule, would stretch under continuous strain.

Surface marking and palpation

The depth and complex articulations involved at this joint make surface marking and palpation impractical.

Accessory movements

With the subject lying prone with the foot overhanging the end of the bed, the talus can be gripped and stabilized by hooking one hand in front of its anterior surface. Firm pressure applied to the back of the heel with the other hand causes the calcaneus to slide forward on the talus.

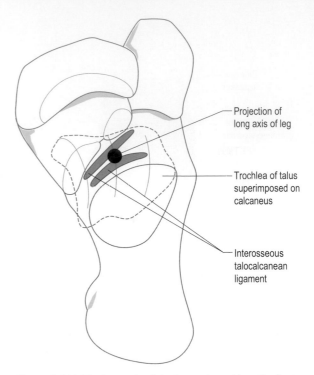

Projection of long axis of leg

Trochlea of talus superimposed on calcaneus

Interosseous talocalcanean ligament

Figure 3.140 The long axis of the leg projected into the foot showing its relationship to the interosseous talocalcanean ligament.

Section summary	
Subtalar joint	
Type	Synovial plane joint
Articular surfaces	Concave facet on under surface of talus; convex
	Facet on upper surface of calcaneus
Capsule	Thin and loose attaching to articular margins
Ligaments	Interosseous (talocalcanean)
	Ligamentum cervicis
	Medial talocalcanean
	Lateral talocalcanean
	Posterior talocalcanean
Stability	Maintained by ligaments especially the interosseous
Movements	Inversion and eversion of the foot

The talocalcaneonavicular joint

Articular surfaces

A synovial joint of the ball and socket variety, the ball being formed by a large continuous facet on the head and lower surface of the neck of the talus (Fig. 3.141A). The articular

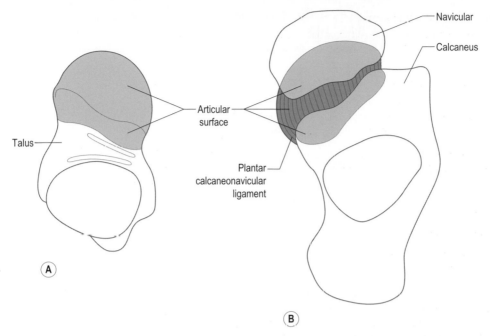

Figure 3.141 (A) Articular surfaces of the talus, (B) the calcaneus, navicular and plantar calcaneonavicular ligament forming the talocalcaneonavicular joint.

surface conforms in shape to the socket and is marked by faint ridges into a facet for the navicular anteriorly, for the sustentaculum tali posteroinferiorly with an anterolateral facet for the anterior part of the calcaneus, and a facet inferomedially for the plantar calcaneonavicular ligament.

The deep and extensive socket is formed partly by bone and partly by ligaments (Fig. 3.141B). Anteriorly the socket is formed by the concave articular surface of the navicular, and posteriorly by the concave upper part of the sustentaculum tali and a concave facet on the anterior end of the upper surface of the calcaneus. These latter two facets may be fused into a single concavity. Between the articular surfaces on the calcaneus and navicular, the head of the talus articulates with the deep surfaces of two ligaments, medially the plantar calcaneonavicular ligament and laterally the calcaneonavicular fibres of the bifurcate ligament.

Joint capsule and synovial membrane

The joint capsule is thin and encloses the common articular cavity. However, because of the nature of the bony and ligamentous socket, a true fibrous capsule is only present on the posterior and dorsal aspects of the joint. One end of the capsular sleeve attaches to the neck of the talus around the articular margin of the head. It extends forwards so that the other end of the sleeve attaches to the upper margin of the navicular, medially with the anterior fibres of the

deltoid ligament that go to the navicular, the upper medial edge of the plantar calcaneonavicular ligament, the floor of the sinus tarsi, the medial limb of the bifurcate ligament and back to the upper surface of the navicular. The posterior part within the sinus tarsi blends with the anterior part of the subtalar joint capsule to form the interosseous (talocalcanean) ligament (p. 360).

Synovial membrane lines all non-articular surfaces, including the fat pad that lies between the bifurcate and plantar calcaneonavicular ligaments on the plantar aspect of the joint. It is thought that this fat pad helps to spread synovial fluid over the moving head of the talus.

Ligaments

Two ligaments are intimately associated with the joint, their deep parts participating in the formation of the articular surfaces.

Plantar calcaneonavicular ligament

A thick, dense fibroelastic ligament of considerable strength extending from the anterior end and medial border of the sustentaculum tali posteriorly, spreading out to attach to the entire width of the inferior surface of the navicular and its medial surface behind the tuberosity (Fig. 3.142). The lower fibres of the ligament lie almost transversely across the foot. The ligament blends with,

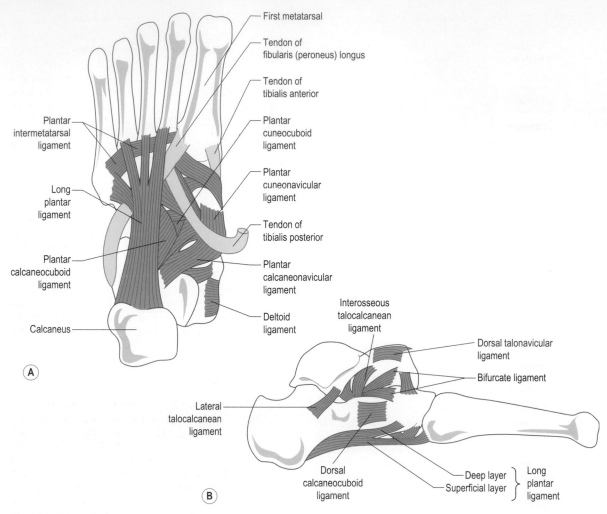

Figure 3.142 Ligaments associated with the talocalcaneonavicular joint viewed (A) from below and (B) laterally.

and is supported by, the deltoid ligament medially. The upper surface of the ligament is smooth and faceted and contains a fibrocartilaginous plate, for articulation with the head of the talus (Fig. 3.141B). Because of its elasticity under the head of the talus, the ligament is also known and referred to as the 'spring ligament'.

Bifurcate ligament

The calcaneonavicular part of the bifurcate ligament, which completes the socket laterally, is composed of short fibres that pass from the upper surface of the anterior end of the calcaneus to the adjacent lateral surface of the navicular (Fig. 3.142).

Dorsal talonavicular ligament

This runs from the neck of the talus to the dorsal surface of the navicular (Fig. 3.142), reinforcing the joint capsule dorsally between the bifurcate and the plantar calcaneonavicular ligaments.

Stability

The major elements contributing to the stability of the joint are the plantar calcaneonavicular and bifurcate ligaments described above, together with the tendon of tibialis posterior. The latter turns into the sole of the foot under the plantar calcaneonavicular ligament and therefore acts as a sling, both for the ligament and for the head of the talus. Tension within the plantar calcaneonavicular ligament, supported by tibialis posterior, resists the tendency of body weight to push the head of the talus downwards between the two bones with which it articulates.

Section summary

Talocalcaneonavicular joint

Type	Synovial ball and socket joint
Articular surfaces	Head and lower surface of neck of talus with posterior concave surface of navicular, anterior calcaneus, plantar calcaneonavicular ligament
Capsule	Attaches to articular margins
Ligaments	Plantar calcaneonavicular ('spring') ligament Calcaneonavicular part of bifurcate ligament Dorsal talonavicular ligament
Stability	Maintained by 'spring' and bifurcate ligaments, tibialis posterior
Movements	Inversion and eversion of the foot

The calcaneocuboid joint

Articular surfaces

The calcaneocuboid joint is the articulation between the facets on the anterior surface of the calcaneus and the posterior surface of the cuboid (Fig. 3.143A). The articular surfaces of the two bones are gently undulating and quadrilateral in shape. The upper part of the calcaneal facet is concave transversely and vertically, while the lower part is convex both transversely and vertically. The articular surface of the cuboid is reciprocally concavoconvex. However, there may be a medial extension of the facet which articulates with the navicular (Fig. 3.143B). If present, this extension is in the form of a plane surface.

Joint capsule and synovial membrane

A simple capsule completely surrounds the joint, separating the joint cavity from adjacent ones. It is thickened above and below by the dorsal and plantar calcaneocuboid ligaments respectively. Synovial membrane lines the inside of the fibrous capsule and attaches to the margins of the articular surfaces.

Ligaments

Dorsal calcaneocuboid ligament

The dorsal calcaneocuboid is a relatively thin broad band which strengthens the dorsal aspect of the capsule (Fig. 3.143C).

Bifurcate ligament

On the dorsomedial aspect of the joint is the calcaneocuboid part of the bifurcate ligament (Fig. 3.143C). It arises, in common with the calcaneonavicular part of the ligament, from the deep hollow on the upper surface of the calcaneus lateral to the anterior articular surface in front of the sinus tarsi undercover of extensor digitorum brevis, and attaches to the adjacent dorsomedial angle of the cuboid. This part of the bifurcate ligament is one of the main connections between the first and second row of tarsal bones.

Plantar calcaneocuboid ligament

On the plantar aspect of the joint are two special ligaments separated by areolar tissue. The deeper plantar calcaneocuboid ligament (Fig. 3.143C) blends with and reinforces the joint capsule. It is a strong, broad band of short fibres which pass forwards from a rounded eminence at the anterior end of the inferior surface of the calcaneus, to the plantar surface of the cuboid behind the ridge that bounds the fibular (peroneal) groove. The plantar calcaneocuboid ligament is sometimes known as the short plantar ligament.

Long plantar ligament

Superficial to the plantar calcaneocuboid ligament is the long plantar ligament, which covers the plantar surface of the calcaneus (Fig. 3.143C). Posteriorly it is attached between the posterior and anterior tubercles of the calcaneus. As it passes forwards the deeper fibres attach to the ridge on the cuboid, while the intermediate fibres bridge the groove on the cuboid to attach to its tuberosity, so forming a fibrous roof over the tendon of fibularis (peroneus) longus. The most superficial of the fibres passes forwards to attach to the bases of the lateral four metatarsals. Most of the ligament is covered by flexor accessorius (quadratus plantae), so that only its posterior part may be visible in the gap between the medial fleshy and lateral tendinous parts of the muscle. The long plantar ligament thus stretches under nearly the whole length of the lateral part of the foot, strengthening the plantar aspect of all of the joints in this region.

Stability

The calcaneocuboid joint receives the weight of the body as it is transmitted to the lateral part of the longitudinal arch of the foot. The essential stability required at this joint is provided by the plantar calcaneocuboid and long plantar ligaments. The tendon of fibularis (peroneus) longus passing anteromedially across the cuboid is an important tie reinforcing the ligaments.

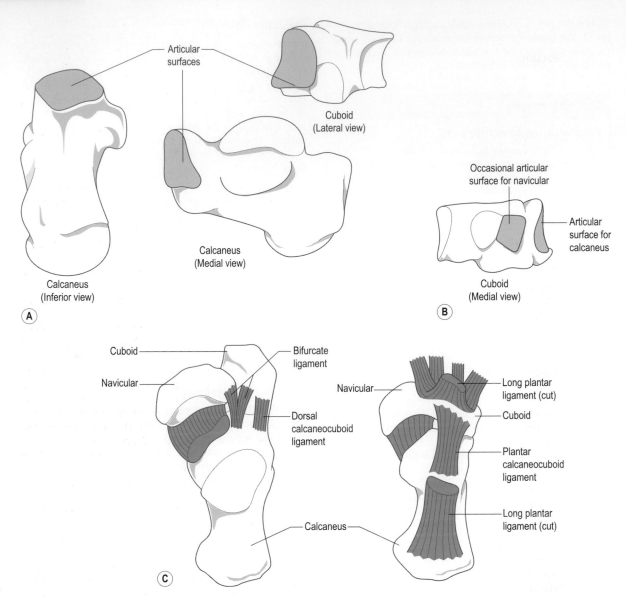

Figure 3.143 (A) Articular surfaces of the calcaneocuboid joint, (B) medial view of the cuboid showing the occasional articular surface for the navicular, (C) ligaments associated with the calcaneocuboid joint.

Surface marking and palpation

The line of the calcaneocuboid joint can be determined by applying pressure to the dorsal surface of the lateral border of the foot just proximal to the tubercle of the fifth metatarsal.

Accessory movements

With the joint line identified, grip the calcaneus firmly between the thumb and index finger of one hand, and the cuboid with the other. Holding the calcaneus still the cuboid can slide up and down against it.

The transverse (mid) tarsal joint

The transverse tarsal joint is the name given to the combined talocalcaneonavicular and calcaneocuboid joints. It provides an irregular plane extending from medial to lateral across the foot, with the talus and calcaneus behind and the navicular and cuboid in front (Fig. 3.144). Although the joints do not communicate they combine in a distinctive movement pattern which is an important contribution to the action of the foot. A small joint cavity frequently exists between the posteromedial angle of the cuboid and the lateral margin of the navicular. When present, the articulation is continuous in front with the cuneonavicular joint.

Ligaments

In addition to the ligaments associated with the component parts of the joint, additional ligaments unite the cuboid and navicular.

The *dorsal* and *plantar cuboideonavicular* ligaments pass between the adjacent parts of the corresponding surfaces of the two bones. A strong *interosseous cuboideonavicular* ligament connects the rough non-articular portions of their adjacent surfaces.

Movements of the subtalar and transverse tarsal joints

In dorsiflexion and plantarflexion the talus moves within the tibiofibular mortise, so that the foot moves as a single unit. In other movements of the foot the calcaneus and navicular move on the talus, and in so doing carry with them the distal tarsal bones and the metatarsals. The movements which occur at the subtalar and transverse tarsal joints produce inversion and eversion of the foot. These movements allow the foot to be placed firmly on slanting or irregular surfaces and yet still provide a firm base of support. In walking across sloping surfaces, for example, the upper foot has to be everted and the lower inverted. Furthermore, when turning at speed the movements of inversion and eversion are essential in order to lean sideways on a foot whose sole is flat on the ground. Most inversion/eversion movements are consequently performed on a foot anchored to the ground, with the leg and body inverting and everting above it.

In inversion, the foot is twisted so that the medial border is raised and the lateral border depressed until the sole faces medially. Inversion is a combined movement of adduction and supination of the foot, accompanied by plantarflexion at the ankle joint; indeed, adduction and supination cannot occur as pure movements, one always accompanies the other. Eversion is the opposite of inversion so that the lateral border of the foot is raised and the medial lowered so that the sole is turned laterally. Similarly, eversion is the combined movement of abduction and pronation of the foot, usually being accompanied by dorsiflexion at the ankle joint. As before, abduction and pronation cannot occur as pure movements; one is always accompanied by the other.

Inversion and eversion occur principally at the subtalar and transverse tarsal joints, although movement does occur at other tarsal and the tarsometatarsal joints, with the amount diminishing the further distal the joint. The composite movements of the calcaneus, navicular and cuboid with respect to the fixed talus are of necessity complex; nevertheless inversion and eversion can be considered to occur around a single axis. This axis runs obliquely upwards, forwards and medially beginning at the posterolateral tubercle of the calcaneus, through the sinus tarsi in the region of the ligamentum cervicis to emerge at the superomedial aspect of the neck of the talus. The subtalar and transverse tarsal joints are thus mechanically linked.

Inversion is produced by tibialis anterior and tibialis posterior, assisted occasionally by extensor and flexor hallucis longus. The dorsiflexion and plantarflexion effects of the tibialis muscles cancel out to produce inversion. Under the action of tibialis posterior, the navicular (and with it the cuboid) is pulled medially (adducted) so that the forefoot moves anteriorly and medially (Fig. 3.145A). This movement is checked by tension in the dorsal talonavicular ligament. At the same time these two bones rotate about an anteroposterior axis passing through the bifurcate ligament, which actively resists torsion and traction stresses. This rotation produces supination because the navicular is raised while the cuboid is lowered (Fig. 3.145A). The elevation and depression of the medial and lateral longitudinal arches respectively cause the sole to face medially.

Eversion is produced by the fibularis (peroneal) muscles longus, brevis and tertius. The plantarflexing actions of fibularis (peroneus) longus and brevis are offset by the

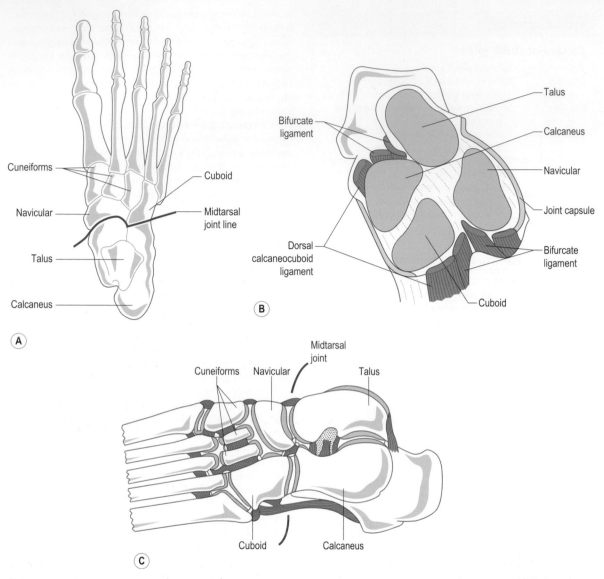

Figure 3.144 (A) Line of the midtarsal joint seen on dorsal view of foot, (B) articular surfaces of the midtarsal joint (opened), (C) oblique section of left foot showing the line of the midtarsal joint.

dorsiflexing action of fibularis (peroneus) tertius and extensor digitorum longus. Under the action of fibularis (peroneal) muscles the navicular and cuboid are pulled laterally (abducted) so that the forefoot moves anteriorly and laterally (Fig. 3.145B). Rotation of the two bones about the same anteroposterior axis as previously produces pronation by raising the cuboid and lowering the navicular (Fig. 3.145B). The effect on the lateral and medial longitudinal arches is to turn the sole to face laterally. Eversion is checked by impact of the talus on the floor of the sinus

tarsi, which consequently closes down. In addition, the upward movement of the cuboid on the calcaneus is limited by the anterior process of the calcaneus and tension developed in the powerful plantar calcaneocuboid ligament, which rapidly stops the joint interspace opening out inferiorly.

During inversion/eversion movements the calcaneus does not remain immobile below the talus. In inversion it is pulled anteriorly by the cuboid, while during eversion it moves posteriorly.

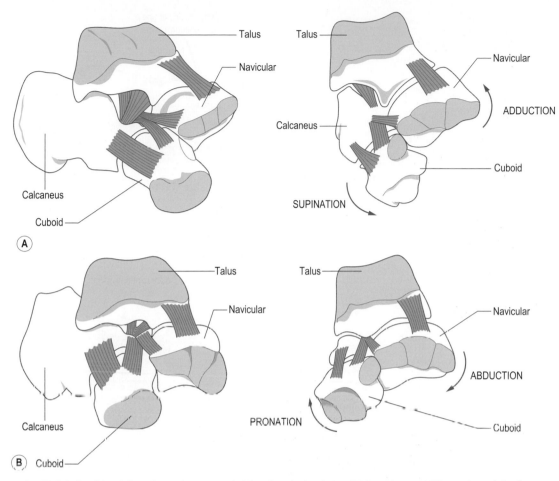

Figure 3.145 Relationship of the talus, calcaneus, cuboid and navicular during (A) inversion and (B) eversion of the foot

---

Section summary

Transverse (mid) tarsal joint

Comprises the talocalcaneonavicular and calcaneocuboid joints acting as a single functional unit.

Ligaments	Dorsal, plantar and interosseous cuboideonavicular ligaments
Movements	Movement at this joint is always accompanied by movement at the subtalar joint and vice versa
	Inversion of forefoot achieved by adduction of foot at subtalar joint and supination of forefoot at midtarsal joint; plantarflexion at ankle increases inversion
	Eversion of forefoot achieved by abduction of foot at subtalar joint and pronation of forefoot at midtarsal joint; dorsiflexion at ankle increases eversion

The cuneonavicular joint

Articular surfaces

The anterior surface of the navicular is generally convex but shows three distinct articular facets (Fig. 3.146A), separated by more or less vertical ridges, for the concave posterior ends of the three cuneiform bones. The articular surfaces glide against each other and move apart so that the inter-space of the joint gapes slightly.

Joint capsule and cavity

A fibrous capsule surrounds the joint, which is distinct on all sides except laterally where it may communicate with the cuneocuboid joint, and always with the cuboideonavicular joint when present. Anteriorly the joint cavity forms recesses between the cuneiforms (Fig. 3.146B), which may communicate with the tarsometatarsal joint between the intermediate cuneiform and the second and

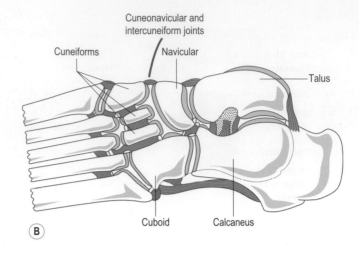

Navicular
(Anterior view)

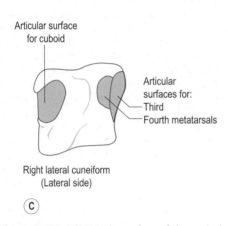

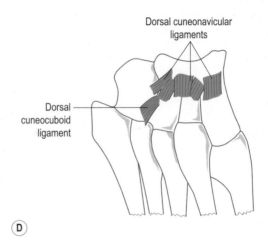

Figure 3.146 (A) Articular surface of the navicular for the cuneiforms, (B) oblique view of the foot showing the cuneonavicular and intercuneiform joints, (C) articular surface on the lateral cuneiform for the cuboid, (D) dorsal cuneonavicular and cuneocuboid ligaments.

third metatarsals, and with the intermetatarsal joints between the second and third, and third and fourth metatarsals.

Ligaments

Dorsal cuneonavicular ligaments

Relatively weak short bands in the upper and medial parts of the capsule passing from the navicular to each cuneiform (Fig. 3.146D).

Plantar cuneonavicular ligaments

Stronger bands on the plantar surface of the joint blending with and inseparable from the slips of insertion of the tibialis posterior tendon.

The intercuneiform joints

The cuneiforms articulate by plane synovial joints on the posterior aspect of their adjacent surfaces (Fig. 3.146B). The three cuneiforms are bound together by the relatively weak, transverse dorsal intercuneiform ligaments, and the much stronger interosseous and plantar intercuneiform ligaments. The latter two ligaments usually form the anterior boundary of the joint cavities.

The cuneocuboid joint

Articular surfaces

The articulation is between a round facet on the posterosuperior surface of the cuboid and a large round articular surface on the posterolateral aspect of the lateral cuneiform (Fig. 3.146C).

Ligaments

Dorsal cuneocuboid ligament

A weak band blending with the capsule passing between the dorsal surfaces of the two bones (Fig. 3.146D).

Plantar cuneocuboid ligament

A weak ligament attaching to the adjacent surfaces of the bones, again blending with the joint capsule.

Interosseous cuneocuboid ligament

This is stronger than the dorsal and plantar ligaments and unites the adjacent surfaces of the bones. It limits the joint cavity anteriorly.

Stability

The cuboid and the three cuneiform bones, being placed side by side across the foot, form the tarsal part of the transverse arch. Stability and integrity of the joints are maintained by the strong interosseous and plantar intercuneiform and cuneocuboid ligaments. The tendon of fibularis (peroneus) longus passing transversely across the foot provides additional support.

Movements

The cuneonavicular, intercuneiform and cuneocuboid joints permit a slight gliding movement between the adjacent bones, which contributes to the flexibility and adaptability of the foot.

Accessory movements

The basic principle of demonstrating accessory movements at the cuneonavicular, intercuneiform and cuneocuboid joints is the same; one of the bones is held steady while the other is moved against it. This is achieved by gripping the two bones between the thumb and index finger, one in each hand. A gliding of one bone against the other is produced by holding one steady and applying pressure to the other. For example, if the navicular is held steady the medial cuneiform can be felt to move up and down against it.

Surface markings and palpation

The tuberosity of the navicular can be readily palpated distal to the head of the talus on the medial border of the foot. With careful and fairly deep palpation the transverse running joint lines of the cuneonavicular joints and the anteroposterior joint lines of the intercuneiform joints can be identified on the dorsum of the foot.

The tarsometatarsal joints

These are between the four anterior tarsal bones (cuboid and the three cuneiforms) and the bases of all five metatarsals (Fig. 3.147A). The 'line of the joint' is irregular and arched, yet it is fairly mobile.

Articular surfaces

The tarsometatarsal joints are all plane synovial joints which overlap one another. The first metatarsal articulates only with the medial cuneiform. The base of the second metatarsal is held in a mortise formed by the three cuneiforms and therefore articulates with all of them – the lateral side of the medial, the distal end of the intermediate, and the medial side of the lateral cuneiform (Fig. 3.147A). Because of this deep mortise, the second metatarsal is the least mobile of the metatarsals. The third metatarsal articulates with the lateral cuneiform only. The fourth metatarsal articulates mainly with the cuboid, but also to a small extent with the lateral cuneiform, while the fifth metatarsal articulates only with the cuboid.

Joint cavities, capsule and synovial membrane

The presence of two strong interosseous tarsometatarsal ligaments divides the joint cavity into three separate parts (Fig. 3.147D). The medial cavity is confined to the articulation between the first metatarsal and the medial cuneiform. The intermediate cavity includes the articulations between the second and third metatarsals and the intermediate and lateral cuneiforms respectively. This cavity is usually prolonged forwards between the bases of the two metatarsals. In addition it may communicate behind with the cuneonavicular joint between the intermediate and medial cuneiforms. The lateral cavity is the articulation between the fourth and fifth metatarsals and the cuboid; it extends forwards between the bases of the two metatarsals.

The joint capsules surrounding the tarsometatarsal joints attach to the articular margins of the various bones. In practice there are very few true capsular fibres as the interosseous, plantar and dorsal tarsometatarsal ligaments close the joints.

Synovial membrane lines the non-articular joint surfaces, attaching to the articular margins.

Ligaments

Dorsal tarsometatarsal ligaments

These are weak, short slips that pass between the adjacent dorsal surfaces of the tarsal bones and metatarsals, each metatarsal receiving a slip from the tarsal bones that it articulates with (Fig. 3.147B).

Plantar tarsometatarsal ligaments

These are similar bands on the plantar surface of joints but are generally less well organized, consisting of both longitudinal and oblique fibres (Fig. 3.147C). Those ligaments associated with the medial two metatarsals are the strongest, with the remaining tarsometatarsal joints being strengthened by fibres from the long plantar ligament.

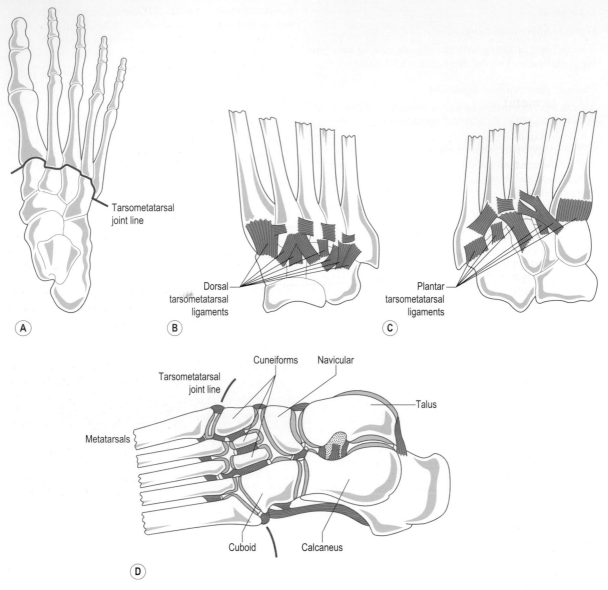

Figure 3.147 (A) Dorsal aspect of foot showing the line of the tarsometatarsal joint, (B) the dorsal tarsometatarsal ligaments, (C) the plantar tarsometatarsal ligaments, (D) oblique view of foot showing the tarsometatarsal and intertarsal joints.

Interosseous tarsometatarsal ligaments

Two interosseous tarsometatarsal ligaments are always present; there may occasionally be a third. The first interosseous ligament passes from the anterolateral surface of the medial cuneiform to the medial side of the base of the second metatarsal. The second interosseous ligament passes from the anterolateral angle of the lateral cuneiform to the medial surface of the base of the fourth metatarsal. Between these two ligaments is found the third, inconstant interosseous

ligament. When present it passes from the lateral side of the second metatarsal to the lateral cuneiform.

Stability

The various ligaments associated with the joints convey a certain degree of stability, particularly medially. Stability is further reinforced on the plantar surface medially by the insertion of muscles. Slips from tibialis posterior tendon reinforce the plantar aspects of the joints of the medial three metatarsals. The insertion of tibialis anterior to the medial sides of the first metatarsal and medial cuneiform, and of fibularis (peroneus) longus laterally, further strengthens and stabilizes the tarsometatarsal joint of the great toe.

The intermetatarsal joints

The bases of the lateral four metatarsals articulate by small synovial joints between facets on their adjacent sides (Fig. 3.147D). There is no joint between the bases of the first and second metatarsals as they are joined together by interosseous fibres only. The joint spaces between the second and third, and third and fourth metatarsals are forward extensions of the intermediate tarsometatarsal joint cavity, while that between the fourth and fifth metatarsals is continuous with the lateral tarsometatarsal joint cavity.

The various joint spaces are closed on their dorsal and plantar aspects by transverse running dorsal and plantar ligaments passing between the adjacent surfaces of the bases of the metatarsals. Anteriorly the joint spaces are limited by the strong interosseous metatarsal ligaments.

Stability

The stability of these joints is due to the dorsal, plantar and interosseous metatarsal ligaments. In particular the strong bands of the interosseous ligaments help to maintain the transverse arch of the foot by holding together the bases of the metatarsals as segments of that arch.

Surface markings and palpation

The transverse joint line between the base of the first metatarsal and the medial cuneiform is readily palpable on the medial side of the foot. Similarly, on the lateral side the joint line between the base of the fifth metatarsal and the

cuboid is easily recognized. However, the joint lines of the second, third and fourth tarsometatarsal joints are not so easy to determine. The lines of the metatarsal joints between the first and second, and fourth and fifth metatarsals can be identified running anteroposteriorly between the bones. Once again, the intervening joint lines are difficult to determine.

Accessory movements

Accessory movements between adjacent bones can be demonstrated by holding one bone steady and moving the other. Perhaps the most impressive of these movements that can be demonstrated in this region is that between the base of the fifth metatarsal and the cuboid.

Blood and nerve supply

The blood supply to the tarsometatarsal and intertarsal joints is from branches of the *dorsalis pedis* on their dorsal aspects and the *medial* and *lateral plantar arteries* of their plantar aspect.

Similarly, the nerve supply is by twigs from the deep fibular (peroneal) nerve dorsally and the medial and lateral plantar nerves on their plantar aspect.

Movements

The interlocking of the bones at the tarsometatarsal joints and the strong intermetatarsal ligaments allows, except at the first tarsometatarsal joint, only a small degree of movement. In spite of the small movement possible the joints contribute to the flexibility of the foot, particularly to inversion and eversion.

As a whole, the line of the tarsometatarsal joints runs obliquely from medial, superior and anterior to lateral, inferior and posterior, so that its medial end lies approximately 2 cm anterior to the lateral end (Fig. 3.148A).

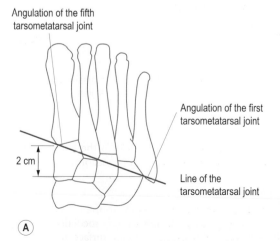

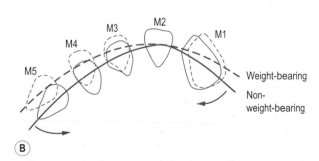

Figure 3.148 (A) The general obliquity of the tarsometatarsal joint axis, showing that the medial end of the joint line is some 2 cm anterior to the lateral end, (B) change in the height of the transverse metatarsal arch upon weight-bearing.

Although the joint line as a whole is oblique, the two ends have opposite obliquity, with the medial joint space running anterolaterally and the lateral running anteromedially. The obliquity of the axis, which essentially only allows flexion and extension movements to occur, is a contributory factor to inversion and eversion. Consequently, the axis of flexion and extension of the lateral tarsometatarsal joints is oblique to the long axis of the metatarsals, so that during plantarflexion the metatarsals move towards the axis of the foot. In other words, flexion of the metatarsals is accompanied by adduction. Similarly, because of the obliquity of the first tarsometatarsal joint, plantarflexion of the first metatarsal is accompanied by about 15° of adduction (Fig. 3.148B).

This plantarflexion and adduction of the first metatarsal is accompanied by a slight degree of rotation. The adduction movements of the metatarsals are assisted by the shapes of the articular surfaces of the cuboid and cuneiforms.

As the heads of the metatarsals move inferiorly and towards the axis of the foot in plantarflexion, this increases the curvature of the transverse arch with a hollowing of the anterior part of the foot (Fig. 3.148B). Conversely, dorsiflexion at the tarsometatarsal joint, in which the metatarsals move superiorly and away from the axis of the foot, causes a flattening of the transverse arch.

The mobility of the first metatarsal in comparison to the remaining metatarsals is necessary during adaptation of the foot to the ground in inversion and eversion, as when walking over rough terrain. In contrast, the immobility of the second metatarsal, wedged between the medial and lateral cuneiforms, together with the slenderness of its shaft are contributory factors in 'spontaneous' fracture of the bone.

The metatarsophalangeal joints

These are synovial condyloid joints between the rounded head of the metatarsal and the cupped base of the proximal phalanx (Fig. 3.149). The convex articular surfaces of the metatarsals cover the dorsal, distal and plantar surfaces, with the plantar surface being more extensive to facilitate plantarflexion at the joint.

Joint capsule and capsular thickenings

The joint capsule is loose and attaches close to the articular margins of the bones (Fig. 3.150A). It is lined by synovial membrane, which also attaches to the articular margins. The capsule is reinforced laterally by strong collateral ligaments, on its plantar surface by the plantar ligament, and dorsally by fibres from the extensor tendons.

Ligaments

Collateral ligaments

The strong collateral ligaments of each joint pass from the tubercles on each side of the head of the metatarsal, fanning out to attach to the sides of the base of the phalanx as well as to the sides of the plantar ligament (Fig. 3.150A). Passing obliquely downwards and forwards, they become tense during flexion of the joint, and therefore serve to restrict this movement.

Plantar ligament

The plantar ligament is a dense fibrocartilaginous plate firmly attached to the plantar border of the base of the proximal phalanx, forming part of the articular surface for the metatarsal head, attaching at the sides to the collateral and deep transverse metatarsal ligaments. In the great

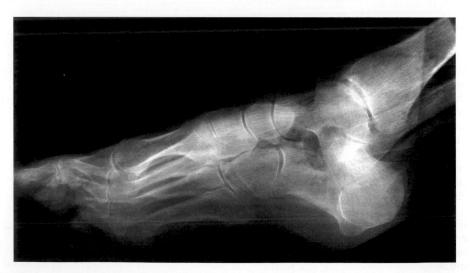

Figure 3.149 Oblique radiograph of the foot showing the metatarsophalangeal and interphalangeal joints.

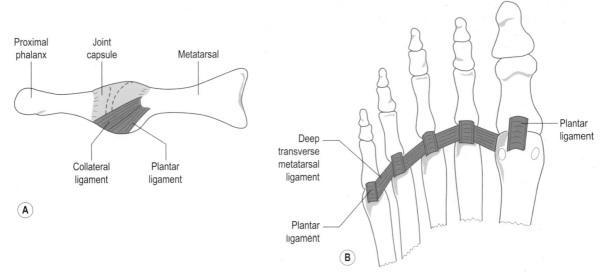

Figure 3.150 (A) An example of a metatarsophalangeal joint showing the arrangement of the collateral and plantar ligaments, (B) the relation of the plantar ligaments to the deep transverse metatarsal ligaments.

toe, the sesamoid bones and their interconnecting ligamentous band almost completely replace the plantar ligament. The sesamoid bones are cartilage covered on their upper surface and articulate with grooves on the undersurface of the head of the first metatarsal. The plantar ligaments are grooved by the long flexor tendons passing to the toes.

Deep transverse metatarsal ligament

The plantar ligaments of all the joints are interconnected by the deep transverse metatarsal ligament, which connects the heads and joint capsules of all the metatarsals (Fig. 3.150B). The ligament is crossed on its plantar surface by the tendons of the lumbricals, and on its dorsal surface by the interosseous tendons.

Stability

The metatarsophalangeal joints have the long extensor and flexor tendons crossing on their dorsal and plantar surfaces respectively. In the lateral four toes these tendons provide some stability for the joints. However, the great toe has no extensor expansion or flexor sheath, the long tendons being held in place by strands of deep fascia. If the phalanges of the great toe become displaced laterally and the fibrous sheaths give way, then the pull of extensor hallucis longus, like that of extensor hallucis brevis, becomes oblique to the long axis of the toe, tending to increase the hallux valgus deformity.

In rheumatoid arthritis, the metatarsophalangeal joints often assume a dorsiflexed position due to the imbalance in muscle tension across the joint. There may be dislocation of the sesamoid bones so that they become repositioned in the first web space. In association with the dorsiflexed

position of the joint, the metatarsal heads become depressed and subcutaneous. If the depression is severe it produces a plantarward convex arch.

Movements

The metatarsophalangeal joints permit dorsiflexion, plantarflexion, abduction, adduction and circumduction.

Dorsiflexion is carried out by the extensors, and is such that the proximal phalanx can be carried beyond the line of the metatarsal (Fig. 3.151A). During dorsiflexion the toes tend to be spread apart and become slightly inclined laterally.

Plantarflexion is performed by the flexor tendons passing to the digits (Fig. 3.151A). During this movement the toes tend to be pulled together.

Because of the arrangement of the interosseous muscles and the immobility of the second metatarsal, abduction and adduction take place about the second toe. Abduction is produced by the dorsal interossei and abductors hallucis and digiti minimi, adduction by the plantar interossei and adductor hallucis.

Accessory movements

With the head of the metatarsal gripped and stabilized between the thumb and index finger of one hand, and the adjacent phalanx held with the other, the phalanx can be slid up and down as well as rotated with respect to the metatarsal.

Surface markings and palpation

With the metatarsophalangeal joints flexed, the heads of the metatarsals stand out and can be easily palpated on the dorsal surface of the foot. If the toes are moved slowly

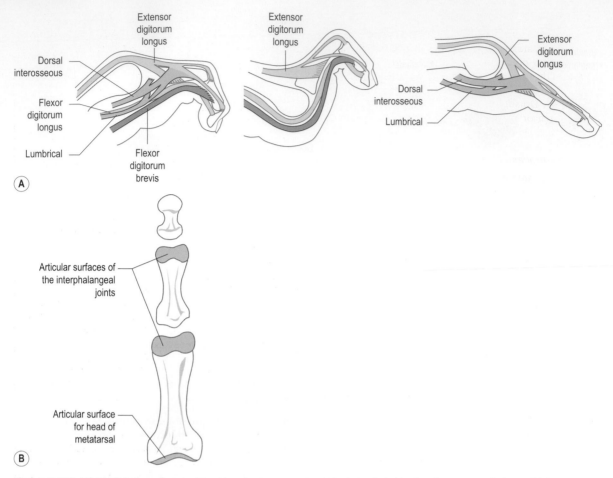

Figure 3.151 (A) Medial view of second toe showing movements at the interphalangeal and metatarsophalangeal joints, (B) articular surfaces of the interphalangeal joints, plantar aspect.

back to the neutral position, the line of each of the metatarsophalangeal joints can be identified on palpation of the dorsal aspect of the foot.

The interphalangeal joints

Articular surfaces

The head of the more proximal phalanx articulates with the base of the next distal phalanx in the series (Fig. 3.149). The heads of the phalanges have a pulley-shaped articular surface, which appears as a double convexity, and articulate with a double concavity on the base of the more distal phalanx (Fig. 3.151B).

Joint capsule and capsular thickenings

As with the metatarsophalangeal joints, the joint capsule attaches close to the articular margins, being reinforced or often replaced by collateral ligaments at the sides, the thick plantar ligament, and the expansion of the extensor

tendon, where appropriate dorsally. The joint capsule is lined by synovial membrane which attaches to the articular margins. The distal interphalangeal joint of the little toe is often obliterated.

Ligaments

Collateral ligaments

The strong collateral ligaments of each joint pass downwards and forwards from the sides of the head of each phalanx, adjacent to the articular surface, to the base of the more distal phalanx, again attaching adjacent to the articular surface (Fig. 3.150). The ligaments also pass to the sides of the plantar ligaments.

Plantar ligaments

Like that of the metatarsophalangeal joint the plantar ligament is a pad of fibrocartilage forming part of the articular surface for the phalangeal head.

Movements

Because of the shape of the articular surfaces the interphalangeal joints are hinge joints permitting dorsiflexion and plantarflexion only (Fig. 3.151A). Plantarflexion towards the sole of the foot is produced by flexors digitorum longus and brevis at the distal and proximal interphalangeal joints respectively. Dorsiflexion away from the sole is produced by the extensor muscles, as well as the lumbricals and interossei.

Accessory movements

These are the same as for the metatarsophalangeal joints except that there is little rotation.

Surface markings and palpation

The larger interphalangeal joints can be palpated on the dorsal surface of the toes, the joint lines running transversely between the bones.

Blood and nerve supply

The blood and nerve supply to both metatarsophalangeal and interphalangeal joints are from branches of the *dorsal* and *plantar digital vessels* and nerves.

Section summary

Joints of the forefoot

Cuneonavicular joint

Type	Synovial plane joint
Articular surfaces	Facets on anterior surface of the navicular with concave posterior surfaces of cuneiforms
Capsule	Surrounds the joint
Ligaments	Dorsal and plantar cuneonavicular ligaments

Intercuneiform joints

Type	Synovial plane joints

Cuneocuboid joint

Type	Synovial plane joint
Articular surfaces	Posterosuperior surface of cuboid; posterolateral surface of lateral cuneiform
Ligaments	Dorsal and plantar cuneocuboid ligaments Interosseous cuneocuboid ligaments
Stability	Maintained by associated ligaments

Tarsometatarsal joints

Type	Synovial plane joints
Articular surfaces	Anterior surfaces of cuboid and cuneiforms with bases of metatarsals
Capsule	Attached to articular margins
Ligaments	Dorsal and plantar tarsometatarsal ligaments Interosseous tarsometatarsal ligaments

Stability	Maintained by ligaments and muscles crossing and/or attaching to the bones
Movements	Dorsiflexion and plantarflexion of toes

Intermetatarsal joints

Type	Synovial plane joints between adjacent surfaces of metatarsal bases

Metatarsophalangeal joints

Type	Synovial condyloid joints
Articular surfaces	Rounded head of metatarsal; base of proximal phalanx
Capsule	Loose, replaced by plantar ligament on plantar surface
Ligaments	Collateral ligaments Plantar ligaments Deep transverse metatarsal ligament
Movements	Plantarflexion and dorsiflexion of toes

Interphalangeal joints

Type	Synovial hinge joint
Articular surfaces	Head of proximal phalanx with base of next distal phalanx
Capsule	Completely encloses joint; reinforced by collateral ligaments; replaced by plantar ligament on plantar surface
Ligaments	Collateral ligaments Plantar ligament
Stability	By ligaments and tendons crossing joint
Movements	Plantarflexion and dorsiflexion of toes

Palpation of the foot

The medial and lateral malleoli are readily palpable (p. 350). Behind each malleolus is a hollow behind which the tendocalcaneus is easily palpated; it can be traced on to the prominent projection of the calcaneus at the heel.

Below the lateral malleolus and slightly anterior to its tip is the fibular (peroneal) tubercle on the lateral side of the calcaneus. In front of the lateral malleolus the talus is palpable for a short distance. If the foot is inverted the head of the talus forms a marked projection on the superolateral aspect of the foot approximately 4 cm anterior to the lateral malleolus. In a slender foot, the distal end of the calcaneus can be felt leading to a hollow over the cuboid to the projection of the base of the fifth metatarsal.

Medially below the malleolus the talus can be felt, especially if the foot is everted, with the sustentaculum tali of the calcaneus below it. Below the sustentaculum tali is the prominence of the heel. Anterior to the medial malleolus the tubercle of the navicular can be palpated, and anterior to that the medial cuneiform, then the base of the first metatarsal. Moving laterally from the medial cuneiform, the remaining cuneiforms and then the cuboid can be identified.

Along the borders of the foot the shafts leading to the heads of the first and fifth metatarsals can be palpated, as can the shafts of all the metatarsals on the dorsum of the foot.

On the sole of the foot, the heads of the metatarsals can be readily palpated, particularly with the toes dorsiflexed. The phalanges may be palpated at their sides along their length.

With care and practice each of the tarsal bones and metatarsals can be identified and palpated. Consequently, in a relaxed position, the integrity and mobility of any of the joints of the foot can be tested by applying firm pressure to the appropriate bones in the direction of movement.

On the dorsum of the foot the tendons of tibialis anterior, extensors hallucis and digitorum longus, and fibularis (peroneus) tertius can be seen and traced from their relative positions in front of the ankle joint to their insertions (Fig. 3.152A). The fleshy belly of extensor

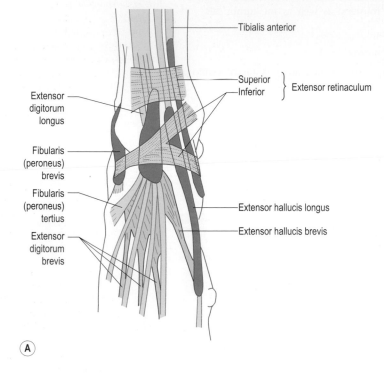

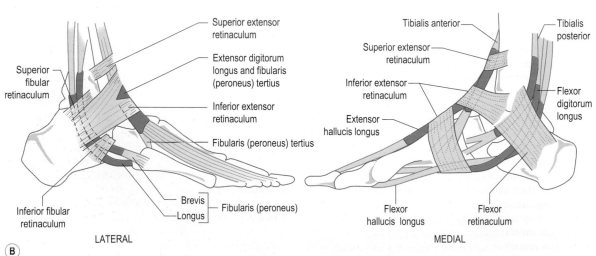

Figure 3.152 (A) Relationship of the tendons on the dorsum of the foot, showing also the extent of their associated synovial sheaths, (B) lateral and medial view showing the long tendons of the calf muscles entering the foot, together with the extent of their synovial sheaths.

digitorum brevis can be seen as a swelling lateral to the digitorum longus tendons, on the dorsal surface of the calcaneus beyond the ankle joint, from which the individual tendons can be traced to the medial four toes.

Running behind the lateral malleolus, the tendons of fibularis (peroneus) longus and brevis can be palpated, separating to pass below and above the fibular (peroneal) tubercle respectively (Fig. 3.152B); fibularis (peroneus) brevis can be traced to its insertion while fibularis (peroneus) longus disappears deep to the plantar aponeurosis.

The tendons of tibialis posterior and flexor digitorum longus can be identified behind the medial malleolus but soon pass into the sole of the foot (Fig. 3.152B). In a slender foot, the tendon of flexor hallucis longus may occasionally be felt in the hollow between the flexor digitorum longus tendon and the tendocalcaneus.

The dorsalis pedis artery, a continuation of the anterior tibial artery, passes into the foot lateral to the tendon of extensor hallucis longus (Fig. 3.152A). Its pulsations can be felt over the medial cuneiform before it passes through the space between the first and second metatarsals. The posterior tibial artery passes behind the medial malleolus midway between the malleolus and the tendocalcaneus (Fig. 3.127A). Its pulsations can be felt by applying light pressure against the lower part of the tibia; it is at this point that the artery can be compressed to stem arterial bleeding.

The dorsal venous arch can usually be seen on the dorsum of the foot, and arising from its medial and lateral ends the long (great) and short (lesser) saphenous veins respectively. The long saphenous vein passes into the leg in front of the medial malleolus, while the short saphenous vein passes behind the lateral malleolus (Fig. 3.127A).

Biomechanics

Owing to its wide variety of functions the foot may be considered as one of the most dynamic structures within the body. It provides physical contact with the environment, being strong to support the weight of the body, yet is also flexible and resilient to absorb shocks transmitted to it and to provide the spring and lift for many activities. Any changes in its structure and/or flexibility will modify its function, resulting in changes in the way the foot is used during activity.

The arched structure of the foot involving a number of bones and their interconnecting joints, together with numerous ligaments and muscle action, give it the stability and flexibility necessary for function. Of the longitudinal arches, the medial has a greater curvature and is characterized by its remarkable elasticity. The more rigid and flatter lateral longitudinal arch makes contact with the ground and thereby provides a firm base for support in the upright position. The factors maintaining the integrity of the arches of the foot are the same as those in any joint of the body,

i.e. ligaments and muscles. However, their relative importance is different in each of the three arches of the foot.

In the medial longitudinal arch, ligaments are important, but by themselves are not capable of maintaining the arch. The plantar aponeurosis is the most important ligament stretched between the supporting pillars of the arch (Fig. 3.153). Extension of the toes tightens the aponeurosis bringing the two pillars together and increases the concavity of the arch. The plantar calcaneonavicular (spring) ligament supports the head of the talus preventing it sinking between the navicular and calcaneus (Fig. 3.153). When this ligament is stretched the medial arch decreases in height. As well as these two major ligamentous components, all the interosseous ligaments assist in maintaining the medial arch. However, muscles are indispensable in the maintenance of this arch, for if they are paralysed or weakened the ligaments alone cannot cope.

In a living foot, the medial arch cannot be flattened because the ligaments are too strong. In static standing the ligaments are the primary support; indeed the muscles are relaxed in this condition. Muscular support of the foot only becomes significant in the larger movements of shifting the weight in the erect posture and in more extensive body movements. If the muscles become weakened the ligaments become progressively stretched and flat-foot results. (Flat-foot is usually accompanied by eversion and adduction of the forefoot.)

The most important and efficient muscular supports are those that run longitudinally beneath the medial arch (Fig. 3.153). Of these, flexor hallucis longus is the most efficient, being the largest of the deep calf muscles. As well as running to the great toe it also gives a slip to the tendons of flexor digitorum longus passing to the second and third toes. During short periods of standing, flexor hallucis longus is not active because body weight is supported

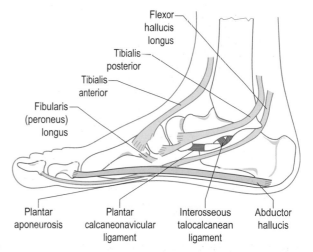

Figure 3.153 The medial longitudinal arch of the foot and muscles associated with its maintenance.

towards the back of the foot with the toe pads off the ground. If standing is prolonged, flexor hallucis longus contracts to press the toe pads against the ground so relieving the ligaments. It is during the toe-off phase of gait, and when landing on the feet, that flexor hallucis longus contracts strongly to take the greater strain placed on the medial arch. Abductor hallucis and the medial part of flexor digitorum brevis also assist in maintaining this arch.

Both tibialis anterior and posterior significantly influence the medial arch (Fig. 3.153), but in a different way to the muscles already mentioned because they have no direct action on the pillars of the arch. Their action of inverting and adducting the foot raises the medial border from the ground. Although of functional importance, these two muscles are less important factors in maintaining the integrity of the arch. Finally, fibularis (peroneus) longus may accentuate the medial arch as it pulls it towards the ground during eversion (Fig. 3.153). Fibularis (peroneal) spasm in young children, however, is usually seen as an everted flat-foot.

In the lateral longitudinal arch, ligaments play a relatively more important role than they do medially. The lateral part of the plantar aponeurosis and the various plantar ligaments are significant in their action. The thicker short plantar ligament joins the calcaneus and cuboid, while the thinner long plantar ligament helps to maintain the concavity of this arch (Fig. 3.154). Several muscles provide support for the lateral arch. Fibularis (peroneus) longus pulls upwards on the lateral arch as its tendon passes under the foot, and in addition provides elastic support for the calcaneus as it passes below the fibular (peroneal) tubercle (Fig. 3.154). It is perhaps the single most important factor maintaining the integrity of the lateral arch. Fibularis (peroneus) brevis and tertius prevent the arch opening

out inferiorly. The tendons of flexor digitorum longus to the fourth and fifth toes, assisted by flexor accessorius (quadratus plantae), together with the lateral half of flexor digitorum brevis and abductor digiti minimi help to maintain the lateral arch by preventing separation of the supporting pillars.

Although the wedge-shape of the cuneiforms, and to some extent the cuboid, suggest that the integrity of the transverse arch is maintained by bony factors, because of the shape of the medial cuneiform, it is clear that bony factors in fact play only a small part. Ligaments binding together the cuneiforms and the bases of the metatarsals are more important. However, the most important factor appears to be fibularis (peroneus) longus, as it pulls the medial and lateral borders of the foot together (Fig. 3.155). At the level of the metatarsal heads, a shallow arch is maintained by the deep transverse ligament of the metatarsal heads, by the transverse fibres binding together the digital slips of the plantar aponeurosis, and by the transverse head of adductor hallucis, which is relatively weak (Fig. 3.155). This shallow arch rests on the ground via soft tissues. The head of the second metatarsal, being the highest above the ground, is the keystone of this arch. The second and third metatarsal heads occupy intermediate positions above the ground, while the first and fifth lie approximately at the same level on the ground. The arch of the metatarsal heads is the area of culmination of the metatarsal rays of the foot. Because of the nature of the arch involving the cuneiforms and the cuboid, each metatarsal forms a different angle with the ground as it runs forwards (Fig. 3.156). The first (medial) ray is the highest, forming an angle between 18° and 25° with the ground. Passing laterally, the angle gradually decreases, being approximately 15° for the second ray, 10° for the third, 8° for the fourth, and only 5° for the fifth (lateral) ray.

Figure 3.154 The lateral longitudinal arch of the foot and muscles associated with its maintenance.

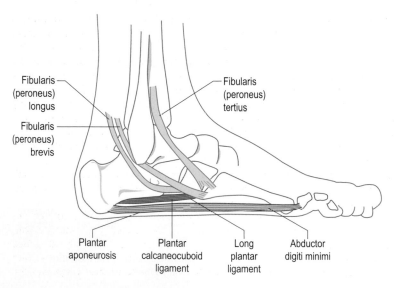

Fibularis (peroneus) longus

Fibularis (peroneus) brevis

Fibularis (peroneus) tertius

Plantar aponeurosis

Plantar calcaneocuboid ligament

Long plantar ligament

Abductor digiti minimi

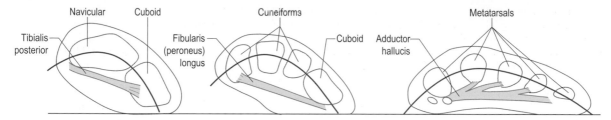

Figure 3.155 The transverse arches of the foot at different levels, together with important muscles involved in their maintenance at each level.

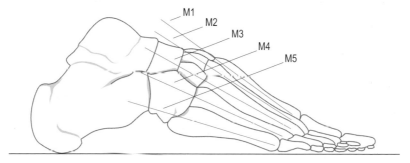

Figure 3.156 Decreasing angulation of each metatarsal with respect to the horizontal, from medial to lateral.

Having considered the various factors which contribute to the integrity of the foot and its arched structure, it is now possible to see how these mechanisms work to enable the foot to act in propulsion of the body and as a shock absorber. A rigid foot could act as a propulsive mechanism because the chief factor responsible for propulsion during walking, running and jumping is contraction of gastrocnemius and soleus, with plantarflexion of the foot at the ankle joint. This mechanism is, however, greatly enhanced and made more efficient by having a flexible foot.

During walking, the weight of the body is taken successively on the heel, the lateral border and then the ball of the foot, with the anterior pillar of the medial longitudinal arch being the last part to leave the ground. In contrast, in sprinting the heel remains clear of the ground, but the anterior pillar is still the last to leave the ground. As the heel leaves the ground during walking the medial toes are extended. Extension of the great toe pulls the plantar aponeurosis around the head of the first metatarsal, and in this way increases the height of the medial longitudinal arch. In addition flexors hallucis longus and digitorum longus are stretched, thereby increasing the force of their subsequent contraction. Contraction of the long and short toe flexors presses the toes against the ground increasing the force of toe-off. By far the most powerful muscle acting this way is flexor hallucis longus. The toes are prevented from buckling under by the action of the lumbricals.

When landing from a jump, the toes and then the forefoot take the strain before the heel touches the ground. In this way much of the damaging high energy and high frequency components associated with ground impact are dissipated. Muscles and ligaments are put under tension and stretched, absorbing as much as 50% of the energy. Even in quiet walking the muscles and ligaments play an essential role in absorbing the shock of landing. Indeed, by stretching the ligaments and muscles the energy stored is released again towards the end of the stance phase of gait, thereby contributing to propulsion and reducing the overall energy cost of walking.

Trabecular arrangement within the foot

The transmission of mechanical forces within and through the foot is reflected, as in other regions, in the arrangement of the bony trabeculae. On the medial side of the foot, following the medial longitudinal arch, trabeculae in the lower end of the tibia can be traced posteriorly and anteriorly (Fig. 3.157B). Those trabeculae arising from the cortex of the anterior surface of the tibia pass obliquely inferiorly and posteriorly through the body of the talus to the calcaneus, where they spread out, eventually reaching the posterior support of the arch, the posterior tubercle of the calcaneus. From the cortex of the posterior surface of the tibia, the trabeculae run inferiorly and anteriorly through the neck and head of the talus, the navicular and medial cuneiform to the first metatarsal.

On the lateral side of the foot, transmission of mechanical forces occurs through the talus and the underlying calcaneus. Two sets of trabeculae can be readily identified (Fig. 3.157A). Arising from the cortex of the anterior surface of the tibia, trabeculae again pass inferiorly through the

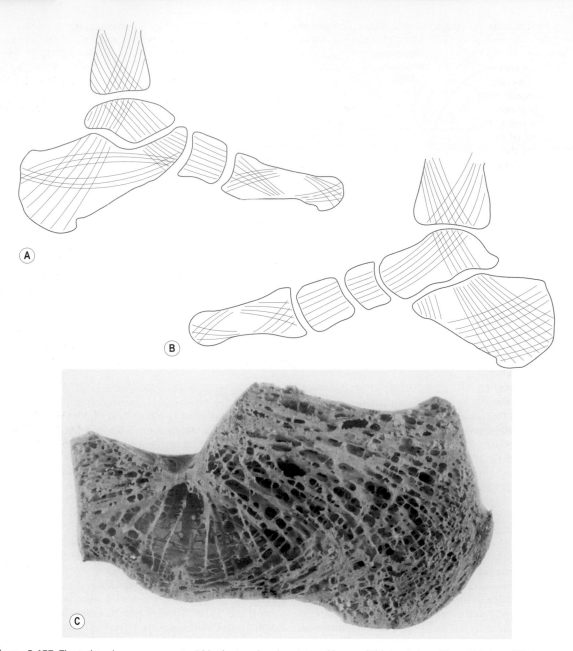

Figure 3.157 The trabecular arrangement within the tarsal and metatarsal bones: (A) lateral view, (B) medial view, (C) photograph of the trabecular arrangement in the calcaneus, sagittal section.

posterior part of the body of the talus and fan out in the calcaneus. Trabeculae from the posterior tibial shaft pass initially through the neck of the talus where it rests on the sustentaculum tali, and then run forwards through the navicular and fifth metatarsal to the anterior part of the lateral longitudinal arch.

Because of the stresses imparted to the calcaneus two additional trabecular systems are seen within it (Fig. 3.157C). A superior arcuate system, concave inferiorly, converges into a dense lamella at the floor of the sinus tarsi. These trabeculae resist compressive stresses. The second system, the inferior arcuate system which is concave superiorly, converges towards the cortical bone of the inferior surface of the calcaneus. This second system resists tension forces. Between these two systems, however, lies an area with relatively few trabeculae, and hence an area of weakness. If a sufficiently violent stress is applied vertically through the talus, injury involving this weak region may result. The long plantar ligament may be able to resist the shock, but the lateral arch gives way at the level of the anterior process of the calcaneus (the keystone of the arch), with the sustentaculum tali being broken off along a vertical line passing through the weak area. The medial tubercle of the calcaneus may also be detached along a line running sagittally. Such fractures of the calcaneus may prove difficult to reduce, as not only has the superior articular surface of the calcaneus to be restored but also the sustentaculum tali, otherwise the medial arch remains collapsed.

Distribution of stresses during static loading

Upon weight-bearing, the forces are distributed in three directions towards the supports of the plantar vault, with the result that each of the arches of the foot is flattened and lengthened.

In the medial arch, the soft tissue under the posterior tubercle of the calcaneus becomes compressed and comes to lie closer to the ground. The sustentaculum tali is lowered, and as a result the talus moves backwards on the calcaneus. As the talus is moving closer to the ground, the navicular rises on the talar head. Both the cuneonavicular and the medial tarsometatarsal joints gape inferiorly, while the angle that the first metatarsal makes with the ground reduces.

The arch is lengthened by a posterior displacement of the heel and a slight anterior displacement of the sesamoid bones under the head of the first metatarsal (Fig. 3.158A).

In the lateral arch, movements of the calcaneus are the same. Both the cuboid and the lateral tubercle of the fifth metatarsal are lowered by similar amounts. The calcaneocuboid and the lateral tarsometatarsal joints gape inferiorly, so that the head of the fifth metatarsal is displaced anteriorly, which together with recession of the calcaneus, lengthens the lateral arch (Fig. 3.158B).

The anterior arch of the foot is flattened so that the foot becomes splayed out either side of the second metatarsal (Fig. 3.158C). The transverse arch involving the cuneiforms and cuboid is also flattened (Fig. 3.158D).

The result of the flattening of these various arches is that the head of the talus and the lateral tubercle of the calcaneus become displaced medially, leading to a twisting of the foot at the midtarsal joint. The hindfoot becomes slightly adducted, pronated and extended, while the forefoot undergoes a relative movement of flexion, abduction and supination.

Because the heels receive about half the body's weight when standing erect, and more momentarily when walking, the wearing of shoes with a small heel area results in indentations being made on floors with a plastic covering.

Foot pathology

The curvature of the various arches of the foot and the orientation of their components depend upon a very delicate balance between the muscles and ligaments involved. An insufficiency or contracture of even a single muscle will

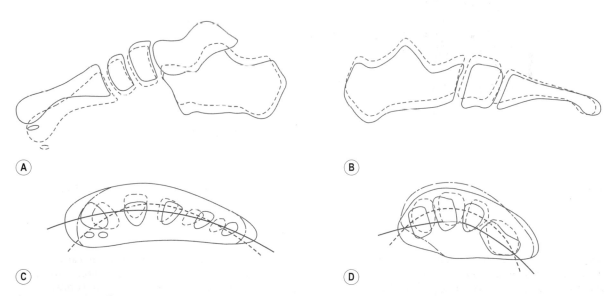

(A)

(B)

(C)

(D)

Figure 3.158 The effects of weight-bearing on (A) the medial and (B) the lateral longitudinal arches, and (C) the anterior and (D) the midfoot transverse arches. The solid line represents the relationship between segments on weight-bearing; the dotted lines show the non-weight-bearing relationships.

disrupt the overall equilibrium of the foot and lead to some form of deformity. The process may be gradual with progressively more muscles becoming involved until the foot assumes an unnatural shape and position. The simple footprint may provide a useful aid to diagnosis. For example, when compared with a normal footprint the various stages in the development of clawfoot (pes cavus) can be identified (Fig. 3.159A). In the first stage, the footprint

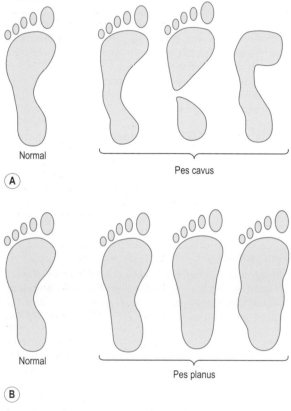

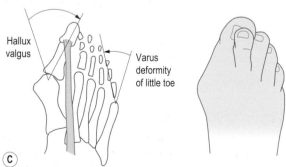

Figure 3.159 The progressive change in the footprint seen in: (A) pes cavus and (B) pes planus, (C) relationship of the skeletal elements of the toes and outward appearance of the foot in hallux valgus with an associated varus of the little toe.

shows a lateral projection on its lateral border, with a deepening of the concavity medially. The next stage shows a divided footprint, and finally the toe prints disappear due to a secondary clawtoe deformity.

In a similar manner, the progression of flat-foot (pes planus) can be seen when compared with a normal footprint (Fig. 3.159B). In this case the medial border of the foot gradually becomes filled in, and may even become convex in long-standing cases.

The muscle imbalance associated with pes cavus may also result in a secondary imbalance involving the anterior arch. The imbalance may be an overloading of the medial or lateral or both supports of the anterior arch, with callosities forming under the appropriate metatarsal heads. Occasionally the anterior arch may become flattened and splayed, with the formation of callosities under each metatarsal head.

If a widely splayed foot is confined within a pointed shoe the great toe becomes displaced laterally (Fig. 3.159C). With time the imbalance soon becomes permanent due to shortening of the capsular ligaments of the joints, lateral dislocation of the sesamoid bones and the tendon, and formation of an exostosis on the medial side of the first metatarsal head. The result is hallux valgus. The intermediate metatarsals are displaced by the great toe, exaggerating the deformity. The fifth toe may undergo a converse deformity, which further enhances the deformity of the intermediate toes (Fig. 3.159C). If the deformity is severe the anterior arch may become convex.

Both normal and pathologic foot function may be assessed clinically by observation of the subject's gait and of the pattern of wear on the soles of shoes. As suggested previously, callosities on the sole of the foot can indicate areas of excessive loading. The magnitude and duration of such loading can be assessed using suitable forms of instrumentation. From such techniques, both minor and major changes in foot function can be determined and evaluated.

NERVE SUPPLY

Introduction

The nerve supply to the lower limb is derived from the ventral rami of the first lumbar to the fourth sacral spinal nerves (Fig. 3.160). Before emerging from their appropriate intervertebral foramina, the roots of these spinal nerves have travelled for some distance in the vertebral canal. The first lumbar nerve roots separate from the spinal cord about two vertebrae above their exit, while those of the fourth sacral nerve separate from the cord at its termination, between the first and second lumbar vertebrae.

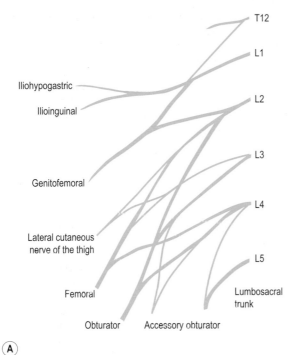

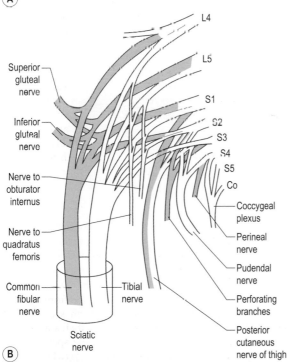

Figure 3.160 Schematic representation of the formation of (A) the lumbar plexus and (B) the lumbosacral plexus.

As each nerve passes out of the intervertebral foramen it receives a grey ramus communicans connecting it to the sympathetic trunk. The first and second lumbar roots are also connected to the trunk by a white ramus communicans.

The ventral rami of the first three and the upper part of the fourth lumbar nerves form the lumbar plexus in the substance of psoas major. The remainder of the fourth lumbar and the ventral ramus of the fifth lumbar nerve form the lumbosacral trunk which passes over the ala of the sacrum to join the ventral rami of the first, second, third and upper part of the fourth sacral nerves to form the lumbosacral plexus. The fourth and fifth sacral nerves form the sacral plexus.

The lumbar plexus

This is formed by the ventral rami of the first, second, third and part of the fourth lumbar nerves; occasionally there is a contribution from the subcostal nerve (T12). The plexus is formed within the substance of psoas major with the rami entering the muscle between its attachments to the body and the transverse process of each vertebra.

The most common arrangement of the plexus is that the first lumbar nerve, with the communication from T12, if present, divides into upper and lower branches. The upper branch again divides into the *iliohypogastric* and the *ilioinguinal* nerves. The lower part of the first lumbar nerve joins the upper part of the second to form the *genitofemoral* nerve (Fig. 3.160A).

The lower part of the second, the third and upper part of the fourth nerves all divide into smaller anterior and larger posterior divisions. The anterior divisions join to form the *obturator* nerve; branches from the third and fourth divisions may occasionally unite to form an *accessory obturator* nerve.

The posterior divisions of all three nerves join to form the *femoral* nerve, with branches from the second and third only forming the *lateral cutaneous nerve of the thigh* (Fig. 3.160A).

Direct muscular branches to psoas major (L1, 2, 3) and quadratus lumborum (L1–4) arise separately from the ventral rami.

Iliohypogastric nerve

The nerve fibres arise from L1 and occasionally T12. The nerve emerges from the upper lateral border of psoas major (Fig. 3.161) passing in front of quadratus lumborum and behind the kidney. Close to the midaxillary line, it pierces the deep surface of transversus abdominis, giving off a lateral cutaneous branch, and continues forwards in the neurovascular plane between transversus abdominis and

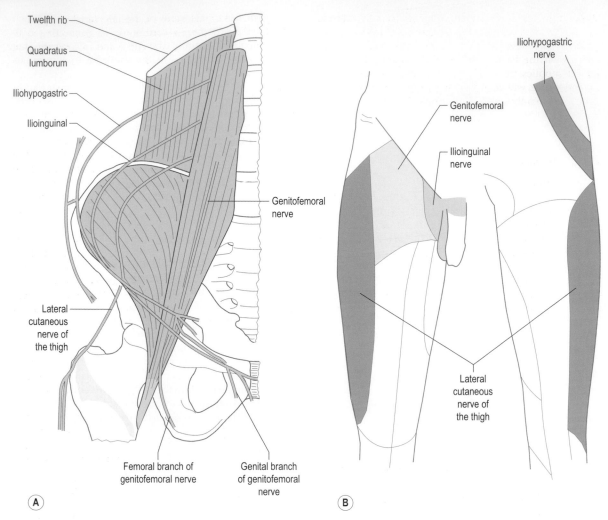

Figure 3.161 (A) Nerves arising from the upper roots of the lumbar plexus and (B) their cutaneous distribution.

internal oblique, which it usually pierces 2 cm medial to the anterior superior iliac spine. It passes medially deep to the external oblique aponeurosis, the nerve becoming cutaneous approximately 4 cm above the superficial inguinal ring. The iliohypogastric nerve supplies the muscles of the lateral abdominal wall, and a strip of skin running from the upper lateral gluteal region to just above the pubis.

Ilioinguinal nerve

The nerve fibres arise from L1 and again occasionally from T12. The nerve emerges from the lateral border of psoas major just below the iliohypogastric nerve and passes obliquely downwards around the inside of the abdomen, lying deep to quadratus lumborum (Fig. 3.161). It pierces

internal oblique passing deep to the external oblique aponeurosis and enters the inguinal canal, reaching the skin through the superficial inguinal ring and external spermatic fascia. The ilioinguinal nerve gives branches to the *internal* and *external oblique* muscles as it passes between them and is sensory to the skin over the pubic symphysis and upper medial part of the femoral triangle, as well as to part of the genitalia.

Lateral cutaneous nerve of the thigh

Formed from the posterior divisions of the second and third lumbar nerves it emerges from the lateral border of psoas major (Fig. 3.161). Passing downwards and forwards onto the pelvic surface of iliacus, it leaves the pelvis just medial to the anterior superior iliac spine, either below or through the inguinal ligament. It passes laterally through, or deep to, sartorius and then the fascia lata to become superficial. The nerve then divides into two branches which pass down the lateral surface of the thigh. The *anterior branch* supplies the anterolateral surface of the thigh as far as the knee, while the *posterior branch* supplies the lateral proximal two-thirds of the thigh below the greater trochanter.

Genitofemoral nerve

Formed within the substance of psoas major from L1 and L2 it emerges from the anterior surface of the muscle close to its medial border (Fig. 3.161), piercing the fascia as it does so. It descends downwards and forwards on the fascia, passing behind the ureter towards the inguinal ligament where it divides into the *genital* and *femoral* branches.

The genital branch enters the deep inguinal ring to pass into the inguinal canal where it supplies *cremaster*, and is then sensory to the skin of the scrotum (or labia major) and the adjacent part of the thigh.

The femoral branch passes under the inguinal ligament lateral to the femoral artery. It becomes superficial by passing through the saphenous opening and supplies the skin over the upper part of the femoral triangle.

Obturator nerve

Formed by the anterior divisions of the second, third and fourth lumbar nerves, which unite within the substance of psoas major, it emerges from the medial border of the muscle on the lateral part of the sacrum (Fig. 3.162). The obturator nerve crosses the sacroiliac joint and obturator internus to enter the obturator canal, below the superior ramus of the pubis above the obturator membrane. On leaving the canal, the nerve lies above obturator externus and divides into *anterior* and *posterior* branches. The anterior branch descends into the thigh anterior to obturator externus and adductor brevis, and posterior to pectineus and adductor longus, with its terminal twigs lying between adductor magnus and the medial intermuscular septum.

The anterior branch sends branches to *adductor longus*, *gracilis*, *adductor brevis* (usually) and *pectineus*

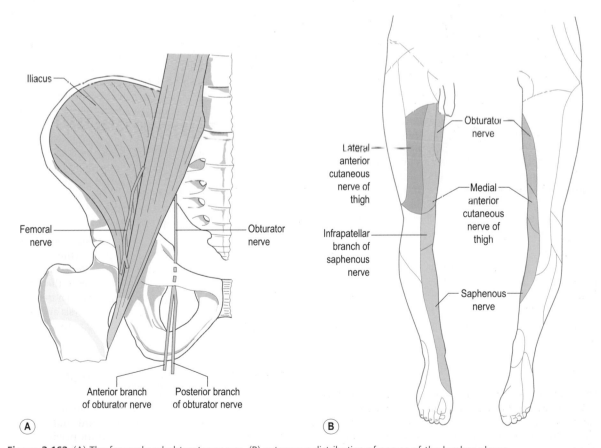

Figure 3.162 (A) The femoral and obturator nerves, (B) cutaneous distribution of nerves of the lumbar plexus.

(occasionally), and is sensory to the skin on the medial side of the thigh, the distal third of which passes via the subsartorial plexus. In addition, an articular branch reaches the hip joint via the acetabular notch.

The posterior branch pierces obturator externus and descends between adductors brevis anteriorly and magnus posteriorly. It passes obliquely through adductor magnus and enters the popliteal fossa. The nerve ends by piercing the oblique popliteal ligament to supply the posterior part of the knee joint including the cruciate ligaments.

The posterior branch supplies *obturator externus* and *adductor magnus*.

Accessory obturator nerve

When present, it arises from the anterior divisions of the third and fourth lumbar nerves (Fig. 3.160A) between the obturator and femoral nerves. It emerges from the medial border of psoas major and descends between the pelvic brim and the external iliac vessels to enter the thigh between the pubic bone and the femoral vessels. Here it usually splits into three branches: one to *pectineus*, one to the hip joint and one which communicates with the anterior branch of the obturator nerve.

Femoral nerve

Formed from the posterior divisions of the second, third and fourth lumbar nerves, it emerges from the lateral border of psoas major to run in the groove between psoas and iliacus deep to the iliac fascia (Fig. 3.162). It passes into the thigh below the inguinal ligament lateral to the femoral sheath to enter the femoral triangle, where it almost immediately divides into a number of branches which can be loosely grouped into *anterior* and *posterior divisions* passing anterior or posterior to the lateral circumflex femoral artery. The anterior division supplies *sartorius* and gives the *medial* and *lateral branches* of the *anterior cutaneous nerve of the thigh*. The posterior division supplies *quadriceps femoris* and sends articular branches to the hip and knee joints as well as giving off the *saphenous nerve*. While in the abdomen the main nerve supplies *iliacus*.

Nerves to the sartorius

The nerves to sartorius arise from the anterior division of the femoral nerve and enter the deep surface of the muscle by long and short fibres.

Anterior cutaneous nerves of the thigh

These arise as lateral and medial branches from the lateral side of the femoral nerve in the upper part of the femoral triangle. After entering the subsartorial canal and crossing to the medial side of the artery, the medial branch divides into anterior and posterior parts, becoming subcutaneous in front of and behind sartorius to

supply an area of skin over the lower part of the medial side of the thigh, knee and upper leg. The two lateral branches pass directly down the thigh, becoming superficial by piercing the fascia over sartorius to supply the skin on the anterior aspect of the thigh as far down as the knee joint.

Nerves to the quadriceps femoris

These pass to all four parts of the muscle. The nerves to *rectus femoris* and *vastus lateralis* pass with branches of the lateral circumflex femoral artery to enter the deep surfaces of the muscles. *Vastus medialis* is supplied by two branches, one of which enters proximally and also supplies *vastus intermedius*. The other branch accompanies the saphenous nerve in the adductor canal to about halfway down the thigh where it enters the medial surface of the muscle. The nerve to vastus intermedius enters the more superficial surface of the muscle, passing through it to supply *articularis genus*. All these branches to the vastus muscles give a supply to the knee joint, while the nerve to rectus femoris sends a branch to the hip joint.

Saphenous nerve

The longest of the branches of the femoral nerve. Beginning about 3 cm below the inguinal ligament, it passes through the femoral triangle to enter the adductor canal on the lateral side of the femoral vessels, giving a branch to the subsartorial plexus. The saphenous nerve pierces the roof of the adductor canal and becomes cutaneous between sartorius and gracilis posteromedial to the knee joint, to which it sends a branch. Passing behind the medial condyles of the femur and tibia, it descends along the medial side of the leg with the long (great) saphenous vein to lie anterior to the medial malleolus. It then passes to the medial side of the foot as far as the head of the first metatarsal. Branches of the nerve supply skin and fascia on the front and side of the knee, leg and foot as far as the base of the great toe.

Section summary
Obturator and femoral nerves

Obturator nerve

From	Anterior divisions of lumbar plexus
Root value	L2, 3, 4
Muscles	Adductors longus, brevis and magnus
supplied	Gracilis
	Obturator externus
	Pectineus (occasionally)

Femoral nerve

From	Posterior divisions of lumbar plexus
Root value	L2, 3, 4

Muscles supplied	Iliacus
	Sartorius
	Quadriceps femoris (rectus femoris, vastus medialis, vastus lateralis, vastus intermedius)
Cutaneous branches	Anterior cutaneous nerve of thigh
	Saphenous nerve

The lumbosacral plexus

Lying on the posterior wall of the pelvis between piriformis and its fascia the lumbosacral plexus (Fig. 3.163) is formed from the ventral rami of the fourth lumbar to fourth sacral nerves. The fifth lumbar and the lower part of the fourth form the lumbosacral trunk, which passes over the ala of the sacrum to join the laterally running ventral rami of the first to fourth sacral nerves. A grey ramus communicans joins each of the rami, while from the second, third and fourth sacral ventral rami, preganglionic parasympathetic fibres (the pelvic splanchnic nerves) pass to join the autonomic plexuses of the pelvis to supply the urogenital organs and lower third of the gastrointestinal tract.

Each ventral ramus divides into anterior and posterior divisions which all converge on the greater sciatic foramen. The following nerves are formed by the union of various anterior divisions: *nerve to quadratus femoris, nerve to obturator internus, pelvic splanchnic nerves, posterior femoral cutaneous nerve* and the *pudendal nerve*. Similarly, from posterior divisions arise the following: branches to *piriformis, coccygeus* and *levator ani*, the *superior and inferior gluteal nerves, posterior femoral cutaneous nerve, perforating cutaneous nerve* and the *perineal branch of S4*. The *sciatic nerve* consists of the medially placed *tibial nerve* (anterior divisions L4 to S3), and the laterally placed *common fibular (peroneal) nerve* (posterior divisions L4 to S2) bound together in a common sheath.

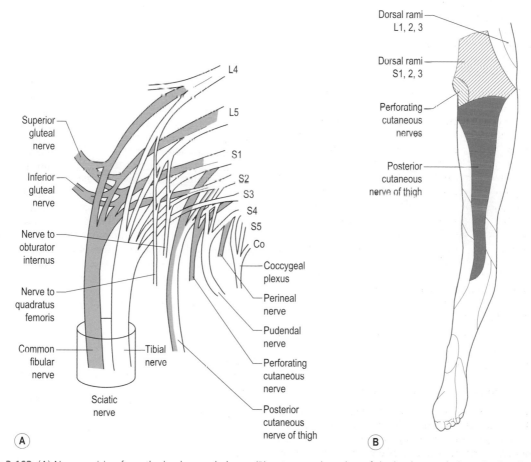

Figure 3.163 (A) Nerves arising from the lumbosacral plexus, (B) cutaneous branches of the lumbosacral plexus in the gluteal region and thigh.

Nerves to piriformis, coccygeus and levator ani

The *nerve* to *piriformis* arises from the posterior division of S2, with an occasional contribution from S1. It passes directly backwards to enter the anterior surface of the muscle. Twigs from S3 and 4 descend to supply *coccygeus* and *levator ani*. The perineal branch of S4 also supplies *coccygeus* and *levator ani*, as well as the *external anal sphincter*, and the overlying skin and fascia (see Fig. 3.166).

Superior gluteal nerve

Formed by the union of the posterior divisions of L4, 5, S1 it passes posterolaterally to leave the pelvis above piriformis with the superior gluteal vessels. It divides into superior and inferior branches, which pass forwards between gluteus medius and minimus. The superior branch supplies *gluteus medius* while the inferior supplies *gluteus medius* and *minimus* and *tensor fascia lata.*

Inferior gluteal nerve

From the posterior divisions of L5, S1, 2 it leaves the pelvis below piriformis superficial to the sciatic nerve, and passes directly into the deep surface of *gluteus maximus* which it supplies.

Posterior cutaneous nerve of the thigh

Formed from the anterior divisions of S2, 3 and the posterior divisions of S1, 2 (Fig. 3.163). It leaves the pelvis through the greater sciatic foramen on the posterior surface of the sciatic nerve below piriformis, and descends down the posterior aspect of the thigh as far as the back of the knee joint deep to the fascia lata, which it then pierces. Branches are given off which supply the skin over the lower part of the buttock, the posterior aspect of the thigh, the popliteal fossa and the upper part of the calf.

Perforating cutaneous nerve

From the dorsal divisions of S2, 3 (Fig. 3.163). It leaves the pelvis by piercing the sacrotuberous ligament, or by passing directly backwards through the medial side of gluteus maximus, to become cutaneous. It supplies the skin covering the lower part of the buttock and medial part of the gluteal fold.

Nerve to quadratus femoris

From the anterior divisions of L4, 5, S1 (Fig. 3.163) it enters the gluteal region through the lower part of the greater sciatic foramen, lying anterior to the sciatic nerve on the posterior surface of the ischium. It supplies *quadratus femoris* and *inferior gemellus* and gives an articular branch to the hip joint.

Nerve to obturator internus

From the anterior divisions of L5, S1, 2 (Fig. 3.163). The nerve passes around the ischial spine, between the sciatic nerve and the pudendal vessels, leaving the greater sciatic foramen and entering the lesser sciatic foramen. It supplies *superior gemellus* and *obturator internus*, the latter from its deep medial surface.

Pudendal nerve

The principal nerve of the perineum, usually being formed from the anterior divisions of S2, 3, 4 (Fig. 3.163). It leaves the pelvis through the inferomedial part of the greater sciatic foramen, with the pudendal vessels and nerve to obturator internus. It passes over the sacrospinous ligament to gain the lesser sciatic foramen and enter the pudendal canal on the deep surface of the obturator fascia. As well as supplying the genitalia, the pudendal nerve also supplies *levator ani* and is cutaneous to the skin around the anus.

Sciatic nerve

Formed by the ventral rami of L4, 5, S1, 2, 3 (Fig. 3.163). The sciatic nerve leaves the pelvis to enter the gluteal region through the greater sciatic foramen, below piriformis (Fig. 3.164). It passes down the back of the thigh deep to biceps femoris lying on, from above downwards, superior gemellus, obturator internus, inferior gemellus, quadratus femoris and adductor magnus. At a point, usually about two-thirds of the way down the thigh, it divides into its terminal branches – the *common fibular (peroneal)* and *tibial nerves*. However, division of the nerve may occur at a higher level in the thigh, or the two components of the nerve may be separate as they leave the pelvis. In the latter case, the tibial nerve leaves below piriformis, while the common fibular (peroneal) may leave above piriformis or pierce the muscle. The nerve can be marked on the surface by a line drawn from a point midway between the ischial tuberosity and the greater trochanter of the femur to a point two-thirds of the way down the back of the thigh where the medial and lateral hamstrings part.

The *sciatic nerve* gives articular branches to the knee joint and muscular branches to *semitendinosus, semimembranosus, biceps femoris* and the *hamstring part* of *adductor magnus*. The short head of biceps femoris is supplied by the common fibular (peroneal) part of the nerve, while the others are all supplied by the tibial part.

Tibial nerve

The medial terminal branch of the sciatic nerve, being formed by the anterior divisions of the ventral rami of L4, 5, S1, 2, 3. It continues the course of the sciatic nerve

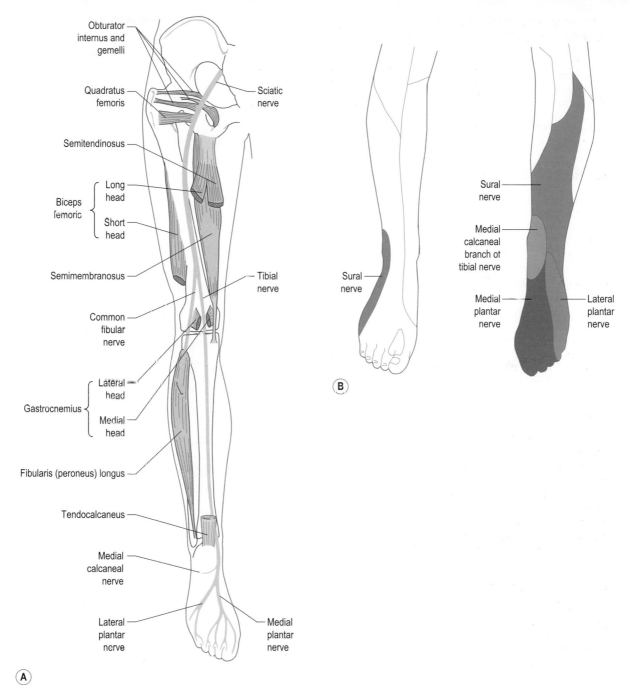

Figure 3.164 (A) The sciatic and tibial nerves and (B) their cutaneous distribution.

through the popliteal fossa (Fig. 3.164), lying at first lateral to the popliteal vessels then crossing superficially to the medial side to enter the leg under the tendinous arch of soleus. The nerve descends obliquely inferomedially between flexor digitorum longus and flexor hallucis longus

to pass behind the medial malleolus deep to the flexor retinaculum between the tendons of these same muscles. As it enters the plantar aspect of the foot the tibial nerve divides into its terminal branches the *medial* and *lateral plantar nerves*.

In the popliteal fossa the tibial nerve gives muscular branches to both heads of *gastrocnemius, soleus, plantaris, popliteus* and *tibialis posterior*. Gastrocnemius is supplied from its deep surface while soleus and plantaris are supplied from their superficial surfaces. The nerve to popliteus descends over the superficial surface of the muscle, bends around its inferior border to enter its deep surface. Articular branches are given off to the ankle, knee and superior tibiofibular joints, and cutaneous branches through the *sural nerve*. The sural nerve (Fig. 3.164) descends between the two heads of gastrocnemius, to pierce the deep fascia in the middle of the calf. It is joined by the *fibular (peroneal) communicating nerve* and passes behind the lateral malleolus and runs forward along the lateral side of the foot. The sural nerve is sensory to skin over the posterior and lateral sides of the lower third of the leg, the lateral border of the foot and fifth toe except over the distal phalanx.

In the leg, the tibial nerve gives muscular branches to the deep surface of *soleus, flexor digitorum longus, flexor hallucis longus* and *tibialis posterior*. It also gives articular branches to the ankle joint and a cutaneous branch to the heel and posterior part of the foot.

Medial plantar nerve

This passes forwards deep to abductor hallucis, accompanied on its medial side by the medial plantar artery, to the interval between the muscle and flexor digitorum brevis. It supplies *abductor hallucis, flexor digitorum brevis, flexor hallucis brevis* and the *first lumbrical*. At the base of the metatarsals, the nerve passes transversely across the foot giving cutaneous branches to the medial side of the sole and great toe and the adjacent sides of the great, second, third and fourth toes, to include the dorsal surface of the distal phalanx and nail bed (Fig. 3.164). Articular branches are given to the tarsal and tarsometatarsal joints.

Lateral plantar nerve

This passes anterolaterally towards the base of the fifth metatarsal (Fig. 3.164) between flexor digitorum brevis and flexor accessorius (quadratus plantae). It then divides into *superficial* and *deep branches*. The superficial branch runs forwards between flexor digitorum brevis and abductor digiti minimi, while the deep branch passes medially with the plantar arch on the plantar surface of the metatarsal bases. The lateral plantar nerve gives muscular branches to *flexor accessorius (quadratus plantae), flexor digiti minimi brevis, abductor digiti minimi, adductor hallucis*, the *lateral three lumbricals* and *all* the *interossei*. It also gives cutaneous branches to the lateral side of the sole and fifth toe and to the adjacent sides of the fourth and fifth toes, including the nail beds and dorsum of the distal phalanx. Articular branches pass to the tarsal and tarsometatarsal joints.

It is worth noting the similarity in the distribution of the lateral plantar nerve in the foot with the ulnar nerve in the hand, and also of the medial plantar nerve with the median nerve in the hand.

Common fibular (peroneal) nerve

The lateral terminal branch of the *sciatic* nerve, being composed of the posterior divisions of L4, 5, S1, 2. It passes along the upper lateral side of the popliteal fossa lying deep to biceps femoris and its tendon until it reaches the posterior part of the head of the fibula (Fig. 3.164). It passes forwards around the neck of the fibula within the substance of fibularis (peroneus) longus, where it ends by dividing into the *superficial* and *deep fibular (peroneal) nerves*. The nerve can be palpated behind the head of the fibula and as it winds around the neck of the fibula. The common fibular (peroneal) nerve gives articular branches to the knee and superior tibiofibular joint. The *lateral cutaneous nerve* of the calf (Fig. 3.165) supplies the posterolateral side of the proximal two-thirds of the leg. It usually arises in common with the *fibular (peroneal) communicating branch* which joins the sural nerve in the middle third of the leg.

Superficial fibular (peroneal) nerve

This descends almost vertically between extensor digitorum longus and fibularis (peroneus) longus anterior to the fibula (Fig. 3.165). About halfway down the leg it becomes superficial by piercing the deep fascia on the anterior surface and divides into its terminal branches, the *medial* and *intermediate dorsal cutaneous nerves*, which pass over the anterolateral aspect of the ankle joint to enter the foot.

The superficial fibular (peroneal) nerve gives muscular branches to *fibularis (peroneus) longus* and *brevis* and cutaneous branches to the anterolateral aspect of the leg and around the lateral malleolus. The medial dorsal cutaneous branch supplies the medial side of the dorsum of the foot, the great toe, and the adjacent sides of the second and third toes. The intermediate dorsal branch supplies the dorsum of the foot and the adjacent sides of the third, fourth and fifth toes. However, the skin over the distal phalanx is supplied by branches from the plantar nerves.

Deep fibular (peroneal) nerve

This passes inferomedially into the anterior compartment of the leg deep to extensor digitorum longus to join the anterior tibial vessels on the anterior surface of the interosseous membrane (Fig. 3.165). It descends on the interosseous membrane deep to extensor hallucis longus and the superior extensor retinaculum. At the ankle joint, the nerve lies deep to the inferior extensor retinaculum and the tendon of extensor hallucis longus which crosses it. As it

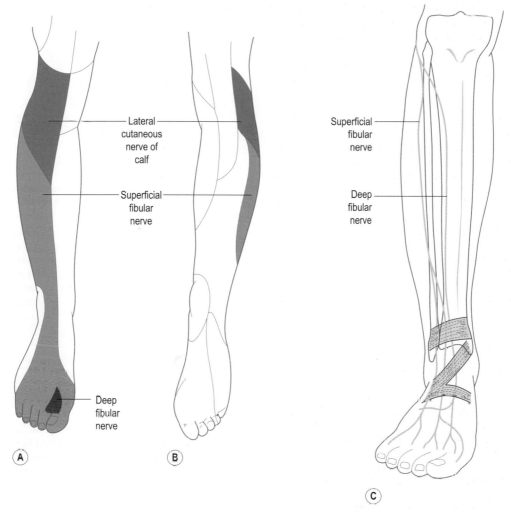

Figure 3.165 Cutaneous distribution of nerves of the leg: (A) anterior and (B) posterior aspects, (C) nerves of the anterior aspect of the leg,

Continued

enters the dorsum of the foot, the nerve lies superficially between the tendons of extensors hallucis and digitorum longus, and divides into the *medial* and *lateral branches*. The medial branch passes to the cleft between the great and second toes, supplying the skin on the adjacent sides as far as the distal interphalangeal joint. The lateral branch supplies *extensor digitorum brevis* and many of the small joints of the foot, particularly on the lateral side.

In the leg, the nerve gives muscular branches to *extensor digitorum longus, tibialis anterior, extensor hallucis longus* and *fibularis (peroneus) tertius*, and sends articular branches to the inferior tibiofibular and ankle joints.

Applied anatomy

The common fibular (peroneal) nerve is vulnerable to injury at the neck of the fibula where it may be crushed from direct trauma (for example, a kick or car bumper), or pressure (a tight immobilizing cast or bandage). This will affect both the superficial and deep fibular (peroneal) nerves with the resulting paralysis of the dorsiflexors causing a 'dropped foot'. This has a profound effect on gait and may require splinting to allow the foot to clear the ground during the swing phase of gait. While extensive, the sensory loss is of less significance.

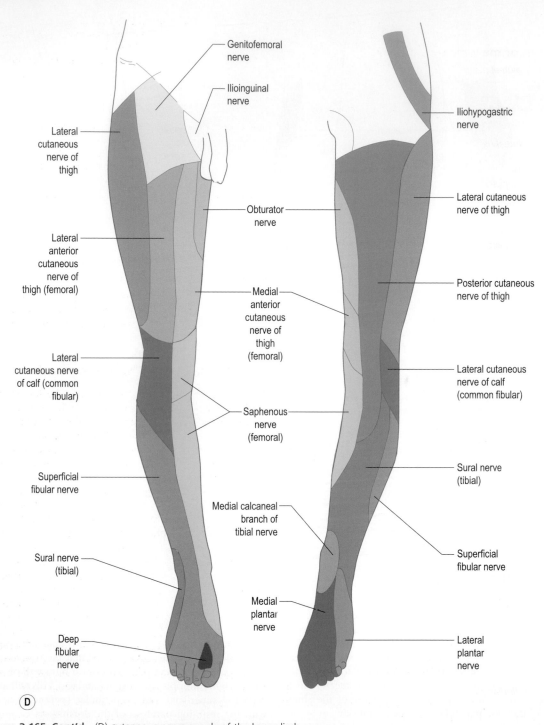

Genitofemoral
nerve

Ilioinguinal
nerve

Iliohypogastric
nerve

Lateral
cutaneous
nerve of
thigh

Lateral cutaneous
nerve of thigh

Obturator
nerve

Lateral
anterior
cutaneous
nerve of
thigh (femoral)

Posterior cutaneous
nerve of thigh

Medial
anterior
cutaneous
nerve of
thigh
(femoral)

Lateral
cutaneous nerve
of calf (common
fibular)

Lateral cutaneous
nerve of calf
(common fibular)

Saphenous
nerve
(femoral)

Sural nerve
(tibial)

Superficial
fibular nerve

Medial calcaneal
branch of
tibial nerve

Superficial
fibular nerve

Sural nerve
(tibial)

Medial
plantar
nerve

Deep
fibular
nerve

Lateral
plantar
nerve

(D)

Figure 3.165, Cont'd   (D) cutaneous nerve supply of the lower limb.

Nerves of the lumbosacral plexus

Superior gluteal nerve
From	Lumbosacral plexus
Root value	L4, 5, S1
Muscles supplied	Gluteus medius and minimus
	Tensor fascia lata

Inferior gluteal nerve
From	Lumbosacral plexus
Root value	L5, S1, 2
Muscles supplied	Gluteus maximus

Sciatic nerve
From	Lumbrosacral plexus
Root value	L4, 5, S1, 2, 3
Muscles supplied	Semitendinosus
	Semimembranosus
	Biceps femoris
	Hamstring part adductor magnus

Tibial nerve
From	Terminal branch sciatic nerve
Root value	L4, 5, S1, 2, 3
Muscles supplied	Gastrocnemius
	Soleus
	Plantaris
	Popliteus
	Tibialis posterior
	Flexor digitorum longus
	Flexor hallucis longus
Cutaneous branches	Sural nerve
	Fibular (peroneal) communicating nerve

Medial plantar nerve
From	Terminal branch tibial nerve
Muscles supplied	Abductor hallucis
	Flexor digitorum brevis
	First lumbrical
	Flexor hallucis brevis

Lateral plantar nerve
From	Terminal branch tibial nerve
Muscles supplied	Flexor accessories
	Flexor digiti minimi brevis
	Abductor digiti minimi
	Adductor hallucis
	Lateral 3 lumbricals
	Plantar and dorsal interossei

Common fibular (peroneal) nerve
From	Terminal branch sciatic nerve
Root value	L4, 5, S1, 2
Cutaneous branches	Lateral cutaneous nerve of calf
	Fibular (peroneal) communicating branch

Superficial fibular (peroneal) nerve
From	Terminal branch common fibular (peroneal) nerve
Muscles supplied	Fibularis (peroneus) longus and brevis

Deep fibular (peroneal) nerve
From	Terminal branch common fibular (peroneal) nerve
Muscles supplied	Extensor digitorum longus
	Tibialis anterior
	Extensor hallucis longus
	Fibularis (peroneus) tertius
	Extensor digitorum brevis

THE SACRAL PLEXUS

This is formed from the ventral rami of S4 and 5. Muscular branches pass to both *coccygeus* and *levator ani*. The plexus is cutaneous to the region next to the coccyx and posterior to the anus.

Dermatomes of the lower limb

Throughout the above account of both the lumbar and lumbosacral plexuses, cutaneous nerves have been given off to supply particular areas of the limb. Gradually the whole of the skin surface is supplied. The various nerves derive their fibres from the rami from which their parent nerves arise. Described in this way they appear to have no particular pattern to them.

A dermatome is an area of skin supplied by one spinal nerve through both its rami. The overlap of these areas, however, is considerable, particularly with the nerves immediately above and below. In some areas the main supply of the nerves is located some distance away from those above and below, and in these situations the overlap is minimal (Fig. 3.166). If these regions are now traced on the surface of the limb, paying little regard to the actual nerve through which the fibres reach the surface, a much clearer pattern of supply emerges. In the development of the lower limb, it should be noted that the section from the knee downwards has completely reversed so that which was the anterior surface of the leg becomes medial and then posterior and that which was posterior becomes lateral and anterior. Consequently, the tibial nerve, which derives its fibres from the anterior divisions of the ventral rami, lies at the back of the limb supplying this area and the plantar aspect of the foot, while the common fibular (peroneal) nerve, which derives its fibres from the posterior divisions of the ventral rami, passes anterolaterally

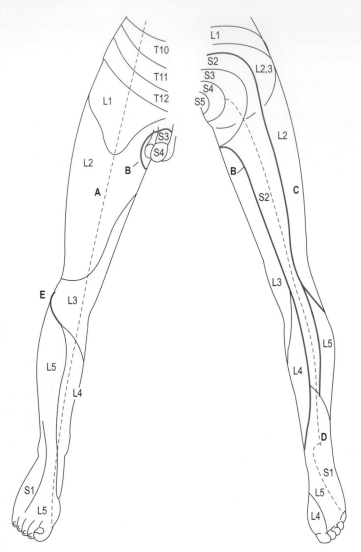

Figure 3.166 Dermatomes of the lower limb; A, preaxial border; B, ventral axial line; C, dorsal axial line; D, postaxial border; E, extension from dorsal axial line.

supplying the areas on the anterior and lateral aspects of the limb.

The dermatomes, therefore, form bands which run from posterior to anterior around the lateral side of the limb, displaying a much simpler distribution than at first imagined. As the cutaneous supply of the posterior primary rami is included in this description, it can be seen that most roots supply a strip of skin, varying in width, from the spinal region to the lower limb. Some strips are so narrow in their upper region that they are merely a line, their main distribution being further down the limb.

BLOOD SUPPLY

The arteries

The main arterial supply to the lower limb is through the *femoral artery*, although there is sufficient collateral circulation to maintain the tissues of the resting limb for a period of several hours. This collateral circulation is an invaluable asset when there is blockage of the femoral artery if surgery needs to be performed on the major vessels.

Femoral artery

A continuation of the external iliac artery, which in turn arises from the abdominal aorta via the common iliac artery. It enters the thigh under the inguinal ligament (Fig. 3.167), being contained in a funnel-shaped prolongation of the abdominal fascia, the *femoral sheath* (p. 267). The femoral vein lies medially within a separate compartment of the sheath, while the femoral nerve is lateral, but outside the sheath. Psoas major lies deep to the artery which can easily be palpated just below the fold of the groin, halfway between the anterior superior iliac spine and the pubic tubercle (the femoral pulse). Within the femoral triangle the artery passes medially across the femoral vein

to enter the subsartorial canal, which continues from the apex of the femoral triangle. It leaves the anterior compartment of the thigh through a fibrous arch in adductor magnus, the adductor hiatus, to enter the popliteal fossa and become the *popliteal artery*.

The femoral artery gives off branches to the lateral iliac and gluteal region through the *superficial circumflex iliac artery*. It supplies the genital region through the *superficial* and *deep external pudendal arteries*; a *descending genicular branch* which participates in the anastomosis around the knee; muscular branches to the surrounding muscles, and through the *profunda femoris* to most of the deep structures in the thigh.

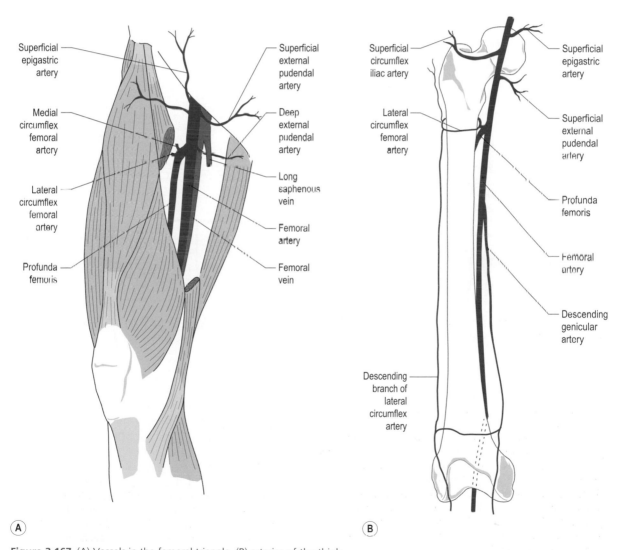

Figure 3.167 (A) Vessels in the femoral triangle, (B) arteries of the thigh.

Profunda femoris

The largest branch of the femoral artery, having approximately the same diameter. It arises from its lateral side approximately 5 cm below the inguinal ligament (Fig. 3.167), passing behind the femoral artery to leave the femoral triangle between pectineus and adductor longus to gain access to the anterior surface of adductors brevis and magnus. In the lower part of the thigh, it passes through adductor magnus as the *fourth perforating artery* contributing to the anastomosis around the knee joint. The profunda femoris gives off several branches soon after its commencement; the *lateral circumflex femoral artery*, which gives *ascending, transverse* and *descending branches* that supply the gluteal region and hip joint, the quadriceps and the knee joint respectively; the *medial circumflex femoral artery* which anastomoses with the *lateral circumflex femoral*, giving branches which supply similar areas.

As well as giving numerous muscular branches, the profunda femoris also gives off the *perforating arteries* which pass either above or through small openings in adductor magnus to link with the descending branch of the lateral circumflex femoral artery (Fig. 3.167B).

Popliteal artery

The continuation of the femoral artery as it emerges through the adductor hiatus (Fig. 3.168). Within the popliteal fossa it runs vertically from the upper medial border and ends by dividing into the *anterior* and *posterior tibial arteries* at the level of the tibial tuberosity.

The popliteal artery is the deepest structure within the popliteal fossa. In the popliteal fossa it gives cutaneous branches to the back of the leg, muscular branches to adductor magnus and the hamstrings, and articular branches to anastomose around the knee joint. The latter are given off in three groups – the upper being the *medial* and *lateral superior genicular arteries* which encircle the lower part of the femur just above the condyles. The *middle genicular artery* is smaller, and pierces the posterior part of the capsule and supplies the cruciate ligaments, and the lower group is formed by the *medial* and *lateral inferior*

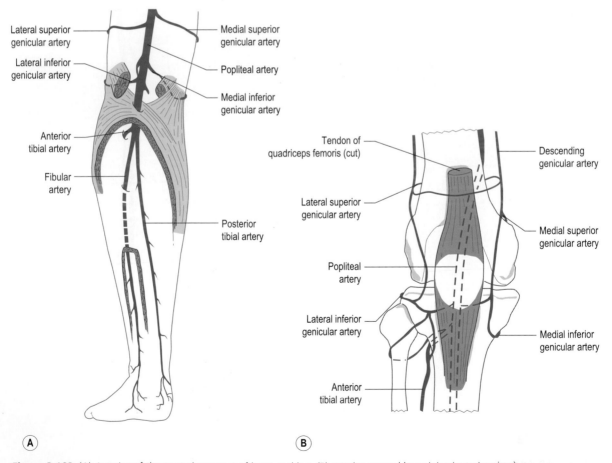

Figure 3.168 (A) Arteries of the posterior aspect of knee and leg, (B) arteries around knee joint (anterior view).

genical arteries which encircle the upper part of the tibial condyles just below the line of the knee joint. The superior and inferior arteries communicate with each other via vertical arteries which pass down either side of the patella (Fig. 3.168). Further details of the blood supply to the knee joint are given on page 317.

Anterior tibial artery

This begins at the distal border of popliteus and ends in front of the ankle joint by becoming the *dorsalis pedis artery* (Figs. 3.168 and 3.169A). It passes forwards through an opening above the interosseous membrane between the tibia and fibula. As the artery passes above the membrane,

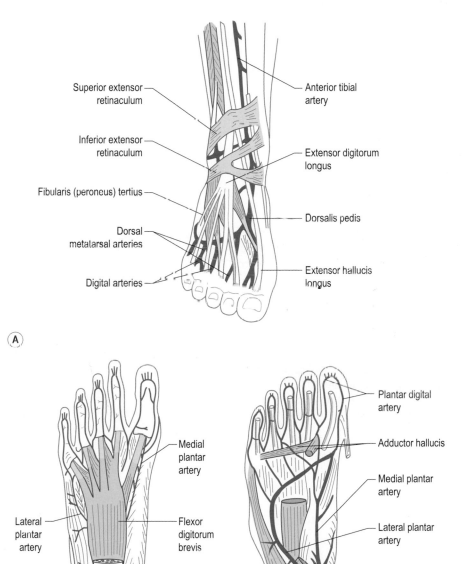

Superior extensor retinaculum

Inferior extensor retinaculum

Fibularis (peroneus) tertius

Dorsal metatarsal arteries

Digital arteries

Anterior tibial artery

Extensor digitorum longus

Dorsalis pedis

Extensor hallucis longus

(A)

Medial plantar artery

Lateral plantar artery

Flexor digitorum brevis

Plantar aponeurosis (cut)

Plantar digital artery

Adductor hallucis

Medial plantar artery

Lateral plantar artery

Flexor accessorius

Flexor digitorum brevis

(B)

Figure 3.169 Arteries and nerves of the foot: (A) dorsal aspect, (B) plantar aspect.

recurrent arteries are given off which pass upwards to join the anastomosis around the knee joint. Running down the anterior surface of the interosseous membrane it becomes superficial and crosses the ankle joint under the extensor retinaculum and between the tendons of extensor hallucis longus and extensor digitorum longus, where it can be easily palpated as the anterior tibial pulse. It enters the dorsum of the foot as the dorsalis pedis artery and runs towards the first interosseous space through which it passes to the plantar aspect of the foot (Fig. 3.169A). In this region the pulse may again be palpated as the dorsalis pedis pulse.

Before leaving the dorsum of the foot the dorsalis pedis gives a large branch, the *arcuate artery*, which passes across the bases of the metatarsals, and becomes continuous laterally with the *fibular (peroneal) artery*. From the arcuate artery arise the *dorsal metatarsal arteries*, which when reaching the cleft of the toe divide, each into two *dorsal digital arteries* to supply the adjacent sides of the toes. The medial side of the great toe and the lateral side of the little toe are usually supplied by separate branches from the arcuate artery.

Posterior tibial artery

The larger of the two terminal branches of the popliteal artery, the posterior tibial artery begins at the distal border of popliteus (Fig. 3.168). It passes down the back of the leg under cover of soleus and gastrocnemius lying on flexor digitorum longus and flexor hallucis longus. About two-thirds of the way down the leg it is only covered by deep fascia and skin, crossing the ankle joint behind the medial malleolus deep to the flexor retinaculum, with the tendon of flexor digitorum longus anterolaterally and the tibial nerve posteromedially. It immediately divides into *medial* and *lateral plantar arteries*. In this region the posterior tibial pulse may be readily palpated.

As the posterior tibial artery passes down the posterior aspect of the leg, it gives off the large *fibular (peroneal) artery* (Fig. 3.168A), which descends between tibialis posterior and flexor hallucis longus. During its course, it gives *malleolar branches* and a *perforating branch* which cross the ankle joint to anastomose with branches of the anterior tibial and dorsalis pedis arteries.

The medial and lateral plantar arteries

The smaller *medial plantar artery* passes forwards along the inner side of the foot (Fig. 3.169B), medial to the medial plantar nerve, between abductor hallucis and flexor digitorum brevis. The artery gives three digital branches, which pass forwards to join the *medial plantar metatarsal arteries*.

The *lateral plantar artery* passes forwards and laterally between flexor digitorum brevis and flexor accessorius (quadratus plantae), on the lateral side of the lateral plantar nerve. At the base of the fifth metatarsal it passes medially and deep to arch across the foot on the metatarsals as the *plantar arch* to become continuous with the dorsalis pedis

artery on the lateral side of the first metatarsal. The arch gives off the *plantar metatarsal arteries*, which each divide when they reach the cleft of the toes, after being joined by the *digital arteries* from the medial plantar artery.

The arteries on the dorsum of the foot communicate with those on the plantar aspect by *perforating arteries* which pass between the metatarsals.

The veins

The venous drainage of the lower limb varies considerably in its arrangement from subject to subject and even from limb to limb. The following description, therefore, is considered to be the most commonly found. The veins are usually described as being superficial and deep, the superficial veins are larger with fewer valves and are situated in the superficial fascia, whereas the deep veins are normally two small vessels accompanying the arteries; these are called the *venae comitantes*. They are situated deep in the limb and possess many valves.

Superficial veins

On the dorsum of the foot the superficial veins are easily seen (Fig. 3.170). They form a network which receives venous blood from the dorsal and plantar aspects of the toes, from either side of the foot, and from the deep plantar areas via the *perforating veins* which pass between the metatarsals. The *dorsal venous network* is drained on either side by the *medial* and *lateral marginal veins*; the medial continues as the *long (great) saphenous vein*, the lateral as the *short (lesser) saphenous vein*.

Long (great) saphenous vein

This passes in front of the medial malleolus and ascends obliquely up the posteromedial aspect of the calf towards the knee (Fig. 3.170). It lies posteromedial to the femoral and tibial condyles, and continues upwards, forwards and laterally in the thigh. It then passes through the cribriform fascia of the saphenous opening, situated just below the centre of the inguinal ligament, to join the *femoral vein* deep in the groin. During its course, the long saphenous vein has between 8 and 20 bicuspid valves.

The long saphenous vein receives many tributaries as it passes up the limb, mainly draining the medial side of the leg and thigh. In the leg, it communicates freely with the *short (lesser) saphenous vein*, and through the deep fascia with the *deep intermuscular veins*, particularly near the knee and ankle joints. Before passing through the saphenous opening, it receives drainage laterally from the iliac region through the *circumflex iliac vein*, from the genital area through the *superficial external pudendal vein*, and from the lower abdominal area through the *superficial epigastric vein*.

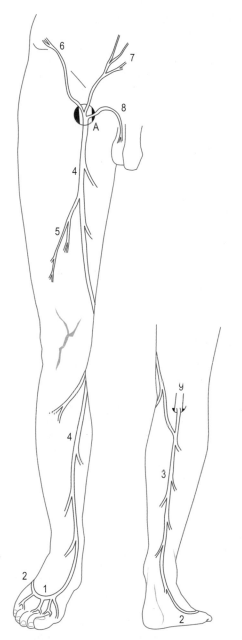

Figure 3.170 Superficial veins of the lower limb. 1, Dorsal venous arch; 2, lateral marginal vein; 3, short saphenous vein; 4, long saphenous vein; 5, lateral accessory vein; 6, superficial circumflex iliac vein; 7, superficial epigastric vein; 8, superficial external pudendal vein; 9, popliteal vein; A, saphenous opening.

Short (lesser) saphenous vein

This passes behind the lateral malleolus, along the lateral side of the tendocalcaneus to the posterior aspect of the calf (Fig. 3.170). It enters the popliteal fossa between the two heads of gastrocnemius by piercing the deep fascia which forms the roof of the fossa, and drains into the *popliteal vein* behind the knee joint. In its course it contains between 6 and 12 bicuspid valves, and is accompanied by the sural nerve. The short saphenous vein receives tributaries from the lateral side of the ankle and leg.

Deep veins

Two veins, the venae comitantes, accompany the smaller arteries of the lower limb and are similarly named. Only the *popliteal*, *femoral* and *profunda femoris veins* are single vessels. The veins possess numerous valves.

Popliteal vein

Formed at the lower border of popliteus by the union of the *anterior* and *posterior tibial veins*, the popliteal vein receives the *genicular* and *short saphenous veins*. In its course through the popliteal fossa, it passes from medial to lateral across the artery. It passes through the adductor hiatus to become the *femoral vein*.

Femoral vein

Ascending in the adductor canal to enter the femoral triangle, the femoral vein ends by becoming the *external iliac vein* as it passes deep to the medial part of the inguinal ligament within the femoral sheath. In its course from the adductor hiatus to the inguinal ligament, the vein passes behind the femoral artery from lateral to medial. During its course it receives the *profunda femoris* and *long saphenous veins*. The long saphenous enters some 3 cm below the inguinal ligament just prior to the femoral vein entering the femoral sheath.

Application

There is abundant communication between the superficial and deep venous systems. The veins deep within the foot tend to drain into the *dorsal venous network*, while in the calf the drainage is from the superficial to the deep via *perforating veins*. Contraction of the calf muscles compresses the local veins within the surrounding fascial compartments, pumping the blood into the deep system because of the arrangement and direction of the valves within the perforating vessels.

Commonly, the valves in the perforating veins become incompetent and allow blood to flow back into the superficial system, the veins of which then become engorged, swollen and tortuous. In such situations the veins appear as large, ugly and often blue swellings on the posteromedial side of the calf and the medial side of the thigh. These are termed 'varicose veins'.

LYMPHATICS

The lymphatic vessels begin in the tissues as a series of blind-ended tubules being composed of a single cell layer. They exist between most tissues, gradually becoming larger in diameter as they pass proximally. In their course, lymph vessels are interrupted by lymph nodes, which serve partly as filters and partly as a source of lymphocytes. The superficial vessels lie in the skin and subcutaneous tissues and frequently accompany the superficial veins. They join the deep vessels at constant sites in the limbs. The deep vessels, draining areas deep to the fascia, accompany blood vessels of the region.

Superficial drainage

The superficial drainage of the lower limb follows the long saphenous vein on the medial side and the short saphenous vein on the lateral side (Fig. 3.171). The vessels on the medial side of the limb converge to the *vertical group* of the *superficial inguinal nodes* around the long saphenous vein near the saphenous opening. The *horizontal group* of superficial inguinal nodes receives lymph from the skin below the level of the umbilicus, as well as from the lower part of the anal canal and external genitalia (excluding the testes in males). The efferent vessels from both groups of superficial nodes pass through the cribriform fascia of the saphenous opening to end mainly in the *external iliac nodes* with some passing to *deep inguinal nodes*.

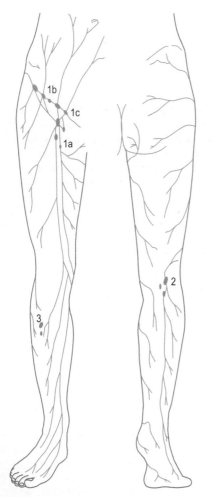
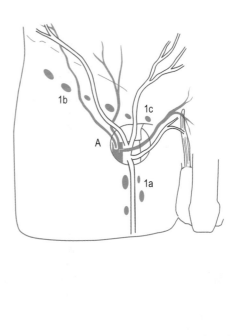

Figure 3.171 Lymphatics of the lower limb. 1, Superficial inguinal lymph nodes (1a, vertical group; 1b, lateral group; 1c, medial group); 2, popliteal nodes; 3, anterior tibial nodes; A, saphenous opening.

The superficial vessels associated with the short saphenous vein pierce the deep fascial roof of the popliteal fossa and drain into the *popliteal nodes*.

Deep drainage

The deep lymph vessels accompany the arteries and venae comitantes, following the same route back to the femoral triangle as the deep veins, gradually growing larger in diameter as they progress up the limb. The few nodes associated with the deep vessels are usually small. An *anterior tibial node* is situated on the upper part of the interosseous membrane. The *popliteal nodes* receive the *anterior* and *posterior tibial vessels* and the lymphatic drainage from the knee as well as the vessels accompanying the short saphenous vein. Efferents from the popliteal nodes pass to the *deep inguinal nodes* which lie on the medial side of the femoral vein. They receive some vessels from the *superficial inguinal nodes*, and all the deep vessels from the territory of the femoral artery.

From the deep and superficial inguinal nodes the lymphatics leave the lower limb by passing under the inguinal ligament through the femoral canal, the most medial compartment of the femoral sheath, to reach the *external iliac nodes*. One of the deep inguinal nodes may be situated within the femoral canal.

Part | 4 |

The trunk and neck

CONTENTS

INTRODUCTION

One of the major features distinguishing human beings from other animals is their bipedal posture and gait. As the hindlimbs progressively took over the locomotor function, the vertebral column assumed a new role. No longer is it held horizontally (Fig. 4.1A), where it is under compression, but it has become a vertical weight-bearing rod, held erect by ligaments and muscles (Fig. 4.1B). This change in function of the vertebral column has been accompanied by changes in its form, as well as by changes in its relationship to the skull and pelvic girdle.

Major differences in the form of the vertebral column are evident among quadrupeds, where it depends mainly on the distribution of mass in the animal. The so-called centre of gravity moves forward or backwards along the vertebral column in relation to there being more weight in the head and forelimbs or hindlimbs and tail respectively. In the evolution of the primates, the centre of gravity moved backwards towards the hindlimbs because both the length and musculature of the hindlimbs increased, in addition to the enlargement of the tail. The first of these changes gave additional power to leaping and grasping, while the latter provided balance when jumping. An important factor in locomotion is that the forward propelling force should pass through the centre of gravity; otherwise the body tends to rotate about the centre of gravity.

The early changes in body form during primate evolution made possible the sitting position, thereby freeing the forelimbs for manipulative activities. It also made possible the later stages of human evolution with the adoption of a bipedal gait. Although the human tail has been lost, the low position of the body's centre of gravity has been maintained because of further development of the hindlimbs, a slightly reduced muscle mass in the forelimbs, and changes in the form and position of the trunk and abdomen.

A prerequisite for efficient arboreal locomotion was the evolution of an extremely flexible vertebral column. The increased range of flexion of the vertebral column, together with that of the lower limbs, gave additional propulsion in jumping as well as the ability to absorb the shock of impact on landing. However, in brachiating primates some flexibility of the vertebral column is lost due to it not having such an important role in locomotion. It seems that human beings could have evolved from a brachiating primate because of the relative stiffness of the vertebral column as well as other morphological features. In addition, four other fundamental changes have occurred in both the vertebral column and the thorax during human evolution, all of which are adaptations to a fully erect posture.

First, the vertebral bodies increase in size towards the lumbar region (Fig. 4.1B) because the compression forces along the trunk are no longer constant but increase progressively from above downwards. The proportions of the vertebral bodies also change from above downwards. Second, the spinous processes are more or less equally developed along the length of the vertebral column because the tendency to bend is no longer restricted to the limb girdles (the points of support), but is more evenly distributed. This is perhaps most clearly seen in the cervical region because the human head is more or less balanced on the vertebral column. Thirdly, the stiffening effect of the sternum and abdominal muscles on the vertebral column has become less important. The sternum has come to lie nearer the vertebral column so that the thorax is wider and less deep (Fig. 4.1C). This has resulted in a less abrupt change in direction of the ribs at their angles. Finally, the increased weight transmission through the pelvis and legs has brought about an enlargement of the sacrum, so that

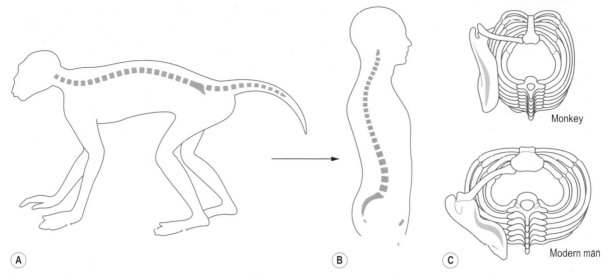

Figure 4.1 Arrangement of the vertebral column in (A) a quadruped monkey and (B) a modern human, (C) changes in the anteroposterior dimensions of the thoracic cage.

the human sacrum is usually composed of five fused segments. It is also relatively wide and more convex on its pelvic surface.

Concomitant with the changes in the vertebral column and trunk, the relation and form of the skull and its associated musculature have also changed dramatically. In quadrupedal animals, the head is supported by the postvertebral (nuchal) muscles and ligaments (these being under tension), which produce compression of the cervical vertebrae. As a more erect posture evolved the strength required in the postvertebral muscles was reduced because more of the weight of the head was carried directly by the vertebrae. This was accompanied by a reduction in the muzzle and an enlargement of the brain (Fig. 4.2A) causing the centre of gravity of the head to move more nearly over the point of support (the occipital condyles). It would be undesirable to have the head perfectly balanced on the occipital condyles because humans possess no powerful prevertebral muscles to support the skull from the front. Consequently, the centre of gravity projection falls just anterior to the occipital condyles, being balanced by the postvertebral muscles (Fig. 4.2B). That the postvertebral muscles have been reduced in importance is evident from the relatively small area of their attachment to the skull in humans compared with other primates (Fig. 4.2C).

In human beings the tail is reduced to between three and five fused coccygeal vertebrae, which curve ventrally and help form the pelvic cavity. Ligaments running from the coccyx to the ischium play an important role in maintaining this relationship, and in so doing contribute to the function of support of the abdominal and pelvic viscera. The abdominal viscera are carried in a sac-like cavity supported behind by the vertebral column, below by the pelvis and anterolaterally by the abdominal muscles (rectus abdominis, the external and internal abdominal obliques and transversus abdominis).

Although the vertebral column has changed in form and orientation during human evolution, it still has to fulfil essentially the same functional requirements as in quadrupedal animals:

1. It carries and supports the thoracic cage, maintaining the balance between it and the abdominal cavity.
2. It gives attachment to many muscles of the pectoral and pelvic girdles.
3. It provides anchorage for many powerful muscles which move the vertebral column; these same muscles maintain the balance and erectness of the human trunk.
4. It surrounds and protects the spinal cord against mechanical injury.
5. It acts as a shock absorber, by virtue of its curvatures and the intervertebral discs, receiving and distributing the impacts associated with the dynamic functioning of the body.
6. It is able, by virtue of its flexibility, to produce and accumulate moments of force as well as to concentrate and transmit forces received from other parts of the body.

The upright posture and independent functioning of the human upper limbs have greatly increased the dynamic demands made on the vertebral column. Nevertheless, the adaptation has been reasonably, although not altogether, successful such that the vertebral column has become a

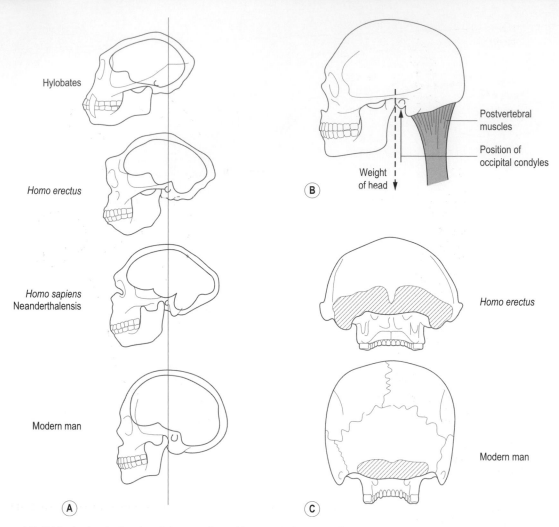

Figure 4.2 (A) Reduction in the size of the muzzle and increase in brain size from lower primates (*Hylobates*) to modern human, (B) relation of the centre of gravity of the head to the occipital condyles, (C) reduction in area of attachment for the postvertebral muscles in modern humans.

complicated and delicate mechanical unit. That the transition from quadrupedal to bipedal has not been entirely successful is witnessed by the fact that low back pain and its associated problems take a heavy toll.

The vertebral column comprises a series of mobile segments held together by ligaments and muscles, each separated from adjacent segments by an intervertebral disc. There are usually 33 bony segments, of which only 24 present as separate bones: the lower nine are fused, five forming the sacrum and four the coccyx. The 24 presacral vertebrae are designated cervical, thoracic and lumbar according to their features and position within the trunk: there are seven cervical, twelve thoracic and five lumbar vertebrae (Fig. 4.3). The length of the vertebral column is between 72 and 75 cm for the majority of individuals, of

which approximately one-quarter is accounted for by the intervertebral discs. About 40% of an individual's height is due to the length of the vertebral column. Variations in height between individuals mainly reflect differences in lower limb length rather than differences in vertebral column length. However, diurnal variations in height, up to 2 cm between early morning and late evening, are due to compression and loss of thickness of the individual intervertebral discs. (It is as well to remember this when charting the change in height of an individual.) Loss of height in elderly individuals is associated with thinning of the discs as a result of age-related changes within them.

The adult vertebral column has four curvatures (Fig. 4.3A), an anterior convexity in the cervical and lumbar regions, and an anterior concavity in the thoracic and sacrococcygeal

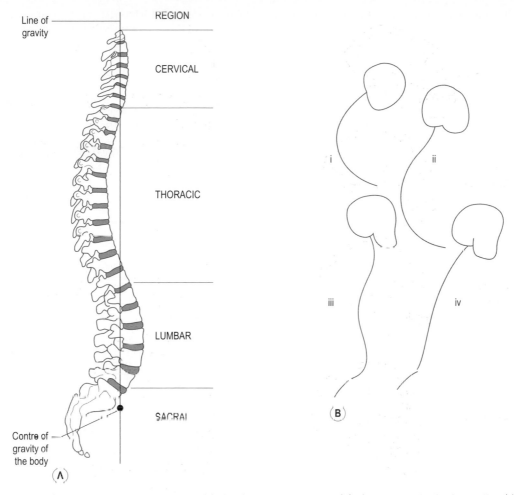

Figure 4.3 (A) Adult vertebral column, lateral view. (B) The changing curvatures of the human vertebral column: (i) at birth; (ii) at 6 months; (iii) adult; (iv) old age.

regions. Both the cervical and lumbar curvatures are acquired, in the sense that they are not present in early fetal development. Until late in fetal development, the vertebral column shows a single curvature concave anteriorly (Fig. 4.3B(i)). Late in fetal life the secondary cervical curvature (Fig. 4.3B(ii)) begins to appear, becoming more accentuated between 6 and 12 weeks after birth as the infant begins to hold up its head to enlarge its visual environment. The secondary lumbar curvature (Fig. 4.3B(iii)) appears when the child begins to sit up at around 6 months, becoming more marked with standing and the onset of walking. It is the extension of the hip that accompanies standing and walking which tilts the pelvis forwards so that the axis of the pelvic cavity is no longer in line with that of the abdominal cavity. The lumbar curvature develops in order to keep the trunk erect when standing. The lumbar curvature is not fully developed until after the age of 2, when a more or less

adult pattern of walking is established. Sadly, in old age the vertebral column tends to assume a gentle C-shaped curve (Fig. 4.3B(iv)) reminiscent of the early fetal curve. The reason being that the shape of the vertebral column is largely determined by the intervertebral discs, and to a much lesser extent by the vertebrae themselves. Consequently, as the discs degenerate and thin with increasing age the secondary curvatures gradually disappear.

The relatively shallow cervical curvature begins in the dens of the axis and ends at the level of the second thoracic vertebra. It can be reduced or obliterated by bending the head forwards. The permanent thoracic curve is due in part to the shape of the thoracic vertebral bodies, the posterior height being greater than that anteriorly. This curvature ends at the 12th thoracic vertebra. An increase in the thoracic curvature is known as *kyphosis*. The lumbar curvature, which tends to be deeper and more prominent in women, ends at the lumbosacral junction. Changes in the orientation of the pelvis (pelvic tilt) are accompanied by changes in the lumbar curvature in an attempt to keep the trunk upright. The wearing of high-heeled shoes throws the pelvis forwards resulting in an increase in the lumbar curvature. A similar situation arises during pregnancy in an attempt to move the centre of gravity backwards and so prevent overbalancing. An increase in lumbar curvature is known as a *lordosis*, although the normal lumbar curve is often referred to as a *lumbar lordosis*. The curvature of the sacrum is permanent because of its fused constituent parts. It should be remembered that the lumbosacral angle is not a part of the vertebral curvatures. However, because it is the region where the mobile and immobile parts of the vertebral column meet, the structures associated with it, e.g. the intervening intervertebral disc and ligaments, are put under considerable stress.

The normal curvatures of the vertebral column make it a flexible support, imparting resilience to axial compressive forces which are absorbed by the giving way and recovery of the various curves.

Seen from the front, the vertebral column appears almost straight and symmetrical with perhaps a very slight right thoracic curve, probably due to the presence of the arch of the aorta. A large lateral curvature is abnormal and is known as a *scoliosis*. Scoliosis also involves rotation of the vertebral column so that the spinous processes of the vertebrae turn towards the concavity of the curvature. Compensatory curves in the reverse direction occur in order to keep the head facing forwards and over the feet. The condition is extremely complex, often appearing in childhood during periods of increased growth (i.e. 6–8 years, and during early puberty).

The vertebral curves pass in front of and behind the line of gravity along which the weight of the head, upper limbs and trunk is projected to the lower limbs. This line is said to pass progressively through the dens, the bodies of the second and 12th thoracic vertebrae and the promontory of the sacrum (Fig. 4.3), with the centre of gravity of the body located just in front of the sacral promontory. However, it must be remembered that the line of gravity is not constant; it is continually changing, both as the body moves and also when standing still. Nevertheless, it is a useful concept in reminding us of the natural balance and beauty of the body. By visualizing changes in the projection of this line, it may be possible to determine the structures put under increased strain in certain postures and pathologies.

Embryological development of the vertebrae

Soon after the formation of each somite it becomes differentiated into the three parts, the ventromedial sclerotome, the medial myotome and the thin lateral dermatome. The sclerotomes come to surround the notochord, followed by the myotomes. Each vertebra is formed from the adjacent parts of two sclerotomes (Fig. 4.4), with the intervening part forming the outer part *(annulus fibrosus)* of the intervertebral disc. The *nucleus pulposus* of the disc is derived from the notochord. As there are eight cervical somites, there are also eight cervical nerves, similarly in the thoracic, lumbar and sacral regions. However, because each vertebra is formed from adjacent somites the cervical nerves come to lie above their correspondingly numbered vertebrae, with the eighth cervical nerve lying below the seventh cervical vertebra. The thoracic, lumbar and sacral spinal nerves all lie below their correspondingly numbered vertebrae.

The dermomyotome breaks up with cells moving both ventrally and dorsally. The original spinal nerve supplying the myotome divides into an anterior and posterior primary ramus supplying the ventral (the *hypomere*) and dorsal (the *epimere*) parts of the myotome respectively. The epimeres come to lie between the transverse and spinous processes, giving rise to the musculature of the trunk and neck. The hypomeres give rise to the prevertebral muscles (the scalenes, quadratus lumborum, psoas and piriformis) and the musculature of the thoracic and abdominal walls.

The occipital myotomes do not participate in the formation of the musculature of the trunk or neck. They give rise to the muscle mass of the tongue. That part of the paraxial mesoderm that does not become segmented eventually becomes incorporated into the branchial (pharyngeal) arches, each of which is innervated by a cranial nerve. The musculature derived from these arches forms the muscles of the face, those around the mandible and those of the larynx and pharynx.

The vertebral column

Extending from the base of the skull to the pelvis, the vertebral column consists of a series of irregularly shaped bones which increase in size from above downwards. The vertebrae are bound together by ligaments and have intervertebral discs between their bodies. In young children, 33 separate vertebrae can be identified; however, by the time adulthood is reached five have fused to form the sacrum and four to form the coccyx. Of the remaining 24, seven are found in the neck (cervical vertebrae), twelve articulate with the ribs (thoracic vertebrae) and five are found in the lower back (lumbar vertebrae). Within each group the vertebrae have similar features, some of which are distinctive, regarding shape and the orientation of their articular processes.

With the exception of the first and second cervical vertebrae, all vertebrae possess a large weight-bearing body anteriorly and vertebral arch posteriorly, which consists of a series of bony processes (Fig. 4.5A). The body varies in shape and size depending on its location in the vertebral column, but is roughly cylindrical with flattened upper

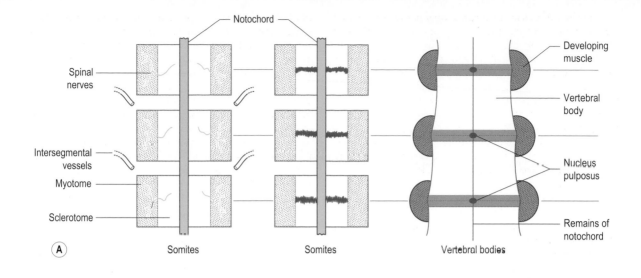

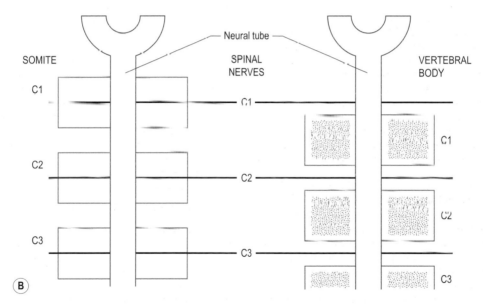

Figure 4.4 (A) Development of vertebrae from the sclerotome component of the somite, (B) the relationship between spinal nerves and the vertebrae.

and lower surfaces. On these surfaces, the markings of the attachment of the intervertebral disc around the periphery can be seen surrounding a roughened central area. The front and sides of the bodies are roughened, particularly at their upper and lower edges, and concave from top to bottom. The posterior surface is fairly smooth and has a large foramen for the passage of the basivertebral vein. The vertebral arch arises from the posterolateral aspect of the body, and with the body surrounds the vertebral foramen. *In situ*

adjacent vertebral foramina, together with the intervertebral discs and the ligamenta flava, form the vertebral canal which houses the spinal cord and its various coverings.

The vertebral arch consists of two *pedicles* and two *laminae*. Each pedicle passes from the upper part of the posterolateral aspect of the vertebral body to join with the anterolateral extremity of the corresponding lamina. The laminae slope backwards to meet in the midline posteriorly, where they are continuous with the posterior

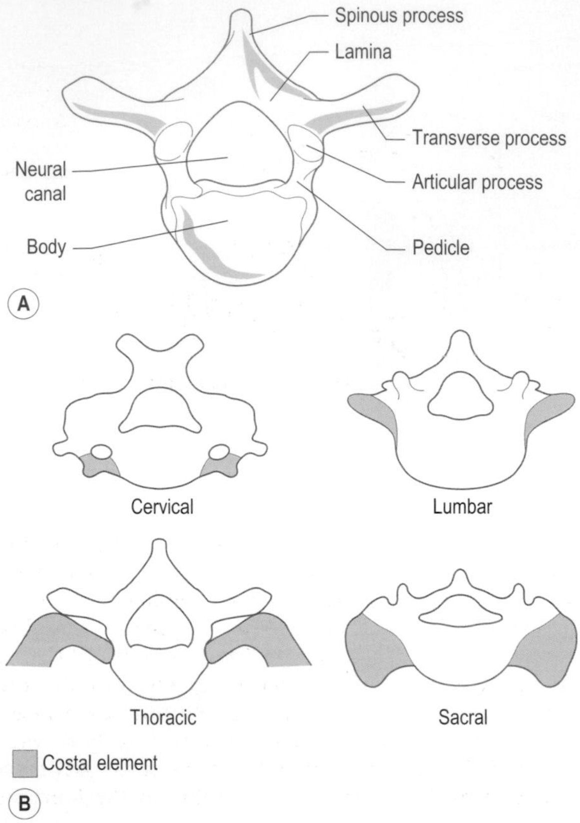

Figure 4.5 (A) Schematic representation of an individual vertebra, (B) fate of the costal element in each of the vertebral regions.

projecting spinous process. Because the pedicles are not as deep as the vertebral bodies, the opening formed between adjacent pedicles (the intervertebral foramen) enables the spinal nerves and supporting blood vessels to leave or enter the vertebral canal. The intervertebral foramen is closed anteroinferiorly by the intervertebral disc, an important relation to note in some pathologies of the vertebral column and spinal nerves.

Arising from the junction of each pedicle and lamina are three processes: the transverse process projecting laterally, and the articular processes, one directed upwards and one directed downwards. In adjacent vertebrae, superior and inferior articular processes make contact with each other by small synovial joints – the *zygapophyseal joints*. It is the shape and orientation of the articular facets on these processes which largely determine the range and type of movement possible between adjacent vertebrae.

With a few exceptions, each part of a vertebra is potentially present in every other vertebra, the main features of which are shown in Fig. 4.5A. The body is the only part

of a vertebra that is represented throughout the whole series of vertebrae; even so, that of the atlas has been displaced as the dens of the axis. The major difference between cervical, thoracic, lumbar and sacral vertebrae is related to the size and fate of their costal elements (Fig. 4.5B). In the cervical region, the medial end of the costal element fuses with the side of the vertebral body. The presence of the vertebral vessels lying anterior to the true transverse process prevents complete fusion of the costal element with the transverse process. Instead, a costotransverse lamella (bar) joins the two parts together laterally, leading to the formation of the *foramen transversarium*. Consequently, the anterior tubercle of the transverse process is the lateral end of the costal element, while the posterior tubercle represents the lateral end of the transverse process. The first and second cervical vertebrae have no anterior tubercles.

In the thoracic region, the costal element remains separate giving rise to the rib, which articulates directly with the body and transverse process of its corresponding vertebra.

In the lumbar region, the costal element has again been incorporated into the vertebra. So complete is its incorporation that apart from the root of the transverse process (lateral process) the remainder is composed of costal element. The true transverse process in the lumbar region gives rise to the accessory and superior processes as well as the lateral process.

Even in the sacrum, the costal elements have been incorporated and form the major part of the lateral masses.

BONES

A description of the sacral and coccygeal vertebrae can be found on pages 211–213.

Lumbar vertebrae

The five lumbar vertebrae are much stouter and stronger than those in either the thoracic or cervical regions (Fig. 4.6). They possess neither foramina transversaria nor articular facets for the ribs. Each has a large, kidney-shaped *body* with almost parallel upper and lower surfaces, except for L5 which is deeper anteriorly than posteriorly. The short strong *pedicles* pass almost directly backwards to join the narrow *laminae* which pass backwards and medially towards the spine. Adjacent laminae are widely separated from each other, leaving diamond-shaped spaces which contain the ligamenta flava.

The *spinous processes* of the lumbar vertebrae project almost horizontally backwards, being level with the lower half of the body. They are wider from above downwards with a thickened posterior edge. That of L5 is frequently rounded.

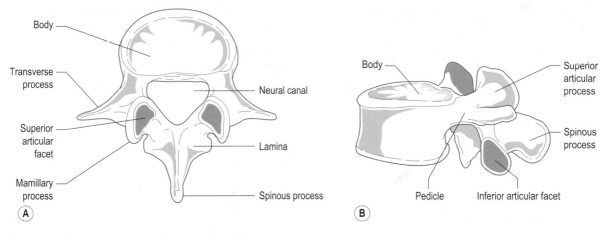

Figure 4.6 A lumbar vertebra: (A) superior view, (B) lateral view.

The *articular processes* project superiorly and inferiorly from the region where the pedicle joins the lamina. The *articular facets* on the superior process are concave transversely and flat vertically: they face posteromedially. On the posterior edge of the superior articular process is the rounded *mamillary process*. The inferior articular processes are set closer together than the superior and have facets which are reciprocally curved to face anterolaterally. The superior facets articulate with the inferior facets of the vertebra immediately above. The inferior articular processes of the fifth lumbar vertebra are more widely set apart and flatter than those of other lumbar vertebrae. Their articular facets face anterolaterally to meet the superior articular facets of the sacrum.

The triangular *vertebral (neural) canal* is larger than that in the thoracic region, but slightly smaller than in the cervical region.

With the exception of the fifth lumbar vertebra the *transverse processes* are short and thin, projecting laterally and slightly backwards from the sides of the vertebral body and base of the pedicles. The third is the longest, while the fourth and fifth are inclined upwards. The transverse processes of L5 are short and stout, and may be fused with the lateral part of the sacrum. From the root of each transverse process, a small tubercle (the accessory process) projects posteriorly. The root of the transverse process is known as the lateral tubercle.

Thoracic vertebrae

The distinctive feature of thoracic vertebrae is the presence of *articular (costal) facets* on the sides of the vertebral body for articulation with the heads of at least one pair of ribs (Fig. 4.7). The *bodies* of thoracic vertebrae are typically heart-shaped, when viewed from above, and on their sides bear articular facets. The bodies of the second

to eighth vertebrae have large, almost complete oval facets near their upper borders, and smaller demifacets near their lower borders. Although similar in basic arrangement, the upper facet of the ninth thoracic vertebra is situated at the base of the pedicle. The body of T1 has a complete oval facet near its upper edge and a demifacet at its lower. The 10th, 11th and 12th vertebrae have single complete facets located at the junction of the body with the pedicle. Between T10 and T12 the facets gradually move from the upper to lower region of the junction (Fig. 4.7B)

The short *pedicles* project almost directly backwards from the upper posterior part of the body. They gradually become larger and stronger from above downwards. The *laminae* are inclined towards the midline, and although narrow from side to side they are deep so that they overlap one another from above.

The *spinous processes* are long and slope downwards, with those in the middle of the series being almost vertical. The upper and lower ones are shorter and generally less sloping. The 12th is almost horizontal, resembling a typical lumbar spine.

The long, thick, rounded *transverse processes* project laterally, backwards and slightly upwards from the junction of the pedicle with the lamina. They have an oval facet on the anterior surface near the tip which faces anterolaterally, for the tubercle of the corresponding rib. In the upper thoracic vertebrae these facets are concave but gradually become flatter from above downwards. The transverse processes of the 11th and 12th thoracic vertebrae have no facets. The transverse process of the 12th thoracic vertebra is very short and shows features similar to the lumbar vertebrae.

From just medial to the base of the transverse process, the articular processes project almost vertically, the superior upwards and the inferior downwards. The articular facets on the superior process are slightly concave

413

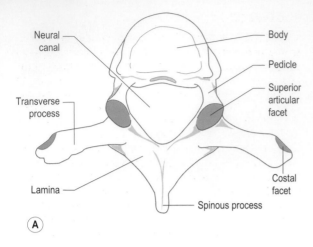

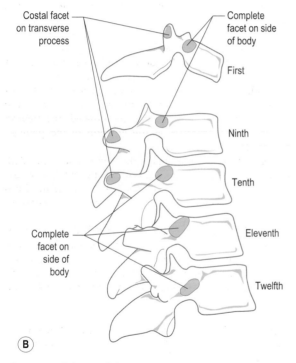

Figure 4.7 (A) Typical thoracic vertebra, superior view, (B) the first, ninth, tenth, eleventh and twelfth thoracic vertebrae, lateral views.

transversely, being flat from above down, and face backwards but also slightly upwards and laterally. The facets on the shorter inferior articular processes are reciprocally curved and face in the opposite direction. The superior articular facets of one vertebra articulate with the inferior facets of the vertebra above. The joints so formed lie on the arc of a circle whose centre lies within or just anterior to the body of the vertebra. Consequently, rotation as well

as flexion and extension are favoured in the thoracic region (p. 463).

The inferior articular processes of T12, although they project vertically, do not lie in the same general plane as the other thoracic articular processes. They are in fact lumbar in type; therefore they are markedly convex transversely and face anterolaterally.

The *vertebral (neural) canal* is smaller than that in either the cervical or the lumbar region, and is nearly circular in appearance.

Cervical vertebrae

The characteristic feature of all cervical vertebrae is the presence of a *foramen transversarium* in each *transverse process* (Fig. 4.8). They tend to be small as they do not carry much weight. The third, fourth, fifth and sixth cervical vertebrae are sufficiently similar to be considered together. The *body* is relatively small, appearing kidney-shaped when viewed from above. Its superior surface projects upwards at the sides, while its inferior surface is correspondingly bevelled. Between the sides of adjacent surfaces of the bodies, each side of the intervertebral disc, are small synovial joints (uncovertebral joints, p. 451). The anterior surface is marked by the attachment of the anterior longitudinal ligament, while laterally the body is hollowed. The posterior surface of the body is flat.

The short *pedicles* pass laterally and backwards, which is why the vertebral canal in the cervical region is triangular and larger than elsewhere. The long narrow *laminae* pass posteromedially, joining together to form the spinous process. The short, bifid *spinous process* projects backwards from the centre of the vertebral arch.

The composite *transverse processes* are stout projections arising from the lateral side of the body and pedicle. Each

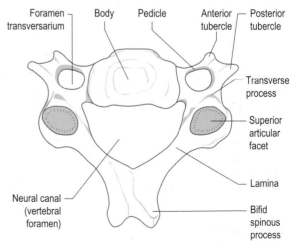

Figure 4.8 Typical cervical vertebra, superior view.

transverse process ends in prominent *anterior* and *posterior tubercles*, joined by the costotransverse lamella. Within each process is a *foramen transversarium*, bound posteriorly by the pedicle, and anteriorly and laterally by the various parts of the transverse process. It transmits the vertebral artery and vein.

The large *superior* and *inferior articular processes* project from the articular mass, which is located at the junction of the pedicle and lamina on each side.

Each process bears an *articular facet*. The slightly convex superior facet faces upwards and backwards while the reciprocally concave inferior facet faces downwards and forwards. The facets become more vertical in the lower part of the cervical spine.

The seventh cervical vertebra

The seventh vertebra (Fig. 4.9), also known as the vertebra prominens, is noted for the length of its non-bifid *spinous process*. It is larger than the preceding cervical vertebra and exhibits similarities to thoracic vertebrae. The body is larger, the pedicles are directed more posteriorly than laterally, the inferior articular facets face more anteriorly than downwards, and the vertebral canal is generally smaller than that of other cervical vertebrae.

The *transverse processes* have a small foramen transversarium, which only transmits the vertebral vein. Occasionally the anterior costal element of the transverse process is much longer than the posterior part, giving the appearance of a rudimentary rib as it extends forwards towards the first rib as either a fibrous or bony strip. This may lead to pressure on the eighth cervical nerve root as it passes forwards over the first rib to take part in the brachial plexus.

The axis (C2)

The axis (C2), the second and strongest of the cervical vertebrae, has many of the features of a typical cervical vertebra (Fig. 4.10). It has, however, the separated body of the atlas fused with the superior surface of its body giving an upward tooth-like projection known as the *odontoid process (dens)*. The dens is slightly pointed at its apex, where the apical ligament attaches it to the anterior margin of the foramen magnum. From the sloping sides of the apex, alar ligaments pass to tubercles on the medial sides of the occipital condyles. Its anterior surface has a small smooth facet, concave from above down and convex transversely for articulation with the facet on the back of the anterior arch of the atlas. Posteriorly is a smooth, constricted horizontal surface where the transverse ligament crosses. Above this constriction there is a slightly expanded area known as the head.

The *large superior articular facets* lie at the junction of the body and pedicle, and transmit the weight of the head to the body of the axis, leaving the dens free to rotate with respect to the atlas. The articular facets are slightly convex, resembling a segment of a dome, and face upwards and laterally. They allow the gliding forwards of one lateral mass while the other glides backwards when the atlas rotates on the axis. The slightly concave inferior facets face downwards and forwards and are situated just behind the transverse processes at the junction of the pedicles and the laminae on either side.

The *transverse processes* are small and rounded, projecting laterally from the sides of the vertebral body. As there are no anterior tubercles, the front of the foramen transversarium is closed by the costotransverse lamella.

The thick and strong laminae pass backwards and medially, joining together to form the strong stout spinous process which is usually bifid. The spinous process projects almost 1 cm further posteriorly than the posterior tubercle of C1 and covers the much thinner spinous process of C3.

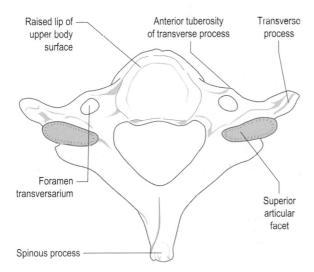

Figure 4.9 The seventh cervical vertebra, superior view.

Labels: Raised lip of upper body surface; Anterior tuberosity of transverse process; Transverse process; Foramen transversarium; Spinous process; Superior articular facet

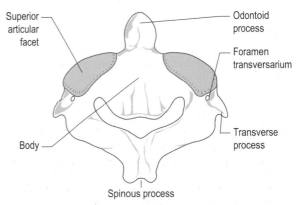

Figure 4.10 The axis, C2; posterosuperior view.

Labels: Superior articular facet; Odontoid process; Foramen transversarium; Body; Transverse process; Spinous process

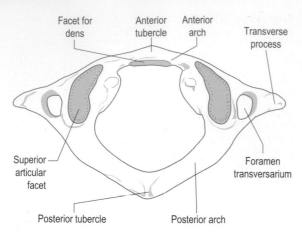

Figure 4.11 The atlas, C1; superior view.

The inferior surface of the axis is very similar to that of a typical cervical vertebra (Fig. 4.8).

The atlas (C1)

This ring of bone bears very little resemblance to a cervical or for that matter any other vertebra (Fig. 4.11). It has no body or spine, but consists of slender *anterior* and *posterior arches* joined on each side by a lateral mass, which bears articular facets superiorly and inferiorly and a *transverse process* laterally. The concave *superior facets* articulate with the condylar processes of the occipital bone, while the concave inferior facets articulate with the second cervical vertebra. The inferior articular facets are segments of a sphere, thereby facilitating rotation between the atlas and axis. Each lateral mass is marked on its medial side by a small tubercle to which the transverse ligament of the atlas attaches.

Passing between the lateral masses anteriorly is the short flattened anterior arch, which has in the middle of its anterior surface the *anterior tubercle* and posteriorly a *facet* for articulation with the dens of the axis. The posterior arch, which represents the pedicles and laminae of typical vertebrae, is a long, curved bar of bone joining the lateral masses and the roots of the transverse processes. Posteriorly, the *posterior tubercle* represents the spinous process. On the superior surface of the posterior arch, running medially from the back of each lateral mass, is a shallow groove for the vertebral artery before it enters the foramen magnum. This groove may be converted into a foramen by cartilaginous or bony tissue.

The strong transverse processes are large and wide. They may also be bifid, even though they represent only the posterior tubercle of a typical cervical vertebra. The *foramen transversarium*, lying close to the lateral mass, transmits the vertebral vessels as well as sympathetic nerves.

Section summary

The vertebrae

Typical lumbar vertebra

- Large, kidney-shaped body.
- Short, strong pedicles arising from upper half of body.
- Horizontally projecting spinous process with thick posterior border.
- Triangular vertebral canal.
- Thin, short transverse process.
- Vertically projecting articular processes; superior transversely concave, facing medially; inferior transversely convex, facing laterally.

Typical thoracic vertebra

- Heart-shaped bodies with demifacets for articulation with head of corresponding rib and rib below.
- Short pedicles arising from upper half of body.
- Small, almost circular vertebral canal.
- Long, downward pointing spinous process.
- Overlapping laminae between vertebrae.
- Long, thick, rounded transverse processes with facet for articulation with tubercle of corresponding rib.
- Vertically projecting flat articular processes; superior facing posteriorly; inferior facing anteriorly.

Typical cervical vertebra

- Small kidney-shaped body with synovial uncovertebral joints at lateral margins.
- Short pedicles projecting almost laterally from body.
- Long, narrow laminae.
- Large, triangular vertebral canal.
- Short, bifid spinous process.
- Composite transverse processes projecting from side of body ending in anterior and posterior tubercles; each process contains foramen transversarium.
- Large superior and inferior articular processes; superior facing superoposteriorly; inferior facing anteroinferiorly.

Atlas (C1)

- Lacks a body; consists of anterior and posterior arches.
- Anterior arch articulates with odontoid process of axis.
- Superior articular facets articulate with occipital condyles of skull.

Axis (C2)

- Strongest cervical vertebra.
- Odontoid process projects superiorly from body; articulates with atlas.

The vertebral column

Ossification of the vertebrae

A typical vertebra ossifies in cartilage from three primary ossification centres and five secondary ossification centres.

The primary centres appear in the vertebral body and in each half of the vertebral arch. That for the centrum, the larger median part of the body, appears between the second and fifth month *in utero*, being present first in the lower thoracic region and then spreading sequentially up and down the column. The centres for the coccygeal vertebrae appear between birth and puberty, with ossification spreading without the formation of secondary centres. Because each centrum is usually ossified from two centres, anomalies may arise in that they may fail to unite and remain as two separate halves, or only one may ossify giving rise to a hemivertebra.

The primary centre in each half of the vertebral arch appears at the junction of the pedicle and lamina at 2 months *in utero* in the upper cervical region, spreading down to the sacrum by the fifth month. From each centre, ossification spreads into the lamina and pedicle where it extends into the centrum to complete the body, and into the root of the transverse process.

At birth, the centrum and each part of the vertebral arch are separated by cartilage. That between the centrum and each arch begins to ossify in the cervical region during the third year, extending to other regions by the seventh year. The laminae begin to unite soon after birth in the lumbar region, spreading to the cervical region by the second year. This process is not complete in the sacrum until the seventh to tenth year. Once the laminae have fused, ossification spreads into the root of the spinous process.

The multiple secondary centres for the upper and lower surfaces of the bodies appear during the ninth year. They fuse to form flat rings of bone around the periphery of these surfaces. Secondary centres appear in the tips of the transverse processes during the 18th year. Fusion of all of these secondary epiphyses with the rest of the vertebra begins at 18, for the bodies, and is complete by age 25.

The lumbar vertebrae also have secondary centres for the mamillary processes. In addition, the first lumbar vertebra may have separate primary centres for its transverse processes, which may remain separate to form a lumbar rib. In the fifth lumbar vertebra there may be two primary centres in each half of the vertebral arch, united by cartilage between the superior and inferior articular processes. Consequently, there is a temporary risk of separation between the two parts.

The upper two cervical vertebrae cannot be considered typical in their pattern of ossification and therefore are dealt with separately. In the atlas a primary centre appears during the second month *in utero* for each half of the vertebral arch and the associated lateral mass, the two halves usually unite during the third or fourth year. Occasionally, a secondary centre may appear in the intervening cartilage. At birth the anterior arch is cartilaginous, but an ossification centre appears by the end of the second year to fuse with the lateral masses, and includes the anterior part of the superior articular surface, during the seventh year.

An epiphysis for each transverse process appears and unites with the rest of the bone during puberty.

In the axis, primary centres for each vertebral arch appear during the second month *in utero*, one for the lower part of the body and two more side by side for the dens and upper part of the body during the fifth month. Those for the dens fuse together 2 months later so that at birth the axis consists of four parts which unite between the third and sixth year. Secondary centres appear in the tip of the dens between 2 and 6 years to fuse with the dens by the age of 12, and for the lower surface of the body which fuses during puberty between 18 and 25. Occasional additional ossification centres may be present.

Bifid cervical spinous processes each have a secondary ossification centre, in addition to which the sixth and seventh cervical vertebrae may have separate primary ossification centres for their costal elements. If present, these usually fuse with the rest of the bone during the fifth year. However, there may be a tendency for that of the seventh to remain separate as a cervical rib.

With so many ossification processes and patterns occurring simultaneously it is not surprising, therefore, to find some variations in the total number of vertebrae present. This is usually due to a reduction, or rarely, an increase in the number of coccygeal vertebrae.

The number of cervical vertebrae is constant at seven. However, the number of thoracic vertebrae may be increased by the presence of ribs associated with L1. Similarly the number of lumbar vertebrae may be reduced as above, or by incorporation of L5 into the sacrum. Such sacralization may be partial or complete. The sacrum may gain or lose additional segments.

Many congenital conditions of the vertebral column are due to incomplete fusion of its constituent parts. Hemivertebrae can cause an abnormal lateral curvature (scoliosis) of the vertebral column. The laminae may fail to fuse or meet in any region, but most commonly in the lumbosacral region, giving rise to *spina bifida*. The spine, laminae and inferior articular processes of L4 and L5 may be joined to the rest of the vertebrae by cartilage and not completely fused. Under certain loading conditions this can lead to a separation whereby the vertebral body, most commonly that of L5, slides forwards. This condition is known as *spondylolisthesis* (see p. 284).

Palpation of the vertebral column

Unfortunately there is considerable variation in the location of bony points from one individual to another and even from side to side within the same individual. This is particularly apparent in the region of the trunk, where the length of the spines can vary considerably, be angled differently or occasionally even be absent. The best way of identifying vertebral spines is therefore to count downwards or upwards from known bony landmarks and cross check with other surface markings. Considerable time and practice is involved in developing palpation techniques, but once mastered it will prove to be an invaluable asset in the future.

The most obvious surface markings are the posteriorly directed spines. However, their palpation can vary considerably, due to the lordosis (convexity forwards) in the cervical and lumbar regions and the kyphosis (concavity forwards) in the thoracic and sacral regions of the vertebral column.

Lumbar region

This region of the vertebral column has a lordosis similar to that of the cervical region. Consequently, it is not easy to identify the spinous processes. With the subject lying prone, and with sufficient support under the abdomen to raise the lumbar region so that it is level, the whole lumbar region can be examined. The spinous processes of the lumbar vertebrae can be palpated in a central cleft down the midline. About 3 cm each side of the midline a small dimple can be seen on the posterior superior aspect of the buttock. This marks the location of the posterior superior iliac spine, which can be easily palpated and acts as an important landmark for the identification of other structures. From these spines the crests of the ilium can be traced upwards and forwards. The spinous process of L5 can be felt in a deep hollow, just above the sacrum, approximately 2 cm above a line drawn between the posterior superior iliac spines. From here the spinous process of L4 is easily recognizable above that of L5. With care, the small gaps between the spinous processes of L4 to T12 can be palpated, and each process identified. The centre of the spinous processes of each vertebra, unlike that in the thoracic region, lies at a level just below the centre of the body of its corresponding vertebra. On either side of the midline is a powerful column of muscle tissue running from the posterior part of the sacrum up towards the thoracic region.

On deep palpation, lateral to this bulk of muscle, small pointed tubercles can be felt, running down either side. These are the tips of the transverse processes, each one being located just above the level of the centre of its corresponding spinous process. Higher up, level with the spinous process of L1, the tip of the 12th rib can be palpated, being level with the ninth costal cartilage in front and lying in the transpyloric plane.

Thoracic region

In this region the spinous processes are much easier to identify, particularly if the subject is sitting with the trunk flexed. The spines of the seventh cervical and the first thoracic vertebrae are even more prominent than when lying prone. It is now easy to identify the spines of individual thoracic vertebrae. Identification is made easier, however, if one finger is placed on the spine above to mark it while that below is determined. Each spinous process appears to

be quite pointed as far as T11. That of the 12th thoracic vertebra, on the other hand, is flattened and similar to that of a lumbar vertebra.

With the subject lying prone, a line of smaller tubercles can be felt approximately 2 cm either side of the spines. These are the posterior aspects of the transverse processes, and as they are usually covered by the long back muscles they are less easy to palpate. The transverse processes pass slightly upwards and laterally, being in line with the upper part of the body of the corresponding vertebra. The spinous processes, however, pass downwards and backwards up to a maximum of 3 cm below the transverse process of the same vertebra. This is important to remember when palpation is being used for diagnostic purposes.

Just beyond the transverse process each rib can be felt passing downwards and laterally around the chest wall.

Cervical region

This region is most easily examined with the subject prone and the forehead supported on the hands with the chin slightly tucked in. Even so, the central area (C3–C5) may be quite difficult to distinguish. There are, however, two unmistakable landmarks, these being the spines of the second and seventh cervical vertebrae.

First find the external occipital protuberance on the occipital bone; this should be almost directly below the most prominent part of the back of the skull. Approximately 2 cm below this, the large prominence of the spine of C2 will be encountered. There is a deep hollow between the two landmarks because C1 has no spinous process, just a relatively small tubercle. Moving down the neck some 10 cm, the long spinous process of C7 (the vertebra prominens), is the next large prominence to be encountered. Usually two prominences can be felt in this region as the spinous process of T1 is just below that of C7. Occasionally, a third prominence, the spinous process of C6, can also be felt. If there is any doubt as to the differential identification of C7, the subject should be asked to raise the forehead from the hands, in which case the spines of C7 and T1 remain under the fingers, whilst that of C6 moves forwards and may not be palpable. If the cervical spine is flexed, the spinous processes may be a little easier to identify, but this depends, to a large extent, on the tautness of the ligamentum nuchae.

Deep palpation lateral to the muscle mass, on either side of the spinous processes, reveals another line of bony projections running up the side of the vertebrae. These are the tips of the transverse processes. They appear blunted because the fingers are feeling both anterior and posterior tubercles at the same time (Fig. 4.8). Near the top of the neck a series of cord-like structures can be felt running downwards and laterally from the tubercles, these are the scalene muscles. The large transverse process of C1 can easily be felt below the mastoid process of the skull. At the lower end of the cervical region, the tubercles become more pronounced with the seventh nearly always projecting further laterally than the rest. It may also be quite tender to the touch. In some subjects the costal element of C7 may be longer than normal and also slightly mobile, giving the appearance of a rudimentary rib; consequently it is termed a 'cervical rib'. Occasionally, this may lead to pressure on the eighth cervical nerve root leading to neurological signs and symptoms in the area of its distribution.

The thoracic cage

A bony and muscular structure (Fig. 4.12) which surrounds, protects and supports the heart and lungs amongst other structures. It is egg-shaped with the narrower end superior towards the neck and the wider end inferior. However, the upper and lower parts of the thoracic cage are cut off obliquely so that the superior opening slopes downwards and forwards at approximately 45°, while the inferior opening slopes downwards and backwards. The openings so formed are known as the *thoracic inlet* and *outlet* respectively.

The thoracic inlet is bounded by the anterior surface of the body of the first thoracic vertebra posteriorly, the medial border of the first rib and its costal cartilage on either side, and the superior surface of the manubrium sterni anteriorly. Through this opening pass the oesophagus and trachea, as well as the vessels and nerves which enter or leave the thorax. In the lateral part of the inlet, the apex of the lung is found supported and covered by the suprapleural membrane. The thoracic outlet is much larger and is bound by the anterior surface of the body of the 12th thoracic vertebra posteriorly, the 12th and the anterior half of the 11th rib on either side, together with the sixth to tenth costal cartilages and the xiphisternal junction anteriorly. The diaphragm covers most of the outlet, but has in it openings for the passage of the aorta, oesophagus and inferior vena cava. Other, smaller structures also pierce the diaphragm to pass between the thorax and abdomen.

The bony components of the thoracic cage are the 12 thoracic vertebrae posteriorly, 12 pairs of ribs laterally and the sternum anteriorly. In general, the ribs pass from behind forwards, connecting the thoracic part of the vertebral column to the sternum. The ribs are so arranged that as they pass forwards they also slope downwards. Consequently, the anterior end of the rib is at a lower level than the posterior part. The space between adjacent ribs, the intercostal space, is filled by muscles, among which are found the intercostal vessels and nerves. The sixth intercostal space is probably the longest, with those above and below gradually decreasing in length. The upper spaces are wider than the lower ones, which tend also to be narrower posteriorly. The articulation of these bony parts is such that the thorax is flattened anteroposteriorly so that its transverse diameter is larger than its anteroposterior diameter. In children the anteroposterior and transverse diameters are more or less equal because the ribs pass more horizontally around the thorax.

The ribs

Long, flat bones which ossify from a cartilage model, part of which persists anteriorly as the costal cartilage of the rib. The costal cartilage of the first rib is very short and often ossified, particularly late in life, whereas the remainder gradually increase in length from above downwards, with the 10th costal cartilage being the longest. The costal cartilages of the upper seven ribs articulate directly with the sternum, while the eighth, ninth and tenth do so via the costal cartilage above. The 11th and 12th ribs merely have cartilage caps anteriorly and end within the abdominal wall musculature. They are often referred to as *floating ribs*. Each rib has a head, neck, angle and shaft. Some ribs exhibit certain features which distinguish them from the rest. These are the first, tenth, eleventh and twelfth ribs. The remaining ribs are known as typical ribs.

A typical rib

Each (Fig. 4.13) has an enlarged *head* at its posterior end, which presents two flattened articular facets with a ridge between them. The facets articulate with the upper border of the body of its own vertebra and the lower border of the body of the vertebra above.

The head is joined to the shaft by a short flattened *neck*, which has upper and lower borders, and anterior and posterior surfaces. The neck continues laterally as the *shaft*, being marked at the junction on its posterior aspect by the *tubercle* of the rib, which has a small oval *articular facet* on its medial part and is roughened laterally. The shaft is flattened having smooth medial and lateral surfaces, a rounded upper and a sharp lower border. The lower border forms the outer margin of the *subcostal groove*, the inner margin being higher on the medial surface of the rib about one-third of the way up. Approximately 3 cm or so lateral to the tubercle, the shaft turns downwards and inwards, giving it a twisted appearance: this is the *angle* of the rib. Only the second rib does not show this twisting of the shaft, in addition to which its outer surface faces upwards and outwards. The anterior end of the rib widens and ends as a roughened hollow where it becomes continuous with its costal cartilage.

The first rib

The first rib is short and sharply curved (Fig. 4.14), giving it a C-shape. Its *head* has only one facet for articulation with the body of the first thoracic vertebra. Joining the head to the *shaft* is a relatively long, narrow *neck*. The *tubercle* of the rib is large and situated on the outer border, and from here the rib slopes downwards and forwards. The shaft has

419

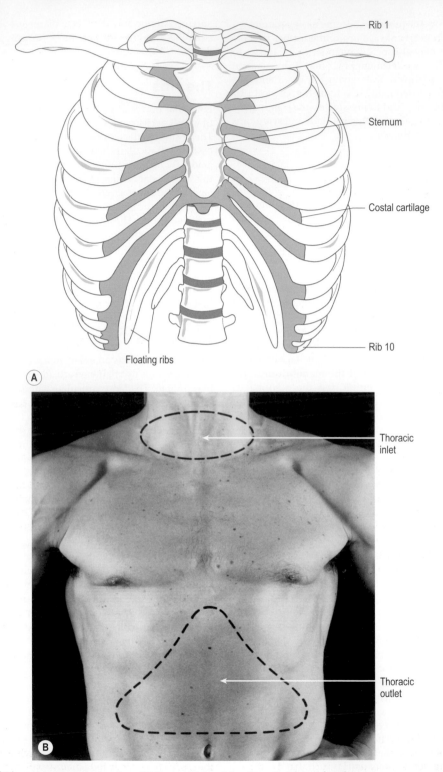

Figure 4.12 (A) Thoracic cage, anterior view, (B) the positions of the thoracic inlet and outlet.

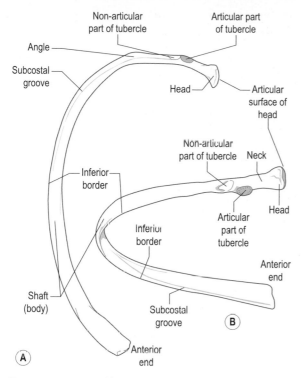

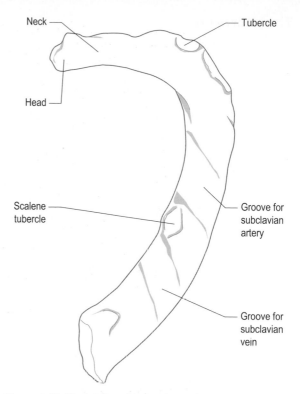

Figure 4.13 A typical left rib viewed (A) inferiorly and (B) posteriorly.

Figure 4.14 The left first rib, superior surface.

superior and inferior surfaces, and medial and thickened lateral borders. It is not angled and remains broad throughout its length. Its lower surface has a shallow subcostal groove running longitudinally while the upper surface has two shallow transverse grooves either side of the *scalene tubercle* which projects from the inner border. The *anterior groove* transmits the *subclavian vein* and the *posterior groove* the *subclavian artery* and the first thoracic nerve, the two vessels being separated by scalenus anterior attaching to the *scalene tubercle*. Anteriorly it articulates with the manubrium sterni by its costal cartilage.

The 10th rib

This shows many of the features of a typical rib, being long with a definite angle. It has, however, only one facet on the head for articulation with the body of the 10th vertebra.

The 11th rib

This is about half the length of the 10th rib, and also possesses only one facet on its head for articulation with its corresponding vertebral body. It does not, however, have an articular facet on its tubercle for articulation with the transverse process.

The 12th rib

This is very short and 'dagger-like' in appearance. Its head has one complete articular facet for articulation with the body of the 12th thoracic vertebra, but there is no tubercle or articular surface for articulation with the transverse process, angle or subcostal groove. Its anterior end is tapered in contrast to the widening of the other ribs.

Costal cartilages

The hyaline costal cartilages are continuous with their respective ribs and connect them either directly (one to seven) or indirectly (eight to ten) to the sternum. The perichondrium of the cartilage is continuous with the periosteum of the rib. It is thus possible to rupture the cartilage within the perichondrium without obvious displacement of the cartilage.

Ossification of the ribs

A primary ossification centre appears near the angle of the rib at six weeks *in utero*. In the second to sixth ribs, secondary centres appear in the head and in each part of the tubercle, while the 11th and 12th ribs only have a secondary centre in the head. All secondary centres appear at puberty, fusing with the remainder of the bone by the age of 25.

The sternum

An elongated, flat bone (Fig. 4.15) situated in the midline of the anterior chest wall. It extends from the root of the neck to the abdominal wall and is composed of three parts: the *manubrium* superiorly, the large *body* and the small irregular *xiphoid process* inferiorly. All three parts have anterior and posterior surfaces, and lateral, superior and inferior borders. Adjacent parts articulate by secondary cartilaginous joints.

The manubrium is the widest part of the sternum, being roughened on its anterior surface and smooth posteriorly. On its superior border is the *jugular (suprasternal) notch*, with the smaller *clavicular notches* either side for articulation with the clavicles. Below the clavicular notch on the lateral border is a *roughened area* for *articulation* with the *first rib costal cartilage*, and lower down a *demifacet* for the *second rib costal cartilage*. The oval inferior surface is roughened for the attachment of the fibrocartilaginous disc of the manubriosternal joint.

The body is composed of four fused segments (sternebrae); again it is roughened on its anterior surface and smooth posteriorly. It does not lie directly in line with the manubrium, but forms an obtuse angle of 160° which can easily be palpated, and as such forms a useful landmark even in obese individuals. The superior surface receives the fibrocartilaginous disc, while the lateral borders show *articular facets* for the *costal cartilages* of the *second* to *seventh ribs*. The second and seventh facets are demifacets, while those in between are full facets. Inferiorly the body is continuous with the irregularly shaped xiphoid process at the secondary cartilaginous xiphisternal joint. The xiphoid process may be perforated or bifid; nevertheless it is thinner than the body of the sternum, being flush with its posterior surface.

The jugular notch is level with the lower border of the body of the second thoracic vertebra, the sternal angle with the lower border of the body of the fourth, and the xiphisternal junction with the ninth thoracic vertebra.

Ossification

Primary ossification centres appear in the manubrium during the fifth month *in utero*, and then in the four sternebrae in the sixth, seventh, eighth and ninth months

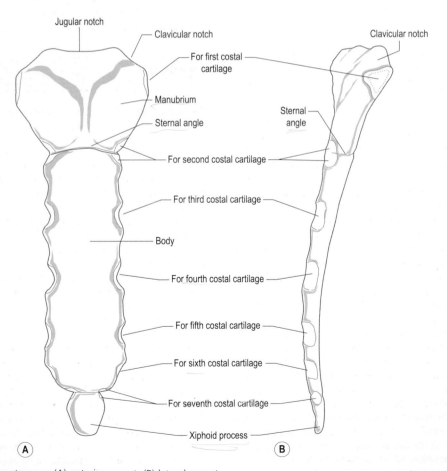

Figure 4.15 The sternum: (A) anterior aspect, (B) lateral aspect.

from above downwards. They fuse in sequence from below upwards in childhood, at puberty and at age 21. The manubrium usually does not fuse with the body until old age. Ossification of the xiphoid process can begin any time after the age of 3; however, it does not fuse with the body until middle age.

Palpation

With the subject seated, a deep hollow can be seen at the base of neck. The inferior margin of this hollow is the jugular (suprasternal) notch, either side of which the medial end of the clavicle can be palpated. Following the clavicle laterally, both its anterior and superior convex surfaces can be easily distinguished. Further laterally the now concave anterior border can be traced to the end of the bone. Beyond this the anterior, lateral and posterior borders and the superior surface of the acromion process can be palpated.

Approximately 2 cm below the suprasternal notch can be felt a ridge of bone beyond which the sternum changes direction. This is the sternal angle at the level of the manubriosternal joint. Either side can be palpated the costal cartilage of the second rib. The space below the second rib and its costal cartilage is the second intercostal space.

At the lower end of the sternum, the pointed xiphoid process can be identified with the costal cartilages of the seventh to tenth ribs running laterally away from it. At the junction of the ninth costal cartilage, there is a marked angle on the anterior rim of the rib cage. This costal angle is level with the tip of the 12th rib and the spinous process of the first lumbar vertebra posteriorly, and the pylorus of the stomach internally. A transverse plane at this level is thus referred to as the transpyloric plane. The lateral border of rectus abdominis also crosses the costal margin at this level. At this junction on the right hand side, the fundus of the gall bladder may be palpated.

The upper two ribs are difficult to palpate. The first is almost completely covered by the clavicle anteriorly, and by thick muscle posteriorly. However, on deep pressure applied to the anteromedial part of the supraclavicular fossa, the superior surface of the first rib can be identified. It should be remembered that this manoeuvre may be painful to the subject because pressure is put on the structures running over the rib at this point.

The second rib is easily identifiable at the sternal angle, but soon becomes lost in thick muscle as it passes posteriorly. The third to eighth ribs are easily identified throughout most of their length, except that they are covered to a variable extent by the scapula. Posterolaterally the angles of the ribs are quite clear, particularly if the scapula is drawn forwards in protraction, lateral to the paravertebral gutter. The anterior half of the 11th rib is quite clear, running anteriorly to the midaxillary line, whilst the tip of the 12th rib can be palpated just beyond the bulk of the long back muscles lying on the transpyloric plane.

Section summary

Ribs and sternum

Typical rib

- Large head with two facets for articulation with body of own vertebra and that above.
- Short, flattened neck.
- Prominent tubercle for articulation with transverse process of corresponding vertebra.
- Long, slender shaft with subcostal groove.
- Articulates anteriorly with its corresponding costal cartilage.

First rib

- Short and sharply curved with superior and inferior surfaces.
- Superior surface has two grooves separated by scalene tubercle.

Sternum

- Elongated flat bone consisting of three parts: manubrium, body, xiphoid process.
- Xiphoid process remains cartilaginous until middle age.

MUSCLES

Muscles producing movements of the trunk and thorax

For descriptive purposes the trunk includes the lumbar and thoracic parts of the vertebral column; therefore movements of these two regions are considered together. The small degree of movement possible between adjacent vertebrae, when combined across several segments, can produce a considerable range of flexion, extension, lateral bending or rotation of the trunk.

Flexion of the trunk occurs when the lumbar and thoracic parts of the vertebral column bend forwards; movement in the opposite direction is extension. These two movements are of considerable functional significance, particularly when they accompany flexion or extension of the hip joint. However, care must be taken when measuring the range of movement of the trunk so that movement at the hip is not included. Lateral flexion (side-bending) of the trunk occurs when the level of the shoulders becomes inclined with respect to that of the pelvis and therefore lateral flexion occurs on both the left and right sides. In this movement adjacent parts of the vertebral bodies on one side come closer together, while those on the opposite side become separated. Rotation of the trunk is produced by the summation of individual vertebral movements which enable the trunk to be twisted to the right or left whilst keeping the shoulders level. Rotation of the pelvis about the hip joint can increase the rotation of the shoulders. Again care must be taken to fix the pelvis when assessing the degree of trunk rotation.

An important structural component of the trunk is the thoracic cage whereby the ribs, articulating with the thoracic vertebrae and sternum, afford protection to the underlying soft tissues. The muscles of inspiration and expiration are therefore considered in this section together with the thoracic cage. The anterior abdominal wall muscles, as well as being involved in producing trunk movements, also raise intra-abdominal pressure; this aids expiration and all straining activities, e.g. micturition, parturition, coughing and vomiting.

As in other sections, a detailed description of any muscle is only given with its first described action.

Muscles flexing the trunk

Rectus abdominis
External oblique
Internal oblique
Psoas minor
Psoas major (p. 233)

Rectus abdominis

Running vertically on the front of the abdomen, rectus abdominis is enclosed within the rectus sheath (Fig. 4.16A). It arises from the front of the *symphysis pubis* and the *pubic crest*, via two tendons, and passes upwards, widening as it does so, to attach to the anterior surfaces of the *xiphoid process* and the *costal cartilages* of the *fifth*, *sixth* and *seventh ribs*. The slightly convex lateral border of the muscle presents as a groove, the *linea semilunaris*, on lean individuals, while the two muscles are separated by the *linea alba*. Transverse tendinous intersections, usually three in number, are found in the anterior part of the muscle and are firmly attached to the rectus sheath. The lowest of these intersections lies at the level of the umbilicus, the highest at the level of the xiphoid, and the third about midway between the other two.

Each rectus abdominis muscle is enclosed in a fibrous sheath formed by the aponeuroses of external and internal oblique and transversus abdominis (Fig. 4.16B), the two sheaths being fused along their medial borders in the region of the linea alba. The formation of the rectus sheath differs at different levels. Above the costal margin it is only present anteriorly and is formed entirely by the aponeurosis of external oblique. Between the costal margin and midway between the umbilicus and symphysis pubis the sheath is formed anteriorly by the aponeuroses of external and internal oblique, while posteriorly it is formed by the aponeuroses of the internal oblique and transversus abdominis (Fig. 4.16B(i)). Below the midpoint between the umbilicus and the symphysis pubis, the aponeuroses of all three muscles

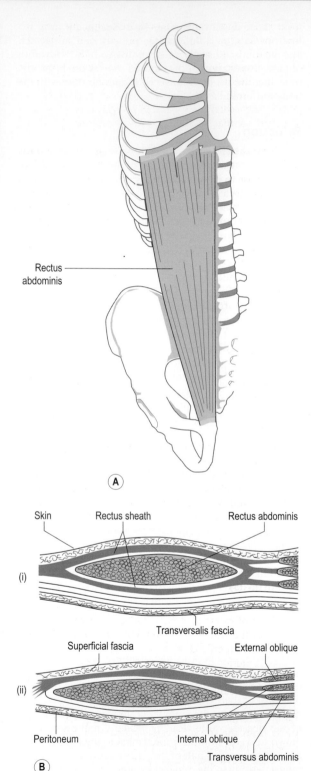

Figure 4.16 (A) The right rectus abdominis, anterior view, (B) formation of the rectus sheath.

pass in front of rectus abdominis so that the posterior wall of the sheath is deficient (Fig. 4.16B(ii)). The inferior limit of the posterior layer of the sheath is marked by a crescentic border, the arcuate line. Below the arcuate line, rectus abdominis lies directly on the transversalis fascia, separating it from extraperitoneal fat. Above the costal margin, the posterior layer is again deficient so that rectus abdominis lies directly on the thoracic wall.

Within the sheath is found pyramidalis, when present. This small muscle, supplied by the subcostal nerve (T12), lies anterior to rectus abdominis, arising from the pubic crest and inserting into the linea alba. It tenses the linea alba, presumably to help provide a stable attachment from which the abdominal muscles can work, particularly when the trunk is flexed.

In the latter stages of pregnancy, the linea alba stretches to increase the distance between the two rectus abdominis muscles, a condition called *divarication* (or *diastasis*) *recti*. A separation of five or more centimetres can occur, but postpartum it returns to normal providing undue strains are avoided.

Nerve supply

By the *anterior primary rami* of the *lower six* or *seven thoracic nerves* (T6/7 to T12). The skin over the muscle is supplied by nerves with root values T4 to L1.

Palpation

The two rectus abdominis muscles can be easily palpated as they run vertically and centrally on the front of the abdomen when the trunk is flexed against resistance. In an athletic subject, it should also be possible to palpate the three transverse tendinous intersections, as well as the linea alba and linea semilunaris at the sides of each muscle.

External oblique (obliquus externus abdominis)

Situated on the anterolateral aspect of the abdominal wall, the fibres of external oblique run downwards and medially from the ribs towards the midline (see Fig. 4.18). It is the most superficial of the three sheets of muscle forming the anterior abdominal wall. The upper attachment is by fleshy slips to the *outer borders* of the *lower eight ribs* and their *costal cartilages*, interdigitating with serratus anterior above and latissimus dorsi below. From here the muscle fibres sweep downwards and medially, with those from the lower two ribs passing almost vertically to attach to the *outer lip* of the *anterior two-thirds* of the *iliac crest*, leaving a free posterior border of the muscle running between the 12th rib and the iliac crest. The remaining fibres give rise to a large aponeurosis which is broader below than above. Each aponeurosis passes

across rectus abdominis, participating in the formation of the *rectus sheath* (p. 424), towards the midline to fuse with that from the opposite side at the linea alba. The *linea alba* is a fibrous raphe which runs from the tip of the xiphoid process of the sternum to the symphysis pubis.

The lower, free border of the aponeurosis stretches between the *pubic tubercle* and the *anterior superior iliac spine*, and forms the *inguinal ligament*. The inguinal ligament folds back on itself so that it is convex downwards – this is caused by the pull of the fascia lata of the thigh which attaches along its length. The medial part of the inguinal ligament is expanded backwards along the pecten pubis, forming the lacunar ligament (a further extension of the lacunar ligament along the pecten is the pectineal ligament). In the anatomical position the lacunar ligament is almost horizontal. Above the pubic tubercle is a triangular cleft in the aponeurosis; this is the superficial inguinal ring (Fig. 4.17). Its base is at the pubic crest and its apex is directed upwards and laterally; the sides are bound by the medial and lateral crura.

Nerve supply

By the *anterior primary rami* of the *lower six thoracic nerves, T7–T12*. The skin over the muscle is supplied by the same nerve roots.

Internal oblique (obliquus internus abdominis)

Lying deep to external oblique, internal oblique (Fig. 4.18) is the middle of the three sheets of abdominal muscles. The muscle fibres arise from the *lateral two-thirds* of the *inguinal ligament*, the *anterior two-thirds* of the *intermediate line* of the *iliac crest*, and from the *thoracolumbar fascia* (see Fig. 4.20C). From this attachment the fibres fan outwards: the most posterior fibres pass almost vertically to attach to the inferior borders of the *lower four ribs*; the more anterior and lower fibres pass upwards and medially, giving way to an aponeurosis along a line extending downwards and medially from the 10th costal cartilage to the body of the pubis. The aponeurosis has a complex involvement in the formation of the *rectus sheath* (p. 424) before interlacing with that of the opposite side at the *linea alba*. That part of the muscle arising from the inguinal ligament passes medially and downwards, blending with the lower part of transversus abdominis to form the *conjoint tendon* which attaches to the *pubic crest* and *pecten pubis*. A few fibres from the inferomedial part of the muscle pass along the spermatic cord to form the cremaster muscle (p. 431).

Nerve supply

By the anterior primary rami of the lower six thoracic nerves (T7–T12) and also the first lumbar nerve (L1).

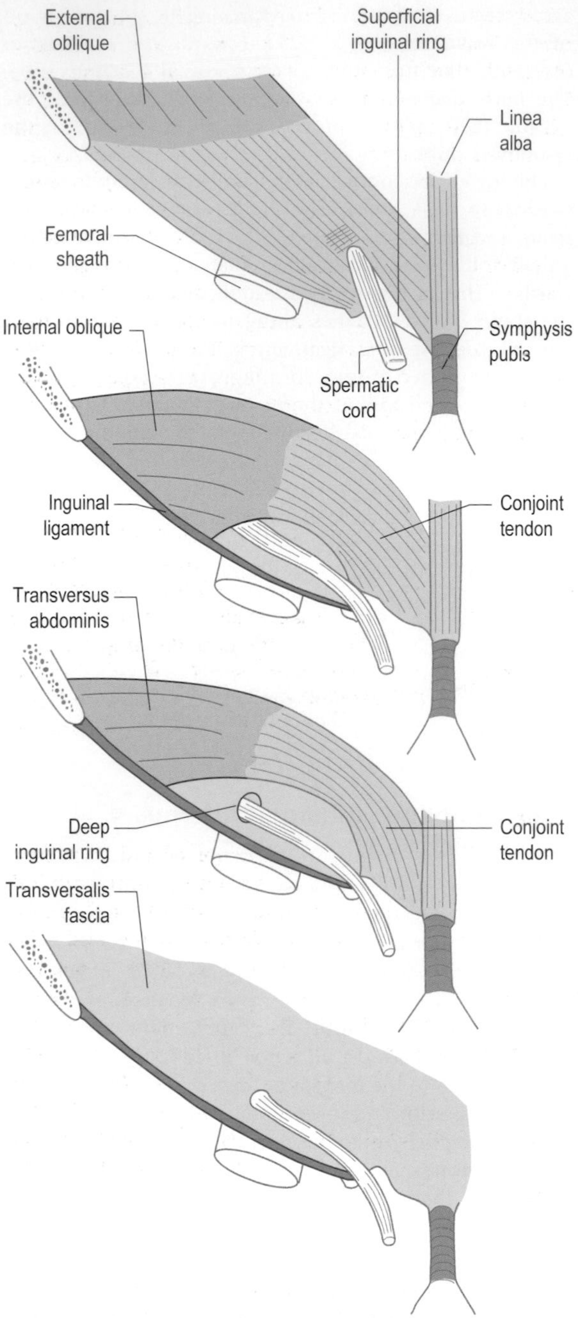

External oblique

Superficial inguinal ring

Linea alba

Femoral sheath

Internal oblique

Inguinal ligament

Transversus abdominis

Deep inguinal ring

Transversalis fascia

Spermatic cord

Symphysis pubis

Conjoint tendon

Conjoint tendon

Figure 4.17 Formation of the inguinal canal.

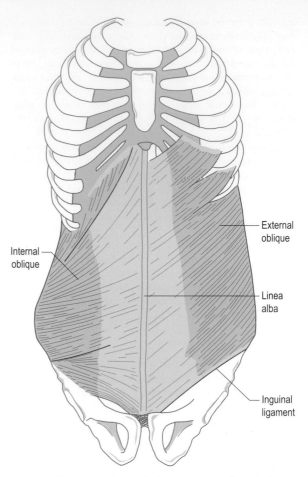

Internal oblique

External oblique

Linea alba

Inguinal ligament

Figure 4.18 The right internal oblique and left external oblique muscles, anterior view.

and alter the degree of pelvic tilt. This latter action has a significant effect in decreasing the lumbar lordosis, and as such is advocated by some in the management of low back pain. These muscles are also involved in rotation and lateral flexion of the trunk, as well as in general functional activities involving the abdomen which are discussed later (p. 439).

Palpation

The flat nature of the oblique muscles makes their palpation difficult in all but muscular subjects. However, a flat hand placed over the lower lateral aspect of the ribs may allow the contraction of external oblique to be felt on resisted flexion. Internal oblique may be similarly palpated if the hand is placed over the lower abdomen just above the anterior half of the iliac crest.

Psoas minor

A weak muscle not always present; however when present it arises from the *sides* of the *bodies* of the *12th thoracic* and

Action

Flexion of the trunk is produced by concentric contraction of external oblique, internal oblique and rectus abdominis of both sides. If the rib cage becomes the fixed point then these same muscles can lift the anterior part of the pelvis *first lumbar vertebrae* and the intervening *intervertebral disc.* The fleshy belly soon gives way to a long tendon which lies on the psoas major and attaches to the *iliopubic eminence* and the *iliac fascia.*

The muscle is supplied by the *anterior primary ramus of L1,* and acts as a weak flexor of the lumbar spine.

Muscles extending the trunk

Quadratus lumborum (p. 430)
Multifidus (p. 428)
Semispinalis (p. 428)
Erector spinae Interspinales

Erector spinae

Erector spinae is a large, complex, powerful mass of muscle consisting of several parts running the length of the vertebral column (Figs. 4.19 and 4.20C).

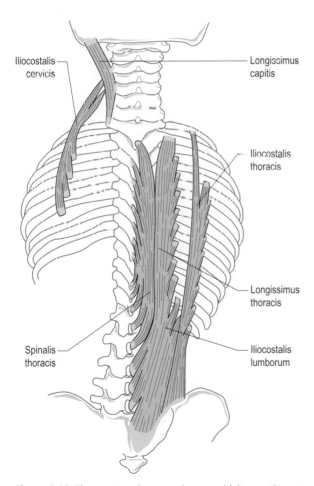

Figure 4.19 The erector spinae muscle mass with its constituent parts shown.

In the lumbar region, it has a broad belly with a well-defined lateral border, but as it extends upwards it divides into three parallel columns, each of which is divided into three parts according to their relative positions (Fig. 4.19).

Erector spinae arises inferiorly from a strong, thick, flat tendon attached along a U-shaped line around the origin of multifidus. The medial limb arises from the *spinous processes of T11 to L5,* spreading onto the *supraspinous ligaments* and associated *median sacral crest.* The lateral limb attaches to the *lateral sacral crest,* the *sacrotuberous, sacrococcygeal* and *posterior sacroiliac ligaments,* and the *posterior part* of the *iliac crest* medial to internal oblique. Deep to the lateral limb erector spinae has a fleshy attachment to the *iliac tuberosity* and the *inner lip* of the *iliac crest.* From this extensive origin the muscle fibres pass upwards deep to latissimus dorsi splitting into three columns.

Iliocostalis

The most lateral of the three columns, iliocostalis can be divided into lumbar, thoracic and cervical parts. Iliocostalis lumborum inserts by six slips into the *inferior borders* of the *lower six ribs* near their *angles.* Medial to each slip arises iliocostalis thoracis which inserts near the angles of the *upper six ribs* and the *transverse process of C7.* Finally, iliocostalis cervicis arises medial to the slips of thoracis to insert into the *posterior tubercles* of the *transverse processes of C4–C7.*

Longissimus

The intermediate column of erector spinae and is the longest and thickest. It can be divided into thoracic, cervical and capitis parts. Longissimus thoracis runs from the *transverse* and *accessory processes* of the *lumbar vertebrae* and adjacent *thoracolumbar fascia* to insert by two sets of slips to the transverse processes of all *12 thoracic vertebrae* and adjacent regions of the *lower 10 ribs.* Longissimus cervicis runs from the *transverse processes of T1–T6,* medial to thoracis, to the *posterior tubercles* of the *transverse processes of C2–C6.* Longissimus capitis arises from the *transverse processes of T1–T5,* in common with longissimus cervicis, and *articular processes of C4–C7,* and inserts into the *posterior aspect* of the *mastoid process.*

Spinalis

The medial relatively insignificant column of erector spinae, again divided into thoracic, cervical and capitis parts. Spinalis thoracis, the most clearly demarcated portion, runs from the *spinous processes of T11 to L2* to those of *T1 to T6.* Spinalis cervicis and capitis are poorly developed, and frequently blend with adjacent muscles.

Nerve supply

All parts of the erector spinae are supplied by adjacent *posterior primary rami* according to their position.

427

Action

When the three muscle columns of both sides of erector spinae act together, they extend the lumbar, thoracic and cervical spines, as well as the head on the neck. Consequently, it is the major extensor of the trunk. However, it is also important in controlling flexion of the trunk. When the three muscle columns of one side act together they produce combined lateral flexion and rotation to the same side. When standing on one leg, the lower part of erector spinae on the non-weight-bearing side works strongly to prevent the pelvis dropping. During walking, erector spinae contracts alternately steadying the vertebral column on the pelvis.

Because the main mass of muscle is situated in the lumbar region, it (particularly longissimus thoracis) is responsible for maintaining the secondary lumbar curvature during sitting and standing.

Palpation

Each erector spinae can be felt and seen as a column of muscle either side of the lumbar spine, particularly during extension. Each muscle mass can also be felt contracting when alternately standing on one leg and then the other.

Interspinales

Short, insignificant muscles of the back, the interspinales (Fig. 4.20B) extend between *adjacent spinous processes*. They are best developed in the cervical and lumbar regions where they consist of bundles of muscle fibres on either side of the interspinous ligament. In the thoracic region they are poorly developed or absent.

Nerve supply

By the *posterior primary rami* of adjacent spinal nerves.

Action

The interspinales can produce extension of the cervical and lumbar spine, but have a more significant role in stabilizing the vertebral column during movement.

Muscles rotating the trunk

Multifidus
Rotatores
Semispinalis
Internal oblique (p. 425)
External oblique (p. 425)

Rotation of the trunk to the left is produced by the simultaneous contraction of the right external and left internal oblique. Conversely, rotation to the right is produced by the left external and right internal oblique. These movements may be accompanied by some flexion of the trunk.

Multifidus

Lies deep to semispinalis and erector spinae in the gutter between the spinous and transverse processes of the vertebrae at all levels. From below upwards the lateral attachment of multifidus is from the *back* of the *sacrum* and the *fascia* covering erector spinae, the *mamillary processes* of the *lumbar vertebrae*, the *transverse processes* of the *thoracic vertebrae*, and the *articular processes* of the *lower four* or *five cervical vertebrae* (Fig. 4.20A). From this extensive origin the muscle fibres are arranged in three layers as they pass upwards and medially to attach to the *spines* of *all vertebrae* from the *fifth lumbar* to the *axis*. The deepest layer attaches to the vertebrae immediately above, the middle layer to the second or third vertebra above, and finally the superficial layer to the third or fourth vertebra above.

Rotatores

These (Fig. 4.20B) are best developed in the thoracic region, being represented by variable bundles in the lumbar and cervical regions. They lie adjacent to the *transverse process* of one *vertebra*, passing upwards to attach to the *lamina* of the *vertebra* above.

Nerve supply

Both the multifidus and the rotatores are supplied by the *posterior primary rami* of adjacent spinal nerves.

Action

Multifidus can produce rotation, as well as extension and lateral flexion, of the vertebral column at all levels, whereas rotatores can only produce rotation in the thoracic region. Both muscles, however, probably have more functional importance in their role as stabilizers of the vertebral column, where they act as extensible ligaments, adjusting their length to stabilize adjacent vertebrae irrespective of the position of the vertebral column.

Semispinalis

Semispinalis (Fig. 4.20A) extends from the lower thoracic region to the base of the skull and consists of three parts. Semispinalis thoracis arises from the *transverse processes* of the *lower thoracic vertebrae* (*T6–T10*) and attaches to the *spinous processes* of the *lower cervical* and *upper thoracic vertebrae* (*C6–T2*). The larger semispinalis cervicis runs from the *transverse processes* of *T1–T6* to the *spinous processes* of *C2–C6*. The largest part of the semispinalis is semispinalis capitis, which runs from the *transverse processes* of *T1–T6* and *articular processes* of *C4–C7* to attach to a medial impression *between* the *superior* and *inferior nuchal lines* on the *base* of the *skull*. The most medial part of semispinalis capitis may be separated from the remainder, and if so is known as spinalis capitis.

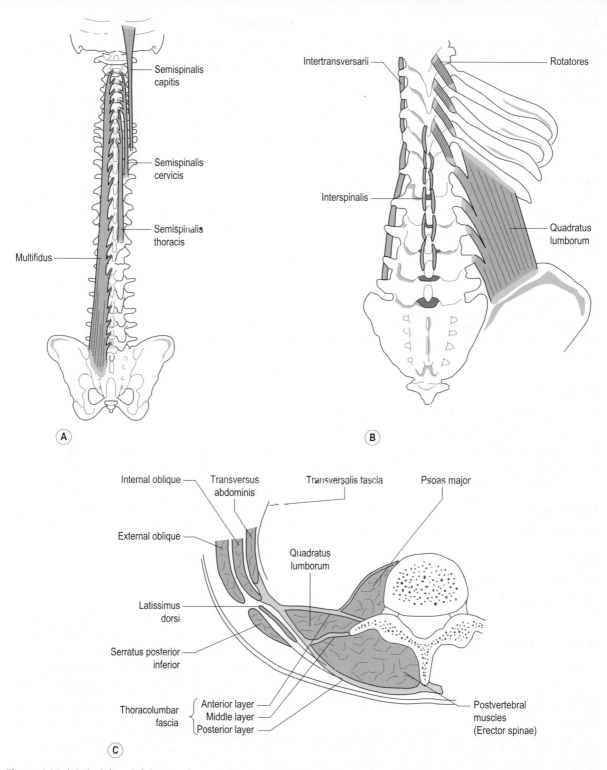

Figure 4.20 (A) The left multifidus muscle, posterior view, (B) right quadratus lumborum, the lower interspinalis, the lower four right rotatores and left intertransversarii muscles, posterior view, (C) transverse section through the region of the upper lumbar spine showing the formation and arrangement of the thoracolumbar fascia.

Nerve supply

By the *posterior primary rami* of adjacent spinal nerves.

Action

Semispinalis, when acting on both sides, produces extension of the thoracic and cervical parts of the vertebral column. When only one side is acting, it produces rotation of the trunk and neck to the opposite side.

Muscles laterally flexing the trunk

Quadratus lumborum
Intertransversarii
External oblique (p. 425)
Internal oblique (p. 425)
Rectus abdominis (p. 424)
Erector spinae (p. 427)
Multifidus (p. 428)

Movement of the trunk to one side is called lateral flexion. Movement to one side is produced by muscles of the same side: rectus abdominis, external and internal oblique, quadratus lumborum and erector spinae.

Quadratus lumborum

A large, flat, quadrilateral muscle of the posterior abdominal wall (Fig. 4.20B) running between the pelvis and the 12th rib, deep to erector spinae. It attaches inferiorly to the *iliolumbar ligament* and *adjacent posterior part* of the *iliac crest*. From here the fibres run upwards and slightly medially to attach to the *medial half* of the *lower border* of the *12th rib*. During its course, the medial border of quadratus lumborum attaches to the *lateral part* of the *anterior surface* of the *transverse processes* of all *lumbar vertebrae*. The muscle is enclosed by the anterior and middle layers of the thoracolumbar fascia (Fig. 4.20C).

Nerve supply

By the anterior primary rami of the subcostal nerve and the upper three or four lumbar nerves (T12, L1, 2, 3 and 4).

Action

Contraction of quadratus lumborum produces lateral flexion of the trunk to the same side. When standing on one leg, it acts strongly on the non-weight-bearing side to stop the pelvis dropping downwards. It also steadies the 12th rib during deep inspiration, so that the origin of the diaphragm is fixed. Both muscles acting together help extend the lumbar vertebral column and also give it lateral stability.

Intertransversarii

Small slips of muscle (Fig. 4.20B) passing between adjacent *transverse processes* in the cervical and lumbar regions. In the cervical region they are reasonably well developed, with each muscle consisting of up to four slips running between adjacent *transverse processes* from the *atlas* (*C1*) to *T1*. In the lumbar region the intertransversarii exist as pairs of muscular slips, the lateral slip running between adjacent *transverse processes*, and the medial slip passing from the *accessory process* of one *vertebra* to the *mamillary process* of the *vertebra* above. The lateral slips of muscle may extend as far as the transverse process of T10. In the cervical region the slips of the intertransversarii lie both anterior and posterior to the emerging ventral ramus of the spinal nerve, whereas in the lumbar region they lie behind the ventral ramus.

Nerve supply

In the cervical region and the lateral part in the lumbar region the intertransversarii are supplied by the *anterior primary rami* of adjacent spinal *nerves*, while the medial lumbar slips are supplied by the *posterior primary rami* of adjacent nerves.

Action

The intertransversarii on one side can produce lateral flexion to the same side in the lumbar and cervical regions. However, their main function is to act as extensile ligaments, thus stabilizing adjacent vertebral segments during movements of the trunk.

Muscles raising intra-abdominal pressure

External oblique (p. 425)
Internal oblique (p. 425)
Rectus abdominis (p. 424)
Transversus abdominis
Cremaster

Transversus abdominis

The deepest of the three sheets of abdominal muscles, and as its name suggests has transversely arranged muscle fibres (Fig. 4.21). It arises from the *lateral third* of the *inguinal ligament* and the *anterior two-thirds* of the *inner lip* of the *iliac crest* inferiorly, the *thoracolumbar fascia* posteriorly, and the *inner surface* of the *costal cartilages* of the *lower six ribs* superiorly. Here it interdigitates with the attachment of the diaphragm (see Fig. 4.23).

The fibres of transversus abdominis pass horizontally around the abdominal wall to end in an aponeurotic sheet which fuses with the posterior layer of the aponeurosis of internal oblique, eventually to reach the linea alba. It is therefore involved in the formation of the rectus sheath (p. 424).

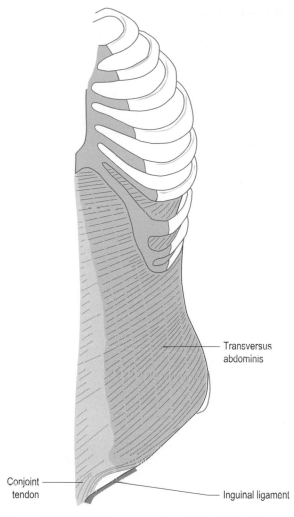

Conjoint tendon

Transversus abdominis

Inguinal ligament

Figure 4.21 The left transversus abdominis, anterior view.

Those fibres arising from the inguinal ligament arch downwards to join with those from internal oblique to form the *conjoint tendon*, which attaches to the *pubic crest* and the *pecten pubis* behind the superficial inguinal ring.

The lower free border of transversus abdominis, between its attachment to the inguinal ligament and the pecten pubis, is concave downwards. Below this border the fascial covering of the muscle (transversalis fascia) comes into contact with both internal and external oblique and the inguinal ligament. In the lateral part of this fascia is a round opening, the *deep inguinal ring* (Fig. 4.17).

Nerve supply

By the anterior primary rami of the lower six thoracic nerves (T7–T12) and the first lumbar nerve (L1).

Action

The actions of all eight muscles of the abdominal wall (four each side) can be related to the function of raising intra-abdominal pressure. This is achieved by the 'sheet' muscles pulling on the rectus sheath via their aponeuroses, flattening the abdomen and so compressing the abdominal viscera. If the diaphragm maintains its tone and resists upward displacement, the subsequent increase in intra-abdominal pressure is important in producing the so-called 'expulsive' acts. Combined with appropriate sphincter relaxation, the increased pressure on the bladder aids micturition; on the rectum it assists in defecation, and on the stomach it helps in vomiting. In the final stages of childbirth, the compressive force produced by these muscles helps to expel the fetus from the uterus.

When the diaphragm is relaxed the increase in intra-abdominal pressure presses the abdominal viscera against its lower surface and so pushes the diaphragm upwards. This increases the pressure inside the thorax so that when the glottis is opened, air is forced from the lungs in a violent, explosive cough or sneeze. The coughing action is further reinforced by the abdominal muscles acting on the lower ribs by pulling them downwards.

Pain resulting from surgical incision of the abdominal wall frequently causes inhibition of these muscles, thereby making coughing very difficult.

The combined action of the abdominal muscles, together with the diaphragm, can also produce a 'muscular corset' which holds the abdominal viscera in place. This action can be increased during activities, such as lifting, whereby a form of pneumatic cushion is formed in front of the vulnerable lumbar spine. This action is frequently seen when people hold their breath, thus anchoring the diaphragm, prior to and when moving a heavy object.

Palpation

A combined contraction of all the abdominal muscles can be easily felt if the hand is placed over the centre of the abdomen while the subject coughs. The increased tension within the anterior abdominal wall can also be appreciated during any of the 'expulsive' acts described.

Cremaster

By loose arrangement of muscle fibres that loop around the spermatic cord and testes. It is continuous with the lower edge of internal oblique and the adjacent part of the inguinal ligament, and attaches to the pubic tubercle. It is usually well developed in males but sparse in females. It is supplied by the *genital branch* of the *genitofemoral nerve* (root value L1 and L2), although voluntary control over the muscle is not possible. A cremasteric reflex is present which raises the testes when the medial side of the thigh is stroked. The reflex is very active in infants, but is much reduced by puberty.

Cremaster, together with dartos, helps to form a mechanism to control the temperature of the testes. Relaxation of these muscles allows the testes to hang well down into the scrotum so that their temperature falls. Contraction of the muscles draws the testes towards the superficial inguinal ring so raising their temperature. Precise regulation of the temperature of the testes is important for the proper formation of spermatozoa, which require a constant temperature approximately 3° lower than core body temperature.

Section summary

Movements of the trunk

Movements of the trunk include those of the thoracic and lumbar spine. Many of the muscles are involved in producing several different movements which are listed below.

Movement	Muscles
Trunk flexion	Rectus abdominis
	External and internal obliques
	Psoas major and minor
Trunk extension	Quadratus lumborum
	Multifidus
	Semispinalis
	Erector spinae
Trunk rotation	Internal oblique
	External oblique
	Multifidus
	Rotatores
	Semispinalis
Lateral flexion	Quadratus lumborum
	Intertransversarii
	External oblique
	Internal oblique
	Rectus abdominis
	Multifidus

- In movements such as flexion and extension the listed muscles on both sides of the trunk work. In lateral flexion muscles on one side work and in rotation it is a combination of some muscles from both sides.
- Many of the muscles listed above, particularly the abdominals, work to raise intra-abdominal pressure for expulsive acts where the diaphragm is fixed, and also forced expiration where the diaphragm moves upwards.

The inguinal canal

An oblique passage through the anterior abdominal wall (Fig. 4.17), approximately 4 cm long, which transmits the spermatic cord in males and the round ligament of the uterus in females, as well as the ilioinguinal nerve in both sexes.

It begins at the *deep inguinal ring* (a round opening in the transversalis fascia approximately 1.5 cm above the midpoint of the inguinal ligament) and ends at the *superficial inguinal ring*, a deficit in the aponeurosis of external oblique above the pubic tubercle and medial end of the inguinal ligament. In the fetus and young children, the deep and superficial rings lie opposite each other, and so facilitate, in the male, the passage of the testes and associated structures into the scrotum from the abdomen. However, with growth the two rings become separated so that in adults the inguinal canal runs downwards and medially from the deep to the superficial rings.

Throughout its course the floor of the canal is formed by the inguinal ligament, with the lacunar ligament in addition medially. The anterior wall is formed by the external oblique aponeurosis throughout, reinforced in its lateral third by muscular fibres of internal oblique. Posteriorly the wall is formed throughout by the transversalis fascia, reinforced by the conjoint tendon in its medial third. These reinforcements of the anterior and posterior canal walls lie opposite the deep and superficial rings respectively. The roof of the canal is formed by the arching fibres of internal oblique and transversus abdominis as they pass from front to back, so forming the conjoint tendon.

As the spermatic cord (or round ligament of the uterus) passes through the canal it acquires coverings from some of the structures forming the canal. On entering the canal, the spermatic cord takes a covering from the margins of the deep inguinal ring – this is the internal spermatic fascia derived from the transversalis fascia. From the lower border of the internal oblique a second covering is acquired – this is the cremasteric fascia containing the cremaster muscle. Finally, as the spermatic cord emerges from the canal the third and last covering is acquired from the margins of the superficial inguinal ring – this is the external spermatic fascia derived from the external oblique aponeurosis.

The inguinal canal is a weak point in the anterior abdominal wall. The reinforcements opposite the deep and superficial inguinal rings afford them some protection. Nevertheless, there is a tendency for contraction of the abdominal muscles to push mobile abdominal contents along the canal. However, contraction of these same muscles also narrows the canal and reduces the size of the inguinal rings. In fact internal oblique is in more or less continuous contraction when standing, so will tend to have a protecting as well as a supporting role as far as the inguinal canal is concerned. Some activities, such as heavy exertion with the trunk rotated, may actually cause the canal to open.

Because of this weakness it is not surprising that part of the mobile abdominal viscera may be squeezed through the superficial inguinal ring, so forming a hernia. For obvious reasons inguinal hernias tend to be twice as common in males as in females. One of two types of inguinal hernia may occur, indirect (oblique) or direct. An indirect hernia follows the course of the testis in its passage through the abdominal wall to appear at the superficial ring; at the deep ring it is lateral to the inferior epigastric vessels. A direct hernia pushes its way through the posterior and sometimes the anterior wall of the canal following no preformed path. It lies medial to the inferior epigastric vessels, and may also appear lateral to the superficial ring.

Both forms of inguinal hernia can be distinguished from a femoral hernia, which emerges through the femoral canal, because its root lies above the inguinal ligament, whereas in a femoral hernia the root is below the inguinal ligament.

Muscles of the pelvic floor

Levator ani
Coccygeus

Levator ani

The two levator ani muscles are broad but thin, stretching to form a gutter-like floor across the lower part of the pelvis, separating the pelvic cavity from the perineum (Fig. 4.22). The two muscles unite in the midline but for most of their extent are separated by the prostate in males and the urethra and vagina in females. Each muscle arises in a continuous manner from the *inner deep surface* of the *pubic body* anteriorly, the *obturator membrane*, and the *pelvic surface* of the *ischial spine* laterally. From this extensive attachment the muscle fibres run backwards, downwards and medially towards the midline, inserting into the *perineal body*, the sides of the *anal canal* and the *anococcygeal raphe* between the anal canal and coccyx. Levator ani usually consists of least two distinct parts: *iliococcygeus* and *pubococcygeus*.

The pubococcygeal part of levator ani is that part arising anterior to the obturator canal. The most anterior fibres pass backwards around either the prostate in males or the vagina in females to the perineal body. More posterior fibres pass to the prostate or vagina to loop around the upper part of the anal canal holding the anorectal junction towards the symphysis pubis. Some of these fibres fuse with the longitudinal muscle of the rectum and via a series of fibroelastic slips pass through the external anal sphincter to attach to the skin around the anus. The most posterior fibres of pubococcygeus insert into the anococcygeal body and sides of the coccyx.

Iliococcygeus arises from the fascia over obturator internus and inserts into the anococcygeal body and sides of the coccyx. It is partly overlapped on its deep surface by the posterior part of pubococcygeus.

Nerve supply

Levator ani has a dual nerve supply, from the *anterior rami of S3 and S4*, and also by a branch from the *perineal branch* of the *pudendal nerve* (root value S4), which enters the muscle on its perineal surface.

Action

Levator ani, together with coccygeus, plays an important role in supporting the pelvic viscera, particularly in females. It is constantly active.

Contraction of the two levator ani has a constricting effect on the openings in the pelvic floor, either reflexly or voluntarily. For example, when intra-abdominal pressure is raised as in coughing, a reflex contraction of levator ani closes the urethra and anus to prevent unwanted micturition or defecation. This action becomes voluntary when these muscles, supplementing the appropriate sphincters, contract to resist an inconvenient urge to micturate or defecate.

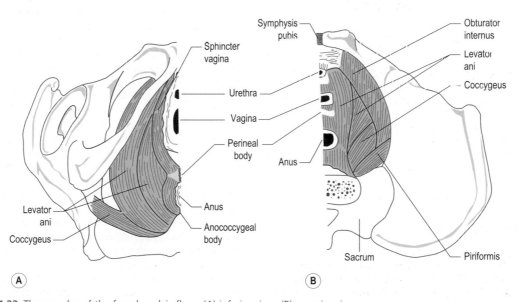

Figure 4.22 The muscles of the female pelvic floor: (A) inferior view, (B) superior view.

In females, the position of levator ani surrounding the vagina is important in supporting the uterus. It is here that the muscle may become excessively stretched during childbirth or surgically traumatized by episiotomy. This stretching may also adversely affect the action of levator ani on the anus and, more commonly, the urethra, possibly leading to stress incontinence. Stress incontinence results in leakage of urine, and possibly faeces, whenever intra-abdominal pressure is raised. Active exercise is necessary to regain normal tone of levator ani, and consequently pelvic floor exercises are taught to restore normal function. Unfortunately, these exercises can be difficult to teach and electrical stimulation may be necessary. It is possible to test and assess the power of contraction of levator ani by inserting the compressible bulb of a 'perineometer' into the vagina. Failing this the strength of contraction can be felt against a gloved finger placed in the vagina.

In males, the problems of stress incontinence are much less common but are sometimes seen following prostatectomy. Pelvic floor exercises are again necessary to restore the tone of levator ani.

Coccygeus

Posterior to, but in the same plane as, levator ani (Fig. 4.22). It is a flat triangular sheet of muscle and fibrous tissue stretching from the *spine* of the *ischium* to the *margin of the coccyx* and *lower two parts* of the *sacrum*. The more fibrous gluteal surface forms the *sacrospinous ligament*.

Nerve supply

By the *anterior primary rami* of *S4*. The skin covering the undersurface of these muscles is supplied by the *anterior primary rami* of *S3* and *S4*.

Action

Forming the posterior part of the pelvic floor, coccygeus assists levator ani in its role in supporting the pelvic viscera and maintaining intra-abdominal pressure. It also pulls the coccyx forwards after it has been pushed backwards during defecation or parturition.

Muscles producing inspiration

Diaphragm
Intercostals
Levatores costorum
Serratus posterior superior

In extreme respiratory distress other muscles, whose primary actions are described elsewhere, may assist in increasing the thoracic dimensions in an attempt to draw more air into the lungs. These muscles are often referred to as accessory muscles of respiration; they include serratus anterior, sternomastoid, scalenes, subclavius, pectoralis minor and pectoralis major. However, to use these muscles effectively their attachment to the ribs and sternum must be free to move, with their other attachment becoming the fixed point. Details of these muscles are given in the sections on the upper limb or neck.

Diaphragm

A musculotendinous sheet (Fig. 4.23) separating the thoracic and abdominal cavities consisting of muscle fibres, attached around the thoracic outlet, which converge to a *central* trefoil-shaped *tendon*. The lumbar part of the diaphragm arises in part from two *crura* which attach to the *anterolateral aspects* of the *bodies* of the *lumbar vertebrae*: the larger *right crus* from the *bodies* and *intervening discs* of *L1–L3*, the smaller *left crus* from the *bodies* and *disc* of *L1* and *L2*. The aorta passes into the abdomen behind the two crura as they cross one another, and in front of the body of T12. At this point the two crura are connected by a tendinous band, the median arcuate ligament. Fibres of the right crus generally pass towards the left, separating to surround the oesophagus before inserting into the central tendon. Fibres of the left crus may also pass behind the oesophageal opening, separating it from the aortic opening. From near the oesophageal opening, the suspensory ligament of the duodenum arises from the right crus to attach to the terminal part of the duodenum.

The remainder of the lumbar part of the diaphragm arises from the *medial* and *lateral arcuate ligaments* which are immediately lateral to the crura. The medial arcuate ligament is a thickening of the fascia covering psoas major and runs from the *side* of the *body* of *L2* to the *transverse process* of *L1*. The lateral arcuate ligament is a thickening of the anterior layer of the thoracolumbar fascia covering quadratus lumborum and runs from the *transverse process* of *L1* to the *tip* of the *12th rib*. Lateral to the arcuate ligaments the costal part of the diaphragm arises from the *inner surface* of the *lower six ribs* and their *costal cartilages*, interdigitating with transversus abdominis, to insert into the anterolateral part of the *central tendon*.

The most anterior sternal part of the diaphragm arises by two slips from the *posterior surface* of the *xiphoid process* of the *sternum*.

All the muscle fibres arch upwards and medially towards their insertion into the central tendon which is situated towards the front of the muscle. Consequently, the short anterior fibres and longer posterior fibres give the appearance of an inverted letter J when viewed from the side. When viewed from the front, two small domes (cupolae) on either side of the central tendon can be seen, that on the right being at a slightly higher level than that on the left; the central part lies opposite the xiphisternal joint.

The upper surface of the diaphragm is covered with parietal pleura, which lines the thoracic cavity. A potential space, the costodiaphragmatic recess, separates the parietal and visceral pleurae, the latter covering the lungs. The

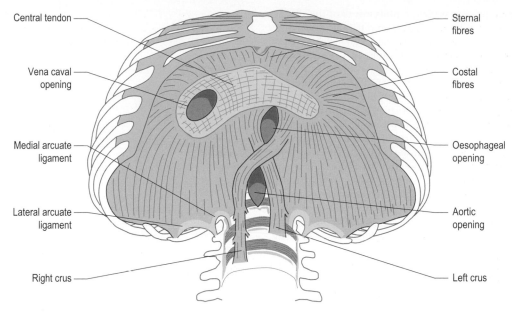

Central tendon

Vena caval
opening

Medial arcuate
ligament

Lateral arcuate
ligament

Right crus

Sternal
fibres

Costal
fibres

Oesophageal
opening

Aortic
opening

Left crus

Figure 4.23 The diaphragm, inferior view.

fibrous pericardium enclosing the heart is firmly attached to the central tendon. The inferior surface of the diaphragm is lined by the parietal layer of peritoneum. This surface is related on the right side to the right lobe of the liver and right kidney, and on the left side to the left lobe of the liver, the fundus of the stomach and left kidney.

Several structures pass between the thorax and abdomen, doing so by either passing through or behind the diaphragm. The major tubular structures (inferior vena cava, oesophagus and aorta) do so by named openings, they may be accompanied by nerves and/or other vessels. The *caval opening* is in the central tendon just to the right of the midline, and transmits the inferior vena cava and right phrenic nerve. The opening is level with the lower border of T8, with the wall of the vena cava firmly adherent to its margin. Consequently, the inferior vena cava is constantly held open.

The *oesophageal opening*, at the level of T10, is to the left of the midline surrounded by fibres of the right and left crura. As well as the oesophagus, the trunks of the vagus nerves (now known as the gastric nerves) and the oesophageal branches of the left gastric vessels pass through this opening. The left phrenic nerve pierces the muscular part of the diaphragm near to the oesophageal opening in front of the left part of the central tendon.

The *aortic opening* lies behind the diaphragm, in front of T12, as the two crura cross each other. The aorta and thoracic duct pass in and out of the abdomen at this opening. The azygos vein is partly covered by the right crus. The greater and lesser splanchnic nerves pierce the crura as they enter the abdomen running to the coeliac ganglion.

Behind the medial and lateral arcuate ligaments pass the sympathetic trunk and subcostal nerve respectively.

Anteriorly, between the sternal and costal attachments of the diaphragm, the superior epigastric artery passes to enter the rectus sheath and supply the upper part of rectus abdominis.

Nerve supply

The diaphragm is supplied with motor and sensory innervation by the *left* and *right phrenic nerves* (root value C3, 4 and 5). Additional sensory fibres to the peripheral part of the diaphragm are supplied by the lower six intercostal (thoracic) nerves.

Action

The diaphragm is the major muscle of inspiration. Downward movement of the diaphragm, elevation of the ribs and forward movement of the sternum all increase the dimensions of the thorax causing air to be drawn into the lungs. From its resting position, sequential contraction of the diaphragm can be described in the following stages. Contraction of the peripheral muscular portion of the diaphragm against the fixed ribs flattens the two cupolae and pulls the central tendon down from the level of T8–T9. Further descent is prevented by compression of the abdominal viscera, which is prevented from bulging outwards by tone in the abdominal muscles and, to a lesser extent, tension on the pericardium. At this point the central tendon becomes the fixed point and further contraction of the muscle fibres causes movement of the ribs and sternum. The lower ribs are lifted upwards and outwards to increase the lateral diameter of the thorax while the upper ribs are raised to push the body of the sternum forwards and so increase the anteroposterior

435

diameter of the thorax. These movements of the ribs are known as 'bucket-handle' and 'pump-handle' respectively (see p. 484) and occur at the costovertebral, costotransverse, sternocostal and interchondral joints. In shallow respiration the descent of the diaphragm can be as little as 1.5 cm, whereas in deep inspiration it can be as much as 10 cm. This descent, together with movement of the ribs and sternum, produces a very efficient mechanism for drawing air into the lungs. From the position of full inspiration the diaphragm relaxes to control the rate of expiration produced by the elastic recoil of the lungs.

The diaphragm also plays an important role in increasing intra-abdominal pressure, where it resists upward movement of the abdominal contents when the abdominal muscles contract, to produce the expulsive acts (defecation, vomiting, micturition, parturition). This action is also important in supporting the lumbar spine during lifting activities by creating a pneumatic cushion to support it. It is likely that compression of the oesophagus by the fibres of the right crus prevents regurgitation of food from the stomach.

The changes in pressure in the thoracic and abdominal cavities, caused by movement of the diaphragm, assist venous and lymphatic drainage from the abdomen to the thorax.

Palpation

The diaphragm is too deep to be directly palpable, but the effects of its contraction can be seen and felt. With the subject sitting and relaxed, the examiner's flat hand should be placed over the subcostal angle. After full expiration, as a breath is taken the abdominal wall can be felt pushing outwards. If the hands are now placed along the lower ribs these will be felt rising upwards and outwards at the same time as the sternum moves forwards.

Frequently, accessory muscles of respiration are used during breathing. In many instances this leads to apical breathing, in which only the upper parts of the lungs are used. Instruction in the correct method of diaphragmatic breathing greatly increases the efficiency of breathing, and is essential in activities which require breathing control, such as singing.

Intercostal muscles (intercostales)

A group of muscles passing between adjacent ribs arranged in three layers (Fig. 4.24). The region between the ribs in which the muscles are situated is known as the intercostal space.

The outer layer is the external intercostal. Its fibres pass obliquely downwards and forwards from the *lower border* of the *rib above* to the *upper border* of the *rib below*. It extends from the tubercles of the ribs posteriorly, becoming thinner anteriorly until it is replaced by, and is continuous with, the external intercostal membrane in the region of the costochondral junction. In the lower part of the thorax the fibres of the external intercostals blend with those of external oblique.

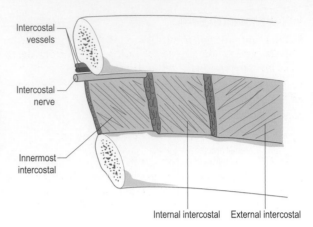

Figure 4.24 The intercostal muscles, lateral view.

The middle layer is the internal intercostal, which extends from the *lower border* of the *costal cartilage* and *costal groove* of the *rib above* to the *upper border* of the *rib below*. It extends from the side of the sternum to the angle of the rib, where it is replaced by the internal intercostal membrane. The muscle fibres pass obliquely downwards and backwards from the costal groove on the rib above to the upper border of the rib below, being at 90° to those of external intercostal. The muscle is thicker anteriorly than posteriorly.

The deepest layer is the innermost intercostal, which although not complete, runs between the *innermost surfaces* of *adjacent ribs*. It is poorly developed in the upper intercostal spaces. The muscle fibres run in a similar direction to those of the internal intercostals, but are separated from them by the intercostal nerve and vessels running in the neurovascular plane between the deep and intermediate muscle layers.

Nerve supply

By the *anterior primary rami* of the adjacent intercostal (thoracic) nerves. The skin over each intercostal space is supplied by the cutaneous branches of the same nerves.

Action

It is generally accepted that some of the external intercostals are active during inspiration, causing elevation of the rib below towards the rib above. The precise role of the internal and innermost intercostals has not been fully established. However, it seems likely that their contraction resists the blowing in and out of the intercostal spaces during respiration and thus produces a more rigid cavity upon which the diaphragm can act. Marked caving in and bulging out of the intercostal spaces occurs when the intercostals are paralysed in quadriplegia. Action in the intercostals has been recorded during many movements involving the trunk. Again, it would appear that their role is one of stabilization of the chest wall.

Levatores costorum

Small strong, triangular muscles found between C7 and T11. Each of the 12 muscles of each side runs from the *tip* of the *transverse process* of the *vertebra above* to the *upper border* of the *rib below*, near its tubercle. The fibres fan out as they pass downwards and laterally.

Nerve supply

By the *dorsal rami* of the adjacent thoracic nerves.

Action

These small muscles elevate the ribs during inspiration. However, their position enables them to also produce a slight degree of rotation and lateral flexion of the trunk.

Serratus posterior superior

A thin, flat muscle (Fig. 4.25) lying anterior to the rhomboids. It arises from the *lower part* of the *ligamentum nuchae* and the *spinous processes* of *C7* to *T3*. The fibres pass downwards and laterally to attach lateral to the *angles* of the *second* to *fifth ribs*.

Nerve supply

By the anterior primary rami of the thoracic nerves T2–T6.

Action

Their attachment to the ribs causes them to be elevated, thus assisting inspiration.

Muscles producing expiration

Transversus thoracis
Subcostals
Serratus posterior inferior
External oblique (p. 425)
Internal oblique (p. 425)
Transversus abdominis (p. 430)
Latissimus dorsi (p. 63)

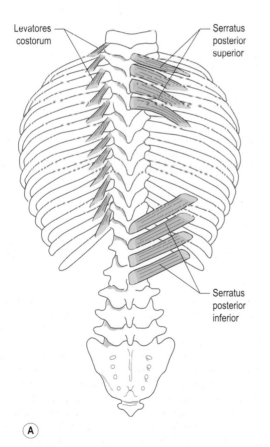

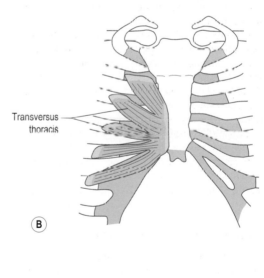

Figure 4.25 (A) The left levatores costorum, and right serratus posterior superior and inferior muscles, posterior view. (B) the left transversus thoracis muscle on the inner surface of the thoracic cage.

The role of the abdominal muscles in forced expiration is considered on page 431.

Transversus thoracis

Found on the inner aspect of the anterior thoracic wall (Fig. 4.25B). It arises from the *posterior surface* of the *xiphoid process*, *lower half* of the *body* of the *sternum*, and *fourth* to *seventh costal cartilages*. The lower fibres pass horizontally while the upper ones pass upwards and laterally to attach to the *inner surface* of the *second* to *sixth costal cartilages*.

Nerve supply

By the *anterior rami* of the adjacent thoracic nerves.

Action

The transversus thoracis pulls the costal cartilages articulating with the sternum downwards and so contributes to expiration.

Subcostals

Irregular slips of muscle extending across one or two intercostal spaces, attaching to the *inner surface* of the *rib* near the *angle*. They are best developed in the lower thoracic region, where their fibres run in the same general direction as those of the innermost intercostal, with which they may be continuous.

Nerve supply

By the *anterior primary rami* of the adjacent thoracic nerves.

Action

The subcostals depress the ribs and so aid expiration.

Serratus posterior inferior

Lies deep to latissimus dorsi (Fig. 4.25A), arising from the *spinous processes* of *T11* to *L2* and the associated *supraspinous ligaments* via the thoracolumbar fascia. The muscle fibres run horizontally to attach to the *lower four ribs* at their *angles*.

Nerve supply

By the anterior primary rami of T9, 10 and 11.

Action

Serratus posterior inferior helps to pull the lower four ribs downwards and backwards and so may assist expiration.

Section summary

Movements of respiration

Respiration is produced by movements of the diaphragm and ribs. These movements are produced by the following muscles:

Movement	Muscles
Inspiration	Diaphragm
	Intercostals
	Levatores costorum
	Serratus posterior superior
Expiration	Transversus thoracis
	Subcostals
	Serratus posterior inferior
	External and internal obliques
	Transversus abdominis
	Latissimus dorsi

- Expiration is normally passive, being produced by the elastic recoil of the lungs.
- In addition to the muscles listed, other muscles attaching to the ribs can assist inspiration when more effort is required to draw air into the lungs, e.g. pectoralis major and minor.

Fasciae of the trunk

Superficial fascia

Over the back, the superficial fascia is thick and contains a large amount of fat held within a meshwork of fibres, whereas at the front and sides of the trunk it contains a variable amount of fat. Laterally it is loosely connected to the skin, but in the midline, particularly in the neck, it holds the skin more firmly to the deep fascia. In the superficial fascia of the anterior abdominal wall fat is commonly deposited in middle age, in the upper part of the abdomen in males and the lower part in females.

As the superficial fascia of the trunk passes down towards the thigh it divides into two layers, between which are found the superficial vessels and nerves. It is continuous with the superficial fascia of the thigh. The deeper of these two layers is a thin elastic membrane, which in the lower part of the abdominal wall is loosely attached to the external oblique aponeurosis, and more firmly to the linea alba and symphysis pubis. In the lower abdomen this deeper membranous layer is a substitute for the deep fascia proper, which is very scant. This membranous layer passes superficial to the inguinal ligament to attach to the deep fascia of the thigh (fascia lata) some 2 cm distal and parallel to the inguinal ligament.

Deep fascia

Over the anterior and lateral parts of the chest and trunk the deep fascia has no special features. It is relatively thin and elastic to allow both the thorax and abdomen to expand. In the lower part of the abdomen it may be replaced by the external oblique aponeurosis and membranous layer of the superficial fascia. It is attached superiorly to the clavicle and side of the sternum, and inferiorly to the iliac crest.

The back, however, is covered by a layer of deep fascia of variable thickness and strength. In the neck it is dense and strong, becoming relatively thin lower down. It covers and encloses the superficial muscles connecting the upper limb to the trunk. It is attached to the spines of the thoracic and lumbar vertebrae, the spine and acromion process of the scapula, the iliac crest and the back of the sacrum. Laterally it is continuous with the deep fascia of the axilla, thorax and abdomen; it also blends with the deep investing fascia of the arm.

Deep to the superficial muscles of the back is found an extremely strong layer of the deep fascia, the thoracolumbar fascia. However, it is really only well developed in the lower thoracic, lumbar and sacral regions.

Thoracolumbar fascia

This (Fig. 4.20C) consists of three separate layers. The posterior layer, which is superficial to erector spinae, is attached medially to the spinous processes of the thoracic, lumbar and sacral vertebrae and the associated supraspinous ligaments. This layer extends from the sacrum and iliac crest to attach to the angles of the ribs, lateral to iliocostalis; latissimus dorsi partly arises from the strong membranous part of this layer in the lower part of the back.

The middle layer of the fascia attaches medially to the tips of the lumbar transverse processes and to the intertransverse ligaments. It extends from the lower border of the 12th rib and lumbocostal ligament above, to the iliac crest and iliolumbar ligament below. It is sandwiched between erector spinae and quadratus lumborum, joining the posterior layer at the lateral border of erector spinae.

The anterior layer of the fascia lies anterior to quadratus lumborum, and attaches to the front of the lumbar transverse processes medially. Laterally it fuses with the middle layer at the lateral border of the quadratus lumborum. It extends from the iliac crest and iliolumbar ligament below, to the lower border of the 12th rib. Superiorly it is thickened between the 12th rib and transverse process of L1 to form the lateral arcuate ligament. It is the thinnest of the three layers.

The single sheet of fascia which is formed laterally acts as the point of attachment for transversus abdominis and internal oblique.

In the lumbar region this thick sheet of fascia is important in filling the gap between the 12th rib and the iliac crest. It acts as a protective membrane and is considered by some to function as a large ligament.

In the thoracic region the fascia is thinner, being sandwiched between erector spinae and latissimus dorsi, and the rhomboids.

Simple activities of the trunk

As in previous sections, this short description is included in order to show how different muscle groups cooperate in producing a desired movement. The sequence of movements involved, together with the responsible joints and muscles are described.

A sit-up

Starting position

The subject lies on their back.

Sequence of movements

The lumbar and thoracic spine is flexed by *rectus abdominis* and *all four abdominal oblique muscles*, each working concentrically. Once the trunk has been raised several centimetres from the ground, a position is reached where *psoas major* can work. With reversed origin and insertion, it pulls on the lumbar spine increasing flexion of the lumbar spine and hips. As this occurs there is a tendency for psoas to lift the legs off the floor unless the feet are fixed. This is opposed by synergic activity in the *hamstrings* which holds the legs against the floor, but which unfortunately tends to flex the knees. This latter action is opposed by the *quadriceps* which keeps the knees extended.

As the body returns to its lying position from sitting, involving extension of the hips and trunk, these same muscles now work eccentrically to control the downward movement which is produced by gravity. If the sit-up is performed with the hands behind the neck, and rotation of the trunk is also included, e.g. so that the right elbow moves to the left knee, then the *left internal* and *right external oblique* muscles will be working strongly to bring about the rotation.

Bending down to touch toes and straightening up again

Starting position

- The anatomical position.

Sequence of movements

The thoracic and lumbar spine and the hip joints are all flexed. From the starting position a brief concentric contraction of the *trunk* and *hip flexors* moves the trunk forwards; it is then that gravity takes over, being the force producing the movement. Following this, flexion of the trunk is controlled by the eccentric contraction of *erector spinae* and *quadratus lumborum*, and that of the hip by eccentric contraction of *gluteus maximus* and the *hamstrings*.

Returning to the upright position, in which the trunk and hips are extended, is produced by concentric contraction of these same muscles.

Sideways bending

Starting position

Standing in full lateral flexion to the left.

Sequence of movements

The trunk is moved from full lateral flexion to the upright position by concentric contraction of the following muscles on the right side of the body: *external* and *internal oblique*, *rectus abdominis*, *quadratus lumborum* and *erector spinae*. Once the trunk has moved past the vertical into right lateral flexion, gravity continues the movement. Corresponding muscles on the left, working eccentrically, now take over to control the movement.

Increasing and decreasing lumbar lordosis

Starting position

Standing with a deep lumbar lordosis.

Sequence of movements

The lumbar spine is moved from its extended to a more neutral position by rotation of the pelvis. The anterior part of the pelvis is raised by concentric contraction of *rectus abdominis*, while at the same time the posterior part is pulled downwards by the *hamstrings* and *gluteus maximus*, also working concentrically. The pelvis is returned to its neutral position by concentric contraction of *erector spinae* and *quadratus lumborum*. If contraction of these muscles continues then the lordosis may be increased as the posterior part of the pelvis is raised.

Muscles of the neck

The neck is one of the most complex areas within the body. Several named regions can be identified with respect to areas bound by specific muscles. Some of these regions contain important structures, e.g. the carotid arteries. However, the regions so described often have no functional basis.

Many of the muscles in the neck are extremely small and impossible to palpate. Nevertheless their actions, either individually or in groups, are important in the alignment and correct posture of the head and neck. Although many of the smaller muscles can be shown to have specific actions, their main role appears to be one of balancing the head on the vertebral column. The following muscles have been described in groups according to the prime movements they produce, without regard to their relative positions within the neck. The movements detailed below include both those of the neck, i.e. the cervical spine, and those of the head on the neck. Consequently, subgroupings of muscles are given depending on whether their action is exclusively on the neck, on the head and neck together, or exclusively on the head with respect to the neck.

Muscles flexing the neck

Longus colli
Sternomastoid
Scalenus anterior (p. 442)

Longus colli

This is in three parts, which lie on the front and sides of the upper thoracic and cervical vertebral bodies (Fig. 4.26A). The lowest part runs obliquely upwards from the *front* of the *bodies* of the *first*, *second* and *third thoracic vertebrae* to the *anterior tubercles* of the *transverse processes* of the *fifth* and *sixth cervical vertebrae*. The middle part runs vertically from the *front* of the *bodies* of the *upper three thoracic* and *lower three cervical vertebrae* to attach to the *front* of the *bodies* of the *second*, *third* and *fourth cervical vertebrae*. The upper part arises from the *anterior tubercles* of the *transverse processes* of the *third*, *fourth* and *fifth cervical vertebrae* to attach to the *anterior tubercle* of the *atlas* (Fig. 4.26A).

Nerve supply

By the anterior primary rami of C3, 4, 5 and 6.

Action

The main action of longus colli, acting either singly or together, is to flex the neck. It is possible that when working singly the lowermost part of longus colli could aid lateral flexion to the same side and rotation to the opposite side, but the central position of both origin and insertion makes this questionable.

Sternomastoid (sternocleidomastoid)

A long strap-like muscle arising from two heads running obliquely around the side of the neck close to the midline anteriorly (Fig. 4.26B). The sternal head is via a narrow, rounded tendon from the *upper part* of the *anterior surface* of the *manubrium sterni*, whereas the broad clavicular head is by a flattened muscular attachment from the *upper surface* of the *medial third* of the *clavicle*. The fibres from the two heads are initially separated by a small gap, with the clavicular head passing deep to the sternal head, but then unite to form a relatively thick muscle belly. Above, sternomastoid inserts by a short, strong, flat tendon into the *lateral surface* of the *mastoid process* of the *temporal bone*, and a thin aponeurosis into the *lateral third* of the *superior nuchal line* of the *occipital bone*.

Nerve supply

The motor supply of sternomastoid is by the *spinal part* of the *accessory* (11th cranial) *nerve*. It receives sensory fibres from the ventral rami of C2 and 3. The skin over the muscle is supplied by roots C2 and 3.

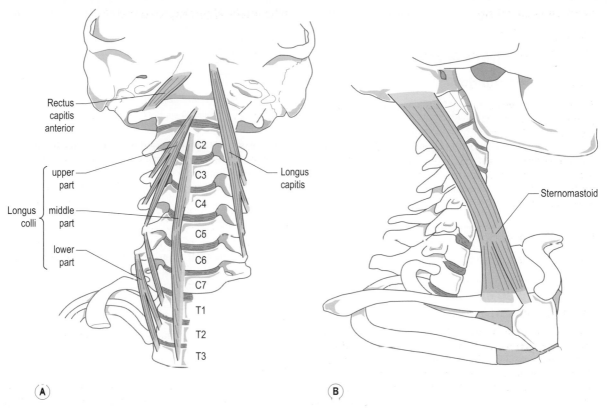

Figure 4.26 (A) The right rectus capitis anterior and longus colli, and the left longus capitis muscles, anterior view, (B) the right sternomastoid muscle, lateral view.

The spinal part of the accessory nerve enters the deep surface of the upper third of sternomastoid, leaving it about halfway down its lateral border. The sensory fibres may join with the accessory nerve.

Action

Contraction of sternomastoid tilts the head to the same side and produces lateral flexion of the neck. At the same time there is rotation of the head to the opposite side. The anterior fibres are said to flex the head on the neck, while the posterior fibres (those to the superior nuchal line) may extend the head at the atlanto-occipital joint.

When both the muscles contract, they produce flexion of the neck by pulling the head forwards. If the head and neck are fixed, then it is possible for the muscles to raise the clavicle and manubrium sterni, and hence the ribs, so that they act as accessory muscles of respiration (see also p. 434).

Palpation

Sternomastoid can be easily palpated if the subject is asked to flex the neck laterally to the same side and then rotate the head to the opposite side against resistance. Both the sternal and clavicular heads can be gripped between the fingers with the gap between them easily identifiable. The round muscle belly is palpable throughout its length, as is the flat tendon at its attachment to the mastoid process.

Muscles flexing the head and neck

Sternomastoid (p. 442)
Longus capitis

Longus capitis

A long narrow muscle (Fig. 4.26A) which runs from the *anterior tubercles* of the *transverse processes of the third, fourth, fifth* and *sixth cervical vertebrae*, passing upwards and medially to attach to the *basilar part* of the *occipital bone* lateral to the pharyngeal tubercle.

Nerve supply

By the *anterior primary rami* of *C1, 2, 3*, and occasionally, *4*.

Action

Longus capitis flexes the head on the neck, as well as flexing the upper cervical spine. However, such a movement is usually achieved by relaxation of the extensor muscles, so that active flexion is only needed against resistance.

Muscles flexing the head on the neck

Rectus capitis anterior

A short strap muscle lying deep to longus capitis (Fig. 4.26A). It passes from the *anterior surface* of the *lateral mass* of the *atlas*, upwards and medially to the *basilar part* of the *occipital bone* between the longus capitis and the occipital condyle.

Nerve supply

By the *anterior primary rami* of C1 and 2.

Action

Rectus capitis anterior flexes the head on the neck. However, its main function is probably to stabilize the atlanto-occipital joint during movement.

Muscles laterally flexing the neck

Scalenus anterior
Scalenus medius
Scalenus posterior
Splenius cervicis
Levator scapulae (p. 58)
Sternomastoid (p. 440)

Scalenus anterior

The anterior of the three scalene muscles (Fig. 4.27B), it lies deep to sternomastoid but in front of the scalenus medius. It arises from the *anterior tubercles* of the *transverse processes* of the *third* to *sixth cervical vertebra*. The muscle fibres run downwards almost vertically, deep to the prevertebral fascia, forming a narrow tendon which attaches to the prominent *scalene tubercle* on the *inner border* of the *first rib* (Fig. 4.14).

Nerve supply

By the anterior primary rami of C4, 5 and 6.

Action

Each scalenus anterior laterally flexes the neck to the same side together with a limited amount of rotation to the opposite side. When both scalenus anterior muscles contract they produce flexion of the neck. If, however, their upper attachment is fixed they act to steady the first rib during respiration, and may assist in its elevation.

Scalenus medius

The middle and largest of the three scalene muscles (Fig. 4.27C). It arises from the *transverse processes* of the *first* and *second cervical vertebrae*, and from the *posterior tubercles* of the *third* to *seventh cervical vertebrae*. The muscle fibres run downwards and laterally to attach to a rough impression on the *upper surface* of the *first rib* behind the groove for the subclavian artery.

Nerve supply

By the *anterior primary rami* of C3–C8 inclusive.

Action

Working singly, scalenus medius produces strong lateral flexion of the neck to the same side. If its upper attachment is fixed, then this strong muscle is very effective in steadying or elevating the first rib during respiration.

Scalenus posterior

The smallest and most posterior of the three scalene muscles (Fig. 4.27C). It arises from the *posterior tubercles* of the *transverse processes* of the *fourth, fifth* and *sixth cervical vertebrae*. Its fibres run downwards and laterally to attach to the *outer surface* of the *second rib* behind the attachment of serratus anterior.

Nerve supply

By the anterior primary rami of C6, 7 and 8.

Action

Scalenus posterior laterally flexes the neck to the same side and may assist in steadying the second rib during respiration.

Splenius cervicis

It arises from the *spinous processes* of the *third* to *sixth cervical vertebrae* from where its fibres run upwards and laterally to attach to the *posterior tubercles* of the *transverse processes* of the *upper three* or *four cervical vertebrae* in front of levator scapulae (Fig. 4.27C).

Nerve supply

By the *posterior primary rami* of C5, 6 and 7.

Action

Acting on its own, splenius cervicis laterally flexes and slightly rotates the neck to the same side. Together, both muscles extend the neck.

Muscles laterally flexing the head and neck

Sternomastoid (see also p. 440)
Splenius capitis
Trapezius (p. 53)
Erector spinae (p. 427)

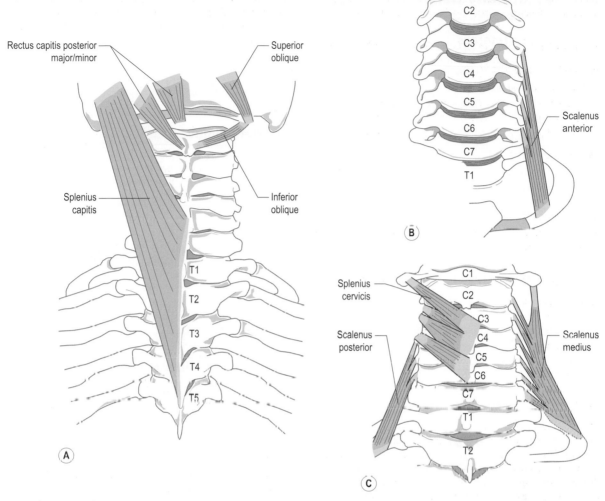

Figure 4.27 (A) The left splenius capitis, rectus capitis posterior minor and major, and the right superior and inferior oblique muscles, posterior view, (B) the left scalenus anterior muscle, anterior view, (C) the right scalenus medius and left scalenus posterior and splenius cervicis muscles, posterior view.

Sternomastoid

Application

Spasm or the contracture of sternomastoid on one side produces the characteristic deformity of lateral flexion of the neck to the same side with rotation of the head to the opposite side. This position is often seen in patients with acute neck pain, or in infants with torticollis (wry neck). Frequently, these conditions require that the muscle is stretched into its longest position by extending and laterally flexing the neck, coupled with rotation of the head in the appropriate direction.

Splenius capitis

Lying deep to the rhomboids, trapezius and sternomastoid, splenius capitis (Fig. 4.27A) arises from the *lower half* of the *ligamentum nuchae* and the *spinous processes* of the *seventh cervical* to the *fourth thoracic vertebrae*. The muscle runs upwards and laterally to attach to the *posterior aspect* of the *mastoid process* of the *temporal bone* and the *lateral third* of the *superior nuchal line* deep to sternomastoid.

Nerve supply

By the *posterior primary rami* of *C3, 4* and *5*.

Action

Acting individually splenius capitis extends the head and neck. However, this action is usually accompanied by lateral flexion of the neck and rotation of the face to the same side. Only when both splenius capitis muscles act together is pure extension of the head and neck achieved.

443

Muscles laterally flexing the head on the neck

Rectus capitis lateralis

A short strap-like muscle running upwards from the *upper surface* of the *transverse process* of the *atlas* to the *jugular process* of the *occipital bone.*

Nerve supply

By the *anterior primary rami* of C1 and 2.

Action

Rectus capitis lateralis produces lateral flexion of the head to the same side. Its main action, however, appears to be stabilizing the atlanto-occipital joint during movement.

Muscles extending the neck

Levator scapulae (p. 58)
Splenius cervicis (p. 442)

Muscles extending the head and neck

Trapezius (p. 53)
Splenius capitis (p. 443)
Erector spinae (p. 427)

Muscles extending the head on the neck

Rectus capitis posterior major
Rectus capitis posterior minor
Superior oblique

Rectus capitis posterior major

A small muscle (Fig. 4.27A) arising from the *spinous process* of the *axis* and runs to attach to the *lateral part* of the *inferior nuchal line* of the *occipital bone*, deep to the superior oblique and semispinalis capitis.

Action

Rectus capitis posterior major extends the head on the neck. Working singly it may produce some rotation of the head to the same side. However, stabilization of the atlanto-occipital joint during movement is probably its main role.

Rectus capitis posterior minor

A small muscle (Fig. 4.27A) arising from the *posterior tubercle* of the *atlas* and inserts into the *medial part* of the *inferior nuchal line* of the *occipital bone*, medial and deep to rectus capitis posterior major.

Action

Rectus capitis posterior minor extends the head on the neck, and like the major muscle, stabilizes the atlanto-occipital joint during movement.

Superior oblique

The superior oblique (Fig. 4.27A) arises from the *upper surface* of the *transverse process* of the *atlas* and inserts between the *superior* and *inferior nuchal lines* of the *occipital bone* lateral to semispinalis capitis.

Action

Superior oblique extends the head on the neck, as well as having an important stabilizing role during movement.

Nerve supply

The innervation of the above three muscles is by the *posterior primary ramus* of *C1.*

Muscles rotating the neck

Semispinalis cervicis (p. 428)
Multifidus (p. 428)
Scalenus anterior (p. 442)
Splenius cervicis (p. 442)

Muscles rotating the head and neck

Sternomastoid (p. 440)
Splenius capitis (p. 443)

Muscles rotating the head on the neck

Inferior oblique
Rectus capitis posterior major (p. 444)

Inferior oblique

The largest (Fig. 4.27A) of the so-called suboccipital muscles. It arises from the *spinous process* of the *axis* and runs upwards and laterally to the *posterior aspect* of the *transverse process* of the *atlas.*

Nerve supply

By the posterior primary ramus of C1.

Action

Inferior oblique turns the face to the same side. Due to the lever arm afforded by the atlas this is quite a strong movement. The muscle also acts as an extensile ligament stabilizing the atlantoaxial joint during movement.

Section summary

Movements of head and neck

In this region movements are described as head on neck at the atlanto-occipital joint and movements of the neck at the joints of the cervical spine. Frequently the same muscles move both of these regions. Movements are produced by the following muscles:

Movement	Muscles
Flexion	Longus colli
	Sternomastoid
	Scalenus anterior
	Longus capitis
	Rectus capitis anterior (head only)
Lateral flexion	Scalenus anterior, medius and posterior
	Levator scapulae
	Sternomastoid
	Splenius capitis
	Trapezius
	Erector spinae
	Rectus capitis lateralis (head only)
Extension	Levator scapulae
	Splenius cervicis
	Trapezius
	Splenius capitis
	Erector spinae
	Rectus capitis posterior, major and minor (head only)
	Superior oblique (head only)
Rotation	Semispinalis cervicis
	Multifidus
	Scalenus anterior
	Splenius cervicis and capitis
	Sternomastoid
	Inferior oblique (head only)
	Rectus capitis posterior major (head only)

JOINTS

Articulations of the vertebral column

Introduction

The function of the human vertebral column, together with a consideration of the adaptations that have occurred in assuming a bipedal posture and gait, have been mentioned in the introductory section to this chapter (p. 406).

Nevertheless it is worth restating that there are 24 free vertebrae in the vertebral column: the five fused sacral and four fused coccygeal vertebrae forming the sacrum and coccyx respectively have already been considered (see pp. 211–213). Of the free vertebrae, seven are termed cervical, twelve thoracic and five lumbar (Fig. 4.28), there being specific differences between the vertebrae in each of these regions. Although in the adult the vertebrae are arranged to give specific curvatures to the column (Fig. 4.28), the joints between adjacent vertebrae have a common plan with the exception of the specialized joints between the atlas and the axis (C1 and C2 respectively).

Anteriorly the bodies of adjacent vertebrae are bound together principally by the strong and important intervertebral discs. The more posterior parts of the vertebral arches are united by synovial joints between the articular processes as well as by ligaments. The joints between the vertebral bodies and the vertebral arches are separated by the intervertebral foramina, through which the spinal nerves emerge. The upper and lower boundaries of these foramina are formed by the pedicles of the arches.

The joints between the vertebral bodies and the connections between the vertebral arches are considered separately. However, it must be remembered that functionally they are both concerned with the structure of the vertebral column, interacting to give it a controlled flexibility.

Joints between vertebral bodies

Between the second cervical and first sacral vertebrae the articulations of adjacent vertebral bodies are cartilaginous joints of the symphysis type. The intervening intervertebral discs are composed of fibrocartilage and are separated from the vertebral bodies by a thin layer of hyaline cartilage.

Intervertebral discs

There are at least 24 intervertebral discs interposed between the vertebral bodies: six in the cervical, twelve in the thoracic and five in the lumbar region, with one between the sacrum and coccyx. (Additional discs may be present between fused sacral segments.) The discs account for approximately one-quarter of the total length of the vertebral column, and are primarily responsible for the presence of the various curvatures. On descending the vertebral column, the discs increase in thickness, being thinnest in the upper cervical region and thickest in the lower lumbar (Fig. 4.28). In the upper thoracic region, however, the discs appear to narrow slightly. In the cervical region the disc is about two-fifths the height of the vertebrae, being approximately 5 mm thick. In the thoracic region the discs average 7 mm in thickness, so that they are one-quarter of the height of the vertebral bodies. The discs in the lumbar regions are at least 10 mm thick, equivalent to one-third of the height of the lumbar vertebral bodies. The relative

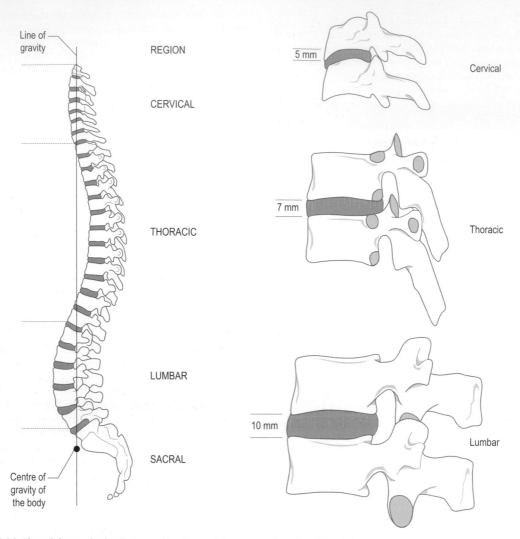

Line of gravity

REGION

CERVICAL

THORACIC

LUMBAR

SACRAL

Centre of gravity of the body

5 mm
Cervical

7 mm
Thoracic

10 mm
Lumbar

Figure 4.28 The adult vertebral column and typical vertebrae in each region, lateral views.

height of the disc to the vertebral bodies is an important factor in determining the mobility of the vertebral column in each of the regions. Individual discs are not of uniform thickness; they are slightly wedge-shaped in conformity with the curvature of the vertebral column in the region of the disc. The curvatures in the cervical and lumbar regions are primarily due to the greater anterior thickness of the discs in these regions.

The overall shape of the discs also varies from one region to another, being similar to the shapes of the adjacent vertebral bodies. Consequently, in the cervical region they tend to be oval, in the thoracic region almost heart-shaped and in the lumbar region kidney-shaped.

It is of considerable practical importance to remember that the intervertebral disc forms one of the anterior boundaries of the intervertebral foramen, and as the spinal

nerves pass through the foramina they lie directly behind the corresponding discs. In addition the discs also form part of the anterior wall of the vertebral canal. Consequently, any posterior bulging of the disc may compress the spinal cord as well as the individual spinal nerves.

Structure

Each disc is structurally characterized by three integrated tissues: the central *nucleus pulposus*, the surrounding *annulus fibrosus* and the limiting *cartilage end plates* (Fig. 4.29A). It is anchored to the vertebral body by the annulus fibrosus fibres and the cartilage end plate.

Nucleus pulposus The soft, highly hydrophilic substance contained within the centre of the disc. There appears to be no clear division between the nucleus pulposus and the surrounding annulus fibrosus, the main difference

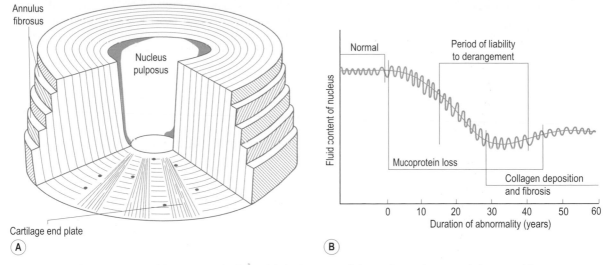

Figure 4.29 (A) Components of the intervertebral disc, (B) Fluid content of the nucleus pulposus and changes with trauma. *(Adapted from Hendry N (1958) The hydration of the nucleus pulposus and its relation to the intervertebral disc derangement.* Journal of Bone and Joint Surgery, *40B, 132–144.)*

being in the density of the fibres contained, with the nucleus having large extrafibrillar spaces containing glycosaminoglycans enabling it to retain fluid. The classical idea, therefore, of a distinct division between the two regions is not true. Furthermore, the concept of the nucleus being round or oval has not been supported by discography, which has shown it to be more rectangular in infants and young children, and anything from oval to multilobed in adults. (Discography is a radiographic technique which allows visualization of the disc in the living subject; clinically it enables the health of the disc to be assessed.) The region between the nucleus and the annulus fibrosus is an area of maximum metabolic activity. It is also sensitive to physical forces as well as to chemical and hormonal regulation of growth processes. Consequently, it may be considered to represent the growth plate of the nucleus pulposus, similar to epiphyseal growth plates, since the nucleus can only increase in size and remodel itself at the expense of the inner part of the annulus fibrosus. The annulus on the other hand increases in horizontal diameter by the addition of new lamellae at the periphery.

The position of the nucleus pulposus within the disc varies regionally, being more centrally located in cervical and thoracic discs and posteriorly located in lumbar discs (Fig. 4.30). Nucleus position is related to certain aspects of function.

The nucleus pulposus consists of a three-dimensional lattice of collagen fibres in which is enmeshed a proteoglycan gel, which is responsible for the hydrophilic nature of the nucleus. Patchy loss and disappearance of this gel occurs with ageing, which lowers the water content until in advanced degeneration the collagen may be devoid of proteoglycan material (Fig. 4.29B). This is the major change

underlying dehydration of the nucleus in later life. In early life, a water content of 80–88% is usual. However, from about the fourth decade onwards this decreases to 70%. These changes in the proteoglycans of the nucleus, both in terms of their loss and composition, change the mechanical behaviour of the disc.

Studies suggest that the nucleus pulposus represents the functional centre of the disc, and that systemic changes within it may be important as a primary cause of pathological change within the disc, and consequently of all pathological change within the intervertebral space. There is, however, the view that in disc degeneration the first morphological change to be observed is the separation of part of the cartilage end plate from the adjacent vertebral body.

Annulus fibrosus A series of annular bands whose geometry varies as a function of vertebral level and intradiscal region. Each annular band has a roughly parallel course, with the directional arrangement of fibres alternating in adjacent bands (Fig. 4.29A), with the obliquity of these bands being greatest in the innermost layer of any given disc. The number of lamellae, as well as their size, thickness and obliquity of arrangement, shows large variations for any given band within different parts of the same disc, for any particular vertebral level, and from individual to individual. Nevertheless, the average number of lamellae is 20 and in general their thickness varies from 200 to 400 μm, increasing from the inside out. Within each lamella the fibrils (0.1–0.2 μm) are uniformly arranged, but their orientation varies considerably from one lamella to another.

Each lamella is composed of obliquely arranged bundles of fibrils, varying in size between 10 and 50 μm. Except for thin fibrils, there is little interconnection between adjacent lamellar sheets; consequently there will be only limited

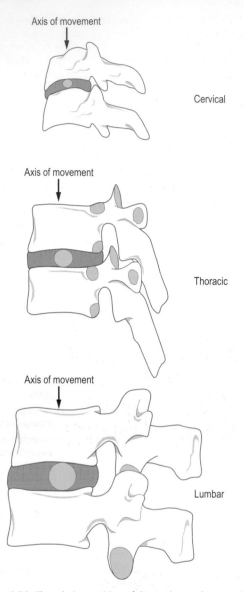

Axis of movement

Cervical

Axis of movement

Thoracic

Axis of movement

Lumbar

Figure 4.30 The relative position of the nucleus pulposus within the intervertebral disc, and its relation to the axis about which movements occur.

restriction to movement during compression and tension. The question arises as to whether orientation of the collagen bundles is predetermined or mechanically induced when movement occurs. There are considerable differences in fibril thickness and lamellar organization in the fetus. It is therefore likely that mechanical phenomena, particularly torsion, are responsible for the arrangement seen in adults.

The density of the fibrocartilaginous lamellae is a function of the annular region, being more closely packed anteriorly and posteriorly than laterally. The lamellar bands do not form complete rings but split intricately or merge

to interlock with other bands. The posterolateral regions of the annulus appear to have marked irregularities and are much less orderly. With ageing, the annulus becomes weakest in these posterolateral regions, thereby predisposing to nucleus herniations.

Elastic fibres are present within both the annulus fibrosus and nucleus pulposus. In the annulus they are circularly, obliquely and vertically arranged, although they are not distributed throughout, but are restricted to the lamellae at the vertebral epiphysis and disc interface. Interlamellar elastic fibres branch and join, freely imparting a dynamic flexibility to the tissue, with obvious implications for function. The intralamellar elastic fibres penetrate the bony vertebrae as perforating fibres.

Within the annulus the total collagen content is not constant, decreasing from the outer layers towards the nucleus. However, the proportion of type I to type II collagen (the principal collagen types within the disc) decreases from the outer layer of the annulus to the nucleus, and also varies from region to region (Fig. 4.31). In other words, type I collagen predominates in the outermost regions of the annulus and type II the innermost; the nucleus pulposus contains type II only. Since type I collagen is typical of tendons and type II of articular cartilage, where large transient compressive forces are generated, the tensile strength of the annulus is probably provided by type I collagen, while the compressive component involves type II. With increasing age, the collagen content of the annulus increases from the inside outwards in the disc, and also downwards from cervical to lumbar regions. However, the proportion of type II collagen does not appear to change with age.

The attachment of the annulus fibrosus to the vertebrae is fairly complicated. The annulus fibres pass over the edges of the cartilage end plate and anchor themselves to and beyond the compact bony zone that forms the outside of the vertebral rim, as well as to the margins of the adjacent vertebral body and its periosteum, thereby forming stable connections between adjacent vertebral bodies. These perforating fibres become interwoven with fibrillar lamellae of the bony trabeculae. This fibrillar anchorage is already present at birth even though the vertebral rim is not ossified.

Cartilage end plate Found on each surface of the vertebral body it represents the anatomical limit of the disc (Fig. 4.29A). It is approximately 1 mm thick at the periphery and decreases towards the centre. It can be considered to have three main functions: (i) it appears to protect the vertebral body from pressure atrophy; (ii) it confines the annulus fibrosus and nucleus pulposus within their anatomical boundaries, and (iii) it acts as a semipermeable membrane to facilitate fluid exchanges between the annulus, the nucleus and the vertebral body via osmotic action. Regarding this third function, however, studies suggest that only the central part of the end plate is permeable.

In the first few years of life, the end plates are loosely attached by a thin layer of calcified material to irregular, radiating, fanshaped ridges and furrows on the vertebral

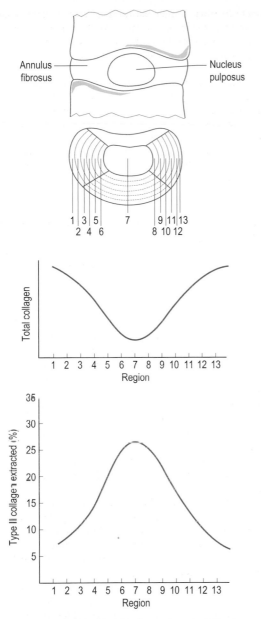

from the vertebral side. However, these channels disappear with increasing age so that by the third decade they are largely obliterated. Following the third decade, retrogressive changes occur in the end plate: it begins to show signs of ossification and there is an increase in calcification. It becomes more brittle, with fimbriation becoming more evident, ranging from thinning to complete destruction of the central end plate zone.

Development of the intervertebral disc The vertebral column begins to develop in the embryonic mesoderm at about 4 weeks, with individual vertebrae developing under the combined inductive influence of the notochord and neural tube. Ablation of either the notochord or neural tube at an early stage results in failure of sclerotomal and myotomal segmentation. The segmental vessels of aortic origin pass between two sclerotomal zones, which then fuse to form the mesenchymal body of the vertebra (Fig. 4.32).

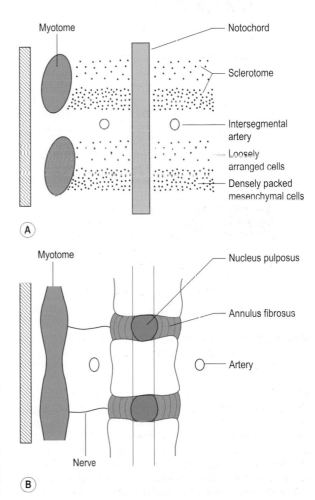

Figure 4.31 Type and ratio of collagen found within the intervertebral disc. *(Adapted from Taylor TKF, Ghosh P and Bushel GR (1981) The contribution of the intervertebral disc to scoliotic deformity.* Clinical Orthopaedics and Related Research, 156, 79–90.)

bodies. Later, a thin layer of calcified material on the end plate firmly adheres to the trabeculae of the porous surface of the vertebral body. It is thought that the end plate is in contact with the bone marrow, through which it receives its nutrients.

In the early part of life, numerous minute vascular channels (cartilage canals) penetrate deeply into the end plate

Figure 4.32 Development of the intervertebral disc, frontal section: (A) at 4 weeks showing sclerotome cells around the notochord and (B) in the adult.

The intervertebral disc therefore develops initially in an environment which contains few blood vessels and is surrounded by a perichondral layer, whose continuity foreshadows the longitudinal vertebral ligaments. The nerves come to lie close to the discs while the intersegmental arteries come to lie either side of the vertebral bodies.

In those regions where the notochord is surrounded by the developing vertebral body it degenerates and disappears. Between the vertebrae, however, the notochord expands as local aggregations of cells within a proteoglycan matrix, forming the gelatinous centre of the disc, the nucleus pulposus. The nucleus is later surrounded by the circularly arranged fibres of the annulus fibrosus, which are derived from the perichordal mesenchyme. The nucleus pulposus and annulus fibrosus constitute the embryonic intervertebral disc. Remnants of notochord may persist in any part of the axial skeleton and give rise to a chordoma. This slow-growing neoplasm occurs most frequently at the base of the skull and in the lumbosacral region.

Following the proliferation and later degeneration of the notochordal cells, there is a fibrocartilaginous invasion of the nucleus pulposus by the orginal mesenchymal intervertebral cells. This invasion occurs at about 6 months *in utero.*

Intervertebral discs lose their embryonic integrity with time, with structural changes occurring in the nucleus throughout adulthood. These normal processes are often considered to be signs of degeneration; they are merely stages in the natural evolution of connective tissue which is subjected to mechanical stress in the form of combined shear and compression forces. The growth of the intervertebral disc, together with the microscopic changes within it, has been correlated with changes associated with weight-bearing in the erect posture. This may be a similar mechanism to that associated with the formation of subcutaneous connective tissue bursae, e.g. housemaid's knee, in which the alternate action of compression forces at right angles to the skin surface and tangential shear stresses induce thickening and delamination of the connective tissue.

Intervertebral discs are subjected to compression, torsion and shear. However, the shear stresses are constantly changing, being dependent on the instantaneous centre of rotation between adjacent vertebrae. This could explain the mechanical delamination of the central region of the disc at different vertebral levels. In other words, the appearance of an irregular cavity in the central region is mechanically induced.

The cartilage end plate also appears to follow this mechanical induction. It is thought to be derived not from the vertebra but from the undifferentiated cells which accumulate in early embryonic life, and develops as an organized structure under mechanical influences. The annular epiphysis of the vertebral body develops in the marginal part of this thin plate of hyaline cartilage, and could therefore be considered to be either part of the disc, or part of the vertebral body.

Ligaments

The bodies of the vertebrae are further held together by longitudinal ligaments which extend the whole length of the vertebral column.

Anterior longitudinal ligament

This supports the anterior aspect of the vertebral column, including the intervertebral discs (Fig. 4.33). It is between 1 and 2 mm thick and consists of three dense layers of

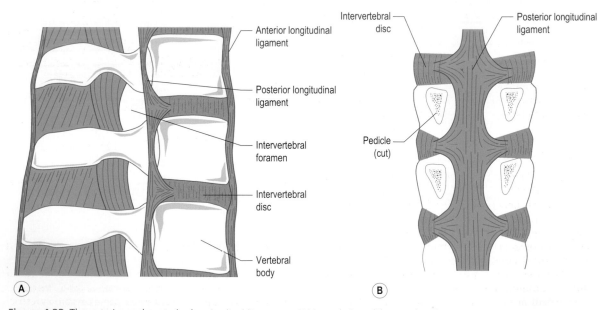

Figure 4.33 The anterior and posterior longitudinal ligaments: (A) lateral view, (B) posterior view.

collagen fibres: those in the superficial layers extend across several vertebrae, while the deepest fibres join adjacent vertebrae. Superiorly, it has a narrow attachment to the anterior tubercle of the atlas. However, as it descends it becomes wider, terminating by spreading over the pelvic surface of the upper part of the sacrum. In the lumbar region the anterior longitudinal ligament is between 20 and 25 mm wide, giving it a cross-sectional area in this region of between 20 and 50 mm^2.

Above the level of the atlas, the anterior longitudinal ligament is continuous with the anterior atlanto-occipital membrane (p. 472).

Posterior longitudinal ligament

This provides the posterior support to the vertebral bodies and is part of the anterior wall of the vertebral canal. It is between 1 and 1.4 mm thick, being composed of two dense layers of collagen fibres: again the more superficial fibres cross several vertebrae, while the deeper ones join adjacent vertebrae. The posterior longitudinal ligament is broader above than below.

Unlike the anterior longitudinal ligament the posterior is attached only to the intervertebral discs and adjacent margins of the vertebral bodies (Fig. 4.33); opposite the middle of each vertebra it is separated from the bone by an interval into which the basivertebral vein passes from the vertebral body. As the ligament narrows in the thoracic and lumbar regions its edges appear serrated (Fig. 4.33). In the lower thoracic and lumbar regions the posterior longitudinal ligament is between 11 and 15 mm wide at the level of the intervertebral disc, whereas at the level of the vertebral body it is only 6–8 mm wide. This gives a cross-sectional area for the ligament of between 3 and 11 mm^2 in the lumbar region, which is considerably less than that of the anterior longitudinal ligament. It extends from the posterior surface of the first sacral segment to the back of the body of the second cervical vertebra, where it becomes continuous with the tectorial membrane (p. 472).

The posterior longitudinal ligament is generally not considered to be as strong as the anterior ligament.

The uncovertebral joints

In the cervical region the intervertebral discs do not extend the full width of the vertebral bodies. Here are found small synovial joints between the lateral parts of adjacent vertebral bodies (Fig. 4.34). The lateral edges of the inferior vertebrae are lipped and fit the bevelled edges of the vertebra above. Consequently, the cartilage-covered superior and inferior articular surfaces face inferolaterally and superomedially respectively. Each joint is surrounded by a capsule which is continuous medially with the intervertebral disc.

Movement at these joints is intimately associated with movements of the cervical spine as a whole, and so, to a

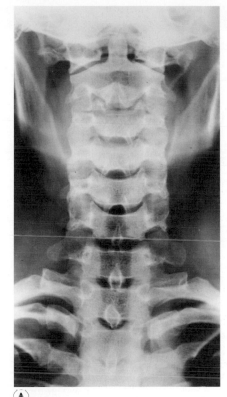

(A)

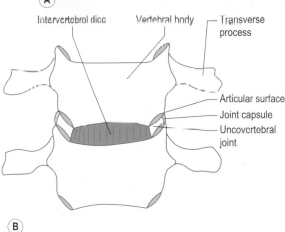

Intervertebral disc — Vertebral body — Transverse process

Articular surface
Joint capsule
Uncovertebral joint

(B)

Figure 4.34 (A) Radiograph of the cervical spine showing the uncovertebral joints, (B) details of the uncovertebral joints.

certain extent, they help to control these movements, and thus stabilize the neck.

The uncovertebral joints lie anterior to the intervertebral foramen. They can, and do, undergo arthritic changes, which can affect the relevant spinal nerves as well as restrict movement between adjacent vertebrae.

Joints between vertebral arches

The vertebral arches are united by synovial joints between their articular processes, as well as by ligaments which pass between the laminae, transverse processes and vertebral spines. In the cervical region, the zygapophyseal joints are behind the transverse processes and form a pillar of bone which is weight-bearing (Fig. 4.35A). In the thoracic region, the joints lie in front of the transverse process, while in the lumbar region they again lie behind them (Fig. 4.35A). Anterior to the zygapophyseal joints in all regions are the intervertebral foramina. Arthritic changes in these joints may give rise to the formation of bony projections (osteophytes) which may compress the spinal nerve within the foramen.

The zygapophyseal joints

Articular surfaces

Although these synovial joints are of the plane variety, the shape and orientation of the joint surfaces on the articular processes vary in the different regions of the vertebral column (Fig. 4.35).

In the cervical region the articular processes form prominent lateral projections arising from the junction of the pedicles and laminae. They carry flat, oval articular facets which lie in an oblique plane. The superior facets face upwards, backwards and slightly medially, while the inferior facets face downwards, forwards and slightly laterally (Fig. 4.35B). The obliquity of the plane of these joints increases slightly from above downwards.

In the thoracic region the articular processes are thin and more or less triangular, and project almost vertically. The articular facet on the superior processes faces mainly backwards but also slightly upwards and laterally, so that it lies on the circumference of a circle whose centre is either just in front of or in the anterior part of the vertebral body (Fig. 4.35A). The inferior processes also have a circumferential arrangement so that their articular

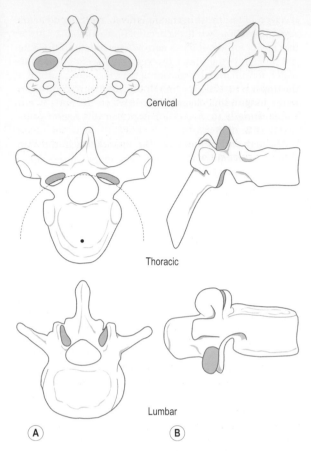

Figure 4.35 The orientation of the articular facets of the zygapophyseal joints in the cervical, thoracic and lumbar regions of the vertebral column: (A) superior view, (B) lateral view.

facets are directed forwards and slightly downwards and medially. They do not project very far below the laminae.

The lumbar articular processes are strong and have a marked upward and downward projection (Fig. 4.35). The articular facets are reciprocally curved in a horizontal plane but virtually straight in a vertical plane. The facets on the superior processes are concave and face medially and backwards, while those on the inferior processes are convex and face forwards and laterally.

All of the articular surfaces, whether in the cervical, thoracic or lumbar region, are covered with hyaline cartilage. The shape and orientation of the articular processes play an important part in determining the type of movement possible within each of the vertebral regions. In the cervical region they are arranged to permit movement in all directions, i.e. flexion, extension, lateral bending and rotation. Those of the thoracic region favour lateral bending and rotation, while in the lumbar region the articular surfaces facilitate flexion, extension and lateral bending. An account of vertebral column movements is given in greater detail on pages 460–470.

Joint capsule and synovial membrane

Each joint is surrounded by a thin fibrous capsule attached to the margins of the articular surfaces. The joint capsule is lax, particularly in the cervical region, to facilitate gliding movements between the two vertebrae. Synovial membrane lines the capsule and also attaches to the margins of the articular surfaces.

Accessory ligaments

While there are no ligaments or thickenings associated directly with the joint capsule, a number of accessory ligaments help to stabilize the vertebral arches.

Ligamentum flavum

This passes between the laminae of adjacent vertebrae from between C1 and C2 down to between L4 and L5 (Fig. 4.36). Its yellowish appearance is due to the presence of a large amount of elastic tissue within it. It is, in fact, the only true elastic ligament in the human body. In each intervertebral interval there are two ligaments, a right and a left. Each is attached to the front of the lower border of the lamina above and passes downwards and backwards to the back of the upper border of the lamina below. The medial borders of the two ligaments meet at the root of the spine; otherwise they are separated by a narrow cleft through which pass veins connecting the internal and external vertebral venous plexuses. Laterally the ligaments extend as far as the joint capsules of the zygapophyseal joints, although they do not blend with them.

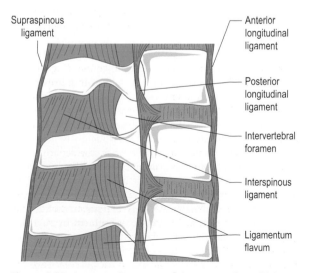

Figure 4.36 Accessory ligaments of the zygapophyseal joints.

Supraspinous ligament

Anterior longitudinal ligament

Posterior longitudinal ligament

Intervertebral foramen

Interspinous ligament

Ligamentum flavum

Studies in non-human primates have shown that the ligamentum flavum has a cross-sectional area greater than that of the anterior longitudinal ligament. There is no reason to suggest that the same is not true in humans, where they assist the postvertebral muscles in maintaining the upright posture, as well as helping to return the trunk to this position following flexion. Because of their elasticity the ligamenta flava permit separation of the laminae during flexion. This same property prevents them forming folds when the vertebral column returns to the upright position: such folds, if they were present, might become caught between the laminae or press upon the dura mater.

Supraspinous ligament

A band of longitudinal fibres running over and connecting the tips of the spinous processes (Fig. 4.36). It is continuous with the posterior edge of the interspinous ligament. The deeper, shorter fibres of the ligament connect adjacent spines, while the more superficial and longer fibres extend over three or four spines. In the cervical region it merges with, and to a large extent becomes replaced by, the ligamentum nuchae.

Ligamentum nuchae

A triangular, midline, fibroelastic septum extending upwards from the spinous process of the seventh cervical vertebra to attach to the external occipital protuberance and crest. The deep part of the ligament attaches to the posterior tubercle of the atlas and the spinous processes of all cervical vertebrae (see Fig. 4.39). In humans the ligamentum nuchae is a rudiment of the well-developed elastic ligament seen in quadrupeds, in which it assists in holding the head erect. Its principal role in humans is perhaps in providing muscle attachments without limiting extension of the neck, as would long cervical spinous processes.

Interspinous ligaments

Thin membranous, relatively weak bands passing between and uniting adjacent vertebral spinous processes (Fig. 4.36). At cervical levels they are insignificant, but at lumbar levels they are longer and stronger.

Intertransverse ligaments

Generally insignificant bands connecting adjacent transverse processes. They tend to be absent at cervical levels and really only become obvious in the lumbar region. In the upper part of the vertebral column they are often replaced by intertransverse muscles.

Blood and nerve supply

The vertebral column receives its arterial supply segmentally from branches of vessels which lie adjacent to it. In the cervical region these are from the *vertebral* and *ascending cervical arteries*, in the thoracic region they are from the *costo-cervical* and *posterior intercostal arteries*, in the lumbar region from the *lumbar* and *iliolumbar arteries*, and in the pelvis from the *lateral sacral arteries*. All of these branches anastomose with and reinforce the *anterior* and *posterior spinal arteries* (p. 497) supplying the spinal cord.

As the branches from the above named arteries pass around the middle of the vertebral body, towards the intervertebral foramen, they give branches to it. Microradiographic studies have shown extensive horizontal and vertical anastomoses between these segmental vessels (Fig. 4.37A).

It is interesting to note that the blood vessels which surrounded the intervertebral disc during the early part of its development subsequently disappear, leaving an essentially avascular structure. Except perhaps for its most peripheral part, which receives a supply from adjacent blood vessels, mature discs are supported by diffusion through the spongy bone of the adjacent vertebral surfaces. Consequently, the blood supply to the vertebral body, as well as its possible disruption, is extremely important.

The blood supply to the vertebral body appears to be zoned (Fig. 4.37B). That part of the body immediately adjacent to the disc is a relatively avascular region, with the more central regions being more vascular. However, this central region can be subdivided into that part supplied by the tortuous *nutrient artery* and that by the *metaphyseal arteries*. The peripheral region of the vertebral body is supplied by short, straight *peripheral arteries*. Diffusion of oxygen and nutrients for disc metabolism is also probably zoned because of the arrangement of the lamellae of the annulus fibrosus. Fluid trapped between lamellae is channelled in a vertical plane. Frequent movement of the lamellae probably speeds up diffusion. One of the consequences of ageing is the gradual narrowing of the lumen of arteries, resulting in a reduced, and eventually obliterated, blood flow. The arteries affected early on are those which have a tortuous course. As far as the vertebral bodies are concerned, the nutrient and metaphyseal arteries are therefore probably the first to suffer. With increasing age, the peripheral arteries apparently become more numerous, so that the outer sleeve of the vertebral body may get a relatively better blood supply than the central region. Disc degeneration and desiccation have been observed to appear at the centre of the disc earlier than at the periphery. Indeed, the periphery may remain healthy and capable of repair long after disintegration and extrusion of the central part of the disc.

Although disc degeneration and low back pain, due to disc hypoxia and starvation (as a secondary effect of reduced blood supply) is an attractive theory, it is difficult to prove. Nevertheless, symptomatic discs differ from normal in their pH and lactic acid concentrations, suggesting that these biochemical changes are due to anoxic respiration of the disc.

Where the segmental arterial supply to the vertebral column is reduced, e.g. by the natural calibre of the vessels, or jeopardized, e.g. by compression of the vessels, then it might be expected that the intervertebral discs adjacent to this vertebra would be liable to degenerative change at an earlier age than elsewhere. One such region is the lower lumbar where the fifth pair of lumbar arteries are small, and may be subjected to compression by a bulging disc. Consequently, the intervertebral discs either side of the fifth lumbar vertebra might be expected to have reduced nutrition and oxygenation. Ninety-five per cent of disc lesions in one study* involved one or both discs either side of L5. Although the evidence is circumstantial, reduced

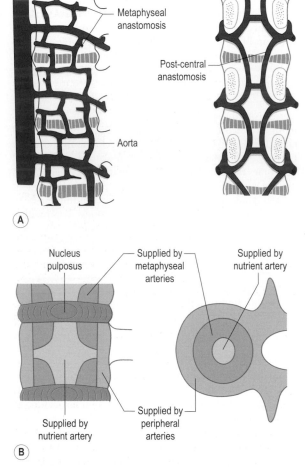

Lateral surface

Metaphyseal anastomosis

Aorta

Dorsal surface

Post-central anastomosis

(A)

Nucleus pulposus

Supplied by metaphyseal arteries

Supplied by nutrient artery

Supplied by nutrient artery

Supplied by peripheral arteries

(B)

Figure 4.37 (A) Anastomoses between the lumbar arteries and their branches, (B) zoning of the blood supply to the intervertebral disc. *(Adapted from Ratcliffe JF (1980) The arterial anatomy of the adult human lumbar vertebral body: a microarteriographic study.* Journal of Anatomy, *131, 57–79.)*

---

*Weinstein PR (1982) Anatomy of the lumbar spine. In: Hardy RW (ed.) *Lumbar Disc Disease*, pp. 5–15. New York: Raven Press.

blood supply to the vertebrae and consequently to the disc are strongly implicated in disc pathology.

The veins of the vertebral column form complex, freely communicating plexuses extending the whole length of the column, both inside and out (Fig. 4.38). The plexuses are drained by a series of intervertebral veins which join the vertebral veins in the neck, the posterior intercostal veins in the chest, lumbar veins in the lower back and lateral sacral veins in the pelvis.

The internal vertebral plexuses form a continuous network between the spinal dura and the walls of the vertebral canal (Fig. 4.38B). The two anterior channels lie either side

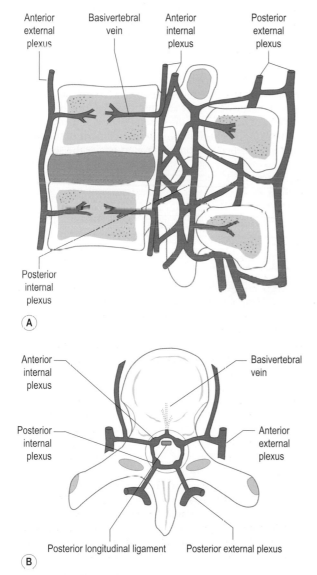

Figure 4.38 Venous drainage of the vertebral column: (A) sagittal section, (B) transverse section.

of the posterior longitudinal ligament, anastomosing across the midline, anterior to the ligament, receiving the basivertebral vein (Fig. 4.38B). They communicate above with the basilar and occipital sinuses and with the sigmoid sinus within the skull. The posterior internal plexuses lie on the inner surfaces of the laminae and ligamenta flava. Again they anastomose across the midline, with some vessels emerging to join the posterior external plexus by passing between the free medial edges of the ligamenta flava. This latter connection may provide a venous pump mechanism as the vessels become squeezed by the ligamenta flava during movements of the vertebral column. The main means of communication between the internal and external plexus, however, is via the intervertebral veins which pass through the intervertebral foramina.

The external plexuses are found along the front of the vertebral column, as well as around the spinous, articular and transverse processes of the vertebrae.

While the precise arrangement of the venous plexuses associated with the vertebral column may not be of interest per se, it has to be remembered that they form a system of great blood-carrying capacity which extends, with few if any valves, from the pelvis to inside the skull, communicating at all levels with the major venous systems of the abdomen, chest, and head and neck. This complex venous network almost certainly plays a significant role in the transfer of metastatic cancer cells to widely separated parts of the body under the influence of differences in venous pressure. Indeed, secondary metastases associated with cancer of the breast or of the prostate invariably involve the vertebrae.

The innervation of the intervertebral disc and the existence of nervous tissue within related structures, e.g. ligaments, are of considerable clinical importance. The sinuvertebral nerve is a recurrent nerve of the vertebral canal which supplies the fibrous connective tissue of the spinal canal, i.e. the posterior longitudinal ligament and the periosteum, together with the venous sinuses and the spinal dura. In many respects the sinuvertebral nerve may be considered equivalent to the recurrent meningeal branch of a cranial nerve. It has been described as having a dual origin: one from the spinal nerve and the other from the sympathetic system. The spinal part arises just distal to the dorsal root ganglion and re-enters the spinal canal to reach the midline, giving branches directly towards the discs above and below each level. It is also likely that some fibres innervate the medial aspect of the zygapophyseal joint capsule.

Numerous fine nerve filaments, as well as encapsulated and non-encapsulated receptors, have been found in the anterior and posterior longitudinal ligaments and the superficial layers of the annulus fibrosus. The innervation of the anterior longitudinal ligament and lateral aspect of the disc appears to be extensive and complex, particularly in the lumbar region, and involves branches from the ventral ramus. The lateral aspect of the zygapophyseal joint capsule probably also receives an innervation from the dorsal ramus as it passes towards the postvertebral muscles.

With both C and A fibres involved in pain transmission, it is clear that structures exist in and around the intervertebral disc whose presence could explain the disc pain caused by mechanical compression of the anterior and posterior nerve fibres in the periphery of the annulus.

Relations

Because the vertebral column extends from the base of the skull to the pelvis, and in doing so is associated with several distinct regions of the body, its relations change as various structures at first approach and then move away from it. Consequently, it is not practical to consider in detail all of these relations. However, an attempt is made to illustrate the major relations in each region.

One of the major functions of the vertebral column is to support and protect the spinal cord against physical trauma. Therefore, within the vertebral canal is found the spinal cord and its meningeal coverings (see p. 494 for further details), as well as the internal vertebral venous plexus which is embedded in loose areolar tissue. As the spinal cord ends at the level of the disc between L1 and L2, below

this level is found the cauda equina, still within the meningeal coverings as far as the level of S2.

In the neck well-defined fascial sheets and membranes enclose the principal muscle masses, the viscera, and main nerves and blood vessels. These fasciae have continuities and bony attachments which form fascial planes and compartments in which the deeper structures of the neck are confined. The cervical vertebrae are located towards the back of the neck. Posteriorly are the postvertebral muscles, enclosed within the superficial layer of cervical fascia. Anterolaterally is the carotid sheath, enclosing the common carotid artery, internal jugular vein and 10th cranial nerve (vagus), with the cervical sympathetic trunk posteromedial to it. The anterior part of the neck comprises essentially two compartments: a superficial muscular compartment, and a deeper visceral compartment enclosing the pharynx, oesophagus, larynx, trachea, thyroid and parathyroid glands. The relationship of the various structures one to another and to the cervical vertebrae at the level of C6 is shown in Fig. 4.39. There are minor variations concerning the precise arrangement of these structures both above and below this level. Nevertheless, the general plan is similar.

In the thoracic region the anterior relations of the vertebrae differ depending on the level at which the section is

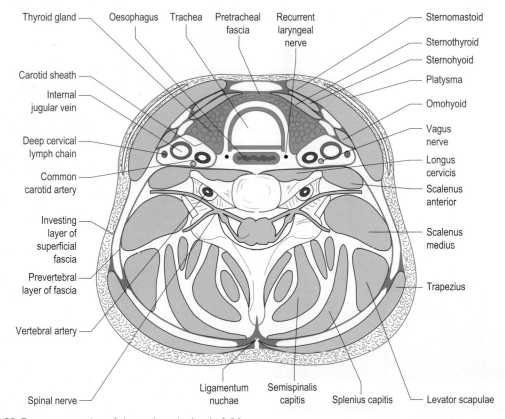

Figure 4.39 Transverse section of the neck at the level of C6.

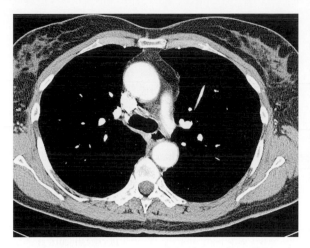

Figure 4.40 Transverse section of the thorax above the level of T8. *(Reproduced with permission from Standring S (2004) Gray's Anatomy 39e, Elsevier Churchill Livingstone.)*

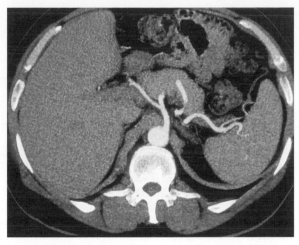

Figure 4.42 Transverse section of the abdomen at the level of L1. *(Reproduced with permission from Standring S (2004) Gray's Anatomy 39e, Elsevier Churchill Livingstone.)*

taken. In general, above the level of T8, thoracic contents only are seen, i.e. lungs, heart, great vessels and the oesophagus (Fig. 4.40). However, below this level, and separated from the thoracic contents by the diaphragm, the liver and stomach become anterior relations (Fig. 4.41). As in the neck, posterior and lateral to the vertebrae are the postvertebral muscles.

Anterior to the lumbar vertebrae lie the major vessels of the abdomen: the aorta and inferior vena cava. Immediately lateral to the upper three lumbar vertebrae lie the kidneys, with various parts of the gastrointestinal tract generally lying anteriorly (Fig. 4.42). Once again the main posterior relation, situated between the transverse and

spinous processes as in other regions, is the postvertebral muscle mass. At lumbar levels it is difficult to differentiate the components of this large muscle mass. Also in this region the thick and strong thoracolumbar fascia can be readily seen (Fig 4.20c).

Stability

Despite its multisegmental composition, the vertebral column is a relatively stable structure. The ease with which the trunk can be held erect with a small amount of muscle activity is in contrast to that required to hold the lower limbs in equilibrium against gravity. Many patients after a long illness are able to sit up, with relatively little effort, long before they can stand. This suggests that the arrangement of the various elements of the vertebral column are such that they provide a high degree of inherent stability which requires little, if any, muscle activity to maintain the position of the vertebral column. Indeed, it has been shown that a vertebral column devoid of all musculature does not collapse but remains erect, and furthermore can support up to 2 kg placed on its upper end without collapsing. How is this possible when the partial centres of gravity of the various links do not lie in the line of the centre of gravity? Furthermore, the centres of the articulations (except for a few) do not fall on this line. Consequently, rotational stresses between vertebrae must be neutralized, otherwise the column would lose its equilibrium and collapse.

What features then contribute to this equilibrium and stability? Of considerable importance is the arrangement and nature of the joints between vertebrae, particularly between their bodies. Being secondary cartilaginous joints they do not permit a large amount of movement. This,

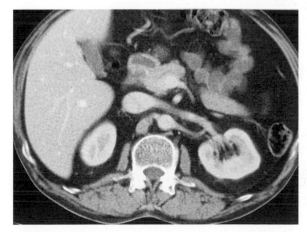

Figure 4.41 Transverse section of the thorax below the level of T8. *(Reproduced with permission from Standring S (2004) Gray's Anatomy 39e, Elsevier Churchill Livingstone.)*

together with the fact that the components of the intervertebral disc interact with the adjacent vertebrae, produces a unit (two vertebrae plus the intervening disc) which is self-stabilizing. Whether loaded or at rest the annulus fibrosus fibres are under tension because of the preloaded state of the disc, brought about by its water-absorbing capacity. The tension developed in these fibres tends to keep the vertebrae aligned. When a force is applied to the upper vertebrae to cause it to move, say in the direction of extension, the nucleus pulposus moves anteriorly, thereby increasing the tension in the anterior part of the annulus, which in turn tends to restore the upper vertebrae to its original position (Fig. 4.43A). Such a mechanism can be seen to work well with asymmetrically applied forces causing movement in the direction of extension, forward flexion and lateral flexion. But, what about axial rotation? During rotation the fibres of the annulus running counter to the direction of movement are stretched, while those

in the opposite direction become relatively relaxed (Fig. 4.43A). Tension is maximum in the central annulus fibres; consequently the nucleus pulposus is compressed, with its internal pressure rising in proportion to the degree of angular rotation. The compression of the nucleus is such that it pushes back against the annulus with a tendency to put the 'relaxed' fibres under tension. This helps to limit movement and also restore the upper vertebra to its original position. It appears then that no matter what force is applied to the intervertebral disc, it always increases the internal pressure of the nucleus pulposus which stretches the annulus fibrosus. This stretching tends to oppose the movement and restore the system to its initial state.

The unique hydrostatic properties of the intervertebral disc result from the interaction of the annulus fibrosus with the water-binding capacity of the proteoglycans in the nucleus pulposus, and from their interaction with respect

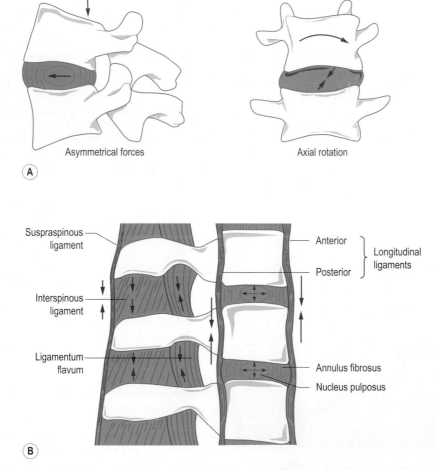

Figure 4.43 (A) Reaction of the intervertebral disc to asymmetrically applied forces and axial rotation, (B) interaction of the intervertebral disc and vertebral ligaments providing an inherent stability of the vertebral column.

to the loading exerted by the longitudinal ligaments binding the vertebrae together. With the intake of water and the subsequent increase in nuclear pressurization, there is an increase in disc height. Not only does this increase in height decrease displacement in all directions, it reduces laxity by pushing the adjacent vertebrae apart, thereby putting the longitudinal ligaments under tension (Fig. 4.43B). However, if the column is deprived of ligaments as well as muscles it automatically collapses. As long as the ligaments and discs are intact it preserves considerable resistance against deformation, which is the result of the elastic tension resistance of the ligaments and the elastic compression resistance of the discs.

As can be seen from Fig. 4.43B, it is not only the longitudinal ligaments which interact with the disc in providing intrinsic stability, the ligamenta flava, interspinous and supraspinous ligaments are equally important. Consequently, for the system to work efficiently and effectively the neural arches must be firmly attached to the bodies. Experiment has shown, however, that if the entire column of arches is separated from that of the bodies, then the column of arches shrinks by 14%. As long as the arches are joined to the bodies they are spread apart, putting the ligaments under tension.

The above points regarding inherent stability of the vertebral column have led to a simple mechanical model (Fig. 4.44), whereby adjacent vertebrae move with respect to each other guided by the zygapophyseal joints, with the elastic properties of the disc and ligaments being replaced by springs. Each vertebra can then be considered to act as a first class lever, with the articular process acting as the fulcrum. Within such a model, the vertebral bodies and intervening disc are essentially the supporting structures and therefore have a static role. The articular and spinous processes, together with adjacent ligaments, have the dynamic role of permitting, guiding and limiting movement.

The presence of the thoracic cage and its articulation posteriorly with the thoracic vertebrae confers a certain degree of stability upon the vertebral column in this region. The ribs themselves are forced into the thoracic cage under considerable stress (tearing or rupture of a costal cartilage through trauma results in the rib springing out slightly). In addition the costal cartilages are constantly changing their shape during respiration. In some respects, therefore, the bodies of the thoracic vertebrae are held between the heads of the ribs. Such an arrangement increases the elastic resistance of the thorax, some of which is used in respiration but the greater part of which is exerted against the vertebral column thus contributing to its intrinsic stability.

That the vertebral column has a certain degree of intrinsic stability is of practical and clinical importance: less muscle activity and therefore energy expenditure is required to maintain it in an erect posture. However, as with all joints, ligaments by themselves are not sufficient to maintain stability. They are of necessity reinforced by muscles. Without the presence and intermittent activity of the associated muscles, the ligaments would gradually stretch and stability would be lost. The large postvertebral muscle mass plays an extremely important role in supporting the ligaments for intrinsic stability. But just as importantly, these muscles are primarily responsible for extrinsic stability, i.e. stability during dynamic movements. However, in explosive dynamic movements the postvertebral muscles by themselves may not be capable of providing the required stability.

With increased dynamic loading of the vertebral column both intra-abdominal and intrathoracic pressures increase (Fig. 4.45). The increase in intra-abdominal pressure, and

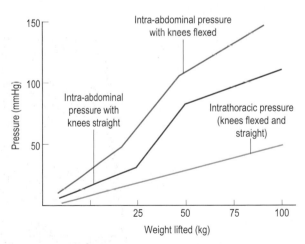

Figure 4.45 Increases in intra-abdominal and intrathoracic pressure with increased dynamic loading of the vertebral column. *(Adapted from Morris JM, Lucas SB and Breslar B (1961) Role of the trunk in stability of the spine.* Journal of Bone and Joint Surgery, *43A, 327–351.)*

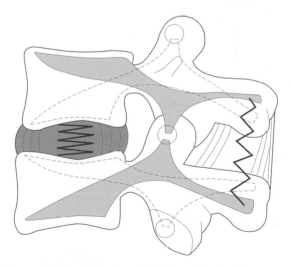

Figure 4.44 Mechanical model of the vertebral unit.

its maintenance, act as a pneumatic cushion which supports the anterior aspect of the lumbar and lower thoracic regions of the vertebral column.

Movements

Movement between adjacent vertebrae occurs because the resilient intervertebral discs are slightly flexible, with the type of movement permitted in each region being largely determined by the shape and orientation of the articular processes. Where the discs are thick in relation to the vertebral bodies, i.e. in the cervical and lumbar regions, the range of movement between adjacent vertebrae is increased. Even so the amount of movement between successive vertebrae is slight, being limited by the arrangement of fibres within the annulus fibrosis of the intervertebral disc. Nevertheless, when added over the whole of the vertebral column the total range of motion becomes considerable.

The basic movements of the vertebral column are flexion (forward bending) and extension (backward bending) about a transverse axis, lateral flexion (bending) about an anteroposterior axis and rotation to the right or left about a vertical axis (Fig. 4.46). Of these, lateral flexion and rotation are always associated movements and neither can take place independently of the other. Indeed, it has been shown for the lumbar region that there is always coupling of movements, so that pure movements in any direction do not occur. Even though the coupled movements are small, they are nevertheless important, and probably serve to help maintain the inherent stability present within the vertebral column during movement. All movements are possible in each region of the vertebral column, with the range of motion determined by disc thickness and articular process geometry. The movements of flexion, extension and lateral flexion involve compression of the intervertebral discs at one edge and stretching at the other. During flexion, the anterior borders of the vertebrae come together, while the posterior borders separate. As full flexion is reached, the anterior part of the disc is compressed, tending to push the nucleus pulposus posteriorly. The posterior part of the disc, the posterior longitudinal ligament and the ligaments of the vertebral arches all become taut; the principal limiting factor to movement is, however, tension in the postvertebral muscles. In extension it is the posterior borders of the vertebrae which become approximated, and the anterior separate. Consequently, the anterior longitudinal ligament becomes increasingly taut while all posterior ligaments are relaxed. Extension is freer and has a wider range of movement than flexion, particularly in the cervical regions. Much of the apparent movement of flexion is in fact due to flexion of the trunk at the hips and also movement of the head at the atlanto-occipital joints.

The importance of the posterior ligaments and intervertebral discs in limiting movements of the vertebral column

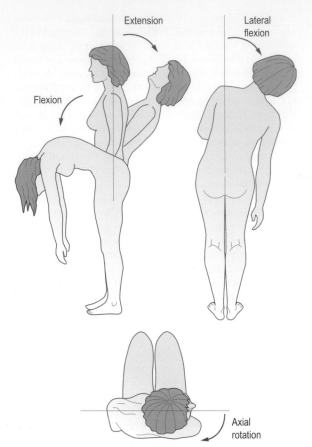

Figure 4.46 Movements of the vertebral column.

has been demonstrated by successive sectioning of these structures and observation and measurement of the movements between adjacent vertebrae. Progressive sectioning of the supra- and interspinous ligaments, the ligamentum flavum, bilateral severing of the articular processes, sectioning of the posterior longitudinal ligament and cutting of the posterior and lateral part of the annulus fibrosus results in a 100% increase in flexion–extension (from 8.4° to 17.3°), a 70% increase in lateral flexion (from 4.9° to 8.3°), and a 650% increase in rotation (from 2.2° to 14.6°) at the thoracolumbar junction. A similar pattern of changes would be expected elsewhere in the vertebral column, although they would be modified by the orientation of the articular processes, particularly regarding the percentage change in rotation.

Lumbosacral junction

Movement here is restricted to flexion and extension, and a small degree of lateral flexion (Fig. 4.47A). In flexion the inferior articular processes of L5 glide upwards over those on the sacrum, with movement being limited by tension

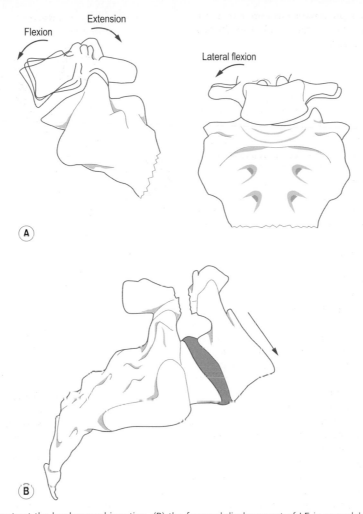

Flexion
Extension
Lateral flexion

A

B

Figure 4.47 (A) Movements at the lumbosacral junction, (B) the forward displacement of L5 in spondylolisthesis.

in the iliolumbar, inter- and supraspinous ligaments and the postvertebral muscles. Backward and downward gliding of the L5 articular processes on those of the sacrum during extension is arrested by apposition of the spines of L5 and S1.

The intervertebral disc between L5 and S1 is markedly wedged, being thicker anteriorly, and is mainly responsible for the angulation between the lumbar and sacral parts of the vertebral column. Because the lumbosacral junction represents the transition between the mobile (lumbar) and immobile (sacral) parts of the vertebral column, it requires considerable support against potentially damaging stresses. To this end the postvertebral muscle mass is extremely thick in this region, reinforced posteriorly by the lower part of the strong thoracolumbar fascia. It is also worth noting that between L5 and S1 the

intertransverse ligaments have been replaced by the much stronger iliolumbar ligaments, which confer additional stability upon this region by restricting lateral flexion (for details of the iliolumbar ligaments see p. 283). The zygapophyseal joints between L5 and S1 have an important weight-bearing function. Occasionally, there is incomplete development of the neural arch of the fifth lumbar vertebra anterior to the inferior articular processes, which may allow separation of the body of L5 from the posterior part of the neural arch. The result is that the body of the fifth lumbar vertebra, together with the trunk above, slides forward and downwards towards the pelvis (Fig. 4.47B), leaving the posterior part of L5 behind articulating with the superior sacral articular processes and tethered by the iliolumbar ligaments. The condition is known as *spondylolisthesis*. The marked angulation at the lumbosacral junction assists this downward and forward movement. The lumbar curvature is accentuated, and the performance of extension exercises is usually extremely painful and dangerous because of the tension on the nerves forming the cauda equina, leading to motor and sensory disturbances along their distribution.

Movements in the lumbar region

Although the intervertebral discs are thick in this region, and in theory would facilitate relatively large movements between adjacent vertebrae, the orientation of the articular processes (Fig. 4.48) tends to confer a certain degree of stability to this region, restricting rotation in particular. Of the lumbar vertebrae, the lower facets of L5 face mainly forwards to articulate with the superior facets on the sacrum. Consequently, movement between L5 and the sacrum is not the same as that between two typical lumbar vertebrae.

Flexion and extension Flexion is relatively free in the lumbar region having a total range of 55°, while extension has a range of 30° (Fig. 4.48A). There is less movement at the thoracolumbar junction than between adjacent

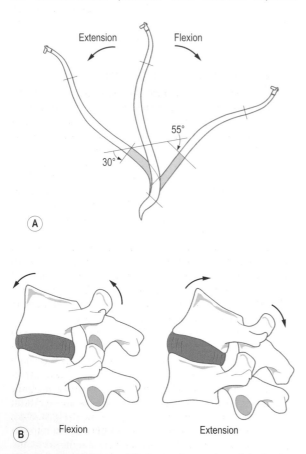

Flexion   Extension

Figure 4.48 (A) Range of flexion and extension of the lumbar region, (B) movements between adjacent vertebrae.

vertebrae, and there is most movement at the lumbosacral junction. The total range of movement decreases with age so that at 65 years it is between one-half and one-third of that at age 10.

In flexion, the vertebrae are tilted forward on each other, so that the lower articular processes of the upper vertebrae glide upwards and forwards on the upper processes of the lower vertebrae (Fig. 4.48B). The spatial orientation of these joints allows the upper vertebrae to move forwards slightly over the lower vertebrae. One result of this forward movement is to increase the superoinferior diameter, while narrowing the anteroposterior diameter of the intervertebral foramen. Flexion from the upright position is controlled by the postvertebral muscles on both sides, being limited by tension in the posterior part of the intervertebral disc, the posterior longitudinal ligament, the ligamenta flava, and the inter- and supraspinous ligaments. Flexion from the supine position is brought about by psoas major and the anterior abdominal muscles, particularly rectus abdominis.

During extension, the inferior articular processes of the upper vertebra glide down into the hollowed superior processes of the lower vertebra, causing the upper vertebra to move backwards slightly on the lower vertebra (Fig. 4.48B). The movement is limited by the anterior longitudinal ligament, the anterior part of the intervertebral disc, apposition of the large lumbar spinous process and the 'close-packing' of the facet joints. Extension from the upright position is controlled by psoas major and the anterior abdominal muscles, while from the prone position it is produced by the postvertebral muscles.

Lateral flexion The range of lateral flexion, as with that of flexion and extension, varies with the individual and with age. During the pre-teenage years the range may be as large as 60° either side of the midline; however, by age 30 this has been halved. On average, the adult range of movement is between 20° and 30° to each side (Fig. 4.49A). Throughout the total age range lateral flexion at the lumbosacral junction is minimal. Because of the small range of lateral flexion in the lumbar region there is very little associated rotation of the vertebrae. Lateral flexion is greatest in the upright position, and is greatly diminished when the vertebral column is flexed and the lumbar curve is lost.

In lateral flexion, the articular processes of the flexed side become close-packed, while the upper process on the opposite side is withdrawn from that below (Fig. 4.49B). Because the articular surfaces also slope slightly, a narrow gap appears between the articular processes on the unflexed side. As a result the intervertebral foramen on the unflexed side enlarges in all directions, while that on the flexed side narrows. Lateral flexion initially requires active muscle contraction; it is not brought about by controlled relaxation (i.e. eccentric contraction) of the muscles of the opposite side. In other words, lateral bending to the right requires contraction of muscles on

Consequently, rotation occurs until all these gaps on one side become obliterated. (Narrowing and eventual obliteration of the gaps on one side produce widening of the gaps on the opposite side.)

The rotation possible, however small, depends on the position of the lumbar column. When it is extended, no rotation at all is possible because of the close-packed position of the zygapophyseal joints. However, the range of possible rotation increases with increasing flexion.

Accessory movements Rotation is really an accumulation of accessory movements occurring at each of the zygapophyseal joints. It consists of a rocking movement of the inferior articular process of the vertebra above within the hollow of the process of the lower vertebra. These same accessory movements can be demonstrated by applying pressure to one side of a lumbar spinous process when the subject is lying prone with the lumbar column slightly flexed.

A downward pressure applied to the mamillary processes or to the spinous process causes a slight forward gliding of the body of the same vertebra with respect to the body above. At the same time the superior articular processes move forward away from the inferior processes of the vertebra above, while the inferior processes are pushed into the superior processes of the vertebra below.

Movements in the thoracic region

In the thoracic region the intervertebral discs are relatively thin with respect to the vertebral bodies, and together with the presence of the ribs and sternum, their movement is limited. The orientation of the articular processes of the thoracic vertebrae, which lie on the arc of a circle with its centre close to the anterior part of the vertebral body, permits flexion, extension, lateral flexion and rotation. However, the inferior processes of the 12th thoracic vertebra resemble those of the lumbar region and therefore movements at the thoracolumbar junction are similar to those between lumbar vertebrae.

Flexion and extension The combined range of flexion and extension in the thoracic part of the vertebral column is between 50° and 70°, with extension being much more limited than flexion (Fig. 4.50A). Flexion is much freer in the lower half of the region as the lower ribs tend to be longer and more flexible because of their longer costal cartilages.

In flexion the inferior articular processes of the vertebra above slide upwards over the superior processes of the lower vertebra (Fig. 4.50B). The interspace between the two vertebrae opens out posteriorly, compressing the anterior part of the intervertebral disc. Flexion is limited by the presence of the thoracic cage, and by the tension developed in the supra- and interspinous ligaments, the ligamenta flava, and posterior longitudinal ligament. From the upright position, flexion is controlled by the postvertebral muscles of both sides, while flexion from the supine position is brought about by the anterior abdominal muscles of

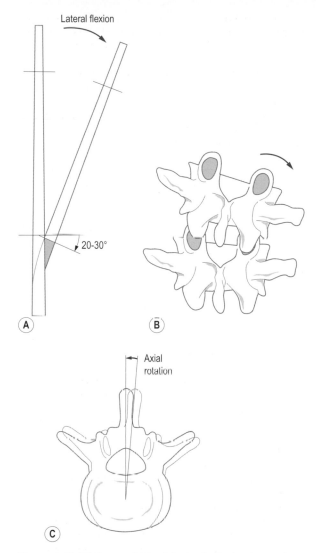

Figure 4.49 (A) Range of lateral flexion in the lumbar region, (B) movement between adjacent vertebrae during lateral flexion, (C) axial rotation.

the right side, i.e. the anterior abdominal muscles and quadratus lumborum. However, once a certain degree of lateral flexion has been reached (about 10°) then the movement is controlled by the eccentric contraction of the muscles of the opposite side. Movement is limited by the intertransverse ligaments of the opposite side as well as opposing muscle groups.

Rotation This is extremely small in the lumbar region, being of the order of a few degrees only (Fig. 4.49C). The limitation to movement is due to the shape and orientation of the lumbar articular facets. In the erect position there is a narrow gap between opposing articular processes.

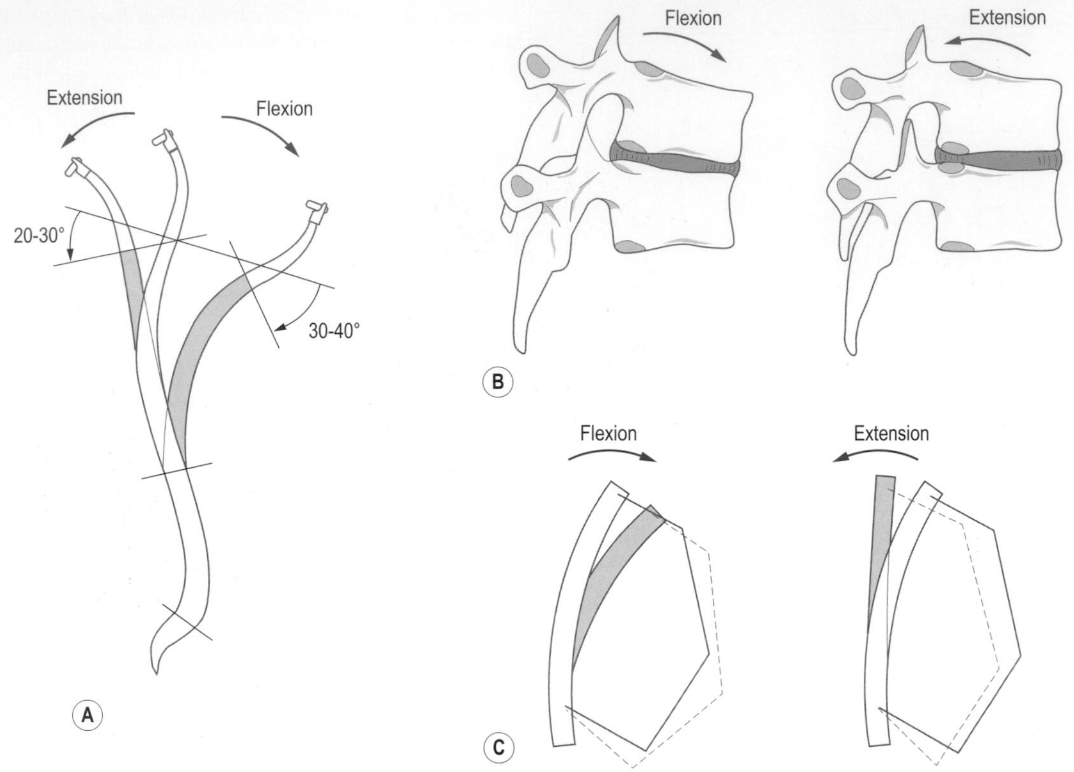

Figure 4.50 (A) Range of flexion and extension of the thoracic region, (B) movements between adjacent vertebrae, (C) effects on the thoracic cage.

both sides, particularly rectus abdominis. During flexion all the angles between the various segments of the thorax and between the thorax and vertebral column increase (Fig. 4.50C).

Extension of the thoracic region approximates the vertebrae posteriorly (Fig. 4.50B). It is limited by impact of the articular and spinous processes between adjacent vertebrae, as well as by tension in the anterior longitudinal ligament. The effect of extension on the thoracic cage is to flatten it by decreasing all angles between the various segments of the thoracic cage and the vertebral column (Fig. 4.50C). Extension from the upright position is controlled by the anterior abdominal wall muscles and from the prone position by the postvertebral muscles.

Lateral flexion In the thoracic region this has a range of 20–25° to each side (Fig. 4.51A), being freer in the lower half of the region. During lateral flexion the articular processes of the two adjacent vertebrae slide relative to each other, so that those on the contralateral side move as during flexion, while those on the ipsilateral side move as during extension (Fig. 4.51B). Lateral flexion is limited by the impact of the articular processes on the side of the movement, and by tension developed in the ligamenta flava and intertransverse ligaments of the opposite side. In the

thoracic region lateral flexion is associated with rotation of the vertebrae. In full lateral flexion the degree of rotation is about 20°. In other words, for each degree of lateral flexion there is almost 1° of accompanying rotation; the rotation being such that the spinous processes of the thoracic vertebrae point towards the concavity of the curve, i.e. they rotate contralaterally.

As with flexion and extension, lateral flexion modifies the shape of the thoracic cage (Fig. 4.51C). On the contralateral side the thorax is elevated, the intercostal spaces widen, the thoracic cage enlarges and the costochondral angle of the 10th rib tends to open out. On the ipsilateral side the reverse occurs: the thoracic cage is lowered and shrinks, the intercostal spaces narrow, and the costochondral angle decreases.

As in the lumbar region, lateral flexion is brought about by the contraction of muscles on the same side.

Rotation The orientation of the articular processes in the thoracic region is such as to promote rotation. This may be a necessary adaptation because of the presence of the ribs. Were it not for the presence of the thoracic cage the range of rotation of the thoracic column would be greater. Even so, it is some 35° in each direction (Fig. 4.52A). During rotation, the inferior processes of the upper vertebrae slide

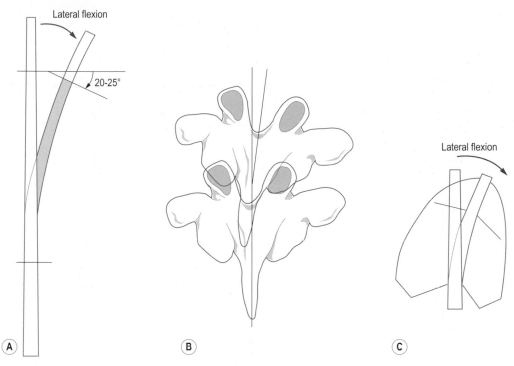

Figure 4.51 (A) Range of lateral flexion of the thoracic region, (B) movements of adjacent vertebrae, (C) effects on the thoracic cage.

sideways outside the superior processes of the lower vertebrae leading to rotation of one vertebral body with respect to the other about a common axis (Fig. 4.52B). This movement of the vertebrae is accompanied by rotation and twisting of the intervertebral disc, which has a tendency to pull the adjacent vertebrae together. A small degree of lateral flexion accompanies this rotation. When the thoracic vertebral column is fully extended both rotation and lateral flexion are greatly reduced, and may be lost completely.

Rotation is limited by tension in the supra- and interspinous ligaments and the ligamenta flava and is brought about largely by the abdominal oblique muscles. For example, rotation of the trunk to the left is achieved by contraction of the right external oblique and left internal oblique.

Because of the articulation of the ribs with the vertebrae, any rotatory movement of the vertebrae will induce a similar movement in the corresponding ribs. However, such movement is limited due to the articulation of the ribs anteriorly with the sternum. Instead there is a distortion of the ribs as follows: an accentuation of the concavity of the rib on the side of rotation, with a flattening of the concavity on the opposite side (Fig. 4.52C). These changes are possible because of the elasticity of the rib and its costal cartilage. These movements of the ribs subject the sternum to shearing forces.

Rotation of the thoracolumbar part of the vertebral column is extremely important during walking. However, the rotation is not simple but rather complex, occurring in opposite directions in the upper and lower parts of the region. As one leg swings through ready for the next heel-strike the pelvis rotates about the hip joint of the supporting leg, carrying with it the trunk. In an attempt to keep the head facing forwards the pectoral girdle rotates in the opposite direction. Studies have shown that the intervertebral disc between the seventh and eighth thoracic vertebrae is not subjected to any rotation, while maximum rotation occurs (in opposite directions) in the discs immediately above and below it (Fig. 4.52D). The degree of rotation decreases towards the pelvic and pectoral girdles.

Accessory movements Accessory movements of the thoracic part of the vertebral column are difficult to demonstrate and usually only used therapeutically. Nevertheless, pressure on a spinous process will, because of its length and angulation, produce a slight rocking of the vertebral body and apposition of the articular surfaces of the zygapophyseal joints associated with its articular processes. A localized pressure applied to the posterior aspect of the base of the transverse process just below the zygapophyseal joint will slightly separate the joint by pushing the upper articular process forwards.

Movements in the cervical region

Taken as a whole, the cervical part of the vertebral column consists of two anatomically and functionally distinct regions. The upper region, known as the suboccipital region, consists of the first and second cervical vertebrae

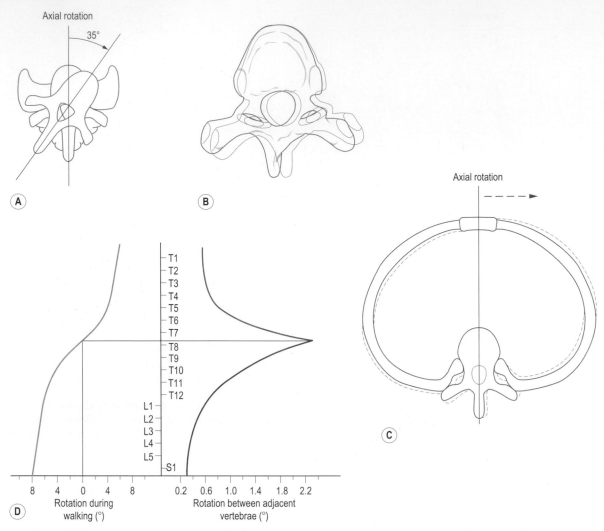

Figure 4.52 (A) Range of axial rotation of the thoracic region, (B) movement between adjacent vertebrae, (C) effects on thoracic cage, (D) rotation in the thoracic region during walking.

and includes the articulation of this unit with the base of the skull. The lower region extends from the lower surface of C2 down to the superior surface of T1. Although each region is distinct functionally, they are complementary to allow movements of flexion and extension, lateral flexion, and rotation of the head and neck. It is easier, however, to consider each of the regions separately.

Lower cervical region

The intervertebral discs in the lower cervical region are relatively thick, which together with the shape and orientation of the articular facets, means that flexion and extension, lateral flexion and rotation are all fairly extensive

movements. As in the thoracic region, lateral flexion and rotation occur as linked movements.

Flexion and extension The total range of flexion and extension in the lower cervical region is approximately 110°, of which about 25° is flexion (Fig. 4.53A; see also Fig. 4.61A). The least movement is between the seventh cervical and first thoracic vertebrae. During flexion the upper vertebral body tilts and slides anteriorly on the lower, compressing the intervertebral space anteriorly (Fig. 4.53B). The tilting of the upper vertebra is facilitated by the anterior ledge on the upper plateau of the lower vertebra, allowing the inferiorly directed projection on the lower plateau of the upper vertebra to move past. Flexion between two vertebrae does not occur about the centre

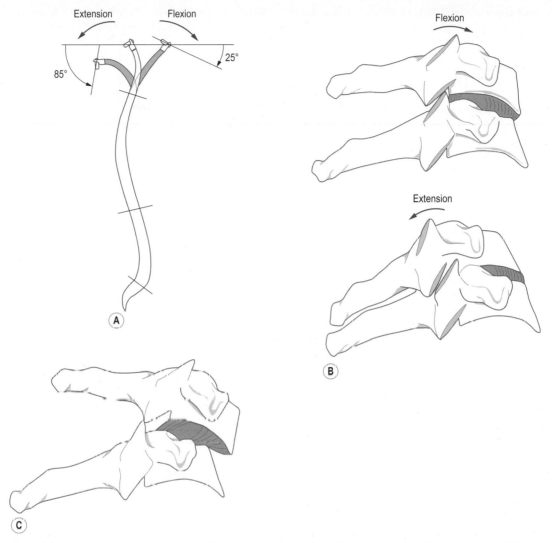

Figure 4.53 (A) Range of flexion and extension of the lower cervical region, (B) movements of adjacent vertebrae, (C) anterior dislocation of the zygapophyseal joints as may occur in severe whiplash injury.

of curvature of the facets on the articular processes. Consequently, the inferior facet of the upper vertebra moves upwards and forwards causing a widening of the joint space posteriorly. Flexion is not limited by bony impact but only by tension developed in the posterior longitudinal ligament, the zygapophyseal joint capsule, the ligamenta flava, the ligamentum nuchae and postvertebral muscles. From the upright position flexion is controlled by the posterior neck muscles of both sides which act against the weight of the head and neck (i.e. trapezius, splenius capitis, longissimus capitis and semispinalis capitis). In the supine position the anterior muscles of both sides flex the cervical column (i.e. the sternomastoid, longus capitis, longus colli and the scalene muscles).

In extension the upper vertebra tilts and slides posteriorly over the lower one (Fig. 4.53B). Again, because the movement does not occur about the centre of curvature of the facets, the superior articular facet moves backwards and downwards causing the anterior part of the joint space to widen and the intervertebral foramen to narrow. Extension is limited by tension developed in the anterior longitudinal ligament and by impact of the superior articular process of the lower vertebra on the transverse process of the upper vertebrae and by impact of the posterior arches through the ligaments. Extension from the upright position is controlled by the anterior neck muscles as above, while in the prone position, the posterior muscles produce extension.

Movements of the vertebral bodies during flexion and extension are guided to some extent by the uncovertebral joints, otherwise the movements would probably not be as pure as they are.

The fact that there is no bony limitation to flexion predisposes the cervical column to a particular form of dislocation. During car accidents, in particular, the vertebral column is strongly extended and then flexed, producing the whiplash injury. Whiplash injuries cause stretching or occasionally tearing of the ligamentous structures limiting flexion. In extreme cases there may be an anterior dislocation of the zygapophyseal joints in which the inferior articular facets of the upper vertebra become hooked on the anterosuperior margin of the superior facet of the lower vertebra (Fig. 4.53C). The dislocation is extremely difficult to reduce as it endangers the medulla oblongata and cervical spinal cord with the risk of paraplegia, quadriplegia or death.

Lateral flexion In the lower cervical region lateral flexion has a range of about 40° to each side (Fig. 4.54A; see also Fig. 4.61B). However, because of the orientation and shape of the articular facets it is impossible to get pure lateral flexion or pure rotation. When the neck is flexed to one side, the lower articular process on that side glides downwards and backwards on the superior process of the vertebra below: on the other side each articular process moves forwards and upwards over the superior process of the vertebra below. Thus flexion to one side is accompanied by a slight rotation to the same side (Fig. 4.54B). If the whole of the lower cervical region is considered rather than the movements between adjacent vertebrae, then there is a slight extension accompanying lateral rotation of the cervical column.

During lateral flexion, the joint spaces of the contralateral uncovertebral joints widen (Fig. 4.54C). The movement, however, is quite complex because it involves not only lateral flexion of one vertebra with respect to another, but also rotation and slight extension.

As in the thoracic and lumbar regions, lateral flexion of the neck requires muscle contraction on the side of flexion. The movement is arrested in part by opposition of the articular facets and also by stretching of the contralateral zygapophyseal joint capsule and compression of the intervertebral disc.

Rotation In the lower cervical region this amounts to 50° in each direction (Fig. 4.55). It cannot occur, however, without there being some associated lateral flexion to the same side. In rotation, the inferior facets on the opposite side of the vertebra above move against the superior facets of the lower vertebra producing a type of camber effect. This results in an increase in the size of the intervertebral foramen on that side. However, on the side to which rotation is occurring, the articular facets separate slightly, thus reducing the size of the intervertebral foramen. The movement is limited by a grinding contact of the opposite side facets, tension within both joint capsules and also torsion of the intervertebral disc. Rotation is produced by posterior

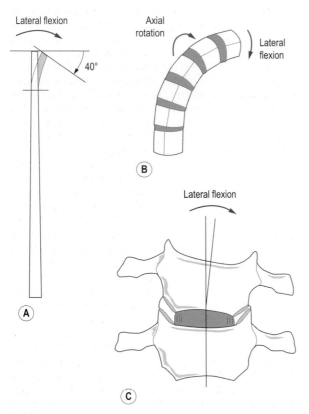

Figure 4.54 (A) Range of lateral flexion of the lower cervical region, (B) schematic representation of linked lateral flexion and rotation, (C) movement between adjacent vertebrae.

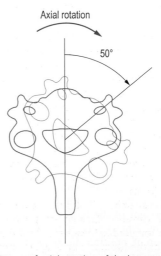

Figure 4.55 Range of axial rotation of the lower cervical region.

muscles which pass upwards and inwards behind the axis of rotation (e.g. semispinalis cervicis), and anterior muscles passing upwards and outwards in front of the axis of rotation (e.g. sternomastoid). This combination of muscle activity produces rotation to the opposite side.

One of the practical consequences of rotation is that it can relieve pressure on the nerve root of the side opposite to the direction of rotation by virtue of the increase in size of the intervertebral foramen. At the same time pressure would be relieved on the same side facet joint, but it would decrease the size of the intervertebral foramen. Practical use is made of these movements in the application of traction in which flexion is combined with rotation to relieve pressure on a nerve root. These movements are also utilized in mobilization and manipulation techniques.

Circumduction A combination of flexion, extension, lateral flexion and rotation, and involves all of the joints in the cervical region. The limiting factors are those which limit the individual movements. Circumduction is a popular movement in keep fit clubs in which 'clicking' sounds may be sought by the untrained practitioner. Combined movements as in circumduction, particularly when performed towards the extremes of the ranges of movement, bring into play certain 'locking mechanisms' in some of the joints. The 'clicking' sounds that may be heard are often due to forcing a joint beyond one of these locking positions. It commonly leads to severe neck pain, which may be referred to the arm and upper chest, usually appearing within a day or two of performing the activity.

Manipulators often make use of these 'locking mechanisms' practically; even so, they must be treated with great care and caution.

Accessory movements These are present in the cervical joints. However, they are not easy to demonstrate. They fall into two groups: (1) general and (2) local.

1. Traction of the cervical spine is the most obvious general movement affecting the neck as a whole, producing distraction of the discs and facet joints throughout the region, causing the nerve roots to be drawn back through the intervertebral foramina. With careful positioning traction can be limited to specific areas. However, skill in location and handling is essential. In the neutral position an elongation of the neck of several millimetres can be achieved depending on the preparation of the subject. Such an elongation will reduce pressure on the intervertebral disc by some 70% (from 30 to 10 kg/cm^2), as well as distraction of the facet joints.

   Traction stretches all of the structures crossing the joints, i.e. capsules, ligamentum nuchae, anterior and posterior longitudinal ligaments, and the muscles. Because of the ligaments passing between C2 and the base of the skull, traction has a negligible effect on the

joints between the base of the skull and C1, and C1 and C2. If traction at these joints is achieved it may have undesirable effects, so care and caution should be exercised.

A side-to-side movement can be produced by keeping the head upright and moving it to the left or right. Fixing the lower part of the cervical column and applying some degree of traction increases this movement. It involves a sideways sliding of the vertebral bodies on each other, with a downwards gliding of the facets of the zygapophyseal and uncovertebral joints of one side and an upward gliding on the other.

2. Local accessory movements can be performed by applying precise pressure to the spinous processes, transverse processes or articular pillars of the vertebrae. Detailed information on the practical application of these procedures is best obtained by consulting the mobilization and manipulation literature.

Application of accessory movements of the vertebral column

A detailed knowledge of the accessory movements of the vertebral column, particularly in the lumbar region, is essential for those using manipulation and movement techniques. There is considerable misunderstanding of how the application of this knowledge can benefit the patient. Although manipulative and movement techniques are not discussed, certain facts are worth stating as they will be of help when consulting the manipulation and movement literature. Because many patients suffer from back problems of one sort or another, comment will be restricted to the lumbar region. Rotation movements of the lumbar spine, i.e. rotation of the thorax on a fixed pelvis, open out the zygapophyseal joints on the side to which the thorax is rotating (the degree of gapping and rotation increasing with flexion of the lumbar column). This is similar to that seen in the cervical region upon rotation. However, unlike the cervical region, where the orientation of the articular facets produces an increase in the size of the intervertebral foramen of the opposite side, in the lumbar region the opposite side joints become compressed preventing further movement.

If, as in most cases when these accessory movements are being applied therapeutically, the trunk is fixed and the lower limbs and pelvis are the mobile segments, then rotation of the pelvis to one side has the same effect on the joints as rotating the thorax to the opposite side. Flexing the lumbar column and applying traction to one leg produces a lateral flexion of the lumbar region to the opposite side, which further increases the separation of the zygapophyseal joints on the side of traction. This stretches the joint capsule and the adjacent part of the intervertebral disc, thereby increasing the size of the intervertebral foramen, releasing trapped nerve roots.

Joints between vertebral arches (zygapophyseal)

Type	Plane synovial joint
Articular surfaces	Facets on superior and inferior articular processes
Capsule	Thin fibrous capsule surrounds joint attaching to articular margins
Ligaments	Ligamentum flavum
	Supraspinous ligaments
	Ligamentum nuchae
	Interspinous ligaments
	Intertransverse ligaments
Stability	Provided by interaction between intervertebral disc and associated ligaments; shape of facet joints
Movements	Lumbar region:
	55° flexion and 30° extension
	20–30° lateral flexion to each side minimal axial rotation
	Thoracic region:
	50–70° combined flexion/extension
	20–25° lateral flexion to each side
	35° axial rotation to each side
	Lower cervical region:
	25° flexion and 110° extension
	40° lateral flexion to each side
	50° axial rotation to each side

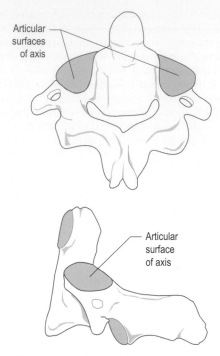

Figure 4.56 Articular surfaces of the axis.

this surface may show a slight convexity. Both articular surfaces are covered with hyaline cartilage.

The atlantoaxial articulations

The joints between the atlas and axis are of two types: bilateral synovial plane joints between the surfaces of their articular processes, and a median synovial pivot joint formed by the articulation of the dens with the anterior arch and the transverse ligament of the atlas. The median joint is supported by a number of accessory ligaments.

The lateral atlantoaxial joints

Articular surfaces

The lateral atlantoaxial joints are between the articular facets on the superior articular processes of the axis and those on the inferior lateral masses of the atlas (Fig. 4.56). The articular surface on the axis is oval with its long axis running from anterosuperomedial to posteroinferolateral, so that it is inclined obliquely laterally and downwards (Fig. 4.56). The articular surface is slightly convex about both axes of the oval, with the convexity being greater about its short axis. The atlantal articular surface is also oval with corresponding long and short axes. However, it is much flatter, showing only a slight concavity if any. Occasionally,

Joint capsule and synovial membrane

Each of the joints is enclosed in a loose fibrous capsule, lined by synovial membrane, which attaches to the margins of the articular surfaces.

Ligaments

Accessory atlantoaxial ligament

This passes obliquely downwards and medially from the back of the lateral mass of the atlas to the back of the body of the axis. In its course it runs in direct contact with the posterior aspect of the joint capsule.

Relations

A membranous expansion from each side of the anterior longitudinal ligament, near its attachment to the anterior tubercle of the atlas, passes to the front of the body of the axis. Posteriorly, bridging the gap between the arch of the atlas and that of the axis, is a membrane which lies in series with the ligamentum flavum attaching to the vertebral arches of lower levels. The second cervical nerve pierces this membrane and passes behind the lateral atlantoaxial joint, which suggests that these joints are of different morphological origin than the joints between the articular processes of lower vertebrae. Because these joints have the same relationship to the nerve as do the uncovertebral joints, they may be considered to be homologous to them, although they are of course much more specialized.

The median atlantoaxial joint

Articular surfaces

The articulation is between a more or less rectangular facet on the front of the dens, which is convex both vertically and transversely, and a reciprocally curved oval facet on the inner aspect of the arch of the atlas (Fig. 4.57). Both surfaces are covered with hyaline cartilage.

On the posterior surface of the dens is a further hyaline cartilage-covered surface, concave superoinferiorly and convex transversely, which articulates with the anterior fibrocartilaginous surface of the transverse ligament of the atlas (Fig. 4.57A).

Joint capsule and synovial membrane

Each of the median joints is enclosed by a thin fibrous capsule, lined by synovial membrane. Of the joint cavities so formed, that between the dens and the transverse ligament is the larger.

Ligaments

Transverse ligament of the atlas

A thick strong band arching behind the dens and attaching to a small tubercle on the medial side of each lateral mass of the atlas (Fig. 4.57A). From the middle of the ligament, a small band of longitudinal fibres ascends to attach to the anterior edge of the foramen magnum: a further band of

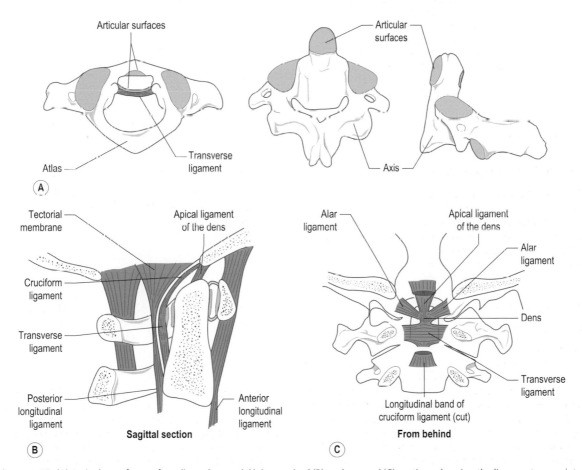

Figure 4.57 (A) Articular surfaces of median atlantoaxial joint; saqittal (B) and coronal (C) sections showing the ligaments associated with the median atlantoaxial joint.

fibres passes downwards to attach to the back of the body of the axis. These two longitudinal bands, together with the transverse ligament, constitute the cruciform ligament of the atlas (Fig. 4.57B).

Accessory ligaments/membranes

A number of accessory ligaments are associated with the median atlantoaxial joints, which unite the axis with the occipital bone. They therefore serve to increase the stability of the suboccipital region.

Tectorial membrane

A broad sheet continuous with the posterior longitudinal ligament, extending from the back of the body of the axis to the occipital bone within the anterior edge of the foramen magnum as far laterally as the hypoglossal canals (Fig. 4.57B). It covers the posterior aspect of the dens, cruciform and alar ligaments; occasionally there is a small synovial bursa between the membrane and the median part of the transverse ligament of the atlas. The lateral parts of the membrane blend with and overlie the accessory atlantoaxial ligament.

As the membrane approaches the foramen magnum it becomes closely applied to and eventually blends with the spinal dura mater.

Alar ligaments

These pass obliquely upwards and laterally from each side of the apex of the dens to the medial side of the respective occipital condyle. Each ligament is a short, strong, rounded band (Fig. 4.57C).

Apical ligament of the dens

A slender band lying immediately in front of the upper longitudinal band of the cruciform ligament (Fig. 4.57B). It stretches from the apex of the dens to the anterior edge of the foramen magnum, lying between the diverging alar ligaments.

Ligamentum nuchae

See page 453.

Ligamentum flavum

See page 453.

The atlanto-occipital joint

Each atlanto-occipital articulation is a synovial joint between the occipital condyle and the facet on the upper surface of the lateral mass of the atlas (Fig. 4.58A). The two joints are symmetrical and can be considered as a single ellipsoid joint because functionally they act together, their movements being mechanically linked.

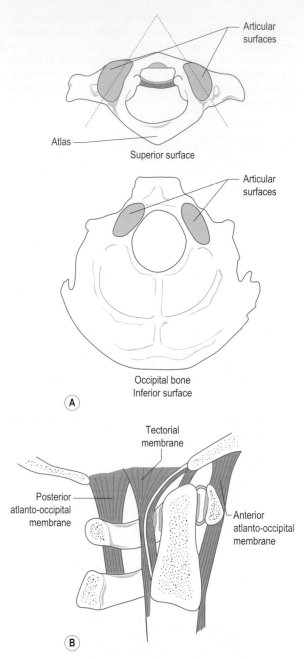

Figure 4.58 Atlanto-occipital joint: (A) the articular surfaces, (B) sagittal section showing the associated ligaments.

Articular surfaces

The oval articular facets on the atlas are concave about both their long and short axes. Occasionally, these facets may be narrowed in their middle part, or even divided into two separate facets. The long axis of the oval runs obliquely from posterolateral to anteromedial, with that of each side converging to meet in the midline in front of the anterior arch of the atlas (Fig. 4.58A).

The articular surfaces of the occipital condyles are reciprocally curved to those on the atlas, i.e. they are elongated and convex, converging anteriorly. Both articular surfaces are covered with hyaline cartilage (Fig. 4.58A).

Joint capsule and synovial membrane

Each joint is enclosed in a thin, loose fibrous capsule which attaches to the margins of the articular surfaces and is lined with synovial membrane.

Ligaments/Membranes

To a large extent the ligaments and membranes passing between the axis and the occipital bone will confer some stability on the atlanto-occipital joints. However, two membranes connect the arches of the atlas with the occipital bone.

Anterior atlanto-occipital membrane

This passes between the anterior arch of the atlas to just in front of the anterior margin of the foramen magnum on the base of the skull (Fig. 4.58B). It consists of densely woven fibres, and is thickened in its central portion by an upward prolongation of fibres from the anterior longitudinal ligament. Its lateral margins blend with the anteromedial part of the joint capsule.

Posterior atlanto-occipital membrane

This is attached above to the posterior margin of the foramen magnum and below to the upper border of the posterior arch of the atlas (Fig. 4.58B). It reaches and blends with the posteromedial aspect of the joint capsule. In the lateral part of its attachment to the atlas, the membrane arches over the vertebral artery and the first cervical nerve as they cross this part of the posterior arch. This lower arched part of the membrane is thickened and may become ossified. (It should be noted that the first cervical nerve passes behind the atlanto-occipital joint.)

Ligamentum nuchae

See page 453.

Movements

Suboccipital region

This rather complex unit allows flexion and extension, lateral flexion, and rotation to occur, but because of the nature of the articulations between the bones involved it is perhaps best to consider the movements between the first and second cervical vertebrae, and between the first cervical vertebra and base of the skull separately.

Flexion and extension A few degrees of flexion and extension are possible at the atlantoaxial joints (Fig. 4.59). The axis about which movement occurs passes more or less through the centre of the dens. Due to the shape of the articular facets of the lateral joints, the lateral mass of the atlas rolls and slides on the superior articular facet of the axis. This movement is only permitted because of the non-rigid transverse ligament, giving some flexibility to the median atlantoaxial joints. During flexion the transverse ligament bends downwards, and during extension it bends upwards.

Atlantoaxial (C1–C2) movements

The movements at these joints are extremely complex because all the joints (i.e. the median and lateral two

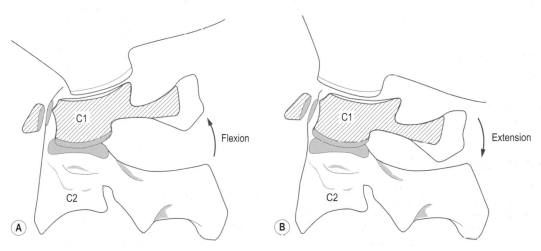

Figure 4.59 (A) Flexion and (B) extension at the atlantoaxial joints.

joints) move together. The principal movement at these joints is one of rotation. However, a few degrees of flexion and extension may be possible.

Rotation During rotation, the head and first cervical vertebra move as a single unit, with the range of movement being 15° to each side. While the anterior arch of the atlas and the transverse ligament pivot around the stationary dens, the lateral masses of the atlas glide over the articular surfaces of the axis – one forwards and the other backwards. Because of the oblique orientation of these lateral joint surfaces, together with the slight convexity of the facets on the axis, rotation of the head is accompanied by a slight vertical descent (approximately 1 mm) of the head. This action slackens the alar ligaments, thus delaying their action in limiting rotation. Rotation can be further increased by tilting the head backwards and to the opposite side.

Rotation is achieved by the action of the suboccipital muscles, sternomastoid and trapezius.

Accessory movements There are very few accessory movements possible at these joints. Forward and backward movement is prevented by the odontoid process being held between the anterior arch of the atlas in front and the transverse ligament behind. Lateral flexion is prevented by the odontoid process abutting against the medial side of the lateral mass of the atlas. Indeed, all of these movements between the two bones are positively discouraged.

Loss of rotation to one side at the atlanto-occipital joint may lead to compensatory rotation of the rest of the cervical spine to the opposite side, often causing pain and discomfort. Restoration of the range can be achieved by fixing the spinous process of the axis and rotating the atlas and the head; the movement can be forced a little beyond the normal. However, it must be stressed that this manoeuvre should only be performed by a qualified practitioner.

Atlanto-occipital movements

Movement at the atlanto-occipital joints occurs primarily in two directions: flexion and extension, and lateral flexion. There is, however, slight slipping between the skull and the atlas which constitutes a form of rotation.

Flexion and extension The total range of flexion and extension is 20° (see Fig. 4.61A), the movement being brought about by the occipital condyles sliding on the lateral masses of the atlas. During flexion, the occipital condyles move backwards on the lateral masses; at the same time the occipital bone moves away from the posterior arch of the atlas (Fig. 4.60A). This latter movement causes a slight flexion at the atlanto-occipital joints as the posterior arches of the atlas and axis become separated. Flexion is limited by the tension developed in the joint capsules, the posterior atlanto-occipital membrane, the ligamentum nuchae and the posterior suboccipital muscles. Flexion of the head from the upright position is controlled by the postvertebral neck muscles of both sides acting against the weight of the head (i.e. trapezius, splenius capitis, longissimus capitis, semispinalis capitis and most of the short suboccipital muscles). From the supine position the head is flexed by the anterior muscles of both sides (i.e. sternomastoid, longus capitis, and the short muscles between the atlas and the occiput).

In extension the occipital condyles slide anteriorly on the lateral masses of the atlas (Fig. 4.60B). This brings the occipital bone close to the posterior arch of the atlas, and tends to cause a slight extension at the atlantoaxial joints so that the posterior arches of the atlas and axis are approximated. Extension is limited by the impact of the occipital bone and the posterior arches of the atlas and axis. In forced extension the posterior arch of the atlas, being caught between the other two bony pieces, may be fractured. Extension is produced in the upright position by the controlled

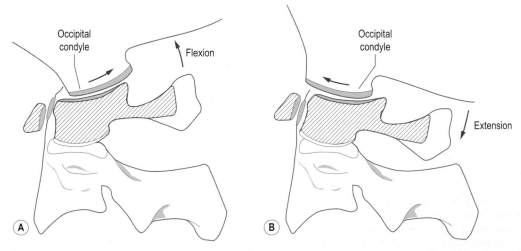

Figure 4.60 (A) Flexion and (B) extension at the atlanto-occipital joints.

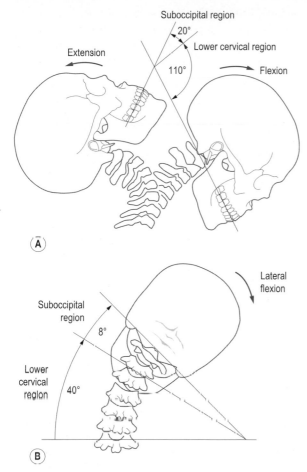

Accessory movements The lateral mass of the atlas can be slid forwards with respect to the occipital bone by applying pressure to the posterior arch when the muscles are relaxed. A forward and backward gliding of the head on the atlas can also be achieved, usually whilst using traction. The head can also be moved sideways producing a sideways gliding of one articular surface against the other. Finally, distraction of the joints can be produced by applying traction to the head; while in distraction a rotatory movement can also be produced.

Figure 4.61 (A) Total range of movement in the cervical region, flexion and extension, (B) lateral flexion.

action of the anterior muscles of both sides, and from the prone position by the action of the postvertebral muscles.

Lateral flexion In the suboccipital region lateral flexion is limited, being no more than 8° to either side (Fig. 4.61B). It consists of movement of the skull against the atlas (5°), and movement of the axis against the third cervical vertebra (3°). The atlanto-occipital movement consists of a slipping of the occipital condyles so that on the side of flexion it moves towards the midline and on the opposite side it moves away from the midline. The movement is limited by tension in the joint capsule and the contralateral alar ligament.

Rotation The small degree of rotation possible at the atlanto-occipital joints is secondary to the rotation of the atlas around the dens. With respect to the lateral masses of the atlas, one occipital condyle moves forwards while the other moves backwards. Rotation to one side is associated with a slight lateral flexion to the opposite side.

Section summary	
Atlantoaxial and atlanto-occipital articulations	
Lateral atlantoaxial joint	
Type	Plane synovial joint
Articular surfaces	Superior facet of axis (C2) with inferior facet of atlas (C1)
Capsule	Loose fibrous capsule enclosing joint attaching to articular margins
Ligaments	Accessory atlantoaxial ligament
Median atlantoaxial joint	
Type	Synovial pivot joint
Articular surfaces	Anterior facet on dens of axis with facet on posterior aspect of arch of atlas; posterior facet on dens with anterior surface of transverse ligament of atlas
Capsule	Thin fibrous capsule encloses each joint
Ligaments	Transverse ligament of atlas
	Tectorial membrane
	Alar ligaments
	Apical ligament of dens
	Ligamentum nuchae
	Ligamentum flavum
Movements	15° axial rotation of atlas (C1) and head around axis (C2)
	Small amount of flexion and extension
Atlanto-occipital joint	
Type	Synovial ellipsoid joint
Articular surfaces	Occipital condyle with facet on superior surface of atlas
Capsule	Thin, loose capsule attaching to articular margins
Ligaments	Anterior atlanto-occipital membrane
	Posterior atlanto-occipital membrane
	Ligamentum nuchae
Stability	From muscles crossing joints and associated ligaments
Movements	20° combined flexion/extension
	8° lateral flexion to each side small amount of rotation

Biomechanics

Trabecular systems

Within the vertebrae are extensive and complex trabecular systems, reflecting the stresses to which each is exposed. Within the body of the vertebra three distinct zones of trabecular bone can be distinguished in a superoinferior direction (Fig. 4.62A). The central zone essentially consists of vertically arranged large-diameter cylinders whose walls are formed by thin solid plates of lamellar bone, with circularly orientated trabeculae arranged around the basivertebral veins. The zones either side, directly beneath the end plate region, are composed of regularly spaced longitudinal and transverse trabeculae. The anterior superior and inferior regions of the vertebral bodies have decreased trabecular bone, and are thus regions of mechanical weakness. To a large extent this explains the wedge-shaped compression fracture of a vertebral body (Fig. 4.62B).

Within each vertebral body there is one principal vertical and several secondary oblique and horizontal systems. Except for the interruption by the intervertebral discs, the vertical system runs throughout the entire column of vertebral bodies from the odontoid process to the sacrum. The oblique accessory systems run in four tracts: a superior and an inferior oblique system on each side. Each superior oblique system runs from the superior articular process of one side downwards through the pedicle to the lower surface of the vertebral body on the opposite side (Fig. 4.62C). The inferior oblique systems run from the inferior articular process of one side upwards via the pedicle to the upper surface of the vertebral body on the opposite side (Fig. 4.62C). These oblique systems do not reach the anterior margin of the vertebral body, so that in this region there are vertical compression trabeculae only (Fig. 4.62A). Posteriorly, behind the articular processes, the oblique systems are continuous with trabeculae within the spinous process (Fig. 4.62C).

Horizontal accessory systems begin in each transverse process and pass into the body of the vertebra where they intersect in the midline (Fig. 4.62D).

In mechanical terms the vertical system sustains body weight and all the jars and shocks which reach the vertebral column in a perpendicular direction. It is the principal trabecular system as stated, and resists atrophy more than the others. In osteoporotic spines the secondary systems atrophy first and their disappearance makes the lines of the vertical system stand out more sharply on radiographs. The spirally wound oblique systems resist torsion, and together

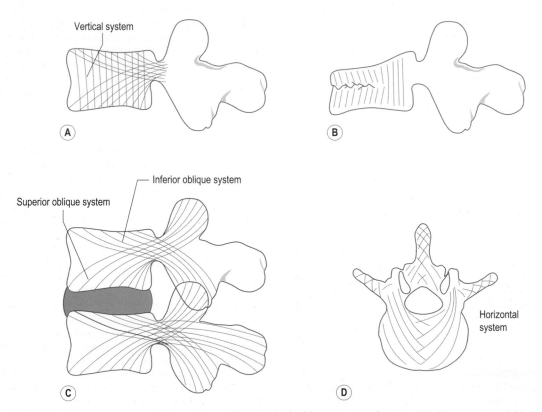

Figure 4.62 Trabecular arrangement within vertebrae: (A) within body, (B) compression fracture, (C) oblique system, (D) horizontal system.

with the vertical system share in the resistance to bending and shear. The horizontal systems are principally tension resistant, as are the minor accessory systems of the transverse and spinous processes, as they resist muscular pull.

It is interesting to note that the compressive breaking strains of lumbar vertebrae, when tested at physiological strain rates, are generally between 7 and 9 kN, with the strength increasing at faster strain rates. However, at least half of the lumbar vertebrae tested in one study were found to have compressive strengths less than the calculated compressive forces.

The vertebral unit

The vertebral unit is a useful concept when modelling of the vertebral column is to be attempted. It consists of two adjacent vertebrae and the intervening intervertebral disc. In terms of its functional components the vertebral column can be considered to consist of an anterior supporting pillar composed of the vertebral bodies and the intervertebral disc, and a posterior pillar comprising the articular processes joined by the vertebral arch. While the anterior pillar plays an essentially static role, the posterior pillar has a dynamic role in vertebral column mechanics.

A further differentiation of the components of the vertebral unit can be distinguished as horizontal layers sandwiched together. The vertebrae themselves form a passive segment, while the intervertebral disc and the posterior ligamentous structures form an active unit, which includes the articular processes. The mobility of this latter segment underlies the movements of the vertebral column.

The oblique trabecular systems described earlier functionally link the anterior and posterior pillars. With these systems as a basis, each vertebra can be compared to a first class lever system with the articular processes acting as the fulcrum (Fig. 4.63). It allows the absorption of axial

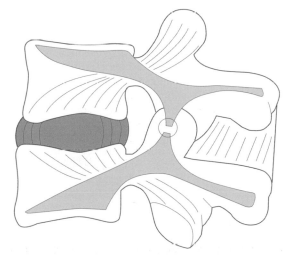

Figure 4.63 Mechanical model of the vertebral unit.

compression forces applied to the vertebral column by direct and passive absorption by the intervertebral disc, and by indirect and active absorption by the posterior ligaments and paravertebral muscles.

Intervertebral disc mechanisms

At the ultrastructural level, the nucleus pulposus, annulus fibrosus and cartilage end plate appear as a close-packed system, which is postulated to act as a buffer against gravity and torsion and so act as a shock absorber of forces transmitted to the vertebral column. The ability to absorb shock is due to the construction of the disc, in which the soft fluid-like nucleus pulposus is encapsulated within the connective tissue lamellae of the annulus fibrosus. Vertical pressure on the vertebral column results in compression of the disc with deformation of the nucleus pulposus so that it expands radially against the surrounding annulus fibrosus; outward expansion of the annulus fibrosus is necessary in order for the disc to absorb shock.

An examination of lateral, posterolateral and posterior disc bulging under compressive loading and applied movements has shown that the disc may bulge up to 2.7 mm beyond its unloaded state. The largest bulges being associated with lateral flexion. Furthermore, no clear relationship was observed between intradiscal pressure and disc bulge. End plate bulge was found to be negligible (0.1 mm) compared with disc bulge. When degenerated discs were tested they exhibited a greater degree of bulging than non-degenerated discs, presumably because they are less efficient. The clinical manifestation of disc bulging, particularly in degenerating discs, may be entrapment of nerve roots as they pass through the narrowed intervertebral foramen.

It has been suggested that the nucleus pulposus plays a major mechanical role in determining the compressive deformation of the disc. However, experiments have shown that the annulus fibrosus appears to be the primary load-bearing structure, capable of performing this function in a near normal fashion even when part or all of the nucleus has been removed. Nevertheless, the efficient functioning of the disc depends to a large extent on the elasticity of the nucleus, which is closely related to its water-binding capacity. Increased pressure within the nucleus (by the imbibition of water) increases the height of the disc, and acts to prestress the disc reducing its laxity, and consequently increasing the vertebral joint stiffness (Fig. 4.64).

Measurements of disc strength and disc pressure show that the tensile strength of the annulus fibrosus is between 15 and 50 kg/cm^2, whilst that of the vertebral bodies varies from 8 to 10 kg/cm^2. In tension, failure is routinely observed at the disc end plate at forces of 850 N at cervical levels and 3000 N at lumbar levels. The vertebral bodies in the cervical and lumbar regions fail in compression at 3000 and 5000 N respectively. The tensile strength of the

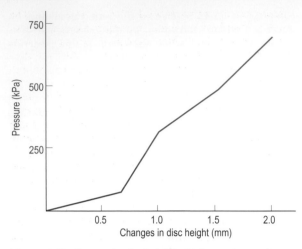

Figure 4.64 Changes in disc height with increasing nuclear pressurization. *(Adapted from Tencer AF and Ahmed AM (1981) The role of secondary variables in the measurements of the mechanical properties of the lumbar intervertebral joint.* Journal of Biomechanical Engineering, *103, 129–137.)*

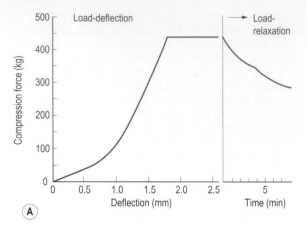

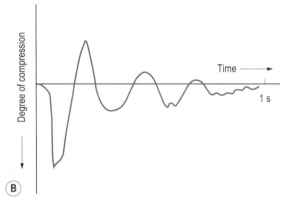

Figure 4.65 (A) Load-deflection and load-relaxation curves of an intervertebral disc. (B) Dynamic loading of the intervertebral disc. *(A, adapted from Markolf K and Morris J (1974) The structural components of the intervertebral disc.* Journal of Bone and Joint Surgery, *56A, 675–687. B, adapted from Hirsch CS (1995) The reaction of the intervertebral disc to compressive forces.* Journal of Bone and Joint Surgery, *37A, 1188–1192.)*

longitudinal ligaments is approximately 200 kg/cm^2, and could therefore provide considerable resistance to disc rupture. The ultimate torsional strength of the intervertebral disc within an intact vertebral column is approximately 40 kg/cm^2. From the foregoing it is possible to appreciate why the vertebral body may fracture without there being any evidence of disc rupture.

Mechanically, the intervertebral disc may be regarded as a viscoelastic structure capable of maintaining very large loads without disintegrating (Fig. 4.65A). The 'end point' of the disc is similar to that of steel. However, when the end point is exceeded, the disc still retains some power of recovery after rest, thereby retaining its properties of elasticity. The mechanical efficiency of the disc appears to be improved with use, and the energy lost during recovery is decreased. This is significant in lifting heavy loads, providing a theoretical basis for the custom of 'taking the strain' before a heavy load is lifted. Complete recovery of a disc after loading is modified by the duration of the force application and by the interaction of the various structural elements.

With static loading, deformation of the disc depends on the duration of the loading, becoming stable after approximately 5 min. With loads of up to 130 kg most of the deformation occurs in the first 30 s following loading; even so no absolute equilibrium is reached even after a few hours, i.e. the disc exhibits creep characteristics. The nucleus pulposus appears to act as an incompressible medium of short duration (approximately 1 s); thereafter the nucleus and annulus interact to redistribute, equilibrate and adapt to the load. The creep characteristic is therefore a mechanism by which the disc can distribute the stress, and

occurs until the disc adapts or reaches a stable state with respect to a particular load, being governed to a large extent by the zygapophyseal articulation. However, when a dynamic load is applied, the intervertebral disc may begin to vibrate (this occurs in less than 1 s and then dies out) (Fig. 4.65B). In this way the disc acts as a shock absorber by damping the oscillations.

The disc therefore has mechanical properties similar to many elastic and semielastic systems. It is important to be aware of these properties in order to understand the mechanisms of disc damage. For example, when a disc is loaded statically to approach its elastic limit, and then dynamically loaded in addition, the vibrations which occur may, at their peak, exceed the tensile limit of the annulus fibrosus, or the attachment to the bone, resulting in disc damage. The strength and the site of failure of the disc appear to be related to the morphology of the nucleus pulposus. However, only when considerable weakness is already present does prolapse occur at a specific site.

In vivo measurements of intradiscal pressures have shown that the force acting across the L5–S1 joint in a subject lifting loads is 30% less than calculated theoretically, and, more impressively, is 50% less in the lower thoracic region. The difference between measured and theoretical findings is thought to be due to absorption of part of the load by the anterior abdominal muscles which contract in response to loading. Increases in disc pressure, compared with quiet standing, have been recorded when coughing (40%), stair climbing (40%) and when walking slowly (15%). The significance of intradiscal pressure is important with respect to disc fluid transport and nutrition. Sustained flexion-loading of a lumbar segment has been shown to produce fluid loss from the disc and a reduction in its height. Therefore sitting postures, which entail flexion of the lumbar column, cause more fluid to be expressed from the lumbar discs than do erect postures, the effect being particularly marked in the nucleus pulposus. The fluid flow in flexed postures is sufficiently large to aid nutrition of the disc. The links between posture, fluid flow and disc nutrition may explain why societies that habitually adopt flexed or squatting postures have a low incidence of lumbar disc degeneration.

With ageing, the mechanical arrangement, and thus the biomechanical response of the intervertebral disc in relation to the zygapophyseal joints, departs from that observed in earlier life. The disc space narrows and the surface contact area of the joints increases. The loss of disc space has been attributed to thinning of the disc. However, it could equally be due to sinking of the disc into the vertebral body, which itself weakens with age due to osteoporosis. Whatever the cause, the result is that the anterior and posterior joints of the vertebral column become less protected by the load-attenuating and load-distributing properties of the nucleus with respect to the annulus, decreasing the range of movement as well as the tolerance to impact. A commonly observed radiographic manifestation in disc/joint degeneration is the appearance of bony spurs, ridges or transverse bars along the superior, inferior and lateral margins of the vertebral bodies, intervertebral foramen and articular surfaces.

What is not certain is whether these changes are the result of purely mechanical factors causing a disturbance in the equilibrium of the surrounding tissues, or whether it is the result of localized disruption of the periosteum causing alteration in the relationship of normal form and bony contour, or indeed whether still other factors are involved. There does appear to be sufficient clinical and radiological evidence to suggest that the changes in the discs occur with changes in the posterior vertebral column at levels where stresses and strains are greatest (i.e. lower cervical and thoracolumbar areas), adversely affecting vertebral column kinematics. For example, when there is a structural change within the disc, there occurs with time an associated alteration in the zygapophyseal joints, or vice versa, thus affecting local biomechanics. The altered biomechanics of one component of the vertebral column may result in asymmetric motion between the column components, which in turn may cause accelerated damage and degeneration that could eventually affect the entire vertebral region.

Changes in the relative position of the vertebral body also involve changes in the relative positions of the zygapophyseal joints. As nucleus turgor is lost and the disc space decreases, these latter joints become partial weight-bearing elements. Although there is convincing radiographic evidence of osteophyte formation, there is no experimental evidence to confirm that osteophyte presence is indicative of degenerative disc disease.

With the intervertebral disc being indispensable for the normal functioning of the vertebral column, it is not surprising that artificial discs are being developed which could replace a degenerated disc. Although some such discs have been developed they, as yet, do not offer the same degree of efficiency as a replacement hip or knee joint does. There are, of course, fundamental differences in the design of joint prostheses and the design of an acceptable viscoelastic component. In addition, there are major problems associated with the process of removal of the degenerated disc, and implantation of the prosthesis, not least of which is the proximity of the spinal cord posteriorly and the emergence of the nerve roots through the intervertebral foramen. Nevertheless, the further development of such prostheses and their implantation offers tremendous benefits over plating and fixation of vertebral segments. It is to be hoped that it will not be too long before intervertebral disc replacement becomes as common as other joint replacements.

Pathology

The result of spinal fusion to counteract the effects of disc degeneration is to create stress concentrations at either end of the fused segment. These increased stresses increase the chance of degeneration or instability in these regions. Obviously, the greater the number of segments that are fused then the more the remaining segments must move to achieve the same overall range of movement.

Rotatory injuries of the vertebral column are associated with fractures of the articular facets, which tend to limit excessive rotation. At lower levels of the vertebral column the common torsional injury is a fracture–dislocation of the thoracolumbar junction. Above this level the thoracic column is relatively stiff by virtue of the presence of the thoracic cage, and the articular processes offer little resistance to rotation. Below this junction the lumbar region gains increased resistance to deformation by the orientation of the articular processes. The T12–L1 area, being a transitional area, has neither the thoracic supplementary protective elements nor the lumbar protective bony geometry, thus torsional deformation and stress are concentrated at the thoracolumbar junction.

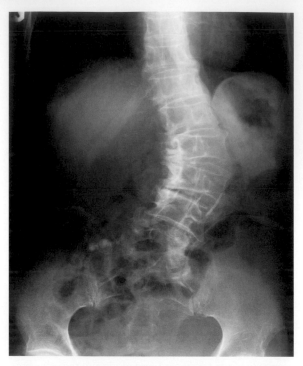

Figure 4.66 Radiograph showing thoracolumbar scoliosis.

The vertebral column possesses natural curvatures in the sagittal plane. A lateral curvature of the column in a frontal plane is termed a *scoliosis* (Fig. 4.66). In reality this is a three-dimensional deformity consisting of lateral flexion and a rotational abnormality. The curves can be broadly subdivided into two types: non-structural and structural. Non-structural curves have no structural abnormality and can easily be reversed by changing posture or by traction. Included in this group are postural curves which are eliminated on forward flexion and show no fixed vertebral rotation, and compensatory curves which are also devoid of vertebral rotation or structural changes. Compensatory curves develop above and below a single structural curve and maintain the alignment of the body. Structural curves, on the other hand, can be identified by demonstrating rib and vertebral rotation to the convexity of the curve on forward flexion. Radiographs of the curve show vertebral rotation with the bodies rotated to the convexity and the spinous processes to the concavity of the curvature (Fig. 4.66).

In theory, scoliotic curves can occur anywhere along the length of the vertebral column. However, in practice they are most commonly found in the thoracic and lumbar regions. Involvement of the thoracic region of the vertebral column is not surprising considering that there is little resistance to rotation in this region. The involvement of the lumbar region is more difficult to account for, although

the importance of lordosis in the development of scoliosis has been stressed by some authorities.

One way of classifying scoliotic curves is in terms of the age of the patient. This is particularly important in *idiopathic scoliosis*, which accounts for some 85% of all scolioses, where the side of the convexity of the curve is related to the age of onset. *Infantile scoliosis* (under the age of 4) is seen most commonly as a left thoracic curve, whereas right thoracic curves are most common in *adolescent scoliosis* (puberty to skeletal maturity). Of these, infantile scoliosis often regresses naturally. *Juvenile scoliosis* (age 4 to puberty) occurs with almost equal frequency to the right and left sides.

A further, and perhaps more important, classification of scoliosis is in terms of its aetiology and pathogenesis. Three classes of curves are considered:

1. Congenital – these arise due to congenital vertebral anomalies which themselves can be classified.
2. Neuromuscular – in which there is some obvious neurological or muscular impairment: this type of curve was very common until recently, frequently developing following poliomyelitis.
3. Idiopathic – in which the aetiology is unclear, although many causes have been postulated: within this class there are four main groups: primary skeletal, neuromuscular, metabolic and hereditary.

Although the aetiology of many of the curves is understood, or at least postulated, the mechanisms by which the deformity develops are less certain.

The incidence of clinically evident scoliosis is of the order of 1–5 per thousand of the population. However, screening programmes have suggested that as many as 15% of the population show some form of lateral curvature of the vertebral column. Presumably many of these are of the non-structural type or never develop to any significance. Nevertheless, scoliosis is more common in the white population than the black and is more prevalent in females than males (ratio of 5:1).

Joints of the thorax

Introduction

The bony components of the thorax articulate with one another in such a way as to provide a rigid, yet slightly mobile thoracic cage (Fig. 4.67). Posteriorly adjacent thoracic vertebrae articulate with one another by both secondary cartilaginous and synovial joints. Anteriorly the various parts of the sternum are joined by secondary cartilaginous joints, while laterally each rib is joined to its costal cartilage by a primary cartilaginous joint. However, it is the articulation of the rib with the vertebral column posteriorly and the costal cartilage with the sternum anteriorly, either directly or indirectly, which provides the mobility necessary during respiration. By the action of

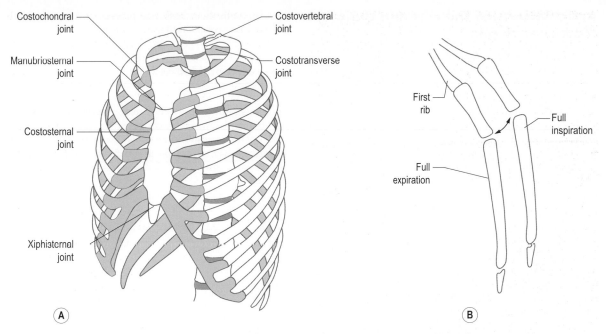

Figure 4.67 (A) Joints of the thorax, (B) movements of the sternum during respiration, viewed laterally.

muscles, the ribs are moved so as to change both the anteroposterior and transverse diameters of the thorax. The precise nature of rib movement differs in different regions of the thorax, being determined by the shape of the articulating surfaces as well as their anterior articulations.

This section considers articulations within a rib (at the costochondral joint) and the sternum (at the manubriosternal and xiphisternal joints) before the articulations of the rib with the vertebrae and sternum. Articulations between vertebrae are considered on page 445.

Costochondral joint

A primary cartilaginous joint between the anterior roughened end of a rib and the lateral end of the costal cartilage (Fig. 4.67A). It is surrounded by perichondrium which is continuous with the periosteum of the rib. Movement at these joints is confined to a slight bending of the cartilage at the junction with the rib. The cartilage, however, may undergo some twisting during the movement of the rib.

The costochondral joints can be stressed by applying continuous pressure to the anterior aspect of the chest wall, as in leaning against a table whilst sitting, or against a fence when standing. This continued stress often results in the joint becoming extremely painful and possibly swollen. Depending on its location, the pain may be misdiagnosed as a cardiac problem, causing the patient further, unnecessary stress. The condition is known as *Tietze's syndrome/chondritis*.

Sternal joints

Manubriosternal joint

A secondary cartilaginous joint between the inferior surface of the manubrium and the upper surface of the body of the sternum (Fig. 4.67A). The opposing surfaces of the two bones are covered with a thin layer of hyaline cartilage, between which is a fibrocartilaginous disc, often hollow at its centre. The joint is strengthened in front and behind by longitudinal fibrous bands and the adjacent sternocostal ligaments. Occasionally the joint resembles a primary cartilaginous joint. In later life it may begin to ossify.

A small amount of movement (7°) is permitted at this joint. During inspiration there is a decrease in the obtuse angle between the manubrium and body of the sternum. There is also a very slight shift of the body upwards (Fig. 4.67B). This is particularly noticeable when pressure is applied to the front of the chest.

Xiphisternal joint

The xiphoid process is an irregularly shaped piece of cartilage joined to the body of the sternum by a secondary cartilaginous joint, supported all round by a fibrous capsule. The xiphoid process ossifies late in life and with it the xiphisternal joint. While it remains cartilaginous, there is a certain amount of flexibility of the xiphoid process at the joint.

481

Articulations of the ribs and their costal cartilages

Posteriorly the typical ribs articulate with the sides of the bodies of two adjacent vertebrae via costovertebral joints and with the transverse process of its corresponding vertebra by a costotransverse joint. Anteriorly, the costal cartilages articulate directly with the sternum by costosternal joints or with each other by interchondral joints. The first, tenth, eleventh and twelfth ribs articulate with only one vertebral body; in addition the 11th and 12th do not articulate with their transverse process or anteriorly with the preceding costal cartilage.

Costovertebral joints

Articular surfaces

The joints are between the convex articular facets on the head of the rib and a large concave demifacet on the upper border of the corresponding vertebra and a smaller demifacet on the lower border of the vertebra above (Fig. 4.68B).

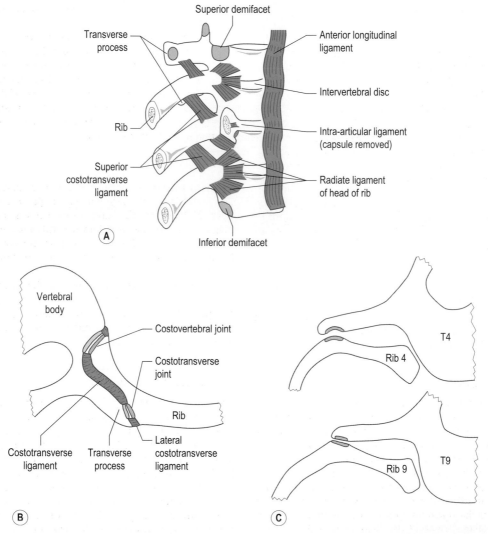

Figure 4.68 (A) The costovertebral joint, lateral view, (B) the costovertebral and costotransverse joints, horizontal section, (C) shape of the costotransverse joint surface at different vertebral levels.

The crest on the head of the rib articulates with a slightly cupped depression on the posterolateral aspect of the intervening intervertebral disc. The joint surfaces are covered with hyaline cartilage.

Joint capsule

A loose fibrous capsule that surrounds the joint. Anteriorly it is thickened to form the radiate ligament of the head of the rib (Fig. 4.68A), while posteriorly it blends with, and is reinforced by, the adjacent denticulation of the posterior longitudinal ligament. An intra-articular ligament completely divides the joint space, each part of which is lined by synovial membrane.

Ligaments

Intra-articular ligament This short, thick band passes from the crest of the head of the rib to the intervertebral disc and divides the joint cavity into two parts (Fig. 4.68A).

Radiate ligament of the head of the rib This passes medially from the front of the head under cover of the lateral part of the anterior longitudinal ligament. Its three bands of fibres radiate upwards to the body of the vertebra above, horizontally to the front of the intervertebral disc and inferiorly to the body of the vertebra below (Fig. 4.68A).

Ribs one, ten, eleven and twelve articulate with their own vertebrae only. Consequently, there is a single joint cavity, no intra-articular ligament and a poorly developed radiate ligament.

Costotransverse joint

These are only present between the tubercles of the upper 10 ribs and the transverse processes of their corresponding vertebrae (Fig. 4.68B).

Articular surfaces

The joint is between the articular facet on the front of the transverse process, near its tip, and an oval facet on the posteromedial aspect of the tubercle of the rib. The shape of the joint surfaces changes from above downwards (Fig. 4.68C). In the upper costotransverse joints, the facet on the rib tubercle is convex and that on the transverse process is reciprocally concave. However, this arrangement differs lower down, with that of the rib facet and of the transverse process becoming flatter. This change in shape is one of the factors responsible for the different movements of the upper and lower ribs during respiration.

Joint capsule

A joint capsule is a thin, fibrous capsule that completely surrounds the joint. It is strengthened posterolaterally by the lateral costotransverse ligament, which although in contact with the capsule does not fuse with it.

Ligaments

Lateral costotransverse ligament This is a stout band passing between the tip of the transverse process, beyond the articular facet, and the roughened lateral part of the costal tubercle (Fig. 4.68B).

Costotransverse ligament Short fibres binding the rib to the transverse process passing from the back of the neck of the rib, to the front of the transverse process, medial to the facet (Fig. 4.68B).

Superior costotransverse ligament This consists of an anterior band which passes from the rib upwards and laterally, and a posterior band which passes upwards and medially from the rib (Fig. 4.68A). These bands attach to the undersurface of the transverse process of the vertebra above and are separated by the external intercostal muscle. The fibres of the anterior band blend laterally with the internal intercostal membrane.

Movements

The movements at costovertebral and costotransverse joints are dealt with together. Although each movement is small, because of the length of the rib it is considerably magnified anteriorly. As the rib is raised or lowered there is a twisting and gliding movement at the costovertebral joints, which is limited by the radiate and intra-articular ligaments.

Concurrently, there is movement at the costotransverse joint. In the lower joints there is gliding and rotation of one plane surface against another. Higher up, because of the curved nature of the joint surfaces, the movement tends to be one of rotation only.

Sternocostal joints

These are between the medial end of the costal cartilage of the first to the seventh ribs and the sternum (Fig. 4.69). That between the first costal cartilage and the sternum is a primary cartilaginous joint (synchondrosis), the cartilage uniting with the socket on the upper part of the lateral border of the manubrium. It prevents any appreciable movement between the two components, an important factor in respiration. The remaining joints are synovial, being surrounded by a fibrous capsule and supported anteriorly and posteriorly by the radiate ligaments. The joint cavity for the second rib is usually divided into two by an intra-articular ligament. With increasing age,

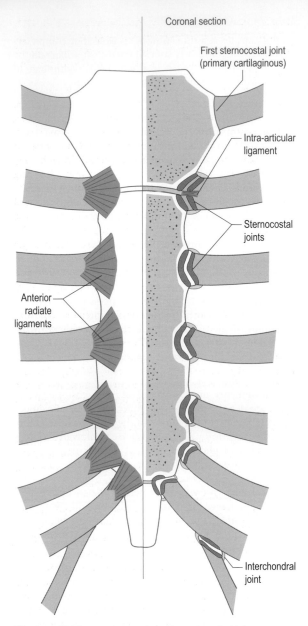

Coronal section

First sternocostal joint
(primary cartilaginous)

Intra-articular
ligament

Sternocostal
joints

Anterior
radiate
ligaments

Interchondral
joint

Figure 4.69 The sternocostal and interchondral joints.

however, these cavities, except for that of the second costal cartilage, become obliterated.

The anterior radiate ligament passes from the medial end of the costal cartilage over the joint to the anterior aspect of the sternum, its highest fibres passing upwards and medially, the middle fibres horizontally, the lower fibres downwards and medially (Fig. 4.69). The fibres interlace with those from the joint above, below and from the opposite side, forming a criss-cross feltwork covering the front of the sternum; this fuses with tendinous fibres of pectoralis major. The posterior radiate ligament of these joints forms a similar pattern over the posterior aspect of the sternum.

The second sternocostal joint is similar to those above except that the anterior and posterior radiate ligaments attach above to the manubrium, horizontally to the fibrocartilaginous pad between the manubrium and the body, and below to the body of the sternum (Fig. 4.69). The intra-articular ligament of the joint restricts its movement.

These synovial joints allow the medial end of the cartilage to glide up and down in the sockets on the lateral border of the sternum when the rib is raised and lowered during respiration.

Interchondral joints

These are formed between the tips of the costal cartilages of the eighth, ninth and tenth ribs and the lower border of the cartilage above (Fig. 4.69). The eighth and ninth costal cartilage joints are synovial, while the tenth is more like a fibrous joint. Small synovial joints are also formed between the adjacent margins of the fifth to ninth costal cartilages.

All these synovial joints are surrounded by fibrous capsules, strengthened anteriorly and posteriorly by oblique ligaments. They provide a slight gliding movement which adds to the general mobility of the region. The interchondral joints can be strained, under similar circumstances as the costochondral joints, leading to pain radiating from the area of the costal cartilage.

Movements of the thoracic cage

During respiration the volume of the thoracic cage changes. This is achieved by movement of the diaphragm, and movements of the ribs and the sternum. The vertical, transverse and anteroposterior diameters of the thorax increase and decrease during inspiration and expiration respectively. Each rib and its costal cartilage can be considered as a lever which moves up and down. The precise nature of the movement depends on several factors including its length and whether it articulates directly with the sternum or not. The ribs that articulate with the sternum gradually increase in length from above down, with their anterior ends lying at a lower level than the posterior ends (Fig. 4.70A). Although the eighth and ninth ribs are shorter, the nature of their attachment to the costal margin means that the transverse diameter of

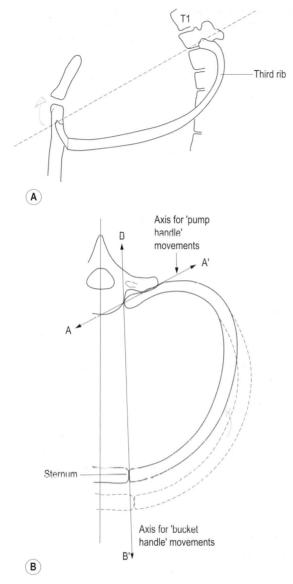

T1

Third rib

(A)

Axis for 'pump handle' movements

D

A'

A

Sternum

Axis for 'bucket handle' movements

B'

(B)

Figure 4.70 (A) The relationship of the anterior and posterior ends of a rib, (B) the axes of movement of the ribs during respiration.

thoracic vertebrae, gradually increasing as the thoracic vertebral column is descended. Consequently, although the axis of movement relative to each rib remains unchanged, the resultant movement of the upper and lower ribs differs.

Movement of the upper ribs (second to fifth) about an axis along their necks causes their anterior ends to be raised, and with them the body of the sternum. Because the first costal cartilage is firmly attached to the manubrium, as well as the first rib being much shorter, movement of its anterior end is slight, the longer second to fifth ribs lifting the body of the sternum upwards and outwards causing a bending of the manubriosternal joint (Fig. 4.67B). In this way the anteroposterior diameter of the thorax is increased (Fig. 4.70B). There is very little lateral movement of these ribs, except perhaps during the terminal part of a deep inspiration, because the axis of movement is less oblique than lower down. The movement has been likened to that of the handle of a pump when drawing water from a well: it is therefore often referred to as the 'pump-handle' movement.

Movement of ribs eight to ten results in an outward and upward movement of their anterior ends. This lateral excursion of the ribs and their costal cartilages causes a widening of the infrasternal angle and a consequent increase in the transverse diameter of the thorax (Fig. 4.70B). Because the shape of the costotransverse joints of these lower ribs is flat, there is both rotation and gliding of one bone against the other (Fig. 4.68C). Consequently, it appears that the axis of the movement of the ribs passes through the costovertebral and sternocostal or interchondral joints (Fig. 4.71B). The upward and outward movement of the shaft of the rib has been likened to raising the handle from the side of a bucket: it is therefore often referred to as the 'bucket-handle' movement.

The intermediate ribs six and seven show both pump-handle and bucket-handle types of movement. Ribs 11 and 12 are not attached anteriorly and thus have very little influence on increasing the transverse diameter of the thorax. They do, however, give attachment to some of the lower fibres of the diaphragm and, with the aid of quadratus lumborum, provide a firm attachment for the diaphragm.

During expiration the reverse movements of the ribs and sternum occur, with a decrease in both the anteroposterior and transverse diameters of the thorax.

As well as these respiratory movements, the ribs also move passively following changes in the thoracic part of the vertebral column. Flexion of the column causes the ribs to move closer together; extension causes them to move further apart; lateral bending causes those on the concave side to come together and those on the convex side to separate; lateral rotation causes a slight relative horizontal displacement of one rib with respect to its neighbour.

the thorax continues to increase as far as the ninth rib, after which it decreases.

The axis about which the rib moves passes along the neck of the rib through the costovertebral and costotransverse joints (Fig. 4.70B). It therefore runs backwards and laterally, following the inclination of the transverse processes of the vertebrae. However, the inclination of the transverse processes is less oblique for the upper

Section summary

Joints of the thorax

Costochondral joints

Type	Primary cartilaginous
Articular surfaces	Anterior end of rib and costal cartilage

Manubriosternal joint

Type	Secondary cartilaginous
Articular surfaces	Inferior surface of manubrium and superior surface of body of sternum
Movements	Small amount of movement associated with inspiration, body carried upwards and outwards

Xiphisternal joint

Type	Secondary cartilaginous
Articular surfaces	Xiphoid process and inferior surface body of sternum; supported by fibrous capsule

Costovertebral joints

Type	Synovial plane joint
Articular surfaces	Facet(s) on head of rib with facet(s) on vertebral bodies; crest of rib head with posterolateral depression on intervertebral disc
Capsule	Loose fibrous capsule surrounds joint; intra-articular ligament divides joint into two compartments
Ligaments	Radiate ligament formed by anterior thickening of capsule posterior longitudinal ligament blends with capsule intra-articular ligament

Costotransverse joints

Type	Synovial plane joint
Articular surfaces	Facet on transverse process of vertebra with facet on tubercle of rib
Capsule	Thin fibrous capsule completely surrounds joint
Ligaments	Lateral costotransverse ligament Costotransverse ligament Superior costotransverse

First sternocostal joint

Type	Primary cartilaginous
Articular surfaces	Anterior end of first costal cartilage with lateral aspect of manubrium

Second to seventh sternocostal joints

Type	Synovial plane joints
Articular surfaces	Anterior end of costal cartilages with facets on side of manubrium (second), body (second to seventh) and xiphoid process (seventh) of sternum
Capsule	Fibrous capsule surround joint
Ligaments	Anterior radiate posterior radiate

Interchondral joints

Type	Synovial plane joints
Articular surfaces	Tips of eighth, ninth and tenth costal cartilages with cartilage above
Capsule	Fibrous capsule surrounds joint, strengthened anteriorly and posteriorly by oblique ligaments

Movement of the ribs

- Raising of the anterior end of the upper ribs and the lateral part of the lower ribs is produced by: twisting and gliding at the costovertebral joints, simultaneous gliding and rotation at the costotransverse joint and gliding at the interchondral joint.
- Upper rib movement increases the anteroposterior diameter of the thorax.
- Lower rib movement increases the transverse diameter of the thorax.

NERVE SUPPLY

The cervical plexus

The cervical plexus (Fig. 4.71) is formed from the anterior primary rami of the upper four cervical nerves and consists of a series of loops between adjacent nerves. The first cervical nerve emerges between rectus capitis anterior and rectus capitis lateralis, descending in front of the transverse process of the atlas to join the ascending branch of the second cervical nerve. As it does so, it sends a large branch to join the hypoglossal nerve. The second, third and fourth nerves divide into upper and lower parts, with adjacent parts joining together. The lower part of the fourth cervical nerve may participate in the brachial plexus. The loop communications of the plexus lie close to the vertebral column in front of levator scapulae and scalenus medius behind the prevertebral muscles. The branches arising from these loops lie posteromedial to the internal jugular vein under cover of sternomastoid. The branches given off can, for descriptive purposes, be grouped into superficial and deep. The former all supply the skin of the lower part of the skull and neck, while the latter separate again into medial and lateral divisions, which go on to supply muscles or communicate with other nerves.

Superficial branches

There are four cutaneous nerves which all appear in the posterior triangle above the midpoint of the posterior border of sternomastoid before piercing the deep fascia to supply the skin of the head and neck (Figs. 4.71A and 4.72).

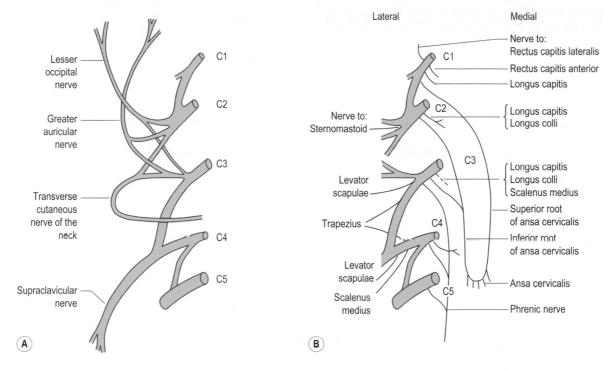

Figure 4.71 (A) Superficial and (B) deep branches of the cervical plexus.

Lesser occipital nerve

This nerve (C2, 3) hooks below the accessory nerve to emerge above the other three and ascends along the posterior border of sternomastoid to pierce the deep fascia at the apex of the posterior triangle. It then divides to supply the skin and fascia on the upper lateral part of the neck, the cranial surface of the auricle and mastoid process, and adjacent part of the scalp (Fig. 4.72). Occasionally, the nerve is double.

Greater auricular nerve

This nerve (C2, 3 or sometimes just C3) is the largest of the cutaneous branches. It emerges below the lesser occipital nerve and passes upwards and forwards over the superficial surface of sternomastoid, deep to platysma, towards the lower part of the auricle. During its course, it divides into a number of branches. Anterior branches pass through the substance of the parotid gland and over the angle of the mandible to supply skin and fascia over the posteroinferior surface of the face (Fig. 4.72). Within the parotid gland these branches communicate with the facial nerve. Intermediate branches supply both surfaces of the lower part of the auricle. Posterior branches supply an area of skin and fascia over the mastoid process. These posterior

branches communicate with the lesser occipital nerve and the posterior auricular branch of the facial nerve.

Transverse cutaneous nerve of neck

This nerve (C2, 3) passes horizontally forwards around the posterior border of sternomastoid, deep to platysma and the external jugular vein. As it passes across sternomastoid it divides into upper and lower branches, which supply the skin and fascia over the anterior triangle of the neck from the mandible to just above the clavicle near the midline (Fig. 4.72).

Supraclavicular nerves

These nerves (C3, 4) appear as one large nerve at the posterior border of sternomastoid just below its midpoint to pass down through the lower part of the posterior triangle, dividing into lateral, intermediate and medial supraclavicular nerves. They all pierce the deep fascia above the clavicle to supply skin and fascia over the lower part of the side of the neck and anterior chest wall as far as the sternal angle (Fig. 4.72). The lateral nerve also supplies the acromioclavicular joint, while the medial supplies the sternoclavicular joint. Branches from the intermediate and lateral nerves may groove or pierce the clavicle.

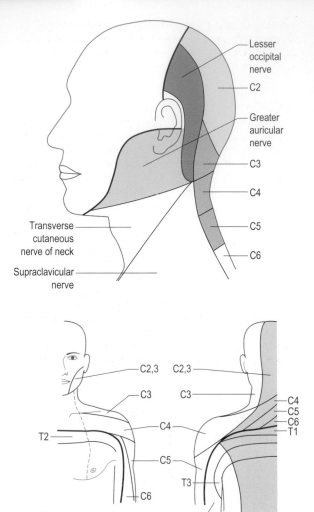

Figure 4.72 (A) Cutaneous distribution of branches from the cervical plexus; (B) dermatome distribution.

Deep branches

Muscular branches

These arise with other branches of the plexus deep to the sternomastoid and can be divided into lateral and medial branches depending on whether they pass into the posterior or anterior triangles respectively.

Lateral branches From the second cervical nerve, a sensory branch enters the deep surface of sternomastoid. From the third and fourth cervical nerves, a sensory branch crosses the posterior triangle to enter the deep surface of trapezius. Separate fibres from C3, 4 innervate levator scapulae, and scalenus medius and posterior (Fig. 4.71B).

Medial branches These innervate the prevertebral muscles rectus capitis lateralis and anterior (C1, 2), longus

capitis (C1, 2, 3, 4) and longus colli (C2, 3, 4) (Fig. 4.71B). From the first cervical nerve a branch joins the hypoglossal nerve. A smaller branch of sensory fibres passes superiorly within the hypoglossal nerve to supply the skull and dura mater of the posterior cranial fossa; this is the meningeal branch. The remaining fibres descend in the hypoglossal nerve to be given off in one of three branches, none of which contain hypoglossal fibres. These are the superior root of the ansa cervicalis, which descends in front of the internal and common carotid arteries to join the inferior root, formed by branches from C2 and 3 passing over the anterior surface of the internal jugular vein, to form the ansa cervicalis (Fig. 4.71B) superficial to the carotid sheath. Branches from the ansa innervate sternohyoid, sternothyroid and omohyoid. The other two branches from the hypoglossal nerve containing C1 fibres are the nerve to thyrohyoid and the nerve to geniohyoid.

Phrenic nerve This arises from C3, 4, 5, although its principal root is C4 (Fig. 4.71B). Branches from C3 reach it either directly or by the nerve to sternohyoid, whereas the C5 fibres may reach it directly or via the accessory phrenic nerve from the nerve to subclavius. The phrenic nerve innervates the diaphragm, being its only motor supply, and has to pass through both the neck and the thorax to reach it.

In the neck each phrenic nerve descends from lateral to medial over the front of scalenus anterior, and lies behind sternomastoid, omohyoid, the internal jugular vein and the suprascapular vessels and is anterior to the cervical fascia. At the base of the neck each phrenic nerve enters the thorax by passing behind the subclavian vein and in front of the subclavian artery.

In the thorax the left and right phrenic nerves are separated from the pleural cavity only by the mediastinal pleura.

Left phrenic nerve This passes in front of the internal thoracic artery, medial to the apex of the left lung and pleura, between the left common carotid and subclavian arteries to cross the aortic arch anterior to the vagus nerve. It then passes anterior to the root of the left lung having the left ventricle and pericardium medially and the pleura of the left lung laterally. It pierces the muscular part of the diaphragm lateral to the oesophageal opening to supply its undersurface mainly on the left.

Right phrenic nerve This is shorter and more vertical, entering the thorax lateral to the right brachiocephalic vein and superior vena cava; it is not in contact with the vagus nerve. It then passes between the pleura and pericardium across the right atrium to reach the inferior vena cava, having passed anterior to the root of the right lung. It passes through the central tendon of the diaphragm through or adjacent to the inferior vena caval opening, to supply the undersurface of the muscle mainly on the right side.

In addition to providing the only motor supply to the diaphragm, the phrenic nerves are also sensory to its central part as well as to the mediastinal and diaphragmatic pleura, fibrous pericardium, diaphragmatic peritoneum and probably also to the liver, gall bladder and inferior vena cava. The phrenic nerves receive fibres from the sympathetic trunk in the neck and the coeliac plexus in the abdomen.

Communicating branches

The cervical plexus sends fibres to the vagus, accessory and hypoglossal nerves, and receives grey rami communicantes from the superior cervical ganglion of the sympathetic trunk.

The spinal cord

Introduction

The central nervous system (CNS) is formed by the aggregation of bundles of axons and clusters of nerve cell bodies. The internal appearance and external form of the central nervous system reflect the manner in which these components are arranged (see also The brain, p. 572).

The central part of the spinal cord is formed by nerve fibres and nerve cell bodies, appearing grey in transverse sections. The outer part of the spinal cord mainly consists of myelinated axons. The myelin endows this material with a white appearance in transverse sections. Since cell bodies are not myelinated, ganglia and nuclei form grey matter, while myelinated axons form white matter. Thus areas in the central nervous system formed mainly by axons are referred to as *white matter* (Figs. 4.73 and 4.74). In contrast, areas formed mainly by cell bodies appear grey and are referred to as *grey matter* (Figs. 4.73 and 4.74).

A collection of cell bodies that forms a prominent, usually rounded swelling is referred to as a *ganglion* (Fig. 4.74). Clusters of axons that form a recognizable bundle within the central nervous system are referred to as a *fasciculus* (Fig. 4.74). When bundles of axons form a raised bump or convex contour on the surface of the central nervous system, typically in the spinal cord, the surface resembles the external surface of a tube, and the elevation is referred to as a *funiculus*.

These terms refer simply to the topographic appearance of collections of axons or cell bodies without regard to their

function. Other terms are used which carry functional implications. A group of axons with a similar function is known as a *tract*, whereas a group of cell bodies with a similar function is known as a *nucleus*.

Glial cells

Nerve cells are not the only constituents of the central nervous system. Interspersed between the nerve cells are several types of cells referred to collectively as glial cells (*glia*: Latin, meaning glue) whose function, in general, is to hold the neurons of the central nervous system together. As a whole, glial cells outnumber nerve cells, and constitute almost half the total volume of the central nervous system.

Four types of glial cells are found in the central nervous system: *oligodendrocytes*, *microglia*, *ependymal cells* and *astrocytes* (Fig. 4.75). Oligodendrocytes are indirectly responsible for myelination in the central nervous system because they myelinate several parallel axons; oligodendrocytes also serve to hold axons together. Microglia are the macrophages of the central nervous system, being responsible for removing foreign matter and cellular debris.

Ependymal cells are exclusively found lining the internal surface of the ventricles of the central nervous system. They

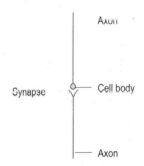

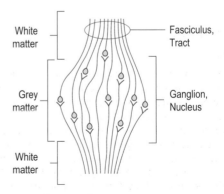

Figure 4.74 In the central nervous system, neurons are connected in series, with the axon of one neuron forming a synapse with the cell body of the next neuron.

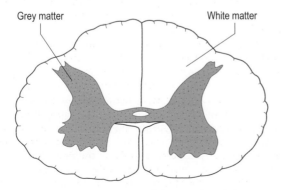

Figure 4.73 Transverse section of the spinal cord showing the arrangement of the white and grey matter.

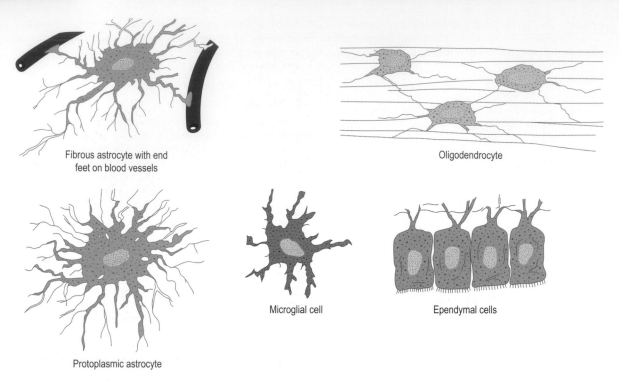

Fibrous astrocyte with end
feet on blood vessels

Oligodendrocyte

Protoplasmic astrocyte

Microglial cell

Ependymal cells

Figure 4.75 The microscopic appearance of neuroglial cells.

are epithelial cells arranged side by side, forming a cellular barrier between the nervous system and the fluid in its cavities.

Astrocytes derive their name from their star-shaped appearance. These cells consist of a cell body with long processes radiating away from it. Two types of astrocyte can be recognized. Microscopically, *protoplasmic astrocytes* occur mainly in the grey matter of the central nervous system and have thick processes that branch repeatedly. They weave between cell bodies, collectively holding them together, with some processes being attached to small blood vessels (Fig. 4.75). In contrast, *fibrous astrocytes* are found mainly in white matter and have fewer but longer processes that weave between the axons. The principal role of astrocytes is to provide a framework for the central nervous system by holding the cell bodies and axons in place relative to one another. However, they also play an important role in the nutrition and metabolic activities of neurons by regulating the exchange of chemicals between themselves and blood vessels.

Development

The central nervous system develops from a single tubular structure, the *neural tube* (Fig. 4.76), which forms along the dorsal surface of the embryo, extending from the site of the future head to the tail. Accordingly the tube has two ends: a head end and a tail end. Because the tube eventually bends

in the sagittal plane during development of the head and brain, the terms 'anterior, posterior, superior and inferior', as used in gross anatomy, cannot be applied to the nervous system without confusion. To circumvent this confusion different terms of reference are therefore used with respect to the central nervous system.

The tail end of the neural tube is called the *caudal end* (Fig. 4.76). The opposite end, regardless of the direction in which it points with respect to the rest of the body, is called the *rostral end* (Fig. 4.76). The surface of the neural tube which faces the belly of the embryo is called its ventral surface, irrespective of whether the tube is straight or curved in the sagittal plane, while the opposite surface is the dorsal surface. These directional terms are equally applicable in descriptions of the adult central nervous system.

The neural tube maintains a narrow cavity along its length, while the thickness of its walls increases as the constituent cells multiply and grow. Most of the caudal part of the neural tube simply grows in length and diameter. Its walls get thicker but its cavity remains narrow. This section of the neural tube forms the future spinal cord, with its cavity becoming the *central canal* of the spinal cord (Fig. 4.76).

Spinal cord

The spinal cord is essentially a long thick cable formed by thousands of longitudinally running axons surrounding a central core of grey matter. It is approximately 45 cm long

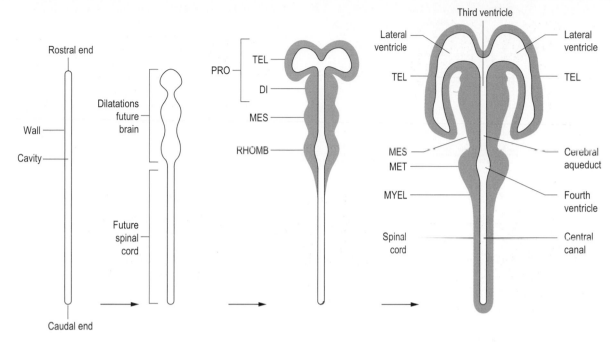

Figure 4.76 Stages in development of the central nervous system, viewed in longitudinal (coronal) sections from a dorsal aspect. (See also pp. 575–576.)

and occupies the vertebral canal at cervical, thoracic and upper lumbar levels.

The external surface of the spinal cord has only a few features that have attracted names (Fig. 4.77). Running along the midline on the ventral surface is a pronounced depression, the *anterior median fissure*, dividing the ventral part of the spinal cord into left and right halves. Dorsally, the spinal cord is marked by several less pronounced longitudinal depressions, one of which lies in the midline, the *posterior median sulcus*, separating the left and right halves of the spinal cord. Another, the *posterior lateral sulcus*, runs along the posterolateral aspect on each side of the entire length of the spinal cord. A further sulcus is present on each side of the rostral half of the spinal cord running longitudinally between the posterior median and posterior lateral sulci: this is the *posterior intermediate sulcus*.

Between the three sulci on each side of the rostral half of the spinal cord lie two slightly rounded prominences that extend longitudinally. The medial prominence, between the posterior median and posterior intermediate sulci, is the fasciculus gracilis, and the lateral, between the posterior intermediate and the posterior lateral sulci, is the fasciculus cuneatus. These fasciculi are formed by bundles of sensory axons destined to synapse in the nucleus gracilis and nucleus cuneatus respectively. Only the fasciculus gracilis is present in the caudal half of the spinal cord; consequently a posterior intermediate sulcus is lacking in this region.

The width of the spinal cord is not uniform along its length. It is expanded a short distance below its rostral

end and again near its caudal end (Fig. 4.77). These expansions occur at levels of the spinal cord responsible for the innervation of the upper and lower limbs. They are formed by the larger number of cell bodies and axons located at these levels to deal with limb functions, and are called respectively the *cervical* and *lumbar enlargement*. The caudal end of the spinal cord tapers to a pointed tip, known as the *conus medullaris*.

Spinal nerves

Short nerves lying within the intervertebral foramina of the vertebral column (Fig. 4.78A). Outside the vertebral column they form elements of the peripheral nervous system, but within the vertebral canal each spinal nerve is connected to the spinal cord by *ventral* and *dorsal* nerve *roots* which are components of the central nervous system.

There are 31 pairs of *spinal nerves*, each pair named according to the vertebra to which they are related. There are eight pairs of cervical spinal nerves, C1–C8, each of which lies above the cervical vertebra with the same segmental number, except for C8 which lies below the C7 vertebra. The remaining spinal nerves take the number of the vertebra that lies above them. Thus, the nerve below the T6 vertebra, for example, is the T6 spinal nerve, and the nerve below the L2 vertebra is the L2. In all, there are 8 cervical (C1–C8), 12 thoracic (T1–T12), 5 lumbar (L1–L5), 5 sacral (S1–S5), and one pair of coccygeal (Co1) spinal nerves.

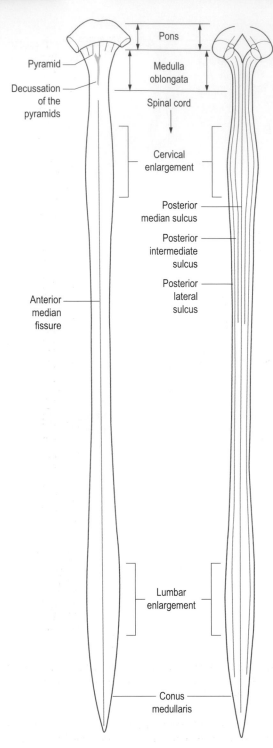

Figure 4.77 Appearance and surface features of the spinal cord.

Labels on figure:
- Pyramid
- Decussation of the pyramids
- Anterior median fissure
- Pons
- Medulla oblongata
- Spinal cord
- Cervical enlargement
- Posterior median sulcus
- Posterior intermediate sulcus
- Posterior lateral sulcus
- Lumbar enlargement
- Conus medullaris

The ventral roots of all the spinal nerves attach in a series along the ventrolateral aspect of the spinal cord on each side (Fig. 4.78B). Similarly, the dorsal roots attach to the posterolateral aspect of the spinal cord along the posterior lateral sulcus. As each dorsal root approaches the spinal cord it divides into a series of small branches called rootlets which form the junction with the spinal cord. Correspondingly, each ventral root is formed by a series of rootlets which emerge from the spinal cord and gather to form the ventral root proper.

The dorsal roots convey only sensory fibres from the spinal nerves to the spinal cord. The ventral roots convey motor fibres from the spinal cord to the spinal nerves, but they also contain some sensory fibres. Near its junction with the ventral root and the spinal nerve, each dorsal root has a swelling, the *dorsal root ganglion* (Fig. 4.78B). This is formed by the cell bodies of all the sensory fibres that run in the related spinal nerve.

Each spinal nerve is attached by its roots to a discrete section of the spinal cord called a *spinal cord segment*. The limits of each segment are demarcated by an imaginary transverse plane drawn through the spinal cord midway between the sites of attachment of the most rostral of the rootlets connected to one particular spinal nerve and the most caudal rootlet of the nerve above. Each spinal cord segment takes the name of the spinal nerve which attaches to it (Fig. 4.78C). Consequently, the spinal cord consists of 8 cervical, 12 thoracic, 5 lumbar, 5 sacral segments and one coccygeal segment.

Internal structure of the spinal cord

The central core of the spinal cord consists of grey matter which in transverse section has a characteristic appearance (Fig. 4.79A), being distributed in a pattern resembling the shape of a butterfly. The area surrounding the central canal of the spinal cord is known as the *central grey matter*. Projecting dorsolaterally and ventrally from this central area are extensions of grey matter known as the *dorsal (posterior)* and *ventral (anterior) horns* (Fig. 4.79B).

The ventral horns on each side are formed by the cell bodies of neurons that innervate voluntary muscles. The axons of these cells leave the spinal cord through the ventral roots. The dorsal horns consist of cells concerned with the processing of sensory information that enters the spinal cord through the dorsal roots. In general, sensory fibres entering the spinal cord synapse on neurons in the dorsal horn, with the axons of these latter cells forming tracts that run upwards or downwards in the spinal cord.

Between the T1–T2 segments of the spinal cord an additional small horn of grey matter is present. It projects laterally in the angle between the dorsal and ventral horns on each side, and is known as the *lateral horn* (Fig. 4.79B). It contains the cell bodies of neurons, the axons of which constitute part of the autonomic nervous system (p. 500).

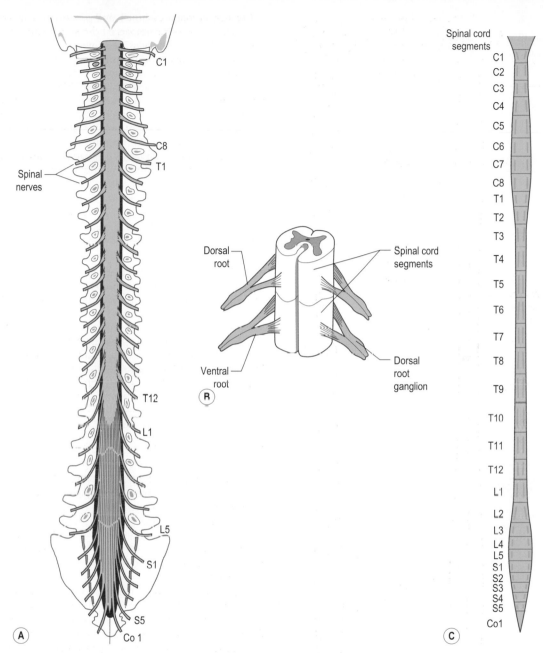

Figure 4.78 (A) The spinal cord and spinal nerves viewed from the back with the laminae of the vertebrae removed to reveal the spinal cord in the vertebral canal, and the spinal nerves in the intervertebral foramina, (B) the two segments of the spinal cord and their spinal nerve roots and rootlets, (C) the spinal cord and its division into segments.

The peripheral part of the spinal cord consists of white matter formed by axons arranged in tracts that convey information to or from the brainstem, or between different segments of the spinal cord. The white matter is divided into sectors by imaginary lines drawn radially through the dorsal and ventral horns (Fig. 4.79C), giving three regions of white matter that can be identified in each half of the spinal cord. That between the two horns laterally is the *lateral funiculus* and is formed by descending tracts conveying motor information and ascending tracts concerned

493

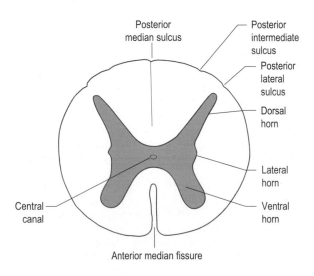

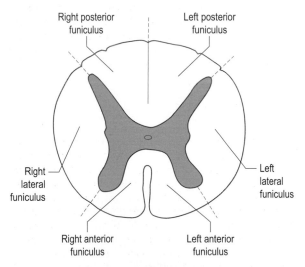

Figure 4.79 The internal structure of the spinal cord and its appearance in transverse section.

with pain, temperature, touch and pressure sensation. The white matter between the dorsal horn and the midline dorsally is the *posterior funiculus*, and consists largely of ascending tracts that convey information on pressure, touch and position sense. Finally, the region between the ventral horn and the midline anteriorly is the *anterior funiculus*, and is formed by a mixture of descending motor and ascending sensory tracts.

Sometimes the region of white matter embracing the anterolateral aspect of the ventral horn is referred to as the *anterolateral funiculus*. This is because the axons in this region, overlapping both the anterior and lateral funiculi, constitute the largest single collection of ascending sensory fibres that convey the sensations of pressure, touch, pain and temperature.

The meninges

Although glial cells hold the neurons of the central nervous system together and provide protection against metabolic insults, the central nervous system is nonetheless a soft, cellular mass.

As such it is vulnerable to external, mechanical insults which might arise were it freely mobile within the skull or vertebral canal. To protect against such insults, the central nervous system is surrounded by three membranes, collectively called the *meninges*, and bathed in *cerebrospinal fluid*.

Meninges of the spinal cord

The outermost of the three meningeal coverings of the central nervous system is a tough fibrous layer: the *dura mater*. In the vertebral canal it forms a long sac, the *dural sac*, which houses the spinal cord (Fig. 4.80). Rostrally, the dural sac attaches to the margins of the foramen magnum, but within the vertebral canal it is relatively mobile, being attached to the vertebral canal by modest fibrous ligaments only. The dural sac is much longer than the spinal cord, reaching as far as the S2 vertebra.

The deepest of the three meningeal coverings is a thin transparent membrane, the *pia mater* (Fig. 4.80). It intimately invests the entire surface of the spinal cord like a fine skin. From the caudal end of the spinal cord it continues as a thread of tissue devoid of any neural elements, piercing the tip of the sac to attach finally to the coccyx (Fig. 4.80).

The middle meningeal covering is a shiny membrane, the *arachnoid* mater; it lines the internal surface of the dura mater (Fig. 4.80). A substantial space persists between the arachnoid and dura mater, and the central nervous system covered by the pia mater. This is the *subarachnoid space*; it is permeated by a network of threads of arachnoid

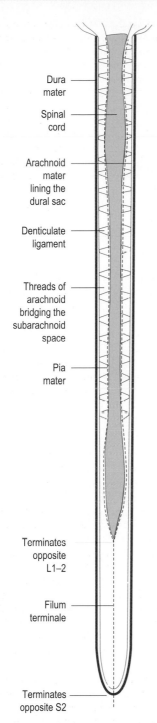

Dura mater

Spinal cord

Arachnoid mater lining the dural sac

Denticulate ligament

Threads of arachnoid bridging the subarachnoid space

Pia mater

Terminates opposite L1–2

Filum terminale

Terminates opposite S2

Figure 4.80 The meningeal coverings of the spinal cord showing the threads of arachnoid mater bridging the subarachnoid space, and the denticulate ligaments holding the spinal cord in place.

mater that connect the arachnoid and dura mater to the pia mater.

Spanning the subarachnoid space is a series of ligaments formed by the pia mater, the *denticulate ligaments* (Fig. 4.80). These triangular, tooth-like extensions of pia mater have their bases attached at regular intervals along the lateral aspect of the spinal cord, and their apices attached to the dural sac. Between the foramen magnum and the tip of the spinal cord, 21 pairs of ligaments project to the dural sac, thereby anchoring the spinal cord and protecting it from injury due to violent contact with the walls of the vertebral canal during movement of the trunk.

Cerebrospinal fluid

Running between the threads of arachnoid mater and filling the subarachnoid space is the *cerebrospinal fluid (CSF)*. The CSF is secreted by strings of capillaries that project into the lateral, third and fourth ventricles of the brain. It fills the cavities of the central nervous system and emerges through apertures in the roof of the fourth ventricle to fill the subarachnoid space surrounding the brain and spinal cord.

CSF is continuously secreted and reabsorbed into the bloodstream. Reabsorption occurs through extensions of the arachnoid mater, the *arachnoid granulations*, which pierce the dura mater of the falx cerebri and project into the superior sagittal sinus (see Fig. 5.38B).

Chemically, the CSF protects the central nervous system by maintaining a constant pH environment; mechanically, it endows the central nervous system with a cushioning fluid environment. Within the skull the brain is buoyant in a pool of CSF, and within the dural sac, the spinal cord is also suspended in a pool of CSF, being held centrally within it by the denticulate ligaments.

Nerve root sheaths

To reach the peripheral nervous system, the spinal nerve roots must penetrate the meninges. As they leave the spinal cord, the proximal ends of the spinal nerve roots are invested by the pia mater (Fig. 4.81); this is one of the morphological reasons for classifying these nerves as part of the central nervous system. The pia mater spreads like a tubular extension over the surface of each nerve, from its point of attachment to the spinal cord as far as where the nerve leaves the vertebral canal. As the spinal nerve roots leave the dural sac, they take a funnel-shaped extension of dura and arachnoid mater with them (Fig. 4.81). These *dural sleeves* extend as far as the spinal nerve, so that the entire lengths of the spinal nerve roots are bathed in CSF.

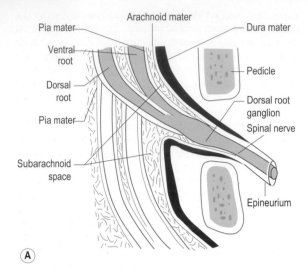

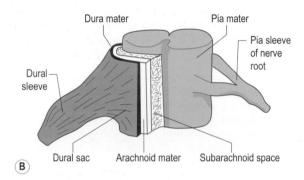

Figure 4.81 The relationship of (A) the meningeal sheaths to the spinal nerves and (B) the meninges to the spinal cord and spinal nerve roots.

Cauda equina

In the early development of the central nervous system, the spinal cord is the same length as the vertebral column, with each spinal nerve running transversely from the spinal cord to its corresponding intervertebral foramen (Fig. 4.82). However, as the fetus develops, differential growth of the vertebral column and spinal cord results in the vertebral column becoming substantially longer than the spinal cord. The dural sac, however, grows to accommodate this difference, so that in the adult, it terminates at the level of S2 while the spinal cord ends opposite the L1–L2 intervertebral disc.

In spite of the differential growth of the spinal cord and vertebral column, the spinal nerves retain their original relationship with their respective intervertebral foramina, and the spinal nerve roots remain attached to their respective spinal cord segments. The difference in length between the spinal cord and vertebral column is accommodated by elongation of the more caudal nerve roots, which become increasingly oblique within the dural sac (Fig. 4.82).

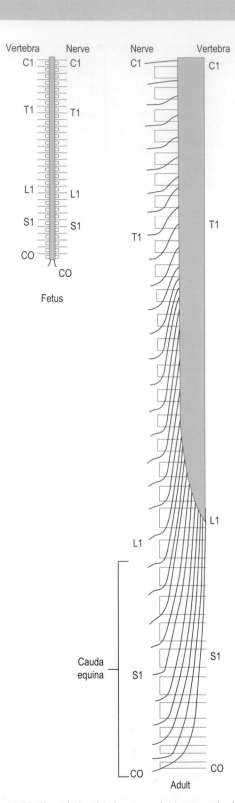

Figure 4.82 The relationship between the intervertebral foramina, spinal nerves and spinal cord in the fetus and adult.

The cervical spinal cord segments remain related to their vertebrae, with the cervical nerve roots more or less retaining their original transverse course. The upper thoracic spinal cord segments are displaced from their respective intervertebral foramina by approximately one segmental level, and progressively lower spinal cord segments are displaced by an increasing margin, until the S5 spinal cord segment lies some 10 vertebral segments short of the S5 intervertebral foramen.

The thoracic, lumbar, sacral and coccygeal nerve roots are therefore progressively longer. Below the caudal tip of the spinal cord, the L2 to Co1 nerve roots form a leash of nerves hanging freely in the dural sac, the *cauda equina* (Fig. 4.82). Each nerve root is ensheathed in its own sleeve of pia mater and is bathed in CSF. As each set of dorsal and ventral nerve roots descends to its respective intervertebral foramen, it passes to the lateral side of the dural sac and penetrates it, as a rule at a level just above the appropriate intervertebral foramen.

Arterial supply

The arterial supply of the central nervous system is derived from the *internal carotid* and *vertebral arteries*.

The vertebral arteries pierce the dural sac and enter the subarachnoid space just above the first cervical vertebra. Each passes forwards and upwards through the foramen magnum to join in the midline in front of the medulla oblongata forming the *basilar artery* which runs upwards in front of the pons (see Fig. 5.40). Branches of the vertebral and basilar arteries pass laterally to supply the medulla oblongata, pons and cerebellum.

Each vertebral artery gives off an *anterior spinal artery* that descends obliquely across the front of the medulla oblongata towards the spinal cord (see Fig. 5.41). The two anterior spinal arteries eventually unite to form a single anterior spinal artery that descends along the anterior median sulcus. Most of the blood supply to the spinal cord is derived from this single vessel (Fig. 4.83). Other smaller branches of the vertebral artery descend along the posterolateral aspect of the spinal cord, the *posterior spinal arteries* supplying only the posterolateral corners of the spinal cord (Fig. 4.83). At variable distances along the spinal cord, the posterior and anterior spinal arteries are reinforced by vessels that pass along the spinal nerve roots. In the cervical region these vessels are derived from the vertebral artery; at thoracic levels they are derived from posterior intercostal arteries, and at lumbar levels from the lumbar arteries (Fig. 4.83).

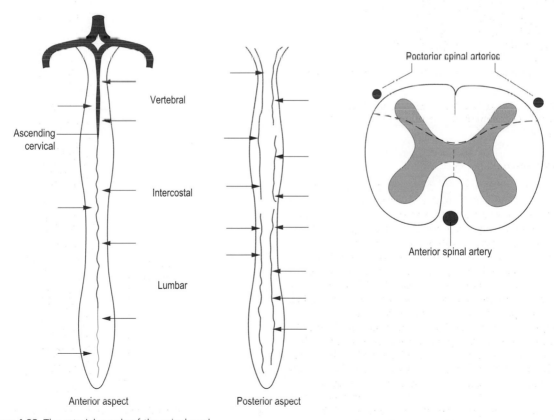

Figure 4.83 The arterial supply of the spinal cord.

The somatic nervous system

This is formed by the peripheral branches of the cranial and spinal nerves. These branches convey motor axons to the muscles of the body and sensory fibres to the central nervous system. That portion of the peripheral nervous system formed by the spinal nerves is described in the following section.

Each spinal nerve is a bundle of sensory and motor axons connected to the spinal cord by the dorsal and ventral nerve roots. Because the ventral and dorsal roots lie within the vertebral canal and within the dural sac, they are classified as part of the central nervous system. The peripheral nervous system is formed by the branches of the spinal nerves outside the vertebral column.

Immediately beyond the intervertebral foramen, each spinal nerve divides into two branches called the *dorsal* and *ventral ramus* (Fig. 4.84). With the exception of the first two cervical spinal nerves, the ventral ramus of each spinal nerve is substantially larger than the dorsal ramus.

All dorsal rami pass backwards into the tissues of the back, innervating principally the back muscles, and the ligaments and joints of the vertebral arches. Most also supply cutaneous branches to the skin over the posterior aspect of the head, neck, trunk and pelvic girdle. The dorsal rami of C1, L4 and 5 lack cutaneous branches, while the cutaneous branches of C4–C6 are inconstant. However, that is not to say that these dorsal rami are entirely motor, for although lacking cutaneous branches they nonetheless convey sensory fibres from the muscles, joints and ligaments that they innervate.

The ventral rami of the spinal nerves supply the sides and anterior parts of the body wall, the limbs and the perineum. In the thoracic region the ventral rami, in general, have a simple arrangement. Except for the first, each thoracic ventral ramus forms a typical intercostal nerve that passes laterally from the intervertebral foramen under the rib of the same segment (Fig. 4.85). The first thoracic ventral ramus forms a small branch that constitutes the first intercostal nerve, but the main branch of the ventral ramus passes over the first rib to join in the formation of the brachial plexus (p. 183). Peripherally, each intercostal nerve innervates the muscles of its intercostal space and the skin overlying that space. The lower six intercostal nerves extend onto the anterior abdominal wall to innervate, in a segmental fashion, the muscles of the anterior abdominal wall and the overlying skin.

At cervical, lumbar and sacral levels, the simple, segmental pattern seen at thoracic levels is modified as the ventral rami form plexuses. A plexus is a network of interconnections between several adjacent ventral rami that allow them to exchange nerve fibres before eventually forming discrete peripheral nerves. The cervical, lumbar and sacral ventral rami form five named plexuses on either side of the vertebral column (Fig. 4.86):

1. The *cervical plexus* is formed by the C1–C4 ventral rami. The peripheral nerves derived from it are distributed to the prevertebral muscles, levator scapulae, sternomastoid, trapezius, the diaphragm and the skin of the anterior and lateral aspects of the neck from the level of the shoulder to the lower jaw and external ear (see pp. 486–489).

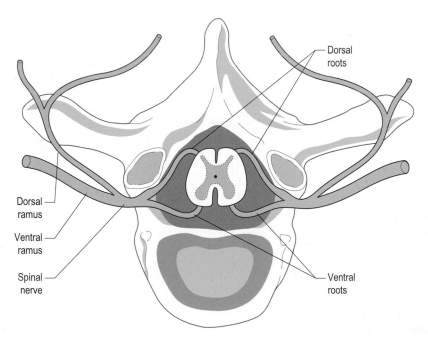

Figure 4.84 The dorsal and ventral rami of a typical spinal nerve.

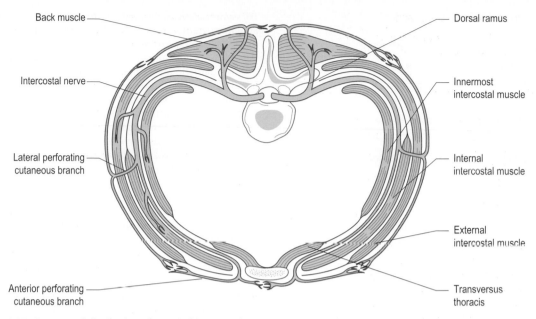

Back muscle

Intercostal nerve

Lateral perforating
cutaneous branch

Anterior perforating
cutaneous branch

Dorsal ramus

Innermost
intercostal muscle

Internal
intercostal muscle

External
intercostal muscle

Transversus
thoracis

Figure 4.85 Course and distribution of a typical intercostal nerve.

2. The *brachial plexus* is formed by the C5 to T1 ventral rami. Nerves derived from this plexus innervate the musculature and joints of the pectoral girdle and upper limb, and the skin of the upper limb (see pp. 183–194).

3. The *lumbar plexus* is formed by the L1–L4 ventral rami with a contribution from the T12 ventral ramus. The smaller nerves derived from this plexus innervate the muscles of the lower anterolateral abdominal wall and the skin of the groin, lateral thigh and anterior parts of the external genitalia. The femoral and obturator nerves are the largest of the lumbar plexus and innervate muscles and skin in the lower limb (see pp. 385–389).

4. The *lumbosacral plexus* is formed by the L4 to S3 ventral rami. Branches of this plexus innervate muscles and skin of the lower limb (pp. 389–395).

5. The *sacral plexus* is formed by the S3–S5 ventral rami and gives rise to nerves that innervate the pelvic floor and the perineum (p. 395).

A knowledge of the distribution of the major peripheral nerves is necessary in clinical practice for the diagnosis and assessment of peripheral nerve injuries and other neurological disorders. The course and distribution of the various peripheral nerves are described in more detail in the appropriate chapters dealing with the regions of the body in which they are found. Their cutaneous distribution follows and complements the segmental dermatome arrangement.

The muscles of the pelvic and shoulder girdles are not grouped into compartments and are not covered by such rules. These muscles usually receive a unique nerve supply or a single nerve may supply two or three muscles. These

relationships are described in the chapters dealing with the muscles of the pelvic and shoulder regions.

Section summary

Spinal cord

- Continuation of the medulla oblongata from the brain to the conus medullaris at the level of L1/L2.
- Long tubular structure lying within the vertebral canal.
- Has cervical and lumbar enlargements.
- Central H-shaped core of grey matter (cell bodies) presenting ventral (anterior) and dorsal (posterior) horns.
- White matter (axons) surrounds grey.

Meninges

- Three coverings: outer fibrous dura mater; middle arachnoid mater; inner pia mater.
- Arachnoid mater separated from pia mater by subarachnoid space containing cerebrospinal fluid.
- Dural sac ends at S2 where pia mater continues as filum terminale to attach to coccyx.

Arterial supply

- From single anterior and paired posterior spinal arteries plus adjacent arteries.

Spinal nerves

- 31 pairs, each pair associated with a spinal cord segment; 8 cervical (C), 12 thoracic (T), 5 lumbar (L), 5 sacral (S), 1 coccygeal (Co).
- Formed within the vertebral canal by ventral (motor) and dorsal (sensory) roots from the spinal cord.

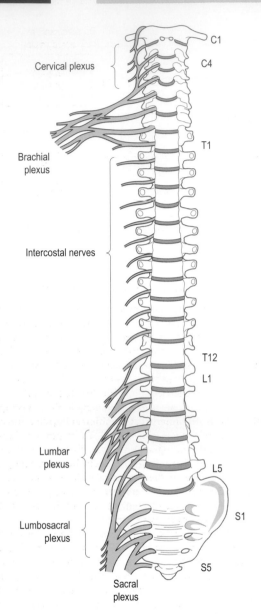

Cervical plexus

C1

C4

Brachial plexus

T1

Intercostal nerves

T12

L1

Lumbar plexus

L5

Lumbosacral plexus

S1

S5

Sacral plexus

Figure 4.86 The major somatic nerve plexuses of the body.

The autonomic nervous system

This consists of those nerves that innervate the viscera of the body, its blood vessels, the salivary and lacrimal glands and the sweat glands and muscles of the hairs (arrectores pilorum). On topographical, anatomical and physiological grounds the autonomic nervous system is divided into two parts: the *sympathetic nervous system* and the *parasympathetic nervous system.*

Topographically, the two systems differ with respect to where they are connected to the central nervous system.

Parasympathetic nerves emerge from the central nervous system in cranial nerves III, VII, IX and X, and in the S2–S4 spinal nerves. Axons of the sympathetic nervous system emerge from the spinal cord in the T1–L2 spinal nerves. Parasympathetic nerves are, therefore described as having a cranial-sacral outflow, and sympathetic nerves as having a thoracolumbar outflow.

Anatomically, both systems are similar in that at the microscopic level any target organ is connected to the central nervous system by two neurons in series (Fig. 4.87). The axon of the first nerve emerges from the central nervous system and synapses with the second neuron, with the collections of the cell bodies of the second neurons forming swellings (ganglia). Axons conveying information from the central nervous system to such ganglia are *preganglionic axons*, while the neurons that form the ganglia and whose axons lead from them to the peripheral target organs are *postganglionic axons*.

The anatomical difference between the sympathetic and parasympathetic nervous systems is that parasympathetic ganglia lie close to the target organ, while sympathetic ganglia lie some distance away. Consequently, postganglionic parasympathetic fibres are quite short, while postganglionic sympathetic fibres are considerably longer.

Pharmacologically, the parasympathetic and sympathetic nervous systems differ in that parasympathetic postganglionic neurons have *acetylcholine* as their transmitter substance, while sympathetic postganglionic neurons have *noradrenaline*. However, preganglionic neurons in both systems have *acetylcholine* as their transmitter (Fig. 4.87).

Physiologically, the autonomic nervous system exerts a variety of effects on different types of tissues; however in general these are all exerted as a result of contraction or relaxation of smooth muscle or of myoepithelial cells in exocrine glands. As a rule, acetylcholine secreted by parasympathetic neurons excites smooth muscle, causing it to contract. Noradrenaline can either excite or inhibit smooth muscle depending on the nature of the molecular receptors found on the smooth muscle membrane. Receptors, known as α-receptors, cause smooth muscle to contract; β-receptors cause it to relax. Different tissues are endowed with either α- or β-receptors, and some tissues contain both types of receptors in different regions or even in the same region. These variations account for the diversity of effects of the sympathetic nervous system.

The sympathetic nervous system stimulates the sinuatrial node of the heart thereby increasing heart rate. It also stimulates cardiac muscle directly to increase the power of cardiac contraction. In the peripheral vascular system the effects of the sympathetic nervous system are variable. Acting on α-receptors, sympathetic nerves cause the smooth muscle of arterioles to contract, resulting in vasoconstriction. Acting on β-receptors, sympathetic nerves cause vasodilatation. In the digestive tract, the sympathetic nervous system relaxes the smooth muscle of the tract, but stimulates its sphincters to contract. In the respiratory tract it

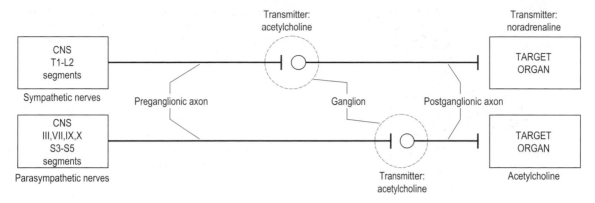

Figure 4.87 Schematic representation of the organization of the sympathetic and parasympathetic nervous systems, also showing the neurotransmitter at each site of synapse.

relaxes the smooth muscle of the bronchi. It causes the sphincter of the bladder to contract, and mediates ejaculation in the male by causing contraction of the smooth muscle in the vas deferens, seminal vesicle and prostate gland. In the eye it causes dilatation of the pupil, and in the skin it causes erection of the hairs and sweating.

These diverse effects may be summarized by an aphorism – the effects of the sympathetic nervous system are those that would be evident in the functions of Fright, Flight, Fight and Fill. The reaction of fright is manifest by pupillary dilatation, hair 'standing on end', sweating and increased heart rate. Flight (running away) and fight involve increased heart rate, increased circulation (vasodilatation) to muscular blood vessels, and diversion (vasoconstriction) of blood away from the digestive tract. All of these actions are mediated by the sympathetic nervous system. The function of 'fill' refers to the relaxation of bronchial smooth muscle to increase air entry, allowing the lungs to fill, and the combination of relaxation of the smooth muscle of the digestive tract and contraction of its sphincters resulting in filling of its lumen.

In a sense, the parasympathetic nervous system exerts opposite effects. It slows the sinuatrial node, reducing heart rate. In the eye, it causes pupillary constriction and accommodation of the lens. It causes contraction of bronchial musculature, and contracts the smooth muscle of the digestive and urinary tracts while relaxing their sphincters. Its action on hollow organs is thus designed to empty them. This emptying effect is further evident in the salivary and lacrimal glands, where the parasympathetic nervous system causes secretion of the respective fluids. Its effect on glands is also evident in the respiratory and digestive tracts where it causes secretion of mucus and acid. The parasympathetic nervous system also has effects on blood vessels. It causes vasodilatation in the genital system to enable erection of the penis or clitoris, and mediates vasodilatation of the internal and external carotid circulations.

The parasympathetic nervous system

The cell bodies of preganglionic parasympathetic neurons are located in special nuclei in the brainstem associated with the nuclei of cranial nerves III, VII, IX and X. The axons of these neurons emerge from the brainstem in these cranial nerves, travelling within them together with their sensory and somatic motor fibres (see pp. 581–584).

Located near the back of the orbit close to the optic nerve is the *ciliary ganglion*, formed by postganglionic parasympathetic neurons that innervate the sphincter pupillae and ciliary muscles of the eye (Fig. 4.88A). These postganglionic neurons receive preganglionic neurons from the oculomotor (III) nerve which synapse on them in the ciliary ganglion. Stimulation of the postganglionic neurons causes the sphincter pupillae to contract, narrowing the pupil, and the ciliary muscle to contract, thereby increasing the refractive (focusing) power of the lens.

Located in the pterygopalatine fossa and associated with the maxillary nerve is the *pterygopalatine ganglion*, formed by postganglionic parasympathetic neurons that innervate the mucous glands of the nose and the lacrimal gland (Fig. 4.88A). It receives preganglionic neurons from the facial (VII) nerve which when stimulated cause lacrimation and nasal and palatine secretion.

The submandibular and sublingual salivary glands are innervated by postganglionic parasympathetic nerves which form a ganglion located in the submandibular region called the *submandibular ganglion* (Fig. 4.88A). This receives preganglionic parasympathetic neurons from the facial nerve, via the chorda tympani.

The parotid gland is innervated by postganglionic parasympathetic neurons whose cell bodies form the *otic ganglion* located just below the foramen ovale associated with the mandibular division of the trigeminal nerve. It receives preganglionic neurons predominantly from the glossopharyngeal (IX) nerve with a small contribution from the facial (VII) nerve (Fig. 4.88A).

501

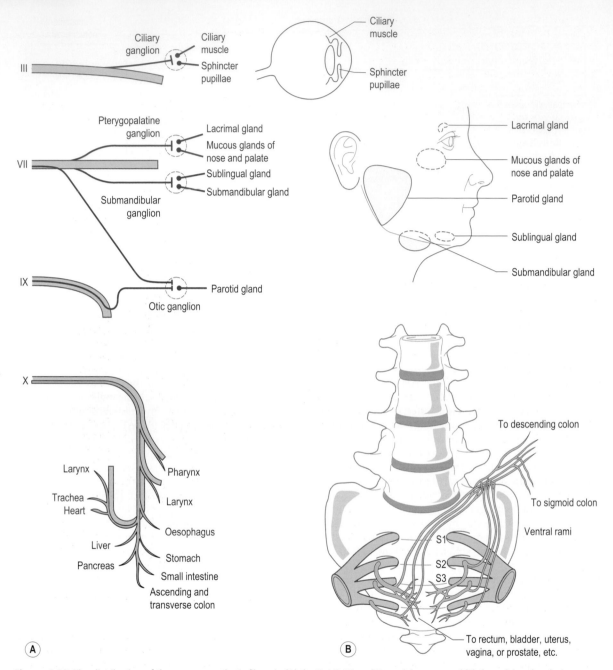

Figure 4.88 The distribution of the parasympathetic fibres in (A) the III, VII, IX and X cranial nerves and (B) the pelvic splanchnic nerves.

The parasympathetic distribution of the vagus nerve includes the heart, the larynx, trachea and bronchi, the pharynx, oesophagus, stomach, small intestine, the large intestine as far as the splenic flexure, as well as the liver, biliary tract and pancreas (Fig. 4.88B). Both vagus nerves convey preganglionic axons to these organs. Postganglionic fibres that innervate the sinuatrial node arise from ganglia lying on the surface of the heart, but the postganglionic fibres of the respiratory, digestive and urinary tracts lie embedded in the walls of these tracts.

The descending colon, sigmoid colon, rectum and pelvic viscera are innervated by parasympathetic neurons that arise from the S2–S4 spinal nerves. These neurons have their cell bodies in the S2, 3 and 4 segments of the spinal cord: their axons travel in the ventral roots of the S2, 3 and 4 spinal nerves. As the ventral rami of these nerves emerge from the anterior sacral foramina, the parasympathetic axons leave them to pass directly to their target viscera. These visceral branches of the S2, 3 and 4 ventral rami are known as the *pelvic splanchnic nerves* (Fig. 4.88B).

Those fibres destined for the pelvic viscera pass directly into them where they synapse on postganglionic parasympathetic neurons located within the walls. Those destined for the descending and sigmoid colon pass upwards out of the pelvis and pass across the left posterior abdominal wall to reach the descending colon, or enter the sigmoid mesocolon to reach the sigmoid colon.

The sympathetic nervous system

Running longitudinally either side of the vertebral column are two chains of nerve fibres called the sympathetic trunks (chains). They extend from the level of C1 to the coccyx crossing the tips of the cervical transverse processes, the heads of the ribs at thoracic levels, and the anterolateral aspects of the vertebral bodies at lumbar and sacral levels (Fig. 4.89). These trunks are formed by preganglionic and postganglionic neurons of the sympathetic nervous system: sites along the trunks where preganglionic and postganglionic neurons synapse are marked by swellings called the *sympathetic ganglia*.

Embryologically, a sympathetic ganglion is formed opposite each spinal nerve, but later in development certain ganglia divide and reform in a variable pattern to give a lesser number of ganglia. Eventually, about 11 thoracic, between one and six, but usually four lumbar, and four sacral ganglia are formed. In front of the coccyx the two sympathetic trunks join to form a single common ganglion known as the ganglion impar. At cervical levels, two main ganglia are formed: a large one at the C1–C3 vertebral level, called the *superior cervical ganglion*, and one at the cervicothoracic level called the cervicothoracic ganglion, or the *stellate ganglion* because of its star-like shape. A third ganglion, the *middle cervical ganglion*, is frequently, but not constantly, formed at the C6 level.

The cell bodies of preganglionic sympathetic neurons are located in the lateral horns of the grey matter in the T1–L2 segments of the spinal cord. Their axons leave the spinal cord in the ventral roots of the spinal nerves at these same levels (Fig. 4.90). Preganglionic sympathetic neurons are not found in spinal nerves or nerve roots at other levels.

After traversing the spinal nerve, preganglionic sympathetic neurons enter the ventral ramus of the spinal nerve, but just beyond the intervertebral foramen they leave the ventral ramus, forming a branch that enters the adjacent sympathetic trunk. This branch conveys preganglionic neurons from the ventral ramus *to* the sympathetic trunk (Fig. 4.90). Because it is a branch, it is called a ramus, and because it communicates with the sympathetic trunk, it is called a ramus communicans. Furthermore, because

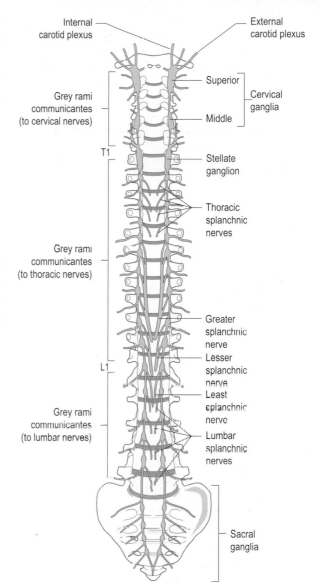

Figure 4.89 The sympathetic trunks and their branches.

preganglionic neurons are myelinated, in fresh specimens they take on a whitish appearance, and so the branch is formally known as a *white ramus communicans* (plural: rami communicantes).

Upon entering the sympathetic trunk, preganglionic sympathetic neurons terminate or pass upwards or downwards within the trunk. Preganglionic neurons from the white rami communicantes, derived from upper thoracic ventral rami as a rule, tend to pass upwards in the sympathetic trunk to reach cervical levels before terminating. Those from lower thoracic and lumbar white rami communicantes tend to pass downwards to lower lumbar and

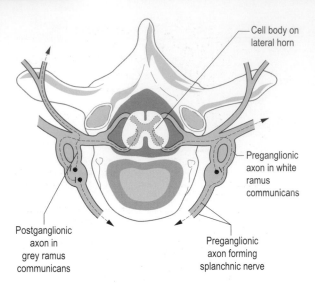

Cell body on lateral horn

Preganglionic axon in white ramus communicans

Postganglionic axon in grey ramus communicans

Preganglionic axon forming splanchnic nerve

Figure 4.90 The arrangement of the white and grey rami communicantes, showing the course of both the pre- and postganglionic sympathetic axons.

sacral levels within the trunk before terminating. Those from mid-thoracic white rami communicantes run only short distances up or down the sympathetic trunk, or terminate at their level of entry into the trunk.

When they terminate, preganglionic sympathetic neurons do so by synapsing with the cell bodies of postganglionic sympathetic neurons located in the sympathetic ganglia. The axons of postganglionic sympathetic neurons then leave the sympathetic trunk, or pass upwards or downwards within it before leaving.

Postganglionic sympathetic neurons leave the sympathetic trunk in any of three ways. Most do so by joining a ventral ramus. Others leave the trunk to follow arteries, or form distinct branches of the sympathetic trunk that pass directly to certain viscera or plexuses (Fig. 4.89).

Postganglionic neurons that join a ventral ramus do so by forming a small branch that passes *from* the sympathetic trunk to the ventral ramus (Fig. 4.90). Such branches are known as rami communicantes, but because postganglionic sympathetic neurons are unmyelinated, they appear greyish, and the branch to the ventral ramus is known as a *grey ramus communicans*. Grey rami communicantes leave the sympathetic trunk at all levels, so that *every* ventral ramus receives a grey ramus communicans. This is distinct from the pattern exhibited by white rami communicantes, which are found only at the T1–L2 levels. Moreover, it is evident that whereas every ventral ramus is connected to the sympathetic trunk by a grey ramus communicans, those at T1–L2 levels have both grey and white rami communicantes. White and grey rami cannot be distinguished by the naked eye, but generally grey rami join the ventral rami slightly proximal to the origins of the white rami.

Once they join the ventral ramus, some of the postganglionic sympathetic neurons assume a short recurrent course to enter the dorsal ramus of the spinal nerve, but the majority pass distally within the ventral ramus. The postganglionic neurons use the course of the dorsal rami and ventral rami to reach their destinations, which are principally the blood vessels in the tissues supplied by the ventral and dorsal rami. In addition, some postganglionic sympathetic neurons follow the course of cutaneous branches of the somatic nerves to reach sweat glands in the skin and the arrectores pilorum muscles.

Because the skin and deep tissues of the head are supplied by cranial nerves and not by spinal nerves, the blood vessels and sweat glands of the head cannot receive their sympathetic nerve supply via grey rami communicantes. A different strategy is used to innervate these structures.

As they run rostrally in the neck, the common, internal and external carotid arteries run close to the cervical portion of the sympathetic trunk. Postganglionic sympathetic neurons destined to supply structures in the head leave the sympathetic trunk and join these arteries. Along the arteries they either form distinct nerves that run parallel to the artery (like the internal carotid nerves), or plexuses that weave around the artery. The plexuses are named according to the artery around which they wind. Thus, internal and external carotid plexuses are formed.

The plexuses and nerves follow the course of their respective arteries, giving rise to branches that follow the branches formed by the artery. Consequently, generally any structure in the head derives its sympathetic innervation from the sympathetic plexus on the nearest available main artery.

A modification to this rule relates to the sympathetic innervation of the eye. The internal carotid nerves accompany the internal carotid artery through the carotid canal and foramen lacerum as far as the cavernous sinus. Here some sympathetic nerves leave the internal carotid artery to pass independently through the superior orbital fissure to reach the ciliary ganglion. Others leave the internal carotid artery to join the ophthalmic division of the trigeminal nerve, eventually entering its nasociliary branch. Sympathetic nerves reach the eye either through the short ciliary branches of the ciliary ganglion or along the long ciliary branches of the nasociliary nerve. Within the eye they are distributed to the dilator pupillae muscle and the blood vessels within the eye.

Sympathetic nerves to thoracic, abdominal and pelvic viscera pass directly from the sympathetic trunk as the *splanchnic nerves*. Splanchnic nerves to the heart and lungs are formed by postganglionic sympathetic axons that arise from the upper four thoracic sympathetic ganglia (Fig. 4.89). They form plexuses around the bronchi and coronary arteries which they follow to reach their respective destinations.

The abdominal viscera are innervated by the *greater, lesser* and *least splanchnic nerves* which have variable origins from the lower seven or eight thoracic segments of the spinal cord (Fig. 4.89). All three nerves enter the abdominal cavity by piercing the ipsilateral crus of the diaphragm. On entering the abdomen they form dense plexuses surrounding the coeliac artery, the renal arteries and the abdominal aorta (Fig. 4.91).

The greater, lesser and least splanchnic nerves are formed by preganglionic sympathetic axons that pass through the sympathetic trunk without synapsing. Instead they synapse with postganglionic neurons whose cell bodies form large ganglia surrounding the coeliac and renal arteries, known respectively as the *coeliac* and *aorticorenal ganglia*. The axons of these postganglionic neurons reach their target organs by forming plexuses that surround and follow the arteries to the abdominal viscera, such that any abdominal organ receives its sympathetic nerve supply along the artery that supplies that organ.

One modification to this pattern relates to the adrenal gland. Embryologically, the medulla of the adrenal gland has the same origin as the sympathetic nervous system, and its cells are equivalent to postganglionic sympathetic neurons. The adrenal gland, therefore, is innervated directly by preganglionic neurons from the coeliac plexus, and not by postganglionic neurons.

The pelvic viscera receive their sympathetic innervation from postganglionic neurons that descend from the coeliac plexus along the aorta, forming the abdominal aortic plexus. This plexus is supplemented by postganglionic neurons from the lumbar splanchnic nerves that arise from the lumbar sympathetic ganglia. At the level of the L4 vertebra, the aorta terminates, but its sympathetic plexus continues on to the anterior surface of L5 and the promontory of the sacrum as the *superior hypogastric plexus* (Fig. 4.91). Leashes of nerves derived from this plexus descend either side of the sacrum with the internal iliac arteries, joining the pelvic splanchnic nerves to form the *inferior hypogastric plexuses*. From these latter plexuses, postganglionic sympathetic neurons enter the bladder, rectum and other pelvic viscera.

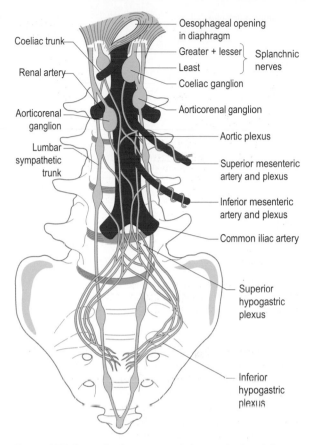

Figure 4.91 Sympathetic nerves and plexuses in the abdomen and pelvis.

Section summary

Autonomic nervous system

Function and organization

- Innervates viscera, blood vessels and glands.
- Preganglionic fibres pass from the central nervous system to synapse in a ganglion.
- Postganglionic fibres pass to target organ/structure.
- Divided into sympathetic and parasympathetic parts.

Parasympathetic nervous system

- Craniosacral outflow with preganglionic fibres in III, VII, IX, X cranial nerves and S2, 3, 4.
- III fibres innervate ciliary muscles and sphincter pupillae (iris) of eye.
- VII fibres innervate nasal, lacrimal, submandibular and sublingual salivary glands.
- IX fibres innervate parotid salivary gland.

- X fibres are distributed to heart, lungs, trachea, bronchi, oesophagus, small intestine, large intestine, liver and pancreas.
- S2, 3, 4 (pelvic splanchnic nerves) are distributed to the descending and sigmoid colon, rectum and pelvic viscera.

Sympathetic nervous system

- Preganglionic fibres emerge from spinal cord between T1 and L2 (thoracolumbar outflow).
- Enter sympathetic chain where they synapse with cell bodies of postganglionic fibres.
- Sympathetic chain extends whole length of and is parallel to vertebral column.
- Postganglionic fibres distributed to all regions and structures within body.
- Some fibres (from T5–T12) form splanchnic nerves which pass into abdomen.

Parasympathetic

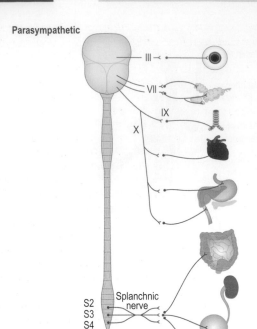

Target organ	Effects
Pupil of eye	Constriction
Lacrimal glands	Increased tear secretion
Salivary glands	Increased secretion of saliva
Trachea and bronchi	Bronchoconstriction
Heart Coronary arteries	Decreased heart rate and contraction Vasoconstriction
Stomach	Increased secretion of gastric juice Increased peristalsis
Liver	Increased bile secretion/blood vessels dilated
Pancreas	Increased secretion of pancreatic juice
Small and large intestine	Increased peristalsis Sphincters relaxed Increased digestion and absorption
Kidney	Increased urinary output
Bladder	Smooth muscle wall contracted Sphincter relaxed
Sex organs/genitalia	Generally vasodilation/erection of tissues in both sexes

Sympathetic

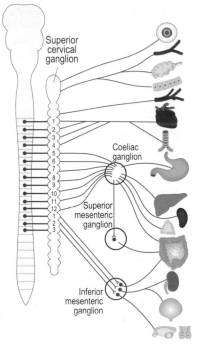

Target organ	Effects
Iris of eye	Pupil dilation (increased visual acuity)
Blood vessels to head	Vasoconstriction
Salivary glands	Inhibition (dry mouth)
Mucosa of nose/mouth	Inhibition (dry mouth and nose)
Vascular supply to muscles	Vasodilation (increased muscle activity)
Heart Coronary arteries	Cardiac output increased (rate and volume) Vasodilation
Trachea and bronchi	Bronchodilation (increased O_2 uptake)
Stomach	Reduced peristalsis/sphincters closed
Liver	Increased conversion of glycogen to glucose
Spleen Adrenal medulla	Contraction Increased secretion of adrenaline/ noradrenaline
Small and large intestine	Decreased peristalsis and tone Sphincters closed/blood vessels constricted
Kidney	Decreased urine output
Bladder	Smooth muscle wall relaxed/sphincter closed
Sex organs/genitalia	Generally vasoconstriction/male ejaculation

THE CARDIOVASCULAR SYSTEM

Introduction

The cardiovascular system comprises the heart and the blood vessels (arteries and veins) conveying blood to and from the various body organs and tissues. Although it appears as a single organ the heart is functionally a double muscular pump, the two parts linked to each other via the pulmonary circulation. The right pump receives deoxygenated blood via the superior and inferior venae cavae and coronary sinus, while the left pump delivers oxygenated blood to the aorta for distribution to the body.

Development

The cardiovascular system begins to develop during the third week with all the component parts (heart, blood vessels and blood cells) being derived from the mesodermal germ layer. It is the first system to function in the embryo with blood beginning to flow and the first heart beats occurring by the end of the third week. The area giving rise to the heart is initially anterior to the *prochordal* and neural plates. However, the central nervous system grows so rapidly following closure of the neural tube that the developing heart and *pericardial cavity* come to lie in front of the future foregut (Fig. 4.92A). It is at this time that the single heart tube is formed, and then elongates, developing alternate constrictions and expansions, giving rise to the future parts of the heart.

The identifiable parts of the heart tube are the *horns* of the *sinus venosus*, the *pulmonary atrium*, the *atrioventricular canal*, the *pulmonary ventricle*, the *bulbus cordis* and the *truncus arteriosus*, the latter being continuous with the aortic sac which gives rise to the aortic arches (Fig. 4.92B). Between the fourth and seventh weeks the heart tube undergoes considerable change, with all except the sinus venosus being divided by septa to give the human four-chambered heart.

The right horn of the sinus venosus is incorporated into the right atrium forming the smooth-walled part, while the left horn forms the coronary sinus. The pulmonary atrium contributes the rough-walled parts to both adult atria; the smooth wall in the left is from the incorporation of the pulmonary veins. The pulmonary ventricle forms the adult left ventricle, while the adult right ventricle arises from the proximal part of the bulbus cordis: the middle part gives rise to the outflow tracts of both ventricles, while the distal part forms the roots of the aorta and pulmonary trunk.

The pulmonary atrium is divided by two septa. The *septum primum* grows from the dorsal wall of the atrium towards the *endocardial cushions:* just prior to fusion perforations appear in its upper part, maintaining a communication between the right and left sides (Fig. 4.92C). At the same time the incomplete *septum secundum* grows to the right of the septum primum from the ventral wall of the atrium forming the *foramen ovale* (Fig. 4.92C). Prior to birth the septa remain separate, allowing blood from the inferior vena cava to pass freely from the right to the left atrium. After birth the septa fuse, forming a complete partition between the two atria.

Endocardial cushions gradually project into the atrioventricular canal; fusion of their intermediate parts gives rise to the right (tricuspid) and left (mitral) atrioventricular openings (Fig. 4.92C).

A spiral septum develops to split the bulbus cordis and truncus arteriosus longitudinally, separating the aorta from the pulmonary trunk, becoming continuous with the developing muscular interventricular septum. This division brings the aorta into communication with the left ventricle and the pulmonary trunk with the right ventricle.

Heart malformations are relatively common, the majority being formed by multiple factors, both genetic and environmental. While some congenital cardiac malformations cause very little disability, others are incompatible with extrauterine life: many, however, can be surgically corrected. Abnormalities may involve the interventricular system, the truncus arteriosus and bulbus cordis, and formation of the valves. The heart may be normal but its position may be on the right rather than the left, and may be associated with a complete or partial transposition of the abdominal viscera.

A pair of aortic arches develop in association with each of the six pairs of branchial arches. The important derivatives from these arches are: the *common carotid* and first part of the *internal carotid arteries* from the third aortic arch; the *aortic arch* from the fourth left and the proximal part of the *right subclavian artery* from the fourth right aortic arch; the proximal parts of the right pulmonary artery from the right sixth and similarly on the left as well as the *ductus arteriosus* (a shunt from the left pulmonary artery to the aorta) from the left sixth aortic arch (Fig. 4.93A)

Fetal circulation

In the fetus blood returns from the placenta in the *umbilical vein*, bypassing the liver via the *ductus venosus* and flows directly into the *inferior vena cava*, to mix with deoxygenated blood returning from the lower limbs, abdomen and pelvis (Fig. 4.93B). From the right atrium blood is directed towards the foramen ovale so that most of it enters the left atrium, then to the aorta to supply the head and upper limbs. Blood, entering the right atrium via the *superior vena cava*, passes to the right ventricle and pulmonary trunk. The high resistance to blood flow in the pulmonary vessels results in the blood passing to the descending aorta via the ductus arteriosus. From the descending aorta blood flows towards the placenta via the two *umbilical arteries* to be reoxygenated to begin the cycle once more.

Changes after birth

At birth important circulatory changes occur due to cessation of blood flow through the placenta and the

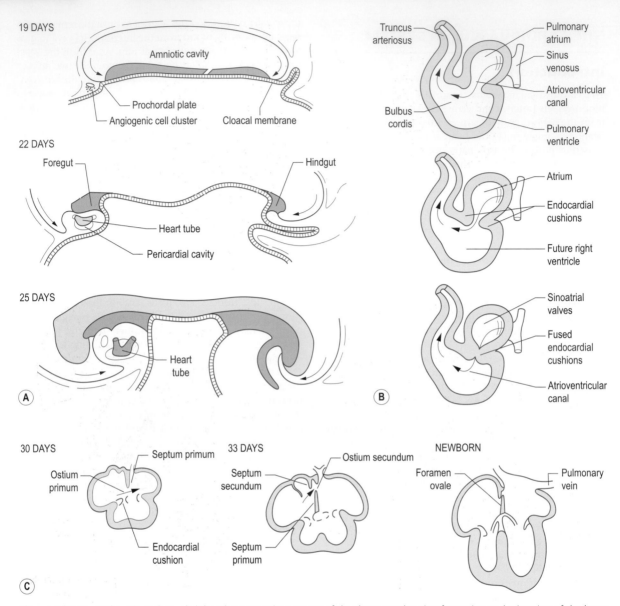

Figure 4.92 Sagittal sections through (A) embryo at various stages of development showing formation and migration of the heart tube and (B) the heart at 4–5 weeks showing the formation of the atrioventricular canal, (C) formation of the interatrial septum.

functioning of the infant's lungs. The foramen ovale closes with the first good breath by the septum primum being pushed against the septum secundum (fusion occurs by the end of the first year). The ductus arteriosus closes with complete anatomical closure by 3 months; the obliterated ductus arteriosus forms the *ligamentum arteriosum* (Fig. 4.93C). The umbilical vein and the ductus venosus both become obliterated after birth, forming the *ligamentum teres* in the free edge of the falciform ligament, and the ligamentum venosum respectively. The distal parts of the umbilical arteries close and become the *medial umbilical ligaments*.

The heart

Surface markings

Although the heart changes its position with respiration and posture, its approximate position within the thorax is as follows. One-third of the heart lies to the right of the midline with the right border extending from the right

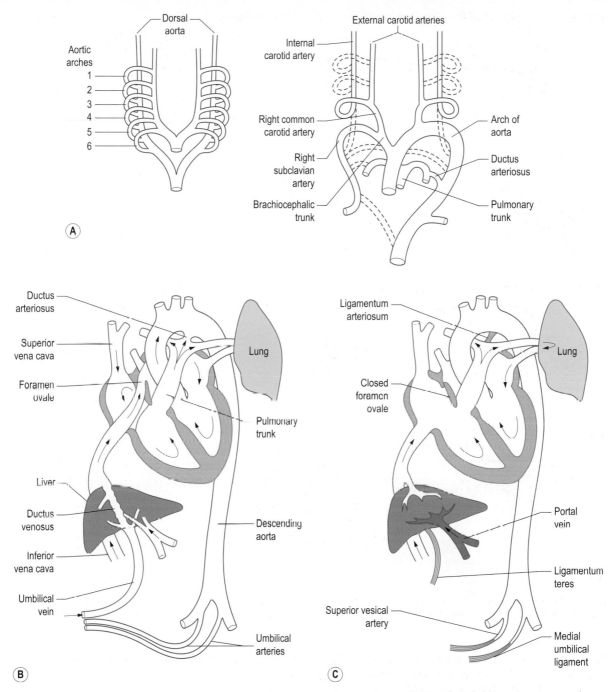

Figure 4.93 (A) The six pairs of aortic arches and their derivatives; pattern of blood flow (B) before birth and (C) after birth.

third to the right sixth costal cartilage, approximately 1 cm to the right of the sternal border. The left border slopes down from the left second intercostal space, 1 cm to the left of the sternal border, to the apex of the heart, which lies in the left fifth intercostal space 9 cm from the midline.

The apex of the heart can usually be palpated in the living subject. The inferior border of the heart sits on the central tendon of the diaphragm and can be indicated by a shallow concavity joining the two inferior points of the right and left borders. Similarly joining the superior points of the right and left borders gives the remaining surface marking (Fig. 4.94A).

The heart itself is pyramidal in shape and presents three surfaces: sternocostal (anterior), diaphragmatic (inferior) and a base (posterior). The sternocostal surface is formed mainly by the right atrium and right ventricle, separated from each other by the atrioventricular groove. The right border of this surface is formed by the right atrium, while the left border is formed by the left ventricle and part of the left atrium. The ventricles are separated by the anterior interventricular groove. The diaphragmatic surface is formed by the right and left ventricles, separated by the posterior interventricular groove, and part of the right atrium. The base of the heart is formed by the left atrium, with a small contribution from the right atrium.

Pericardium

The heart lies within the *middle mediastinum* of the thorax (Fig. 4.94B) surrounded by a double fold of serous membrane *(serous pericardium)* contained within a dense connective tissue sac *(fibrous pericardium)* itself attached to the central tendon of the diaphragm (Fig. 4.94B). The outer (parietal) layer of the serous pericardium is attached to the inner surface of the fibrous pericardium, while the inner (visceral) layer is adherent to the heart wall. These two layers are continuous at the roots of the great vessels (Fig. 4.94C) thus forming a closed space. This pericardial space may become filled with blood or other fluids (cardiac tamponade) which may interfere with the normal functioning of the heart.

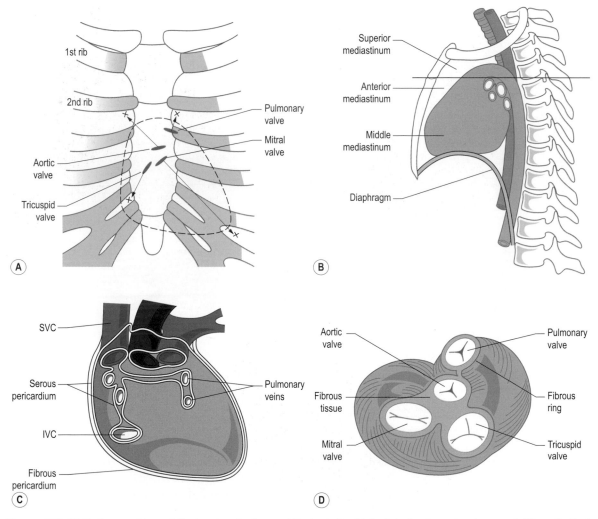

Figure 4.94 (A) Surface markings of the heart and valves and the best sites (x) for listening to each valve sound, (B) mediastinal divisions, (C) posterior aspect of the heart showing the reflection of the serous and fibrous pericardium, (D) fibrous framework of the heart.

Structure

The structure of the greater part of the heart wall is cardiac muscle (myocardium). The serous pericardium together with a thin subserous layer of connective tissue is the epicardium, while the chambers of the heart are lined by endocardium. Within the heart wall is a connective tissue skeleton composed of four firmly connected rings of fibrous tissue providing a relatively rigid attachment for the valves and myocardium (Fig. 4.94D). The myocardium consists of two separate systems of spiralling and looping bundles of fibres, one associated with the atria and the other with the ventricles. Nowhere are the two systems continuous with each other, hence the need for a specialized atrioventricular conducting system.

Blood supply

The arterial supply to the heart is provided by the right and left coronary arteries, branches of the ascending aorta just above its origin from the left ventricle (Fig. 4.95A). The *right coronary artery* arises from the right aortic sinus and passes forward between the pulmonary trunk and the right auricle to enter the atrioventricular groove between the right atrium and ventricle (Fig. 4.95A). It continues in this groove around the inferior margin of the heart, giving off the *right marginal artery*, towards the posterior interventricular groove where it anastomoses with the *circumflex branch* of the *left coronary artery* and gives off the *posterior interventricular artery*, which runs in the posterior interventricular groove (Fig. 4.95B). The right coronary artery gives branches to supply both atria

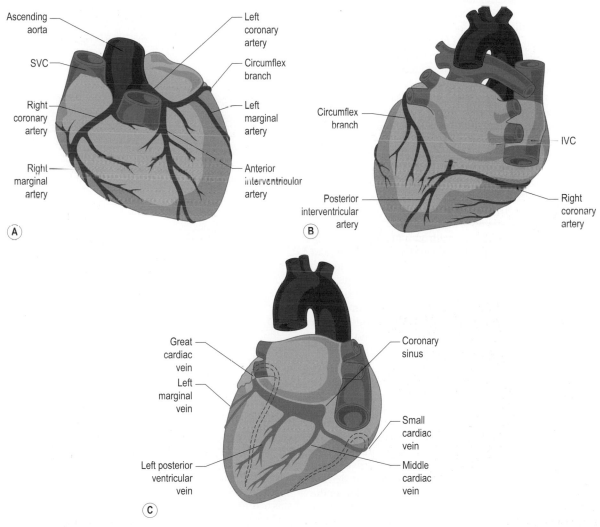

Figure 4.95 Arterial blood supply of the heart: (A) anterior aspect, (B) posterior aspect, (C) venous drainage of the heart viewed posteriorly. IVC, inferior vena cava; SVC, superior vena cava.

and ventricles, as well as providing the major blood supply to the conducting system. The sinuatrial node is supplied by a sinuatrial branch from the right coronary artery in 60–70% of individuals, while the atrioventricular node artery is given off close to the posterior interventricular artery, which in turn supplies the atrioventricular bundle and its branches.

The left coronary artery arises from the left aortic sinus and passes forwards between the pulmonary trunk and left auricle towards the atrioventricular groove. Here it divides into a circumflex branch, which continues around the atrioventricular groove, giving off the *left marginal artery*, towards the posterior of the heart to anastomose with the right coronary artery, and the *anterior interventricular artery*, which passes down in the anterior interventricular groove towards the apex of the heart supplying both ventricles and the interventricular septum (Fig. 4.95A).

The coronary arteries and their larger branches are functional end-arteries; the anastomoses are not large enough to provide an alternative blood supply should the main artery become blocked. Sudden blockage of a coronary artery or one of the large branches is almost invariably fatal. Blockage of smaller vessels leads to necrosis of the heart tissue supplied by that vessel (heart attack). Narrowing of the lumen of the coronary arteries or major branches results in a reduction in the blood supply to the muscle (ischaemia), with the patient complaining of pain behind the sternum; the pain often radiates down the left arm (angina pectoris).

The greater part of the venous drainage of the heart is by a system of veins that drain into the *coronary sinus*, which lies in the posterior part of the atrioventricular groove and empties into the right atrium (Fig. 4.95C). The larger veins follow the course of the arteries; however, their names differ from those of the arteries. The *great cardiac vein* lies in the anterior interventricular groove beginning at the apex of the heart. It receives the *left posterior ventricular vein* and the *left marginal vein* to form the coronary sinus. Close to its termination the coronary sinus is joined by the *small* and *middle cardiac veins* (Fig. 4.95C). There are in addition two or three anterior cardiac veins draining from the surface of the right ventricle directly into the right atrium, and a number of very small veins draining directly into the atria from the myocardium (venae cordae minimae).

Chambers of the heart

The heart consists of four chambers, the right and left atria and the right and left ventricles. The right atrium receives deoxygenated blood from the systemic circulation via the superior and inferior venae cavae and the coronary sinus, which open into the smooth posterior part derived from the sinus venosus (p. 507). The rough anterior part, derived from the pulmonary atrium (p. 507), is separated from the posterior part by the *crista terminalis* (Fig. 4.96A): the thickenings are the musculi pectinati. The superior end of the crista terminalis surrounds the *sinuatrial node*. On the interatrial septum a shallow oval depression *(fossa ovalis)* indicates the site of the fetal foramen ovale that allowed blood to

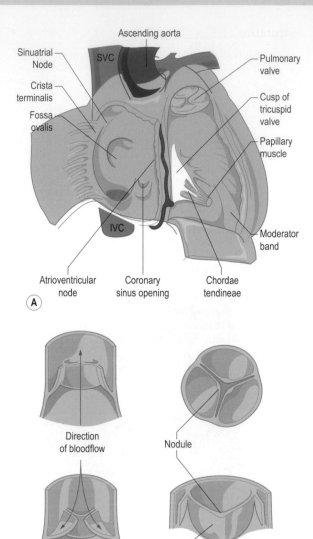

Figure 4.96 (A) The right atrium and ventricle opened showing internal features of each chamber, (B) structure of the pulmonary and aortic valves; their opening and closing is dependent on the direction of blood flow.

enter the left atrium and bypass the pulmonary circulation (p. 507). From the right atrium blood enters the right ventricle via the right atrioventricular (tricuspid) valve.

The right ventricle is separated from the left ventricle by the interventricular septum. The internal surface shows irregular muscle projections (trabeculae carnae), some of which are attached to the ventricular wall along their whole length, while others are free at one end. The latter pass into the ventricular cavity and become attached to the *chordae tendineae* of the *cusps* of the *tricuspid valve*: these are the *papillary muscles* (Fig. 4.96A). The *moderator band* lies partly free in the ventricle and carries the right branch of the atrioventricular bundle of the conducting system. The pulmonary opening, leading to the pulmonary trunk, is guarded by the *pulmonary valve*. During ventricular systole (contraction) the papillary muscles contract and pull on the chordae tendineae and so prevent the cusps of the tricuspid valve from inverting and allowing blood to flow back into the right atrium.

The left atrium has two pulmonary veins opening into it on each side, bringing freshly oxygenated blood from the lungs. Most of the lining is smooth, due to incorporation of the terminal parts of the pulmonary veins. Only the auricle develops from the pulmonary atrium of the fetus. Blood passes into the left ventricle via the left atrioventricular (mitral) valve.

The left ventricle has the thickest walls of all the chambers of the heart, due to the extremely high pressures it has to develop in order to force blood into the systemic circulation. The wall also possesses trabeculae carnae; three papillary muscles attach to the free margins of the cusps of the mitral valve. The aortic opening leading to the aorta is guarded by the aortic valve; above each semilunar valvule is an aortic sinus with the right and left giving rise to the right and left coronary arteries respectively.

The pulmonary and aortic valves each consist of three semilunar valvules attached to the vessel wall, the free border projecting upwards into the vessel lumen. In the middle of the free border a thickened nodule assists in approximating the central areas of the edges of the valvule (Fig. 4.96B). The concavities formed by the valvules face away from the direction of blood flow so that during ventricular systole they lie against the vessel wall. However, in diastole when interventricular pressure falls and there is a tendency for blood to return to the ventricle the valvules fill and approximate their free borders, thereby preventing further blood flow (Fig. 4.96B). The surface markings of the tricuspid, mitral, pulmonary and aortic valves are shown in Fig. 4.94A, which also shows the most appropriate site for listening to each valve with a stethoscope.

Conducting system

The conducting system of the heart consists of specialized cardiac muscle fibres arranged as groups of cells (nodes) or bundles of fibres, which are responsible for the rhythmic initiation and propagation of the impulse associated with the heart beat and the coordinated contraction of the atria and ventricles. Contraction is initiated at the *sinuatrial node* in the right atrium: the impulse spreading throughout the two atria and to the *atrioventricular node* (Fig. 4.97A),

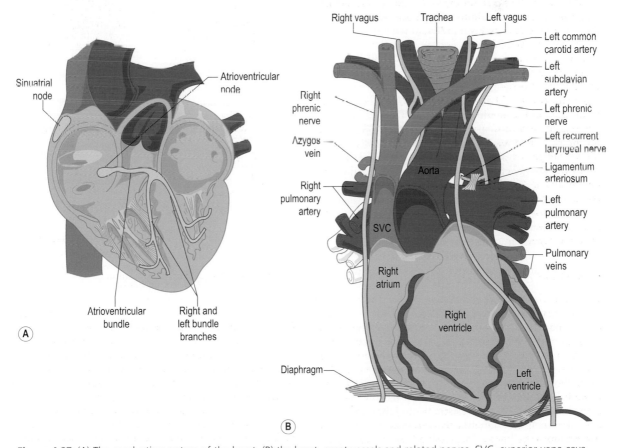

Figure 4.97 (A) The conducting system of the heart, (B) the heart, great vessels and related nerves. SVC, superior vena cava.

which lies in the interatrial septum close to the opening of the coronary sinus. The atrioventricular node passes the impulse to the *atrioventricular bundle* located in the interventricular septum. Within the septum the bundle divides into the *right* and *left bundle branches*. The right bundle branch runs towards the apex of the heart and enters the moderator band to supply the papillary muscles and then breaks up into fine fibres (Purkinje fibres) to supply the remainder of the right ventricle. The left bundle branch breaks up into two or more strands as it passes towards the apex of the heart. If the atrioventricular bundle is interrupted, for example by lack of blood supply, total heart block results in which the ventricles beat slowly and rhythmically at their own rate independent of the atria, which continue to contract at a rate determined by the sinuatrial node.

The nerve supply to the heart is from both the sympathetic and parasympathetic parts of the autonomic nervous system and reaches the sinuatrial node via cardiac plexuses of nerves associated with the arch of the aorta and bifurcation of the trachea. The sympathetic fibres have their cell bodies in the first four or five thoracic spinal nerve segments and reach the heart by the superior and middle cervical sympathetic ganglia; the parasympathetic fibres come from the vagus. Increased sympathetic stimulation increases heart rate while increased parasympathetic stimulation slows heart rate.

The great vessels

The great vessels (Fig. 4.97B) connect the heart to the pulmonary and systemic circulations. The *superior vena cava* (SVC) carries blood from the thorax, upper limbs and the head and neck; while the inferior (IVC) carries blood from the abdomen, pelvis and lower limbs. The superior vena cava is formed by the union of the left and right brachiocephalic veins behind the first costal cartilage. The inferior vena cava passes through the central tendon of the diaphragm and almost immediately enters the right atrium.

The pulmonary trunk divides into the *right* and *left pulmonary arteries* just below the level of the sternal angle and pass towards the lung root. Close to the bifurcation the left pulmonary artery is joined to the underside of the aortic arch by the *ligamentum arteriosum*, the remnant of the fetal ductus arteriosus (p. 508).

The *aorta* leaves the left ventricle and passes upwards, backwards and to the right as far as the level of the sternal angle (the ascending aorta) it then curves over to the front of the vertebral column at the level of the fourth thoracic vertebra (the aortic arch) to pass down through the thorax anterior to the vertebral column (the descending aorta). The aorta passes behind the diaphragm at the level of the 12th thoracic vertebra to enter the abdomen.

Section summary

Heart

A muscular double pump with the right side receiving and pumping deoxygenated blood to the lungs, whilst the left side receives oxygenated blood from the lungs and pumps it to all tissues in the body.

Structure

- Four chambers, two atria (right and left) and two ventricles (right and left).
- Contained within a dense sac of connective tissue (fibrous pericardium) lined with serous pericardium.

Right side

- Right atrium receives venous blood from the superior and inferior venae cavae and the coronary sinus.
- Tricuspid valve connects the right atrium and ventricle.
- Right ventricle pumps blood to the lungs via the pulmonary trunk which has a semilunar valve.

Left side

- Left atrium receives oxygenated blood from the lungs by pulmonary veins.
- Mitral (bicuspid) valve connects the left atrium and ventricle.
- Left ventricle has thickest walls; pumps blood into the aorta via the aortic valve.

Conducting system

- Contraction is initiated at the sinuatrial node, spreads through the atria to the atrioventricular node, then via the atrioventricular bundle which divides and forms Purkinje fibres.
- Sympathetic and parasympathetic stimulation increases and decreases heart rate respectively.

Blood supply

- Arterial supply from right and left coronary arteries from ascending aorta.
- Venous drainage from system of veins running into coronary sinus which drains into right atrium.

THE RESPIRATORY SYSTEM

Introduction

The repiratory system is concerned with the exchange of gases (oxygen, carbon dioxide) between the air in the lungs and the blood in the capillaries of the pulmonary circulation: this is sometimes referred to as external respiration. (Internal respiration occurs between the blood in the systemic capillaries and the cells and tissues of the body organs.) In inspiration, muscle action increases the thoracic diameters so that the pressures within the pleural cavity and lung spaces become less than that of the atmosphere; consequently the lung expands and air rushes in. Expiration, unless forced, is relatively passive being due to the recoil of the lungs, muscular relaxation and atmospheric pressure acting on the chest wall. The maximum amount of air that can be exhaled (vital capacity) following the deepest inspiration is 3000/3500 ml (females/males): the amount of air breathed in and out (tidal volume) during quiet respiration is 500 ml. The total lung surface available for respiratory exchange is 70 m^2.

Rigid or semi-rigid walls in the nasal cavity, pharynx, larynx, trachea and larger bronchi keep the upper airways open at all times. The soft, flexible and elastic lungs respond passively to changes in thoracic diameter, having more room for expansion laterally, inferiorly and anteriorly, where chest enlargement is not limited, than in the apical, upper anterior and posterior mediastinal regions. Respiration predominantly takes place in the peripheral (superficial) zone of the lung most immediately expanded by thoracic enlargement, being some 5 mm deep in quiet and up to 30 mm in forced respiration.

Development

During the fourth week of development the respiratory system (larynx, trachea, bronchi and lungs) appears as an outgrowth from the anterior wall of the primitive pharynx (Fig. 4.98A), the epithelium and glands being of endodermal origin and the connective tissue and muscular components being derived from splanchnic mesoderm surrounding the developing foregut. With expansion of the distal end of the *respiratory diverticulum* to form the *lung buds* a septum forms so that the foregut becomes divided into the oesophagus and the trachea and lung buds (Fig. 4.98A).

The two developing lung buds expand and grow into primitive pleural cavities, which become separated from the pericardial and peritoneal cavities. The *parietal* and *visceral pleura* develop from the mesoderm lining the body wall and that covering the outside of the lungs respectively; they are continuous at the lung root. During the fifth week the right lung bud divides into three main bronchi and the left into two, foreshadowing the adult pattern (Fig. 4.98B). The secondary bronchi undergo progressive branching, giving rise to 10 segmental bronchi in each lung, representing the future bronchopulmonary segments, each of which is surrounded by mesenchymal tissue.

The main bronchi continue to divide repeatedly until the end of the sixth month when there have been some 17 divisions (Fig. 4.98B): an additional six divisions occur postnatally. Until the seventh month the bronchioles continue to divide into smaller and smaller canals, while the vascular supply to the lungs increases. However, respiration is not possible until the cells of the respiratory bronchioles become flat and thin (Fig. 4.98C). By the seventh month

an adequate gas exchange is possible so that a premature infant is capable of surviving. There is considerable growth of the lungs, both of respiratory bronchioles and alveoli, after birth.

In addition to alveolar epithelial cells surrounding the capillaries, other cells produce surfactant, which lowers the surface tension at the air–blood barrier. Prior to birth the lungs are filled with fluid, but with the beginning of respiration most of this fluid is rapidly resorbed by the blood and lymph capillaries, with a small amount being expelled during parturition. However, the surfactant remains in the lungs as a thin lining on the alveolar cell membrane.

Upper respiratory tract

The upper respiratory tract (Fig. 4.99A) is lined by pseudostratified, ciliated, columnar respiratory epithelium beneath which lies lymphoid tissue, further mucous and serous glands and a rich vascular plexus: hence the cleansing, warming and moistening of the inspired air that occurs. In forced respiration or respiratory distress the oral cavity also becomes part of the upper respiratory tract, but the function of the respiratory epithelium is lost.

Nasal cavity

This communicates externally by the nostrils (nares) of the largely cartilaginous external nose and posteriorly with the *nasopharynx* via the choanae, as well as with the paranasal sinuses, which, although they have no respiratory function, are important in vocalization. The nasal septum divides the cavity into two parts, each having a roof, floor, lateral and medial walls, the latter being the septum. Each irregular lateral wall has medial and downward projections (*superior*, *middle* and *inferior conchae*, Fig. 4.99B) thought to cause turbulence of the inspired air. In infection swelling of the thick lining may block the nasal cavity.

Pharynx

A fibromuscular tube extending from the base of the skull to the cricoid cartilage opposite C6, lying behind and communicating with the nasal cavity (nasopharynx), oral cavity (*oropharynx*) and larynx (*laryngopharynx*) (Fig. 4.99C). Its inner epithelial lining (respiratory in the nasopharynx, stratified squamous in the oro- and laryngopharynx) is separated from an incomplete muscular layer (constrictor muscles) by the pharyngobasilar fascia, and surrounded by the buccopharyngeal membrane. The nasopharynx is part of the respiratory system, the oropharynx part of both the respiratory and digestive systems, while the laryngopharynx is part of the digestive system. During swallowing respiration is temporarily suspended as the nasopharynx is closed by raising the soft palate and the laryngeal inlet is partly closed by the epiglottis as the bolus of food passes towards the oesophagus.

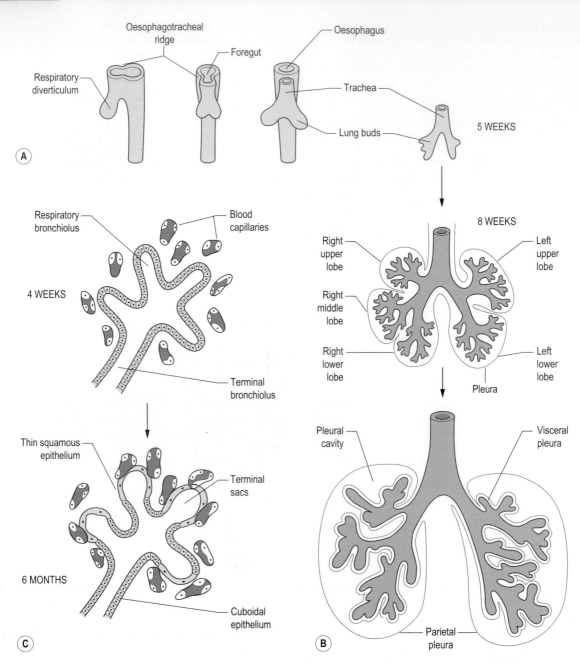

Figure 4.98 Successive stages in the development of (A) the respiratory diverticulum and (B) the trachea and lungs, (C) histological and functional development of the lungs.

Larynx

The framework of cartilages forming the larynx is maintained by membranes and muscles (Fig. 4.100A), with the muscles acting primarily as sphincters for protection of the lower respiratory tract or for increasing intrathoracic pressure, as well as moving the vocal ligaments during respiration and phonation. The larynx narrows at the *vestibular folds*, widens between the vestibular and *vocal folds* and narrows again at the vocal folds (the *rima glottidis*). Below the vocal folds the larynx again widens to become continuous with the *trachea* (Fig. 4.100B).

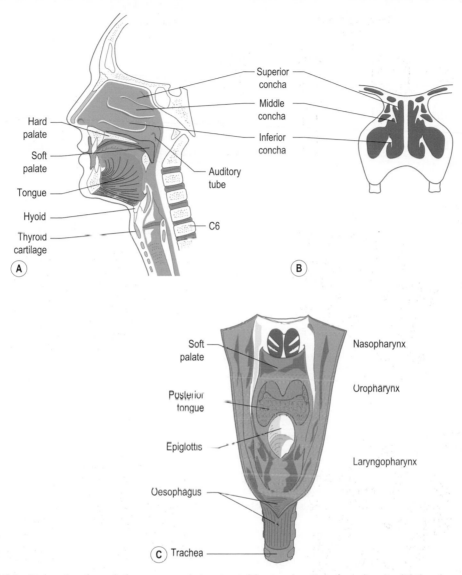

Figure 4.99 (A) Sagittal section through the upper respiratory tract, (B) coronal section of nasal cavity, (C) the pharynx opened and viewed from behind showing the naso-, oro- and laryngopharynx.

Trachea and principal bronchi

The trachea extends from the *cricoid cartilage* to its bifurcation into *right* and *left main bronchi* (Fig. 4.100C) at the level of T4/T5 in the thorax: in full inspiration the bifurcation (carina) extends as low as T6. The fibromembranous trachea and principal bronchi are reinforced by incomplete cartilage rings completed posteriorly by smooth muscle (trachealis). The right bronchus is shorter, straighter, larger and a more direct continuation of the trachea than the longer, narrower left bronchus. inhaled foreign bodies thus tend to pass into the right lung. The principal bronchi divide into secondary bronchi, one for each lobe of the lung, which in turn divide into tertiary bronchi, one for each bronchopulmonary segment.

The lungs and pleura

Lungs

The lungs occupy most of the space in the thoracic cavity, each lung lying free within its pleural cavity, attached only by its root to the mediastinum. Conforming to the outline of the thoracic cage, each lung has an apex, base, costal and mediastinal surfaces, and anterior, inferior and posterior borders (Fig. 4.101). The anterior border of each lung is

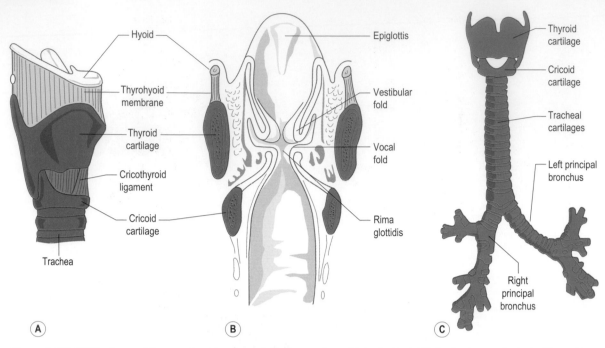

Figure 4.100 (A) The larynx, (B) coronal section through the larynx viewed from behind, (C) the trachea and principal bronchi.

sharp, with that of the left lung having a shallow notch (cardiac notch) below which is the lingula: the posterior border is rounded and lies in the paravertebral gutter. The shorter, wider and heavier right lung is divided into three lobes *(superior, middle* and *inferior)* by two fissures *(oblique* and *horizontal)*, while the smaller left lung has only two lobes *(superior,* including the lingula, and *inferior)* separated by the *oblique fissure* (Fig. 4.101).

The rounded apex of the lung projects into the neck due to the obliquity of the thoracic inlet while the concave base is separated from the liver, on the right, and the liver, spleen and stomach, on the left, by the diaphragm. The medial surface of the right lung is related to the right atrium, oesophagus, azygos vein, trachea, superior vena cava and subclavian artery; that of the left to the left ventricle, oesophagus, arch and descending aorta, subclavian and common carotid arteries (Fig. 4.101).

The lung root is the only part of the lung not covered with pleura being the entry and exit site for structures to and from the lung (Fig. 4.101): the *bronchus* is posterior, with the *pulmonary artery* anterosuperior and the *pulmonary veins* anteroinferior. Also present are bronchial vessels lying close to the bronchi and lymph nodes. A downward extension of the pleura surrounding the lung root forms the pulmonary ligament. Passing anterior to the lung root is the phrenic nerve and the anterior pulmonary plexus and posteriorly is the vagus and posterior pulmonary plexus.

Pleura

Each lung is surrounded by a pleural sac which has two serous membranous layers *(visceral* and *parietal pleura)*, the layers being continuous with each other at the *lung root (hilum)*. The visceral pleura is firmly adherent to the lung tissue extending deep into the fissures, while the parietal pleura lines the thoracic wall (costal pleura), covers the upper surface of the diaphragm (diaphragmatic pleura), and the mediastinum (mediastinal pleura) (Fig. 4.102). At the thoracic inlet the parietal pleura arches over the apex of the lung (cervical pleura), being reinforced superiorly by dense connective tissue (the suprapleural membrane).

The potential space *(pleural cavity)* between the visceral and parietal pleura has a thin film of fluid to reduce friction during movement. Inflammation of the pleura elicits pain and may give rise to audible and palpable friction. Excess fluid within the pleural cavity (hydrothorax) produces a dullness on percussion and a reduction in breath sounds, while air (pneumothorax) produces a resonance and an absence of breath sounds.

The parietal pleura is supplied with blood from the intercostal and internal thoracic arteries; venous drainage is into corresponding veins while the lymphatics pass to nodes on the inner chest wall: innervation is by the intercostal (mediastinal pleura) and phrenic (diaphragmatic pleura) nerves; pain may be referred to the thoracic and/or abdominal walls, or to the neck and shoulder (from the

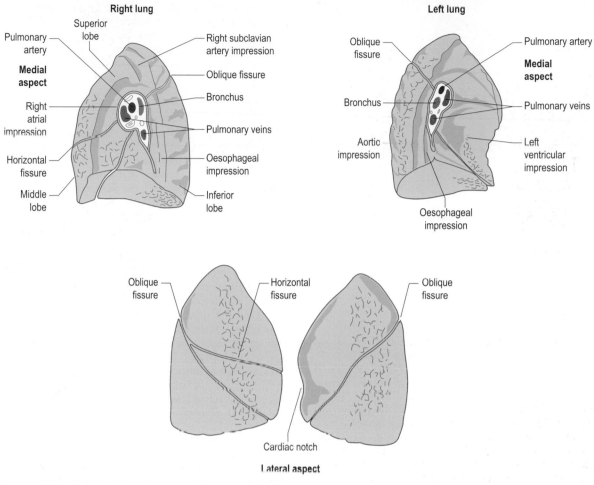

Right lung

Pulmonary artery

Superior lobe

Right subclavian artery impression

Medial aspect

Oblique fissure

Right atrial impression

Bronchus

Pulmonary veins

Horizontal fissure

Oesophageal impression

Middle lobe

Inferior lobe

Left lung

Oblique fissure

Pulmonary artery

Medial aspect

Bronchus

Pulmonary veins

Aortic impression

Left ventricular impression

Oesophageal impression

Oblique fissure

Horizontal fissure

Oblique fissure

Cardiac notch

Lateral aspect

Figure 4.101 Medial and lateral aspects of the right and left lungs.

mediastinal pleura). The visceral pleura is supplied by bronchial arteries and its venous and lymphatic drainage is similar to that of the lung: innervation is from the autonomic nervous system and is not sensitive to pain.

Surface markings

The surface markings of the lungs and pleura are important in any physical examination of the chest (Fig. 4.103). On both sides the parietal pleura and lung run down from the apex 3 cm above the medial third of the clavicle, passing behind the sternoclavicular joint to the sternal angle close to the midline. On the right side, both the parietal pleura and lung run down to the sixth costal cartilage, where they then diverge, while on the left the pleura and lung deviate to the left some 3 cm at the fourth costal cartilage before passing to the sixth costal cartilage. Passing laterally, the lower border of the lungs cross the midclavicular line at the sixth costal cartilage and the parietal pleura at the eighth rib. In the midaxillary line the lungs are at the eighth rib and the pleura at the 10th, while posteriorly the lungs cross the 10th rib and the pleura at the 12th, both 5 cm from the midline. Posteriorly the lower border of the parietal pleura on each side dips below the costal margin. The posterior borders of both lung and parietal pleura pass upwards towards the apex of the lung.

Since the lung does not extend as far downwards as the parietal pleura a potential space, the costodiaphragmatic recess, exists; the lung does not enter this recess except in deep respiration.

With the arm abducted to 90° the oblique fissure can be marked on a line which runs from the spinous process of T3 (level with the spine of the scapula) along the medial border of the scapula as far as the midaxillary line and then along the sixth rib and costal cartilage anteriorly. The horizontal fissure of the right lung runs laterally along the

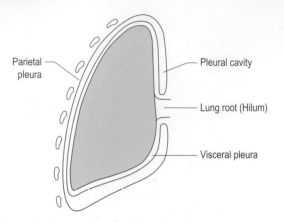

Figure 4.102 Organization of the parietal and visceral pleura and formation of the pleural cavity.

lower margin of the fourth costal cartilage and rib to meet the oblique fissure in the midaxillary line.

Bronchopulmonary segments

Each bronchopulmonary segment is a functionally independent unit that can be defined radiographically and often removed surgically. Their location and relation to the relevant tertiary bronchus are important when drainage of a particular lung segment is required. Within a bronchopulmonary segment tertiary bronchi repeatedly subdivide, becoming the bronchioles when cartilage is no longer present in their walls. The bronchioles further subdivide, eventually becoming distributed as the alveoli where respiratory exchange occurs.

In the right lung the superior lobe bronchus arises from the main bronchus 2.5 cm from the carina and soon divides into tertiary segmental bronchi (apical, posterior and anterior), which pass to similarly named bronchopulmonary segments in the superior lobe (Fig. 4.104). Five centimetres from the carina the main bronchus divides into the middle and inferior lobe bronchi passing to the middle and inferior lobes of the lung respectively. The middle lobe bronchus divides into lateral and medial segmental bronchi supplying corresponding bronchopulmonary segments in the middle lobe, while the inferior lobe bronchus gives the apical basal segmental bronchus opposite the origin of the middle lobe bronchus and then divides into the medial basal, anterior basal, lateral basal and posterior basal segmental bronchi (Fig. 4.104).

In the left lung the superior lobe bronchus arises 4–5 cm from the carina but then divides into superior and inferior (lingular) parts to supply the superior lobe and lingula respectively (Fig. 4.104). The superior division gives three segmental bronchi (apical, posterior and anterior), as in the right lung, passing to similarly named bronchopulmonary segments. The lingular division gives the superior and inferior segmental bronchi. (The lingula of the left lung is comparable to the middle lobe of the right lung.) The inferior lobe bronchus continues for a further 1.5 cm before giving rise to the apical basal segmental bronchus and then divides into the medial basal, anterior basal, lateral basal and posterior basal segmental bronchi (Fig. 4.104).

It should be noted that variations in the pattern of branching of the tertiary bronchi have been observed, which may give rise to fewer bronchopulmonary segments in either of the lungs than have been described above.

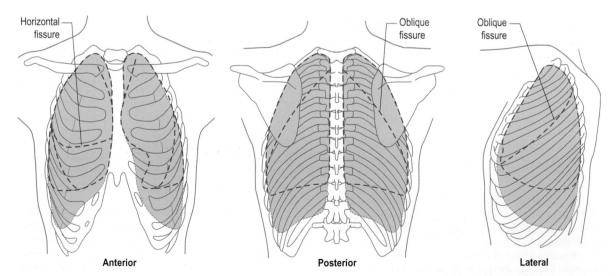

Figure 4.103 Surface projections of the lungs (and visceral pleura) and fissures (– – –): parietal pleura (shaded).

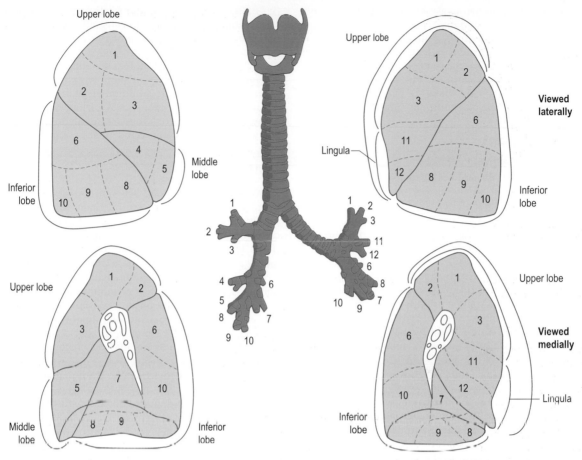

Figure 4.104 The trachea and distribution of the principal and segmental bronchi. Bronchopulmonary segments: 1, apical; 2, posterior; 3, anterior; 4, lateral; 5, medial; 6, apical basal; 7, medial basal; 8, anterior basal; 9, lateral basal; 10, posterior basal; 11, superior; 12, inferior.

Section summary

Respiratory system

The respiratory system is concerned with the exchange of oxygen and carbon dioxide between air in the lungs and blood in the pulmonary circulation (external respiration) and between the blood, cells and tissues (internal respiration).

Upper respiratory tract

- Comprises: nasal cavity, pharynx, larynx, trachea, main bronchi.
- Trachea bifurcates into right and left principal bronchi.
- Principal bronchi divide into secondary bronchi (one for each lobe), which divide into tertiary bronchi (one for each bronchopulmonary segment).

Lungs

- Right lung has three lobes (superior, middle, inferior) separated by two fissures (oblique, horizontal).
- Left lung has two lobes (superior, inferior) separated by the oblique fissure.
- Bronchus and pulmonary vein enter and pulmonary arteries leave at hilum of lung.
- Lungs wrapped in inner visceral pleura, separated from parietal pleura, which lines thoracic cavity, by pleural cavity.

THE DIGESTIVE SYSTEM

Introduction

The digestive tract is essentially a long tube extending from the oral cavity to the anus and comprises the mouth, pharynx, oesophagus, stomach, and small and large intestines. Ingested food is broken down, some of which is absorbed into the body through the wall of the digestive tract while indigestible material is excreted. It has a moist epithelial lining containing numerous glands secreting either mucous or digestive enzymes: some enzymes are produced by associated organs of the tract (salivary glands, liver, and pancreas). In specific regions the epithelium has a specialized surface for absorption of the simpler chemical compounds. Relatively thick inner circular and outer longitudinal muscle layers surround the epithelium, keeping the mixed food and enzymes moving along the tract (peristalsis).

Development

The primitive gut (foregut, midgut, hindgut) forms during the fourth week as the result of lateral folding of the embryo and incorporation of the dorsal part of the yolk sac (see Fig. 1.25): the rest of the yolk sac and allantois remain outside the embryo. The endoderm of the primitive gut gives rise to the epithelial lining of most of the digestive tract and its derivatives (biliary system and parenchyma of the liver and pancreas), except at the cranial and caudal extremities where it is derived from the ectoderm of the stomodeum and proctodeum respectively. The splanchnic mesoderm surrounding the primitive gut gives rise to the muscular and connective tissue (peritoneal) components.

The foregut gives rise to the oesophagus, trachea and lung buds (Fig. 4.98A), stomach, proximal duodenum, liver and biliary system, and pancreas; the midgut forms the primary intestinal loop giving rise to the distal duodenum, jejunum, ileum, caecum, vermiform appendix, ascending and proximal transverse colon; while the hindgut forms the distal transverse, descending and sigmoid colon, rectum and upper part of the anal canal: the lower anal canal is derived from the proctodeum.

The stomach appears as a dilatation beyond the oesophagus in the fourth week of development. Due to differential growth the stomach rotates 90° clockwise about its long axis and acquires *greater* and *lesser curvatures;* the dorsal mesentery is carried to the left, forming the *omental bursa* and lesser sac behind the stomach (Fig. 4.105A).

The *liver*, *gall bladder* and biliary system appear as an outgrowth *(liver bud)* from the proximal duodenum during the fourth week (Fig. 4.105B), which divides into two parts, enlarges and grows between the layers of the ventral mesentery. The liver grows so rapidly that it fills most of the abdominal cavity by the ninth week: part of the liver (the bare area) outgrows the ventral mesentery and comes to lie in direct contact with the diaphragm (Fig. 4.105B).

Elongation of the midgut forms the primary intestinal loop, which because of the rapid growth of the liver and the kidneys, projects into the umbilical cord during the sixth to tenth weeks (Fig. 4.105B).

The endoderm of the hindgut, as well as giving rise to the remainder of the digestive tract, also forms the internal lining of the bladder and urethra. The terminal part of the hindgut is an endoderm-lined cavity (the *cloaca*) in direct contact with the surface ectoderm at the *cloacal membrane*. The *urorectal septum* divides the cloaca into anterior (urogenital) and posterior (anorectal) parts, with the cloacal membrane dividing into the *urogenital membrane* anteriorly and the *anal membrane* posteriorly (Fig. 4.105). The anal membrane becomes surrounded by mesenchymal swellings so that at 8 weeks it is at the bottom of the proctodeum (anal pit). During the ninth week the anal membrane ruptures, allowing a communication between the rectum and the outside (Fig. 4.105C). Thus the upper part of the anal canal is derived from the endoderm of the hindgut, while the lower part is ectodermal in origin.

Oral cavity

Divided into the vestibule, between the lips and cheeks externally and gums and teeth internally, which receives the secretions of the parotid salivary gland, and the oral cavity proper (mouth) which is internal to the teeth (Fig. 4.106A). The *hard* and *soft palates* form the roof of the mouth and mylohyoid the floor: it contains the *tongue* and receives the secretions of the submandibular and sublingual salivary glands. The mouth is continuous posteriorly with the oropharynx by the oropharyngeal isthmus.

The highly mobile upper and lower lips help to retain saliva and food within the mouth, as well as change their shape to produce different sounds in phonation. The cheek has buccinator as the principal muscle and acts to prevent food accumulating in the vestibule. The lining of the cheek is continuous with that covering the gums (gingivae), which in turn is continuous with the periosteum lining the alveolar sockets.

Teeth

Each tooth has a *crown* above and a *root* below the gum margin (Fig. 4.106B), with the greater part being formed by *dentine*, which is covered by *enamel* over the crown and *cementum* over the root. Within the dentine is the *pulp cavity* containing vessels and nerves. Each tooth is anchored to the surrounding alveolar bone by the *periodontal membrane* (Fig. 4.106B). The teeth of the maxilla are innervated by anterior, middle and posterior superior alveolar nerves, from the maxillary division of the trigeminal, while those of the mandible are innervated by the inferior alveolar nerve from the mandibular division of the trigeminal. Superior and inferior alveolar arteries from the maxillary artery accompany the nerves.

There are two sets of teeth *(deciduous* and *permanent)* whose components (incisor, canine, premolar, molar) erupt at specific times (Fig. 4.106C). The deciduous dentition begins to appear by 6 months and is completed by the eruption of the second molar at 2 years. From age 6 the deciduous dentition is gradually replaced by the permanent dentition, which is usually complete by age 20; the third molar may remain permanently unerupted. As the permanent teeth are lost the surrounding alveolar bone is gradually resorbed, reducing the depth of the mandible (p. 559).

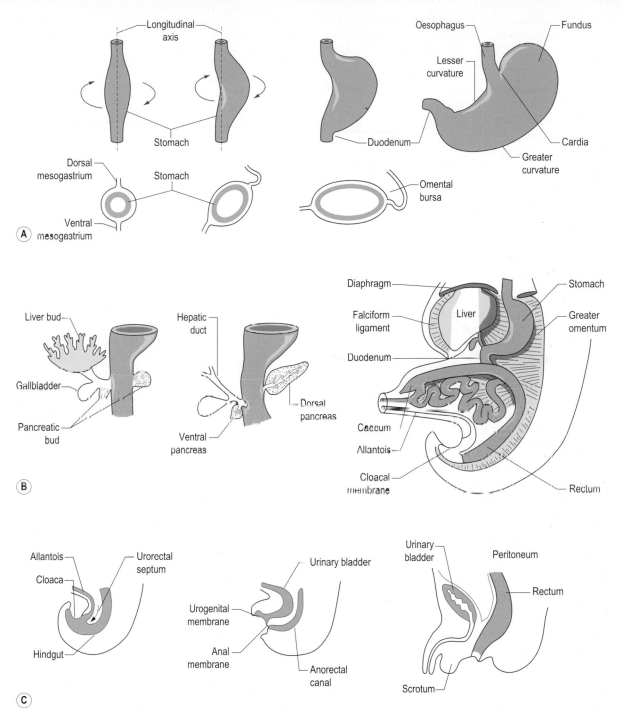

Figure 4.105 (A) Rotation of the stomach and formation of the omental bursa, (B) development of the liver and pancreas also showing herniation of the midgut, (C) development of the urinary bladder and anorectal canal.

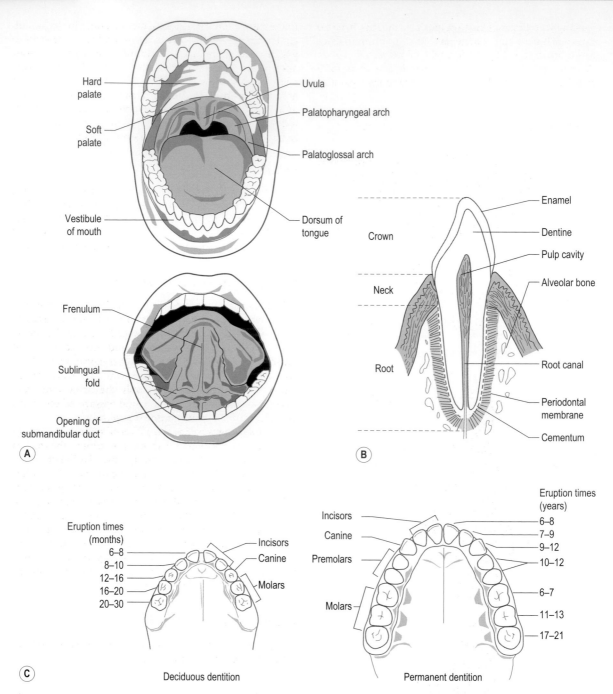

Figure 4.106 (A) Boundaries of the oral cavity, (B) longitudinal section through a tooth, (C) the deciduous and permanent dentition, showing the average eruption times for each tooth.

Tongue

The mobile muscular tongue is divided into anterior two-thirds, within the mouth, and posterior third, in the oropharynx, by the *sulcus terminalis* at the apex of which is

the *foramen caecum*, the origin of the thyroglossal duct from which the thyroid developed (Fig. 4.107A): the upper surface of the tongue is the dorsum. The extrinsic muscles (genioglossus, hyoglossus, styloglossus, palatoglossus) attach it to the skull and mandible and are responsible for changing its position, while the intrinsic muscles have no bony attachment and change its shape. All except palatoglossus are supplied by the hypoglossal nerve.

The roughness of the anterior part is due to filiform (white conical) and fungiform (red globular) papillae under the covering mucous membrane. Anterior to the sulcus terminalis are 8–12 *circumvallate papillae*, in the walls of which are found numerous taste buds. The nodular appearance of the posterior part is due to the underlying *lingual tonsil*. The mucous membrane is richly innervated: the anterior two-thirds by the lingual nerve (general sensation) and chorda tympani of the facial nerve (taste), and the posterior third by the glossopharyngeal nerve for both general sensation and taste, including the circumvallate papillae.

Palate

The palate is divided into the larger anterior hard palate, formed by the maxilla and palatine bones, which gives attachment posteriorly to the soft, mobile (muscular) palate (Fig. 4.106A). During swallowing and phonation the soft palate is tensed and elevated to close off the nasopharynx.

Pharynx and oesophagus

The pharynx extends from the base of the skull to the level of the cricoid cartilage at the level of C6, and consists of three overlapping circularly arranged muscles (superior, middle and inferior constrictors) and three longitudinal muscles (stylopharyngeus, palatopharyngeus and salpingopharyngeus). It communicates with the nasal cavity, mouth and larynx anteriorly; only the oropharynx and laryngopharynx are involved in the passage of the bolus of food from the mouth to the oesophagus.

The oesophagus begins at the level of the cricoid cartilage (C6), descends through the thorax posterior to the trachea and left atrium and passes through the diaphragm at the level of T10 to the left of the midline to enter the cardiac region of the stomach (Fig. 4.107). The oesophagus is constricted at its origin, at T4/5 where the *arch of the aorta* and the *left bronchus* cross it anteriorly, and where it pierces the diaphragm: these are the commonest sites of cancer and stricture. Although the latter constriction does not constitute an anatomical sphincter it is an important functional sphincter controlling the entry of food into the stomach and preventing the reflux of stomach contents.

Abdomen and pelvis

The greater part of the remainder of the digestive tract lies within the abdomen and pelvis.

Stomach

The stomach lies free within the abdominal cavity anchored only at its proximal and distal ends. It is divided into the *fundus* above the oesophageal opening, the *body* below the fundus and the *pylorus* separated from the body by the angular notch *(incisura angularis)* (Fig. 4.108A): the pyloric antrum lies next to the body while the thickened *pyloric sphincter* controls the passage of gastric contents into

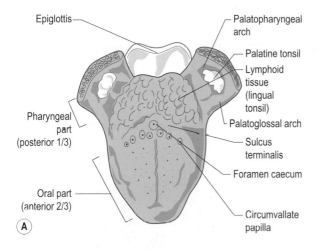

Epiglottis

Palatopharyngeal arch

Palatine tonsil

Lymphoid tissue (lingual tonsil)

Pharyngeal part (posterior 1/3)

Palatoglossal arch

Sulcus terminalis

Foramen caecum

Oral part (anterior 2/3)

Circumvallate papilla

(A)

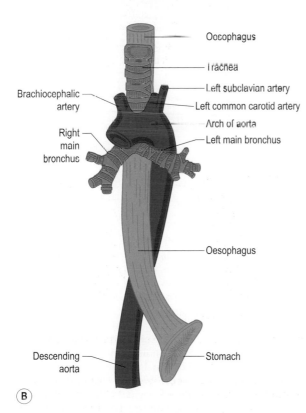

Oesophagus

Trachea

Left subclavian artery

Left common carotid artery

Brachiocephalic artery

Arch of aorta

Right main bronchus

Left main bronchus

Oesophagus

Descending aorta

Stomach

(B)

Figure 4.107 (A) Dorsum of the tongue, (B) oesophagus and its relations in the thorax.

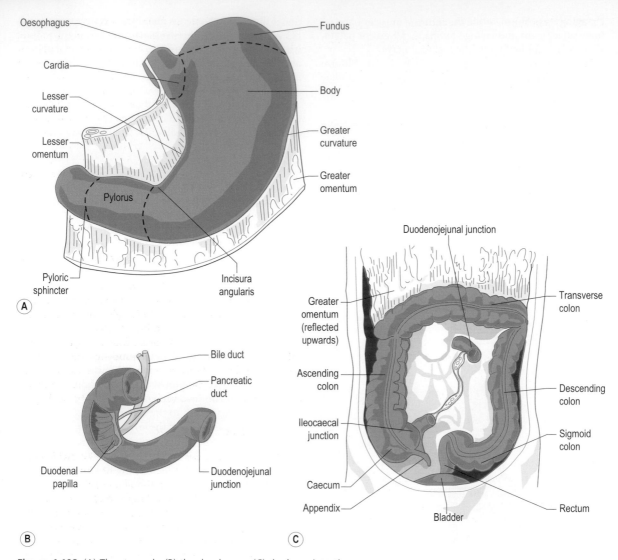

Figure 4.108 (A) The stomach, (B) the duodenum, (C) the large intestine.

the duodenum. The muscular wall of the stomach contains an internal oblique layer in addition to the longitudinal and circular layers: its interior is thrown into a series of longitudinal folds.

The stomach produces 2500 ml of secretion per day; thus it has a rich blood supply from the coeliac trunk (artery of the foregut), with blood draining into the portal vein and thence to the liver.

Small intestine

The c-shaped duodenum (first part of the small intestine) (Fig. 4.108B) lies directly on the posterior abdominal wall overlying the vertebral column; it has a thick muscular wall with the interior thrown into irregular circular folds.

The duodenum receives the secretions of both the pancreas (pancreatic juice for the digestion of proteins, fats and carbohydrates) and the liver via the gall bladder (bile for the digestion of fats); the two ducts have a common site of entry at the *major duodenal papilla*.

The jejunum and ileum (remainder of the small intestine) are relatively mobile and fill any available space in the abdominopelvic cavity, being the most common parts of the digestive tract found within hernias. The jejunum is fixed to the posterior abdominal wall at the *duodenojejunal junction*, and the ileum at the *ileocaecal junction*. The blood supply to the small intestine beyond the first part of the duodenum as far as the transverse colon of the large intestine is by branches from the superior mesenteric artery, the artery of the midgut.

Large intestine

The *caecum,* the first part of the large intestine is a blind ending pouch below the ileocaecal junction, and has the *appendix* attached to its medial surface (Fig. 4.108C). The extremely mobile appendix is of variable length (3–20 cm) and may be found lying behind the caecum or over the pelvic brim.

The *ascending colon* lies directly on the muscles of the posterior abdominal wall on the right of the abdomen from the caecum to just below the liver, where it bends sharply to become the *transverse colon.* The transverse colon is suspended from the posterior abdominal wall below and behind the stomach; on reaching the spleen it bends sharply downwards to become the *descending colon.* The descending colon also lies directly on the posterior abdominal wall descending as far as the pelvic brim (Fig. 4.108C).

The *sigmoid colon* joins the descending colon to the *rectum,* which begins at the level of the third sacral vertebra and follows the curve of the lower sacrum and coccyx to the pelvic floor, where it is supported by levator ani. The rectum pierces the pelvic floor, becoming the anal canal at the anorectal junction. The anal canal extends downwards and backwards to end at the anus.

From partway along the transverse colon to halfway along the anal canal the blood supply to the digestive tract is from branches of the inferior mesenteric artery, the artery of the hindgut. The lower half of the anal canal is supplied by branches from the internal iliac artery.

Venous drainage of the digestive tract below the diaphragm to halfway along the anal canal is via the portal vein to the liver. The transition in the anal canal between systemic and portal drainage represents the site of a portosystemic anastomosis; a similar arrangement exists in the lower third of the oesophagus. In cases of portal hypertension the venous vessels at the sites of these anastomoses may become distended (varices) and rupture, causing bleeding: in the anal canal the varices are known as haemorrhoids.

Pancreas

Both an endocrine (insulin secretion) and exocrine (pancreatic juice secretion) gland the *pancreas* has a rich blood supply as well as a substantial duct system. It lies obliquely across the posterior abdominal wall at the level of L1 and is divided into the head, neck, body and tail (Fig. 4.109A). The expanded head is within the C-shape of the *duodenum,* the body extends to the left as far as the hilus of the *left kidney,* and the tail ends at the hilus of the *spleen.*

Liver and biliary tract

A large solid organ about 1/40th of adult body weight (1/20th at birth) the liver lies under cover of the costal margin, being directly related to the inferior surface of the diaphragm. It has main *right* and *left lobes* and smaller *caudate* and *quadrate lobes* (Fig. 4.109B), which are both functionally part of the left lobe. The liver receives a large blood supply, 30% from the hepatic artery and 70% from the portal vein, and produces bile; venous drainage is by the hepatic veins to the inferior vena cava. The region where the bile ducts leave and the hepatic artery and portal vein enter the liver is known as the *porta hepatis.*

The *right* and *left hepatic bile ducts* join to form the common hepatic duct just below the liver. The *cystic duct* from the gall bladder joins the common hepatic to form the *common bile duct* (Fig. 4.109C), which empties into the *duodenum* in conjunction with the *pancreatic duct.* The liver produces 1500 ml of bile each day: it is stored and concentrated (by a factor of 10) in the *gall bladder,* an accessory organ of the biliary tract, and periodically released.

Spleen

Although not part of the digestive system the spleen is closely related to the stomach and pancreas, as well as the kidney, transverse and descending colon. It is part of the reticuloendothelial system and concerned with haematopoiesis in the fetus and the reutilization of iron from the haemoglobin of destroyed red blood cells.

The spleen lies under cover of the thoracic cage on the left side of the body between the ninth and eleventh ribs. Rupture of the spleen following a violent blow to the lower left ribs may produce massive intraperitoneal haemorrhage, which if left untreated or undiagnosed may be fatal.

Abdominal regions

For descriptive purposes the abdomen can be divided into nine regions, projected onto the anterior abdominal wall, by the intersection of the horizontal *transpyloric* and *transtubercular* planes and the vertical right and left *midclavicular* *planes* (Fig. 4.110A). The transpyloric plane lies midway between the jugular notch (sternum) and the upper border of the symphysis pubis; the transtubercular plane passes through the iliac tubercles. Alternatively the abdomen can be divided into quadrants centred on the umbilicus (Fig. 4.110B). Even though the position of the umbilicus is variable the quadrants are extensively used in clinical practice.

Palpation

In adults virtually no normal abdominal viscus is palpable. The anteroinferior edge of the liver may be palpated in the infant below the right costal margin extending out of the right hypochondrium. Since the liver moves with the diaphragm during respiration it may become palpable below the right costal margin in adults on full inspiration. The gall bladder is not normally palpable. However, its fundus lies at the intersection of the right costal margin with the linea semilunaris. The spleen can be palpated in the infant, moving with respiration in the left hypochondrium:

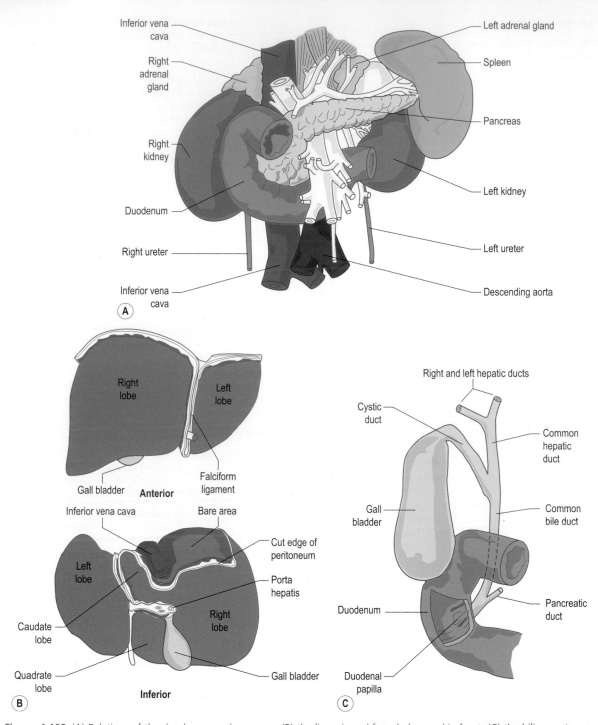

Figure 4.109 (A) Relations of the duodenum and pancreas, (B) the liver viewed from below and in front, (C) the biliary system.

in the adult it cannot be felt until it is about three times its normal size. The digestive tract is not normally palpable; an enlarged caecum or sigmoid colon may be felt in the right or left iliac fossae respectively. In thin subjects, the pulsations of the abdominal aorta are transmitted to the anterior abdominal wall and are easily seen. Movements of the digestive tract are rarely visible; if seen such movements are abnormal.

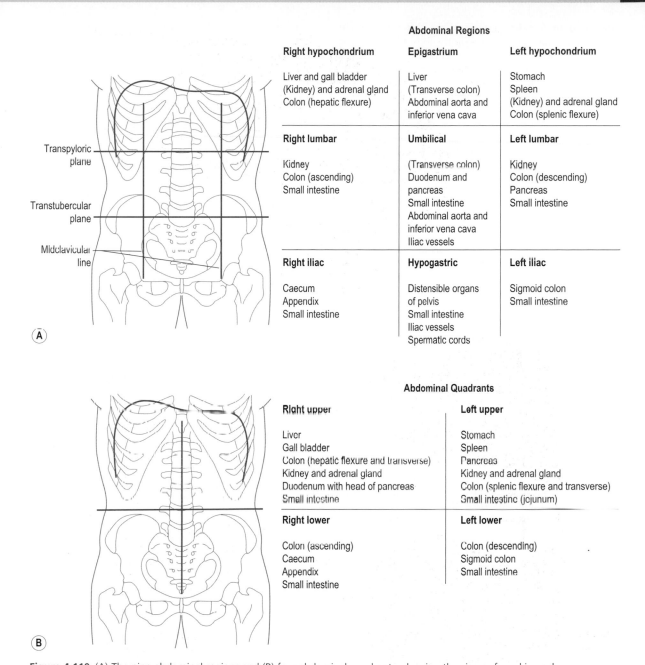

Abdominal Regions

Right hypochondrium	Epigastrium	Left hypochondrium
Liver and gall bladder	Liver	Stomach
(Kidney) and adrenal gland	(Transverse colon)	Spleen
Colon (hepatic flexure)	Abdominal aorta and	(Kidney) and adrenal gland
	inferior vena cava	Colon (splenic flexure)
Right lumbar	**Umbilical**	**Left lumbar**
Kidney	(Transverse colon)	Kidney
Colon (ascending)	Duodenum and	Colon (descending)
Small intestine	pancreas	Pancreas
	Small intestine	Small intestine
	Abdominal aorta and	
	inferior vena cava	
	Iliac vessels	
Right iliac	**Hypogastric**	**Left iliac**
Caecum	Distensible organs	Sigmoid colon
Appendix	of pelvis	Small intestine
Small intestine	Small intestine	
	Iliac vessels	
	Spermatic cords	

Transpyloric plane
Transtubercular plane
Midclavicular line

(A)

Abdominal Quadrants

Right upper	Left upper
Liver	Stomach
Gall bladder	Spleen
Colon (hepatic flexure and transverse)	Pancreas
Kidney and adrenal gland	Kidney and adrenal gland
Duodenum with head of pancreas	Colon (splenic flexure and transverse)
Small intestine	Small intestine (jejunum)
Right lower	**Left lower**
Colon (ascending)	Colon (descending)
Caecum	Sigmoid colon
Appendix	Small intestine
Small intestine	

(B)

Figure 4.110 (A) The nine abdominal regions and (B) four abdominal quadrants, showing the viscera found in each.

Hernias

An abdominal hernia is a protrusion of gut and its covering peritoneum into a space where it is not normally found: they may be internal (impalpable) or external, providing an obvious palpable protrusion. Hiatus hernia is the commonest form of internal hernia in which either the gastrointestinal junction and upper stomach passes through the oesophageal opening in the diaphragm into the posterior mediastinum, or part of the stomach passes through to lie adjacent to the oesophagus.

The commonest type of external hernia is inguinal hernia (75%) and then femoral hernia, both of which need to be distinguished from other 'lumps in the groin' (enlarged superficial inguinal lymph nodes). Factors predisposing to external hernias include: sex, males are more commonly

529

affected than females because of the differing contents of the inguinal canal; raised intra-abdominal pressure through obesity, coughing, straining or lifting; patent processus vaginalis and abdominal muscle weakness (p. 432).

Most inguinal hernias are indirect (and predominantly occur in males), in which a loop of small intestine follows the course of the spermatic cord through the abdominal wall such that a bulge appears above the inguinal ligament medial to the pubic tubercle. Once through the superficial ring the hernia enters the scrotum (or rarely the labium major). In direct inguinal hernias the protruding gut pushes through the posterior wall of the inguinal canal, usually deep to or slightly lateral to the superficial ring.

Femoral hernias are twice as common in females as in males, particularly in women who have had children. Here a loop of gut pushes through the femoral ring into the femoral sheath, and therefore appears below the inguinal ligament lateral to the pubic tubercle: the hernia is limited by the fascia lata except at the saphenous opening, through which the hernia may progress. With all external hernias there is a danger of strangulation, in which the intestine becomes twisted and its blood supply is cut off. Prompt surgical intervention is required to prevent gangrenous necrosis.

Chewing and swallowing

Chewing is the process of breaking food up into smaller components, during which time digestion begins by the action of the salivary enzymes. Chewing is achieved by alternate protraction and retraction of the mandible on each side, coupled with elevation on the side of protraction. The tongue and buccinator work to keep the food between the teeth and prevent it from accumulating either in the vestibule or in the mouth respectively. The resulting mass is then formed into a bolus in preparation for swallowing.

In the voluntary stage of swallowing the bolus of food is pushed backwards between the tongue and hard palate towards the oropharynx; the floor of the mouth is raised by the contraction of mylohyoid which also elevates the hyoid. At the same time the soft palate is tensed and elevated to prevent food entering the nasopharynx and nasal cavity. When the bolus has passed the palatoglossal folds they are brought together to close the oropharyngeal isthmus and prevent food returning to the mouth: contraction of the palatoglossal muscles also helps push the bolus into the oropharynx.

Once the bolus makes contact with the pharyngeal wall the swallowing reflex is initiated. Respiration is temporarily suspended, by closure of the nasopharynx above and the laryngeal inlet below, as the bolus is propelled towards the oesophagus by the serial contraction of the constrictor muscles (peristalsis). Simultaneous contraction of the longitudinal pharyngeal muscles pulls the laryngeal walls upwards and inwards, thus elevating the larynx and pharynx to receive the bolus. In this way the food is forced into the oesophagus.

When swallowing fluids the tongue forms a longitudinal midline furrow along which the fluid flows to pass either side of the epiglottis into the piriform fossae to enter the oesophagus.

Section summary

Digestive system

The digestive tract is a long tube extending from the oral cavity to the anus: it comprises the mouth, pharynx, oesophagus, stomach, small and large intestines, liver and pancreas. Ingested food is broken down, some absorbed through the wall of the digestive tract and undigested material excreted.

Oral cavity
- Contains the teeth, tongue and receives secretions from salivary glands.

Pharynx
- Fibromuscular tube.
- Communicates with the nasal and oral cavities, and larynx; continues as the oesophagus.

Stomach
- Lies free within abdominal cavity; comprises fundus, body and pylorus.
- Pyloric sphincter controls the release of contents into duodenum.

Small intestine
- Connects stomach and large intestine.
- Has three parts: duodenum, jejunum, ileum.
- Duodenum receives secretions from liver and pancreas.

Large intestine
- Has six parts: ascending, transverse, descending and sigmoid colon; rectum and anal canal terminating at anus.

Pancreas
- An endocrine and exocrine gland lying behind stomach.
- Endocrine secretions enter bloodstream directly, exocrine secretions enter duodenum.

Liver
- Large organ lying below diaphragm (mainly right side); divided into right, left, caudate and quadrate lobes.
- Receives blood from digestive tract.
- Produces bile, which is secreted into duodenum or stored and concentrated in gall bladder.

THE UROGENITAL SYSTEM

Introduction

The urinary part of the urogenital system is concerned with the formation, concentration, storage and excretion of urine. It lies partly in the abdomen (kidneys, ureters), partly in the pelvis (bladder), with the remainder passing through the perineum (urethra).

Development of the urinary system

On the basis of function the urogenital system can be considered as two different components: the urinary system and the genital system. However, they are embryologically and anatomically intimately related, developing from a common mesodermal ridge (the intermediate mesoderm) along the entire length of the posterior body

wall, with the excretory ducts of both systems initially entering a common cavity, the *cloaca*.

Urinary system

During development three different kidney systems are formed: the pronephros, the *mesonephros*, and the *metanephros*, which forms the permanent kidney. The pronephros appear during the fourth week and are merely a few cell clusters in the cervical region which soon degenerate (Fig. 4.111A). However, most of their ducts are used by

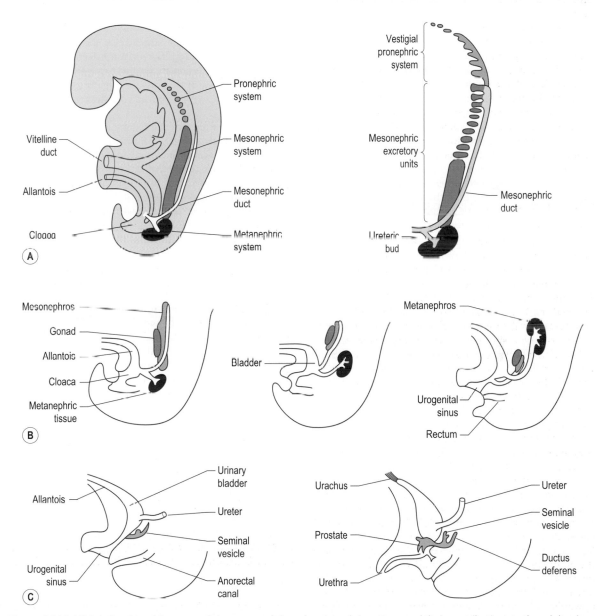

Figure 4.111 (A) Relationship of the pronephric, mesonephric and metanephric systems and their contributions to the adult urinary system, (B) ascent of the kidney, (C) development of the urogenital sinus into the urinary bladder.

the developing mesonephros which appear later in the fourth week, caudal to the pronephros, forming large ovoid structures either side of the midline (Fig. 4.111A). At the end of the ninth week all that remains of the mesonephros are a few of the caudal tubules in both sexes and the mesonephric duct in the male.

The metanephros begins to develop in the fifth week, becoming functional 6 weeks later. The permanent kidneys develop from two sources: the metanephric diverticulum (the ureteric bud) giving rise to the ureter, renal pelvis, calyces and collecting tubules, and the metanephric mesoderm giving rise to the nephrons (the excretory units) of the kidney (Fig. 4.111B). No new nephrons are formed after birth.

The kidneys 'migrate' from the pelvis into the abdomen, attaining their adult position by week nine (Fig. 4.111B). Eventually they come to lie close to the vertebral column at the level of L1. The kidneys become fully functional during the second half of pregnancy, the urine passing into the amniotic cavity and forming the major part of the amniotic fluid. However, since the placenta eliminates metabolic waste products from the fetal blood the kidneys do not need to be functional prior to birth.

The epithelium of the bladder arises from the urogenital sinus, the other layers developing from adjacent splanchnic mesoderm. The distal parts of the mesonephric ducts become incorporated into the bladder: their proximal parts degenerate in the female, but persist in the male to form the epididymus, vas deferens and ejaculatory duct when the mesonephros degenerates. In the male the pelvic part of the *urogenital sinus* forms the prostatic and membranous parts of the *urethra* (Fig. 4.111C). The epithelium of both the male and female urethra is of endodermal origin, the surrounding connective tissue and smooth muscle being derived from adjacent splanchnic mesoderm. In males at the end of the 12th week the epithelium of the prostatic urethra proliferates and penetrates the surrounding mesenchyme to form the prostate gland: in females the proximal part of the urethra gives rise to the urethral and paraurethral glands.

The urinary system

Kidneys and ureters

The kidneys lie on the posterior abdominal wall in the paravertebral gutters opposite the upper three lumbar vertebrae, the left slightly higher than the right: their exact position varies with respiration and body posture. Their anterior relations are (Figs. 4.109A and 4.112): the medial aspect of the upper poles to the adrenal (suprarenal) glands; the hilar region of the left kidney to the pancreas and the right to the duodenum; part of the lower pole of each kidney to the flexures of the large intestine, the remainder to small intestine; the liver overlies the right kidney and the stomach and spleen the left. Posteriorly

each kidney lies on psoas, quadratus lumborum and transversus abdominis (from medial to lateral), with the diaphragm superiorly (Fig. 4.112A). Except in thin individuals, the kidneys are not palpable.

Each kidney is covered by a fibrous *capsule* which is surrounded by perirenal fat, in turn enclosed by the anterior and posterior layers of the perirenal fascia; the layers fuse superiorly but remain separate inferiorly. In wasting diseases the fat disappears and the kidney moves downward. The kidney has outer convex and inner concave borders; the hilus, where the vessels and ureter enter and leave, is on the concave border. At the hilus the *renal vein* is anterior to the *renal artery*, which is anterior to the *pelvis* of the *ureter* (Fig. 4.112B).

When cut, the internal structure of the kidney can be clearly seen (Fig. 4.112C). The *cortex* lies deep to the capsule and extends inwards *(renal columns)* between the *medullary pyramids*. The apex of each pyramid leads to *minor calyces* which drain into *major calyces* and thence to the *pelvis* and *ureter*.

The functional unit of the kidney is the nephron (over 1 million in each kidney) comprising glomerulus (filtration of blood plasma), proximal convoluted tubule, loop of Henle (straight tubule), distal convoluted tubule (together with the proximal tubule resorbs water and dissolved materials back into the circulation) and collecting ducts. Both the cortex and outer medulla contain glomeruli; some convoluted tubules are found in the cortex but the majority are in the medulla; the loops of Henle and collecting ducts are in the medulla.

The ureter (continuation of the renal pelvis) passes down over psoas and the pelvic brim, anterior to the bifurcation of the common iliac artery, to enter the *bladder* obliquely at the upper lateral angle of the trigone (Fig. 4.113B). Close to the bladder it is crossed by the vas deferens (males), while in females it lies below the broad ligament (see Fig. 4.115B) lateral to the cervix, where it can be palpated. Along its length the ureter has three constrictions (at the junction with the renal pelvis, as it crosses the pelvic brim and where it enters the bladder); renal stones may become impacted at these sites, causing intense pain.

Bladder

A hollow muscular organ, the *bladder* varies in shape and size depending on the amount of urine contained: when empty it lies in the pelvis, but as it fills (up to 500 ml) it enlarges upward into the abdominal cavity. It has outer longitudinal, middle circular and inner longitudinal muscle layers (detrusor muscle). The mucous membrane lining (transitional epithelium) is thrown into folds due to its loose attachment to the underlying tissue, except at the *trigone* (triangular area between the openings of the ureters and the urethra) where it is firmly attached and thus appears smooth (Fig. 4.113B).

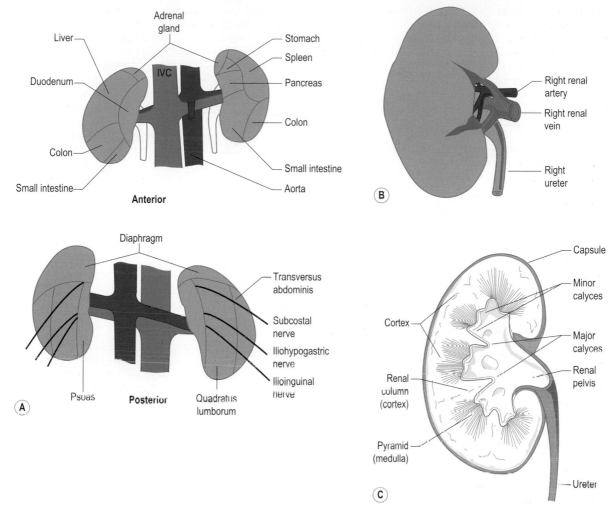

Figure 4.112 (A) Schematic representation of the anterior and posterior relations of the kidneys, (B) the right kidney showing the arrangement of structures at the hilum, (C) coronal section of the kidney showing the internal organization. IVC, inferior vena cava.

When empty the bladder has four pyramidal sides; the *base* is posterior and the *superior* and *inferolateral surfaces* meet at the apex (Fig. 4.113A). Anteriorly, the *apex* lies behind the symphysis pubis; the superior surface has coils of small intestine resting on it; the inferolateral surfaces rest on levator ani (pelvic floor) and against obturator internus, while posteriorly the base is related to the *uterus* (females) or seminal vesicles and vas deferens (males); further posteriorly is the rectum (Fig. 4.113C). The bladder is relatively immobile where the urethra leaves (the neck), being fixed by ligaments in both sexes; it is also firmly adherent to the *prostate* in males.

In infants the bladder is an abdominal organ and therefore readily palpable in the hypogastrium when full. A completely full bladder in adults is theoretically palpable above the pubis; however, rectus abdominis makes this difficult.

Urethra

The *urethra* is much longer in males than females (20 cm compared with 4 cm). In males it passes from the neck of the bladder through the prostate *(prostatic urethra)*, pelvic floor and perineal membrane *(membranous urethra)* and penis *(penile or spongy urethra)* to end at the external urethral opening at the tip of the *glans penis* (Fig. 4.113D). The prostatic urethra has the ejaculatory and prostatic ducts opening into it.

The female urethra runs from the neck of the bladder through the pelvic floor and perineal membrane to open

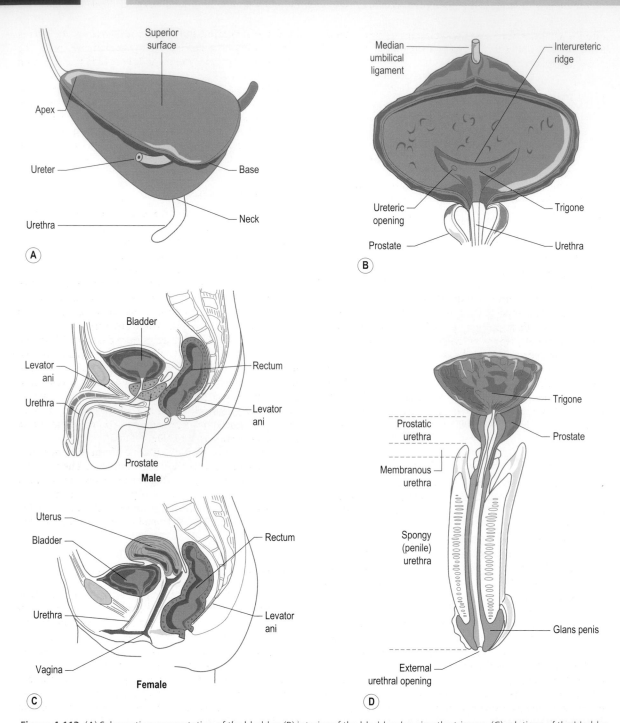

Figure 4.113 (A) Schematic representation of the bladder, (B) interior of the bladder showing the trigone, (C) relations of the bladder in the male and female, (D) the uretha in the male.

into the vestibule (between the labia minor) just anterior to the *vagina*, being firmly attached to the anterior wall of the vagina.

Micturition

When the bladder contains more than 300–400 ml of urine the desire to micturate increases: although in some individuals it can hold 600 ml there is a point at which it will empty involuntarily. The voluntary act of micturition initially involves contraction of the abdominal muscles and diaphragm to raise intra-abdominal pressure, followed by contraction of detrusor and relaxation of the sphincter mechanisms in the proximal part of the urethra in males and along most of the urethra in females. Towards the end of micturition the abdominal muscles contract again, then relax; detrusor relaxes and the sphincters contract.

During micturition the neck of the bladder becomes funnel-shaped due to relaxation of the pelvic floor. If these muscles are torn during childbirth, the ability to lift the neck of the bladder becomes impaired; the sphincter mechanism cannot then prevent urine being expelled when intra-abdominal pressure increases (stress incontinence). In males hypertrophy of the middle lobe of the prostate may act as a valve at the internal urethral opening forming a recess in which urine collects and remains after micturition: in some cases the amount of urine retained may be sufficient that the individual has the constant urge to micturate, yet is only able to void a relatively small amount.

Development of genital system

The early genital systems of both sexes are similar, but later under the influence of the testis-determining factor gene on the Y chromosome a series of events occur which determines the fate of the rudimentary sexual organs.

The gonads appear as a pair of longitudinal (genital) ridges formed by mesodermal epithelium and the underlying mesenchyme from which develop the primary sex cords, and after the sixth week the primordial germ cells. Up to the seventh week the gonads of both sexes are identical and are known as the indifferent gonads. In males the primary sex cords form the *testes* and become connected to the epididymus. In females the primary sex cords degenerate; however, secondary sex cords extend from the surface epithelium of the developing ovary into the underlying mesenchyme and incorporate the primary germ cells. As the *ovary* separates from the regressing mesonephros it becomes suspended by its own ligament, the mesovarium.

Both males and females initially have two pairs of genital ducts: the *mesonephric* and the *paramesonephric ducts*. The fate of the mesonephric duct in the male has already been discussed (p. 531); the paramesonephric ducts degenerate. In females the paramesonephric ducts form the main genital duct, with three parts being recognizable: the first two parts form the uterine (fallopian) tubes while the fused parts become the body and *cervix* of the uterus (Fig. 4.114A). The surrounding mesenchyme forms the myometrium of the uterus and its peritoneal covering (perimetrium).

The fibromuscular wall of the vagina develops from mesenchyme with the epithelial lining being derived from the urogenital sinus. Until late in fetal life the lumen of the vagina is separated from the cavity of the urogenital sinus by a membrane (the *hymen*), (Fig. 4.114A) which usually ruptures during the perinatal period.

Descent of the gonads

As the mesonephros degenerates the *gubernaculum* descends from the inferior pole of each gonad and passes obliquely from the anterior abdominal wall at the site of the future inguinal canal, and attaches to the labioscrotal swelling. The processus vaginalis, an evagination of peritoneum, develops anterior to the gubernaculum and passes through the abdominal wall along the same path pushing before it layers of the abdominal wall. In males these layers become the coverings of the spermatic cord.

In males by week 28 the testes have descended from the posterior abdominal wall to the deep inguinal ring, the internal opening of the inguinal canal. Their passage through the inguinal canal usually takes 2 or 3 days, passing outside the peritoneum and processus vaginalis (Fig. 4.114B). By week 32 the testes have entered the scrotum, after which the inguinal canal contracts and closes around the spermatic cord (Fig. 4.114C).

In females the ovaries descend from the posterior abdominal wall to just below the pelvic brim. The proximal gubernaculum attaching to the uterus below the fallopian tubes becomes the *round ligament of the ovary*, while the distal part becomes the *round ligament of the uterus*, which passes through the inguinal canal to terminate in the labia major.

Female reproductive system

The greater part of the female reproductive system (ovaries, uterine tubes, uterus and upper part of the vagina) lies in the pelvis; only a small part lies in the perineum (lower part of the vagina). The ovaries are the female gonads, analogous to the testes in males; the uterine tubes convey the ovum (egg) to the uterus, where, if fertilization has occurred, the zygote becomes implanted; development into the embryo and fetus takes place in the uterus. The vagina connects the uterus to the exterior and serves for the introduction of sperm as well as the birth canal.

The peritoneal lining of the pelvic cavity extends laterally as a horizontal fold over the uterus and uterine tubes dividing the cavity into anteroinferior and posterosuperior compartments, containing the bladder and rectum respectively (Fig. 4.113C); the anterior and posterior layers of the fold either side of the uterus form the broad ligaments.

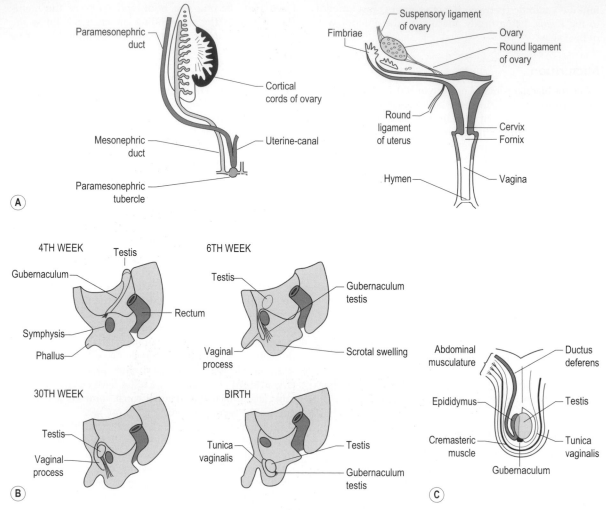

Figure 4.114 (A) Representation of the female genital ducts at 8 weeks and their contribution to the female genital system, (B) descent of the testes, (C) adult relationship of the testes and their coverings.

Ovaries and uterine tubes

Each *ovary* (3 cm long, 2 cm wide, 1 cm thick) lies in the posterior layer of the *broad ligament* closely related to the opening *(infundibulum)* of the *uterine tube* (Fig. 4.115A). The lower pole is connected to the uterus by the round ligament of the ovary, continuous with the round ligament of the uterus which passes through the inguinal canal to the labia major. At approximately monthly intervals a single ovum (Graafian follicle) is shed by one ovary; this enters the peritoneal cavity and is immediately engulfed by the infundibulum of the uterine tube to be conveyed to the uterus, which takes 4 days.

The ovary develops on the posterior abdominal wall and lies in the iliac fossa until about the age of 6 years, after which it moves into the pelvis. During pregnancy the ovary is pulled upwards by the enlarging uterus, returning to the pelvis after childbirth; but its position is more variable than before and no longer vertical. The surface of the ovary is smooth in the child but becomes more irregular due to successive ovulations resulting in the formation of areas of fibrous tissue; after menopause it becomes smaller.

The uterine tube extends laterally from the junction of the fundus and body of the uterus, its lumen communicating with that of the uterus (Fig. 4.115B). The narrowest part *(isthmus)* is as it joins the uterus, widening (the ampulla) as it passes laterally to end as the funnel-shaped infundibulum fringed with *fimbriae*, one of which *(ovarian fimbria)* is attached to the superior pole of the ovary.

Fertilization takes place in the uterine tube, with the fertilized ovum being moved by contraction of the smooth

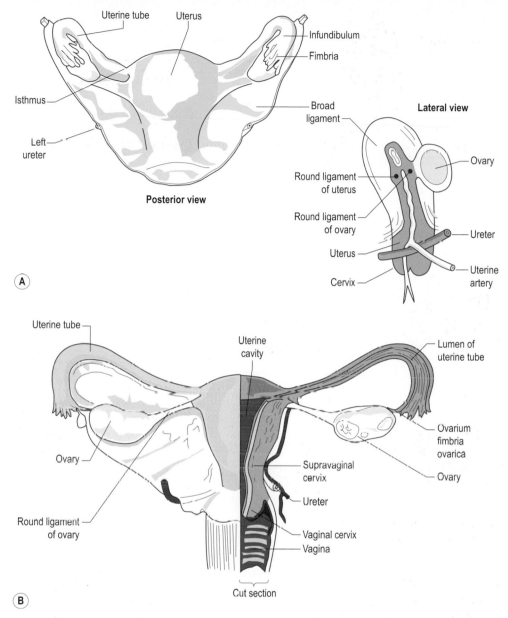

Figure 4.115 (A) The uterus and broad ligament, viewed posteriorly and laterally, (B) the uterus shown without the broad ligament.

muscular wall of the uterine tube as well as by the action of cilia towards the uterus. Delay in the passage may result in the dividing ovum becoming lodged in the uterine tube (tubal pregnancy); however, rupture of the tube with severe haemorrhage usually follows within 4–6 weeks: occasionally this is fatal.

The hormonal changes during the menstrual cycle produce changes in the vascularity of the lining of the uterine tubes.

Uterus

A thick-walled muscular organ lying in the pelvic cavity, the uterus can be divided into three parts: the *fundus* above the opening of the *uterine tubes*, a *body* which narrows inferiorly and a neck *(cervix)* projecting into the *vagina*. The cavity of the uterus (triangular mediolaterally but merely a slit anteroposteriorly) is continuous with the cervical canal through the *isthmus*, the canal opening into the vagina through the external os (Fig. 4.116A).

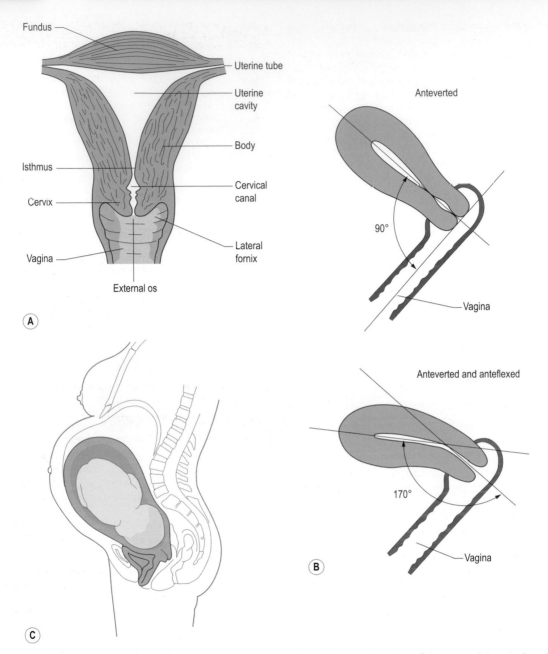

Figure 4.116 (A) Schematic representation of the uterus (coronal section), (B) 'normal' position of the uterus, (C) sagittal section of pregnant woman near full term showing the enlarged uterus occupying the majority of the abdominal and pelvic cavities.

The mucous membrane lining of the uterus (endometrium) undergoes extensive changes during the menstrual cycle in response to circulating levels of ovarian hormones and is shed each month if fertilization of the ovum does not occur. If fertilization occurs the ovum becomes embedded in the uterine wall through which it acquires its nutrients for growth to a full-term fetus.

Apart from the vaginal part of the cervix the uterus is relatively free and mobile. In the majority of women the body overlies the superior surface of the empty bladder, the normal *anteverted* position (Fig. 4.116B); the uterus is also bent forward along its own axis *(anteflexion)* (Fig. 4.116B). The uterus can also slope backwards (retroversion) and/or be bent backwards along its own axis (retroflexion).

Although the uterus lies above the bladder it is separated from it by the uterovesical pouch; posteriorly it is separated from the middle third of the rectum by the rectouterine pouch (of Douglas). Inferiorly the cervix is supported by the muscular pelvic diaphragm (levator ani) and condensations of pelvic fascia, which form three ligaments (transverse cervical, pubocervical and sacrocervical). The fibromuscular perineal body, to which part of levator ani attaches, is important in maintaining the pelvic floor, and should it become damaged during childbirth, prolapse of the pelvic viscera may occur, particularly the uterus, but in severe cases also the bladder (see p. 433).

Pregnancy

During pregnancy the fetus and uterus enlarge, occupying more and more of the abdominal cavity (Fig. 4.116C); from being a pelvic organ the uterus enlarges upwards reaching the symphysis pubis by 12 weeks, the umbilicus by 24 weeks and the xiphisternum by 36 weeks. In the last 4 weeks the fetal head may descend into the pelvis (engagement of the head) and the fundus may also descend to some extent. At the end of labour the uterus almost returns to its former size (7.5 cm long, 5 cm broad, 2.5 cm thick) in about 6 weeks.

Vagina

The terminal part of the female reproductive tract, being continuous with the uterine cavity at the external os and opening into the vestibule between the labia minor, lying parallel with the pelvic inlet. The anterior and posterior walls are usually opposed but become widely distended and elongated during parturition. The space between the cervix and the vagina is the *fornix*, being divided into anterior, posterior and lateral parts; the posterior wall of the vagina is longer than the anterior so that the posterior fornix is deeper than the anterior.

Anteriorly the vagina is related to the base of the bladder and the urethra, which is bound to it; posteriorly is the rectouterine pouch, the ampulla of the rectum and inferiorly the perineal body. Laterally the upper part of the vagina is related to the broad ligament containing the ureter and uterine artery. Levator ani surrounds the vagina.

The mucous membrane lining of the vagina is firmly attached to the underlying muscular coat, being thick and corrugated by transverse elevations (vaginal rugae); longitudinal ridges form the anterior and posterior rugal columns in the corresponding walls. The musculature is arranged longitudinally and has a thin layer of erectile tissue between the mucosal and muscular coats.

Section summary

Urinary system

The urinary system is concerned with the formation, concentration, storage and excretion of urine. It comprises the kidneys, ureters, bladder and urethra.

Kidneys

- Lie on the posterior abdominal wall, covered by fibrous capsule and surrounded by fat.
- Consists of cortex and medulla.
- Renal artery enters and renal vein and ureter leave at hilum.
- Ureter runs from hilum to bladder.

Bladder

- Hollow muscular organ resting on levator ani behind the symphysis pubis.
- Receives ureter from each kidney; drained by the urethra.
- In males the urethra passes through prostate gland, perineal membrane, and penis.
- In females the urethra pierces the pelvic floor and perineal membrane, opening into the vestibule between the labia minor anterior to vagina.

Genital system

Male reproductive system

Comprises the testes, epididymus, vas (ductus) deferens, seminal vesicles and urethra.

Female reproductive system

Ovaries and uterine (Fallopian) tubes

Each ovary lies in the posterior layer of the broad ligament close to the opening (infundibulum) of the uterine tubes.
Ovary connected to the uterus by the round ligament of the ovary, which is continuous with the round ligament of the uterus.
Uterine tubes extend laterally from the sides of the uterus being narrow (isthmus) where they join the uterus.

Uterus

Thick walled muscular organ lying in pelvic cavity
Has fundus above opening of uterine tubes, body narrowing inferiorly to the cervix which projects into the vagina.
Cervix is supported by levator ani and transverse cervical, pubocervical and sacrocervical ligaments (condensations of pelvic fascia).
Is usually anteverted and anteflexed so that body lies superior to bladder.
Separated from bladder and rectum by uterovesical and rectouterine pouches respectively.

Vagina

Superior part surrounds cervix, lower part opens into vestibule between labia minor.
Space between vagina and cervix is fornix (anterior, lateral and posterior).

THE ENDOCRINE SYSTEM

A series of ductless glands (Fig. 4.117) collectively form the endocrine system, a major communicating system which, with the nervous system, regulates and coordinates body functions. The system comprises the pituitary (hypophysis), thyroid, parathyroid and adrenal (suprarenal) glands, pancreas (islets of Langerhans) and gonads; other regions (hypothalamus, kidneys, digestive tract, thymus and pineal gland) also have endocrine functions. Disorders of the endocrine glands are numerous and may result in disturbances of growth and development, metabolism and reproduction.

Glands

Pituitary gland

A small gland (1 cm diameter) in two parts situated within the skull; the anterior lobe (adenohypophysis) develops from the ectoderm of the oral cavity, while the posterior lobe (neurohypophysis) develops from the brain, being connected to the hypophysis by the infundibulum. It acts mainly by controlling the activities of all other endocrine glands.

Thyroid gland

The largest of the endocrine glands, it consists of two lobes joined across the midline by an isthmus; it lies in the neck

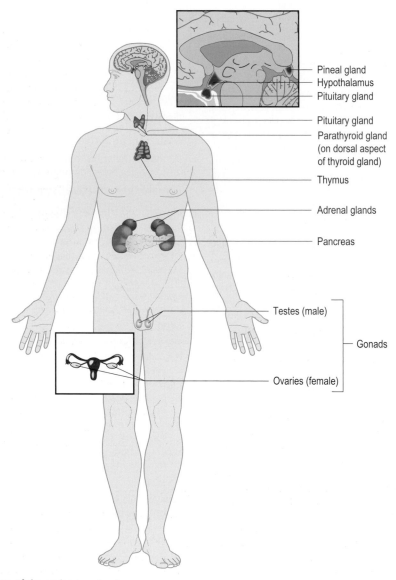

Pineal gland
Hypothalamus
Pituitary gland

Pituitary gland
Parathyroid gland (on dorsal aspect of thyroid gland)

Thymus

Adrenal glands

Pancreas

Testes (male)

Gonads

Ovaries (female)

Figure 4.117 Location of the endocrine glands.

with the lobes covering the lower part of the larynx and upper part of the trachea and the isthmus overlying the second, third and fourth tracheal rings. It secretes the hormones thyroxine and tri-iodothyronine, which are important in controlling the rate of oxidation (metabolic rate) in the body, and calcitonin (from C-cells), which lowers both calcium and phosphate levels in the plasma. Unlike other endocrine glands, the thyroid maintains a large store of hormones.

Because of its attachment to the larynx the gland moves upwards during swallowing. Its size varies with age, sex and general nutritional status, increasing slightly in women during menstruation and pregnancy. Pathological enlargement produces a goitre.

Parathyroid glands

Four small glands (3 mm wide, 1 mm deep, 6 mm long) lying on or embedded in the back of the lobes of the thyroid gland. Each gland consists of chief cells producing parathyroid hormone (parathormone), which maintains the normal relation between blood and skeletal calcium levels, and oxyphil cells whose function is unknown. Removal of the parathyroids is followed by increased neuromuscular excitability and muscle spasms, eventually leading to death within a few days.

Adrenal glands

Each adrenal gland lies on the upper pole of its respective kidney; the right is triangular and the left crescentic in shape. Each gland consists of an outer cortex and an inner medulla which function as separate endocrine glands; the medulla develops from the same cells that give rise to the sympathetic nervous system. The cortex consists of three zones which produce specific steroid hormones: the outer zone produces mineralocorticoids (e.g. aldosterone) whereas the intermediate and inner zones produce glucocorticoids (e.g. cortisol) and sex hormones (e.g. androgens). The cortex is essential to life, with interference in its function causing disruption of fluid and electrolyte balance in the body; it is also concerned with carbohydrate metabolism and is important in normal bodily reactions to stress.

The cells of the medulla secrete a specific catecholamine, either adrenaline or noradrenaline. Adrenaline is released in response to stress, it increases heart rate, raises blood pressure, and causes the release of sugar into the bloodstream from the liver; it may also be a neurotransmitter in the brain, where it is associated with many functions including cardiovascular and respiratory responses. Noradrenaline has widespread actions which include cardiac stimulation, blood vessel constriction, and relaxation of the bronchioles and the digestive tract. Within the brain it is involved in the regulation of body temperature, food and water intake, and cardiovascular and respiratory control.

Islets of Langerhans

These are scattered throughout the pancreas and produce glucagon (secreted by A-cells) and insulin (secreted by B-cells). Glucagon is secreted in response to low blood glucose concentrations and acts to raise blood glucose levels by stimulating the conversion of liver glycogen into glucose. It also stimulates the secretion of insulin, pancreatic somatostatin (inhibits insulin and glucagon release) and growth hormone. Insulin is released in response to increased blood glucose levels (e.g. after a meal) and acts to lower the level by accelerating glucose uptake by most tissues (except the brain) and promoting its conversion into glycogen and fat.

Gonads

Steroid hormones are produced and secreted mainly by the gonads (testes in males, ovaries in females) which are necessary for sexual development and the control of reproductive function. The most important are certain androgens (testosterone, dihydrotestosterone) found predominantly in males and progesterones (progesterone) and certain oestrogens (oestradiol, oestrone, oestriol) found predominantly in females. They probably act on the brain to influence sexual and other behaviour.

Androgens are necessary for the development of male genitalia in the fetus; during puberty they promote the development of secondary sexual characteristics (growth of the penis and testes, appearance of pubic, facial and body hair, increase in muscle strength, deepening of the voice); in adults they are required for the production of sperm and the maintenance of libido.

Oestrogens are responsible for the development of female secondary sexual characteristics as well as promoting sexual readiness and preparing the uterus for implantation of the embryo. Progesterone is required for the maintenance of pregnancy, and preparation of the uterus for implantation; it also inhibits ovulation during pregnancy and prepares the breasts for lactation.

Hypothalamus

The part of the brain which has an important role in regulating the internal environment (food intake, water balance, body temperature) as well as controlling the release of hormones from the pituitary. It is also involved in the control of the emotions by the limbic system.

Thymus

A lymphoid gland of variable size and shape found in the thorax; it is present at birth and continues to grow until puberty after which it usually regresses. Its presence is essential in the newborn for the development of lymphoid tissue and immunological competence; in adults it is concerned with lymphocyte production, most of which are destroyed

541

by the gland itself with only a few being released into the circulation.

Pineal gland

A small gland found within the brain above the third ventricle but separated by the blood–brain barrier; it synthesizes melatonin from serotonin. The secretion of melatonin and its concentration within blood both fluctuate, being highest during darkness. It may be associated with the synchronization of circadian rhythms (sleep and waking, the daily rise and fall of cortisol production).

Part | 5 |

Head and brain

CONTENTS

BONES

Introduction

An awesome looking structure, commonly used to portray death and instil terror, the skull is in reality a complex arrangement of many individual bones which form the skeleton of the head and face. It consists of two main elements: the large, hollow cranial cavity (the walls of which enclose the brain) and the bones of the face anteroinferiorly, which complete the walls of the orbits and nasal cavity and also forms the roof of the mouth. The mandible, although a separate bone, completes the bony framework of the face.

The discovery of many fossilized vertebrate remains, particularly skull fragments and teeth, has led to a detailed study of the skull revealing a fascinating history of the evolutionary development of the human skeleton. The size and shape of the cranial, orbital and nasal cavities, the jaw and the teeth have all provided a detailed catalogue of the probable evolutionary history of vertebrates. Estimates of relative brain size from the size and shape of the cranial cavity reveal information regarding possible intelligence. The size and shape of the jaws and teeth give clues as to the type of food eaten, and indirectly provide information regarding changes in body posture with respect to adaptation to the gathering and provision of food (e.g. running and hunting). The size and direction of the orbits, together with changes in the upper cervical spine, indicate the increasing importance of the power and range of vision. This, together with the bipedal posture adopted during the same period, gives an insight into the defence, protection and manipulation of tools necessary for survival.

The skull

The cranial cavity of the human skull is relatively large, and projects anteriorly over the facial skeleton (Fig. 5.1). Its walls are composed of plates of bone arranged as two layers of compact bone enclosing a central layer of cancellous bone. These bones ossify in the membrane (intramembranous ossification) and are joined edge to edge by fibrous interlocking joints known as *sutures* (p. 17).

The top and most of the sides of the skull are formed by the two large *parietal bones*, which articulate along their medial borders at the *sagittal suture* (Fig. 5.1A). Anteriorly the skull is formed by the *frontal bone*, which joins the two parietal bones at the *coronal suture*. The point where the coronal and sagittal sutures meet is termed the bregma. At birth, the region of the bregma is not ossified and appears as an easily felt diamond-shaped area of connective tissue known as the *anterior fontanelle*. This gap gradually decreases in size and closes about 18 months after birth.

The frontal bone forms the forehead and separates the orbital cavities from the anterior cranial fossa. It therefore forms the roof of each orbital cavity and the floor of the anterior cranial fossa.

The remaining part of each side of the cranial cavity is formed by the squamous part of the *temporal bone* and part of the *greater wing of the sphenoid*. The region where the frontal, parietal, temporal and sphenoid bones almost meet is termed the pterion; it has a small sphenoidal fontanelle which closes within 3 months of birth. Passing medially and slightly forwards from the squamous part of the temporal bone is the petrosal part of the temporal bone, which forms most of the floor of the middle cranial fossa; it contains and protects the organs of hearing and balance.

The most posterior section of the cranial cavity is formed by the *occipital bone*, which meets the posterior border of each parietal bone at the *lambdoid suture*. The meeting point of the lambdoid and sagittal sutures is termed the lambda. Again, at birth, this region is not ossified and appears as a

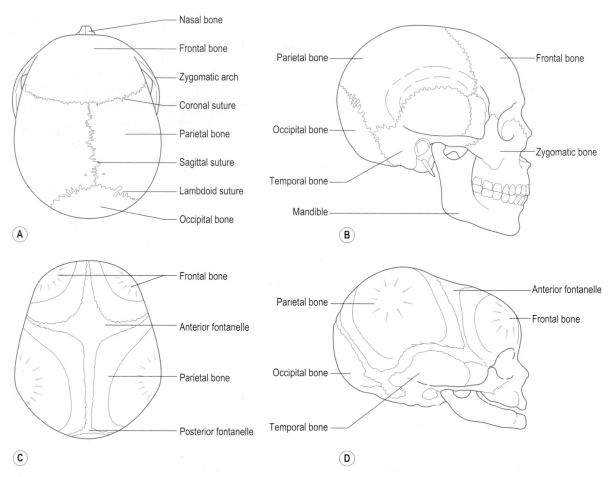

Figure 5.1 Superior and lateral views of the adult (A, B) and fetal skull (C, D).

triangular area of connective tissue known as the *posterior fontanelle*. This gap rapidly decreases in size and closes between the third and sixth month after birth. The occipital bone curves forwards to surround the foramen magnum and projects anteriorly as the basilar part of the occipital bone, which joins the body of the sphenoid.

The outside of the top of the skull needs little description as the parietal bones and the relevant parts of the frontal and occipital bones show few bony landmarks except for the sutures where they join. It is, however, worth noting that the skull is widest towards the back. Its roundness means that the effects of a blow to the head can be distributed and minimized. On the inner aspect of the skull, in the region of the sagittal suture, is a faint groove formed by the superior sagittal sinus as it passes posteriorly between the two layers of the dura mater as far as the internal occipital protuberance. On either side of this groove deeply pitted areas can usually be seen. They are formed by the underlying arachnoid granulations, which themselves are involved in the regulation of the flow of cerebrospinal fluid (CSF) into the systemic circulation. The base of the skull, however, shows many features and openings which allow blood vessels and nerves to enter and leave the cranial cavity.

Growth of the skull

At birth the cranial cavity is relatively large but the face is small (Fig. 5.1D), being approximately one-eighth of the whole skull compared with one-third in the adult. The teeth are not fully formed and the paranasal sinuses are rudimentary, consequently both the jaws and nasal cavities are small. The individual bones are joined by cartilage; ossification progressing with age. There is no mastoid process so the styloid process and the stylomastoid foramen are closer to the side of the head. Undue pressure applied during a forceps delivery in the region of the developing mastoid may therefore damage the facial nerve. The maxilla is shallow because it has no sinus, and consists mainly of alveolar processes and developing teeth. The nasal bones are flat, so that the infant has no bridge to the nose, which together with the absence of superciliary arches gives the forehead its prominent appearance. The orbits are relatively large and have the nasal cavity lying almost completely between them.

After birth the skull grows rapidly until the seventh year, with the greatest increase in the size of the cranial cavity occurring during the first year. During the second year, the styloid process and the stylomastoid foramen come to lie deeper as the mastoid process begins to grow. By the seventh year the orbits are almost adult-sized; the petrous part of the temporal bone, the body of the sphenoid and the foramen magnum have, however reached full-size. The jaws have enlarged in preparation for the eruption of the permanent teeth. Growth of the skull after age 7 is slower than it was before, except for during puberty when a rapid growth

in all directions occurs, particularly in the frontal and facial regions accompanying the increasing size of the paranasal sinuses.

From the early twenties to middle age there is gradual fusion of the various sutures between the individual bones of the skull, beginning with the sagittal suture.

The skull viewed anteriorly

When viewed from the front (Fig. 5.2) the upper third of the skull consists of the cranial cavity and is formed by the frontal bone; this is the forehead. Parts of the parietal, sphenoid and temporal bones can also be seen in this region. Below the forehead are the two large orbital cavities, whose margins are formed by the frontal bone above, the *zygomatic bone* inferolaterally and the *maxilla* inferomedially. The walls of the orbital cavity are formed by the maxilla inferiorly, the zygomatic bone and *greater wing* of the *sphenoid* laterally, the frontal bone and the *lesser wing* of the *sphenoid* superiorly, and the *ethmoid* and *lacrimal* bones medially. Deep within the cavity can be seen the *superior* and *inferior orbital fissures* and the *optic canal*. Along the medial part of the superior margin is the *supraorbital notch*, transmitting the corresponding vessels and nerves. Towards the front of the medial wall is the opening for the nasolacrimal duct, which conveys secretions from the eye (i.e. tears) to the nose for eventual swallowing. Below the inferior orbital margin on the front of the face is the large *infraorbital foramen*, through which pass the infraorbital nerve and vessels onto the face.

Below and medial to the orbits is the pear-shaped nasal aperture, bound almost entirely by the maxillae, with only the superior part being bound by the two *nasal bones*. Within the nasal cavity a more-or-less midline septum can be seen; this is formed by the *vomer* and the *perpendicular plate* of the *ethmoid*. Projecting medially into the cavity from its lateral walls can be seen the *inferior* and the *middle conchae* (the two superior conchae lie deeply behind the nasal bones). The inferior alveolar margin of the maxilla, supporting the teeth, forms an almost horizontal convex arch, which fits snugly with the corresponding arch of the mandible.

The skull viewed from the lateral side

The temporal bone is central to the lateral aspect of the skull (Fig. 5.1B). Above it joins the parietal bone posteriorly and the greater wing of the sphenoid anteriorly, while posteroinferiorly it joins the occipital bone. It is marked just behind and below its centre by the external acoustic meatus, above which is a bony ramus passing forwards to meet a similar backwardly projecting process from the zygomatic bone to form the zygomatic arch. Just below the posterior part of this arch the mandibular fossa can be seen. The *styloid process* is seen projecting downwards from deep to the external acoustic meatus.

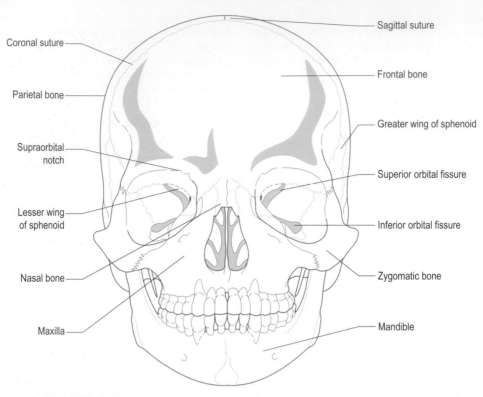

Figure 5.2 Anterior view of the skull.

The skull viewed inferiorly

The skeleton of the face attaches to the undersurface of the anterior half of the cranial cavity. Consequently, a view from below reveals the mandible, *maxilla* and *palatine bones*. With the mandible removed the maxilla and the palatine bones can be clearly seen (Fig. 5.3A).

The anterior and lateral margins of the maxilla are marked by the sockets for the teeth, being somewhat smaller anteriorly, increasing in size as they pass posteriorly around the alveolar margin. The area between these bony margins is formed by the palatine part of the maxilla anteriorly and the horizontal plate of the palatine bone posteriorly. Projecting backwards in the midline from the posterior border of the palatine bones is the *nasal spine*, while laterally are the *medial* and *lateral pterygoid plates* of the *sphenoid* directed inferiorly, the medial with the hook-like *pterygoid hamulus*. At the lateral edges of the palatine bones are the *greater* and *lesser palatine foramina* transmitting the corresponding vessels and nerves. At the front of the hard palate is the *incisive canal*, through which pass terminal branches of the greater palatine and sphenopalatine vessels, and the nasopalatine nerves.

The central section of the undersurface of the skull is formed mainly by the sphenoid with its many bony processes and the foramina. Laterally lies the temporal bone marked by the *mandibular fossa* which receives the condyle of the mandible. Posterolaterally is the *external acoustic meatus*, while posteromedially is the *styloid process* of the temporal bone – a long and slender projection passing downwards and anteromedially (Fig. 5.3A). Medial to the styloid process anteriorly is the *carotid canal* and posteriorly the *jugular foramen*.

In front of the carotid canal is the *foramen spinosum*, separated from it by the *spine* of the *sphenoid*. Behind the external acoustic meatus is the large breast-shaped mastoid process of the temporal bone, with the *stylomastoid foramen* lying medially between it and the styloid process. It is here that the facial nerve (seventh cranial nerve) emerges from the skull. Centrally the large *foramen magnum* is unmistakable with the *occipital condyles* on its anterolateral borders. The *hypoglossal (anterior condylar) canal* lies anterior to the condyle, while posterior is the *posterior condylar canal*. Posterior to the foramen magnum, and being joined to it by the external occipital crest, is the *external occipital protuberance*. Passing laterally towards the mastoid process is the *superior nuchal line*; between this line and the foramen magnum are less distinct lines, the *inferior nuchal lines*, which also curve laterally. The area between the two nuchal lines and the foramen magnum gives attachment to the postvertebral muscles.

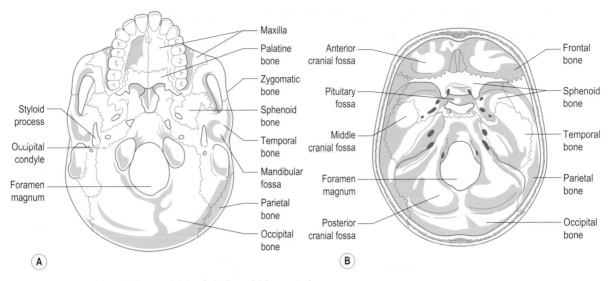

Figure 5.3 Base of the skull viewed (A) inferiorly and (B) superiorly.

The floor of the cranial cavity

The floor of the cranial cavity is formed by the base of the skull, with both the right and left sides having similar features (Fig. 5.3B). The two halves are divided in the sagittal plane by the *crista galli* anteriorly, the *body* of the *sphenoid* centrally, and the foramen magnum and *internal occipital protuberance* posteriorly. The floor is further divided into *anterior, middle* and *posterior cranial fossae* by prominent ridges of bone, with each fossa lying at a different level, the posterior being at the lowest level and the anterior the highest.

Anterior cranial fossa. This lies above and in front of the middle cranial fossa (Fig. 5.3B), being separated from it by the posterior concave edge of the lesser wings of the sphenoid. It contains the lower part of the frontal lobes of the brain. The walls and most of the floor of the fossa are formed by the frontal bone, with the posterior part of the floor being formed by the lesser wings of the sphenoid. Between the two sides is a sagittally directed elongated hollow. Running centrally in this hollow is a crest known as the crista galli on either side of which is a perforated horizontal plate, the *cribriform plate*. It is through this plate that the olfactory nerves from the upper part of the nasal cavity pass to the olfactory bulb above the cribriform plate.

Middle cranial fossa. This lies behind and below the anterior cranial fossa (Fig. 5.3B), being formed by the temporal and sphenoid bones. It consists of a median and two lateral parts. The lateral parts contain the temporal lobes of the brain, while the raised median part is formed by the body of the sphenoid. Passing laterally from the body of the sphenoid are the greater wings of the sphenoid,

which, together with the squamous part of the temporal bone, turn superiorly, forming the lateral wall of the skull in the temporal region. The floor of the middle cranial fossa is formed by part of the greater wings of the sphenoid anteriorly and the gently sloping superior surface of the petrous part of the temporal bone posteriorly. Anteriorly each greater wing of the sphenoid turns upwards forming the anterior wall of the fossa. The superior part of this wall is overlapped by the posterior edge of the anterior cranial fossa, i.e. the lesser wings of the sphenoid.

The body of the sphenoid has a smooth, hollowed depression superiorly, the *pituitary* (hypophyseal) *fossa*, in which sits the pituitary gland. Small horn-like (clinoid) processes project on either side from the front and back of the hollow. They give attachment anteriorly to a horizontal fold of the dura mater (the tentorium cerebelli) which passes between the cerebral and cerebellar hemispheres. The cavernous sinuses (p. 591), part of the intracranial venous sinus system, are formed between the two layers of the dura mater either side of the body of the sphenoid. Within the walls, or through these sinuses, pass many of the nerves destined for the orbit, as well as the internal carotid artery. The internal carotid artery emerges from the sinus anterolateral to the anterior clinoid process. Anterior to this region, between the roots of the lesser wings of the sphenoid, is the optic canal running forwards and laterally conveying the optic nerve and ophthalmic artery to the orbit. Passing transversely between the two optic canals is a groove called the sulcus chiasmatis; it does *not* lodge the optic chiasma. Between the greater and lesser wings of the sphenoid is a gap which becomes narrower as it passes laterally. This is the superior orbital fissure and opens

547

directly into the back of the orbit. It transmits many structures to and from the orbit – the third, fourth, ophthalmic division of the fifth, and sixth cranial nerves, and the ophthalmic veins. Below the superior orbital fissure in the anterior wall of the middle cranial fossa close to the body of the sphenoid is the rounded *foramen rotundum*, through which passes the maxillary division of the fifth cranial nerve into the pterygopalatine fossa.

In the floor of the middle cranial fossa between the greater wing of the sphenoid and the petrous temporal bone is the *foramen lacerum*. In life, its edges are connected by fibrous tissue, which supports the internal carotid artery as it passes medially from the carotid canal to the side of the body of the sphenoid. Lateral to the foramen lacerum are two openings in the greater wing of the sphenoid. The larger, medial one is the *foramen ovale*, which transmits the mandibular division of the trigeminal nerve as well as the lesser petrosal nerve and small blood vessels. The more lateral opening is the *foramen spinosum*, through which passes the middle meningeal artery to supply the meninges of the brain.

Posterior cranial fossa. The largest and deepest of all three fossae (Fig. 5.3B); in it lodges the cerebellum. The temporal bone is seen as a hard ridge passing backwards and laterally from the body of the sphenoid. Along its superior border runs a longitudinal groove for the superior petrosal sinus.

The floor and most of the posterior wall of the fossa are formed by the concave surface of the occipital bone; a small part of the posterior wall being formed by the parietal bones. The anterolateral wall is formed by the posterior surface of the petrous part of the temporal bone, with the body of the sphenoid and the basilar part of the occipital bone forming the anterior wall.

The most obvious feature of the posterior fossa is the large oval opening (foramen magnum), slightly narrower transversely than from front to back. It transmits the spinal cord, being at the junction of the spinal cord and brainstem, as well as a number of blood vessels and the spinal part of the 11th cranial nerve. Passing upwards from the posterior margin of the foramen magnum is the internal occipital crest ending at the internal occipital protuberance. Running down towards this protuberance is the continuation of the groove of the superior sagittal sinus. As it approaches the protuberance, the groove passes to the right to run transversely around the posterolateral wall of the fossa as the groove for the transverse sinus. On reaching the petrous part of the temporal bone, the groove turns downwards and continues as an S-shaped towards the jugular foramen. This latter groove is known as the groove of the sigmoid sinus. A similar arrangement of transverse and sigmoid grooves can be observed in the left-hand side of the posterior cranial fossa. The sigmoid sinus passes through the jugular foramen to become the internal jugular vein. As it does so it is joined by the inferior petrosal sinus,

which runs in the groove between the petrous temporal and the basioccipital bones towards the jugular foramen. As well as these two venous channels passing through the jugular foramen, it also transmits the ninth, tenth and eleventh cranial nerves.

Anterior to the jugular foramen is the carotid canal through which passes the internal carotid artery and its associated plexus of sympathetic nerves. In the posterior surface of the petrous part of the temporal bone is the opening of the *internal auditory meatus*; both the seventh and eighth cranial nerves enter this canal.

Running forwards and upwards from the anterior margin of the foramen magnum is the clivus, which has the basilar artery separating it from the brainstem. On the lateral part of the clivus, medial to the groove for the inferior petrosal sinus, runs the sixth cranial nerve before piercing the dura to run in the sinus. Above the lateral margin of the foramen magnum is the hypoglossal canal, which runs anterolaterally and transmits the 12th cranial nerve.

Palpation

Virtually the whole of the outer surface of the cranium is palpable, being either subcutaneous or lying just below a thin sheet of muscle. The top of the skull composed of parietal and frontal bones can be taken between the examiner's hands as in giving a blessing. On the upper part of the occipital bone, the external occipital protuberance can be readily determined by the finger tips. On the temporal bone, to the side of the skull, the external acoustic meatus is visible, and behind the pinna of the ear the mastoid process can be identified becoming pointed at the level of and behind the lobe of the ear. From just above the external acoustic meatus the posterior part of the zygomatic arch can be identified running forwards to join the zygomatic bone, which presents as the point of the cheek and is thus easily palpable.

Anterosuperior to the cheek the margins of the orbit can be palpated, being particularly marked deep to the eyebrow.

The mandible

The mandible (Fig. 5.4) or lower jaw completes the skull. It consists of a horizontal convex *body*, with two upwardly projecting *rami* from the posterior ends of the body. The rami and body all present an outer and an inner surface, with the body having superior and inferior borders, while the rami have anterior and posterior borders. The outer surface of the body is slightly concave from top to bottom and markedly convex from side to side, and may show a vertical line in the midline where the two halves have fused. Towards the front of the body on each side is the *mental foramen* transmitting the mental nerve and vessels. The inner surface of the body is divided by a slightly raised ridge of bone, the *mylohyoid line*, into upper and lower areas. Behind the front of the mandible, above this line, is the

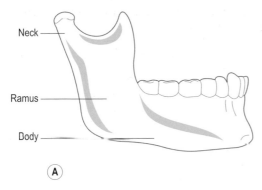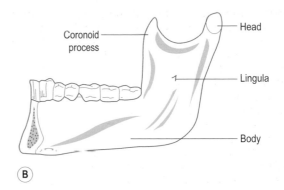

Figure 5.4 (A) Lateral and (B) medial views of the mandible.

sublingual fossa which lodges the sublingual salivary gland, while below this line, in the middle third of the body, is the *submandibular fossa* for the submandibular gland. On the inner surface of the anterior midline region are two pairs of bony projections, the *genial (mental) spines*. The upper pair lie above the mylohyoid line and give rise to part of the tongue (genioglossus), while the lower pair lie below the line and give attachment to geniohyoid. The superior border of the body consists of a series of sockets for the teeth. The bone behind the third molar is thickened as it joins the anterior border of the ramus. The lower border of the body is thick and rounded, but shows two fossae anteriorly, one each side of the midline. These are the *digastric fossae*, each giving attachment to the anterior belly of digastric. Posteriorly the body meets the lower posterior part of the ramus at the *angle*, which tends to be everted in males and inverted in females. Just in front of the angle the lower border is notched where the facial artery crosses it to enter the face.

Each ramus is continuous with the body, but is flatter. The inner surface is marked by the *mandibular foramen*, which is partially covered by a small bony process, the *lingula*. The inferior alveolar nerve and vessels enter the mandible at the mandibular foramen, giving off mylohyoid branches before they do so. These pass in the downward and forward running mylohyoid groove, also seen on the inner surface of the ramus. The area behind the mylohyoid groove towards the angle of the mandible is roughened for the attachment of medial pterygoid. The sphenomandibular ligament attaches to the lingula. The outer surface of the ramus is roughened for the attachment of masseter. The posterior border is thick and rounded, particularly at its lower end, and may show a shallow fossa for part of the parotid gland. The anterior border is also thick inferiorly but becomes a thin pointed projection superiorly, the *coronoid process*. The concave superior border passes between the coronoid process anteriorly and the *condylar process (head)* posteriorly. The head, which articulates with the mandibular fossa of the temporal bone, is much broader transversely than from front to back, and is marked at its lateral and medial ends by prominent tubercles.

Palpation

The entire length and depth of the mandible are palpable with the angle posteriorly, being prominent, particularly in males, even though the posterior border and lateral surface of the ramus is mainly covered by muscle and part of the parotid gland. The lateral tubercle on the condyle can be palpated anterior to the tragus of the ear below the posterior part of the zygomatic arch. Identification is easier with the subject opening and closing the mouth, when the condyle can be felt gliding forwards on the articular surface of the temporal bone.

The hyoid bone

The U-shaped hyoid bone (Fig. 5.5) is deficient posteriorly and suspended in the neck by muscle attachments below the tongue and above the larynx (see Fig. 5.8). Being attached to the tongue, it moves up and down with the tongue during swallowing. Because the hyoid is also firmly attached to the larynx by the thyrohyoid membrane, when the hyoid moves upwards it carries the larynx with it. The hyoid consists of a *body* anteriorly, a pair of *greater horns* which project upwards and backwards from the body, and a pair of *lesser horns* which also project upwards and backwards from the junction of the body with the greater horns. The lesser horns and upper part of the body of the hyoid are derived from the second pharyngeal arch, while the greater horns and lower part of the body are from the third pharyngeal arch. The stylohyoid ligament, also from the second arch, attaches to the lesser horn. The hyoid is connected by muscles and ligaments to the tongue, the mandible, base of the skull (styloid process), thyroid cartilage and sternum.

Ossification

A pair of ossification centres, which soon unite, appear in the body shortly before birth, as well as a single centre for each greater horn. During the first year centres appear for the lesser horns. Not until middle age do the body and

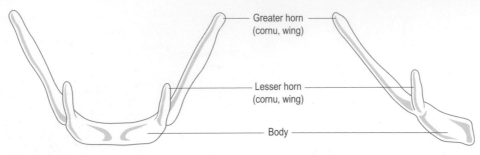

Figure 5.5 Superior (left) and lateral (right) views of the hyoid bone.

greater horns fuse, and only in old age do the lesser horns fuse with the remainder.

Palpation

The two greater horns can be felt through the skin below the mandible if firm pressure is applied between the finger and the thumb above the thyroid cartilage of the larynx. In this position the bone can be moved from side to side. The body can be palpated in the front of the neck about 2 cm above the laryngeal prominence with the chin elevated.

Section summary

Bones of the head

Skull

- The skeleton of the head and face consist of a large hollow cranial cavity (housing the brain) and bones of the face anteroinferiorly.
- The cranial cavity is formed by the parietal, frontal, temporal, sphenoid and occipital bones.
- The facial skeleton is formed by the frontal and zygomatic bones and the maxillae, with the mandible inferiorly.
- The floor of the cranial cavity is divided into anterior, middle and posterior cranial fossae.

Hyoid

- Is suspended by muscles connecting it to the tongue, mandible, styloid process, thyroid cartilage and sternum.

MUSCLES

Muscles which change the shape of the face

The muscles which change the shape of the face are commonly known as the muscles of facial expression (Fig. 5.6). Although with the contraction of individual or groups of muscles, human beings can smile, frown, look happy or sad, and generally convey many emotions, it must be remembered that the primary action of the majority of these muscles is to dilate (open) or constrict (close) the eyes, nose and mouth. In addition, one of these muscles, buccinator, plays an important role in mastication (p. 552), while others are involved in forming and shaping the sounds produced by the larynx into recognizable words.

To be able to perform these many complex actions it is not surprising that the majority of the muscles have only one attachment to bone, the other being to the superficial fascia and skin. Consequently, when the facial muscles contract to dilate or constrict the eyes, nose or mouth, their secondary action has a profound effect on facial expression, through which an individual can convey their emotions to the world at large, and recognize the feelings of others.

All of the muscles of facial expression are derived from the second pharyngeal arch, and consequently are supplied by branches of the *facial* (seventh cranial) *nerve*. Damage to branches of the facial nerve will result in a loss of tone in the muscles supplied by that branch, with a consequent sagging of that part of the face. The skin of the face is supplied by the ophthalmic, maxillary and mandibular branches of the *trigeminal* (fifth cranial) *nerve*.

These muscles are best considered in groups in relation to whether they act upon the eyes, nose or mouth. Those muscles which do not adhere to this plan are considered individually.

Movements of the eyebrows

Occipitofrontalis
Corrugator supercilii
Procerus

Occipitofrontalis

Raising the eyebrows, as in an expression of surprise, is produced by occipitofrontalis (Fig. 5.6), which is in two parts united by the strong aponeurosis of the scalp. The posterior part arises from the *outer part* of the *superior nuchal line* and the *mastoid part* of the *temporal bone* from where it passes to

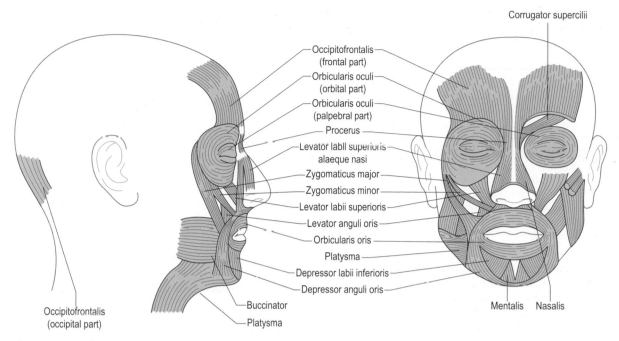

Corrugator supercilii

Occipitofrontalis
(frontal part)

Orbicularis oculi
(orbital part)

Orbicularis oculi
(palpebral part)

Procerus

Levator labii superioris
alaeque nasi

Zygomaticus major

Zygomaticus minor

Levator labii superioris

Levator anguli oris

Orbicularis oris

Platysma

Depressor labii inferioris

Depressor anguli oris

Buccinator

Platysma

Occipitofrontalis
(occipital part)

Mentalis   Nasalis

Figure 5.6 Facial muscles.

attach to the aponeurosis. The larger anterior part of the muscle runs from the aponeurosis to the superficial fascia of the forehead and eyebrow, interlacing with orbicularis oculi.

Action

The contraction of the posterior or anterior bellies of occipitofrontalis pulls the scalp backwards or forwards respectively. The anterior bellies acting from the aponeurosis raise the eyebrows either together in surprise, or individually as in a quizzical expression. This action produces the deep transverse lines on the forehead.

Corrugator supercilii

This small muscle (Fig. 5.6) arises from the *medial end* of the *superciliary arch* (eyebrow), blending with orbicularis oculi. Its fibres run upwards and laterally to attach to the skin of the eyebrows. On contraction they pull the two eyebrows together medially and downwards as in a frown, so producing the deep vertical wrinkles between the eyebrows.

Procerus

These small muscles, are continuous with each other across the midline. They run from the *lower part* of the *nasal bone* to the skin over the lower part of the forehead, intermingling with the anterior belly of occipitofrontalis (Fig. 5.6). On contraction they pull down the medial part of the eyebrow producing transverse wrinkles at the root of the nose.

Muscles around the eye

Orbicularis oculi

Orbicularis oculi

One of the most complex muscles of the face and by far the most important around the eye (Fig. 5.6). It consists of three parts – an orbital part, a palpebral part and a lacrimal part. The orbital part surrounds the orbit, spreading onto the forehead, temple and cheek, taking its origin from the *medial margin* of the *orbit* and the *medial palpebral ligament*. The fibres arising from above the ligament run elliptically around the orbit to attach again below the ligament.

The palpebral part of the muscle is much thinner and lies within the eyelids. These fibres arise from the medial palpebral ligament and run outwards within the eyelids to unite at the lateral palpebral raphe.

The lacrimal part of orbicularis oculi arises from fascia behind the lacrimal sac and *crest* of the *lacrimal bone*, passing laterally to attach to the tarsal plates of each eyelid.

Action

Orbicularis oculi plays an important part in protecting the eye by washing the eyeball with tears and by firmly closing to protect the eye against insult and injury. The orbital part of the muscle, which can act independently or with other parts, draws the skin of the forehead and cheek towards the medial angle of the orbit, so that the eye is screwed up tightly as when a bright light is shone into the eyes or when dust blows towards them. This action produces the characteristic 'crow's feet' at the lateral corner of the eye. The palpebral part of the muscle exerts much finer control over the individual eyelids, and by reflex or voluntary contraction pulls down the upper and raises the lower eyelid to close the eye. By pulling the eyelids medially during regular blinking, tears produced by the lacrimal gland are wiped over the surface of the eyeball to keep it moist and wash particles towards the lacrimal puncta. The small movements of the eyelids are very important in non-verbal communication between humans.

The lacrimal part of the muscle dilates the lacrimal sac.

Application

When orbicularis oculi is paralysed there is an inability to close the eye. The resultant failure of this mechanism for blinking and washing the eyeball means that the eyeball tends to become red and inflamed.

Muscles around the nose

Nasalis
Levator labii superioris alaeque nasi
Depressor septi

These small muscles around the nose (Fig. 5.6) act to dilate or constrict the nasal apertures. These actions are rudimentary in humans, but are of obvious importance in some animals, for example, camels. The dilators are nasalis and levator labii superioris alaeque nasi, which also acts on the upper lip, while the constrictor is depressor septi.

Muscles around the mouth

Orbicularis oris
Buccinator
Levator labii superioris alaeque nasi
Levator labii superioris
Zygomaticus major and minor
Levator anguli oris
Depressor anguli oris and depressor labii inferioris
Risorius and mentalis

Orbicularis oris

Surrounding the mouth (Fig. 5.6) orbicularis oris is a composite sphincter muscle with deep and superficial parts. Vertically it extends from the nasal septum to midway between the chin and lower lip. The deep fibres are continuous with buccinator, while the superficial fibres are all derived from other muscles. The deep fibres attach to both the *maxilla* and *mandible* near the lateral incisor. Between the deep and superficial layers of the muscle, the intrinsic fibres of orbicularis oris run elliptically around the mouth, having no bony attachment. Some of the superficial fibres decussate like those of buccinator.

Action

Orbicularis oris produces movements of the lips as in 'puckering' or whistling. It has an important role in speech by changing the shape of the mouth and lips to help form recognizable sounds. It is also important in mastication as its contraction against the teeth helps to keep food between the teeth during chewing. Contractions of the various muscles which insert into orbicularis oris change its shape.

Buccinator

Buccinator is continuous with the deep part of orbicularis oris and forms the substance of the cheek (Fig. 5.6). It arises from both the *maxilla* and the *mandible* opposite the molar teeth, and from the *pterygomandibular raphe*, which stretches from the *pterygoid hamulus* to the *posterior end* of the *mylohyoid line*. The fibres run forwards to blend with those of orbicularis oris. The medial fibres decussate posterolateral to the angle of the mouth, so that the lower fibres run to the upper lip and the upper fibres to the lower lip. The interlacing of deep buccinator fibres, with some of the superficial fibres of orbicularis oris, forms an easily felt nodule at the angle of the mouth known as the modiolus.

Action

Buccinator presses the cheek against the teeth or resists outward pressure against the cheek. The former action is important in mastication as it prevents the accumulation of food in the vestibule of the mouth – a common problem when buccinator is paralysed. When blowing up a balloon, buccinator can be felt resisting the outward pressure of air against the cheek. When the lips are protruded by orbicularis oris, buccinator causes the cheek to 'cave in', thereby producing a sucking action. Buccinator can also be used to pull orbicularis oris posteriorly against the teeth, or to retract the angle of the mouth to expose the premolar and molar teeth.

Palpation

If a finger is placed between the cheek and teeth, buccinator can be felt contracting as the finger is pressed against the teeth.

Levator labii superioris alaeque nasi

This muscle (Fig. 5.6) arises from the *maxilla* and attaches to the *ala* of the *nose* and skin, and the muscle of the upper lip.

Levator labii superioris

This muscle (Fig. 5.6) arises from the *maxilla* above the infraorbital foramen and attaches to the upper lip towards its lateral end.

Zygomaticus major and minor

These both arise from the *zygomatic bone* and attach to the upper lip near its angle (Fig. 5.6).

Levator anguli oris

This arises from the *maxilla* below the infraorbital foramen (Fig. 5.6), inserting into skin and muscle at the angle of the mouth. A few of its fibres may pass to the lower lip.

Action

All of these muscles when contracting on both sides raise the upper lip as in smiling; contraction of one side produces a sneer. The tone of these muscles is important in maintaining the normal horizontal position of the mouth. When paralysed, there is a gradual drooping of the angle of the mouth on the affected side leading to leakage of saliva and a constant dribble from the corner of the mouth. Zygomaticus major, assisted by levator anguli oris, raises and pulls the angle of the mouth laterally, as in laughing.

The following facial muscles have their effect on the lower lip.

Depressor anguli oris and depressor labii inferioris

Depressor anguli oris arises from the *front* of the *mandible* below the mental foramen and blends with the muscles of the lower lip at the angle of the mouth (Fig. 5.6), as well as inserting into skin. Some of its fibres may pass into the upper lip. It is continuous with platysma and overlaps depressor labii inferioris which also arises from the *mandible* below the mental foramen. Its fibres attach to the skin and muscle of the lower lip, the fibres from each side blending together (Fig. 5.6).

Risorius and mentalis

These two muscles, risorius and mentalis although considered as part of the lower lip musculature, usually only have connections to the skin of the lower lip. Risorius arises from the fascial covering of the parotid gland and inserts into skin at the angle of the mouth. Mentalis arises from the *mandible* below the incisors and passes to the skin of the chin (Fig. 5.6). Platysma (Fig. 5.6), with which risorius may be completely fused, is considered later (p. 557).

Action

Depressor anguli oris and depressor labii inferioris pull the mouth downwards and laterally, giving an expression of sadness. By itself depressor labii inferioris curls the lower lip downwards. Mentalis on the other hand, pulls the skin of the chin upwards, causing protrusion of the lower lip as in pouting. Risorius merely pulls the angle of the mouth laterally.

Application

If the muscles on one side of the face are paralysed, as in Bell's palsy, the typical problems encountered are an inability to close the eye; a tendency for food to accumulate in the vestibule of the mouth, i.e. between the cheek, lips and teeth of the paralysed side; the corner of the mouth droops downwards on the affected side, and on smiling, the mouth is pulled towards the sound side by the unopposed action of the intact muscles.

Section summary

Muscles of the face

The muscles of the face (facial expression) principally act to open or close the orbit, nose and mouth. However, because they insert into the skin they are also intimately involved in changing facial expressions.

Movement of the eyebrows

Occipitofrontalis
Corrugator supercilii
Procerus

Muscles around the eye

Orbicularis oculi

Muscles around the nose

Nasalis
Levator labii superioris alaeque nasi
Depressor septi

Muscles around the mouth

Orbicularis oris
Buccinator
Levator labii superioris alaeque nasi
Levator labii superioris
Zygomaticus major
Zygomaticus minor
Levator anguli oris
Depressor anguli oris
Depressor labii inferioris
Risorius
Mentalis

Muscles moving the mandible

The movements of the mandible are complex, involving the coordinated action of the muscles attached to it. Because of the nature of the temporomandibular joint (p. 559), four basic movements of the mandible can be identified; these being:

1. *protraction* – pulling the mandible forwards so that its head articulates indirectly with the articular eminence of the temporal bone
2. *retraction* – pulling the mandible backwards so that its head moves back into the mandibular fossa
3. *elevation* – closing the mouth
4. *depression* – opening the mouth.

Combinations of these basic movements occur in chewing and grinding. Opening the mouth, particularly against resistance, as when eating sticky foods, will involve the infrahyoid group of muscles in order to stabilize the hyoid bone and provide a firm base against which mandibular movements can be made.

Muscles elevating the mandible

Masseter
Medial pterygoid
Temporalis (p. 555)

Masseter

A flat quadrilateral muscle with deep and superficial parts (Fig. 5.7A). The superficial part arises from the *zygomatic process* of the *maxilla* and the *anterior two-thirds* of the *zygomatic arch*, the muscle fibres running downwards and backwards to attach to the *outer surface* of the *angle* of the *mandible*, extending onto the *lower half of the outer surface* of the *ramus*. The deeper part arises from the *deep*

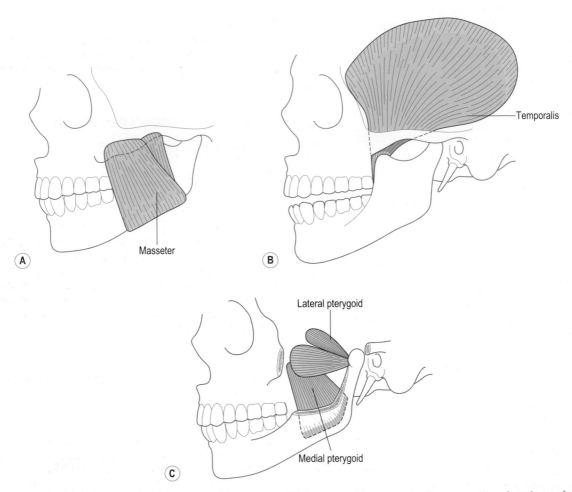

Figure 5.7 Muscles of mastication: (A) masseter, (B) temporalis, (C) the pterygoid muscles with the zygomatic arch and part of the mandible removed.

surface of the *zygomatic arch*, the fibres of which run downwards and backwards to attach to the *ramus* and *coronoid process of the mandible*.

Nerve supply

The *mandibular division* of the *trigeminal* (fifth cranial) *nerve*, with the skin over the muscle supplied mainly by the anterior rami of C2 and 3, and partly by the mandibular branch of the trigeminal nerve.

Action

Masseter elevates the mandible so approximating the upper and lower teeth. The superficial fibres help pull the mandible forwards during protraction.

Palpation

Masseter can be easily palpated by applying pressure through the skin of the cheek below the zygomatic arch when the teeth are clenched together.

Medial pterygoid

A thick quadrilateral muscle (Fig. 5.7C; see also Fig. 5.14) arising from the *medial side* of the *lateral pterygoid plate* and the *pyramidal process* of the *palatine bone*. A smaller head arises from the *maxillary tubercle*. From these two origins, which surround the lower fibres of lateral pterygoid, the muscle fibres run downwards, backwards and laterally to attach to a rough *triangular impression* on the *inner surface* of the *mandible* between the angle and the mylohyoid line.

Nerve supply

By the *mandibular division* of the *trigeminal* (fifth cranial) *nerve*.

Action

Medial pterygoid elevates the mandible and so closes the mouth. Because of the direction of the muscle fibres, it also pulls the mandible forwards. When the medial and the lateral pterygoid muscles of one side contract together, the chin swings to the opposite side. Such movements are important in chewing.

Muscles retracting the mandible

Temporalis
Digastric (p. 556)
Geniohyoid (p. 556)

Temporalis

A large, flat, fan-shaped muscle arising from the *temporal fossa* of the *temporal bone* and the fascia covering it (Fig. 5.7B). The anterior fibres run almost vertically, while the most posterior are almost horizontal. All of the muscle

fibres, however, converge to a thick tendon which passes deep to the zygomatic arch to insert into the *apex* and *deep surface* of the *coronoid process* and *anterior border* of the *ramus* of the *mandible*. A few fibres may become continuous with buccinator, while other more superficial fibres may fuse with masseter.

Nerve supply

The *mandibular division* of the *trigeminal* (fifth cranial) *nerve*, with the skin over the muscle supplied mainly by the mandibular branch of the trigeminal nerve and the anterior rami of C2 and 3.

Action

The posterior horizontal fibres retract the mandible after it has been protruded. The anterior vertical fibres elevate the mandible and close the mouth; they are constantly active to counteract the effects of gravity.

Palpation

Temporalis can be palpated by applying firm pressure above the zygomatic arch through the temporal fascia over the temporal fossa, particularly when the teeth are firmly clenched together.

Muscles protracting the mandible

Lateral pterygoid
Medial pterygoid (p. 555)
Masseter (p. 554)

Lateral pterygoid

The lateral pterygoid (Fig. 5.7C; see also Fig. 5.14) has two heads, an upper head which arises from the *inferior surface* of the *greater wing* of the *sphenoid*, and a lower head which arises from the *lateral surface* of the *lateral pterygoid plate*. From this extensive origin the fibres pass backwards and slightly laterally to insert into the *front* of the *neck* of the *mandible* and the *capsule* and *intra-articular disc* of the temporomandibular joint.

Nerve supply

By the *mandibular division* of the *trigeminal* (fifth cranial) *nerve*.

Action

Contraction of lateral pterygoid pulls the head of the mandible, the intra-articular disc and the joint capsule forwards onto the articular eminence – a movement which occurs during protraction of the mandible and opening the mouth. Working with medial pterygoid of the same side, lateral pterygoid produces a slight rotation of the jaw so that the chin swings to the opposite side. When all four pterygoid muscles contract, the mandible is protruded so that the lower incisors are carried in front of the upper ones (see Fig. 5.16B). When only both lateral pterygoids are working, the head of the mandible is carried forwards; there is an accompanying slight rotation of the head against the intra-articular disc so that the mouth opens. This is essentially what happens when sleeping with the mouth open. Lateral pterygoid is the main antagonist to retraction; its eccentric contraction therefore helps to control this movement.

Muscles depressing the mandible

Digastric
Mylohyoid
Geniohyoid
Platysma

Digastric

Digastric as its name suggests, has two bellies, anterior and posterior, united by an intermediate tendon (Fig. 5.8). The posterior belly passes downwards and forwards from the *medial surface* of the *mastoid process* in close association with stylohyoid, to an intermediate tendon which passes through the insertion of stylohyoid and is held by an aponeurotic sling to the *upper surface* of the *hyoid*. The anterior belly of the muscle runs upwards and forwards to attach to the *digastric fossa* on the *lower border* of the *mandible* close to the symphysis menti.

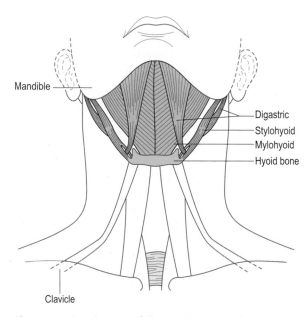

Figure 5.8 Anterior view of the suprahyoid muscles.

Mandible

Digastric
Stylohyoid
Mylohyoid
Hyoid bone

Clavicle

Nerve supply

The posterior belly is supplied by the *facial* (seventh cranial) *nerve*, and the anterior belly by the *nerve to mylohyoid* from the inferior alveolar branch of the *mandibular division* of the *trigeminal* (fifth cranial) *nerve*.

Action

If the hyoid is fixed, digastric can depress the mandible thereby opening the mouth, an action required when acting against resistance. With the mandible elevated, digastric also helps in retraction of the mandible. With the mandible fixed, digastric raises the hyoid and with it the larynx, an important action in swallowing.

Mylohyoid

The mylohyoid (Fig. 5.8) arises from the *mylohyoid line* on the *inner surface* of the *body* of the *mandible*. Its fibres run downwards and medially towards the midline to insert into a *median fibrous raphe* which extends from the symphysis menti to the upper surface of the body of the hyoid bone. The two mylohyoid muscles thus give rise to a muscular sheet which forms the floor of the mouth.

Nerve supply

By the nerve supply to the mylohyoid from the inferior alveolar branch of the *mandibular division* of the *trigeminal* (fifth cranial) *nerve*.

Action

Mylohyoid can depress the mandible against resistance, elevate the hyoid, and also raise the floor of the mouth. In elevating and fixing the hyoid, it helps to press the tongue against the roof of the mouth and so is important in swallowing.

Geniohyoid

A small muscle lying deep to mylohyoid. It runs from the *inferior genial (mental) spine* on the *posterior surface* of the *symphysis menti* to *the front* of the *body of* the *hyoid bone*.

Nerve supply

By the fibres from the *anterior primary ramus* of *C1*, which reach the muscle by travelling with the hypoglossal nerve.

Action

Depending upon which end of the muscle is fixed, geniohyoid can elevate the hyoid or depress the mandible. When the hyoid is pulled upwards and forwards, the floor of the mouth is shortened and the pharynx is widened ready to receive food.

Platysma

A broad, flat sheet of muscle lying in the superficial fascia over the front of the neck (Fig. 5.6). It is variably developed in different people, arising from the skin and superficial part of the chest and shoulder. The fibres cross the clavicle running upwards over the front and sides of the neck to attach to the *lower border* of the *body* of the *mandible* and fascia of the lower part of the face. In the neck the platysma of each side is separated by a gap in the midline: just below the chin the most medial fibres decussate with those of the opposite side.

Nerve supply

By the *facial* (seventh cranial) *nerve*.

Action

Platysma can depress the mandible and therefore help to open the mouth. However, as a muscle of facial expression, it can produce expressions of horror by depressing the angle of the mouth and lower lip. When a supreme effort is being made, as in weight lifting, platysma can be seen standing out in the neck. Similarly, platysma can be seen to stand out in runners after strenuous exertions. In these circumstances platysma may be acting as an accessory muscle of respiration by pulling on the chest wall. However, it is more probable that the muscle acts to prevent compression of the great veins and the sucking in of the soft tissues of the neck due to the violent respiratory efforts being made. Paralysis of the muscle causes the skin of the neck to fall away in slack folds.

Section summary

Movements of the mandible

Movement of the mandible occurs at the temporomandibular joint.

Movement	Muscles
Elevation	Masseter
	Medial pterygoid
	Temporalis
Retraction	Temporalis
	Digastric
	Geniohyoid
Protraction	Lateral pterygoid
	Medial pterygoid
	Masseter
Depression	Digastric
	Mylohyoid
	Geniohyoid
	Platysma

Muscles depressing the hyoid bone

Sternohyoid
Sternothyroid
Thyrohyoid
Omohyoid

Sternohyoid

A strap muscle (Fig. 5.9) arising from the *medial end* of the *clavicle*, the *posterior sternoclavicular ligament*, and the *posterior surface* of the *manubrium sterni*. The fibres run upwards and slightly medially to attach to the *lower border* of the *body* of the *hyoid*.

Nerve supply

By the *anterior primary rami of C1, 2 and 3* via the *ansa cervicalis*.

Sternothyroid

The sternothyroid (Fig. 5.9) is smaller than, and is situated posterior to, sternohyoid. It arises from the *posterior surface* of the *manubrium sterni*, below the attachment of sternohyoid, and from the *first costal cartilage*. Being broader than sternohyoid the fibres run upwards and laterally over the trachea and the thyroid gland to attach to the *oblique line* of the *thyroid cartilage*

Nerve supply

By the *anterior primary rami of C1, 2 and 3* via the *ansa cervicalis*.

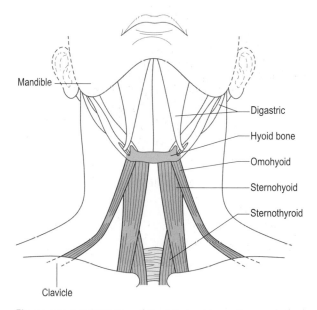

Figure 5.9 Anterior view of the infrahyoid muscles.

Mandible

Digastric
Hyoid bone
Omohyoid
Sternohyoid
Sternothyroid

Clavicle

Thyrohyoid

A short muscle running between the *oblique line* of the *thyroid cartilage* as a continuation of sternothyroid, and the *lower border* of the *body* and *greater horn* of the *hyoid*.

Nerve supply

By the *anterior primary ramus* of *C1*, which reaches the muscle via the *hypoglossal nerve*.

Omohyoid

This (Fig. 5.9) resembles digastric in that it consists of two bellies united by an intermediate tendon. The inferior belly arises from the *transverse scapular ligament* and adjacent *margins* of the *suprascapular notch* on the *upper border* of the *scapula*, and runs forwards and slightly upwards to an intermediate tendon. From the tendon the superior belly runs upwards and medially to attach to the *lower border* of the *body* of the *hyoid*.

Under cover of sternomastoid the intermediate tendon is bound down to the clavicle and the first rib by a sling of deep cervical fascia, thereby maintaining the angled appearance of the muscle.

Nerve supply

By the *anterior primary rami of C1, 2 and 3* via the *ansa cervicalis*.

Action of infrahyoid muscles

The so-called infrahyoid (strap) muscles act to depress the hyoid after it has been elevated during swallowing. Alternatively, by fixing the hyoid, the suprahyoid muscles can act on the mandible to depress it against resistance.

Thyrohyoid acting by itself raises the thyroid cartilage towards the hyoid, thereby pulling the larynx upwards under the root of the tongue. In this way thyrohyoid is responsible for closing the laryngeal inlet preventing food from entering the larynx during swallowing. Acting by itself, sternothyroid opens the laryngeal inlet by pulling the thyroid cartilage away from the hyoid. During forced inspiration sternothyroid probably acts to keep the laryngeal inlet open.

Muscles elevating the hyoid bone

Stylohyoid
Digastric (p. 556)
Mylohyoid (p. 556)
Geniohyoid (p. 556)

Stylohyoid

A small thin muscle (Fig. 5.8) arising from the *posterior part* of the *styloid process* of the *temporal bone* near its root. It runs downwards above the posterior belly of digastric, splitting into two slips enclosing the intermediate tendon of digastric. These slips unite before attaching to the *root of the greater horn* of the *hyoid*.

Nerve supply

By the *facial* (seventh cranial) *nerve*.

Action

Stylohyoid elevates and retracts the hyoid, taking with it the tongue and lengthening the floor of the mouth.

Mastication and swallowing (deglutition)

Mastication is the process involved in chewing and grinding food between the molar and the premolar teeth in order to break it down into smaller fragments so that it can be moulded into a softer, more manageable bolus and swallowed. The requisite movements of the mandible occur at the temporomandibular joints (p. 559) and involve opening and closing of the mouth, together with protraction and retraction of the mandible.

Initially, food is bitten off by the incisor teeth by means of the contraction of masseter, medial pterygoid and temporalis. Chewing movements are essentially produced by the alternate contraction of the pterygoid muscles first on one side and then the other. The elevators (masseter and medial pterygoid) of the side on which the pterygoids are functioning are also active to keep the teeth opposed. On the opposite side, temporalis is active to retract that side of the mandible, particularly if the food is sticky. There may also be activity in digastric, mylohyoid and geniohyoid to help separate the teeth momentarily when the bolus is sticky. During these swinging movements of the mandible the food is prevented from escaping between the teeth by the action of other muscles intimately involved in the masticatory process. If food passes into the vestibule of the mouth it is returned by the contraction of buccinator. If it escapes medially into the oral cavity it is ground against the hard palate and pushed back between the teeth by the action of the tongue. Contraction of mylohyoid helps keep the tongue against the hard palate, while orbicularis oris around the mouth prevents food escaping through the lips.

Once a suitable consistency has been achieved the bolus of food is collected from the anterior part of the mouth by the tip of the tongue and pressed back towards the soft palate by mylohyoid, digastric and stylohyoid. At the same time the elevators of the mandible and mylohyoid contract, and the hyoid is elevated carrying with it the larynx. Geniohyoid then pulls the hyoid forwards to widen the pharynx in anticipation of receiving the bolus. The soft palate is raised so closing off the nasopharynx and preventing food entering the nose. Elevation of the tongue and the hyoid closes the laryngeal inlet so that food does not enter the airways. The food bolus slides over the surface of the epiglottis into the laryngopharynx. Respiration is reflexly inhibited as the constrictor muscles of the pharynx contract successively to push the bolus of food towards the oesophagus. When the bolus has passed the laryngeal inlet, the hyoid and the larynx are pulled down to their resting positions by contraction of sternohyoid, sternothyroid and omohyoid. Breathing is once again resumed and another mouthful can be dealt with.

Swallowing fluids is essentially similar, except that in the initial stages the tongue forms a tube. The fluid is then forced backwards by the tongue, flowing down over the sides of the epiglottis, thereby avoiding the laryngeal inlet. See also page 516.

Application

Painful spasm of the muscles of mastication is frequently seen following traumatic overstretching, a situation which can arise during dental extraction. The resulting 'trismus' responds well to gentle heat and soft tissue techniques applied to the muscles.

JOINTS

The temporomandibular joint

Introduction

The temporomandibular joint is the means whereby the mandible (lower jaw) articulates indirectly with the skull. It is one of the few joints in the body which contains a complete intra-articular disc dividing the joint space into upper and lower compartments. This feature facilitates the movements which are possible at the joint in such a way that the combined gliding and hinge actions produce rotation of the mandible about a transverse axis which passes between the two lingulae. The position of this axis means that during mandibular movements the structures entering the mandibular foramen, i.e. the inferior alveolar nerve and vessels, are not put under tension.

The articular surfaces of the two components of the joint are not congruent. Indeed, they are highly incongruent; the intra-articular disc improves congruence between the articular surfaces. However, it is probably a functional advantage to have highly incongruent joint surfaces at the temporomandibular joint because of the growth changes associated with the mandible, and therefore the changing demands made upon the joint.

At birth the two halves of the mandible are separate, with the body being a shell of alveolar bone enclosing the developing teeth. The ramus is short and the angle it makes with the body is obtuse, about 175°, so that the condyle lies

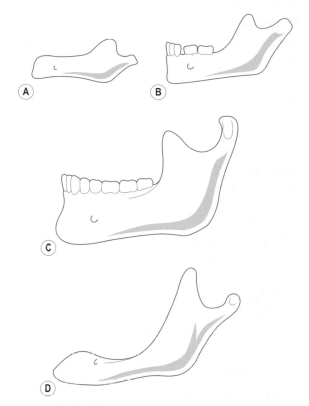

Figure 5.10 Age changes in the mandible: (A) at birth, (B) age 4, (C) adult, and (D) old age.

nearly in line with the body (Fig. 5.10A). During the first year the two bodies begin to fuse at the symphysis menti, with union being complete by the end of the second year. The ramus gradually enlarges, with a corresponding decrease in the angle between the body and ramus so that by about age 4 it has decreased to 140° (Fig. 5.10B). Progressive changes within the mandible continue so that in the adult this angle has decreased to 110° or less (Fig. 5.10C). With the loss of the teeth in old age, the alveolar sockets are absorbed and the angle between body and ramus now increases, due to bony remodelling, to 140° (Fig. 5.10D). The condyle becomes bent backwards so that the mandibular notch is widened.

The mandible grows in width between its angles, as well as in length, height and thickness. The condylar process grows not only upwards, but also backwards and sideways, thus contributing to its increase in total length, as well as height and width. Modelling maintains the shape of the condyle and the curves of the anterior margin of the ramus and coronoid process.

In adults, the long axis of the condyle is directed medially and backwards, forming an angle of about 30° with the frontal plane (Fig. 5.11). This angle is not, however, always consistent from side to side. If prolonged, the two

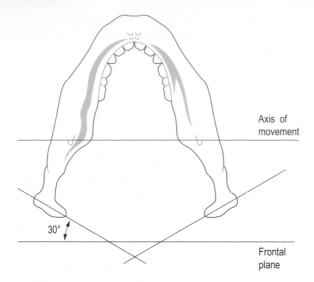

Figure 5.11 Inferior aspect of the mandible.

Axis of movement

Frontal plane

30°

axes would meet in the median plane at the anterior margin of the foramen magnum.

The temporomandibular joint, together with its associated ligaments and muscles, as well as the occluding surfaces of the teeth limits the forcible opposition of the jaws during biting. However, this is only one aspect of normal function which includes mastication, suckling, swallowing, yawning and speaking, although this latter activity involves little force at the joint.

Articular surfaces

Each temporomandibular joint is a synovial condyloid joint between the mandibular fossa of the temporal bone on the base of the skull and the corresponding condyle of the mandible. Between each pair of joint surfaces is a complete intra-articular disc. The bony articular surfaces are covered by fibrocartilage, with the superficial part having a large fibrous element with only the deeper parts containing the cartilage.

Mandibular fossa

The temporal articular surface extends from the squamotympanic fissure to the anterior margin of the articular eminence and is concavoconvex from behind forwards (Fig. 5.12). The fossa is oval, being wider mediolaterally than anteroposteriorly (average values being 23 and 19 mm respectively). At birth this surface is more-or-less plane, with the lateral part higher than the medial. Changes during childhood result in the sinuous surface shape seen in sagittal section. Mediolaterally the surface becomes more horizontal. In adults the fossa consists of a continuous plate of compact bone with no underlying spongy bone.

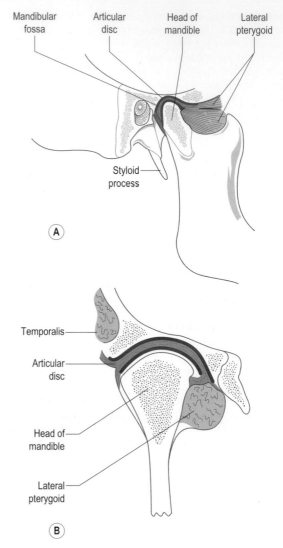

Figure 5.12 Sections of the temporomandibular joint: (A) sagittal and (B) coronal.

Remodelling occurs throughout life in this bony plate, and may lead to deviations from the usual joint shape.

Condyle of the mandible

The spindle-shaped condyle lies at right angles to the ramus of the mandible, with its long axis passing posteromedially. It is slightly smaller than the mandibular fossa, being on average 20 mm long mediolaterally and 10 mm wide anteroposteriorly. The articular surface of the condyle is curved and is displaced slightly posteriorly with respect to the highest point of the condyle (Fig. 5.12). Mediolaterally there is a great degree of symmetry.

Intra-articular disc

An articular disc separates the incongruent bony articular surfaces, moulding itself to them during movements at the joint (Fig. 5.12). The disc is thought to be derived from the tendon of lateral pterygoid, which in the embryo passes posteriorly between the head of the mandible and the temporal bone to attach to the malleus. That part of the tendon compressed between the two bones is said to form the disc. In adults when viewed from above it is an avascular oval plate of fibrous tissue. In the newborn it is relatively flat, but with development of the bony constituents of the joint the disc becomes shaped rather like a 'jockey's cap', so that its upper surface is concavoconvex to fit the mandibular fossa and the articular eminence, and its undersurface concave over the condyle.

The central region of the disc, although the most dense, is the thinnest, about 1.1 mm, while anteriorly and posteriorly its thickness varies up to 2.0 and 2.8 mm respectively. Very rarely is the disc perforated in its central region. Although essentially avascular the periphery of the disc, which is attached to the inside of the capsule, is highly vascularized and innervated. Anteriorly the disc attaches near the condyle, and receives, together with the adjoining capsule, the attachment of the upper fibres of lateral pterygoid. Posteriorly it attaches nearer to the temporal bone than the mandible, and is therefore not as freely mobile as elsewhere. This region of the disc is less dense than other parts, with the fibres having a crinkled arrangement enabling a small degree of elongation and recoil.

At birth the central region consists of relatively soft collagenous tissue, which becomes coarser with age and use. The functional wear that occurs is thought to be a major factor in the conversion of the region into a disc of fibrocartilage.

Joint capsule and synovial membrane

The strong, fibrous joint capsule is thin and loose, surrounding the articular surfaces and enclosing the intra-articular disc. It is attached above to the margins of the mandibular fossa as far anteriorly as the transverse prominence of the articular eminence. Below, the capsule attaches to the neck of the mandible (Fig. 5.13). To the inner aspect of the capsule is attached the intra-articular disc (Fig. 5.12). Above this attachment, the capsule is loose, thereby permitting free movement of the disc with respect to the mandibular fossa; below, the capsule is much more taut with the fibres of the medial and lateral parts of the capsule being short so that the disc appears to be attached directly to the medial and lateral ends of the condyle. Part of the tendon of lateral pterygoid inserts into the front of the joint capsule, and therefore indirectly to the intra-articular disc.

The capsular attachments are such that they allow a rotatory movement to occur between the condyle and the undersurface of the disc, as well as enabling the disc and condyle to move backwards and forwards as one unit against the mandibular fossa.

Two synovial membranes are associated with the joint, one for each compartment, i.e. above and below the disc. Each membrane lines all the non-articular surfaces and fuses with the periphery of the disc.

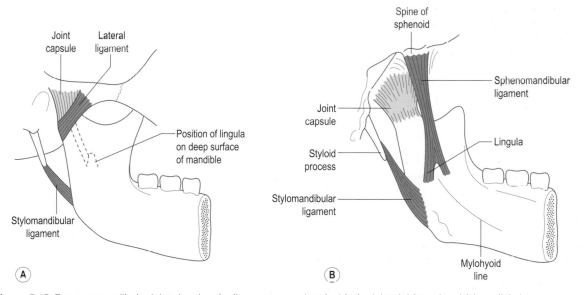

Figure 5.13 Temporomandibular joint showing the ligaments associated with the joint: (A) lateral and (B) medial views.

Ligaments

Three ligaments are associated with the joint – the lateral ligament blends with the capsule, while the sphenomandibular and the stylomandibular ligaments are accessory to the joint. These accessory ligaments, nevertheless have an important role in guiding the movement of the mandible against the base of the skull during associated muscle contraction. To a large extent the pterygomandibular raphe also acts to guide movement of the mandible.

Lateral ligament

This has a broad attachment to the lower border and the tubercle of the zygomatic bone. From here the fibres pass downwards and backwards, blending with the joint capsule to attach to the lateral and posterior parts of the neck of the mandible (Fig. 5.13A). It therefore reinforces the lateral aspect of the joint capsule.

Sphenomandibular ligament

A strong, thin flat band lying on the medial side of the joint passing downwards and forwards from the spine of the sphenoid to the lingula and adjacent area on the medial side of the ramus of the mandible (Fig. 5.13B). The sphenomandibular ligament develops from the sheath of that part of the cartilage of the first pharyngeal arch (Meckel's cartilage) lying between the base of the skull and the mandibular foramen.

Stylomandibular ligament

A thickening of the deep cervical fascia extending from near the apex of the styloid process to the lower part of the posterior border of the ramus of the mandible near the angle (Fig. 5.13). It separates the parotid (posterior) and the submandibular (anterior) salivary glands.

Blood and nerve supply

The arterial supply to the temporomandibular joint is from branches of the middle meningeal and anterior tympanic branches of the maxillary artery, and from the superficial temporal and ascending pharyngeal arteries. The superficial temporal and maxillary arteries are the terminal branches of the external carotid artery, while the ascending pharyngeal arises from the external carotid lower down in the neck. Venous drainage is to the retromandibular vein and thence to the jugular system within the neck. The lymphatic drainage of the joint is to the deep parotid lymph nodes.

The nerve supply to the joint is by twigs from the auriculotemporal and masseteric branches of the mandibular division of the trigeminal (fifth cranial) nerve.

Relations

The joint is relatively superficial, being covered by skin and subcutaneous tissue only. The condyle can be palpated immediately below the middle third of the zygomatic arch. Posteriorly crossing the zygomatic arch, but anterior to the external auditory meatus, are the superficial temporal vessels (Fig. 5.14). Pressure applied in this region reveals the pulse.

Anteriorly is the tendon of lateral pterygoid as it passes from its anterior and deep attachments to insert partly into the anterior joint capsule and neck of the mandible. Also anteriorly is the tendon of temporalis (Fig. 5.14), lying superficial to lateral pterygoid, as it passes deep to the zygomatic arch from the temporal fossa to attach to the anterior and medial aspects of the coronoid process.

The root of the styloid process is posteromedial to the mandibular fossa, which has the carotid sheath and its contents medial to it.

Stability

Stability at the joint depends to a large extent on whether the mouth is open or closed, being greater with the mouth closed and the teeth occluded.

With the mouth closed the teeth themselves stabilize the mandible on the maxilla and take any strain if a blow is received on the chin. Furthermore, the forward movement of the condyle is discouraged by the articular eminence and contraction of the posterior fibres of temporalis. Backward movement is prevented by the lateral ligament and by contraction of lateral pterygoid.

With the mouth open the joint is less stable, because the condyle has moved forward onto the articular eminence as well as rotating against the intra-articular disc. Backward dislocation is opposed again by the lateral ligament

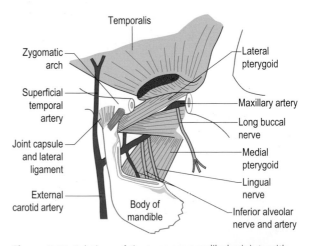

Figure 5.14 Relations of the temporomandibular joint, with part of the mandible removed.

and lateral pterygoid, while forward dislocation is restricted by the articular eminence, tension in the lateral ligament and contraction of masseter, temporalis and medial pterygoid. However, with the mouth open there is little to prevent an upward dislocation through the thin roof of the mandibular fossa. In practice this injury is extremely rare as a blow to the drooping mandible usually causes the mouth either to close or open more fully.

The most common dislocation is one in which one or both condyles pass forward beyond the articular eminences, i.e. a forward dislocation (Fig. 5.15A). This may occur in yawning or opening the mouth too widely. Reduction of the dislocation is often prevented by spasm of the posterior deep fibres of masseter, which tend to hold the dislocated jaw open because these fibres now pass down behind the axis of rotation due to the forward position of the condyles. To reduce the dislocation the spasm must be overcome, with or without an anaesthetic. If the dislocation is on one side only it can be reduced by placing the thumb on the back teeth of the dislocated side and pushing downwards and backwards. This releases the condyle and the mandible springs back into place; care must be taken quickly to remove the thumb. If the dislocation is bilateral then the angles of the mandible have to be gripped tightly with one hand and pulled downwards and backwards, while the other hand pushes the chin upwards (Fig. 5.15A). The mandible should, on release of the condyles, spring back into position (Fig. 5.15B). Occasionally

the dislocation tends to recur, particularly on yawning. With loss of permanent dentition the joint is less stable and anterior dislocation occurs more readily. It is also more easily reduced because of the relative lengthening of the ligaments of the joint due to tooth loss.

Movements

The movements of the mandible can be described as lowering (depression) and raising (elevation) of the jaw, protraction (forward movement), retraction (backward movement) and side-to-side movements. All of these movements are used to some extent during chewing. However, to fully understand and appreciate the movements which occur at the temporomandibular joint, some knowledge of the relationship between the upper and lower teeth with the mouth closed is necessary. With the teeth in contact (occluded) the upper incisors lie in front of the lower ones, there usually being some contact between their opposing surfaces (Fig. 5.16A). When the mouth is opened from this position the edges of the lower incisors pass downwards and forwards until the edges of both sets of incisors are directed towards each other. The downward and forward movement of the mandible, as in initiating opening the mouth, indicates the compound movement of forward gliding and rotation that occurs at the temporomandibular joint.

The form of the articular surfaces of the joint enables a hinge like rotation between the condyle and the inferior surface of the articular disc in the lower compartment of the joint, while at the same time the condyle and disc glide forwards (and slightly downwards) against the mandibular fossa as a single unit in the upper compartment. It is the sinuous form of the mandibular fossa and articular eminence which ensure that the teeth become separated by direct downward movement of the mandible as well as by its hinge action.

Throughout this combined gliding and hinge action the mandible tends to rotate about a transverse axis through the two lingulae. The position of this axis is determined by the sphenomandibular and the stylomandibular ligaments acting as stays, while various muscles are contracting to produce the movements.

Protraction and retraction

Protraction and retraction are the actions whereby the condyle and disc move as one unit against the mandibular fossa. In protraction the condyle and disc glide forwards so that the condyle rides on the articular eminence (Fig. 5.16B). Retraction is the opposite movement whereby the condyle is relocated in the mandibular fossa (Fig. 5.16C). Protraction is primarily produced by contraction of lateral pterygoid pulling on the disc and neck of the mandible. It is helped by medial pterygoid and the superficial fibres of masseter. Retraction is produced by the posterior almost horizontal fibres of temporalis,

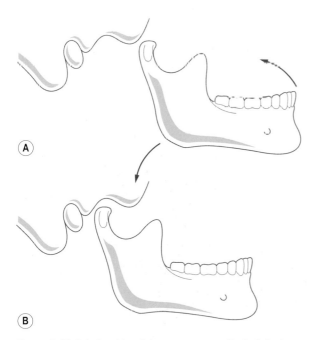

Figure 5.15 Relationship of the temporomandibular joint in (A) bilateral dislocation and (B) with the mouth closed.

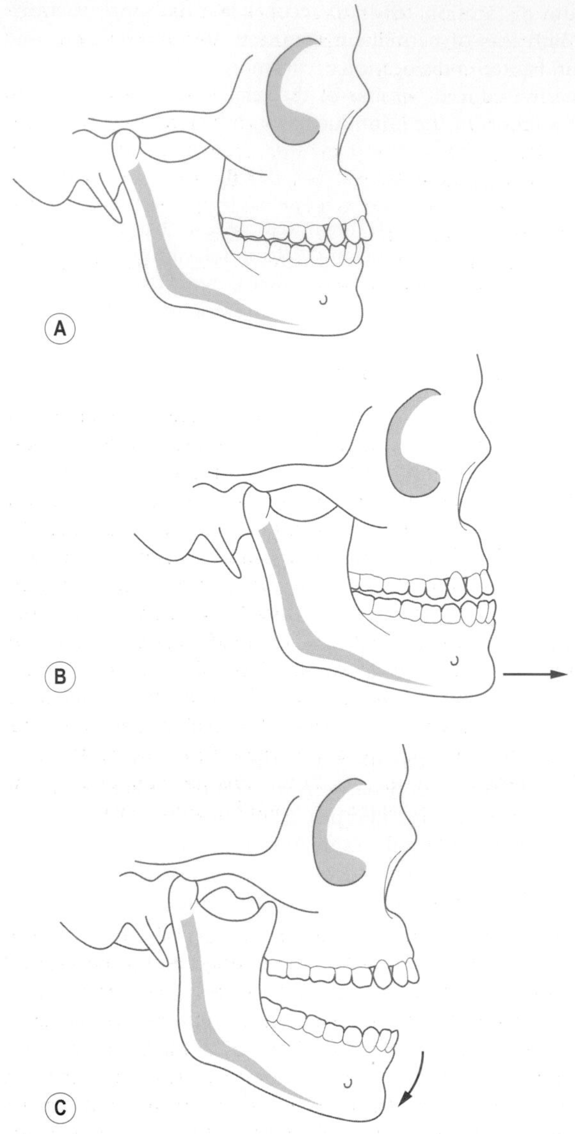

Figure 5.16 Movements of the mandible: (A) the closed position of the mouth, i.e. retracted and elevated, (B) protraction, (C) depression.

temporomandibular joint. Depression of the mandible usually involves the eccentric contraction of temporalis to give a controlled opening of the mouth under the influence of gravity. If there is resistance to opening, as when chewing sticky buns, then geniohyoid, mylohyoid and digastric contract pulling on the mandible from a fixed hyoid bone.

Closing the mouth (elevation) is a very powerful action produced by the combined action of masseter, medial pterygoid and temporalis, the two former holding the mandible in a sling. Temporalis is considered to be the antigravity muscle contracting to keep the mouth closed under normal conditions.

Side-to-side (grinding) movements

In chewing and grinding movements, the mandible is alternately protracted and retracted with the two sides usually moving in opposite directions so that one side is protracted while the opposite side is retracted. These movements are combined with elevation and depression with the result that the mandibular teeth move diagonally across the maxillary teeth; the intervening food is therefore crushed and ground. Consequently, all of the so-called 'muscles of mastication' are called into action, working rhythmically and alternately. Medial pterygoid is of particular importance in this respect as it plays a major role in producing the oblique movement of the mandible. However, to keep the food between the teeth and to prevent it lodging in the vestibule of the mouth or returning to the oral cavity proper, buccinator and the muscles of the tongue respectively are also working hard and so must therefore be included in any discussion of mastication.

Accessory movements

A number of accessory movements are possible at the temporomandibular joint; however, only two are described here. With the subject lying supine and the head turned through 90° so that the face looks laterally, a downward directed pressure on the mandibular condyle (using the thumb) causes it to move transversely within the mandibular fossa. Keeping the subject's head in the same position, pressure applied behind the lobe of the ear to the back of the condyle, and directed forwards, produces the forward movement of the condyle.

Surface marking and palpation

The line of the temporomandibular joint can be clearly and easily identified by placing the tip of the finger immediately in front of the tragus of the ear. As the subject opens their mouth the mandibular condyle moves forwards, and a large depression can then be felt; this is the cavity of the joint.

assisted by the suprahyoid muscles. Lateral pterygoid may contract eccentrically during retraction to control the movement.

Elevation and depression

Elevation and depression (Fig. 5.16A and C) of the mandible involve the hinge-like rotation of the condyle against the intra-articular disc in the lower compartment of the

Temporomandibular joint

Type
- Synovial condyloid joint.

Articular surfaces
- Head (condyle) of mandible with mandibular fossa of temporal bone.
- Fibrous intra-articular disc between head of mandible and mandibular fossa.

Capsule
- Strong, thin loose fibrous capsule completely surrounds joint attaching to articular margins of mandibular fossa and neck of mandible.

Ligaments
- Lateral ligament blends with joint capsule.
- Accessory sphenomandibular and stylomandibular ligaments.

Stability
- Provided by adjacent muscles and ligaments: depends on whether mouth is open or closed.
- Dislocation can occur anteriorly when one or both condyles pass beyond articular eminences.

Movements
- Protraction and retraction.
- Elevation and depression.
- Side-to-side movement.

THE EAR, EYE AND BRAIN

The ear

Development

The ear consists of three parts which function together but have different origins. The membranous part of the internal ear originates from the otic vesicle (surface ectoderm origin) during the fourth week. The otic vesicle divides into an anterior part which forms the saccule and the cochlear duct, and a dorsal part forming the utricle, the semicircular canals and the endolymphatic duct. The surrounding bony labyrinth develops from the adjacent mesenchyme. Except for the cochlear duct, from which develops the organ of Corti (spiral organ), the membranous labyrinth is concerned with maintaining balance.

The epithelial lining of the middle ear (tympanic cavity, mastoid antrum and auditory) tube is derived from the endoderm of the tubotympanic recess of the first branchial pouch. The auditory ossicles develop from the dorsal ends of the cartilages of the first (malleus and incus) and second (stapes) branchial arches.

The external auditory (acoustic) meatus develops from the first branchial cleft, and is separated from the tympanic cavity by the tympanic membrane, which is derived from three sources – the ectoderm of the first branchial cleft, intermediate mesodermal layer, and the endoderm of the first branchial pouch. The external ear (auricle) develops from six mesenchymal swellings around the margin of the first branchial pouch.

Components of the ear

The three parts of the ear (external, middle and internal) are all, except for the auricle, found within the temporal bone; the auricle is attached to the tympanic part of the temporal bone. The external ear collects the sounds and conveys these to the tympanic membrane, which separates the external and the middle ear, causing it to vibrate. Vibration of the tympanic membrane is transmitted across the middle ear by the three auditory ossicles (malleus, incus and stapes) to the internal ear. The middle ear communicates with the nasopharynx via the Eustachian (auditory) tube. The internal ear consists of two functionally distinct parts; that concerned with hearing (the cochlear part) and that with balance and position sense (the vestibular part). The sensory endings of both parts are supplied by the vestibulocochlear nerve, the eighth cranial nerve.

External ear

This consists of the auricle and the external acoustic (auditory) meatus (Fig. 5.17A) which collect and convey sound respectively towards the tympanic membrane. The auricle projects backwards and laterally from the side of the head, being connected to the fascia by three small, insignificant muscles. It is a single piece of elastic cartilage, except for the fibrofatty lobule, covered with skin; the named parts are shown in Fig. 5.17A. In adults its shape is extremely variable, increasing threefold in length from birth to adulthood; it also tends to increase in size and thickness in old age.

The external auditory meatus is 25 mm long. It is cartilaginous in its outer third, being continuous with the cartilage of the auricle, and bony in its medial two-thirds, being formed by the tympanic part of the temporal bone. The meatus curves upwards and backwards as it passes medially, its inferior wall being 5 mm longer than the superior wall because of the obliquity of the tympanic membrane. The skin lining the meatus is firmly attached to the underlying bone and the outer third of the canal contains numerous ceruminous (wax secreting) cells and hairs. The meatus lies behind the temporomandibular joint, with the mastoid air cells being immediately posterior.

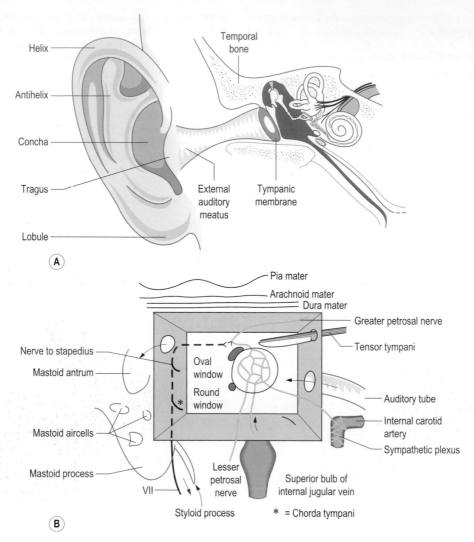

Figure 5.17 Schematic diagrams showing (A) the external, middle and internal ear, and (B) the middle ear with the tympanic membrane and auditory ossicles removed.

Middle ear

A narrow, irregular cavity containing the auditory ossicles immediately medial to the tympanic membrane (Fig. 5.17A). The middle ear can be conveniently viewed as a six-sided space, with that part above the tympanic membrane being the epitympanic recess. The cavity communicates with the nasopharynx via the auditory (Eustachian) tube which opens into the anterior wall, and with the mastoid air cells via the aditus in the posterior wall (Fig. 5.17B). The auditory tube enables the pressure on both sides of the tympanic membrane to be equalized; it is opened during swallowing.

The tympanic membrane is circular and concave laterally, and consists of three layers – modified skin externally,

an intermediate fibrous layer and a mucous membrane internally. The majority of the membrane is tense; however, there is a small flaccid area anterosuperiorly. Between the internal and external layers runs the *chorda tympani* branch of the facial nerve conveying taste sensations from the anterior two-thirds of the tongue and carrying parasympathetic fibres to the submandibular and the sublingual salivary glands.

The auditory ossicles articulate by the synovial joints transmit the vibrations of the tympanic membrane to the inner ear. The malleus attaches to the inner surface of the membrane and articulates with the incus, which in turn articulates with the stapes, the oval base of which lies in the *oval window*. Movements of the malleus and stapes are controlled and reflexly dampened down by contraction of *tensor tympani* and stapedius respectively, both of which are found within the middle ear. Tensor tympani is innervated by the mandibular division of the trigeminal nerve and stapedius by the facial nerve.

Internal ear

Situated within the petrous part of the temporal bone it consists of a complex series of fluid-filled spaces, known as the membranous labyrinth, occupying a similarly shaped cavity, the bony labyrinth. Displacement of the fluid in these spaces stimulates the sensory endings of the lining epithelium.

The bony labyrinth consists of three parts, the vestibule (containing the *utricle* and *saccule* of the membranous labyrinth), the *semicircular canals (anterior, posterior and horizontal)* and the cochlea (Fig. 5.18A). The anterior and posterior semicircular canals are at right angles to each other and lie 45° to the sagittal plane, with the anterior being anterior and lateral, and the posterior, posterior and lateral. The lateral semicircular canal lies horizontally. The membranous semicircular ducts are dilated at one end (the ampulla) (Fig. 5.18A) in which is a thickening (the ampullary crest) where endings of the vestibulocochlear nerve terminate. The three ducts open into the utricle, which communicates with the saccule, which in turn communicates with the cochlea. Thickenings in both the utricle and saccule are known as the maculae and contain terminations of the vestibulocochlear nerve. The ampullary crests of the semicircular canals convey information about rotatory and angular movements of the head, while the maculae convey information about linear and tilting movements. Disease of the semicircular ducts, the utricle and saccule gives rise to giddiness of varying degrees.

The bony cochlea consists of two and three-quarter turns of a spiral, resembling a shell lying on its side. It has a central supporting column of bone (the modiolus) to which is attached a thin lamina of bone partially dividing the spiral into two parts, the *scala vestibuli* above and the *scala tympani*

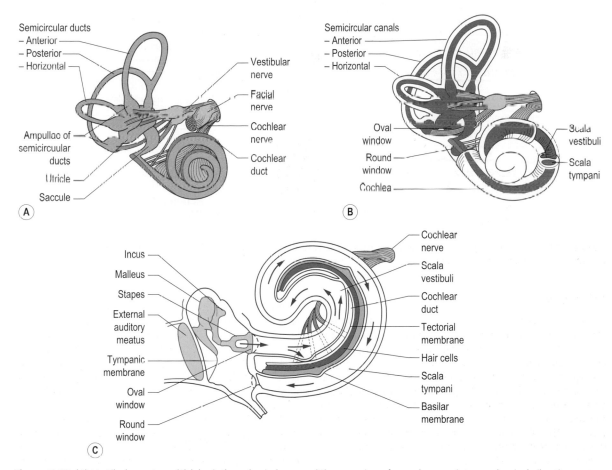

Figure 5.18 (A) Vestibular system, (B) labyrinth and spiral organ, (C) conversion of sound waves into mechanical vibrations (schematic diagram).

below (Fig. 5.18B). The membranous cochlear duct lines the bony cochlea and is triangular in cross-section. The outer wall of the triangle is thickened to form the spiral ligament, the lower part of which is the *basilar membrane* and the upper part the vestibular membrane. The thickened and highly specialized spiral organ (organ of Corti) lies on the basilar membrane. Pulsations transmitted to the perilymph within the membranous cochlea by movement of the *stapes* in the *oval window* pass through the scala tympani, and are transmitted to the fluid in the scala vestibuli, being adjusted by compensatory movements of the *round window*, thus causing movement of the basilar membrane, thereby stimulating the hair cells of the spiral organ (Fig. 5.18C); the end result is auditory perception. Low frequency sounds cause maximum activity in the basilar membrane; high frequency sounds are limited to the basal portion of the cochlea.

Section summary

The ear

The ear consists of three parts, external, middle and inner, each of which has a specific function. As well as being associated with hearing, the inner ear is also important in maintaining balance and equilibrium.

External ear

- Comprises the auricle and external auditory (acoustic) meatus which collect and convey sound towards the tympanic membrane.
- Chorda tympani (branch of VII) crosses the tympanic membrane between its external and internal layers.

Middle ear

- Narrow irregular cavity medial to tympanic membrane containing auditory ossicles (malleus, incus, stapes).
- Ossicles transmit vibrations of tympanic membrane to inner ear.
- Communicates with nasopharynx via the auditory tube and mastoid air cells via aditus.

Inner ear

- Located within petrous temporal bone comprising bony labyrinth (balance) and bony cochlea (hearing) parts.
- Bony labyrinth consists of vestibule (utricle, saccule) and semicircular canals (anterior, posterior, and horizontal): movement of perilymph in semicircular ducts detects angular and rotational movements; movement of maculae in utricle and saccule detect linear and tilting movements of head.
- Bony cochlea consists of $2^{3}/_{4}$ turns of a spiral with a central supporting column of bone: movement of stapes in oval window transmits pulsation to perilymph causing movement of basilar membrane, stimulating cells of spiral organ (auditory perception).

The eye

Development

The eyes begin to develop either side of the developing forebrain as optic vesicles by the end of the fourth week. Continuous with the forebrain the optic vesicles contact the surface ectoderm and induce development of the lens placode. When the optic vesicle invaginates to form the pigmented and neural layers of the retina, the lens placode also invaginates forming the lens pit and lens vesicle.

The retina, the optic nerve, the muscles and the epithelium of the iris, and the ciliary body are all derived from the neuroectoderm of the forebrain, while the lens and epithelium of the lacrimal glands, eyelids, conjunctiva and cornea all arise from the surface ectoderm. The extraocular muscles and all of the connective and vascular tissue of the cornea, the iris, the ciliary body, the choroid and the sclera are of mesodermal origin.

The eyeball

This consists of three concentric layers, an outer fibrous supporting layer comprising the *sclera* and *cornea*; a middle vascular, pigmented layer comprising the *choroid*, the ciliary body and the iris; and an inner layer of nerve elements, the *retina*. The interior of the eyeball contains fluid under pressure and is divided into anterior and posterior compartments by the lens and its attachments, and contain aqueous humour and the vitreous body respectively (Fig. 5.19A).

A thin fibrous sheet surrounds the sclera forming a socket for the eyeball, separating it from the other contents of the orbit. The eyeball is supported inferiorly by the *suspensory ligament*, and is surrounded and protected by extraocular fat. In adults the eyeball is almost spherical, having a diameter of 25 mm; however, the anteroposterior diameter may be greater or less than normal, giving rise to *myopia* (short-sightedness) or *hypermetropia* (long-sightedness) respectively (Fig. 5.19B). Relative to body size the eyeball is much larger in infants and children since it completes the majority of its growth in the antenatal period; it is also slightly larger in women than in men.

The two eyes look forwards; an imaginary line connecting the centre of the corneal curvature (the anterior pole) to the centre of the scleral curvature (the posterior pole) is the *optic axis* (Fig. 5.19C). The *visual axis* is, however more important and joins the centre of the cornea to the *fovea of the retina*, it represents the course taken by light from the centre point of vision. When looking at distant objects the visual axes of the two eyes are parallel, the optic axes are slightly and the optic nerves markedly convergent posteriorly.

Outer fibrous layer

The sclera is the posterior opaque part of the fibrous layer and forms about five-sixths of the circumference of the eyeball, the remainder being cornea. It is 1 mm thick

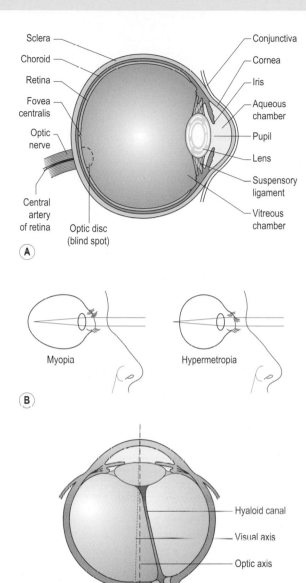

Figure 5.19 (A) Horizontal section through the eye, (B) myopia (short-sightedness) and hypermetropia (long-sightedness), (C) visual and optic axes of the eye.

1 mm thick, being covered by conjunctiva. The cornea is avascular but richly innervated by the ophthalmic nerve with abrasion of its surface being extremely painful, its sensitivity to touch forms the basis of the corneal reflex which results in reflex contraction of orbicularis oculi and closure of the eye.

The majority of the refraction of the eye takes place at the surface of the cornea and not at the lens. Irregularities in the curvature of the cornea, which ideally should correspond to a section of a perfect sphere, interfere with the ability to form sharp images on the retina. When the cornea is more curved in one direction than the other the condition is known as astigmatism.

The surface conjunctiva and the cornea are kept moist and clean by a watery fluid secreted by the lacrimal gland. Constant blinking is an important part of the mechanism of fluid flow across the cornea; drying of the cornea causes serious damage to its surface cells.

Middle vascular layer

Often called the uvea it consists of three parts, the choroid, the ciliary body and the *iris* (Fig. 5.19A). The choroid is a thin membrane lining the sclera as far as the corneoscleral junction, being loosely connected to the sclera except near where the optic nerve pierces when it is firmly attached. It consists of two parts, an outer pigmented (brown) layer, which prevents light passing through the sclera and the scattering of light entering via the pupil, and an inner vascular layer which is nutritive to the outer layer of the retina.

The ciliary body is a wedge-shaped ring connecting the choroid to the iris, and contains the *ciliary muscle* and the *ciliary processes*, being lined by the ciliary part of the retina (Fig. 5.20A). The inwardly projecting part of the wedge is directed towards the lens, being connected to it by fibres of the suspensory ligament of the lens. The ciliary muscle consists of two sets of smooth muscle fibres – an inner oblique set and an outer radial set – with both sets being under parasympathetic control. Contraction of the ciliary muscle reduces tension in the suspensory ligament of the lens, allowing its natural elasticity to increase its curvature so that the eye can focus on near objects, the process of accommodation. The ciliary processes are 60–80 radiating projections, 2 mm in length, and also give attachment to the suspensory ligament of the lens.

The iris is a thin contractile membrane, firmly attached at its periphery to the ciliary body, lying in front of the lens with a central opening, the pupil. It contains smooth muscle fibres organized into an inner circular sphincter pupillae and an outer radially arranged dilator pupillae. These two muscles control the size of the pupil and thus the amount of light entering the eye. Both muscles are under autonomic control, the sphincter pupillae being parasympathetic and the dilator pupillae sympathetic. The colour of the iris is due to the pigment cells in its posterior layer. In individuals with few pigment cells, and because of the

posteriorly and 0.5 mm thick anteriorly and gives attachment to the tendons of the extraocular muscles. The anterior part of the sclera is covered by *conjunctiva* and forms the 'white of the eye'. Posteriorly the sclera is pierced, 3 mm medial to the fovea, by the *optic nerve* and accompanying vessels (Fig. 5.19A).

The forward-bulging cornea is continuous with the sclera at the corneoscleral junction. It is dense and uniformly

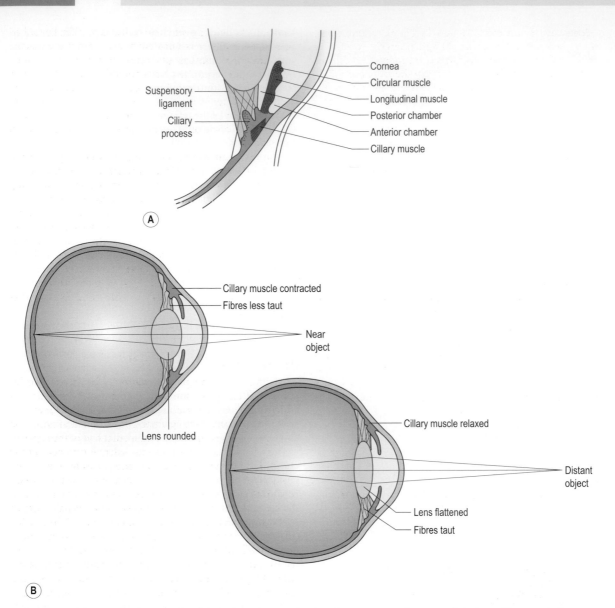

Figure 5.20 (A) Horizontal section through the anterior eye showing the suspensory ligament of the lens, (B) process of accommodation.

way other elements of the iris absorb and reflect light, the iris appears pale blue; with increasing numbers of pigment cells the iris darkens and may become dark brown.

Refracting media

The iris partly divides the region in front of the lens into anterior (aqueous) and posterior (vitreous) chambers, both of which contain aqueous humour (Fig. 5.19A). This is a clear watery solution formed by the epithelium of the ciliary processes, the fluid is resorbed at the iridocorneal angle into the sinus venosus to re-enter the circulation. Interference with the process of resorption results in an increased intraocular pressure, glaucoma, which affects the peripheral part of the visual field due to the pressure on the retina. Behind the lens and ciliary body is the posterior chamber containing the vitreous body, a transparent, colourless semi-gelatinous material (Fig. 5.19A).

The lens is biconvex, 10 mm in diameter and 4 mm thick, becoming thinner in old age. It largely consists of concentric lamellae of lens fibres surrounded by a capsule firmly attached to the ciliary body by the suspensory ligament of the lens. Both the capsule and lens are transparent and elastic. The shape of the lens is modified by the ciliary muscle as the eye focuses on objects at different distances (Fig. 5.20B). After middle age the lens becomes less elastic and the ability to accommodate is gradually lost (presbyopia), so that glasses are required for close work. The lens may also become less transparent with increasing age, giving rise to cataract.

Inner nervous layer

The light-sensitive layer, more commonly known as the retina, and extends onto the ciliary body and iris but in this region contains no nerve elements so is non-functioning. The retina is 0.5 mm thick posteriorly thinning to 0.1 mm anteriorly; however, both the optic disc and the fovea centralis are much thinner areas. It comprises two parts, an outer pigmented epithelial layer and an inner transparent layer containing the light receptors (*rods* and *cones*) (Fig. 5.21). The region where the fibres forming the optic nerve converge to pass through the choroid and sclera is the optic disc. It contains no light receptors and is therefore insensitive to light (the blind spot). Three mm lateral to the optic disc is the macula, which has at its centre a depression, the fovea centralis, where vision is most acute (Fig. 5.19A).

Within the inner transparent layer the rods and cones lie closest to the choroid, so that light has to pass through most of the retina before reaching them (Fig. 5.21). The cones are used in bright light as well as for colour discrimination. The macula contains only cones so that it functions

in detailed vision, i.e. when an object is specifically looked at it is always focused onto the macula. From the macula outwards the number of cones in the retina rapidly decreases, however, the number of rods increases; the rods are used in dim light. The pigment contained in the rods is bleached out in bright light but reforms in dim light so that objects previously not visible are seen (dark adaptation). There are six times as many rods as cones in the retina.

The blood supply to the retina is essentially from the central artery of the retina, a branch of the ophthalmic artery, which divides into four branches each supplying a separate quadrant of the retina. Because each branch is an end-artery, blockage results in blindness in the associated quadrant. The retina may also become detached from the choroid, either spontaneously or as a result of a blow to the eye, and vision is impaired. If the retina is torn, fluid passes outside the layer of rods and cones with vision again being lost. In both cases laser treatment can be used to reattach the retina in order to prevent further separation.

Movements of the eyeball

The direction of the gaze is controlled by the extraocular muscles, these being the four *rectus* muscles (*superior, medial, inferior, lateral*) and the two *oblique* muscles (*superior, inferior*). The recti all attach posteriorly to a tendinous ring surrounding the optic canal and medial part of the superior orbital fissure, and insert into the sclera 6 mm behind the edge of the cornea. Superior oblique passes forwards from the medial wall of the orbit to hook around the trochlea on the frontal bone, then backwards to attach to the upper surface of the sclera behind the equator of the eyeball. Inferior oblique passes laterally from the medial part of the maxilla below inferior rectus to attach to the sclera, again behind the equator of the eyeball. Of these muscles lateral rectus is innervated by the abducens (sixth cranial) nerve, superior oblique by the trochlear (fourth cranial) nerve and the remainder by the oculomotor (third cranial) nerve.

Medial and lateral rectus cause the eye to look horizontally, medially and laterally respectively (Fig. 5.22). However, because of the oblique course of the superior and inferior rectus within the orbit they tend to pull the eye medially in addition to turning it to look up and down respectively (Fig. 5.22). The two oblique muscles on the other hand tend to pull the eye laterally as well as moving it up and down (Fig. 5.22). Because the oblique muscles attach behind the equator they pull on the back of the eyeball. Consequently, superior oblique turns the eye to look downwards and laterally, while inferior oblique turns it to look upwards and laterally. In reality almost every movement of the eyeball involves at least three muscles. Coordinated movement between the two eyes is controlled by the brain.

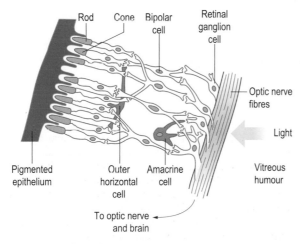

Figure 5.21 Organization of the retina.

Section summary

The eye

The eye consists of three layers, the outer fibrous (sclera and cornea), the middle vascular (choroid, ciliary body, iris) and the inner nervous (retina), surrounding two fluid-filled cavities (anterior and posterior chambers). The visual axis passes between the centre of the cornea and the fovea of the retina.

Outer fibrous layer

- Sclera – anterior part covered by conjunctiva forming 'white of the eye'.
- Cornea – continuous with sclera, is avascular but richly innervated.

Middle vascular layer

- Choroid is a thin membrane lining the sclera with outer pigmented and inner vascular layers.
- Ciliary body connects the choroid to the iris: contains ciliary muscle and processes.
- Iris is a thin contractile membrane attached peripherally to the ciliary body; lies anterior to lens.

Inner nervous layer

- Retina comprises outer pigmented and inner transparent layers, the latter containing light receptors (rods, cones).
- Optic disc (blind spot) is where optic nerve passes through choroid and sclera; contains no light receptors.

Chambers

- Contain clear watery solution within the anterior (aqueous) and posterior (vitreous) chambers.
- Lens alters curvature to facilitate changes in focus.

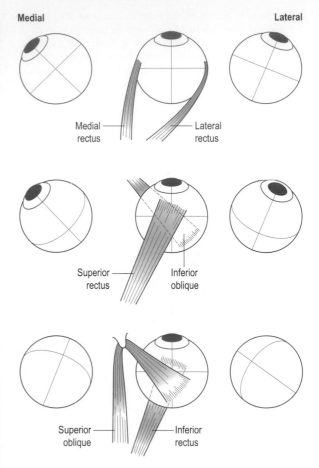

Figure 5.22 Diagrammatic representation of the extraocular muscles of the right eye and their actions.

The brain

Introduction

The central nervous system (CNS) is formed by the aggregation of bundles of axons and clusters of nerve cell bodies. The internal appearance and external form of the CNS reflect the manner in which these components are aggregated (see also The spinal cord p. 489).

The outer part of the brain is formed by nerve fibres and nerve cell bodies, such material appears grey on the transverse sections. The inner part of the brain consists mainly of myelinated axons, the myelin endows this material with a white appearance in transverse sections. Since cell bodies are not myelinated, the ganglia and nuclei form grey matter, while myelinated axons form white matter. Thus areas in the CNS formed mainly by axons are referred to as *white matter* (Figs. 5.23 and 5.24). In contrast, the cell bodies are not myelinated, so that areas formed mainly by cell bodies appear grey, and are referred to as *grey matter* (Figs. 5.23 and 5.24).

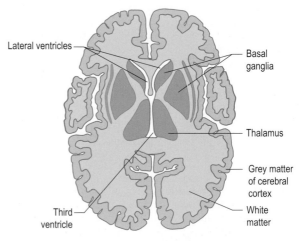

Figure 5.23 Transverse section of the brain showing the arrangement of the white and grey matter.

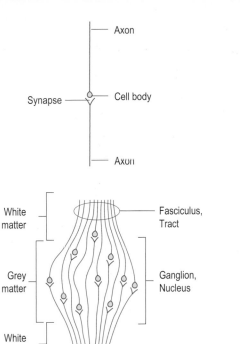

Figure 5.24 In the CNS, neurons are connected in series, with the axon of one neuron forming a synapse with the cell body of the next neuron.

A collection of cell bodies that forms a prominent, usually rounded swelling is referred to as a *ganglion* (Fig. 5.24). Clusters of axons that form a recognizable bundle within the CNS are referred to as a *fasciculus* (Fig. 5.24). When bundles of axons form a raised bump or convex contour on the surface of the CNS, typically in the spinal cord, the surface resembles the external surface of a tube, and the elevation is referred to as a *funiculus*.

These terms refer simply to the topographic appearance of collections of axons or cell bodies without regard to their function. Other terms are used which carry functional implications. A cluster of axons with a similar function is known as a *tract*, whereas a cluster of cell bodies with a similar function is known as a *nucleus*.

Glial cells

Nerve cells are not the only constituents of the CNS. Interspersed between the nerve cells are several cell types referred to collectively as glial cells where their function, in general, is to hold the neurons of the CNS together. As a whole, glial cells outnumber nerve cells and constitute almost half the total volume of the CNS.

Four types of the glial cells are found in the CNS – *oligodendrocytes*, *microglia*, *ependymal cells* and *astrocytes* (Fig. 5.25). Oligodendrocytes are responsible for myelination in the CNS; in addition, because they myelinate several parallel axons, they also serve to hold axons together. Microglia are the macrophages of the CNS, being responsible for removing foreign matter and cellular debris.

Ependymal cells are exclusively found lining the internal surface of the ventricles of the CNS. They are epithelial cells arranged side by side forming a cellular barrier between the nervous system and the fluid in its cavities.

Astrocytes derive their name from their star-shaped appearance. These cells consist of a cell body with long processes radiating away from it. Two types of astrocyte can be recognized. Microscopically, *protoplasmic astrocytes* occur chiefly in the grey matter of the CNS and have thick processes that branch repeatedly. They weave between cell bodies, collectively holding them together, with some processes being attached to small blood vessels (Fig. 5.25). In contrast, *fibrous astrocytes* are found chiefly in white matter and have fewer but longer processes that weave between axons. The principal role of astrocytes is to provide a framework for the CNS by holding the cell bodies and axons in place relative to one another. However, they also play an important role in the nutrition and metabolic activities of neurons by regulating the exchange of chemicals between themselves and blood vessels.

Development

The CNS develops from a single tubular structure, the *neural tube* (Fig. 5.26), which forms along the dorsal surface of the embryo, extending from the site of the future head to the tail. The tube therefore has two ends – a head end and a tail end. Because the tube eventually bends in the sagittal plane during development of the head and brain, the terms 'anterior, posterior, superior and inferior', as used in gross anatomy, cannot be applied to the nervous system without confusion. To overcome this different terms of reference are therefore used with respect to the CNS.

The tail end of the neural tube is called the *caudal end* (Fig. 5.26). The opposite end, regardless of the direction in which it points with respect to the rest of the body, is called the *rostral end* (Fig. 5.26). The surface of the neural tube which faces the belly of the embryo is called its ventral surface, irrespective of whether the tube is straight or curved in the sagittal plane, while the opposite surface is the dorsal surface. These directional terms are equally applicable to the descriptions of the adult CNS.

The neural tube maintains a narrow cavity along its length, while the thickness of its walls increases as the constituent cells multiply and grow. Most of the caudal part of the neural tube simply grows in length and diameter. Its walls get thicker but its cavity remains narrow. The rostral end of the neural tube undergoes several changes,

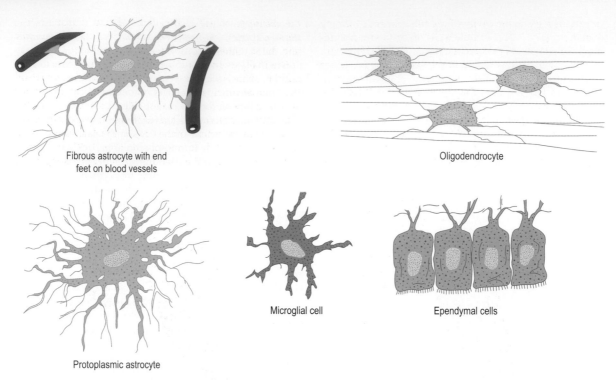

Figure 5.25 Microscopic appearance of neuroglial cells.

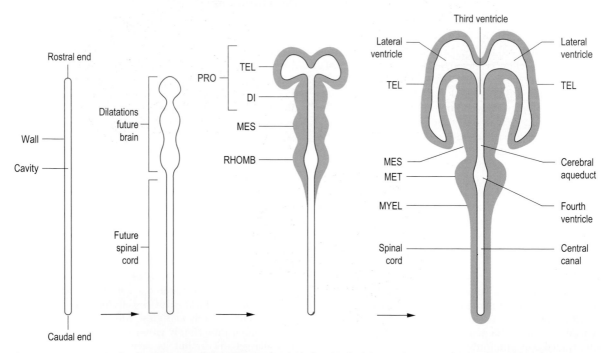

Figure 5.26 Stages in the development of the CNS, as viewed in longitudinal (coronal) sections from a dorsal aspect.

In various regions the cavity of the tube expands forming dilatations (Fig. 5.26), and the walls of the tube surrounding these dilatations thicken as a result of nerve cell proliferation. The most rostral end of the neural tube is referred to as the *prosencephalon* (PRO), more caudally is the *mesencephalon* (MES) and then the *rhombencephalon* (RHOMB), which subsequently divides into the *metencephalon* (MET) and the *myelencephalon* (MYEL). As the walls of the prosencephalon thicken they expand laterally as the *telencephalon* (TEL), the contained cavity being the *lateral ventricle* (Fig. 5.26). The remaining part of the prosencephalon is the *diencephalon* (DI). The diencephalon surrounds the *third ventricle*, the mesencephalon the *cerebral aqueduct* and the rhombencephalon the *fourth ventricle*.

Gradually, the telencephalon enlarges and grows in a set pattern. There is some growth rostrally on each side, beyond the original rostral limit of the neural tube, but the major growth is laterally and caudally (Fig. 5.26). Cells over the caudolateral aspect of the telencephalon eventually grow to form an extension that overlaps the lateral aspect of the main mass of the telencephalon (Fig. 5.27). The lateral ventricles participate in all this growth and elongate in

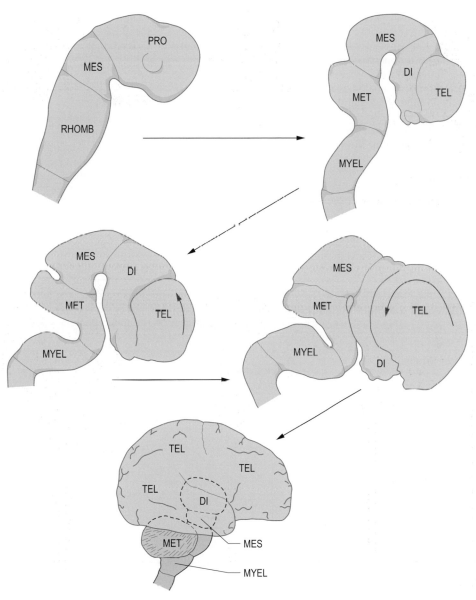

Figure 5.27 Successive stages in the development of the rostral end of the CNS, viewed from the right side

the directions along which the cells of the telencephalon proliferate. Overall, the telencephalon and lateral ventricles assume a curved shape extending laterally, then caudally, and eventually curving laterally, ventrally and rostrally.

As the telencephalon on each side develops, it covers and buries the diencephalon (Figs. 5.26 and 5.27).

Caudal to the diencephalon, the cavity of the neural tube expands to form the fourth (IV) ventricle, which assumes a diamond shape when viewed from behind (Fig. 5.26). The tissue surrounding this ventricle proliferates to form the rhombencephalon (referring to the rhomboid shape of the fourth ventricle). The rostral part of the rhombencephalon develops into a separate section called the metencephalon (Fig. 5.27). The dorsal portion (that part behind the fourth ventricle) differentiates into the *cerebellum*, while the ventral portion of the metencephalon (in front of the fourth ventricle) forms the *pons*, a bundle of neurons that arch around the ventral surface of the neural trube. The caudal portion of the rhombencephalon, the myelencephalon, remains relatively undifferentiated and connects the pons to the spinal cord (Fig. 5.27).

The mesencephalon (Fig. 5.27), between the diencephalon and the rhombencephalon, does not undergo great proliferation. Its walls simply thicken and its cavity remains tubular and narrow, its cavity is the cerebral aqueduct (Fig. 5.26).

Collectively, the mesencephalon, the pons and the myelencephalon constitute a single structural unit, the *brainstem* that connects the spinal cord to the diencephalon, with the cerebellum attached to it dorsally.

Nomenclature

The terminology used above with respect to the various subdivisions of the CNS is the most formal nomenclature that is in use. However, other systems of nomenclature based on Latin or English are also used. Unfortunately, in practice no single system is used consistently, and certain terms are used only in the most formal of circumstances. The various terminologies and their equivalents are given in Table 5.1, with the most commonly used terms in italics. In general, Latin or English forms are used when nouns are required to refer to various parts, but when adjectives are required it is usually the Greek forms that are used.

The term 'prosencephalon' is rarely used; neither is 'metencephalon', which refers collectively to the pons and the cerebellum, while 'myelencephalon' is virtually never used outside an embryological context. Otherwise, this region is referred to as the *medulla oblongata*, which alludes to the tapering shape of this part of the CNS as it narrows to the diameter of the spinal cord. The terms 'mesencephalon' and 'midbrain' are equivalent and are used interchangeably and with equal frequency.

Adult morphology

In the fully developed CNS most of its subdivisions are clearly recognizable. The spinal cord is long and cylindrical, occupying the rostral three-quarters of the vertebral column. The remainder of the CNS lies within the skull (Fig. 5.28). The cerebellum fills the posterior cranial fossa while the brainstem lies behind the clivus. The telencephalon and the diencephalon occupy the remainder of the cranial cavity, i.e. the middle and anterior cranial fossae.

Diencephalon

The diencephalon is not readily apparent in the intact adult brain as it lies deep to the telencephalon. It can be demonstrated in dissections in which the telencephalon and the cerebellum have been removed (Fig. 5.29). The largest component of the diencephalon is the *thalamus*, a large

Table 5.1 Nomenclature of the division of the CNS

GREEK SYSTEM	LATIN SYSTEM	COMMON ENGLISH	RELATED CAVITY
Prosencephalon:		*Forebrain*	
telencephalon	Cerebrum		Lateral ventricle
diencephalon			Third ventricle
Mesencephalon		*Midbrain*	Cerebral aqueduct
Rhombencephalon:		*Hindbrain*	Fourth ventricle
metencephalon	*Cerebellum*		
	Pons		
myelencephalon	*Medulla oblongata*		
	Medulla	*Spinal cord*	Central canal

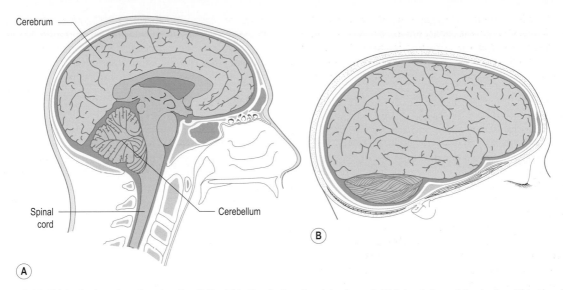

Figure 5.28 (A) Sagittal section showing the CNS within the skull and vertebral canal, (B) lateral view of the brain within the skull.

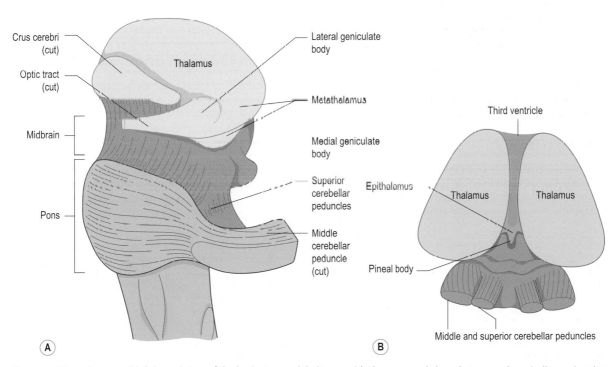

Figure 5.29 Thalamus: (A) left lateral view of the brainstem and thalamus with the crus cerebri, optic tract and cerebellar peduncles cut (see also Fig. 5.31), (B) the two thalami astride the rostral end of the brainstem, viewed from above.

collection of nuclei. Each thalamus is continuous caudally with the midbrain, and is separated from the opposite thalamus by the cavity of the third ventricle. The functions of the thalamus are diverse and include various motor, sensory and emotional processes.

Other components of the diencephalon include the *metathalamus*, *epithalamus* and *hypothalamus*. The metathalamus consists of two prominences projecting from the posterior, lateral, inferior surface of each thalamus (Fig. 5.29A). These are the *medial* and *lateral geniculate*

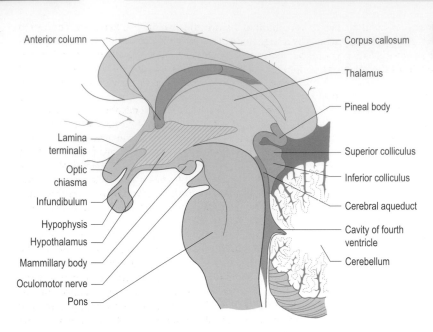

Figure 5.30 Sagittal section of the brain and brainstem showing the location of the hypothalamus.

Labels on figure:
Anterior column
Lamina terminalis
Optic chiasma
Infundibulum
Hypophysis
Hypothalamus
Mammillary body
Oculomotor nerve
Pons
Corpus callosum
Thalamus
Pineal body
Superior colliculus
Inferior colliculus
Cerebral aqueduct
Cavity of fourth ventricle
Cerebellum

bodies which are involved in certain auditory and visual processes respectively.

The epithalamus consists of a strip of neural tissue that bridges the posterior medial ends of the two thalami (Fig. 5.29B). Projecting from the posterior edge of the epithalamus in the midline is the *pineal body (gland)*, a structure concerned with hormonal regulation, particularly of the reproductive hormones.

The hypothalamus forms the rostroventral part of the lateral wall of the third ventricle and is best seen in sagittal sections (Fig. 5.30). It is a thin layer of tissue that hangs from the ventromedial surface of the thalamus and blends with the opposite hypothalamus in the midline, thereby forming a sling of neural tissue across the floor of the third

ventricle. Embryologically, the hypothalamus is formed by the ventral and ventrolateral walls of the rostral end of the neural tube while the lateral walls form the thalamus and the telencephalon. The rostral end of the hypothalamus is marked by the *lamina terminalis*, a strip of neural tissue bridging the midline and derived from the rostral end of the original neural tube.

Brainstem and cerebellum

The most obvious feature of the ventral surface of the brainstem is the thick bundle of transversely running nerve fibres that constitute the *pons* (Fig. 5.31). Rostral to the pons is the *midbrain*, caudal is the *medulla oblongata*.

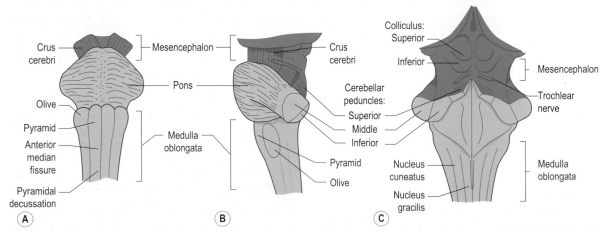

Labels on figure:
(A) Crus cerebri, Olive, Pyramid, Anterior median fissure, Pyramidal decussation, Mesencephalon, Pons, Medulla oblongata
(B) Crus cerebri, Cerebellar peduncles: Superior, Middle, Inferior, Pyramid, Olive
(C) Colliculus: Superior, Inferior, Cerebellar peduncles: Superior, Middle, Inferior, Nucleus cuneatus, Nucleus gracilis, Mesencephalon, Trochlear nerve, Medulla oblongata

Figure 5.31 Features of the brainstem: (A) anterior, (B) lateral and (C) posterior aspects.

The ventral surface of the midbrain is marked by two obliquely orientated, semicylindrical prominences, each known as a *crus cerebri* or *cerebral peduncle* (Fig. 5.31). The crura cerebri convey motor fibres from the cerebrum to the brainstem.

The fibres of the crura cerebri appear to run under the pons and emerge caudal to it on the ventral surface of the medulla oblongata as two longitudinal prominences either side of the midline. Each of these is known as a *pyramid* (Fig. 5.31); they convey motor fibres to the spinal cord. A shallow midline cleft, the *anterior median fissure*, separates the two pyramids, but caudally this fissure becomes obliterated for a short distance as the fibres in the pyramids cross the midline before passing into the substance of the spinal cord. The region where the pyramids cross is known as the *pyramidal decussation*. The caudal end of the decussation marks the junction of the medulla oblongata and spinal cord (Fig. 5.31).

Lateral to each pyramid is a rounded prominence, the *olive*, on the surface of the medulla oblongata. Cells within the olive are involved in the functions of the cerebellum.

The posterior surface of the midbrain is marked by four rounded projections, the colliculi, which are arranged in pairs with a *superior* and an *inferior colliculus* on each side (Fig. 5.31C). The superior colliculi mediate certain reflexes involved in turning the head and eyes in response to visual stimuli, while the inferior colliculi mediate similar responses to auditory stimuli.

The caudal end of the medulla oblongata is marked posteriorly by two rounded elevations on each side of the midline, the medial of which is the *nucleus gracilis* and the lateral the *nucleus cuneatus* (Fig. 5.31C). These nuclei are responsible for processing sensory information concerning pressure, touch and position sense that passes from the spinal cord to the brainstem and thalamus.

The *cerebellum* is a mass of neural tissue (Fig. 5.30), the principal function of which is the control of posture, repetitive movements and the geometric accuracy of voluntary movements. It is connected to the brainstem by three pairs of structures known as *peduncles* (Fig. 5.31B and C). The largest of these are the *middle cerebellar peduncles*, which run around the lateral aspect of the brainstem and are directly continuous with the pons. Passing rostrally from the cerebellum into the dorsal aspect of the midbrain are the two *superior cerebellar peduncles* connecting the cerebellum to the rostral portion of the brainstem and thalamus. Passing caudally from the cerebellum into the dorsolateral aspects of the medulla oblongata are the two *inferior cerebellar peduncles* which connect the cerebellum with the medulla oblongata and spinal cord.

Cerebrum

The cerebrum is divided into left and right halves by a large cleft, the *longitudinal fissure* (Fig. 5.32A); each half is referred to as a *cerebral hemisphere*. Deep within the longitudinal fissure the two hemispheres are connected by a large body of transversely running nerve fibres, the *corpus callosum* (Fig. 5.32A). When the two cerebral hemispheres are separated by a midline incision, their medial surfaces are revealed and the transected surface of the corpus callosum is seen (Fig. 5.32B).

The surface of each cerebral hemisphere is formed by a layer of grey matter, the cerebral cortex, which varies in thickness between 2 and 5 mm. It contains the cell bodies of the neurons responsible for the various functions of the cerebrum. Deep to the cortex, the cerebral hemispheres are formed by a large mass of white matter that consists of the axons of the neurons in the cerebral cortex, together with those that enter the cerebrum from the brainstem and diencephalon.

For descriptive purposes, each cerebral hemisphere is divided into four lobes that are named according to the bones of the skull to which they are most closely related (Fig. 5.32C). The anterior and posterior end of each cerebral hemisphere tapers to a single prominence known as the *frontal* and the *occipital pole* respectively. Similarly the anterior end of the temporal lobe is known as the *temporal pole*.

The surface of the cerebral cortex is thrown into ridges separated from one another by valley-like depressions, the ridges are referred to as *gyri* (singular: gyrus) and the depressions as *sulci* (singular: sulcus). Some sulci and gyri are significant because they subdivide the cerebral hemispheres both topographically and functionally (Fig. 5.32D).

The most obvious sulcus is that which separates the temporal lobe from the parietal and frontal lobes; this is the *lateral sulcus*. From the medial aspect of the temporal pole it runs backwards and upwards medial to the temporal lobe, ending at the junction of the temporal lobe with the parietal lobe.

Various transverse sulci cross the frontal and parietal lobes, located near the middle of the cerebral hemisphere is the *central sulcus* (Fig. 5.32D). It extends from the medial to the lateral surface of each hemisphere, reaching almost as far as the lateral sulcus. Whereas other sulci may extend onto the medial aspect of the cerebral hemisphere, or may reach almost as far as the lateral sulcus, the central sulcus is the only continuous sulcus that does both, and on these grounds it can be easily recognized.

The central sulcus serves to demarcate the frontal lobe from the parietal lobe. All the tissue anterior to the central sulcus therefore constitutes the frontal lobe, with the parietal lobe lying behind it. The sulcus running parallel to the central sulcus and in front of it is the *precentral sulcus*, with the gyrus between them being the *precentral gyrus* (Fig. 5.32D). Stimulation of the precentral gyrus results in contraction of voluntary muscles, and for this reason this gyrus is referred to as the *motor cortex*.

Similarly, the sulcus immediately behind the central sulcus is the *postcentral sulcus*, with the gyrus between the two being the *postcentral gyrus* (Fig. 5.32D). This

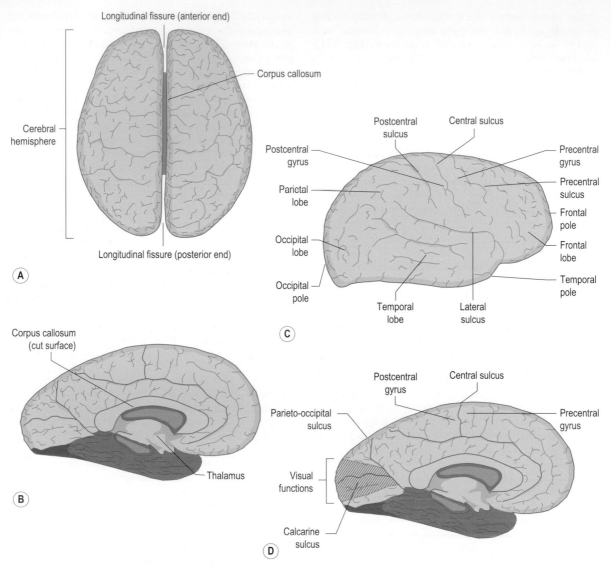

Figure 5.32 Brain: (A) superior view of brain showing its division into right and left cerebral hemispheres by the longitudinal fissure, (B) medial view of the left cerebral hemisphere showing the cut surface of the corpus callosum, (C) right cerebral hemisphere and its subdivision into lobes and principal sulci of the lateral surface, (D) medial surface of the left cerebral hemisphere showing the principal sulci and gyri.

gyrus receives sensory information from the brainstem and thalamus, and is therefore often referred to as the *sensory cortex*.

Behind the postcentral sulcus, the medial surface of the cerebral hemisphere is crossed by a clearly marked transverse sulcus that extends onto the upper part of the lateral surface of the hemisphere. This is the *parieto-occipital sulcus* which demarcates the parietal lobe from the occipital lobe (Fig. 5.32D). Anteriorly the parietal lobe is separated from the temporal lobe by the posterior end of the lateral sulcus, but posteriorly no major topographical feature delineates

the junction between the temporal lobe and the parietal and occipital lobes.

On the medial aspect of the cerebral hemisphere, the lower end of the parieto-occipital sulcus merges with the *calcarine sulcus* (Fig. 5.32D). Either side of the calcarine sulcus the cortex of the cerebrum is responsible for vision.

The occipital lobe is involved in the function of vision. The parietal lobe is, in general, responsible for sensory functions, such as the perception of touch, pressure, and the position of the body and limbs. However, it is also responsible for more sophisticated functions such as three-dimensional perception, analysis of visual images, language, geometry and calculations (Fig. 5.33). The frontal lobe subserves motor functions and governs the expression of intellect and personality. The upper part of the temporal lobe opposite the postcentral gyrus is involved in the perception of sound, while the remainder is concerned with memory and various emotional functions.

Base of the brain

The base of the brain is formed largely by the ventral surfaces of the frontal, temporal and occipital lobes of the cerebral hemispheres (Fig. 5.34). The medial edge of the most rostral end of the temporal lobe is the *uncus* and subserves the perception of smell.

Anteriorly neural tissue is absent between the two frontal lobes where they are separated by the longitudinal fissure. Caudal to this space is the floor of the third ventricle, formed by the junction of the hypothalamus from each side. Most posteriorly the brainstem projects caudally from the inferior or caudal surface of the diencephalon, which lies deep to the cerebral hemispheres.

Projecting in the midline from the floor of the third ventricle is a funnel-shaped extension of neural tissue, the *infundibulum* (Figs. 5.30 and 5.34). It connects the hypothalamus to the pituitary gland and consists of axons from cells in the hypothalamus which regulate the secretion of oxytocin and antidiuretic hormone by the pituitary gland. Caudal to the infundibulum are two small prominences, the *mammillary bodies*, the functions of which are not fully understood (Fig. 5.34).

On the ventral aspect of the diencephalon is an X-shaped structure consisting of two rostral and two caudal limbs;

the caudal limbs pass around the crura cerebri ending in the metathalamus (Fig. 5.34). The rostral limbs are the terminal ends of the *optic nerves*, which transmit visual information from the retina of the eye. The intersecting region is the *optic chiasma* in which the medial fibres from each optic nerve cross the midline. The caudal limbs are the *optic tracts*, each of which conveys visual information from one half of the visual field to the ipsilateral lateral geniculate body (Fig. 5.29A).

Cranial nerves

Certain nerves project from the cranial portion of the CNS, i.e. that portion within the skull. They connect the brainstem and diencephalon with various structures and tissues in the head and neck, as well as with various thoracic and abdominal viscera. While inside the skull, these nerves constitute part of the CNS and, as a rule, become peripheral nerves only once they leave the skull. There are 12 pairs of cranial nerves that are known by number or name, the name usually reflects the form, function or distribution of the nerve.

The first cranial nerve *(I)*, the *olfactory nerve*, mediates the perception of smell. It is composed of a series of short nerves that project from the olfactory epithelium in the roof of the nose through the cribriform plate of the ethmoid bone to the *olfactory bulb*. Because of their small size they can only be seen in microscopic dissections or histological preparations of the cribriform plate. Each olfactory nerve synapses on cell bodies that form the ipsilateral olfactory bulb, the axons of which constitute the *olfactory tract* (Fig. 5.34). The olfactory tract ramifies in the anterior

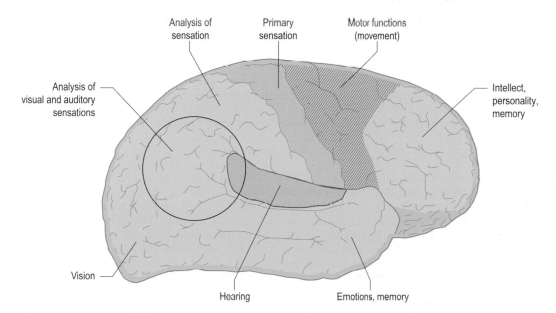

Figure 5.33 Lateral aspect of right hemisphere showing the principal functions of various regions of the cerebrum.

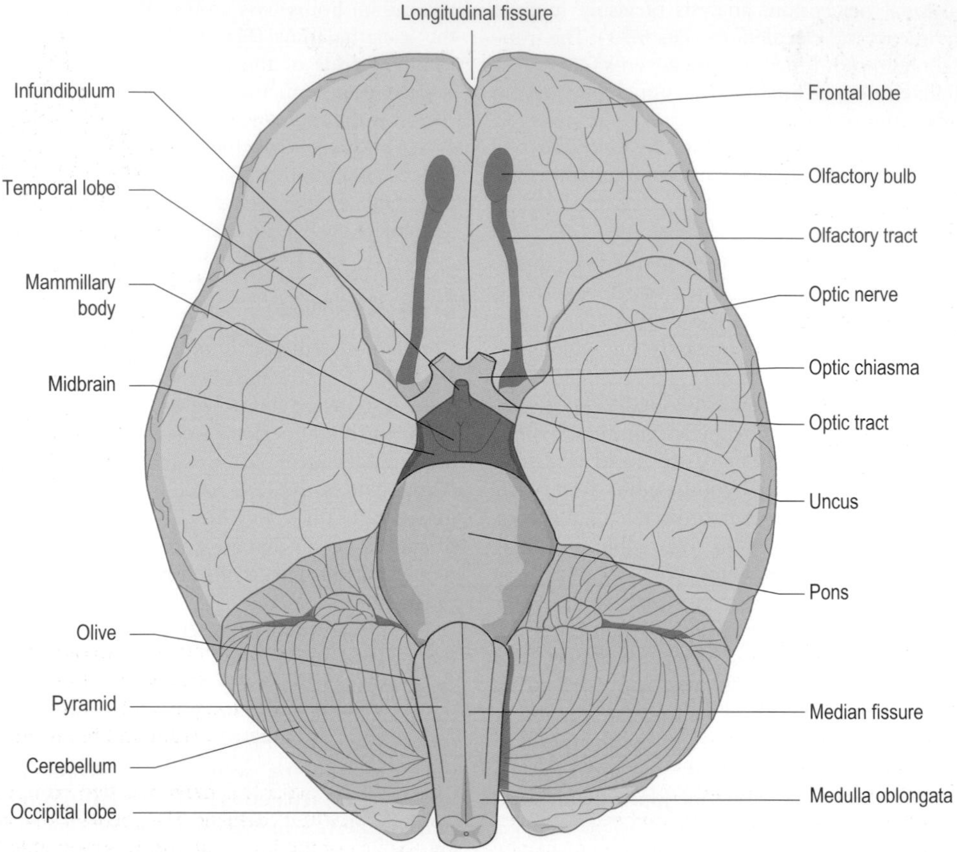

Figure 5.34 Base (inferior surface) of the brain.

end of the hypothalamus where it has diverse connections, the majority of which eventually pass to the uncus.

The second cranial nerve (II), the *optic nerve*, transmits visual information from the ipsilateral retina to the optic chiasma (Fig. 5.34). Information from both eyes is transmitted by each optic tract to the lateral geniculate body on each side. After processing this information, cells in the lateral geniculate body eventually relay it to the occipital lobes of the ipsilateral cerebral hemisphere through axons that run in the white matter of the cerebral hemisphere.

The third cranial nerve (III), the *oculomotor nerve*, innervates certain muscles of the eye. It emerges from the ventral surface of the midbrain at the level of the superior colliculus immediately medial to the crus cerebri on each side (Fig. 5.35A and B). It reaches the orbit by passing through the cavernous sinus and then the superior orbital fissure.

The fourth cranial nerve (IV), the *trochlear nerve*, innervates the superior oblique muscle of the eye. It emerges from the dorsal aspect of the midbrain immediately below the inferior colliculus (Fig. 5.35C), and then winds around the crus cerebri to appear on the ventral aspect of the

brainstem (Fig. 5.35A and B). It enters the orbit by passing through the cavernous sinus and then the superior orbital fissure.

The fifth cranial nerve (V), the *trigeminal nerve*, divides into three large branches, the ophthalmic, maxillary and mandibular divisions. The nerve is mainly sensory, with the cell bodies of the sensory fibres located in a swelling on the nerve, the *trigeminal ganglion*, formed at the junction of its three divisions (Fig. 5.35A and B). The part of the nerve between the ganglion and the brainstem is known as the *sensory root* of the trigeminal nerve. It is attached to the lateral aspect of the pons, and demarcates the junction of the pons with the middle cerebellar peduncle.

Although the pons and middle cerebellar peduncle are continuous with one another, being formed essentially by the same nerves, they are by convention demarcated by an imaginary line passing longitudinally through the point of attachment of the trigeminal root to the brainstem. The middle cerebellar peduncle lies dorsal to this line and pons ventral to it.

The trigeminal nerve projects ventrally onto the superior surface of the petrous temporal bone, with the trigeminal

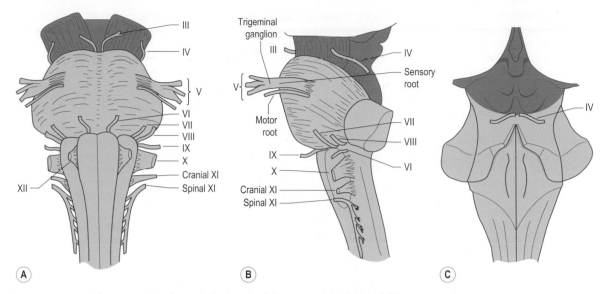

Figure 5.35 Cranial nerves arising from the brainstem: (A) anterior, (B) left lateral, (C) posterior views.

ganglion resting in a shallow depression on its surface. Beyond the ganglion the three divisions leave the skull to supply particular regions of the face and skull.

The first or *ophthalmic division (V1)* enters the orbit through the superior orbital fissure, its branches supply the ethmoidal air cells, the walls of the nasal cavity, the cornea and conjunctiva and the skin of the forehead. The second or *maxillary division (V2)* leaves the cranial cavity through the foramen rotundum, its branches are distributed to the nasal cavity, the upper teeth, the hard and soft palates and the skin of the face covering the maxilla and zygoma. The third or *mandibular division (V3)* passes through the foramen ovale and supplies the lower teeth, the temporomandibular joint, the tongue and the skin overlying the lower jaw and the temporal regions.

Although largely sensory, the trigeminal nerve is accompanied by a small nerve known as the *motor root* of the trigeminal nerve. This root emerges from the pons just below the sensory root (Fig. 5.35A and B) and joins the mandibular division in the foramen ovale to supply the muscles of mastication.

The sixth cranial nerve (VI), the *abducens nerve*, innervates the lateral rectus muscle of the eye which is responsible for abducting the eye. It emerges from the brainstem at the caudal border of the pons between it and the beginning of the pyramid (Fig. 5.35A and B), entering the orbit through the superior orbital fissure.

The seventh and eighth cranial nerves (VII and VIII), *facial* and *vestibulocochlear* respectively, emerge as a pair from the lateral aspect of the medulla oblongata immediately below the caudal border of the pons (Fig. 5.35A and B). They both enter the internal auditory meatus. The facial nerve is the more medial of the pair, and after emerging

from the stylomastoid foramen innervates the muscles of facial expression. It also contains taste fibres from the anterior two-thirds of the tongue, and parasympathetic fibres destined for the lacrimal, submandibular and sublingual glands.

The vestibulocochlear nerve has two components – the vestibular and the cochlear. The cochlear division arises in the cochlea of the inner ear and is responsible for hearing. The vestibular division innervates the vestibular apparatus of the inner ear and is responsible for the sense of balance.

The ninth, tenth and eleventh cranial nerves (IX, X and XI) appear as a series of tiny rootlets aligned along the lateral surface of the medulla oblongata (Fig. 5.35A and B). The uppermost of these rootlets converge to form the ninth cranial nerve, the *glossopharyngeal nerve*; the middle rootlets form the tenth cranial nerve, the *vagus nerve*, and the lowest rootlets form the cranial part of the 11th cranial nerve, the *accessory nerve*. All three nerves leave the skull through the jugular foramen.

The glossopharyngeal nerve supplies stylopharyngeus, but principally is sensory to the pharynx and posterior one-third of the tongue. The vagus nerve has a diverse distribution including the muscles of the pharynx and the larynx, and the mucosa of the larynx, the viscera of the chest and much of the abdomen. The ninth and tenth cranial nerves also innervate the carotid body and the carotid sinus and are therefore involved in cardiovascular reflexes.

The *accessory nerve* has two components – the cranial and the spinal. The *cranial accessory nerve* is formed by the lowest rootlets from the lateral aspect of the brainstem. The *spinal accessory nerve* is formed by rootlets that emerge from the lateral aspect of the C1–C5 segments of the spinal cord. They converge to form a single trunk that enters the skull through the foramen magnum (Fig. 5.35A and B). The spinal and cranial parts of the accessory nerve join for a short distance, but the fibres of the cranial part are rapidly transferred to the vagus nerve through which they are distributed to the muscles of the pharynx, larynx and soft palate. The spinal accessory nerve leaves the skull independently but in company with the glossopharyngeal and vagus nerves pass through the jugular foramen, innervating the sternomastoid and trapezius.

The 12th cranial nerve (XII), the *hypoglossal nerve,* supplies the muscles of the tongue. It emerges from the brainstem as a series of rootlets attached along the depression between the pyramid and the olive on each side (Fig. 5.35A and B), and leaves the skull through the hypoglossal canal in the occipital bone.

Nerve root sheaths

To reach the peripheral nervous system, the cranial nerves must penetrate the meninges. As they leave the brainstem the proximal ends of the cranial nerves and the spinal nerve roots are invested by the pia mater (Fig. 5.36); this is one of the morphological reasons for classifying these nerves as part of the CNS. The pia mater spreads like a tubular extension over the surface of each nerve, from its point of attachment to the brainstem as far as where the nerve leaves the skull or the vertebral canal.

Generally, the cranial nerves pierce the arachnoid and dura mater at the foramina through which they leave the skull. Here the dura and arachnoid mater are drawn into the foramen by the exiting nerve, and eventually the meningeal tissues blend with the fibrous sheath of the peripheral part of the cranial nerve (Fig. 5.36). Thus, cranial nerves are covered by pia mater and are bathed in CSF as far as their respective foramina.

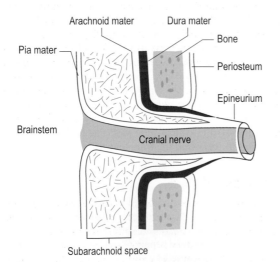

Figure 5.36 Relationship of the meningeal sheaths to the cranial nerves.

Exceptions to this rule apply to the second to the fifth cranial nerves. A dural and arachnoid sheath accompanies the optic nerve as far as the back of the eye; the nerve is therefore covered by pia mater and is bathed in CSF up to this point. The third, fourth and fifth cranial nerves pierce the dura mater that forms the cavernous sinus, and run outside the dura along the walls of the cavernous sinus before passing through the superior orbital fissure. The fifth cranial nerve penetrates the dura mater over the apex of the petrous temporal bone. Its three divisions then run extradurally before entering their respective foramina. The sensory and motor roots are, however, covered by pia mater and bathed in CSF as far as the middle of the trigeminal ganglion, which is where the dura and arachnoid mater blend with the nerve.

Internal structure of the brain

For introductory purposes, it is sufficient to examine the appearance of simple transections of the CNS.

Brainstem

In general the brainstem consists of a central core of grey matter, the *reticular formation,* surrounded by white matter formed by various motor and sensory tracts, and in which are interspersed various nuclei associated with the cranial nerves (Fig. 5.37A). Transverse sections have a diverse appearance because of the variations in external shape and internal contents at different levels. Details of this internal structure are better addressed when the CNS is studied in detail.

Forebrain

The cerebral hemispheres have a characteristic appearance irrespective of whether they are sectioned coronally, horizontally or sagittally. The cerebral cortex is a layer of grey matter extending over the entire external surface of each hemisphere (Fig. 5.37B), the cells of which are responsible for processing information related to the functions of the lobe in which they lie.

Deep to the cortex, white matter forms the main bulk of each hemisphere. It consists of axons of cells in the cerebral cortex that communicate with cells in adjacent, distant and contralateral regions of the cerebrum, and also the axons that pass out of or into the cerebrum.

Sections through the central parts of the cerebral hemisphere reveal the cavity of the lateral ventricle and large collections of grey matter surrounding the lateral and third ventricles (Fig. 5.37). The large mass adjacent to the third ventricle is the thalamus; the remaining masses are collectively known as the *basal ganglia* – a group of nuclei involved in the control of movement.

The meninges

Although the glial cells hold the neurons of the CNS together and afford them protection against metabolic insults, the CNS is nonetheless a soft, cellular mass. As such

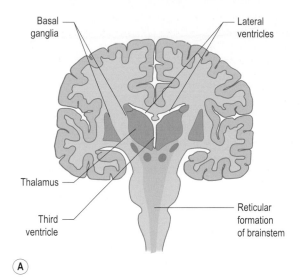

Basal ganglia

Lateral ventricles

Thalamus

Third ventricle

Reticular formation of brainstem

(A)

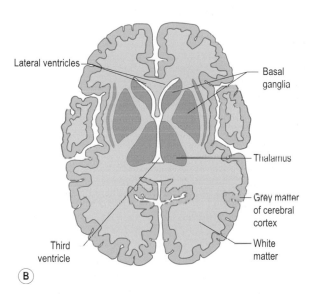

Lateral ventricles

Basal ganglia

Thalamus

Grey matter of cerebral cortex

Third ventricle

White matter

(B)

Figure 5.37 Internal appearance of the brain and brainstem: (A) coronal and (B) transverse sections.

it is vulnerable to external, mechanical insults which might arise were it freely mobile within the skull or vertebral canal. To protect against such insults, the CNS is surrounded by three membranes, collectively called the *meninges*, and is bathed in *cerebrospinal fluid*.

Intracranial meninges

The meningeal layers that surround the spinal cord also surround the brain and the brainstem. The pia mater follows every convolution of its surface, extending into the depths of any sulcus or fissure (Fig. 5.38). It is adherent to the underlying neural tissue and endows the CNS with a smooth surface. Specimens in which the pia mater has been stripped have a granulated surface. The arachnoid mater lines the deep surface of the dura mater and forms a network of fine threads bridging the pia mater and dura (Fig. 5.38).

Whereas the dura mater forms a mobile sac around the spinal cord, in the skull it is applied to the internal surface of the cranial cavity where it fuses with the endosteum of the cranium (Fig. 5.38). However, at certain sites, the dura mater separates from the adjacent bone to form large folds that project towards the centre of the cranial cavity, and which enclose vascular channels called venous sinuses. These transmit venous blood draining from the CNS, the cranium and the orbit (see p. 589).

The major reflections of the dura mater within the skull are the *falx cerebri* and *tentorium cerebelli* (Fig. 5.39). Each fold consists of two layers of dura mater with a free edge and two surfaces. At the attached edge of the fold the two layers are reflected onto the inner surface of the cranium, being continuous with the rest of the dural lining of the cranial bones.

The falx cerebri projects from the roof of the cranium in the median plane, extending from the crista galli in front to the internal occipital protuberance behind (Fig. 5.39), in profile it is sickle-shaped. The falx projects into the longitudinal fissure and thus separates the two cerebral hemispheres. Along its attached margin the falx cerebri encloses the superior sagittal sinus, while its free edge encloses the inferior sagittal sinus (Fig. 5.38B).

The tentorium cerebelli lies in a transverse plane projecting from the anterior, lateral and posterior walls of the posterior cranial fossa, forming a partial roof to it (Fig. 5.39). Its peripheral attachments are diverse. Anteriorly, on each side, it attaches to the posterior clinoid process and the posterosuperior margin of the petrous temporal bone. Laterally and posteriorly it is attached to the margins of the bony sulcus of the transverse sinus. From these attachments the tentorium cerebelli projects centrally, lying under the basal surface of the occipital and temporal lobes, and overlying the cerebellum. Centrally the tentorium forms a U-shaped free edge which extends from one anterior clinoid process to the other. The notch formed transmits the midbrain from the posterior cranial fossa to the diencephalon. Posteriorly in the midline, the posterior end of the falx cerebri fuses with the upper surface of the tentorium cerebelli and encloses the straight sinus.

Cerebrospinal fluid

Running between the threads of arachnoid mater and filling the subarachnoid space is the *cerebrospinal fluid*. The CSF is secreted by strings of capillaries that project into the lateral, third and fourth ventricles of the brain. It fills the cavities of the CNS and emerges through the apertures in the roof of the fourth ventricle to fill the subarachnoid space surrounding the brain and spinal cord.

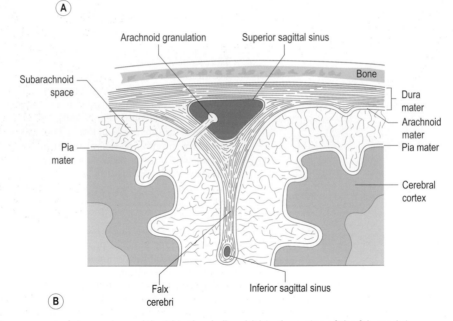

Figure 5.38 Arrangement of the meninges: (A) within the skull and (B) in the region of the falx cerebri.

CSF is secreted continuously and reabsorbed into the bloodstream through extensions of the arachnoid mater, the *arachnoid granulations* that pierce the dura mater of the falx cerebri and project into the superior sagittal sinus (Fig. 5.38B).

Chemically, the CSF protects the CNS by maintaining a constant pH environment, while mechanically, it endows the CNS with a cushioning fluid environment. Within the skull the brain is buoyant in a pool of CSF, and within the dural sac, the spinal cord is also suspended in a pool of CSF, being held centrally within it by the denticulate ligaments.

Arterial supply

The arterial supply of the CNS is derived from the *internal carotid* and *vertebral arteries*. The internal carotid artery enters the skull through the carotid canal, and then traverses the foramen lacerum and cavernous sinus (Fig. 5.40). It penetrates the dura and arachnoid mater medial to the

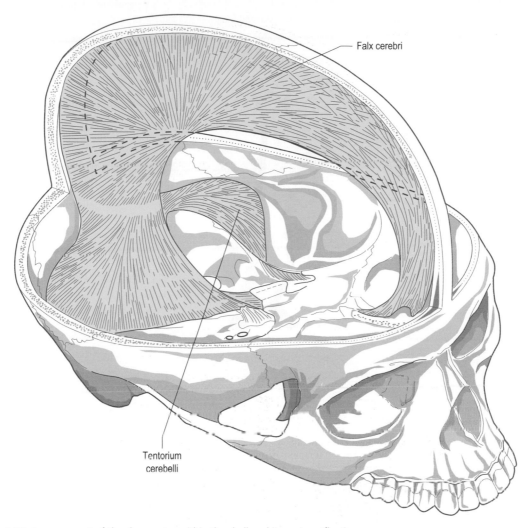

Falx cerebri

Tentorium
cerebelli

Figure 5.39 Arrangement of the dura mater within the skull and its major reflections.

anterior clinoid process and enters the subarachnoid space over the ventral surface of the brain where it divides into its terminal branches – the *anterior cerebral* and *middle cerebral arteries* (Fig. 5.41).

Each anterior cerebral artery curves medially, crossing the optic nerve, to enter the longitudinal fissure (Fig. 5.41). Before doing so, the two anterior cerebral arteries anastomose with one another via the *anterior communicating artery*. The main trunk of the anterior cerebral artery follows the contour of the corpus callosum and distributes branches to the medial aspect of the cerebral hemisphere (Fig. 5.41).

The middle cerebral artery passes laterally to gain the floor of the lateral sulcus, and then runs backwards along it. Deep, penetrating branches arise from its proximal part and enter the cerebral hemisphere from below to supply the region of the basal ganglia. Superficial branches arise from the middle cerebral artery as it runs in the lateral sulcus. These pass in all directions onto the external surface of the cerebral hemisphere, supplying most notably the motor and sensory regions of the cerebral cortex (Fig. 5.42).

The vertebral arteries pierce the dural sac and enter the subarachnoid space just above the first cervical vertebra. Each passes forwards and upwards through the foramen magnum to join in the midline in front of the medulla oblongata to form the *basilar artery*, which runs upwards in front of the pons (Fig. 5.41). Branches of the vertebral and basilar arteries pass laterally to supply the medulla oblongata, pons and cerebellum.

The terminal branches of the basilar artery are the *posterior cerebral arteries* (Fig. 5.41), each of which passes around the lateral surface of the midbrain to gain the inferior aspect

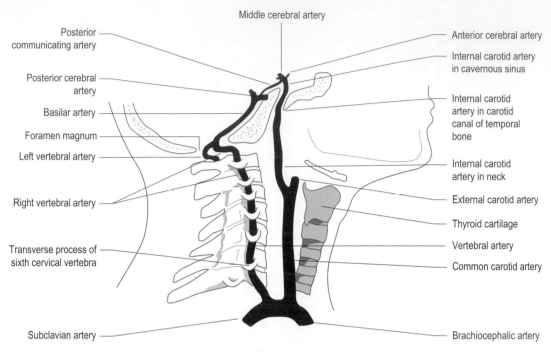

Figure 5.40 Right internal carotid and vertebral arteries entering the skull to supply the brain.

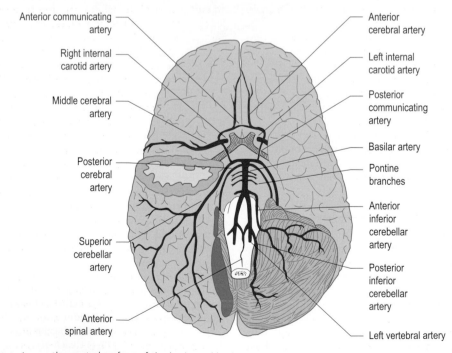

Figure 5.41 Arteries on the ventral surface of the brain and brainstem.

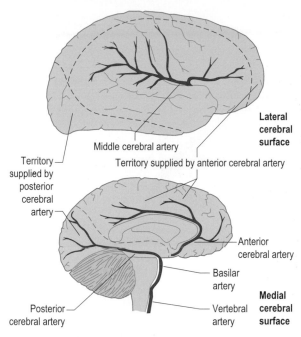

Figure 5.42 Distribution of the anterior, middle and posterior cerebral arteries over the cerebral cortex.

of the temporal lobe of the cerebral hemisphere (Fig. 5.42). Each supplies the posterior parts of the hemisphere, notably the occipital lobe. Branches from the proximal part of the posterior cerebral artery supply the midbrain, thalamus, metathalamus and epithalamus.

On the ventral surface of the brain, each posterior cerebral artery anastomoses with the ipsilateral internal carotid artery through the *posterior communicating artery* (Fig. 5.41). These latter vessels allow the vertebrobasilar circulation to communicate with that of the internal carotid arteries. Similarly, the anterior communicating artery permits flow between the two internal carotid arteries. The ring of vessels formed by the anterior communicating, the anterior cerebral, the internal carotid, the posterior communicating, and the posterior cerebral arteries and the end of the basilar artery is known as the *circle of Willis*.

Venous drainage and dural venous sinuses

The veins that drain the internal aspects of the cerebral hemispheres, the basal ganglia and thalamus converge to form a large vessel called the *great cerebral vein* that passes posteriorly to enter the straight sinus. Veins from the external surface of the cerebral hemispheres drain upwards and downwards towards the superior sagittal and transverse sinuses.

The superior vertebral veins may rupture during blows to the head where they penetrate the arachnoid and dura mater to enter the *superior sagittal sinus* (Fig. 5.43A). If they

are torn at these sites, blood may accumulate between the arachnoid and dura mater cleaving the two membranes apart, forming a progressively enlarging collection of blood that slowly compresses the underlying brain. Because of the location of the blood, this condition is known as a subdural haematoma.

The dural venous sinuses are vascular channels that are located at constant sites along the internal surface of the cranium. Most are enclosed by bone on their external aspect and by a fold of the dura mater internally. Others are located wholly within specific folds of dura mater. Some sinuses are small, resembling a small diameter vein threaded between the dura mater and skull, others are much larger channels. The difference between a sinus and a vein is that a sinus lacks the adventitial and muscular layers normally found in the wall of a vein. These are replaced by the bone and dura mater that surrounds the sinus. Consequently, only a layer of endothelial cells separates the blood inside a sinus from its surrounding tissues.

The major venous sinuses of the skull are the paired cavernous, transverse and sigmoid sinuses, and the single superior and inferior sagittal sinuses, and the straight sinus (Fig. 5.43A).

The *superior sagittal sinus* runs posteriorly along the attached edge of the falx cerebri (Figs. 5.38B and 5.43A). It drains blood from the external surfaces of the cerebral hemispheres, and reabsorbs CSF through the arachnoid granulations. The *inferior sagittal sinus* is enclosed by the free edge of the falx cerebri.

The *straight sinus* is enclosed by the dura mater along the line of fusion of the falx cerebri and tentorium cerebelli (Fig. 5.43A). It receives the great cerebral vein and the *inferior sagittal sinus*. The superior sagittal sinus, the straight sinus and the occipital sinus may meet at the internal occipital protuberance, where the transverse sinuses begin. This region is known as the *confluence of sinuses* (Fig. 5.43A).

From the internal occipital protuberance each *transverse sinus* runs around the posterior and lateral walls of the posterior cranial fossa in the attached margin of the tentorium cerebelli (Fig. 5.43A). The right transverse sinus is usually a continuation of the superior sagittal sinus, while the left is a continuation of the straight sinus. They receive veins from the surfaces of the cerebral hemisphere and cerebellum as well as the *superior petrosal sinus*. The transverse sinus is directly continuous with the *sigmoid sinus* located in a groove on the mastoid portion of the temporal bone. The sigmoid sinus joins with the *inferior petrosal sinus* to exit the skull via the jugular foramen and forms the internal jugular vein.

The smaller venous sinuses of the skull are located along the edges of the bones that project into the middle and posterior cranial fossae (Fig. 5.43A). The superior and inferior petrosal sinuses run along the respective edges of the petrous temporal bone; the *sphenoparietal sinus* runs along the lesser wing of the sphenoid and the occipital sinus runs along the internal occipital crest.

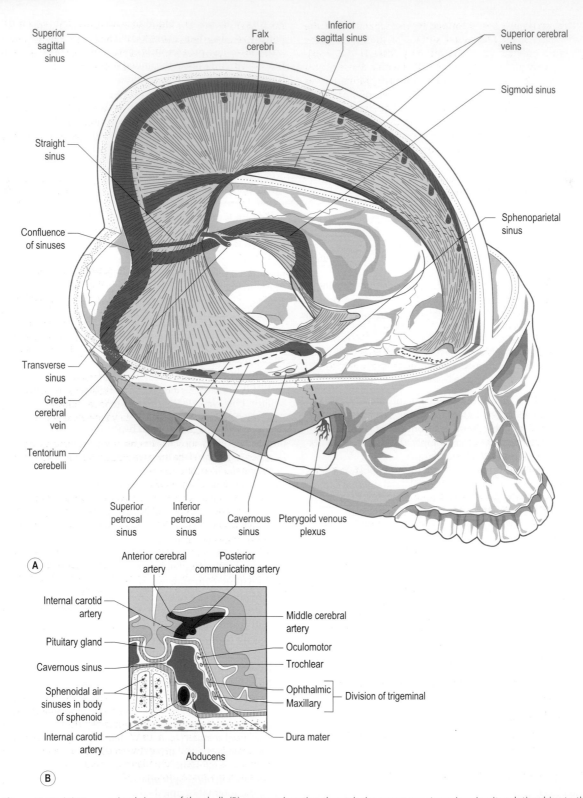

Figure 5.43 (A) Venous dural sinuses of the skull, (B) a coronal section through the cavernous sinus showing its relationships to the sphenoid, internal carotid artery and III (oculomotor), IV (trochlear), V (trigeminal) and VI (abducens) cranial nerves.

The *cavernous sinus* is formed by the separation of the dura mater from the lateral surface of the body of the sphenoid either side of the pituitary fossa (Fig. 5.43B). It is a narrow trabeculated space filled with a dense plexus of venous channels. The third and fourth cranial nerves and the ophthalmic and maxillary divisions of the trigeminal nerve pass along the lateral wall of the sinus, running deep to the dura mater but external to the endothelium of the sinus. The sixth cranial nerve runs with the internal carotid artery between the endothelium and the bony floor of the sinus. The sinus receives venous blood principally from the sphenoparietal sinus and the superior ophthalmic vein, which drains the orbit, as well as from the middle meningeal vein and veins from the cerebral hemisphere.

Ultimately the majority of the blood in the dural venous sinuses drains to one or other of the internal jugular veins, although some may drain through small emissary veins that penetrate the skull at specific sites to communicate with veins outside the skull.

Section summary

The brain

Features

- Expanded rostral end of neural tube located within the skull.
- Comprises cerebrum, cerebellum and brainstem (pons, midbrain, medulla oblongata); continuous with spinal cord beyond the medulla oblongata.
- Contains ventricles (lateral, third, fourth), thalamus and hypothalamus.
- Medulla oblongata has anterior median fissure with pyramids either side; posteriorly nucleus gracilis medially and nucleus cuneatus laterally lie either side of midline.

Cerebrum

- Separated by longitudinal fissure into right and left cerebral hemispheres, joined by corpus callosum.
- Has outer cortex of grey matter (cell bodies) and large inner mass of white matter (axons).
- Cerebral cortex thrown into folds (gyri) and depressions (sulci).
- Transverse central sulcus separates precentral gyrus (motor cortex) and postcentral gyrus (sensory cortex).
- Each hemisphere divided into four lobes (frontal, parietal, temporal, occipital).
- Temporal and parietal lobes separated by lateral sulcus.
- Parietal and temporal lobes separated by parieto-occipital sulcus.
- Frontal lobe subserves motor functions and is responsible for expressions of intellect and personality.
- Parietal lobe responsible for sensory functions (touch, pressure, position sense), three-dimensional perception, analysis of visual images, language, geometry and calculations.
- Occipital lobe is involved in vision.
- Temporal lobe is involved in perception of sound, memory and emotion.

Cranial nerves

- 12 pairs inside skull, form part of CNS; once outside skull are part of peripheral nervous system.
- (I) Olfactory nerve mediates sense of smell.
- (II) Optic nerve transmits visual information.
- (III) Oculomotor nerve innervates extraocular muscles (superior, medial and inferior rectus, inferior oblique): conveys parasympathetic fibres to the ciliary ganglion.
- (IV) The trochlear nerve innervates superior oblique.
- (V) The trigeminal nerve divides into three parts: ophthalmic division (V1) distributed to nasal cavity, cornea, conjunctiva, skin of forehead; maxillary division (V$_2$) distributed to nasal cavity, upper teeth, hard and soft palates, skin of face; mandibular division (V$_3$) distributed to lower teeth, tongue, temporomandibular joint, skin over mandible and muscles of mastication.
- (VI) The abducens nerve innervates lateral rectus.
- (VII) The facial nerve innervates the facial muscles, conveys taste from anterior two-thirds of tongue; conveys parasympathetic fibres to the pterygopalatine and the submandibular ganglia.
- (VIII) The vestibulocochlear nerve transmits information regarding sense of balance (vestibular part) and auditory information (cochlear part).
- (IX) The glossopharyngeal nerve innervates stylopharyngeus, sensory to pharynx and posterior third of tongue, innervates carotid body and sinus: conveys taste from posterior third of tongue; conveys parasympathetic fibres to otic ganglion.
- (X) Vagus nerve innervates muscles of the pharynx and the larynx, innervates carotid body and sinus: conveys parasympathetic fibres to viscera of chest and abdomen.
- (XI) Accessory nerve: cranial fibres distributed with vagus to muscles of pharynx and larynx; spinal fibres innervate sternomastoid and trapezius.
- (XII) Hypoglossal nerve innervates the muscles of the tongue.

Meninges

- Three coverings: outer fibrous dura mater, middle arachnoid mater and inner transparent pia mater.
- Dura reflected within skull as falx cerebri (in sagittal plane) and tentorium cerebelli (in transverse plane).
- Arachnoid lines deep surface of dura forming network of fine threads bridging gap between dura and pia mater.
- Pia mater is adherent to underlying neural tissue.
- Subarachnoid space contains CSF.

Cerebrospinal fluid

- Produced by choroid plexus in roof of lateral and third ventricles.
- Reabsorbed into venous system through arachnoid granulations.

Blood supply

- Arterial supply from right and left internal carotid and vertebral arteries.
- Internal carotid divides into anterior and middle cerebral arteries: anterior communicating artery connects anterior cerebral arteries.
- Vertebral arteries join forming basilar artery, which terminates as posterior cerebral arteries: posterior communicating artery connects posterior and middle cerebral arteries.
- Venous drainage by system of intracranial venous sinuses (superior and inferior sagittal, transverse, cavernous, superior and inferior petrosal, sigmoid) draining into internal jugular vein.

Index

Note: Page numbers followed by *b* indicate boxes, *f* indicate figures and *t* indicate tables.

Index